CLASSIC TOPICS ON THE HISTORY OF MODERN MATHEMATICAL STATISTICS

CLASSIC TOPICS ON THE HISTORY OF MODERN MATHEMATICAL STATISTICS

From Laplace to More Recent Times

PRAKASH GORROOCHURN

Department of Biostatistics, Mailman School of Public Health,
Columbia University, New York, NY

Published by John Wiley & Sons, Inc., Hoboken, New Jersey
Published simultaneously in Canada

For general information on our other products and services or for technical support, please contact our Customer Care Department within the United States at (800) 762-2974, outside the United States at (317) 572-3993 or fax (317) 572-4002.

Wiley also publishes its books in a variety of electronic formats. Some content that appears in print may not be available in electronic formats. For more information about Wiley products, visit our web site at www.wiley.com.

Library of Congress Cataloging in Publication data can be found on file at the Library of Congress.

ISBN: 9781119127925

Set in 10/12pt Times by SPi Global, Pondicherry, India

10 9 8 7 6 5 4 3 2

To Nishi and Premal

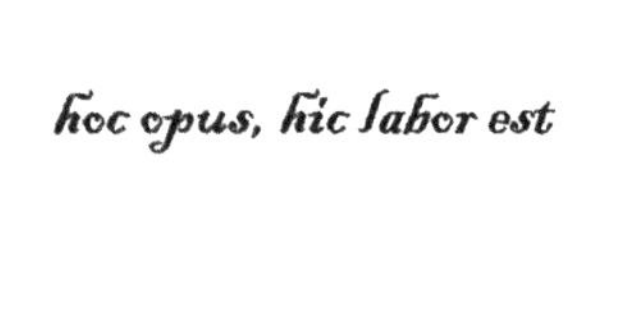
hoc opus, hic labor est

CONTENTS

PREFACE

This book describes the works of some of the more recent founders of the subject of mathematical statistics. By more recent, I have taken it to mean from the Laplacean to the immediate post-Fisherian period. The choice is, of course, entirely subjective, but there are certain strong reasons for starting a work on the history of modern mathematical statistics from Laplace. With the latter, a definite form was given to the discipline. Although Laplace's levels of rigor were certainly not up to the standards expected today, he brought more sophistication to statistics compared to his predecessors. His systematic work on probability, error theory, and large sample methods laid the foundations for later investigations. One of his most important results, the Central Limit Theorem, was the prototype that was worked and improved upon by generations of succeeding statisticians.

This book may be viewed as a continuation of my previous book *Classic Problems of Probability* (Gorroochurn, 2012a). However, apart from the fact that this book deals with mathematical statistics rather than probability, I have now endeavored to go into a more detailed and complete coverage of the topics. That does not mean that I have treated every single historical topic. Rather, what I have treated, I have tried to do so in a thorough way. Thus, the reader who wishes to know exactly how Laplace first proved the Central Limit Theorem, how Fisher developed ANOVA, or how Robbins first developed the empirical Bayes method may find the book helpful. In my demonstration of the mathematical results, I have purposely used (as much as possible) the same notation as the original writers so that readers can have a much better feel for the original works. I have also included page numbers referring to the original derivations so readers can easily check against the original works. I really hope readers will be encouraged to read the original papers as these are often treasure troves of statistics in the making.

I also hope that readers will find the book a useful addition to the few books on the history of statistics that have hitherto been written. Two of the major books I ought to mention are Stigler's *The History of Statistics: The Measurement of Uncertainty Before 1900* (Stigler, 1986b) and Hald's *A History of Mathematical Statistics from 1750 to 1930* (Hald, 1998).* The first is a superlative and engaging essay on the history of statistics, but

*Other books on the history of statistics include Hald (1990), Pearson (1978), Pearson and Kendall (1970), Kendall and Plackett (1977), Hald (2007), MacKenzie (1978), Benzecri (1982), Cowles (2001), Chatterjee (2003), Porter (1986), and Westergaard (1932).

since it does not treat the post-1900 period, none of Fisher's work is treated. The second book is a comprehensive treatment of the pre-1930 period, written in modern mathematical notation. However, it devotes so much space to Laplace and Gauss that, to my taste, the coverage of Fisher is not as complete as those of the two other mathematicians. Moreover, Hald's book has no coverage of the Neyman–Pearson theory. This, of course, is no fault of Hald since the book is perfectly suited for the aims the author had in writing it.

But why do I insist on the coverage of Fisher's statistical work? My simple answer is as follows: *virtually the whole of modern mathematical statistics sprang either from Fisher's original work or from the extensions (and sometimes corrections) others made to his original work.* Much of the modern statistical vocabulary, especially in the important field of estimation, originated from Fisher. The word "statistic" itself, in its technical sense, is Fisher's creation. Almost single-handedly, Fisher introduced sufficiency, efficiency, consistency, likelihood, ANOVA, ANCOVA, and countless other statistical concepts and techniques. Many of the current branches of statistics, such as multivariate analysis and nonparametric statistics, had an important original contribution made by Fisher. Unfortunately, it was not possible for me to treat all the different branches of statistics: my main concern here is what may be termed parametric statistical inference.

While going through the book, readers will find that I often quote the original authors. My purpose in doing so is for readers not to be influenced by my own paraphrasing of the original author's thoughts and intentions, but to hear it from the "horse's mouth" itself. Thus, I believe there is a different effect on readers when I say "Fisher was somewhat indifferent to unbiasedness as a desideratum for an estimator because unbiased estimators are not invariant to transformations" than when Fisher himself said that "…lack of bias, which since it is not invariant for functional transformation of parameters has never had the least interest for me" (Bennett, 1990, p. 196).* For the most part, though, my quotation of the original author has been accompanied by my own interpretation and comments, so that readers are not left in the dark as to what the original author really meant. The reader will find a similar approach in the late Lehmann's book *Fisher, Neyman, and the Creation of Classical Statistics* (Lehmann, 2011). I hope this strategy will bring a sense of authenticity to the subject of history I am writing about.

This book may be divided into three parts. The first part deals mainly with Laplace's contributions to mathematical statistics. In this part, readers will find a detailed analysis of Laplace's papers and books. Readers will also learn about Laplace's definition of probability, his philosophy of universal determinism and how it shaped his statistical research, his early preference for methods of inverse probability (based on the principle of indifference), his investigation of various laws of error, his powerful method of asymptotic approximation, his introduction of characteristic functions, his later use of methods based on direct probability, his proofs of the Central Limit Theorem, and his development of least squares. The first part also contains important related work of other scholars such as Fourier, Lagrange, Adrain, Poisson, Gauss, Hagen, and Lyapunov.

The second part of this book deals mainly with Galton's discovery of regression and correlation, Pearson's invention of the X^2 goodness-of-fit statistic, and Student's innovation of the small-sample t-statistic, culminating in Fisher's creation of new statistical methods. In the section devoted to Galton, readers will learn about Galton's

*See also p. 376 of the current book.

observation of an interesting phenomenon, which he thought was unidirectional and which he first termed reversion (which is later changed to regression), and the first appearance of correlation. Extensions of Galton's work by Weldon, Edgeworth, Yule, Pearson, and Fisher are also considered, as are related works prior to Galton including those of Lagrange, Adrain, Gauss, Laplace, Plana, Bravais, and Bertrand. In the section devoted to Pearson, readers will learn how Pearson developed the X^2 goodness-of-fit statistic and clashed with Fisher when the latter pointed out an error in Pearson's development. Derivations of the χ^2-distribution prior to Pearson is also considered, including those of Bienaymé, Abbe, and Helmert. In the section devoted to Student, readers will learn about his original, but somewhat faulty, derivation of the t-statistic (which was called z and slightly different from today's t) and Fisher's later consolidation of Student's result. In this section, previous derivations of the t-distribution, based on inverse probability, by Lüroth and Edgeworth are also examined. Finally, in the section devoted to Fisher, readers will find a detailed account of Fisher's development of estimation theory, significance testing, ANOVA and related techniques, and fiducial probability. Also included in this section are Fisher's (in)famous disputes with Jeffreys and Neyman–Pearson, and the use of maximum likelihood and significance testing prior to Fisher.

The third and final part of this book deals with post-Fisherian developments. In this part, readers will learn about extensions to Fisher's theory of estimation (by Darmois, Koopman, Pitman, Aitken, Fréchet, Cramér, Rao, Blackwell, Lehmann, Scheffé, and Basu), Wald's powerful statistical decision theory, and the Bayesian revival ushered by Ramsey, de Finetti, Savage, and Robbins in the first half of the twentieth century.

A few of the sections in the book are non-historical in nature and have been denoted by an asterisk (*). Moreover, a handful of historical topics that are found in the "Further Extensions" sections have been given without detailed demonstration.

As in my previous book, I have strived for simplicity of exposition. I hope my efforts will inculcate love for statistics and its history in the reader.

Prakash Gorroochurn
pg2113@columbia.edu
April 22, 2015

ACKNOWLEDGMENTS

Warren Ewens and Bruce Levin each read large portions of the original manuscript. Both encouraged me throughout the writing of the book. Various sections of earlier drafts were read by Bernard Bru, Bin Cheng, Daniel Courgeau, Andrew Dale, Sukumar Desai, Stephen Fienberg, Hans Fischer, Dominique Fourdrinier, David Hitchcock, Vesa Kuusela, Eugenio Regazzini, Nancy Reid, Christian Robert, Thomas Severini, Glenn Shafer, Craig Smorynski, Aris Spanos, Veronica Vieland, Alan Welsh, and Sandy Zabell. Jessica Overbey, Bei Wang, and Julia Wrobel helped verify the references. Finally, Ji Liao, Lynn Petukhova, Jiashi Wang, Gary Yu, Wenbin Zhu, and many students from my probability class assisted with the proofreading. To all these people I express my deepest gratitude. All remaining errors and omissions are, of course, my sole responsibility.

Susanne Steitz-Filler, Sari Friedman, and Allison McGinniss from Wiley were all very helpful in realizing this project.

My final and continued gratitude goes to my mother and late father.

INTRODUCTION: LANDMARKS IN PRE-LAPLACEAN STATISTICS

The word "statistics" is derived from the modern Latin *statisticus* ("state affairs"). *Statisticus* itself originates from the classic Latin *status*, from which word "state" is derived. In the eighteenth century,* the German political scientist Gottfried Achenwall (1719–1779) brought *statistisch* into general usage as the collection and evaluation of data relating to the functioning of the state. The English word "statistics" is thus derived from the German *statistisch*.

Following Achenwall, the next landmark was the creation of the science of political arithmetic in England, in the eighteenth century. Political arithmetic was a set of techniques of classification and calculation on data obtained from birth and death records, trade records, taxes, credit, and so on. It was initiated in England by John Graunt (1620–1674) and then further developed by William Petty (1623–1687). In the nineteenth century, political arithmetic developed into the field of statistics, now dealing with the analysis of all kinds of data. Statistics gradually became an increasingly sophisticated discipline, mainly because of the powerful mathematical techniques of analysis that were infused into it.

The recognition that the data available to the statistician were often the result of chance mechanisms also meant that some notion of probability was essential both for the statistical analysis of data and the subsequent interpretation of the results. The calculus of probability had its origins well before the eighteenth century. In the sixteenth century, the physician and mathematician Gerolamo Cardano (1501–1575) made some forays into chance calculations, many of which were erroneous. His 15-page book entitled *Liber de ludo aleae* (Cardano, 1663) was written in the 1520s but published only in 1663. However, the official start of the calculus of probability took place in 1654 through the correspondence between Blaise Pascal (1623–1662) and Pierre de Fermat (1601–1665) concerning various games of chances, most notably the problem of points. Meanwhile, having heard of the exchange between the two Frenchmen, the Dutch mathematician Christiaan Huygens (1629–1695) wrote a small manual on probability, *De Ratiociniis in ludo aleae* (Huygens, 1657), which came out in 1657 as the first published book on probability. Thereupon, a brief period of inactivity in probability followed until Pierre Rémond de Montmort

*The early history of statistics is described in detail in the books by Pearson (1978) and Westergaard (1932).

(1678–1719) published his book *Essay d'Analyse sur les Jeux de Hazard* in 1708 (Montmort, 1708). But the real breakthrough was to come through James Bernoulli's (1654–1705) posthumous *Ars Conjectandi* (Bernoulli, 1713), where Bernoulli enunciated and rigorously proved the *law of large numbers*. This law took probability from mere games of chance and extended its applications to all kinds of world phenomena, such as births, deaths, accidents, and so on. The law of large numbers showed that, viewed microscopically (over short time intervals), measurable phenomena exhibited the utmost irregularity, but when viewed macroscopically (over an extended period of time), they all exhibited a deep underlying structure and constancy. It is no exaggeration then to say that Bernoulli's *Ars Conjectandi* revolutionized the world of probability by showing that chance phenomena were indeed amenable to some form of rigorous treatment. The law of large numbers was to receive a further boost in 1730 through its refinement in the hands of Abraham de Moivre (1667–1754), resulting in the first derivation of the normal distribution.

In the meantime, two years before the release of the *Ars Conjectandi*, the Englishman John Arbuthnot (1667–1735) explicitly applied the calculus of probability to the problem of sex ratio in births and argued for divine providence. This was the first published test of a statistical hypothesis. Further works in demography were conducted by the Comte de Buffon (1707–1788), Daniel Bernoulli (1700–1782), and Jean le Rond d'Alembert (1717–1783).

Although *Ars Conjectandi* was duly recognized for its revolutionary value, James Bernoulli was not able to bring the book to its full completion before he passed away. One aspect of the problem not treated by Bernoulli was the issue of the probability of hypotheses (or causes), also known as *inverse probability*. This remained a thorny problem until it was addressed by Thomas Bayes (1701–1761) in the famous "An Essay towards solving a problem in the Doctrine of Chances" (Bayes, 1764). In the essay, again published posthumously, Bayes attacked the inverse problem addressed by Bernoulli. In the latter's framework, the probability of an event was a known quantity; in the former's scheme, the probability of an event was an unknown quantity and probabilistic statements were made on it through what is now known as *Bayes' theorem*. The importance of this theorem cannot be overstated. But inasmuch as it was recognized for its revolutionary value, it also arose controversy because of a particular assumption made in its implementation (concerning the prior distribution to be used).

In addition to the aforementioned works on probability, another major area of investigation for the statistician was the investigation of errors made in observations and the particular laws such errors were subject to. One of the first such studies was the one performed by the English mathematician Thomas Simpson (1710–1761), who assumed a triangular error distribution in some of his investigations. Other mathematicians involved in this field were Daniel Bernoulli, Joseph-Louis Lagrange (1736–1813), Carl Friedrich Gauss (1777–1855), and especially Adrien-Marie Legendre (1752–1833), who was the first to publish the method of least squares.

PART ONE: LAPLACE

1

THE LAPLACEAN REVOLUTION

1.1 PIERRE-SIMON DE LAPLACE (1749–1827)

Laplace was to France what Newton had been to England. Pierre-Simon de Laplace* (Fig. 1.1) was born in Beaumont-en-Auge, Normandy, on March 23, 1749. He belonged to a bourgeois family. Laplace at first enrolled as a theology student at the University of Caen and seemed destined for the church. At the age of 16, he entered the College of Arts at the University of Caen for two years of philosophy before his degree in Theology. There, he discovered not only the mathematical writings of such greats as Euler and Daniel Bernoulli, but also his own aptitude for mathematical analysis. He moved to Paris in 1769 and, through the patronage of d'Alembert, became Professor in Mathematics at the *École Royale Militaire* in 1771.

Laplace lived through tumultuous political times: the French revolution took place in 1789, Robespierre came in power in a coup in 1793, Louis XVI was executed in the same year followed by that of Robespierre himself the next year, and Napoléon Bonaparte came to power in 1799 but fell in 1815 when the monarchy was restored by Louis XVIII. Laplace was made Minister of the Interior when Napoléon came in power but was then dismissed only after 6 weeks for attempting to "carry the spirit of the infinitesimal into administration."†

But Napoléon continued to retain the services of Laplace in other capacities and bestowed several honors on him (senator and vice president of the senate in 1803, Count

*A description of Laplace's life and work can be found in Gillispie (2000), Andoyer (1922), Stigler (1975), and Hahn (2005).

† In the *Correspondance de Napoléon 1er*, the following exact description of the situation can be found: "In the Interior, minister Quinette was replaced by Laplace, a first-rate mathematician, but who quickly proved to be a more than mediocre administrator. From his very first work, the officials realized they had made a mistake; Laplace never took a question from its real point of view; he looked for subtleties everywhere, only had problematic ideas, and finally brought the spirit of the infinitesimal into administration." (Napoléon [Bonaparte], 1870, p. 330).

Classic Topics on the History of Modern Mathematical Statistics: From Laplace to More Recent Times, First Edition. Prakash Gorroochurn.

FIGURE 1.1 Pierre-Simon de Laplace (1749–1827). Wikimedia Commons (Public Domain), http://commons.wikimedia.org/wiki/File:Pierre-Simon_Laplace.jpg

of the Empire in 1806, Order of the Reunion in 1813). Nevertheless, Laplace voted Napoléon out in 1814, was elected to the French Academy in 1816 under Louis XVIII, and then made Marquis in 1817.

Thus, throughout the turbulent political periods, Laplace was able to adapt and even prosper unlike many of his contemporaries such as the Marquis de Condorcet and Antoine Lavoisier, who both died. Laplace continued to publish seminal papers over several years, culminating in the two major books, *Mécanique Céleste* (Laplace, 1799) and *Théorie Analytique des Probabilités* (Laplace, 1812). These were highly sophisticated works and were accompanied by the easier books, *Exposition du Système du Monde* (Laplace, 1796) and *Essai Philosophique sur les Probabilités* (Laplace, 1814) aimed at a much wider audience.

From the very start, Laplace's research branched into two main directions: applied probability and mathematical astronomy. However, underlying each branch, Laplace espoused one unifying philosophy, namely, universal determinism. This philosophy was vindicated to a great extent in so far as celestial mechanics was concerned. By using Newton's law of universal gravitation, Laplace was able to mathematically resolve the remaining anomalies in the theory of the Solar System. In particular, he triumphantly settled the issue of the "great inequality of Jupiter and Saturn." In the *Exposition du Système du Monde*, we can read:

> We shall see that this great law [of universal gravitation]...represents all celestial phenomena even in their minutest details, that there is not one single inequality of their motions which is not derived from it, with the most admirable precisions, and that it explains the cause of several singular motions, just perceived by astronomers, and which were too slow for them to recognize their law. (Laplace, 1796, Vol. 2, pp. 2–3)

Laplace appealed to a "vast intelligence," dubbed Laplace's demon, to explain his philosophy of universal determinism*:

> All events, even those that on account of their smallness seem not to obey the great laws of nature, are as necessary a consequence of these laws as the revolutions of the sun. An intelligence which at a given instant would know all the forces that move matter, as well as the position and speed of each of its molecules; if on the other hand it was so vast as to analyse these data, it would contain in the same formula, the motion of the largest celestial bodies and that of the lightest atom. For such an intelligence, nothing would be irregular, and the curve described by a simple air or vapor molecule, would seem regulated as certainly as the orbit of the sun is for us. (Laplace, 1812, p. 177)

However, we are told by Laplace, ignorance of the underlying laws makes us ascribe events to chance:

> ...But owing to our ignorance regarding the immensity of the data necessary to solve this great problem, and owing to the impossibility, given our weakness, to subject to calculation

*Similar statements to the three ones that follow were later repeated in the *Essai Philosophique sur les Probabilités* (Laplace, 1814, pp. 1–4).

> those data which are known to us, even though their numbers are quite limited; we attribute phenomena which seem to occur and succeed each other without any order, to variable or hidden causes, who action has been designated by the word *hazard*, a word that is really only the expression of our ignorance. (*ibidem*)

Probability is then a relative measure of our ignorance:

> Probability is relative, in part to this ignorance, in part to our knowledge. (*ibidem*)

It is perhaps no accident that Laplace's research into probability started in the early 1770s, for it was in this period that interest in probability was renewed among many mathematicians due to work in political arithmetic and astronomy (Bru, 2001b, p. 8379). Laplace's work in probability was truly revolutionary because his command of the powerful techniques of analysis enabled him to break new ground in virtually every aspect of the subject. The advances Laplace made in probability and the extent to which he applied them were truly unprecedented. While he was still alive, Laplace thus reached the forefront of the probability scene and commanded immense respect. Laplace passed away in Paris on March 5, 1827, exactly 100 years after Newton's death.

Throughout his academic career, Laplace seldom got entangled in disputes with his contemporaries. One notable exception was his public dissent with Roger Boscovich (1711–1787) over the calculation of the path of a comet given three close observations. More details can be found in Gillispie (2000, Chapter 13) and Hahn (2005, pp. 67–68).

Laplace has often been accused of incorporating the works of others into his own without giving due credit. The situation was aptly described by Auguste de Morgan* hundreds of years ago. The following extract is worth reading if only for its rhetorical value:

> The French school of writers on mathematical subjects has for a long time been wedded to the reprehensible habit of omitting all notice of their predecessors, and Laplace is the most striking instance of this practice, which he carried to the utmost extent. In that part of the "Mecanique Celeste" in which he revels in the results of Lagrange, there is no mention of the name of the latter. The reader who has studied the works of preceding writers will find him, in the "Théorie des Probabilités," anticipated by De Moivre, James Bernoulli, &c, on certain points. But there is not a hint that any one had previously given those results from which perhaps his sagacity led him to his own more general method. The reader of the "Mecanique Celeste" will find that, for any thing he can see to the contrary, Euler, Clairaut, D'Alembert, and above all Lagrange, need never have existed. The reader of the "Système du Monde" finds Laplace referring to himself in almost every page, while now and then, perhaps not twenty times in all, his predecessors in theory are mentioned with a scanty reference to what they have done; while the names of observers, between whom and himself there could be no rivalry, occur in many places. To such an absurd pitch is this suppression carried, that even Taylor's name is not mentioned in connexion with his celebrated theorem; but Laplace gravely informs his readers, "Nous donnerons quelques théorêmes généraux qui nous seront utiles dans la suite," those general theorems being known all over Europe by the names of

*De Morgan was the first to have used the phrase "inverse probability" in his *An Essay on Probabilities* (De Morgan, 1838, Chapter III), although Dale (1999, p. 315) has traced the appearance of "inverse" to an earlier review by De Morgan (1837, p. 239) of Laplace's *Théorie Analytique des Probabilités*, and also to an even earlier set of lectures given at the *École Polytechnique* in the late eighteenth or early nineteenth century.

> Maclaurin, Taylor, and Lagrange. And even in his Theory of Probabilities *Lagrange's theorem* is only "la formule (p) du numéro 21 du second livre de la Mécanique Céleste." It is true that at the end of the Mecanique Celéste he gives historical accounts, in a condensed form, of the discoveries of others; but these accounts never in any one instance answer the question—Which pages of the preceding part of the work contain the original matter of Laplace, and in which is he only following the track of his predecessor? (De Morgan, 1839, Vol. XXX, p. 326)

Against such charges, recent writers like Stigler (1978) and Zabell (1988) have come to Laplace's defense on the grounds that the latter's citation rate was no worse than those of his contemporaries. That might be the case, but the two studies also show that the citation rates of Laplace as well as his contemporaries were all very low. This is hardly a practice that can be condoned, especially when we know these mathematicians jealously guarded their own discoveries. Newton and Leibniz clashed fiercely over priority on the Calculus, as did Gauss and Legendre on least squares, though to a lesser extent. If mathematicians were so concerned that their priority over discoveries be acknowledged, then surely it was incumbent upon them to acknowledge the priority of others on work that was not their own.

1.2 LAPLACE'S WORK IN PROBABILITY AND STATISTICS

1.2.1 "Mémoire sur les suites récurro-récurrentes" (1774): Definition of Probability

This memoir (Laplace, 1774b) is among the first of Laplace's published works and also his first paper on probability (Fig. 1.2). Here, for the first time, Laplace enunciated the definition of probability, which he called a *Principe* (Principle):

> *The probability of an event is equal to the product of each favorable case by its probability divided by the product if each possible case by its probability, and if each case is equally likely, the probability of the event is equal to the number of favorable cases divided by the number of all possible cases.** (Laplace, 1774b, OC 8, pp. 10–11)

The above is the classical (or mathematical) definition of probability that is still used today, although several other mathematicians provided similar definitions earlier. For example†:

- Gerolamo Cardano's definition in Chapter 14 of the *Liber de ludo aleae*:

> So there is one general rule, namely, that we should consider the whole circuit, and the number of those casts which represents in how many ways the favorable result can occur, and compare that number to the rest of the circuit, and according to that proportion should the mutual wagers be laid so that one may contend on equal terms. (Cardano, 1663)

* Italics are Laplace's.

† See also Gorroochurn (2012b).

MÉMOIRE

SUR

LES SUITES RÉCURRO-RÉCURRENTES

ET SUR LEURS USAGES

DANS LA THÉORIE DES HASARDS (1).

Mémoires de l'Académie Royale des Sciences de Paris (*Savants étrangers*),
Tome VI, p. 353; 1774.

1. On peut concevoir ainsi la formation des suites récurrentes : si φ exprime une fonction quelconque de x, et que l'on y substitue successivement, au lieu de x, 1, 2, 3, ..., on formera une suite de termes dont je désigne par y_x celui qui répond au nombre x; cela posé, si dans cette suite chaque terme est égal à un nombre quelconque de termes précédents, multipliés chacun par une fonction de x à volonté, la suite est alors récurrente.

Telle est l'idée la plus générale que l'on puisse s'en former, et c'est sous ce point de vue que je les ai considérées dans un Mémoire antérieur présenté à l'Académie (2).

Je suppose maintenant que φ est une fonction de x et de n, et que l'on y substitue successivement au lieu de x et de n les nombres 1, 2, 3, ... : on formera pour chaque valeur de n une suite dont je désigne le terme répondant aux nombres x et n par ${}_n y_x$; or, si ${}_n y_x$ est égal à un nombre quelconque de termes précédents pris dans un nombre quel-

(1) Par M. de la Place, Professeur à l'École Royale militaire

(2) *Voir* le Tome IV des *Mémoires de Turin*.

FIGURE 1.2 First page of Laplace's "Mémoire sur les suites récurro-récurrentes" (Laplace, 1774b)

- Gottfried Wilhelm Leibniz's definition in the *Théodicée*:

> If a situation can lead to different advantageous results ruling out each other, the estimation of the expectation will be the sum of the possible advantages for the set of all these results, divided into the total number of results. (Leibniz, 1710, 1969 edition, p. 161)

- James (Jacob) Bernoulli's statement from the *Ars Conjectandi*:

> ... if complete and absolute certainty, which we represent by the letter a or by 1, is supposed, for the sake of argument, to be composed of five parts or probabilities, of which three argue for the existence or future existence of some outcome and the others argue against it, then that outcome will be said to have $3a/5$ or 3/5 of certainty. (Bernoulli, 1713, English edition, pp. 315–316)

- Abraham de Moivre's definition from the *De Mensura Sortis*:

> If p is the number of chances by which a certain event may happen, & q is the number of chances by which it may fail; the happenings as much as the failings have their degree of probability: But if all the chances by which the event may happen or fail were equally easy; the probability of happening to the probability of failing will be p to q. (de Moivre, 1711, p. 215)

Although Laplace's *Principe* was an objective definition, Laplace gave it a subjective overtone by later redefining mathematical probability as follows:

> The probability of an event is thus just the ratio of the number of cases favorable to it, to the number of possible cases, *when there is nothing to make us believe that one case should occur rather than any other.** (Laplace, 1776b, OC 8, p. 146)

In the above, Laplace appealed to the *principle of indifference*† and his definition of probability relates to our beliefs. It is thus a subjective interpretation of the classical definition of probability.

1.2.2 "Mémoire sur la probabilité des causes par les événements" (1774)

1.2.2.1 Bayes' Theorem The "Mémoire sur la probabilité des causes par les événements" (Laplace, 1774a) (Fig. 1.3) is a landmark paper of Laplace because it introduced most of the fundamental principles that he first used and would stick to for the rest of his career.‡ Bayes' theorem was stated and inverse probability was used as a general method for dealing with all kinds of problems. The asymptotic method was introduced as a powerful tool for approximating certain types of integrals, and an inverse version of the Central Limit Theorem was also presented. Finally the double exponential distribution

* Italics are ours.

† See Section 1.3 for more on the principle of indifference.

‡ Laplace's 1774 memoir is also described in Stigler (1986a), which also contains a full English translation of the paper.

MÉMOIRE

SUR

LA PROBABILITÉ DES CAUSES

PAR LES ÉVÉNEMENTS (¹).

Mémoires de l'Académie royale des Sciences de Paris (Savants étrangers),
Tome VI, p. 621; 1774.

I.

La théorie des hasards est une des parties les plus curieuses et les plus délicates de l'Analyse, par la finesse des combinaisons qu'elle exige et par la difficulté de les soumettre au calcul; celui qui paraît l'avoir traitée avec le plus de succès est M. Moivre, dans un excellent Ouvrage qui a pour titre : *Theory of chances*; nous devons à cet habile géomètre les premières recherches que l'on ait faites sur l'intégration des équations différentielles aux différences finies; la méthode qu'il a imaginée pour cet objet est fort ingénieuse et il l'a très heureusement appliquée à la solution de plusieurs problèmes sur les Probabilités; on doit convenir cependant que le point de vue sous lequel il a envisagé cette matière est indirect. Les équations aux différences finies sont susceptibles des mêmes considérations que celles aux différences infiniment petites, et doivent être traitées d'une manière analogue; la seule différence qui s'y rencontre est que, dans le cas des différences infiniment petites, on peut négliger certaines quantités qu'il n'est pas

(¹) Par M. de la Place, Professeur à l'École royale militaire.

FIGURE 1.3 First page of Laplace's "Mémoire sur la probabilité des causes par les événements" (Laplace, 1774a)

was introduced as a general law of error. Laplace here presented many of the problems that he would later come back to again, each time refining and perfecting his previous solutions.

In Article II of the memoir, Laplace distinguished between two classes of probability problems:

> The uncertainty of human knowledge bears on events or the causes of events; if one is certain, for example, that a ballot contains only white and black tickets in a given ratio, and one asks the probability that a randomly chosen ticket will be white, the event is then uncertain, but the cause on which depends the existence of the probability, that is the ratio of white to black tickets, is known.
>
> In the following problem: *A ballot is assumed to contain a given number of white and black tickets in an unknown ratio, if one draws a white ticket, determine the probability that the ratio of white to black tickets in the ballot is p:q*; the event is known and the cause unknown.
>
> One can reduce to these two classes of problems all those that depend on the doctrine of chances. (Laplace, 1774a, OC 8, p. 29)

In the above, Laplace distinguished between problems that require the calculation of direct probabilities and those that require the calculation of inverse probabilities. The latter depended on the powerful theorem first adduced by Bayes and which Laplace immediately enunciated as a *Principe* as follows:

> PRINCIPE—*If an event can be produced by a number n of different causes, the probabilities of the existence of these causes calculated from the event are to each other as the probabilities of the event calculated from the causes, and the probability of the existence of each cause is equal to the probability of the event calculated from that cause, divided by the sum of all the probabilities of the event calculated from each of the causes.** (*ibidem*)

Laplace's first statement in the above can be written mathematically as follows: if $C_1, C_2, \ldots, C_n$ are n exhaustive events ("causes") and E is another event, then

$$\frac{\Pr\{C_i \mid E\}}{\Pr\{C_j \mid E\}} = \frac{\Pr\{E \mid C_i\}}{\Pr\{E \mid C_j\}} \quad \text{for } i, j = 1, 2, \ldots, n; i \neq j \tag{1.1}$$

Equation (1.1) implies that

$$\Pr\{C_j \mid E\} = \frac{\Pr\{C_i \mid E\}\Pr\{E \mid C_j\}}{\Pr\{E \mid C_i\}}$$

$$\sum_{j=1}^{n}\Pr\{C_j \mid E\} = \frac{\Pr\{C_i \mid E\}\sum_{j=1}^{n}\Pr\{E \mid C_j\}}{\Pr\{E \mid C_i\}}$$

$$1 = \frac{\Pr\{C_i \mid E\}\sum_{j=1}^{n}\Pr\{E \mid C_j\}}{\Pr\{E \mid C_i\}}$$

* Italics are Laplace's.

$$\therefore \quad \Pr\{C_i \mid E\} = \frac{\Pr\{E \mid C_i\}}{\sum_{j=1}^{n} \Pr\{E \mid C_j\}}. \tag{1.2}$$

Equation (1.2) is Laplace's second statement in the previous quotation. It is a *restricted* version of Bayes' theorem because it assumes a discrete uniform prior, that is, each of the "causes" $C_1, C_2, \ldots, C_n$ is equally likely: $\Pr\{C_i\} = 1/n$ for $i = 1, 2, \ldots, n$.

It should be noted that Laplace's enunciation of the theorem in Eq. (1.2) in 1774 made no mention of Bayes' publication 10 years earlier (Bayes, 1764), and it is very likely that Laplace was unaware of the latter's work. However, the 1778 volume of the *Histoire de l'Académie Royale des Sciences*, which appeared in 1781, contained a summary by the Marquis de Condorcet (1743–1794) of Laplace's "Mémoire sur les Probabilités," which also appeared in that volume (Laplace, 1781). Laplace's article made no mention of Bayes or Price,* but Condorcet's summary explicitly acknowledged the two Englishmen:

> These questions [on inverse probability] about which it seems that Messrs. Bernoulli and Moivre had thought, have been since then examined by Messrs. Bayes and Price; but they have limited themselves to exposing the principles that can be used to solve them. M. de Laplace has expanded on them.... (Condorcet, 1781, p. 43)

As for Laplace himself, his acknowledgment of Bayes' priority on the theorem came much later in the *Essai Philosophique Sur les Probabilités*:

> Bayes, in the *Transactions philosophiques* of the year 1763, sought directly the probability that the possibilities indicated by past experiences are comprised within given limits; and he has arrived at this in a refined and very ingenious manner, although a little perplexing. This subject is connected with the theory of the probability of causes and future events, concluded from events observed. Some years later I expounded the principles of this theory.... (Laplace, 1814, English edition, p. 189)

It also in the *Essai Philosophique* that Laplace first gave the general (discrete) version of Bayes' theorem:

> The probability of the existence of anyone of these causes is then a fraction whose numerator is the probability of the event resulting from this cause and whose denominator is the sum of the similar probabilities relative to all the causes; if these various causes, considered *a priori*, are unequally probable it is necessary, in place of the probability of the event resulting from each cause, to employ the product of this probability by the possibility of the cause itself. (*ibid.*, pp. 15–16)

*Richard Price (1723–1791) was Bayes' friend and published Bayes' paper after the latter's death. The paper was augmented by Price with an introduction and an appendix.

Equation (1.2) can thus be written in general form as

$$\Pr\{C_i \mid E\} = \frac{\Pr\{E \mid C_i\}\Pr\{C_i\}}{\sum_{j=1}^{n} \Pr\{E \mid C_j\}\Pr\{C_j\}} \tag{1.3}$$

which is the form in which Bayes' theorem is used today. The continuous version of Eq. (1.3) may be written as

$$f(\theta \mid \mathbf{x}) = \frac{f(\mathbf{x} \mid \theta) f(\theta)}{\int_{-\infty}^{\infty} f(\mathbf{x} \mid \theta) f(\theta) d\theta} \propto f(\mathbf{x} \mid \theta) f(\theta), \tag{1.4}$$

where θ is a parameter, $f(\theta)$ is the prior density of θ, $f(\mathbf{x} \mid \theta)$ is joint density* of the observations $\mathbf{x}$, and $f(\theta \mid \mathbf{x})$ is the posterior density of θ. It is interesting that neither of the above two forms (and not even those assuming a uniform prior) by which we recognize Bayes' theorem today can be found explicitly in Bayes' paper.

Laplace almost always used (1.4) in the form

$$f(\theta \mid \mathbf{x}) = \frac{f(\mathbf{x} \mid \theta)}{\int_{-\infty}^{\infty} f(\mathbf{x} \mid \theta) d\theta} \propto f(\mathbf{x} \mid \theta).$$

Evidently, Laplace assumed a uniform prior in Bayes' theorem, though he seldom, if ever, bothered to explicitly state this important assumption. The Laplacian assumption of a uniform prior stems from Laplace's somewhat cavalier adoption of the principle of indifference and will be discussed in more detail in Section 1.3. However, for the time being, we should note that statistical inference based on Bayes' theorem used with a uniform prior came to be known soon after as the *method of inverse probability* and that inferences based on the general version of Bayes' theorem came to be known as *Bayesian methods* much later in the 1950s (Fienberg, 2006).

1.2.2.2 Rule of Succession In Article III of the "Mémoire sur la probabilité des causes par les événements," Laplace set out to use inverse probability to solve the following problem:

> PROBLEM 1—*If a ballot contains an infinite number of white and black tickets in an unknown ratio, and if we draw p+q tickets of which p are white and q are black; we ask the probability that a next ticket drawn from this ballot will be white.*† (Laplace, 1774a, OC 8, p. 30)

To solve the above problem, Laplace used an argument which boils down to saying that the posterior distribution of the (unknown) proportion x of white tickets is proportional to the likelihood of the observed balls. In accordance with our previous remark, this statement is not true because one needs to multiply the likelihood by the prior density of x, but

* The joint density $f(\mathbf{x} \mid \theta)$ is also the likelihood of θ, that is, $f(\mathbf{x} \mid \theta) \equiv L(\theta \mid \mathbf{x})$.

† Italics are Laplace's.

Laplace had tacitly assumed that the prior is uniform on [0, 1]. Therefore, under the latter assumption, the posterior distribution of x is

$$\frac{x^p(1-x)^q}{\int_0^1 x^p(1-x)^q\,dx}. \tag{1.5}$$

The probability $\tilde{E}$ that the next ticket is white is then the expectation of the above distribution, so Laplace wrote

$$\tilde{E} = \frac{\int_0^1 x^{p+1}(1-x)^q\,dx}{\int_0^1 x^p(1-x)^q\,dx}.$$

Since

$$\int_0^1 x^p(1-x)^q\,dx = \frac{1\cdot 2\cdot 3\cdots q}{(p+1)\cdots(p+q+1)},$$

Laplace was finally able to obtain

$$\tilde{E} = \frac{p+1}{p+q+2} \tag{1.6}$$

(*ibid.*, p. 31). The above formula is known as *Laplace's rule of succession*. It is a perfectly valid formula as long as the assumptions it makes are all tenable. However, it has been much discussed in the literature because of its dependence on the principle of indifference and of the controversial way in which it has often been applied.*

1.2.2.3 Proof of Inverse Bernoulli Law. Method of Asymptotic Approximation. Central Limit Theorem for Posterior Distribution. Indirect Evaluation of $\int_0^\infty e^{-t^2}dt$

In Article III of the "Mémoire sur la probabilité des causes par les événements," we also see an interesting proof of the inverse of Bernoulli's law,† which Laplace enunciated as follows:

> *One can assume that the numbers p and q are so large, that it becomes as close to certainty as we wish that the ratio of the number of white tickets to the total number of tickets in the urn lies between the two limits $\frac{p}{p+q}-\omega$ and $\frac{p}{p+q}+\omega$, ω being arbitrarily small.*‡ (*ibid.*, p. 33)

In modern notation the above can be written as

$$\Pr\left\{\frac{p}{p+q}-\omega < x < \frac{p}{p+q}+\omega\right\} \to 1 \quad \text{as} \quad p,q\to\infty,$$

* See Section 1.3 for more on the principle of indifference and the law of succession.

† The inverse of Bernoulli's law refers to Bernoulli's law with an *unknown* probability of success and should not be confused with inverse probability (see Gorroochurn, 2012a, p. 140).

‡ Italics are Laplace's.

where p is the number of white tickets, q is the number of black tickets, and x is the (unknown) proportion of white tickets.

Laplace's proof was based on inverse probability, and used what has become known as *Laplace's method of asymptotic approximation.** The basic idea behind this method here is to approximate an integral whose integrand involves high powers by the integral of an appropriate Gaussian distribution, the approximation being more accurate for points close to the maximum of the integrand. We now describe Laplace's proof. Using the posterior distribution of x in Eq. (1.5), we have

$$\begin{aligned}\Pr\left\{\frac{p}{p+q}-\omega<x<\frac{p}{p+q}+\omega\right\}&=\frac{(p+1)\cdots(p+q+1)}{1\cdot 2\cdot 3\cdots q}\int_{-\omega+p/(p+q)}^{\omega+p/(p+q)} x^p(1-x)^q\,dx\\&=\frac{(p+1)\cdots(p+q+1)}{1\cdot 2\cdot 3\cdots q}\left\{\int_{-\omega+p/(p+q)}^{p/(p+q)} x^p(1-x)^q\,dx+\int_{p/(p+q)}^{\omega+p/(p+q)} x^p(1-x)^q\,dx\right\}\\&=E.\end{aligned} \tag{1.7}$$

Laplace wished to expand E around the point at which the maximum of $x^p(1-x)^q$ occurs, that is, around $x=p/(p+q)$. By making the substitution $z=x-p/(p+q)$, the integral E above can be written as

$$\begin{aligned}E&=\frac{(p+1)\cdots(p+q+1)}{1\cdot 2\cdot 3\cdots q}\cdot\frac{p^p q^q}{(p+q)^{p+q}}\\&\times\left\{\int_0^\omega\left(1-\frac{p+q}{p}z\right)^p\left(1+\frac{p+q}{q}z\right)^q dz+\int_0^\omega\left(1+\frac{p+q}{p}z\right)^p\left(1-\frac{p+q}{q}z\right)^q dz\right\}.\end{aligned} \tag{1.8}$$

Laplace set

$$\omega=\frac{1}{(p+q)^{1/n}}\quad\text{for}\quad 2<n<3.$$

Laplace's next task was to approximate each of the expressions inside the integrals in Eq. (1.8) above. For example, consider $\{1-(p+q)z/p\}^p$ and set†

$$\begin{aligned}u&=\log\left(1-\frac{p+q}{p}z\right)^p\\&=p\log\left(1-\frac{p+q}{p}z\right)\\&=-p\left\{\frac{p+q}{p}z+\frac{(p+q)^2z^2}{2p^2}+\frac{(p+q)^3z^3}{3p^3}+\cdots\right\}\\&=-(p+q)z-\frac{(p+q)^2z^2}{2p}-\frac{(p+q)^3z^3}{3p^2}-\cdots.\end{aligned} \tag{1.9}$$

*A modern treatment of Laplace's method of asymptotic approximation can be found in the book by Small (2010, Chapter 6).

†All logarithms (log) used in this book are to base e, unless otherwise stated.

From the limits of the integral in (1.8), it is seen that $0 \le z \le \omega$ where $\omega = (p+q)^{-1/n}$ for $2 < n < 3$. For the last shown term on the right side of (1.9), we therefore have, for $2 < n < 3$,

$$0 \ge -\frac{(p+q)^3 z^3}{3p^2} \ge -\frac{(p+q)^{3-3/n}}{3p^2} \to 0 \quad \text{as} \quad p, q \to \infty.$$

Thus, $(p+q)^3 z^3 / (3p^2) \to 0$ and similarly for the later terms in (1.9). Hence, for infinitely large p and q, Laplace was able to write Eq. (1.9) as

$$u = \log\left(1 - \frac{p+q}{p} z\right)^p = -(p+q)z - \frac{(p+q)^2 z^2}{2p} - \cdots$$

$$\therefore \left(1 - \frac{p+q}{p} z\right)^p = \exp\left\{-(p+q)z - \frac{(p+q)^2 z^2}{2p} - \cdots\right\}$$

(*ibid.*, p. 34). By using a similar reasoning for the other terms in (1.8), namely, for $\{1+(p+q)z/p\}^p$, $\{1+(p+q)z/q\}^p$, and $\{1-(p+q)z/q\}^p$, Laplace obtained

$$E = \frac{(p+1)\cdots(p+q+1)}{1\cdot 2\cdot 3\cdots q} \cdot \frac{p^p q^q}{(p+q)^{p+q}} \int_0^\omega 2\exp\left\{-\frac{(p+q)^3 z^2}{2pq}\right\} dz.$$

Using Stirling's formula $\nu! \sim \sqrt{2\pi\nu}(\nu/e)^\nu$ (where ν is large),

$$E = \frac{(p+q)^{3/2}}{\sqrt{2\pi}\sqrt{pq}} \int_0^\omega 2\exp\left\{-\frac{(p+q)^3 z^2}{2pq}\right\} dz. \tag{1.10}$$

Laplace next made the substitution $\log \mu = -(p+q)^3 z^2 / (2pq)$. When $z = 0$ then $\mu = 1$; when $z = \omega = (p+q)^{-1/n}$ for $2 < n < 3$ and p, q are infinitely large, then $\mu = 0$. Thus, Laplace obtained

$$\int_0^\omega 2\exp\left\{-\frac{(p+q)^3 z^2}{2pq}\right\} dz = -\frac{\sqrt{2pq}}{(p+q)^{3/2}} \int_1^0 \frac{d\mu}{\sqrt{-\log \mu}} d\mu.$$

By appealing to Euler's *Calcul Intégral*, he demonstrated that $\int_0^1 d\mu / \sqrt{-\log \mu} = \sqrt{\pi}$ so that the above becomes

$$\int_0^\omega 2\exp\left\{-\frac{(p+q)^3 z^2}{2pq}\right\} dz = \frac{\sqrt{2\pi}\sqrt{pq}}{(p+q)^{3/2}}.$$

Hence, from Eq. (1.10) above, Laplace finally obtained

$$E = \Pr\left\{\frac{p}{p+q} - \omega < x < \frac{p}{p+q} + \omega\right\} = \frac{(p+q)^{3/2}}{\sqrt{2\pi}\sqrt{pq}} \cdot \frac{\sqrt{2\pi pq}}{(p+q)^{3/2}} = 1$$

(*ibid.*, p. 36) for infinitely large p and q. This completes Laplace's proof, obtained, as Todhunter has rightly noted, by "a rude process of approximation" (Todhunter, 1865, p. 468).

Two further important results can also be found in Laplace's proof. First, from Eq. (1.10), it is seen that

$$E=\Pr\left\{\frac{p}{p+q}-\omega<x<\frac{p}{p+q}+\omega\right\}\approx\frac{2(p+q)^{3/2}}{\sqrt{2\pi}\sqrt{pq}}\int_0^{\omega}\exp\left\{-\frac{(p+q)^3z^2}{2pq}\right\}dz$$

for large values of p and q. This can be rewritten as

$$\Pr\left\{\left|x-\frac{p}{p+q}\right|<\omega\right\}=\frac{2}{\sigma\sqrt{2\pi}}\int_0^{\omega}\exp\left(-\frac{z^2}{2\sigma^2}\right)dz \tag{1.11}$$

where $\sigma=\sqrt{pq}/(p+q)^{3/2}$ and $z=x-p/(p+q)$. The above result is the Central Limit Theorem for the posterior distribution of x, and later came to be known as the *Bernstein–von Mises theorem* due to the works of Bernstein (1917) and von Mises (1931).

Second, from his proof Laplace had effectively, albeit indirectly, obtained the important identity:

$$\int_0^{\infty}e^{-t^2}dt=\frac{\sqrt{\pi}}{2}. \tag{1.12}$$

To see this, recall that we mentioned earlier that Laplace had obtained the result

$$\int_0^1\frac{d\mu}{\sqrt{-\log\mu}}=\sqrt{\pi} \tag{1.13}$$

by using a theorem from Euler's *Calcul Intégral*. This theorem states that

$$\int_0^1\frac{\mu^n d\mu}{\sqrt{(1-\mu^{2k})}}\int_0^1\frac{\mu^{n+k}d\mu}{\sqrt{(1-\mu^{2k})}}=\frac{1}{k(n+1)}\cdot\frac{\pi}{2}\quad\text{for all}\quad n,k. \tag{1.14}$$

In the above, Laplace then took $n=0$ and $k\to 0$. Since

$$\lim_{k\to 0}\frac{1-\mu^{2k}}{2k}=-\log\mu,$$

Equation (1.14) becomes

$$\int_0^1\frac{d\mu}{\sqrt{2k}\sqrt{-\log\mu}}\int_0^1\frac{d\mu}{\sqrt{2k}\sqrt{-\log\mu}}=\frac{\pi}{2k}$$

whence Eq. (1.13) follows by taking $k=1$. Now Eq. (1.13) is identical to Eq. (1.12) as can be seen by making the substitution $\mu=e^{-t^2}$ in Eq. (1.13):

$$\int_0^1\frac{d\mu}{\sqrt{-\log\mu}}=\int_{\infty}^0\frac{-2te^{-t^2}}{t}dt=\sqrt{\pi}.$$

$$\therefore\quad\int_0^{\infty}e^{-t^2}dt=\frac{\sqrt{\pi}}{2}.$$

1.2.2.4 Problem of Points As a further application of the principle of inverse probability, in Article IV Laplace next solved a modified version of the Problem of Points* (Laplace, 1774a, OC 8, pp. 39–41):

> PROBLEM II—*Two players A and B, whose skills are respectively unknown, play a certain game, for example, piquet, with the condition that the one who wins a number n of rounds, will obtain an amount a deposed at the start of the game; I assume that the two players are forced to quit the game, when A is short of f rounds and B is short of h rounds; this given, we ask for how the amount should be divided between the two players.*† (*ibid.*, p. 39)

In the usual version of the problem, the skills (i.e., probabilities of winning one round) of the two players are assumed to be known. However, in Laplace's modified version, they are *unknown*.

Laplace's solution was as follows. If A's and B's skills are assumed to be known to be p and q $(=1-p)$, respectively, then B should receive an expected amount given by

$$aq^{f+h-1}\left\{1+\frac{p}{q}(f+h-1)+\frac{p^2}{q^2}\frac{(f+h-1)(f+h-2)}{1\cdot 2}+\cdots+\frac{p^{f-1}}{q^{f-1}}\frac{(f+h-1)\cdots(h+1)}{1\cdot 2\cdot 3\cdots(f-1)}\right\} \quad (1.15)$$

(*ibid.*, p. 40). The above is equivalent to saying that B's *probability* of winning the game is

$$\sum_{j=0}^{f-1}\binom{f+h-1}{j}p^j q^{f+h-1-j}. \quad (1.16)$$

Although Laplace did not provide the details of the derivation, the above formula can be obtained as follows. If the game had continued, the maximum number of possible more rounds would have been $f+h-1$. Player B will win the game if and only if A can win only 0, 1, …, $f-1$ rounds (since A is f rounds short) out of $f+h-1$. Now A can win j rounds out of $f+h-1$ rounds in $\binom{f+h-1}{j}$ ways. By using the binomial distribution, the required probability of B winning the game can be obtained.‡

However, A's skill is unknown. Laplace denoted this probability by x and then assigned it a Unif (0, 1) distribution ("*nous pouvons la supposer un des nombres quelconques, compris depuis 0 jusqu'à 1*"). Then the probability that A wins $n-f$ rounds and B wins $n-h$ rounds is proportional to $x^{n-f}(1-x)^{n-h}$. By using Eq. (1.5), the posterior distribution of x is

$$\frac{x^{n-f}(1-x)^{n-h}}{\int_0^1 x^{n-f}(1-x)^{n-h}\,dx}. \quad (1.17)$$

*The Problem of Points is also discussed in Edwards (1982), Hald (1990), Todhunter (1865, Chap. II), Smorynski (2012, pp. 4–25), and Gorroochurn (2012a, Prob. 4 and end of Prob. 9;2014)

† Italics are Laplace's.

‡ Equation (1.15) first appeared in the second edition of Pierre Rémond de Montmort's (1678–1719) *Essay d'Analyse sur les Jeux de Hazard* (Montmort, 1713, pp. 244–245) as the first formula for the Problem of Points. The solution had been communicated to Montmort by John Bernoulli (1667–1748) in a letter that is reproduced in the *Essay* (*ibid.*, p. 295). Laplace referred to Montmort in his memoir and also asserted that the case of three players had not been solved yet. However, this is not correct as de Moivre had in fact solved the general case of n players in the *Miscellanea Analytica* (de Moivre, 1730, p. 210, Prob. VI).

Moreover, for a given x the expected amount B should receive is, from Eq. (1.15),

$$a(1-x)^{f+h-1}\left\{1+\frac{x}{1-x}(f+h-1)+\frac{x^2}{(1-x)^2}\frac{(f+h-1)(f+h-2)}{1\cdot 2}+\cdots\right.$$
$$\left.+\frac{x^{f-1}}{(1-x)^{f-1}}\frac{(f+h-1)\cdots(h+1)}{1\cdot 2\cdot 3\cdots(f-1)}\right\}. \tag{1.18}$$

Therefore, B's posterior expected amount is obtained by multiplying Eqs.(1.17) and (1.18) and integrating from 0 to 1:

$$\frac{a\int_0^1(1-x)^{f+h-1}\left\{1+\frac{x}{1-x}(f+h-1)+\frac{x^2}{(1-x)^2}\frac{(f+h-1)(f+h-2)}{1\cdot 2}+\cdots+\frac{x^{f-1}}{(1-x)^{f-1}}\frac{(f+h-1)\cdots(h+1)}{1\cdot 2\cdot 3\cdots(f-1)}\right\}dx}{\int_0^1 x^{n-f}(1-x)^{n-h}dx}.$$

By using

$$\int_0^1 x^p(1-x)^q\,dx=\frac{1\cdot 2\cdot 3\cdots q}{(p+1)\cdots(p+q+1)},$$

B's posterior expected amount becomes

$$\frac{a(n-h+1)(n-h+2)\cdots(n+f-1)}{(2n-f-h)(2n-f-h+1)\cdots(2n)}$$
$$\times\left\{1+\frac{f+h-1}{1}\cdot\frac{n-f+1}{f+n-1}+\frac{(f+h-1)(f+h-2)}{1\cdot 2}\cdot\frac{(n-f+1)(n-f+2)}{(f+n-1)(f+n-2)}+\cdots\right.$$
$$\left.+\frac{(f+h-1)(f+h-2)\cdots(h+1)}{1\cdot 2\cdot 3\cdots(f-1)}\cdot\frac{(n-f+1)(n-f+2)\cdots(n-1)}{(f+n-1)(f+n-2)\cdots(n+1)}\right\}$$

(*ibid*., p. 41). Note that the posterior *probability* of B winning the game is the expression above divided by a. Note also that, using the compact form in (1.16) rather than (1.15) for the probability of B winning (when the latter's skill is known), the posterior probability of B can be written as

$$\frac{1}{\binom{2n}{f+h-1}}\sum_{j=0}^{f-1}\binom{n+j}{n-h}\binom{n-j-1}{n-f}.$$

1.2.2.5 First Law of Error The "Mémoire sur la probabilité des causes par les événements" also contains a justification for Laplace's choice of the double exponential distribution (also known as the first law of error of Laplace) as a general law of error. This distribution arose in the context of finding the type of center that should be chosen from several given observations of the same phenomenon. We here describe both the origin of the double exponential distribution and Laplace's subsequent investigation, based on inverse probability, of an appropriate center. In Article V, Laplace first explained:

> We can, by means of the preceding theory [i.e. by using inverse probability], solve the problem of determining the center that one should take among many given observations of the same phenomenon. Two years ago I presented such a solution to the Academy, as a sequel to the Memoir "Sur les séries récurro-récurrentes" printed in this volume, but it appeared to me to be of such little usefulness that I suppressed it before it was printed. I have since learned from Jean Bernoulli's astronomical journal that Daniel Bernoulli and Lagrange have considered the same problem in two manuscript memoirs that I have not seen. This announcement both added to the usefulness of the material and reminded me of my ideas on this topic. I have no doubt that these two illustrious geometers have treated the subject more successfully than I; however, I shall present my reflections here, persuaded as I am that through the consideration of different approaches, we may produce a less hypothetical and more certain method for determining the mean that one should take among many observations. (*ibid.*, pp. 41–42)

Laplace's investigation was limited to the case of three observations as follows. Let the time axis be represented by the line AB in Figure 1.4. Suppose the instant a given phenomenon occurs three observations are made on it and give the time as a, b, and c. Let $p=b-a$ and $q=c-b$, and suppose the true instant of the phenomenon is V. The question is, how should V be estimated? Laplace's answer was to first represent the probability distribution of deviations (or errors) x from V by the (density) function $\phi(x)$.

Laplace stated (*ibid.*, p. 43) that the density $\phi(x)$ must satisfy the following three conditions:

1. C1: It must admit a vertical axis of symmetry "because it is as probable that the observation deviates from the truth to the right as to the left."
2. C2: It must have the horizontal axis as asymptote "because the probability that the observation differs from the truth by an infinite amount is evidently zero."
3. C3: The area under the curve must be unity "because it is certain that the observation will fall on one of the points" on the horizontal axis.

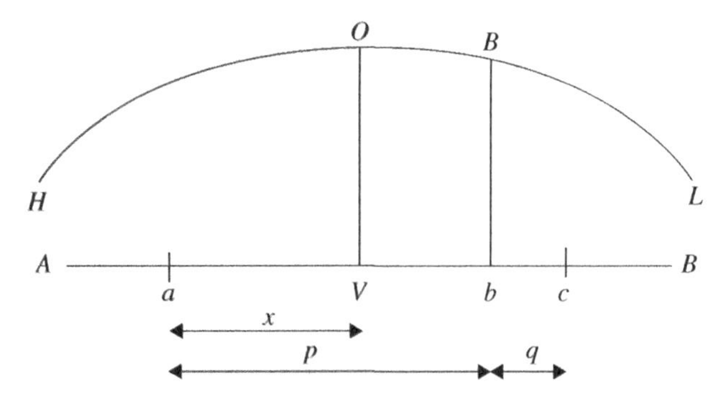

FIGURE 1.4 Laplace's determination of an appropriate center for three observations a, b, and c

As for the center, there are two possibilities, namely:

1. *The center of probability* ("*milieu de probabilité*"), which is such that the true value is equally likely to be above or below the center.
2. *The center of error* or *astronomical mean* ("*milieu d'erreur* ou *milieu astronomique*"), which is such that the sum of absolute errors multiplied by their probabilities is a minimum.

In an analysis based on inverse probability, the first center is also the posterior median, and the second center is one that minimizes the posterior expected error. However, through a straightforward proof, Laplace showed that both centers were the same, that is, the posterior median also minimized the posterior expected error.

Coming back to the choice for $\phi(x)$, the three conditions C1–C3 were still not sufficient to identify $\phi(x)$ uniquely. Laplace stated:

> But of infinite number of possible functions, which choice is to be preferred? The following considerations can determine a choice...Now, as we have no reason to suppose a different law of the ordinates than for their differences, it follows that we must, subject to the rules of probabilities, suppose that the ratio of two infinitely small consecutive differences to be equal to that of the corresponding ordinates. We thus will have
>
> $$\frac{d\phi(x+dx)}{d\phi(x)} = \frac{\phi(x+dx)}{\phi(x)}.$$
>
> Therefore
>
> $$\frac{d\phi(x)}{dx} = -m\phi(x)$$
>
> which gives $\phi(x) = \zeta e^{-mx}$ ... the constant ζ needs to be determined from the assumption that the total area of the curve...is unity...which gives
>
> $$\zeta = \frac{1}{2}m, \quad \text{so that} \quad \phi(x) = \frac{m}{2}e^{-mx}$$
>
> (*ibid.*, pp. 45–46)

In the above, Laplace had appealed to the principle of indifference and stated that, for two different deviation x_1 and x_2,

$$\frac{d\phi(x_2)}{d\phi(x_1)} = \frac{\phi(x_2)}{\phi(x_1)}.$$

From the above we have

$$\frac{\phi'(x_1)}{\phi(x_1)} = \frac{\phi'(x_2)}{\phi(x_2)} \quad \Rightarrow \quad \frac{\phi'(x)}{\phi(x)} = k_1,$$

where k_1 is a constant. Upon integration we obtain the general solution $\phi(x)=k_2e^{k_1x}$, where k_2 is another constant. By applying conditions C1–C3 to this solution, Laplace thus obtained $\phi(x)=(me^{-mx})/2$. Since $\phi(x)$ should be symmetric about the vertical axis, it is more appropriate to represent Laplace's density as

$$\phi(x)=\frac{1}{2}me^{-m|x|} \quad \text{for} \quad m>0, \tag{1.19}$$

which is the double exponential distribution.

Laplace's aim was now to calculate the posterior median of the deviations subject to the error distribution in (1.19). His analysis was as follows. If the distance *Va* in Figure 1.4. is denoted by x, then $Vb=p-x$ and $Vc=p+q-x$. Then the equation of the curve *HOL* is $y=\phi(x)\phi(p-x)\phi(p+q-x)$. In Laplace's analysis, y should be viewed as being proportional to the posterior distribution of x based on a uniform prior. From a to b, Laplace obtained

$$y=\frac{1}{2}me^{-mx}\cdot\frac{1}{2}me^{-m(p-x)}\cdot\frac{1}{2}me^{-m(p+q-x)}=\frac{m^3}{8}e^{-m(2p+q-x)}. \tag{1.20}$$

From b to c, the curve *HOL* becomes

$$y=\frac{1}{2}me^{-mp}\cdot\frac{1}{2}me^{-m(x-p)}\cdot\frac{1}{2}me^{-mq}=\frac{m^3}{8}e^{-m(x+q)}.$$

Similarly, Laplace found the equations for other sections of the curve *HOL*, and hence the areas under these sections. By adding these areas and integrating with respect to x, he obtained the total area under *HOL* as

$$\frac{m^2}{4}e^{-m(p+q)}\left(1-\frac{1}{3}e^{-mp}-\frac{1}{3}e^{-mq}\right)^{*} \tag{1.21}$$

(*ibid*., p. 47). Laplace now determined the area to the left of *bB* as

$$\frac{m^2}{8}e^{-m(p+q)}\left(1-\frac{2}{3}e^{-mp}\right)$$

and this area is greater or less than half the total area in (1.21) depending on whether $p>q$ or $p<q$. Laplace assumed that the former was the case so that the posterior median lies in the interval (a, b). By equating the area to the left of the posterior median x, where $a<x<b$, with half the total area, he eventually obtained

$$\begin{aligned} m^2e^{-m(2p+q-x)} &= m^2e^{-m(p+q)}\left(1+\frac{1}{3}e^{-mp}-\frac{1}{3}e^{-mq}\right) \\ \Rightarrow x &= p+\frac{1}{m}\log\left(1+\frac{1}{3}e^{-mp}-\frac{1}{3}e^{-mq}\right) \end{aligned} \tag{1.22}$$

*The divisor of m^2 is wrongly given as 8 in Laplace's paper, but from what follows in the paper it is apparent that it should be 4.

(*ibidem*). Now if the arithmetic mean of the three observations is used, then (Fig. 1.4)

$$\frac{(x)+(p-x)+(p+q-x)}{3}=0 \quad \Rightarrow \quad x=\frac{2p+q}{3}$$

is the correction that should be used. However, when $m \to 0$, Eq. (1.22) implies

$$x \to \frac{2p+q}{3};$$

that is, as $m \to 0$, the median and mean give the same correction. But Laplace dismissed the $m \to 0$ case because:

> ...the assumption that m is infinitesimally small implies that all points...[are] equally probable; which is out of any likelihood by the very nature of things.... (*ibidem*)

Continuing with the analysis, Laplace observed that the true value of m is unknown, and "recourse to other means of obtaining this value [i.e. the value of the posterior median] is required." Laplace used inverse probability again. He reasoned that since Eq. (1.21) was obtained by integrating $y=\phi(x)\phi(p-x)\phi(p+q-x)$ with respect to x for $\phi(x)=(me^{-mx})/2$, it was also proportional to the posterior density of m given p and q, that is, the latter posterior density was proportional to

$$m^2 e^{-m(p+q)}\left(1-\frac{1}{3}e^{-mp}-\frac{1}{3}e^{-mq}\right).$$

Then using the density y given in (1.20) at the point x for a given m, he wrote the "total" density at x for *any* m as

$$\int_0^\infty y m^2 e^{-m(p+q)}\left(1-\frac{1}{3}e^{-mp}-\frac{1}{3}e^{-mq}\right)dm.$$

(*ibid.*, p. 49). The aim is now to find the median of the above function. To do this one needs to split the area under it into two halves. In analogy with the left equation in (1.22), one obtains for the posterior median x,

$$\begin{aligned}
\int_0^\infty m^2 e^{-m(2p+q-x)} \times m^2 e^{-m(p+q)}\left(1-\frac{1}{3}e^{-mp}-\frac{1}{3}e^{-mq}\right)dm &= \int_0^m m^2 e^{-m(p+q)}\left(1+\frac{1}{3}e^{-mp}-\frac{1}{3}e^{-mq}\right) \\
&\quad \times m^2 e^{-m(p+q)}\left(1-\frac{1}{3}e^{-mp}-\frac{1}{3}e^{-mq}\right)dm \\
\int_0^\infty m^4 e^{-m(3p+2q-x)}\left(1-\frac{1}{3}e^{-mp}-\frac{1}{3}e^{-mq}\right)dm &= \int_0^m m^4 e^{-m(2p+2q)}\left(1+\frac{1}{3}e^{-mp}-\frac{1}{3}e^{-mq}\right) \\
&\quad \times\left(1-\frac{1}{3}e^{-mp}-\frac{1}{3}e^{-mq}\right)dm
\end{aligned} \tag{1.23}$$

(*ibidem*).

However, as Stigler has pointed out (Stigler, 1986b, pp. 113–115), Laplace's equation in (1.23) is incorrect. This is because the posterior distribution of x is proportional to y in Eq. (1.20), and the constant of proportionality involves m. This fact needs to be taken into account in going from the first equation in (1.22) to (1.23).

In any case, by performing the integration in (1.23), Laplace obtained the following 15th degree equation for x:

$$\frac{1}{(3p+2q-x)^5}-\frac{1}{3(4p+2q-x)^5}-\frac{1}{3(3p+3q-x)^5}=\frac{1}{(2p+2q)^5}-\frac{1}{3(2p+3q)^5}-\frac{1}{9(4p+2q)^5}+\frac{1}{9(2p+4q)^5} \tag{1.24}$$

(Laplace, 1774a, OC 8, p. 50). For the above equation, Laplace was able to show that there was one value of x for which $0<x<p$ and evaluated the root for $q/p=0,.1,.2,\ldots,1.0$.

Laplace must have been discouraged by the complicated equation in (1.24). Perhaps this could explain why he did not push his investigations on the subject any further and did not try other forms of $\phi(x)$ such as the normal distribution.

Following Stigler (1986b, p. 116), we now show how Laplace's error should be corrected in going from the first equation in (1.22) to (1.23). From Eq. (1.20), by taking the proportionality constant depending on m into account, the posterior distribution of x is

$$f(x \mid p,q,m) \propto \frac{f(p,q \mid x,m)}{f(p,q \mid m)} \propto \frac{f(p,q \mid x,m)}{f(m \mid p,q)}$$

where the f's denote posterior densities when conditioned on p and q, and direct densities otherwise. Then Eq. (1.23) becomes

$$\int_0^\infty m^2 e^{-m(2p+q-x)}dm = \int_0^m m^2 e^{-m(p+q)}\left(1+\frac{1}{3}e^{-mp}-\frac{1}{3}e^{-mq}\right)dm.$$

This is the equation Laplace should have obtained instead of the incorrect (1.23). Upon integration of the above equation, we are led to a cubic in x:

$$(2p+q-x)^3 = \frac{1}{(p+q)^{-3}+\frac{1}{3}(2p+q)^{-3}-\frac{1}{3}(p+2q)^{-3}}.$$

This is a simpler equation than (1.24).

1.2.2.6 Principle of Insufficient Reason (Indifference) Near the end of the "Mémoire sur la probabilité des causes par les événements," we find an interesting paragraph where Laplace explicitly stated and endorsed the principle of indifference:

> We suppose in the theory that the different ways in which an event can occur are equally probable, or where they are not, that their probabilities are in a given ratio. When we wish then to make use of this theory, we regard two events as equally probable when we see no

> reason that makes one more probable than the other, because if they were unequally possible, since we are ignorant of which side is the greater, this uncertainty makes us regard them as equally probable. (Laplace, 1774a, OC 8, p. 61)

The principle would prove to be a major workhorse in Laplace's work in probability and statistics, and will be discussed in more detail in Section 1.3.

1.2.2.7 Conclusion We have described the major mathematical topics covered in the "Mémoire sur la probabilité des causes par les événements." However, scarcely any of the analysis we have considered in this section is reproduced in Laplace's *Théorie Analytique des Probabilités* (Laplace, 1812). This is because the material was superseded later by Laplace's more refined analysis. Nevertheless, as we mentioned before, the memoir is truly a superlative work, not least because of the sophistication of the analytical methods employed by Laplace. In the words of Andoyer:

> …as can be seen from the previous analysis we can repeat that the first works of Laplace immediately put him in the first rank amongst mathematicians; in addition, all his future work already appears here: it will develop and flourish, but the principles are fixed right from the start and will remain unchanged. (Andoyer, 1922, p. 105)

1.2.3 "Recherches sur l'intégration des équations différentielles aux différences finis" (1776)

1.2.3.1 Integration of Difference Equations. Problem of Points The "Recherches sur l'intégration des équations différentielles aux différences finis" (Laplace, 1776b) is devoted to a large extent to techniques of "integration" for difference equations. This method is hardly ever used nowadays not only because it is cumbersome but also because of the existence of much simpler techniques (such as generating functions, which were later developed extensively by Laplace himself). However, we shall illustrate it with a simple example. Suppose we wish to solve for u_n in

$$u_n = a_n u_{n-1} + b_n \quad \text{for} \quad n = 1, 2, \ldots. \tag{1.25}$$

The method of integration consists of defining a new variable

$$v_n \equiv \frac{u_n}{\prod_{i=1}^{n} a_i}. \tag{1.26}$$

Substituting for (1.25) in (1.26), we obtain

$$v_n \prod_{i=1}^{n} a_i = a_n v_{n-1} \prod_{i=1}^{n-1} a_i + b_n$$
$$v_n = v_{n-1} + \frac{b_n}{\prod_{i=1}^{n} a_i}.$$

We now define $\Delta v_n = v_{n+1} - v_n$ so that the above becomes

$$\Delta v_n = \frac{b_{n+1}}{\prod_{i=1}^{n+1} a_i}.$$

This now needs to be "integrated," resulting in

$$v_n = \sum \left(\frac{b_{n+1}}{\prod_{i=1}^{n+1} a_i} \right),$$

whence u_n can be obtained. Here, "integration" (Σ) of a function f_n is used in the sense that

$$\sum \Delta f_n = f_n. \tag{1.27}$$

Thus, if $u_n = 2u_{n-1} + 1$, then we define $v_n \equiv u_n / 2^n$, resulting in $\Delta v_n = 1/2^{n+1}$. In view of (1.27), the "integral" must be of the form $v_n = \alpha + \beta / 2^n$ where α and β are constants. This must satisfy $\Delta v_n = 1/2^{n+1}$ so that $\beta = -1$. Hence, $v_n = \alpha - 1/2^n$ and $u_n = \alpha 2^n - 1$.

Laplace proposed to solve the Problem of Points by using the above method:

> PROBLEM XIV *Two players A and B whose respective skills (probabilities of winning one round) are in the ratio p to q, respectively, play a game such that for x needed rounds, A is n rounds short and B is x-n rounds short of winning; it is required to determine the respective probabilities [of winning the game] of these two players.** (*ibid*., OC 8, p. 160)

Let ${}_n y_x$ be B's probability of winning the game. Then Laplace reasoned as follows: if B loses the next round then her probability of winning the game will be ${}_{n-1} y_{x-1}$; otherwise, the probability will be ${}_n y_{x-1}$. Laplace was thus able to write the following partial difference equation:

$${}_n y_x = \frac{q}{p+q} {}_n y_{x-1} + \frac{p}{p+q} {}_{n-1} y_{x-1} \quad n = 1, 2, \cdots, x,$$

subject to the boundary conditions ${}_n y_n = 1$ and ${}_0 y_x = 0$. Laplace's lengthy solution using the method of "integration" proceeded by first converting the above partial difference equation into an ordinary one and then applying the technique outlined at the start of this section. His final solution was

$${}_n y_x = \frac{1}{\left(\frac{p}{q}+1\right)^{x-1}} \left\{ 1 + \frac{p}{q}(x-1) + \frac{p^2}{q^2}\frac{(x-1)(x-2)}{1 \cdot 2} + \frac{p^3}{q^3}\frac{(x-1)(x-2)(x-3)}{1 \cdot 2 \cdot 3} + \cdots + \frac{p^{n-1}}{q^{n-1}}\frac{(x-1)(x-2)\cdots(x-n+1)}{1 \cdot 2 \cdots (n-1)} \right\}$$

(*ibid*., p. 162).

1.2.3.2 *Moral Expectation. On d'Alembert* Toward the middle of the memoir (Article XXV), Laplace engaged in some philosophical thoughts regarding probability. He

* Italics are Laplace's

reiterated his endorsement of the principle of indifference (which he first explicitly stated in the "Mémoire sur la probabilité des causes par les événements"*) and gave a slightly different definition of mathematical probability from the one first enunciated in the "Mémoire sur les suites récurro-récurrentes."† Laplace's definition here made use of the principle of indifference. He next considered the issue of mathematical versus moral expectation. Mathematical expectation is simply the "product of the amount at stake and the probability of obtaining it." On the other hand:

> Moral expectation depends…in the amount at stake and the probability of obtaining it; but it is not always proportional to the product of these two quantities; it depends on countless variable circumstances…one can regard the moral expectation itself as the product of an advantage and the probability of obtaining it; but one must distinguish, in the advantage wished for, its relative value from its absolute value. (Laplace, 1776b, OC 8, p. 148)

Moreover, an "ingenious" rule for the relative value of an amount had been provided by Daniel Bernoulli in the context of the St Petersburg problem:

> The relative value of a very small amount is…proportional to this absolute value divided by the total wealth of the interested person. (*ibidem*)

Thus, if a player's fortune changes from x to dx, then the change in the relative value (or utility) of the player's fortune is given as

$$dU \propto \frac{dx}{x}.$$

The moral expectation (or mean utility) is then the probability of obtaining an amount multiplied by the relative value of that amount.

Laplace would later devote an entire chapter on the topic of moral expectation in the *Théorie Analytique des Probabilités* (Laplace, 1812, Book II, Chapter X).

Laplace next clarified an important misconception that was made by none other than his mentor, d'Alembert. On several occasions, the latter had questioned the validity of several principles in the calculus of probabilities. In a well-documented instance, d'Alembert had claimed that if the toss of a fair coin resulted in a succession of, say, heads, then the next toss would more likely be a tail:

> Let's look at other examples which I promised in the previous Article, which show the lack of exactitude in the ordinary calculus of probabilities.
>
> In this calculus, by combining all possible events, we make two assumptions which can, it seems to me, be contested.
>
> The first of these assumptions is that, if an event has occurred several times successively, for example, if in the game of heads and tails, heads has occurred three times in a row, it is equally likely that head or tail will occur on the fourth time? However I ask if this assumption is really true, & if the number of times that heads has already successively occurred by the hypothesis, does not make it more likely the occurrence of tails on the fourth time? Because

* See pp. 24–25

† See p. 7

after all it is not possible, it is even physically impossible that tails never occurs. Therefore the more heads occurs successively, the more it is likely tail will occur the next time. If this is the case, as it seems to me one will not disagree, the rule of combination of possible events is thus still deficient in this respect. (d'Alembert, 1761, pp. 13–14)

Laplace clearly saw the error in d'Alembert's reasoning, but it is interesting to see the tact with which he approached the subject. Referring to the issue of moral expectation, Laplace first said:

Most of those who have written about chances have seemed to confuse expectation and moral probability with expectation and mathematical probability, or settle at least one them with the other; they have thus wanted to extend their theories beyond what they are susceptible to, which has made them obscure and incapable of satisfying the minds accustomed to the rigorous clarity of Mathematics. M. d'Alembert has proposed against them very fine objections, which has arisen the attention of mathematicians and has made feel the absurdity to which one could be lead, in many circumstances, from the results of the Calculus of Probabilities, and, therefore, the need to establish in this matter a distinction between the mathematical and the moral.... (Laplace, 1776b, OC 8, pp. 148–149)

After the initial praise, Laplace was now ready to correct his mentor's error (but without mentioning d'Alembert's name*):

The Doctrine of chances assumes that if *heads* and *tails* are equally likely, then so will be all the combinations *(head, head, head, etc.)*, *(head, tail, head, etc.)*, *etc.* Several philosophers have thought that this assumption is wrong, and that the combinations in which an event happens several times consecutively are less likely than others; but one should assume for this to be the case that the past events have an influence on future ones, which is inadmissible. (*ibid.*, p. 151)

In the above, Laplace correctly stated that, by the property of independence, all sequences of heads and tails are equally likely when a fair coin is tossed.

1.2.4 "Mémoire sur l'inclinaison moyenne des orbites" (1776): Distribution of Finite Sums, Test of Significance

At the start of this memoir (Laplace, 1776a), Laplace recalled Daniel Bernoulli's investigations regarding the planes of planetary orbits (cf. Section 5.8.4). Bernoulli had used probabilistic calculations to show that the inclinations of the planes of different planets were so small that these slight deviations could not be attributed to pure chance. Laplace now wished to investigate the inclinations of the orbits of comets.† He noted that his senior colleague du Séjour had previously studied the subject:

*According to Lecat (1935, p. 2), d'Alembert's error was first explicitly pointed out in Bertrand's book *Calculs des Probabilités* (Bertrand, 1889, p. 2).

†Laplace's work is also described in Gillispie (2000, Chapter 5), Sheynin (1973, pp. 286–291), and Hald (1998, pp. 51–55).

> [Mr. du Séjour] has found that the average inclination of 63 comets observed so far was 46°16′, which differs little from 45°, and that the ratio of direct to retrograde comets was 5/4, which deviates little from unity. Hence he concluded, rightly, that there is for comets no cause which makes move in one direction rather than the other, and approximately in the same plane, and thus that which determines the movement of the planets is entirely independent of the general system of the universe. (Laplace, 1776a, OC 8, pp. 280–281)

In analogy with the motion of planets, Laplace assumed that it was equally likely for a comet to be in direct and in retrograde motions. Assuming the motion of each of n comets is independent of the other, the probability that $n-\mu$ are in direct motion, and therefore μ are in retrograde motion, is given by the term

$$\frac{n(n-1)\cdots(n-\mu+1)}{1\cdot 2\cdot 3\cdots \mu}\left(\frac{1}{2}\right)^{n-\mu}\left(\frac{1}{2}\right)^{\mu}=\frac{n(n-1)\cdots(n-\mu+1)}{1\cdot 2\cdot 3\cdots \mu}\left(\frac{1}{2}\right)^{n}$$

in the binomial expansion of $\left(\frac{1}{2}+\frac{1}{2}\right)^n$. One can recognize the above as the probability of μ successes in a Bino (n, 1/2) distribution.

On the other hand, suppose one wished the ratio of retrograde to direct comets to be between $(n-\mu)/\mu$ and $(n-\mu')/\mu'$. Then the probability of the latter event can be calculated by adding all terms between

$$\frac{n(n-1)\cdots(n-\mu+1)}{1\cdot 2\cdot 3\cdots \mu}\left(\frac{1}{2}\right)^{n}$$

and

$$\frac{n(n-1)\cdots(n-\mu'+1)}{1\cdot 2\cdot 3\cdots \mu'}\left(\frac{1}{2}\right)^{n}$$

in the binomial expansion of $\left(\frac{1}{2}+\frac{1}{2}\right)^n$. However, these examples were too simple and in Article II Laplace wished to solve a more difficult problem:

> Given that an indefinite number of bodies are randomly thrown into space and orbiting the sun, the probability is required that the mean inclination of their orbits with respect to a given plane, such as the ecliptic, is between two limits, for example 40° and 50°? (*ibid.*, p. 282)

Laplace first considered the case of $n=2$ comets. In Figure 1.5, he drew a "curve of probability" *AZMzB* such that the ordinate of this curve is proportional to the probability of the mean inclination on the x-axis (line *AB*). His aim was to determine the equation of the probability curve so that he could integrate it and obtain the ratio of the area between any two given limits and the total area. This ratio would then be equal to the probability that the mean inclination is between the two above-mentioned limits. Laplace assumed the mean inclination is equally likely to take any value between zero and its maximum value

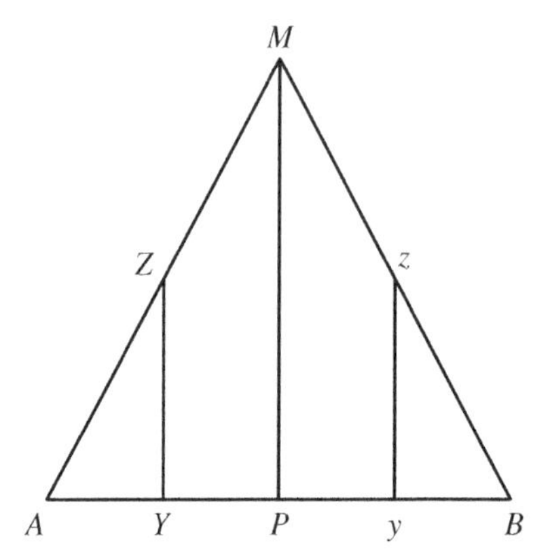

FIGURE 1.5 Laplace's curve of probability for $n=2$ comets

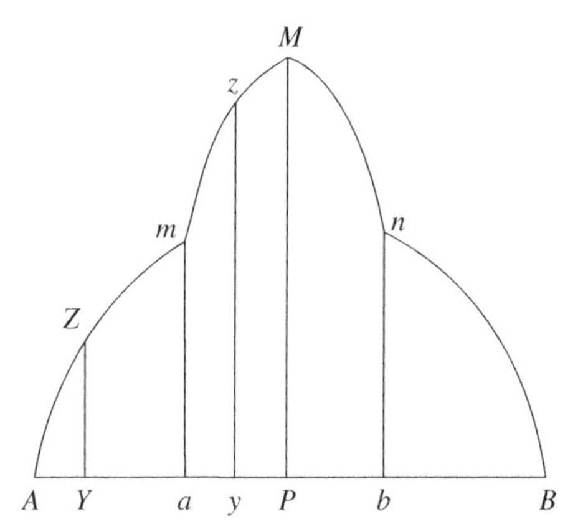

FIGURE 1.6 Laplace's curve of probability for $n=3$ comets

of 90°. He denoted the latter value by α and represented it by the point B on the x-axis. The ordinates were measured in units of $1/\alpha$, and Laplace assumed that the ordinate of the point M is α. MP bisects the line AB and the equation of AM is

$$y=\frac{2x}{\alpha}. \tag{1.28}$$

Laplace next considered the case $n=3$ (Fig. 1.6). The line AB (of length α) is now divided into three equal segments Aa, ab, and bB. First, consider the case of the curve AZm (i.e., when the mean inclination x satisfies $x<\alpha/3$). Assuming the inclination of one comet is f, the mean inclination of the other two comets is

$$\frac{3x-f}{2}<\frac{\alpha}{2} \quad \text{for} \quad 0\leq f\leq 3x.$$

Laplace now related these two comets with those of the $n=2$ case studied previously. From (1.28), the number of cases (which is proportional to the probability on the y-axis) with mean inclination $(3x-f)/2$ is $(3x-f)$. Therefore, the corresponding *total* number of cases is $\int_0^{3x}(3x-f)df = 9x^2/2$ and hence

$$y = \frac{9x^2}{2\alpha}$$

(*ibid.*, p. 284) is the equation of the curve *AZm*. We thus see that Laplace had wrongly drawn this curve in Figure 1.6: it should concave *up*, not *down*.

For the case $n=3$, Laplace next considered the curve *mMn* (Fig. 1.6), and denoted the distance between a and the mean inclination of the three comets by z. By using a similar reasoning as before, Laplace was able to write the equation of *mMn* as

$$y = \frac{1}{2}\alpha + 3z - \frac{9}{\alpha}z^2.$$

From the above cases, Laplace was now able to describe the general case when the interval *AB* is divided into n equal parts. For the rth part, that is, when $x \in [\alpha(r-1)/n, \alpha r/n]$ for $r = 1, 2, \ldots, n$, he obtained the following recursive equation for ${}_r y_{n,z}$, the ordinate of the curve at $x = \{\alpha(r-1)/n\} + z$:

$$ {}_r y_{n,z} = \int_0^{nz} {}_r y_{n-1,\frac{nz-f}{n-1}} df + \int_0^{\alpha-nz} {}_{r-1}y_{n-1,\frac{\alpha-s}{n-1}} ds, \tag{1.29}$$

(*ibid.*, p. 288). After some heavy algebra, Laplace was able to solve Eq. (1.29):

$$\begin{aligned}
{}_r y_{n,z} &= \pm \frac{(nz)^{n-1}}{(r-1)!(n-r)!} \mp \frac{\alpha(nz)^{n-2}}{(r-2)!(n-r)!} \mp \frac{\alpha^2(nz)^{n-3}}{1\cdot 2} \\
&\quad \times \left\{ \frac{1}{(r-2)!(n-r-1)!} - \frac{1}{(r-3)!(n-r)!} \right\} + \cdots \\
&\quad \mp \frac{\alpha^{q-1}(nz)^{n-q}}{(q-1)!} \left\{ \frac{1}{(r-2)!(n-r-q+2)!} - \frac{2^{q-1}-q}{(r-3)!(n-r-q+3)!} + \cdots \right. \\
&\quad \left. \mp \frac{(q-1)^{q-1} - (q-2)^{q-1}q + \cdots}{(r-q)!(n-r)!} \right\} + \cdots \\
&\quad + \frac{\alpha^{n-1}}{(n-1)!} \left\{ (r-1)^{n-1} - n(r-2)^{n-1} + \frac{n(n-1)}{2}(r-3)^{n-1} - \cdots \right\}
\end{aligned}$$

(*ibid.*, p. 298). In the above, the upper (lower) sign should be used when r is odd (even), except for the term

$$\mp \frac{(q-1)^{q-1} - (q-2)^{q-1}q + \cdots}{(r-q)!(n-r)!},$$

for which the upper (lower) sign should be used when q is odd (even).

To finish off the analysis, Laplace needed to determine one important quantity, namely, the area (${}_rK_n$) under the curve ${}_ry_{n,z}$ from $x=\alpha(r-1)/n$ to $x=\alpha r/n$. He was able to show that

$$
\begin{aligned}
{}_rK_n &= \int_{\alpha(r-1)/n}^{\alpha r/n} {}_ry_{n,z}\,dx \\
&= \int_0^{\alpha/n} {}_ry_{n,z}\,dz \\
&= \frac{\alpha^2}{n!n}\left\{r^n - \frac{n+1}{1}(r-1)^n + \frac{(n+1)n}{2!}(r-2)^n - \cdots \right. \\
&\quad \left. +(-1)^{r-1}\frac{(n+1)(n)(n-1)\cdots(n-r+3)}{(r-1)!}\right\}
\end{aligned} \tag{1.30}
$$

(*ibid*., p. 300), where $x=\{\alpha(r-1)/n\}+z$. From the above, he was also able to calculate the total area under the curve ${}_ry_{n,z}$ from $x=0$ to $x=\alpha$ as

$$S = {}_nK_n + {}_{n-1}K_n + \cdots = \frac{\alpha^2}{n}. \tag{1.31}$$

The ratio ${}_rK_n/S$ then gives the probability that the mean inclination X lies within the limits $x=\alpha(r-1)/n$ and $x=\alpha r/n$:

$$
\begin{aligned}
\Pr\left\{\frac{\alpha(r-1)}{n} \le X \le \frac{\alpha r}{n}\right\} &= \frac{{}_rK_n}{S} \\
&= \frac{1}{n!}\left\{r^n - \frac{n+1}{1}(r-1)^n + \frac{(n+1)n}{2!}(r-2)^n - \cdots\cdot \right. \\
&\quad \left. +(-1)^{r-1}\frac{(n+1)(n)(n-1)\cdots(n-r+3)}{(r-1)!}\right\}
\end{aligned} \tag{1.32}
$$

As an illustration, in Article IX Laplace considered the 63 comets whose orbits had been studied at that time. However, he noted that the calculations would be too laborious and instead chose to work with $n=12$. He next performed a test of significance, indeed the first based on the arithmetic mean. The line AB (cf. Figs. 1.5 and 1.6) is first divided into 12 equal parts, each representing $\alpha/12 = 90°/12 = 7°30'$. Using Eq. (1.32), with $r=6$,

$$
\begin{aligned}
\Pr\{45° - 7°30' \le X \le 45°\} &= \Pr\left\{\frac{5\alpha}{12} \le X \le \frac{\alpha}{2}\right\} \\
&= \frac{6^{12}}{12!}\left\{1 - 13\left(\frac{5}{6}\right)^{12} + \frac{13\cdot12}{1\cdot2}\left(\frac{4}{6}\right)^{12} - \frac{13\cdot12\cdot11}{1\cdot2\cdot3}\left(\frac{3}{6}\right)^{12} + \right. \\
&\quad \left. \frac{13\cdot12\cdot11\cdot10}{1\cdot2\cdot3\cdot4}\left(\frac{2}{6}\right)^{12} - \frac{13\cdot12\cdot11\cdot10\cdot9}{1\cdot2\cdot3\cdot4\cdot5}\left(\frac{1}{6}\right)^{12}\right\} \\
&= .339
\end{aligned}
$$

(*ibid.*, p. 301). From this number, Laplace made three conclusions:

1. The probability that the mean inclination of the 12 comets is more than 37°30′ is $.5+.339=.839$.
2. The probability that the mean inclination of the 12 comets is less than 52°30′ is the same as above.
3. The probability that the mean inclination of the 12 comets lies between 37°30′ and 52°30′ is $2\times.339=.678$.

For the 12 most recently observed comets, Laplace calculated a mean inclination of 42°31′ and thus concluded:

> For us to suspect that these comets have a cause which tends to make them move in the ecliptic, there should have been very large odds that, if they were thrown at random, their mean inclination would exceed 42°30′; however we have just found that the odds are 839 to 161, which is less than 6 to 1, that it will exceed 37°30′, and the odds are considerably less that it will exceed 42°30′. (*ibid.*, p. 302)

Laplace thus rejected the hypothesis of a "cause" governing the motion of comets.

A modern treatment of Laplace's problem can be found in Wilks (1962, p. 204). Assume

$$Y_n = U_1 + U_2 + \cdots + U_n,$$

where the U_i's are IID* with density functions $f_{U_i}(u_i)$. Now, the above can be re-written as $Y_n = Y_{n-1} + U_n$. Also, let $V = Y_{n-1}$. Therefore,

$$\begin{aligned} f_{Y_n,V}(y_n,v) &= f_{Y_{n-1},U_n}(y_{n-1},u_n)\left|J\left(\frac{y_n,v}{y_{n-1},u_n}\right)\right|^{-1} \\ &= f_{Y_{n-1}}(y_{n-1})f_{U_n}(u_n)\cdot 1 \\ f_{Y_n}(y_n) &= \int_{-\infty}^{\infty} f_{Y_{n-1}}(y_{n-1})f_{U_n}(u_n)dv \\ &= \int_{-\infty}^{\infty} f_{Y_{n-1}}(y_{n-1})f_{U_n}(y_n - y_{n-1})dy_{n-1}. \end{aligned} \tag{1.33}$$

Assuming the U_i's are Unif (0, 1), the density of Y_1 is $f_{Y_1}(y_1)=1$ for $0 \le y_1 \le 1$. This can be written in terms of the indicator function $I(\cdot)$ as

$$f_{Y_1}(y_1) = I_{\{0\le y_1 \le 1\}},$$

where

$$I_{\{y\in A\}} = \begin{cases} 1 & \text{for } y \in A, \\ 0 & \text{otherwise} \end{cases}.$$

From Eq. (1.33),

* Independent and identically distributed

$$
\begin{aligned}
f_{Y_2}(y_2) &= \int_{-\infty}^{\infty} f_{Y_1}(y_1) f_{U_n}(y_2 - y_1)\,dy_1 \\
&= \int_{-\infty}^{\infty} I_{\{0 \le y_1 \le 1\}} I_{\{0 \le y_2 - y_1 \le 1\}}\,dy_1 \\
&= \int_{-\infty}^{\infty} I_{\{0 \le y_1 \le 1\} \cap \{0 \le y_2 - y_1 \le 1\}}\,dy_1 \\
&= \begin{cases} \int_0^{y_2} dy_1 & \text{for } 0 < y_2 < 1 \\ \int_{y_2 - 1}^{1} dy_1 & \text{for } 1 < y_2 < 2 \end{cases}.
\end{aligned}
$$

Hence,

$$
f_{Y_2}(y_2) = \begin{cases} y_2 & \text{for } 0 < y_2 < 1 \\ 2 - y_2 & \text{for } 1 < y_2 < 2 \end{cases}.
$$

In general, by using (1.33), it can be shown through mathematical induction that

$$
f_{Y_n}(y_n) = \frac{1}{(n-1)!} \left\{ y_n^{n-1} - \binom{n}{1}(y_n - 1)^{n-1} + \binom{n}{2}(y_n - 2)^{n-1} - \cdots + (-1)^k \binom{n}{k}(y_n - k)^{n-1} \right\}
$$

for $k < y_n \le k+1$, $k = 0, 1, \ldots, n-1$. Since the mean X is related to the sum Y_n through

$$
X = \frac{Y_n}{n},
$$

the density of X is

$$
\begin{aligned}
f_X(x) &= f_{Y_n}(y_n) \left| \frac{dy_n}{dx} \right| \\
&= n f_{Y_n}(nx) \\
&= \frac{n}{(n-1)!} \left\{ (nx)^{n-1} - \binom{n}{1}(nx - 1)^{n-1} + \binom{n}{2}(nx - 2)^{n-1} - \cdots + (-1)^k \binom{n}{k}(nx - k)^{n-1} \right\}
\end{aligned}
$$

for $k/n < x \le (k+1)/n$, $k = 0, 1, \ldots, n-1$. The probability that the mean X lies between $x = \alpha(r-1)/n$ and $x = \alpha r/n$ is then

$$\Pr\left\{\frac{\alpha(r-1)}{n} \le X \le \frac{\alpha r}{n}\right\} = \int_{\alpha(r-1)/n}^{\alpha r/n} f_X(x)dx$$

$$= \frac{n}{(n-1)!}\int_{\alpha(r-1)/n}^{\alpha r/n}\left\{(nx)^{n-1} - \binom{n}{1}(nx-1)^{n-1} + \binom{n}{2}(nx-2)^{n-1} - \cdots\right.$$

$$\left. + (-1)^k\binom{n}{k}(nx-k)^{n-1}\right\}dx,$$

which is the same as (1.32).

In Articles XI–XIII, Laplace proved some elegant results on the calculus of operations obtained by Lagrange (1774). Laplace later refined his proofs in 1780 (Laplace, 1780) and then again in 1782 by the use of generating functions (Laplace, 1782). We shall consider Lagrange's results and Laplace's 1782 proofs in Section 1.2.7.

1.2.5 "Recherches sur le milieu qu'il faut choisir entre les resultants de plusieurs observations" (1777): Derivation of Double Logarithmic Law of Error

This memoir was read to the *Académie des Sciences* by Laplace on March 8, 1777, but remained unpublished until 1979, when it was discovered by Charles Gillispie (1979). It was Laplace's second article read to the *Académie* on the theory of errors, the first being the "Mémoire sur la probabilité des causes par les événements" (Laplace, 1774a). It was followed by the "Mémoire sur les probabilités" (Laplace, 1781), where many of the current results were also incorporated.

Laplace's 1777 memoir was a comment on Lagrange's previous "Mémoire sur l'utilité de la méthode de prendre le milieu entre les résultats de plusieurs observations" (Lagrange, 1776). In the latter memoir, Lagrange had obtained the arithmetic mean as the single correction to use when several observed errors were given by applying the techniques of direct probability (see Section 5.7.2). However, Laplace objected to Lagrange's method:

> ...This great mathematician [Lagrange] has determined through an ingenious analysis the error incurred by taking an arithmetic mean between the results of several observations, where the law of facility of each error is given; ... but it seems to me that this research requires other principles than those used by this learned writer. (Gillispie, 1979, p. 228)

The "other principles" Laplace referred to in the above were those of inverse probability:

> The problem under consideration can be looked at from two different viewpoints, depending on whether the observations are considered before or after they are made; in the first case, the search for the center to be taken between the observations consists of determining *a priori* which function of the results of the observations is most advantageous for the center of the observations; in the second case, the search for the center consists of determining a similar function *a posteriori*, that is by consideration of the respective distances between the observations. One can easily see that these two ways to consider the problem must lead to different results; but it is at the same time clear that the second is the only one which must be used. (*ibid.*, p. 229)

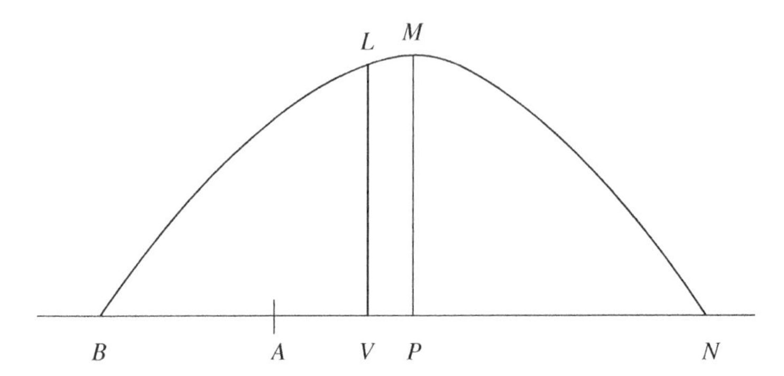

FIGURE 1.7 Laplace's probability curve for several observations made on V when the law of error ($BLMN$) is assumed to be known

Thereupon, Laplace stated that the center that ought to be chosen was the one that minimized the sum of absolute errors multiplied by their probabilities. This criterion was what he had earlier (Laplace, 1774a, p. 43) called the center of error (=posterior expected error), and which he had shown was minimum when it coincided with the center of probability (=posterior median).

To fix ideas, Laplace first considered the case when the observations had not been made yet, but the law of error was known. Suppose a phenomenon takes place at an instant V (Fig. 1.7) and observations $p, p^{(1)}, \ldots, p^{(n-1)}$ are made on V. Starting from the point A, let AV denote the error made in taking the sample mean $(p+p^{(1)}+\cdots+p^{(n-1)})/n$ as the true instant of the phenomenon, and let VL denote the corresponding value of the probability density. Also, the correction that needs to be applied to the arithmetic mean is $AP=\mu$, where PM is the line that divides the area under the density into two halves. Laplace's aim was to determine the probability that the arithmetic mean of the errors will be x, that is, the sum of errors will be nx.

Let the density of the error made on the first observation be $\phi(t)$, where $-b \leq t \leq x$. So as to deal with only positive values, let $z=t+h$, where h is the magnitude of the largest possible negative error, so that z lies between $h-b=q$ and $h+x$.

Now let $\phi(t)$ be piecewise continuous such that from $z=q$ to $z=q^{(1)}$ it is $\psi(z)$, from $z=q^{(1)}$ to $z=q^{(2)}$ it is $\psi^{(1)}(z)+\psi(z)$, ..., and from $z=q^{(r)}$ to $z=\infty$ it is $\psi^{(r)}(z)+\psi^{(r-1)}(z)+\cdots+\psi(z)$. Laplace thus denoted the density of the error made on the first observation by

$$\mathbf{l}\cdot\psi(z)+\mathbf{l}^{(1)}\cdot\psi^{(1)}(z)+\cdots+\mathbf{l}^{(r)}\cdot\psi^{(r)}(z) \tag{1.34}$$

(Gillispie, 1979, p. 232), where $\mathbf{l}$ means that $\psi(z)$ needs to be taken from $z=q$ to $z=\infty$, $\mathbf{l}^{(1)}$ means that $\psi^{(1)}(z)$ needs to be taken from $z=q_1^{(1)}$ to $z=\infty$, and so on. Similarly, the density of the second observation is

$$\mathbf{l}_1\cdot\psi_1(z)+\mathbf{l}_1^{(1)}\cdot\psi_1^{(1)}(z)+\cdots+\mathbf{l}_1^{(r)}\cdot\psi_1^{(r)}(z), \tag{1.35}$$

where $\mathbf{l}_1$ means that $\psi_1(z)$ needs to be taken from $z=q_1$ to $z=\infty$, $\mathbf{l}_1^{(1)}$ means that $\psi_1^{(1)}(z)$ needs to be taken from $z=q_1^{(1)}$ to $z=\infty$, and so on. Similarly the density of the other observations can be defined.

Laplace next considered the case of two observations such that the sum of the respective errors t and $t^{(1)}$ is $2x$. Then using $z = t + h$ and $z^{(1)} = t^{(1)} + h$, we have $z + z^{(1)} = 2x + 2h = s$, say, so that $z^{(1)} = s - z$. Now the density of the error t is given in (1.34) and that of the error $t^{(1)}$ can be written from (1.35) as

$$\mathbf{l}_1 \cdot \psi_1(s-z) + \mathbf{l}_1^{(1)} \cdot \psi_1^{(1)}(s-z) + \cdots + \mathbf{l}_1^{(r)} \cdot \psi_1^{(r)}(s-z).$$

Therefore the density of $s = z^{(1)} + z$ is the product of these two densities:

$$\left\{\mathbf{l} \cdot \psi(z) + \mathbf{l}^{(1)} \cdot \psi^{(1)}(z) + \cdots + \mathbf{l}^{(r)} \cdot \psi^{(r)}(z)\right\}\left\{\mathbf{l}_1 \cdot \psi_1(s-z) + \mathbf{l}_1^{(1)} \cdot \psi_1^{(1)}(s-z) + \cdots + \mathbf{l}_1^{(r)} \cdot \psi_1^{(r)}(s-z)\right\}.$$

To find the density of the sum of errors, the above density is integrated over all possible values of z:

$$\int\left\{\mathbf{l} \cdot \psi(z) + \mathbf{l}^{(1)} \cdot \psi^{(1)}(z) + \cdots + \mathbf{l}^{(r)} \cdot \psi^{(r)}(z)\right\}\left\{\mathbf{l}_1 \cdot \psi_1(s-z) + \mathbf{l}_1^{(1)} \cdot \psi_1^{(1)}(s-z) + \cdots + \mathbf{l}_1^{(r)} \cdot \psi_1^{(r)}(s-z)\right\}dz.$$

Upon replacing z by $q + u$ in $\psi(z)$ and writing the latter as $\pi(u)$, replacing z by $q^{(1)} + u$ in $\psi^{(1)}(z)$ and writing the latter as $\pi^{(1)}(u)$, and so on, Laplace wrote the above as

$$\begin{aligned}\Gamma(s) = \int_0^s &\left\{\mathbf{l} \cdot \pi(u) + \mathbf{l}^{(1)} \cdot \pi^{(1)}(u) + \cdots + \mathbf{l}^{(r)} \cdot \pi^{(r)}(u)\right\} \\ &\times \left\{\mathbf{l}_1 \cdot \pi_1(s-u) + \mathbf{l}_1^{(1)} \cdot \pi_1^{(1)}(s-u) + \cdots + \mathbf{l}_1^{(r)} \cdot \pi_1^{(r)}(s-u)\right\}du\end{aligned}$$

(*ibid.*, p. 234). Similarly, the density ${}^{n-2}\Gamma(s)$ of the sum of n errors can be determined.

Laplace noted that considerable simplification occurred in the above formulas when $\pi, \pi^{(1)}, \ldots; \pi_1, \pi_1^{(1)}, \ldots$ were expressed as polynomials in u, for example,

$$\begin{aligned}&\mathbf{l} \cdot \pi(u) + \mathbf{l}^{(1)} \cdot \pi^{(1)}(u) + \cdots = f + f^{(1)}u + f^{(2)}u^2 + \cdots, \\ &\mathbf{l}_1 \cdot \pi_1(u) + \mathbf{l}_1^{(1)} \cdot \pi_1^{(1)}(u) + \cdots = f_1 + f_1^{(1)}u + f_1^{(2)}u^2 + \cdots \\ &\cdots\end{aligned} \tag{1.36}$$

Then,

$$\begin{aligned}{}^{n-2}\Gamma(s) = \varepsilon^{n-1} &\left\{f + f^{(1)}\varepsilon + 1 \cdot 2 \cdot f^{(2)}\varepsilon^2 + 1 \cdot 2 \cdot 3 f^{(3)} \cdot \varepsilon^3 + \cdots\right\} \\ &\times \left\{f_1 + f_1^{(1)}\varepsilon + 1 \cdot 2 f_1^{(2)}\varepsilon^2 + \cdots\right\}\cdots\left\{f_{n-1} + f_{n-1}^{(1)}\varepsilon + 1 \cdot 2 f_{n-1}^{(2)}\varepsilon^2 + \cdots\right\}\end{aligned}$$

(*ibid.*, p. 237), where $\varepsilon^r = s^r/(1 \cdot 2 \cdot 3 \cdots r)$. As an example, in Art. V Laplace considered the case when each observation had the same constant density K for $-h \le t \le g$. This is the same case considered previously by Lagrange (see Section 1.4.4). Now, $q = 0$, and let $q^{(1)} = mh$, $q^{(2)} = m^{(1)}h, \cdots$ Then, $\mathbf{l}, \mathbf{l}^{(1)}, \mathbf{l}_1^{(2)}, \cdots$ can be replaced by $1, i^m, i^{m^{(1)}}, \cdots$ and then $F(s)i^{\mu}$ replaced by $F(s - \mu n)$ for any function F of s, where $h + g = p = mn$.

Now, using the formula for ${}^{n-2}\Gamma(s)$ above,

$${}^{n-2}\Gamma(s) = \varepsilon^{n-1}\left\{f + f^{(1)}\varepsilon + 1 \cdot 2 \cdot f^{(2)}\varepsilon^2 + 1 \cdot 2 \cdot 3 f^{(3)}\varepsilon^3 + \cdots\right\}^n. \tag{1.37}$$

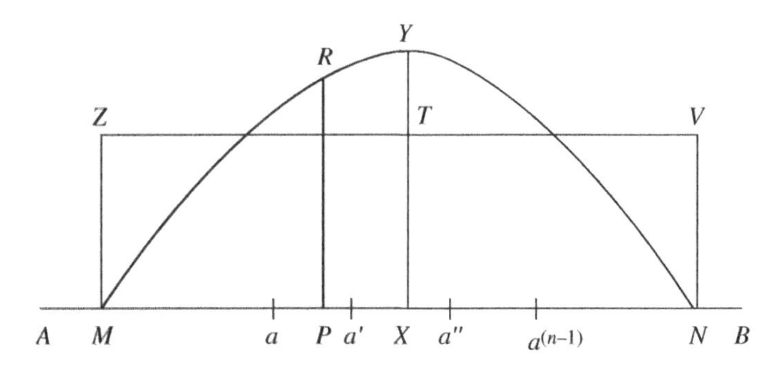

FIGURE 1.8 Laplace's probability curve for several observations when the law of error (*MRYN*) is assumed to be unknown

Since $z = h + g$, we have $\psi(s) = K$ for $0 \le z \le g + h$. For $g + h \le z < \infty$, we have $\psi^{(1)}(s) + \psi(s) = 0$ so that $\psi^{(1)}(s) = -\psi(s) = -K$. Then

$$\mathbf{l} \cdot \pi(u) + \mathbf{l}^{(1)} \cdot \pi^{(1)}(u) = K\left(\mathbf{l} - \mathbf{l}^{(1)}\right) = K\left(1 - i^{m}\right),$$

since $\mathbf{l} = 1$ and $\mathbf{l}^{(1)} = i^{m}$. From Eq. (1.36), $f = K(1 - i^{m})$ and $f^{(1)} = 0$ so that Eq. (1.37) gives

$$^{n-2}\Gamma(s) = \varepsilon^{n-1} \cdot K^{n}\left(1 - i^{m}\right)^{n} = \frac{K^{n} s^{n-1}}{1 \cdot 2 \cdots (n-1)}\left(1 - i^{m}\right)^{n}.$$

Upon replacing $F(s)i^{\mu}$ by $F(s - \mu n)$ for any function F of s, Laplace thus obtained from the above the probability that the sum of the errors is nx as

$$\frac{K^{n}}{1 \cdot 2 \cdot 3 \cdots (n-1)}\left\{s^{n-1} - n(s-p)^{n-1} + \frac{n(n-1)}{2}(s-2p)^{n-1} - \frac{n(n-1)(n-2)}{2 \cdot 3}(s-3p)^{n-1} + \cdots\right\}$$

(*ibid.*, p. 238), where $p = mn = h + g$ and $s = nh + nx$. Note that the above is the same as Lagrange's (1.148).

Laplace next (Art. VI.)* considered the case when the observations had already been made. Suppose (Fig. 1.8) that the observations $a, a', a'', \ldots$ have been made of the true instant of the phenomenon. Let the origin be at a, and let $aP = x$, $aa' = q^{(1)}$, $a'a'' = q^{(2)}, \ldots$ so that $a'P = q^{(1)} - x$, $a''P = q^{(2)} - x, \ldots$. Then Laplace observed that the true instant of the phenomenon was more likely to be at the point P than at the point X if, assuming the true point is P, the probability of the distances aa', $a'a'', \ldots$ was higher than the corresponding probability by assuming the true point is X. This observation, he stated, was a corollary to Bayes' theorem, which he then enunciated under the uniform prior assumption. The ordinate PR was then proportional to the probability that the true instant is P. Laplace's reasoning here can be written as

*The derivation that follows is also described in Dale (1999, pp. 185–190).

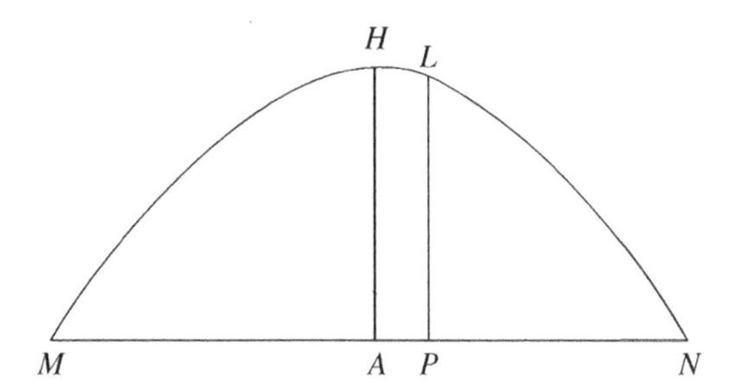

FIGURE 1.9 The line AH divides the area under MHN into two equal halves

$$\Pr\{\text{true instant is } P \mid aa', a'a'', \ldots\} = \frac{\Pr\{aa', a'a'', \ldots \mid \text{true instant is } P\}}{\sum \Pr\{aa', a'a'', \ldots\}} \tag{1.38}$$

Laplace next assumed the true instant to be P so that the errors made in the observations $a, a', a'', \ldots$ were $-x, q^{(1)} - x, q^{(2)} - x, \ldots$ and the probabilities of the respective errors were $\phi(x), \phi(q^{(1)} - x), \phi(q^{(2)} - x), \ldots$ Now, let the law of error be unknown (Art. VII), let the errors of observations lie in $[-h, h]$, and let the points M and N in Figure 1.8 be such that $a^{(n-1)}N = h$ and $aM = h$. Laplace now reasoned that the probability that the true instant was P was the ratio of PR to the total area under MRN. His reasoning was based on Eq. (1.38), which can be written as

$$\begin{aligned}\Pr\{\text{true instant is } P \mid aa', a'a'', \ldots\} &= \frac{\Pr\{aa', a'a'', \ldots \mid \text{true instant is } P\}}{\sum \Pr\{aa', a'a'', \ldots\}} \\ &= \frac{PR}{\text{total area under } MRN}.\end{aligned}$$

As can be seen from the above, PR is also the product of the probabilities that the error in the first observation is $-x$, the error in the second observation is $q^{(1)} - x$, the error in the third observation is $q^{(2)} - x$, and so on. Laplace next argued that PR was the probability for a given system ("*un sistême [sic] donné*"). To take into account all possible systems, one needed the *sum* of all such probabilities that the error in an observation is f.

To find the above sum of probabilities, Laplace argued as follows: In Figure 1.9, let A denote the origin, $AP = f$, and the curve MHN be $\phi(f)$. Further, let AH divide the total area under the curve MHN into two halves. If $\phi(f)$ is symmetric about AH, this also means that $AM = AN = h$. Now, the area under HAN is 1/2, and Laplace imagined that this number was divided into a very large number m of infinitesimally small parts* which needed to be distributed over a finite number h of points on AN. If A is the only point, then all m parts need to be above A, and there is only one combination of parts possible. If there are two

*The exact phrase used by Laplace was "an infinite number m of parts."

points A and B, then there are only $m/2$ combinations.* If there are three points A, B, and C, let z be the number of parts above C. Now each of the three points must have at least z parts so that $(m-3z)$ parts need to be distributed between A and B. Using the previous case, the number of combinations for a given z is thus $(m-3z)/2$ and the total number of combinations for any z is

$$\int_0^{m/3} \frac{m-3z}{2}dz = \frac{m^2}{(1\cdot 2)^2 3}.$$

Using the same argument, Laplace wrote the number of combinations when there are h points on AN as

$$\frac{m^{h-1}}{\{1\cdot 2\cdot 3\cdots(h-1)\}^2 h} \tag{1.39}$$

(*ibid.*, p. 245). For the case when the number of parts above A cannot exceed p, Laplace obtained the number of combinations as

$$\Gamma(h,m,p) = \frac{1}{\{1\cdot 2\cdot 3\cdots(h-1)\}^2 h}\left\{m^{h-1} - h(m-p)^{h-1} + \frac{h(h-1)}{1\cdot 2}(m-2p)^{h-1} - \cdots\right\}$$

(*ibid.*, p. 246), the terms in the above continuing until one of $(m-p)$, $(m-2p)$, ... becomes negative.

Laplace next considered a situation in which PL (Fig. 1.9) contains p parts and the number of points preceding the point P is n. Each of the latter points will contain at least p parts, so let the corresponding total number of parts be $np+\mu$. Then the number of parts to the right of PL is $m-(np+\mu)-p = m-p(n+1)-\mu$ and these are distributed over $h-n-1$ points. Now:

1. The number of combinations for the $np+\mu$ parts over the n points before P is

$$\frac{\mu^{n-1}}{\{1\cdot 2\cdot 3\cdots(n-1)\}^2 n},$$

since each of the n points must have p parts and there are only μ remaining parts to be distributed.

2. The number of combinations for the $m-p(n+1)-\mu$ parts over the $h-n-1$ points after P is $\Gamma(h-n-1, m-p(n+1)-\mu, p)$, since each these points cannot contain more than p parts.

*To understand this, suppose m is small with $m=4$. Then there can only be one or two parts above A. The case of three parts above A cannot be counted as it corresponds to the one-part case. Thus, with $m=4$, there are only $m/2=2$ combinations.

Then the total number of combinations that result in *PL* having *p* parts can be obtained by multiplying the number of combinations in 1· and 2· previously and integrating over μ:

$$\int \frac{\mu^{n-1}\Gamma(h-n-1, m-p(n+1)-\mu, p)}{\{1\cdot 2\cdot 3\cdots(n-1)\}^2 n} d\mu = \int \frac{\mu^{n-1}}{\{1\cdot 2\cdot 3\cdots(n-1)\}^2 n\{1\cdot 2\cdot 3\cdots(h-n-2)\}^2 (h-n-1)} \times \left[\{m-(n+1)p-\mu\}^{h-n-2} - (h-n-1)\times \{m-(n+2)p\}^{h-n-2} + \cdots\right] d\mu \tag{1.40}$$

(*ibid.*, p. 247). For the integral $\int \mu^{n-1}\{m-(n+1)p-\mu\}^{h-n-2}\, d\mu$ above, Laplace argued that μ should be taken from zero to $m-(n+1)p$; for the integral $\int \mu^{n-1}\{m-(n+2)p-\mu\}^{h-n-2}\, d\mu$ above, μ should be taken from zero to $m-(n+2)p$; and so on. The integral in (1.40) then becomes

$$\frac{1}{(1\cdot 2\cdot 3\cdots n)\{1\cdot 2\cdot 3\cdots(h-n-1)\}\{1\cdot 2\cdot 3\cdots(h-2)\}}\left[\{m-(n+1)p\}^{h-2} - (h-n-1) \times\{m-(n+2)p\}^{h-2} + \frac{(h-n-1)(h-n-2)}{1\cdot 2}\{m-(n+3)p\}^{h-2} - \cdots\right].$$

Recall that Laplace wished to find the sum of *PR* across all possible systems. This sum can be obtained by multiplying the above with *p* and integrating with respect to the latter:

$$\int \frac{1}{(1\cdot 2\cdot 3\cdots n)\{1\cdot 2\cdot 3\cdots(h-n-1)\}\{1\cdot 2\cdot 3\cdots(h-2)\}}\left[\{m-(n+1)p\}^{h-2} - (h-n-1) \times\{m-(n+2)p\}^{h-2} + \frac{(h-n-1)(h-n-2)}{1\cdot 2}\{m-(n+3)p\}^{h-2} - \cdots\right] pdp.$$

For the integral $\int \{m-(n+1)p\}^{h-2}\, pdp$, *p* should be taken from zero to $m/(n+1)$; for the integral $\int \{m-(n+2)p\}^{h-2}\, pdp$, *p* should be taken from zero to $m/(n+2)$; and so on. The sum of *PR*, that is, the sum of all ordinates relative to the point *P*, can then be obtained from the last equation as

$$\frac{m^h}{(1\cdot 2\cdot 3\cdots n)\{1\cdot 2\cdot 3\cdots(h-n-1)\}\{1\cdot 2\cdot 3\cdots h\}}\left\{\frac{1}{(n+1)^2} - (h-n-1)\frac{1}{(n+2)^2} + \frac{(h-n-1)(h-n-2)}{1\cdot 2}\frac{1}{(n+3)^2} - \cdots\right\}$$

(*ibid.*, p. 249). After simplification, Laplace wrote the above formula as

$$\frac{m^h}{\{1\cdot 2\cdot 3\cdots(h-1)\}^2 h^2}\left(\frac{1}{n+1} + \frac{1}{n+2} + \cdots + \frac{1}{h}\right).$$

Now, to obtain the probability that the error in an observation is AP (Fig. 1.9), one needs to divide the last expression by the total number of combinations in (1.39). Thus, the probability that the error of observation is AP is

$$\frac{m}{h}\left(\frac{1}{n+1}+\frac{1}{n+2}+\cdots+\frac{1}{h}\right).$$

Let $AP = n+1 = x$, $x/h = z$ so that $1/h = dz$. As h becomes infinitely large (so that $h=m$), the above becomes

$$\begin{aligned}\frac{m}{h^2}\left(\frac{1}{\frac{n+1}{h}}+\frac{1}{\frac{n+2}{h}}+\cdots+\frac{1}{\frac{h}{h}}\right)&=\frac{m}{h^2}\left(\frac{1}{z}+\frac{1}{z+dz}+\frac{1}{z+2dz}+\cdots\right)\\&=\frac{m}{h^2dz}\left(\frac{dz}{z}+\frac{dz}{z+dz}+\frac{dz}{z+2dz}+\cdots\right)\\&=\frac{m}{h^2dz}\log z^{-1}\\&=\log z^{-1}\end{aligned}$$

(*ibid.*, p. 250). Thus, the probability of the error $AP=x$ is $\log z^{-1}=\log(h/x)$, and Laplace obtained a preliminary version of the double logarithmic law of error. He later (*ibid.*, p. 253) also showed that this law of error resulted in a center that coincided with the arithmetic mean.

1.2.6 "Mémoire sur les probabilités" (1781)

1.2.6.1 Introduction In "Mémoire sur les probabilités" (Laplace, 1781) (Fig. 1.10), we find a maturing of Laplace's treatment of probability and a deeper analysis of problems. At the very start, Laplace stated the two main issues he was going to deal with:

> In this Memoir I propose to deal with two important points in the analysis of chances which do not appear to have yet been sufficiently explored: the purpose of the first is the way to calculate the probability of compound events from simple events whose respective probabilities are unknown; the purpose of the second is the influence of past events on the probability of future events, and the law according to which, by its development, they inform us of the cause which make them happen. (Laplace, 1781, OC 9, p. 383)

Laplace thus wished to calculate compound and conditional probabilities, and correctly noted that the two types of probabilities were related. He stated that the simple probabilities that needed to be used in the computations were unknown. How could these simple probabilities be determined? In Article II, Laplace noted:

MÉMOIRE SUR LES PROBABILITÉS[1]

Mémoires de l'Académie royale des Sciences de Paris, année 1778; 1781.

I.

Je me propose de traiter dans ce Mémoire deux points importants de l'analyse des hasards qui ne paraissent point avoir encore été suffisamment approfondis : le premier a pour objet la manière de calculer la probabilité des événements composés d'événements simples dont on ignore les possibilités respectives; l'objet du second est l'influence des événements passés sur la probabilité des événements futurs, et la loi suivant laquelle, en se développant, ils nous font connaitre les causes qui les ont produits. Ces deux objets, qui ont beaucoup d'analogie entre eux, tiennent à une métaphysique très délicate, et la solution des problèmes qui leur sont relatifs exige des artifices nouveaux d'analyse; ils forment une nouvelle branche de la théorie des probabilités, dont l'usage est indispensable lorsqu'on veut appliquer cette théorie à la vie civile. Je donne, relativement au premier, une méthode générale pour déterminer la probabilité d'un événement quelconque, lorsqu'on ne connait que la loi de possibilité des événements simples, et, dans le cas où cette loi est inconnue, je détermine celle dont on doit faire usage. La considération du second objet me conduit à parler des naissances : comme cette matière est une des plus intéressantes auxquelles on puisse appliquer le Calcul des probabilités, je fais en sorte de la traiter avec tout le soin dû à son impor-

(1) Remis le 19 juillet 1780.

FIGURE 1.10 First page of "Mémoire sur les probabilités" (Laplace, 1781)

> The latter [simple probabilities] may be determined in these three ways: 1° *a priori*, when, by the very nature of the events, one sees that they are possible in a given ratio; thus, in the game of heads and tails, if the coin thrown in the air is homogeneous and if its two sides are completely alike, one deems *head* and *tail* equally likely; 2° *a posteriori*, by repeating a large number of times the experiment which can bring about the event concerned, and by examining how many times it happens; 3° finally, by considering the motives which can bring us to decide on the existence of the event; if, for example, the respective skills of two players A and B are unknown, since there is no reason to assume that A is stronger than B, one concludes that the probability that A wins is ½. The first way gives the absolute probability of the events; the second gives an approximation, and the third only gives their possibilities relative to our state of our knowledge. (*ibid*., pp. 384–385)

In the above, Laplace had given a succinct distinction between three kinds of probability: classical (mathematical), frequentist, and epistemic (based on an appeal to the principle of indifference).

As an illustration, in Article II Laplace solved a variation of the *Gambler's Ruin Problem**: Two players A and B play a game such that the one who wins the first n rounds wins the game. It is assumed that the stronger player has a probability $(1+\alpha)/2$ (where $-1\le\alpha\le1$) of winning one round (so that the weaker player has a corresponding probability of $(1-\alpha)/2$), but it is ***unknown*** whom of A or B is the stronger player. Laplace solved the problem as follows. If A is the stronger player, then the probability that she wins the game is $\{(1+\alpha)/2\}^n$. Likewise, her probability is $\{(1-\alpha)/2\}^n$ if she is the weaker player. However, there is no reason suppose that A is the stronger or weaker player so that each possibility is given a probability of 1/2 by the principle of indifference. Hence, the (unconditional) probability that A will win the game is

$$P=\frac{1}{2}\left\{\frac{(1+\alpha)}{2}\right\}^n+\frac{1}{2}\left\{\frac{(1-\alpha)}{2}\right\}^n=\frac{(1+\alpha)^n+(1-\alpha)^n}{2^{n+1}}$$

(*ibid*., p. 387).

1.2.6.2 Double Logarithmic Law of Error In Article XII of the "Mémoire sur les probabilités," Laplace gave a mathematical justification for his use of the double logarithmic distribution as a law of error:

$$f_X(x)=\frac{1}{2a}\log\frac{a}{|x|}\quad\text{for}\quad -a\le x\le a. \tag{1.41}$$

Like the case of double exponential distribution, Laplace here again made use of the principle of indifference. Using his own notation as much as possible, we now describe his justification. Consider a game where the probability of winning (of skill) of a player is treated as a random variable.† Let x (where $-a\le x\le a$) denote the excess skill of the player relative to the mean. Divide the interval [0, a] into n equal intervals and let the ordinate at some x be denoted by $y(x)$, where

* See Gorroochurn (2012a, Prob. 5)

† Laplace, of course, did not think in terms of a random variable, the concept being first introduced later by his disciple, Poisson.

$$\begin{aligned} y(0) &= z + z_1 + \cdots + z_{n-1}, \\ y(1/n) &= z + z_1 + \cdots + z_{n-2}, \\ &\vdots \\ y(a) &= z. \end{aligned}$$

In Laplace's derivation, $y(x)$ was the probability of x, but the probabilities were not normalized because he set

$$y(0) + y(1/n) + \cdots + y(a) = s.$$
$$\therefore \quad nz + (n-1)z_1 + \cdots + z_{n-1} = s.$$

By letting $nz = t, (n-1)z_1 = t_1, \ldots, z_{n-1} = t_{n-1}$, the above becomes

$$t + t_1 + \cdots + t_{n-1} = s.$$

In terms of the $t_i's$, the value of $y(r/n)$ is

$$\frac{t}{n} + \frac{t_1}{n-1} + \frac{t_2}{n-2} + \cdots + \frac{t_{n-r}}{r}.$$

Laplace now appealed to the principle of indifference and assumed that $t, t_1, \ldots, t_{n-r}$ were each Unif (0, 1/b), where b is a constant. Using this assumption, he showed that the average of $y(r/n)$ is

$$N\left(\frac{1}{n} + \frac{1}{n-1} + \frac{1}{n-2} + \cdots + \frac{1}{r}\right),$$

where N is a function of n. But, as n becomes infinitely large, we have

$$\begin{aligned} \frac{1}{n} + \frac{1}{n-1} + \frac{1}{n-2} + \cdots + \frac{1}{r} &= \log n - \log r \\ &= \log\left(\frac{n}{r}\right) \\ &= \log\left(\frac{a}{x}\right). \end{aligned}$$

Thus, the average value of y for a given x is $N\log(a/x)$.

Note that in the above, x must satisfy $0 < x \leq a$. A similar reasoning as above can be applied to the interval $[-a, 0]$, resulting in an average limiting value of $N\log(a/x)$ for $y(r/n)$. The density of Y, the deviation of the player's skill from the mean, is thus

$$N\log\left(\frac{a}{x}\right) \quad \text{for} \quad -a \leq x \leq a,$$

where x is replaced by $-x$ in the logarithm when $x<0$. Since the above is symmetric about the vertical axis, we need

$$\int_0^a N \log\left(\frac{a}{x}\right) dx = \frac{1}{2} \quad \Rightarrow \quad N = \frac{1}{2a}.$$

Hence, Laplace was able to obtain the law of error in Eq. (9.30):

$$y = \frac{1}{2a} \log\left(\frac{a}{x}\right) \quad \text{for} \quad -a \le x \le a$$

(*ibid.*, p. 412), where x is replaced by $-x$ in the logarithm when $x<0$. The above is more appropriately written as

$$y = \frac{1}{2a} \log\left(\frac{a}{|x|}\right) \quad \text{for} \quad -a \le x \le a.$$

The above simple derivation of the double logarithmic distribution should be compared with the more convoluted one in Sec 1.2.5. In Article XIII, Laplace also said:

> This is the equation that one must use when one has, relatively to the possible values of x, no information, except it is as likely to be less as it is to be more: now this is what happens in many situations. Let us assume, for example, that we have the true instant of a phenomenon observed by several people; each one of them can easily fix the largest error his observation is susceptible to, either in excess, or less, by taking in the limit half of the largest interval that he can assume between two similar observations, with discarding them as bad; this interval is what we have called $2a$...Now, it is natural to think that the same errors, either more or less, are equally likely and that their probability is as likely to be less or more; if we have no information, relative to their probability, one of course falls to the preceding case; one must thus assume that the probability, of the positive error x, as well as the negative error $-x$, is equal to $\frac{1}{2a}\log\frac{a}{x}$; and this is the law of probability that must be used, when looking for the mean that must be chosen between several observations. (*ibid.*, p. 413)

1.2.6.3 Definition of Conditional Probability. Proof of Bayes' Theorem In Article XIV, Laplace motivated his investigation of future contingencies based on past events as follows:

> The previous theory assumes we have no reason to attribute to one of the players more skill than the others, which is true when the game starts; but as the game progresses and the events of the game happen, we acquire new insights on their respective strengths, so that they would be exactly known if the number of rounds was infinite, as we show in the following: the players' skills and, more generally, the different causes of events are thus related to their existence by laws which are very important to know well, and from this point of view, there is no doubt that past events affect the probability of future events. (*ibid.*, p. 414)

Now, let E be a past event, e a future event, and $E+e$ the compound event that E happens first followed by e. Laplace then argued that

$$\Pr\{e \mid E\} = \frac{\Pr\{E+e\}}{\Pr\{E\}}. \tag{1.42}$$

The above is an explicit definition of conditional probability.

Continuing Article XV, for the first time Laplace provided a proof of Bayes' theorem. Recall that, independently of Bayes, this theorem was enunciated by Laplace in the "Mémoire sur la probabilité des causes par les événements" (Laplace, 1774a).* Both there and here, Laplace assumed equally likely *a priori* probabilities by appealing to the principle of indifference. We now give Laplace's proof. Laplace considered an event E, and also n exhaustive causes $A_1, A_2, \ldots, A_n$ such that

$$x_i = \Pr\{A_i \mid E\},$$
$$a_i = \Pr\{E \mid A_i\}.$$

Then the probability of a second event E' similar to E is

$$\Pr\{E' \mid E\} = a_1 x_1 + a_2 x_2 + \cdots + a_n x_n. \tag{1.43}$$

Although Laplace did not give the details, the above formula can be obtained as follows:

$$\begin{aligned}\Pr\{E' | E\} &= \sum_{i=1}^{n} \Pr\{E' \mid A_i, E\} \Pr\{A_i \mid E\} \\ &= \sum_{i=1}^{n} \Pr\{E' \mid A_i\} \Pr\{A_i \mid E\} \\ &= \sum_{i=1}^{n} a_i x_i.\end{aligned}$$

In Eq. (1.43), Laplace had thus assumed that $\Pr\{E' \mid A_i, E\} = \Pr\{E' \mid A_i\}$, that is, E' and E are conditionally independent given any A_i. Continuing from Eq. (1.43), the conditional probability of *two* consecutive events similar to E is

$$a_1^2 x_1 + a_2^2 x_2 + \cdots + a_n^2 x_n \tag{1.44}$$

and so on. Next, Laplace noted that the probability $\Pr\{E' \mid E\}$ in Eq. (1.43) could be obtained by using the right side of Eq. (1.42) as follows:

$$\begin{aligned}\frac{\Pr\{E' + E\}}{\Pr\{E\}} &= \frac{\sum_{i=1}^{n} \Pr\{(E' + E) \mid A_i)\} \Pr\{A_i\}}{\sum_{i=1}^{n} \Pr\{E \mid A_i\} \Pr\{A_i\}} \\ &= \frac{\left(a_1^2 + a_2^2 + \cdots + a_n^2\right) / n}{\left(a_1 + a_2 + \cdots + a_n\right) / n} \\ &= \frac{a_1^2 + a_2^2 + \cdots + a_n^2}{a_1 + a_2 + \cdots + a_n}.\end{aligned} \tag{1.45}$$

* See Section 1.2.2.1.

In the last formula, Laplace had assumed that $\Pr\{A_i\} = 1/n$ for $i = 1, 2, \ldots, n$, that is, all possible causes were equally likely. Equating (1.43) and (1.44), we have

$$a_1 x_1 + a_2 x_2 + \cdots + a_n x_n = \frac{a_1^2 + a_2^2 + \cdots + a_n^2}{a_1 + a_2 + \cdots + a_n}. \tag{1.46}$$

Similarly, the conditional probability of two consecutive events similar to E in Eq. (1.44) can also be derived by using the right side of (1.42), resulting in

$$a_1^2 x_1 + a_2^2 x_2 + \cdots + a_n^2 x_n = \frac{a_1^3 + a_2^3 + \cdots + a_n^3}{a_1 + a_2 + \cdots + a_n}. \tag{1.47}$$

Likewise, the conditional probability of i consecutive events similar to E can also be derived by using the right side of (1.42), resulting in

$$a_1^i x_1 + a_2^i x_2 + \cdots + a_n^i x_n = \frac{a_1^{i+1} + a_2^{i+1} + \cdots + a_n^{i+1}}{a_1 + a_2 + \cdots + a_n} \quad \text{for} \quad i = 1, \ldots, n-1.$$

By solving for $x_1, x_2, \ldots, x_n$ in the $(n-1)$ equations above, in conjunction with $x_1 + x_2 + \cdots + x_n = 1$, Laplace thus obtained

$$\begin{aligned} x_1 &= \frac{a_1}{a_1 + a_2 + \cdots + a_n}, \\ x_2 &= \frac{a_2}{a_1 + a_2 + \cdots + a_n}, \\ &\vdots \\ x_n &= \frac{a_n}{a_1 + a_2 + \cdots + a_n}, \end{aligned}$$

(*ibid*., p. 417). This completes Laplace's proof of Bayes' theorem (under the assumption of a discrete uniform prior), which in modern notation can be written as

$$\Pr\{A_i \mid E\} = \frac{\Pr\{E \mid A_i\}}{\sum_{j=1}^{n} \Pr\{E \mid A_j\}}.$$

As was mentioned in Section 1.2.2.1, Laplace's "Mémoire sur les probabilités" made no mention of Bayes. However, a summary of the article that appeared with the complete article in the 1778 volume of the *Histoire de l'Académie Royale des Sciences* made explicit mention of both Bayes and Price. The summary had been written by the Marquis de Condorcet (Fig. 1.11), who was then the permanent secretary of the *Académie des*

FIGURE 1.11 Marquis de Condorcet (1743–1794). Wikimedia Commons (Public Domain), http://commons.wikimedia.org/wiki/Marquis_de_Condorcet#/media/File:Nicolas_de_Condorcet.PNG

Sciences, and the 1778 volume itself appeared in 1781. The summary thus contained the first explicit acknowledgment of Bayes as being the initial discoverer of the theorem.*

1.2.6.4 Proof of Inverse Bernoulli Law Refined In Article XVIII Laplace gave a superior solution† to a problem he previously considered in Article III‡ of the "Mémoire sur la probabilité des causes par les événements" (Laplace, 1774a). Suppose, out of $(p+q)$ children, p boys and q girls are born. Let the unknown probability of a boy be denoted by X. Under the assumption that $X \sim \text{Unif}\,[0, 1]$, the posterior distribution of X is

$$\frac{x^p(1-x)^q}{\int_0^1 x^p(1-x)^q\,dx}$$

(cf. Eq. 1.5). Laplace wished to show that

$$\Pr\left\{\frac{p}{p+q}-\theta < x < \frac{p}{p+q}+\theta\right\} \to 1 \quad \text{as} \quad p,q \to \infty \tag{1.48}$$

where $\theta > 0$ is arbitrarily small and proceeded by writing the probability on the left side above as

$$P = \frac{\int_{-\theta+p/(p+q)}^{\theta+p/(p+q)} x^p(1-x)^q\,dx}{\int_0^1 x^p(1-x)^q\,dx} \tag{1.49}$$

(cf. Eq. 1.7). In the above integral, define

$$y = x^p(1-x)^q. \tag{1.50}$$

Also set $p = 1/\alpha$, $q = \mu/\alpha$, and

$$y = \alpha z \frac{dy}{dx}. \tag{1.51}$$

By integrating Eq. (1.51) by parts successively,

* In an article probably written a few years before 1780, d'Alembert also provided a similar acknowledgment: "...the search for the probability of causes by events, which has been dealt with by several mathematicians. See in the *Philosophical Transactions* of 1763 & 1764 the works of MM. Bayes & Price on this subject..." (d'Alembert, 1780, p. 60).

† The derivation that follows is also described in Dale (1999, pp. 198–200).

‡ See also p. 13, where the same problem was considered using an urn model.

$$\begin{aligned}
\int y dx &= \alpha \int z dy \\
&= C + \alpha zy - \alpha \int y dz \\
&= C + \alpha yz - \alpha^2 \int z \frac{dy}{dx} dz \\
&= C + \alpha yz - \alpha^2 \int z \frac{dz}{dx} dy \\
&= C + \alpha yz - \alpha^2 \left\{ z \frac{dz}{dx} y - \int y d\left(z \frac{dz}{dx} \right) \right\} \\
&= C + \alpha yz - \alpha^2 \left\{ z \frac{dz}{dx} y - \int y \frac{d(zdx)}{dx} \right\} \\
&= C + \alpha yz - \alpha^2 \left\{ z \frac{dz}{dx} y - \int \alpha z \frac{dy}{dx} \frac{d(zdx)}{dx} \right\} \\
&= C + \alpha yz - \alpha^2 \left\{ z \frac{dz}{dx} y - \alpha \int z \frac{d(zdx)}{dx^2} dy \right\} \\
&\cdots \\
&= C + ayz \left[1 - \alpha \frac{dz}{dx} + \alpha^2 \frac{d(zdz)}{dx^2} - \alpha^3 \frac{d\{zd(zdz)\}}{dx^3} + \cdots \right]
\end{aligned} \tag{1.52}$$

(Laplace, 1781, OC 9, p. 424), where C is a constant of integration. Now, since $y = x^p(1-x)^q$, where $p = 1/\alpha$ and $q = \mu/\alpha$, and using Eq. (1.51), we have $z = \{x(1-x)\}/\{1-(1+\mu)x\}$. Therefore, (1.52) can be used to obtain "by a quick approximation"

$$\int_0^{-\theta+p/(p+q)} y dx = \int_0^{-\theta+1/(1+\mu)} y dx = \frac{\alpha \mu^{q+1} \{1-(1+\mu)\theta\}^{p+1} \left(1 + \frac{1+\mu}{\mu}\theta\right)^{q+1}}{\theta(1+\mu)^{p+q+3}} \times \left[1 - \frac{\alpha\{\mu + (1+\mu)^2 \theta\}}{(1+\mu)^3 \theta^2} + \cdots \right]$$

for $\alpha \ll \theta^2$. Laplace commented that:

> This [the above] series has the advantage of giving the limits between which the value of $\int y dx$ is contained; indeed, this value is less than the first term of the series and greater than the second term of the series. (*ibid.*, p. 425)

That is, for $\xi < 1/(1+\mu)$, in Eq. (1.52):

$$\alpha yz \left(1 - \alpha \frac{dz}{dx} \right) < \int_0^{\xi} y dx < \alpha yz.$$

Laplace next gave a more accurate approximation to $\int ydx$ and hence to P in Eq. (1.49). By integrating Eq. (1.52) from $x=-\theta+p/(p+q)=-\theta+1/(1+\mu)$ to $x=\theta+p/(p+q)=\theta+1/(1+\mu)$, he showed that Eq. (1.49) could be written as

$$
\begin{aligned}
P &= \frac{1}{\int_0^1 x^p(1-x)^q\,dx}\int_{-\theta+p/(p+q)}^{\theta+p/(p+q)} ydx \\
&= 1-\frac{\alpha^{1/2}\mu^{1/2}}{\sqrt{2\pi}(1+\mu)^{3/2}\theta}\left[1-\alpha\frac{\{12\mu^2+(1+\mu)^2(1+\mu+\mu^2)\theta^2\}}{12\mu(1+\mu)^3\theta^2}\right]\times \\
&\left[\{1-(1+\mu)\theta\}^{p+1}\left(1+\frac{1+\mu}{\mu}\theta\right)^{q+1}+\{1+(1+\mu)\theta\}^{p+1}\left(1-\frac{1+\mu}{\mu}\theta\right)^{q+1}\right]
\end{aligned}
\tag{1.53}
$$

(*ibid.*, p. 427). In the above, Laplace had made use of Stirling's formula to approximate the factorials, and terms which were $O(\alpha^{5/2})$ had been neglected. To finish off the proof of (1.48), Laplace noted that as p and q both increase, the expression in the last pair of square brackets decreases considerably. Moreover, since $p=1/\alpha$, $q=\mu/\alpha$, the quantity α also decreases. Hence, Laplace was able to conclude that

$$P\to 1 \quad \text{as} \quad p,q\to\infty$$

(*ibid.*, p. 429), which is what he wished to prove.

As an application, in Article XIX Laplace considered a variation of Arbuthnot's problem (cf. Section 5.8.1). From 1745 to 1770, the births in Paris comprised 251,527 boys and 241,945 girls. Laplace wished to determine the probability that "the possibility of the birth of a boy was less than or equal to ½." He took $1/(1+\mu)-\theta=1/2$ so that $\theta=(1-\mu)/\{2(1+\mu)\}$. Then, with $p=251{,}527$ and $q=241{,}945$, we have $\mu=q/p=.9619047$and $\alpha=1/p=1/251{,}527$. The required probability is $\Pr\{x\le 1/2\}=1-\Pr\{x>1/2\}=1-P$, where

$$
\begin{aligned}
1-P &= \frac{\sqrt{\dfrac{2\alpha\mu}{(1+\mu)\pi}}}{1-\mu}\left(\frac{1+\mu}{2}\right)^{p+1}\left(\frac{1+\mu}{2\mu}\right)^{q+1} \\
&\times\left\{1-\alpha\frac{48\mu^2+3\mu(1-\mu)^2+(1-\mu)^4}{12\mu(1+\mu)(1-\mu)^2}+\alpha^2\cdots\right\}<1.1521\times 10^{-42}.
\end{aligned}
$$

Laplace concluded:

> ...one can consider as truthful as any other moral truth, that the difference observed in Paris between the births of boys and girls is due to a greater possibility in the births of boys. (*ibid.*, pp. 431–432)

Laplace also commented on the births in London between 1664 till 1757. A total of 737,629 boys and 698,958 girls were born. Since the ratio of boys to girls was slightly larger and since there was a larger number of births in total, Laplace surmised that $1-P$ would be even smaller here.

1.2.6.5 Method of Asymptotic Approximation Refined The integration in Eq. (1.49) involves the evaluation of the integral $\int_0^1 ydx$ where

$$y = x^p (1-x)^q,$$

and both p and q are large positive integers. This integration is elementary since

$$\int_0^1 x^p (1-x)^q \, dx = \frac{p!q!}{(p+q+1)!} = \frac{1 \cdot 2 \cdots p \times 1 \cdot 2 \cdots q}{1 \cdot 2 \cdot 3 \cdots (p+q+1)}$$

and this can be simplified by using Stirling's formula $\nu! \sim \sqrt{2\pi\nu}(\nu/e)^\nu$ (where is ν large). But how shall we calculate $\int_0^1 ydx$ when y is an *arbitrary* function involving *large* powers of x?

This was essentially the problem that Laplace proceeded to solve in Article XXIII of the "Mémoire sur les probabilités." In his own words:

> We have been able to do this [i.e. evaluate $\int_0^1 ydx$] when $y = x^p(1-x)^q$ by means of the beautiful theorem of M. Stirling on the value of the product 1.2.3…u, when u is a very large number. But this procedure is indirect, and it is natural to think that there exists a method to directly determine k [i.e. the integral $\int_0^1 ydx$], for arbitrary y, and to which the theorem is a corollary: that which I shall present has seemed to me to fulfill this objective in the most general way. (Laplace, 1781, OC 9, p. 445)

Laplace's procedure used the *method of asymptotic approximation*, which he now refined, having introduced it previously in the "Mémoire sur la probabilité des causes par les événements"* (Laplace, 1774a, OC 8). We now describe Laplace's procedure. Let y be an arbitrary function of x satisfying $(y)_{x=0} = (y)_{x=1} = 0$, α be a very small constant, a be the value of x at which the maximum of y occurs such that $(y)_{x=a} = A$, and $\theta = x - a$. Then a Taylor series expansion of $g(x) = \alpha \log y$ about $x = a$ gives

$$\begin{aligned}
\alpha \log y &= \alpha \log A + (x-a)g'(a) + \frac{(x-a)^2}{2!} g''(a) + \frac{(x-a)^3}{3!} g'''(a) + \frac{(x-a)^4}{4!} g''''(a) + \cdots \\
&= \alpha \log A + \frac{\theta^2}{2!} g''(a) + \frac{\theta^3}{3!} g'''(a) + \frac{\theta^4}{4!} g''''(a) + \cdots \\
&= \alpha \log A - \theta^2 (f + f'\theta + f''\theta^2 + \cdots).
\end{aligned} \tag{1.54}$$

In the above, we have used $g'(a) = 0$, which follows from $(y')_{x=a} = 0$, and the $f^{(i)}$'s are constants defined by $f^{(i)} = -\theta^{i+1} g^{(i+1)}(a)/(i+1)!$ From Eq. (1.54), Laplace obtained

$$y = A \exp\left\{-\frac{\theta^2}{\alpha}(f + f'\theta + f''\theta^2 + \cdots)\right\}$$

* See Section 1.2.2.3.

so that

$$\begin{aligned}\int_0^1 y\,dx &= A\int_0^1 \exp\left\{-\frac{\theta^2}{\alpha}\left(f+f'\theta+f''\theta^2+\cdots\right)\right\}dx \\ &= A\int_{-a}^{1-a} \exp\left\{-\frac{\theta^2}{\alpha}\left(f+f'\theta+f''\theta^2+\cdots\right)\right\}d\theta,\end{aligned} \tag{1.55}$$

since $\theta = x - a$. Next, Laplace set

$$\frac{\theta^2}{\alpha}\left(f+f'\theta+f''\theta^2+\cdots\right)=t^2. \tag{1.56}$$

From Eq. (1.54), it is seen that Eq. (1.56) is obtained by making the transformation

$$\alpha \log y = \alpha \log A - \alpha t^2 \quad \Rightarrow \quad y = Ae^{-t^2}. \tag{1.57}$$

Also, Laplace used Lagrange's inversion formula* to invert Eq. (1.56) and wrote

$$\begin{aligned}\theta &= t\sqrt{\alpha}\,(f+f'\theta+f''\theta^2+\cdots)^{-1/2} \\ &= t\sqrt{\alpha}\left\{f+f't\sqrt{\alpha}\,(f+f'\theta+f''\theta^2+\cdots)^{-1/2}+f't^2\alpha(f+f'\theta+f''\theta^2+\cdots)^{-3/2}+\cdots\right\} \\ &= \alpha^{1/2}t\left\{h+h^{(1)}\alpha^{1/2}t+h^{(2)}\alpha t^2+\cdots\right\}.\end{aligned} \tag{1.58}$$

From the above,

$$d\theta = \alpha^{1/2}\left\{h+2h^{(1)}\alpha^{1/2}t+3h^{(2)}\alpha t^2+\cdots\right\}dt. \tag{1.59}$$

Thus, using Eqs. (1.58) and (1.59), Eq. (1.55) becomes

$$\int_0^1 y\,dx = A\alpha^{1/2}\int_{-\infty}^{\infty} e^{-t^2}\left\{h+2h^{(1)}\alpha^{1/2}t+3h^{(2)}\alpha t^2+\cdots\right\}dt. \tag{1.60}$$

In the above, the limits $\pm\infty$ are obtained from Eq. (1.57), namely, $y = Ae^{-t^2}$, and by using $(y)_{x=0} = (y)_{x=1} = 0$. Laplace next used, for $n = 1, 2, \ldots,$

$$\begin{aligned}&\int_{-\infty}^{\infty} t^{2n-1}e^{-t^2}dt = 0, \\ &\int_{-\infty}^{\infty} t^{2n}e^{-t^2}dt = 2\frac{(1)(3)\cdots(2n-1)}{2^n}\int_{-\infty}^{\infty} e^{-t^2}dt.\end{aligned}$$

Thus, Eq. (1.60) becomes

$$\int_0^1 y\,dx = k = 2\alpha^{1/2}A\left\{h+\frac{1\cdot 3}{2}h^{(2)}\alpha+\frac{1\cdot 3\cdot 5}{2^2}h^{(4)}\alpha^2+\frac{1\cdot 3\cdot 5\cdot 7}{2^3}h^{(6)}\alpha^3+\cdots\right\}\int_{-\infty}^{\infty} e^{-t^2}dt \tag{1.61}$$

*Lagrange's inversion formula states: suppose $t = (x-a)/v(x)$ where $v(x)$ is analytic in a neighborhood of $x = a$ such that $v(a) \neq 0$. Then a power series of x in terms of t is given by $x = a + \sum_{j=1}^{\infty} b_j t^j$ where $b_j = (d^{j-1}\{v(x)\}^j / dx^{j-1})_{x=0} / j!$

(Laplace, 1781, OC 9, p. 447). Therefore, "it only remains to have the integral $\int e^{-t^2}dt$ from $t=0$ to $t=\infty$." Unaware that he had already indirectly obtained this integral in his 1774 paper "Mémoire sur la probabilité des causes par les événements" (Laplace, 1774a),* Laplace devised a new technique† and obtained

$$\int_0^\infty e^{-t^2}dt = \frac{\sqrt{\pi}}{2} \tag{1.62}$$

(*ibid.*, pp. 447–448). Equation (1.61) then becomes

$$k = \int_0^1 ydx = A\sqrt{\pi\alpha}\left\{h + \frac{1\cdot 3}{2}h^{(2)}\alpha + \frac{1\cdot 3\cdot 5}{2^2}h^{(4)}\alpha^2 + \frac{1\cdot 3\cdot 5\cdot 7}{2^3}h^{(6)}\alpha^3 + \cdots\right\}. \tag{1.63}$$

It now remained to find the constants $h, h^{(2)}, h^{(4)}, \ldots$ From Eq. (1.57), Laplace wrote

$$\theta = t\sqrt{\frac{\theta^2}{\log(A/y)}}.$$

By using a result in his "Mémoire sur l'usage du calcul aux différence partielles" (Laplace, 1780), Laplace was able to find the coefficients $h, h^{(2)}, h^{(4)}, \ldots$ in Eq. (1.58):

$$h^{(2n)} = \frac{1}{1\cdot 2\cdot 3\cdots(2n+1)\alpha^{n+1/2}}\left[\frac{d^{2n}}{d\theta^{2n}}\left\{\theta^{2n+1}\left(\log\frac{A}{y}\right)^{-n-1/2}\right\}\right]_{\theta=0} \tag{1.64}$$

(Laplace, 1781, OC 9, p. 448). Since α had been assumed to be a very small constant, Laplace was concerned with only the first term of Eq. (1.63), that is,

$$k = \int_0^1 ydx = A\sqrt{\pi\alpha}\,h. \tag{1.65}$$

Next a Taylor series expansion of $g(y) = \log y$ about $x = a$ (the point at which the maximum A of y occurs) gives

$$g(y) = g(a) + (x-a)g'(a) + \frac{(x-a)^2}{2!}g''(a) + \cdots.$$

Let $\theta = x - a$. Since $g'(a) = 0$, the above becomes

$$\log y = \log A + \frac{\theta^2}{2!}\left[\frac{d^2y}{yd\theta^2} - \frac{1}{y^2}\left(\frac{dy}{d\theta}\right)^2\right]_{\theta=0} + \cdots = \log A + \frac{\theta^2}{2A}\left(\frac{d^2y}{d\theta^2}\right)_{\theta=0} + \cdots$$

$$\Rightarrow \quad \log\left(\frac{A}{y}\right) = -\frac{\theta^2}{2A}\left(\frac{d^2y}{d\theta^2}\right)_{\theta=0}.$$

* See Eq. (1.12).

† We shall describe Laplace's technique next.

Substituting for log(A/y) in Eq. (1.64),

$$h=\left[\frac{\theta}{\alpha^{1/2}}\left\{-\frac{\theta^2}{2A}\left(\frac{d^2y}{d\theta^2}\right)_{\theta=0}\right\}^{-1/2}\right]_{\theta=0}=\sqrt{\frac{-2A}{\alpha\left(\frac{d^2y}{d\theta^2}\right)_{\theta=0}}}.$$

Using Eq. (1.65), Laplace finally obtained the formula

$$\int_0^1 ydx=A\sqrt{\pi\alpha}\sqrt{\frac{-2A}{\alpha\left(\frac{d^2y}{d\theta^2}\right)_{\theta=0}}}=\left(\sqrt{\frac{2\pi y^3}{-\frac{d^2y}{dx^2}}}\right)_{x=a} \tag{1.66}$$

(*ibid.*, p. 449). This describes Laplace's method of integration by using asymptotic approximations.

As an application of his method, in Article XXV Laplace referred to Article XVIII of his current memoir where he had to evaluate the integral $\int_0^1 ydx$, with

$$y=x^p(1-x)^q$$

(p and q are positive large integers), by making use of Stirling's formula.* Laplace had then obtained

$$\begin{aligned}\int_0^1 ydx&=\int_0^1 x^p(1-x)^q\,dx\\&=\frac{1\cdot 2\cdots p\times 1\cdot 2\cdots q}{1\cdot 2\cdot 3\cdots(p+q+1)}\\&=\frac{\sqrt{2\pi\alpha}\,\mu^{q+1/2}}{(1+\mu)^{p+q+3/2}}\left[1+\alpha\frac{\left\{(1+\mu)^2-13\mu\right\}}{12\mu(1+\mu)}+\cdots\right],\end{aligned} \tag{1.67}$$

where $p=1/\alpha$, $q=\mu/\alpha$, and α is a very small quantity (*ibid.*, p. 427). Laplace now wished to reproduce the formula in Eq. (1.67) by using his method of asymptotic approximation. First, he noted that the maximum of y occurred at $x=a$ where

$$a=\frac{p}{p+q}=\frac{1}{1+\mu}.$$

Then

$$\begin{aligned}(y)_{x=a}&=A=\frac{\mu^q}{(1+\mu)^{p+q}}=\frac{\mu^{\mu/\alpha}}{(1+\mu)^{(1+\mu)/\alpha}}.\\ \Rightarrow\quad \alpha\log A&=\mu\log\left(\frac{\mu}{1+\mu}\right)+\log\left(\frac{1}{1+\mu}\right).\end{aligned} \tag{1.68}$$

*Stirling's formula: $\nu!\sim\sqrt{2\pi\nu}(\nu/e)^\nu$ (where is ν large).

Also,

$$\begin{aligned} y &= x^p(1-x)^q = x^{1/\alpha}(1-x)^{\mu/\alpha}. \\ \Rightarrow \quad \alpha \log y &= \log x + \mu \log(1-x) \\ &= \log(a+\theta) + \mu \log(1-a-\theta) \\ &= \mu \log\left(\frac{\mu}{1+\mu} - \theta\right) + \log\left(\frac{1}{1+\mu} + \theta\right), \end{aligned} \tag{1.69}$$

where $x = a + \theta$. From Eqs. (1.68) and (1.69), Laplace obtained

$$\begin{aligned} \log A - \log y &= -\frac{1}{\alpha}\log\{1+(1+\mu)\theta\} - \frac{\mu}{\theta}\log\left(1 - \frac{1+\mu}{\mu}\theta\right) \\ &= \frac{(1+\mu)^2(1+\mu)}{2\alpha\mu}\theta^2 + \frac{(1+\mu)^3(1-\mu^2)}{3\alpha\mu^2}\theta^3 + \frac{(1+\mu)^4(1+\mu^3)}{4\alpha\mu^3}\theta^4 + \cdots. \end{aligned} \tag{1.70}$$

From Eq. (1.64),

$$\begin{aligned} h &= \left[\frac{\theta}{\alpha^{1/2}}\left(\log\frac{A}{y}\right)^{-1/2}\right]_{\theta=0} \\ &= \frac{\sqrt{2\mu\alpha}}{\alpha^{1/2}(1+\mu)^{3/2}}, \\ h^{(2)} &= \frac{1}{(3\cdot 2\cdot 1)\alpha^{3/2}}\left[\frac{d^2}{d\theta^2}\left\{\theta^3\left(\log\frac{A}{y}\right)^{-3/2}\right\}\right]_{\theta=0} \\ &= \frac{\alpha\sqrt{2\mu\alpha}}{\{\alpha(1+\mu)\}^{3/2}} \frac{\{(1+\mu)^2 - 13\mu\}}{18\mu(1+\mu)}, \\ &\dots. \end{aligned}$$

Hence, Laplace used Eq. (1.63) to obtain

$$\begin{aligned} k &= \int_0^1 y\,dx \\ &= A\sqrt{\pi\alpha}\left\{L_1 + \frac{(1)(3)}{2}L_3\alpha + \cdots\right\} \\ &= \frac{\sqrt{2\pi\alpha}\,\mu^{q+1/2}}{(1+\mu)^{p+q+3/2}}\left[1 + \alpha\frac{\{(1+\mu)^2 - 13\mu\}}{12\mu(1+\mu)} + \cdots\right], \end{aligned}$$

(*ibid.*, p. 455), which is the same as his previous solution in Eq. (1.67). Note that Laplace's method here does not depend on Stirling's formula. Note also that, if Laplace had simply used (1.66) he would have obtained the less accurate, but much faster, approximation

$$\int_0^1 y dx = \frac{\sqrt{2\pi\alpha}\,\mu^{q+1/2}}{(1+\mu)^{p+q+3/2}},$$

which is the first term of the series in (1.67).

1.2.6.6 Stirling's Formula The reader may remember that Laplace wished to develop a general method to which Stirling's formula would be a corollary.* In Article XXV, Laplace proved that indeed his method of asymptotic approximation could be used to derive Stirling's formula. We now describe his derivation. First Laplace let

$$y = x^p e^{-x}$$

(*ibid.*, p. 456), where p is a positive large integer. Since $\int_0^\infty x^p e^{-x}dx = p\int_0^\infty x^{p-1}e^{-x}dx$, he wrote

$$\int_0^\infty x^p e^{-x} dx = 1\cdot 2\cdot 3\cdots p. \tag{1.71}$$

Now the maximum of y occurs at $x = a$ where

$$a = p.$$

Then

$$(y)_{x=a} = A = p^p e^{-p} = \left(\frac{1}{\alpha}\right)^{1/\alpha} e^{-1/\alpha}.$$

$$\Rightarrow \quad \alpha \log A = -\log\alpha - 1,$$

where $\alpha = 1/p$. Also,

$$\begin{aligned} y &= x^p e^{-x} = x^{1/\alpha} e^{-x} \\ \Rightarrow \quad \alpha \log y &= \log x - \alpha x \\ &= \log(a+\theta) - \alpha(a+\theta) \\ &= \log\left(\frac{1}{\alpha}+\theta\right) - \alpha\left(\frac{1}{\alpha}+\theta\right) \\ &= \log(1+\alpha\theta) - \log\alpha - 1 - \alpha\theta. \end{aligned}$$

Therefore,

$$\log A - \log y = \theta - \frac{1}{\alpha}\log(1+\alpha\theta).$$

Now, from (1.64),

$$h = \frac{\sqrt{2}}{\alpha}, \; h^{(2)} = \frac{\sqrt{2}}{18\alpha}.$$

* See quotation from Laplace above Eq. (1.54).

Hence, from Eq. (1.63),

$$\begin{aligned}\int_0^1 y\,dx &= A\sqrt{\pi\alpha}\left\{\frac{\sqrt{2}}{\alpha}+\frac{(1)(3)}{2}\cdot\frac{\sqrt{2}}{18\alpha}\cdot\alpha+\cdots\right\}\\ &= p^p e^{-p}\sqrt{\frac{\pi}{p}}\left(p\sqrt{2}+\frac{1}{6\sqrt{2}}+\cdots\right)\\ &= p^{p+1/2}e^{-p}\sqrt{2\pi}\left(1+\frac{1}{12p}+\cdots\right).\end{aligned}$$

Since $\int_0^\infty x^p e^{-x}dx = p!$ (cf. Eq. 1.71), Laplace thus established Stirling's formula

$$1\cdot 2\cdot 3\cdots p = p^{p+1/2}e^{-p}\sqrt{2\pi}\left(1+\frac{1}{12p}+\cdots\right)\quad \text{for large } p$$

(*ibid.*, p. 458).

1.2.6.7 Direct Evaluation of $\int_0^\infty e^{-t^2}dt$ We mentioned a few paragraphs ago that in the "Mémoire sur les Probabilités" (Laplace, 1781, OC 9, pp. 447–448) Laplace obtained the important result

$$\int_0^\infty e^{-t^2}dt = \frac{\sqrt{\pi}}{2}$$

(cf. Eq. 1.62). In fact, he had already indirectly obtained this integral in his 1774 paper "Mémoire sur la probabilité des causes par les événements" (Laplace, 1774a).* Let us now see how he derived this result in 1781. First, Laplace considered the double integral

$$\begin{aligned}\int_0^\infty\int_0^\infty e^{-s(1+u^2)}ds\,du &= \int_0^\infty\left[-\frac{1}{1+u^2}e^{-s(1+u^2)}\right]_0^\infty du\\ &= \int_0^\infty \frac{1}{1+u^2}du\\ &= \left[\tan^{-1}u\right]_0^\infty\\ &= \frac{\pi}{2}.\end{aligned}$$

By substituting $u\sqrt{s}=t$ and reversing the order of integration, the above becomes:

$$\begin{aligned}\int_0^\infty\int_0^\infty e^{-s}e^{-t^2}du\,ds &= \int_0^\infty\int_0^\infty \frac{e^{-s}e^{-t^2}}{\sqrt{s}}dt\,ds\\ &= B\int_0^\infty \frac{e^{-s}}{\sqrt{s}}ds,\end{aligned}\tag{1.72}$$

* See also Eq. (1.12).

where $B=\int_0^\infty e^{-t^2}dt$. Now let $s=s'^2$ so that

$$\int_0^\infty \frac{e^{-s}}{\sqrt{s}}ds = 2\int_0^\infty e^{-s'^2}ds' = 2B.$$

Therefore, Eq. (1.72) becomes

$$\frac{\pi}{2} = B(2B) = 2B^2,$$

whence

$$B = \int_0^\infty e^{-t^2}dt = \frac{\sqrt{\pi}}{2}$$

(Laplace, 1781, OC 9, p. 448).

1.2.6.8 Theory of Errors In Articles XXX–XXXIII of the "Mémoire sur les probabilités," Laplace turned his attention to the theory of errors having already visited the topic briefly in Article XII.* He repeated some of the remarks he made in the "Mémoire de la probabilité des causes par les événements" concerning the works of Lagrange, Daniel Bernoulli, and Euler (Laplace, 1774a, OC 8, pp. 41–42).†

Laplace assumed that the *facilité* (probability) of an error x is $\phi(x)$ for the first observer, $\phi'(x)$ for the second observer, $\phi''(x)$ for the third observer, and so on. All probability functions are assumed to be even functions of x. Suppose now that the intervals between the first and second, second and third, ... observations be the observed quantities $p, p', \ldots$. Then the errors in the successive observations are $0 \le z \le c$ and the joint probability of $p, p', \ldots$ is $\phi(x)\phi'(p-x)\phi''(p'-x)\ldots$. The latter is also the posterior distribution of x, given the $p, p', \ldots$ and assuming a uniform prior for x, and can be written as

$$y = \phi(x)\phi'(p-x)\phi''(p'-x)\ldots, \tag{1.73}$$

where $-f \le x \le c-f$. Laplace called (1.73) the *courbe des probabilités* (probability curve).

Next, Laplace commented that, for a given set of observations, there was an infinite number of possible means depending on the conditions one imposed. Laplace gave four such possible conditions:

> [1] the sum of the positive errors should be equal to the sum of negative errors; ...[2] the sum of positive errors, multiplied by their respective probabilities, should be equal to the sum of the negative errors, multiplied by their respective probabilities. [3] One can also force the mean to be the point where the true instance of the phenomenon is most probable, as M. Daniel Bernoulli has done... [4] There is one which belongs to the nature of the problem and which should serve to fix the mean to be chosen from several observations; this condition is that, by fixing the instant of the phenomenon to this point, the resulting error should be a minimum; now since, in the usual theory of chances, one evaluates the advantage by taking

*Section 1.2.6.2.

†See also p. 20.

> the sum of the product of each possible advantage, multiplied by the probability of obtaining it, so also here the error should be estimated by the sum of the products of each possible error, multiplied by its probability; the mean to be chosen should thus be such that the sum of the products should be less than for any other instant. (Laplace, 1781, OC 9, p. 477)

Laplace's four conditions can be written as (Sheynin, 1977, pp. 9–10)

1. $\int_{-N}^{0} ydx = \int_{0}^{N} ydx,$

where N is the maximum possible error;

2. $\int_{-N}^{0} xydx = \int_{0}^{N} xydx;$
3. maximum likelihood;
4. $\min_{x} \int_{x} xydx.$

Laplace's preference was thus for the fourth condition. He made the substitution $x = z - f$ so that $0 \le z \le c$. However, because of the assumption of evenness, Laplace actually used $-c \le z \le c$. If h is the value of z that should be chosen according to condition (4), then the quantity

$$k\int_{0}^{c}|z-h|ydz = k\int_{0}^{h}(h-z)ydz + k\int_{h}^{c}(z-h)ydz \tag{1.74}$$

is a minimum, where k is a constant that Laplace introduced to normalize the *courbe des probabilités*. Laplace now varied the quantity h by a small amount δh so that the variation on the right side of (1.74) is

$$k\delta h\int_{0}^{h} ydz - k\delta h\int_{h}^{c} ydz,$$

which should be set to zero, (1.74) being a minimum. Therefore,

$$\int_{0}^{h} ydz = \int_{h}^{c} ydz, \tag{1.75}$$

which is the same condition as (1), and results in h being the median of y.

The above presupposes that the *courbe des probabilités* y is known. But most of the time, we are told by Laplace, this is not the case, and "the calculus of probability must provide the means to compensate for this ignorance." Referring to his Article XIII,* Laplace then stated:

> …if, in that case, $\pm a$, $\pm a'$, $\pm a''$, … are the limits of the errors of the first, second,…observation, one must assume
>
> $$\phi(x) = \tfrac{1}{2a}\log\tfrac{a}{x},\quad \phi'(x) = \tfrac{1}{2a'}\log\tfrac{a'}{x},\quad \ldots$$
>
> (Laplace, 1781, OC 9, p. 479)

* See p. 46.

But Laplace was fully cognizant that the last distribution did not lend itself to easy analysis:

> But one must agree that they make the preceding method very difficult to apply: but in exposing it my aim has rather been to inform of everything that the doctrine of chances can shed light on this matter.... (*ibidem*)

It is unfortunate that Laplace did not try the Gaussian form $\phi(x) \propto e^{-x^2}$ for he would have found the analysis much more tractable. But then it is hard to conceive how he could have reached such a form from the principle of indifference, devoted as he was to the latter.

1.2.7 "Mémoire sur les suites" (1782)

1.2.7.1 De Moivre and Generating Functions This memoir is important because it is here that Laplace introduced and named the method of *generating functions*. The technique had first been used by Abraham de Moivre in the *Miscellanea Analytica* (de Moivre, 1730, Book VII, pp. 194–197) (Fig. 1.12) in the context of finding the number of ways of obtaining a given sum when n fair dice with f faces each are thrown.* The problem was reproduced in the second edition of de Moivre's *Doctrine of Chances* (de Moivre, 1738):

> To find how many Chances there are upon any number of Dice, each of them of the same number of Faces, to throw any given number of points. (*ibid.*, p. 35)

Before moving on to Laplace's paper, let us briefly examine de Moivre's reasoning in terms of generating functions through an example. Suppose we wanted to find the total number of ways of obtaining a sum of 15 with five throws of a fair six-faced die. De Moivre's genius consisted of instead considering five throws of a second regular die that had t faces marked 1, t^2 faces marked 2, ..., and t^6 faces marked 6. Then, for a sum of 15 with the second die, one possible way would be to obtain the five-tuple (1, 1, 1, 6, 6), which can be obtained in $t \times t \times t \times t^6 \times t^6 = t^{15}$ ways, by using the product rule; another possibility is to obtain the five-tuple $(1, 2, 3, 6, 3)$, which can be obtained in $t \times t^2 \times t^3 \times t^6 \times t^3 = t^{15}$ ways. Similarly, by considering all other possibilities, it can be seen that the total number of ways of obtaining a sum of 15 with the second die is given by the term involving t^{15} in

$$\underbrace{\left(t+t^2+t^3+t^4+t^5+t^6\right)\times\left(t+t^2+t^3+t^4+t^5+t^6\right)\times\cdots\times\left(t+t^2+t^3+t^4+t^5+t^6\right)}_{5\text{ times}}.$$

Upon expansion, it will be seen that the total number of ways for the second die is $651t^{15}$. Now the die becomes the same as the first one by setting $t = 1$. Hence, the total number of ways for the first type of die is 651 (note that this is the coefficient of t^{15} in the expansion of $(t + t^2 + \cdots + t^6)^5$). Note, however, that solving the same problem without introducing the second die would be much less easy, as the reader can verify, since it would involve finding *all* five-tuples having a sum of 15.

*See also Seal (1949).

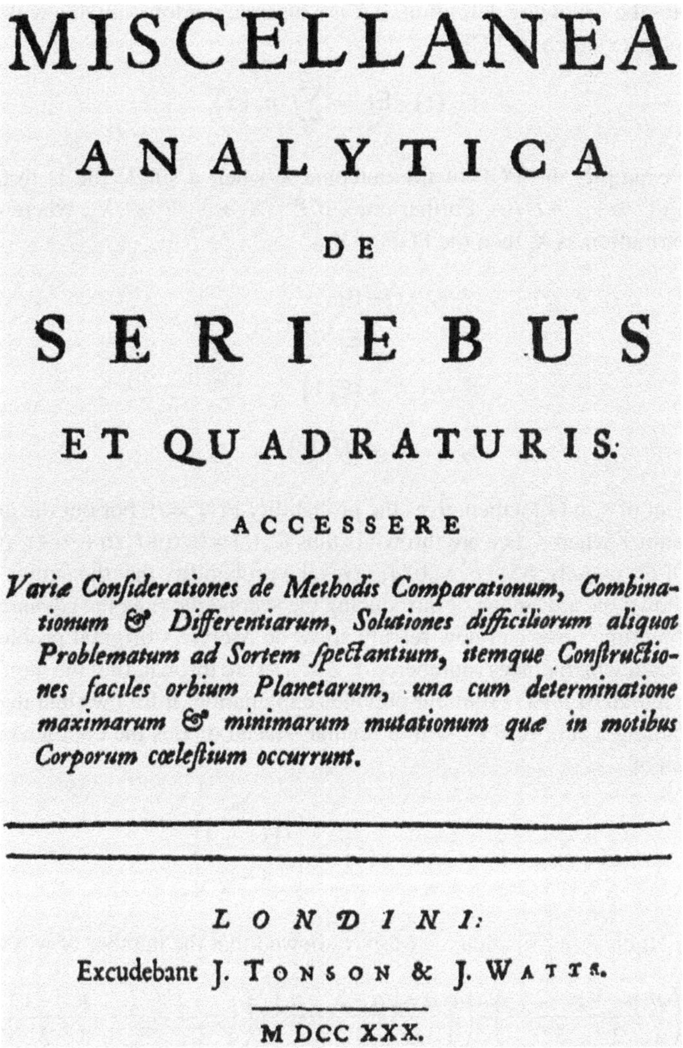

MISCELLANEA

ANALYTICA

DE

SERIEBUS

ET QUADRATURIS:

ACCESSERE

Variæ Considerationes de Methodis Comparationum, Combinationum & Differentiarum, Solutiones difficiliorum aliquot Problematum ad Sortem spectantium, itemque Constructiones faciles orbium Planetarum, una cum determinatione maximarum & minimarum mutationum quæ in motibus Corporum cœlestium occurrunt.

LONDINI:

Excudebant J. TONSON & J. WATTS.

M DCC XXX.

FIGURE 1.12 Title page of de Moivre's Miscellanea Analytica (de Moivre, 1730)

Introducing the second die leads to the expression $(t+t^2+\cdots+t^6)$ for the total number of faces for that die. This expression is its associated *generating function*. From this initial generating function, it is not difficult to reach another type of generating function, namely, the *probability generating function (PGF)*. In the earlier example, suppose we wanted to find the *probability* of obtaining a sum of 15 by throwing $n=5$ of the first type of die, assuming $f=6$. To do this we need to take the coefficient of t^{15} in the expansion of $(t+t^2+\cdots+t^6)^5$ and then divide by 6^5, which is the same as taking the coefficient of t^{15} in the expansion of

$$\left(\frac{1}{6}t+\frac{1}{6}t^2+\frac{1}{6}t^3+\frac{1}{6}t^4+\frac{1}{6}t^5+\frac{1}{6}t^6\right)^5.$$

This motivates the following definition. If X is a discrete random variable with probability mass function $p_X(x)$, then its PGF is

$$G_X(t) = \mathsf{E}t^X = \sum_x t^x p_X(x).$$

For our die example, the PGF of the outcome X when a single die is thrown is thus $G_X(t) = (t/6 + t^2/6 + \cdots + t^6/6)$. Furthermore, if $Y = X_1 + X_2 + \cdots + X_n$, where each X_i has the same distribution as X, then the PGF of Y is

$$\begin{aligned} G_Y(t) &= \mathsf{E}t^Y \\ &= \mathsf{E}t^{X_1+X_2+\cdots+X_n} \\ &= \left(\mathsf{E}t^X\right)^n \\ &= \{G_X(t)\}^n. \end{aligned}$$

The coefficient of t^i in $G_Y(t)$ then gives the probability $\Pr\{Y = i\}$. For our die example, the PGF of the sum Y when 5 dice are thrown is thus $G_Y(t) = (t/6 + t^2/6 + \cdots + t^6/6)^5$, and the coefficient of t^{15}, namely, $651/6^5 \approx .084$, gives the probability that the sum is 15.

Having shown the advantage of introducing the second die (i.e., the advantage of using generating functions), we can now readily solve de Moivre's original problem: given n regular dice each which f faces (numbered $1, 2, \ldots, f$) are thrown, find the number of ways of obtaining a total of $p+1$. From our previous explanation, it follows that the number of ways of obtaining a sum of $b+1$ with n regular f-faced dice is the coefficient of $p+1$ in the expansion of

$$\left(t + t^2 + \cdots + t^f\right)^n = \left\{\frac{t\left(1-t^f\right)}{1-t}\right\}^n$$

where $|t| < 1$. Upon simplification, de Moivre showed that the number of ways is

$$\left(\frac{p}{1}\frac{p-1}{2}\frac{p-2}{3}\cdots\right) - \left(\frac{q}{1}\frac{q-1}{2}\frac{q-2}{3}\cdots\frac{n}{1}\right) + \left(\frac{r}{1}\frac{r-1}{2}\frac{r-2}{3}\cdots\frac{n}{1}\frac{n-1}{2}\right)$$
$$-\left(\frac{s}{1}\frac{s-1}{2}\frac{s-2}{3}\cdots\frac{n}{1}\frac{n-1}{2}\frac{n-2}{3}\right) + \cdots$$

where $p - f = q$, $q - f = r$, $r - f = s$, etc. (de Moivre, 1730, p. 194; 1738, p. 35). The above formula needs to be continued until a zero or negative factor is obtained. In modern notation the formula can be written as

$$\sum_{i=0}^{\infty}(-1)^i\binom{n}{i}\binom{p-fi}{n-1}.$$

As we have thus shown, de Moivre was the first to introduce the method of generating function. However, it was in the able hands of Laplace that the method of generating functions was to blossom and find applications in all kinds of probability problems, which would have otherwise been almost intractable. Indeed, Laplace regarded generating functions so central to his probability program that Book I of his later major book *Théorie Analytique des Probabilités* (Laplace, 1812) would be completely devoted to the technique (spanning about 170 pages in the first edition of the book). Moreover, he also extended this technique to the more general method of *characteristic functions* (CF).

1.2.7.2 Lagrange's Calculus of Operations as an Impetus for Laplace's Generating Functions In the "Mémoire sur les Suites" Laplace made these important remarks:

> ...the relationship of generating functions to the corresponding variables leads to me to the singular analogy of positive powers with differentials and of negative powers with integrals, which was first observed by Leibnitz, and since then put into greater daylight by M. de la Grange (Mémoires de Berlin, année 1772). (Laplace, 1782, OC 10, p. 2)

Indeed, Laplace's search for a proof of an important result by Lagrange on the calculus of operations was very probably the genesis of his idea of generating functions. We now explain Lagrange's work before moving on to Laplace.

Lagrange's work on the calculus of operations appeared in a 1774 paper entitled "Sur une nouvelle espèce de calcul rélatif à la différentiation & à l'intégration des quantités variables" (Lagrange, 1774) (Fig. 1.13).* In this memoir, Lagrange proved that, for any function $u = u(x)$,

$$\Delta u = \exp\left(\xi \frac{du}{dx}\right) - 1 \qquad (1.76)$$

(Lagrange, 1774, Oeuvres 3, p. 450), where ξ is a small change in x, and $\Delta u = u(x+\xi) - u(x)$. From Eq. (1.76), he provided the following generalization:

$$\Delta^{\lambda} u = \left\{\exp\left(\xi \frac{du}{dx}\right) - 1\right\}^{\lambda} \qquad (1.77)$$

(*ibidem*), where λ is a positive integer, and Δ^{λ} denotes Δ iterated λ times.

Not only does Eq. (1.76) shows the link between calculus and finite differences, but it also formalizes an observation Leibniz had made previously (Leibniz, 1971) concerning the analogy between differentiation and exponentiation:

$$\text{differentiation}: \quad d^n(fg) = \sum_{i=0}^{n} \binom{n}{i} (d^i f)(d^{n-i} g),$$

$$\text{exponentiation}: \quad (f+g)^n = \sum_{i=0}^{n} \binom{n}{i} f^i g^{n-i},$$

where n is a positive integer. Lagrange's proof of (1.76) was relatively straightforward and as follows. First he established the Taylor series expansion:

$$u(x+\xi) = u(x) + \xi \frac{du}{dx} + \frac{\xi^2}{2} \frac{d^2u}{dx^2} + \frac{\xi^3}{2 \cdot 3} \frac{d^3u}{dx^3} + \cdots. \qquad (1.78)$$

* See also Gillispie (2000, Chapters 9 and 11) and Lubet (2004, pp. 204–211).

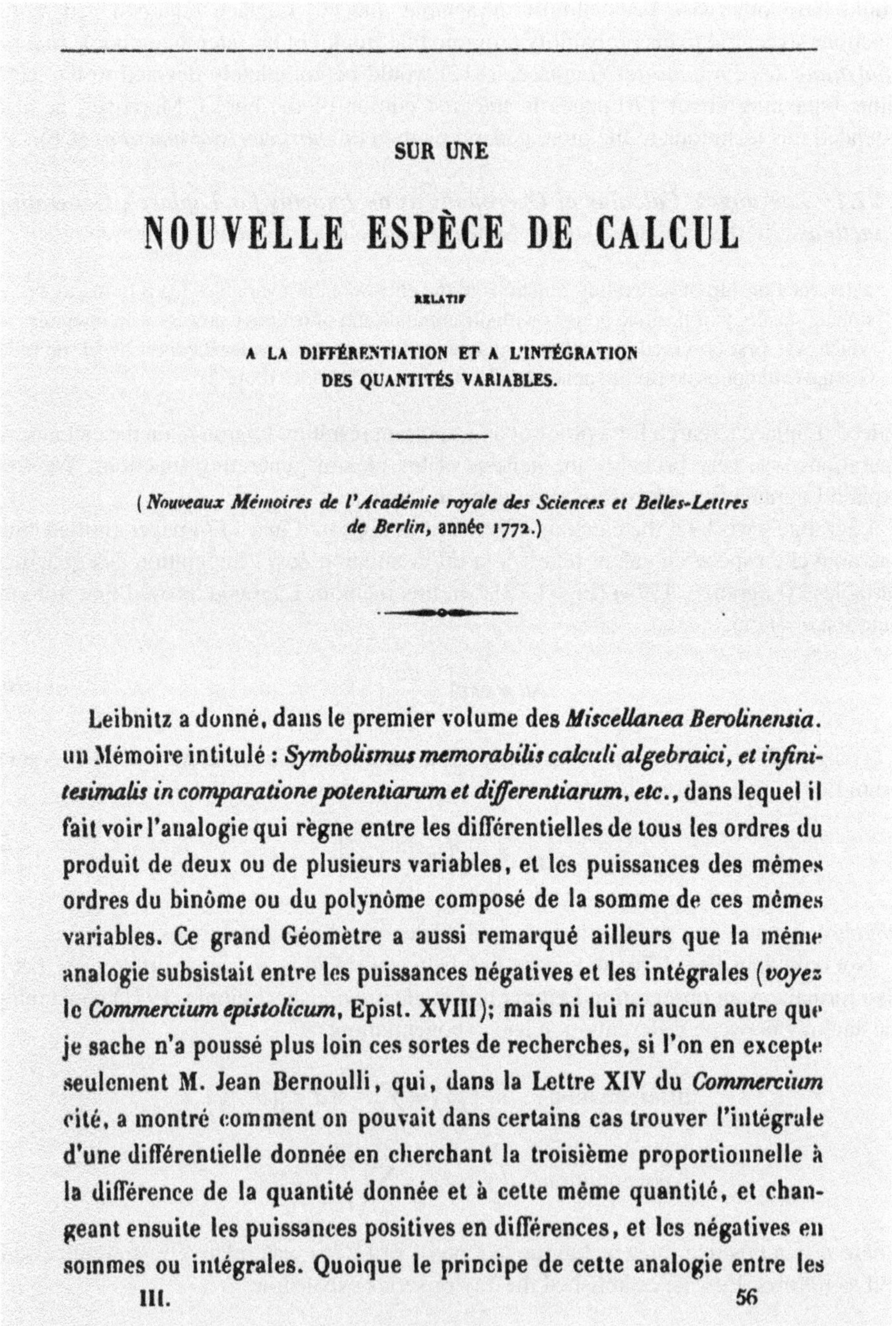

SUR UNE

NOUVELLE ESPÈCE DE CALCUL

RELATIF

A LA DIFFÉRENTIATION ET A L'INTÉGRATION
DES QUANTITÉS VARIABLES.

(*Nouveaux Mémoires de l'Académie royale des Sciences et Belles-Lettres de Berlin*, année 1772.)

Leibnitz a donné, dans le premier volume des *Miscellanea Berolinensia*, un Mémoire intitulé : *Symbolismus memorabilis calculi algebraici, et infinitesimalis in comparatione potentiarum et differentiarum, etc.*, dans lequel il fait voir l'analogie qui règne entre les différentielles de tous les ordres du produit de deux ou de plusieurs variables, et les puissances des mêmes ordres du binôme ou du polynôme composé de la somme de ces mêmes variables. Ce grand Géomètre a aussi remarqué ailleurs que la même analogie subsistait entre les puissances négatives et les intégrales (*voyez* le *Commercium epistolicum*, Epist. XVIII); mais ni lui ni aucun autre que je sache n'a poussé plus loin ces sortes de recherches, si l'on en excepte seulement M. Jean Bernoulli, qui, dans la Lettre XIV du *Commercium* cité, a montré comment on pouvait dans certains cas trouver l'intégrale d'une différentielle donnée en cherchant la troisième proportionnelle à la différence de la quantité donnée et à cette même quantité, et changeant ensuite les puissances positives en différences, et les négatives en sommes ou intégrales. Quoique le principe de cette analogie entre les

III. 56

FIGURE 1.13 First page of Lagrange's "Sur une nouvelle espèce de calcul relatif à la différentiation et à l'intégration des quantités variables" (Lagrange, 1774)

Then he expanded $\exp(\xi du/dx)-1$ in powers of du as follows:

$$\exp\left(\xi\frac{du}{dx}\right)-1=1+\frac{\xi}{dx}du+\frac{1}{2}\frac{\xi^2}{dx^2}du^2+\frac{1}{2\cdot 3}\frac{\xi^3}{dx^3}du^3+\cdots-1$$
$$=\frac{\xi}{dx}du+\frac{1}{2}\frac{\xi^2}{dx^2}du^2+\frac{1}{2\cdot 3}\frac{\xi^3}{dx^3}du^3+\cdots.$$

Lagrange next replaced du^j by d^ju so that the above becomes

$$\exp\left(\xi\frac{du}{dx}\right)-1=\xi\frac{du}{dx}+\frac{1}{2}\xi^2\frac{d^2u}{dx^2}+\frac{1}{2\cdot 3}\xi^3\frac{d^3u}{dx^3}+\cdots.$$

Using (1.78), the last expression becomes

$$\exp\left(\xi\frac{du}{dx}\right)-1=u(x+\xi)-u(x)=\Delta u$$

and Eq. (1.76) is established.

We make the following comment regarding Lagrange's proof. Of course we cannot just replace du^j by d^ju, as Lagrange did, because these two quantities are not equal. Indeed, Eq. (1.76) is incorrect and should instead be written as

$$\Delta u=\exp\left(\xi\frac{d}{dx}\right)u-1. \tag{1.79}$$

The above can also be written in terms of the operators Δ and d/dx as follows:

$$\Delta=\exp\left(\xi\frac{d}{dx}\right)-1.$$

If this form is used, then all the results that follow hold true without having to make any changes in the differentials.

Let us also consider some other related results obtained by Lagrange. When $\mu=-\lambda$ $(\lambda>0)$ is a negative integer, Lagrange wrote $\Delta^\mu=\Sigma^\lambda$ and Eq. (1.77) becomes

$$\sum^\lambda u=\left\{\exp\left(\xi\frac{du}{dx}\right)-1\right\}^{-\lambda} \tag{1.80}$$

(Lagrange, 1774, Oeuvres 3, p. 451). In the above, the right side needs to be expanded in powers of du, then $(du)^{-j}$ $(j>0)$ needs to be replaced by $d^{-j}u$, and finally d^{-j} needs to be replaced by $\int^j$, that is, $\int$ iterated j times. Furthermore, if $u=u(x,y,z,\ldots)$ then Eqs. (1.77) and (1.80) are both extended naturally as follows, for λ a positive integer:

$$\Delta^\lambda u=\left[\exp\left\{\xi\left(\frac{\partial u}{\partial x}+\frac{\partial u}{\partial y}+\frac{\partial u}{\partial z}+\cdots\right)\right\}-1\right]^\lambda,$$

$$\Sigma^\lambda u=\left[\exp\left\{\xi\left(\frac{\partial u}{\partial x}+\frac{\partial u}{\partial y}+\frac{\partial u}{\partial z}+\cdots\right)\right\}-1\right]^{-\lambda}.$$

Finally, Lagrange also set $\omega = \xi du/dx$ and wrote

$$\left(e^{\omega}-1\right)^{\lambda} = \omega^{\lambda}\left(1 + A\omega + B\omega^2 + C\omega^3 + D\omega^4 + \cdots\right). \tag{1.81}$$

Taking logarithms on both sides,

$$\lambda \log\left(e^{\omega}-1\right) - \lambda \log \omega = \log\left(1 + A\omega + B\omega^2 + C\omega^3 + D\omega^4 + \cdots\right).$$

Differentiating,

$$\lambda\left(\frac{e^{\omega}}{e^{\omega}-1} - \frac{1}{\omega}\right) = \frac{A + 2B\omega + 3C\omega^2 + 4D\omega^3 + \cdots}{1 + A\omega + B\omega^2 + C\omega^3 + D\omega^4 + \cdots}. \tag{1.82}$$

Now since

$$\frac{e^{\omega}}{e^{\omega}-1} = \frac{1}{1-e^{-\omega}} = \frac{1}{\omega - \dfrac{\omega^2}{2} + \dfrac{\omega^3}{2\cdot 3} - \dfrac{\omega^4}{2\cdot 3\cdot 4} + \cdots},$$

Equation (1.81) can be cross multiplied, and upon comparing like terms in ω^i, Lagrange obtained

$$\begin{aligned}
A &= \frac{\lambda}{2},\\
2B &= \frac{(\lambda+1)A}{2} - \frac{\lambda}{2\cdot 3},\\
3C &= \frac{(\lambda+2)B}{2} - \frac{(\lambda+1)A}{2\cdot 3} + \frac{\lambda}{2\cdot 3\cdot 4},\\
4D &= \frac{(\lambda+3)C}{2} - \frac{(\lambda+2)B}{2\cdot 3} + \frac{(\lambda+1)A}{2\cdot 3\cdot 4} - \frac{\lambda}{2\cdot 3\cdot 4\cdot 5},\\
&\ldots\ldots
\end{aligned}$$

With these values, and using Eq. (1.81), he was thus able to write Eq. (1.77) as

$$\Delta^{\lambda} u = \frac{d^{\lambda}u}{dx^{\lambda}}\xi^{\lambda} + A\frac{d^{\lambda+1}u}{dx^{\lambda+1}}\xi^{\lambda+1} + C\frac{d^{\lambda+2}u}{dx^{\lambda+2}}\xi^{\lambda+2} + D\frac{d^{\lambda+3}u}{dx^{\lambda+3}}\xi^{\lambda+3} + \cdots \tag{1.83}$$

(*ibid.*, p. 453).

Having thus previously "proved" Eq. (1.76),* Lagrange went on to give (1.77) (and thus also Eq. 1.83) without any proof because he thought such a proof would be quite difficult. In spite of his incomplete proofs, Lagrange's "nouvelle espèce de calcul" was very attractive, and Laplace set out to provide the missing links just a few years later (Laplace, 1776a; 1780; 1782). It is the third memoir that concerns us here, namely, the "Mémoire sur les Suites" (Laplace, 1782), for it contains his first use of the method of generating functions. In fact, most of the memoir was later reproduced in the *Théorie*

*See p. 67.

Analytique des Probabilités (Laplace, 1812, Book I, Part I). Here is how Laplace motivated the method of generating functions:

> The theory of series is one of the most important subjects of Analysis; all problems that can be reduced to approximations, and consequently almost all the applications of Mathematics to nature, depend on this theory; thus we can see that it has mainly taken the attention of geometers; they have found a great number of beautiful theorems and ingenious methods, either to develop functions as series, or to sum series exactly or approximately; but they have succeeded only by indirect and particular methods, and there can be doubt that, in this branch of Analysis, as in all others, there is a general and simple method of consideration, from which truths already known follow, and which leads to several new truths. The search of a similar method is the subject of this Memoir; that which I have reached is founded on the consideration of what I name *generating functions*: it is a new calculus which can be called *calculus of generating functions*, and which has appeared to me as deserving to be learned by geometers. (Laplace, 1782, OC 10, p. 1)

Laplace next proved Eq. (1.77) by using generating functions as follows. Let y_x be a function of x and suppose its generating function (G) is

$$u = y_0 + y_1 t + y_2 t^2 + \cdots$$

(*ibid.*, p. 5). That is, $\mathrm{G}y_x = u$. To find $\mathrm{G}y_{x+r}$, Laplace reasoned that y_{x+r} is the coefficient of t^{x+r} in u and is therefore also the coefficient of t^x in $t^{-r}u$. Therefore,

$$\mathrm{G}y_{x+r} = t^{-r}u.$$

This implies that

$$\mathrm{G}\Delta y_x = \mathrm{G}y_{x+1} - \mathrm{G}y_x = \frac{u}{t} - u = \left(\frac{1}{t} - 1\right)u,$$

and, by iteration,

$$\mathrm{G}\Delta^j y_x = \left(\frac{1}{t} - 1\right)^j u, \tag{1.84}$$

where j is a positive integer. If we define ${}^i\Delta y_x \equiv y_{x+i} - y_x$, where i is a positive integer, then

$$\mathrm{G}\left({}^i\Delta y_x\right) = \mathrm{G}y_{x+i} - \mathrm{G}y_x = \frac{u}{t^i} - u = \left(\frac{1}{t^i} - 1\right)u,$$

and by iteration,

$$\mathrm{G}\left({}^i\Delta^j y_x\right) = \left(\frac{1}{t^i} - 1\right)^j u. \tag{1.85}$$

Next Laplace expanded as follows:

$$\left(\frac{1}{t^i} - 1\right)^n u = \left\{\left(1 + \frac{1}{t} - 1\right)^i - 1\right\}^n u,$$

where n is a positive integer. By equating the coefficients of t^i on both sides of the last expression, and by using Eqs. (1.84) and (1.85), Laplace was thus able to obtain

$$ {}^{i}\Delta^{n} y_x = \left\{ \left(1+\Delta y_x\right)^i - 1 \right\}^n \tag{1.86}$$

(*ibid.*, p. 35). Laplace next let Δy_x become infinitesimally small so that $\Delta y_x = dy$. Also, he set $idx = a$, where i is infinitely large and a is a finite quantity. Then Eq. (1.86) becomes

$$\begin{aligned} {}^{i}\Delta^{n} y_x &= \left\{ \left(1+dy\right)^i - 1 \right\}^n \\ &= \left\{ \left(1+\frac{dy}{dx}dx\right)^i - 1 \right\}^n \\ &= \left\{ \left(1+\frac{a\,dy/dx}{i}\right)^i - 1 \right\}^n \\ &= \left\{ \exp\left(a\frac{dy}{dx}\right) - 1 \right\}^n \end{aligned}$$

(*ibid.*, p. 37). By taking $idx = a = \xi$, it is seen that the above is the same as Eq. (1.77). This completes Laplace's proof.

1.2.8 "Mémoire sur les approximations des formules qui sont fonctions de très grands nombres" (1785)

1.2.8.1 Method of Asymptotic Approximation Revisited This memoir (Laplace, 1785) is an important paper since it contains Laplace's introduction of the method of characteristic functions (CF). Before introducing these, Laplace spent some time refining his method of asymptotic approximation, having previously dealt with this method in the "Mémoire sur la probabilité des causes par les événements" (Laplace, 1774a) and the "Mémoire sur les probabilités" (Laplace, 1781).* We now consider these topics.

Laplace wished to determine the integral $\int_{\theta}^{\theta'} y\,dx$, where y is an arbitrary function of x such that

$$Y = (y)_{x=a}$$

is its maximum value. He let $(x-a)^{\mu+1}$ be the factor of smallest order of $\log(Y/y)$, where μ is a nonnegative integer. By writing

$$y = Ye^{-t^{\mu+1}} \tag{1.87}$$

and

$$v(x) = \frac{x-a}{\left\{\log(Y/y)\right\}^{1/(\mu+1)}}, \tag{1.88}$$

* See Sections 1.2.2.3 and 1.2.6.5.

Laplace obtained

$$x-a=\nu(x)t.$$

By applying Lagrange's inversion formula, the above becomes

$$x=a+t\nu(a)+\frac{t^2}{1\cdot2}\left(\frac{d\nu^2}{dx}\right)_{x=a}+\frac{t^3}{1\cdot2\cdot3}\left(\frac{d^2\nu^3}{dx^2}\right)_{x=a}+\cdots.$$

Hence, from Eq. (1.87), assuming that at $x=\theta,\theta'$ the values of t are $-\infty,\infty$, respectively,

$$\int_\theta^{\theta'} ydx=\int_{-\infty}^{\infty} y\frac{dx}{dt}dt=Y\int_{-\infty}^{\infty} e^{-t^{\mu+1}}\left\{\nu(a)+t\left(\frac{d\nu^2}{dx}\right)_{x=a}+\frac{t^2}{1\cdot2}\left(\frac{d^2\nu^3}{dx^2}\right)_{x=a}+\cdots\right\}dt \quad (1.89)$$

(Laplace, 1785, OC 10, p. 217). It remains to obtain the quantities $\nu(a),(d\nu^2/dx)_{x=a}$, $(d^2\nu^2/dx^2)_{x=a},\ldots$. Since $(x-a)^{\mu+1}$ is assumed to be a factor of $\log(Y/y)$, Laplace wrote

$$\log\left(\frac{Y}{y}\right)=(x-a)^{\mu+1}\left\{A+B(x-a)+C(x-a)^2+\cdots\right\}. \quad (1.90)$$

By differentiating with respect to x and then setting $x=a$,

$$A=-\frac{1}{1\cdot2\cdot3\cdots(\mu+1)}\frac{d^{\mu+1}\log y}{dx^{\mu+1}},$$

$$B=-\frac{1}{1\cdot2\cdot3\cdots(\mu+2)}\frac{d^{\mu+2}\log y}{dx^{\mu+2}},$$

$$\ldots$$

Using Eqs. (1.88) and (1.90), then

$$\nu=\left\{A+B(x-a)+C(x-a)^2+\cdots\right\}^{-1/(\mu+1)} \quad (1.91)$$

and

$$\begin{aligned}\nu^r&=\left\{A+B(x-a)+C(x-a)^2+\cdots\right\}^{-r/(\mu+1)}\\&=A^{-r/(\mu+1)}-\frac{r}{\mu+1}A^{-(r+\mu+1)/(\mu+1)}B(x-a)+\left\{\frac{r(r+\mu+1)}{1\cdot2(\mu+1)^2}A^{-\frac{r+2\mu+2}{\mu+1}}B^2-\frac{r}{\mu+1}A^{-\frac{r+\mu+1}{\mu+1}}C\right\}\\&\quad\times(x-a)^2+\cdots\end{aligned}$$

for $r=1, 2, 3, \ldots$ (*ibidem*). From the above expression, the quantities $\nu(a),(d\nu^2/dx)_{x=a}$, $(d^2\nu^3/dx^2)_{x=a},\ldots$ can be obtained and hence (1.89) can be evaluated.

There was also a simple case that Laplace considered. He stated that the case $\mu=1$ was the most common,* that is, the factor of smallest order of $\log(Y/y)$ would be $(x-a)^2$

*Laplace used $\mu+1=2i$ and then set $i=1$ (Laplace, 1785, OC 10, p. 227).

most of the time. Moreover, the formula in (1.89) converged very rapidly as long as $dy/dx \gg y$, which takes place when y contains high powers, and the limits θ and θ' are far from $x = a$. Then the integral in (1.89) becomes

$$\int_{\theta}^{\theta'} y dx = Y \int_{-\infty}^{\infty} e^{-t^2} \left\{ v(a) + t \left(\frac{dv^2}{dx} \right)_{x=a} + \frac{t^2}{1 \cdot 2} \left(\frac{d^2 v^3}{dx^2} \right)_{x=a} + \cdots \right\} dt.$$

Since

$$\int_{-\infty}^{\infty} t^k e^{-t^2} dt = \begin{cases} \frac{1 \cdot 3 \cdots (k-1)}{2^{k/2}} \sqrt{\pi}, & k \text{ even} \\ 0, & k \text{ odd} \end{cases} \tag{1.92}$$

we have

$$\int_{\theta}^{\theta'} y dx = Y \sqrt{\pi} \left\{ v(a) + \frac{1}{2} \frac{1}{1 \cdot 2} \left(\frac{d^2 v^3}{dx^2} \right)_{x=a} + \frac{1 \cdot 3}{2^2} \frac{1}{1 \cdot 2 \cdot 3 \cdot 4} \left(\frac{d^4 v^5}{dx^4} \right)_{x=a} + \cdots \right\} \tag{1.93}$$

(*ibid.*, p. 229). The above represents Laplace's formula for the integral $\int_{\theta}^{\theta'} y dx$ by using his method of asymptotic approximation.

Note that, from (1.91), $\nu(a)$ in the above is given by, for $\mu = 1$,

$$\begin{aligned} v(a) &= \frac{1}{\left(\sqrt{A}\right)_{x=a}} \\ &= \frac{1}{\left(\sqrt{-\frac{1}{2} \frac{d^2}{dx^2} \log y} \right)_{x=a}} \\ &= \frac{1}{\sqrt{-\frac{1}{2} \left(\frac{1}{y} \frac{d^2 y}{dx^2} - \frac{1}{y^2} \left[\frac{dy}{dx} \right]^2 \right)_{x=a}}} \\ &= \left(\sqrt{\frac{-2y}{\frac{d^2 y}{dx^2}}} \right)_{x=a}. \end{aligned}$$

Hence, if we consider only the first term on the right side of Eq. (1.93), we obtain

$$\int_{\theta}^{\theta'} y dx = Y \sqrt{\pi} \left(\sqrt{\frac{-2y}{\frac{d^2 y}{dx^2}}} \right)_{x=a} = \left(\frac{y^{3/2} \sqrt{2\pi}}{-\frac{d^2 y}{dx^2}} \right)_{x=a}$$

(*ibidem*). This is the same expression as (1.66), assuming that the limits in both are far from $x = a$.

1.2.8.2 Stirling's Formula Revisited From his refined method of asymptotic approximation, as just described, Laplace was able to prove Stirling's formula again, having already done so before in the "Mémoire sur les probabilités" (Laplace, 1781, OC 9, p. 458).* He set $y = x^s e^{-x}$. Then the maximum of y occurs at $x = s$ so that, remembering that $\mu = 1$, Eq. (1.87) becomes

$$y = Ye^{-t^2} = s^s e^{-s} e^{-t^2}.$$

Laplace next set

$$x - s = \theta$$

so that $y = x^s e^{-x} = s^s e^{-s} e^{-t^2}$ becomes

$$(s+\theta)^s e^{-(s+\theta)} = s^s e^{-s} e^{-t^2}$$
$$\left(1+\frac{\theta}{s}\right)^s e^{-\theta} = e^{-t^2}.$$

The above implies

$$t^2 = -s\log\left(1+\frac{\theta}{s}\right)+\theta = \frac{\theta^2}{2s} - \frac{\theta^3}{3s^2} + \frac{\theta^4}{4s^3} - \cdots.$$

By using Lagrange's inversion formula,

$$\theta = t\sqrt{2s} + \frac{2t^2}{3} + \frac{t^3}{9\sqrt{2s}} + \cdots.$$

Then

$$\begin{aligned}
\int_{-\infty}^{\infty} y\,dx &= \int_{-\infty}^{\infty} x^s e^{-s}\,dx \\
&= \int_{-\infty}^{\infty} s^s e^{-s} e^{-t^2}\,dx \\
&= s^s e^{-s} \int_{-\infty}^{\infty} e^{-t^2}\,d\theta \\
&= s^s e^{-s} \int_{-\infty}^{\infty} e^{-t^2} \frac{d\theta}{dt}\,dt \\
&= s^s e^{-s} \int_{-\infty}^{\infty} e^{-t^2} \left(\sqrt{2s} + \frac{4t}{3} + \frac{t^2}{3\sqrt{2s}} + \cdots\right) dt \\
&= s^s e^{-s} \left\{\sqrt{2\pi s} + 0 + \frac{1}{3\sqrt{2s}} \cdot \frac{\sqrt{\pi}}{2} + \cdots\right\} \\
&= s^{s+1/2} e^{-s} \sqrt{2\pi} \left(1 + \frac{1}{12s} + \cdots\right).
\end{aligned}$$

* See also p. 58 for a description of that derivation.

Hence, Laplace established Stirling's formula

$$\int_{-\infty}^{\infty} x^s e^{-s} dx = s! = s^{s+1/2} e^{-s} \sqrt{2\pi} \left(1 + \frac{1}{12s} + \cdots\right)$$

(Laplace, 1785, OC 10, p. 259).

1.2.8.3 Genesis of Characteristic Functions We now describe how Laplace introduced the concept of characteristic functions (CF).* This was motivated by his desire to determine the middle term of the series expansion of the multinomial expression $(1+1+\cdots+1)^n$, where n is a large positive integer.

At the end of Article XXII, Laplace was able to determine

$$\frac{\text{Middle term in expansion of } (1+1)^{2s}}{2^{2s}} \approx \frac{1}{\sqrt{\pi s}} \quad (s \text{ large}) \tag{1.94}$$

(*ibid.*, p. 270) without the use of CF. The importance of this result is as follows. Suppose an error X_i can take two values -1 and $+1$ each with probability 1/2, and let $T_{2s} = X_1 + X_2 + \cdots + X_{2s}$ be the total error when $2s$ observations are made. Then $T_{2s} = 0$ if and only if s of the observations are $+1$s and the other s are -1s. From the binomial distribution,

$$\Pr\{T_{2s} = 0\} = \frac{\binom{2s}{s}}{2^{2s}}.$$

But $\binom{2s}{s}$ is the middle term in the binomial expansion of $(1+1)^{2s}$. Therefore, using (1.94),

$$\Pr\{T_{2s} = 0\} = \frac{\text{Middle term in expansion of } (1+1)^{2s}}{2^{2s}} \approx \frac{1}{\sqrt{\pi s}} \quad (s \text{ large}).$$

This shows the importance of the result in (1.94), which in fact had first been obtained by de Moivre (1730; 1733; 1738).

Now, suppose X_j can take the three values -1, 0, and 1 each with equal probability 1/3, and s observations are made. Then a total error of zero occurs if and only if there are as many $+1$s as -1s, and the remaining are 0s. In the same way as before, the latter event occurs with a probability equal to

$$\frac{\text{Middle term(s) in expansion of } (1+1+1)^{s}}{3^{s}}.$$

However, how shall we determine the middle term(s) of $(1+1+1)^s$? It was with the aim to answer such questions that Laplace introduced the CF in Article XXIII. In his own words:

> One can reach the preceding results [Eq. (1.94) above] more simply in the following manner: for this, let y_s be the middle term of the binomial $(1+1)^{2s}$; it can be seen that this term is equal to the term independent of $e^{\omega\sqrt{-1}}$ in the development of the binomial $(e^{\omega\sqrt{-1}} + e^{-\omega\sqrt{-1}})^{2s}$; but, if we multiply this binomial by $d\omega$, and if we then take the integral from $\omega = 0$ to $\omega = 180°$, it is clear that this integral will be equal to πy_s; by substituting $2\cos\omega$ for $e^{\omega\sqrt{-1}} + e^{-\omega\sqrt{-1}}$; one therefore obtains

*The term was first used by Poincaré (1912, p. 206) for the moment-generating function (David, 1995). It was later used appropriately by Lévy in his book *Calculs des Probabilités* (Lévy, 1925, p. 161).

$$y_s = \frac{2^{2s}}{\pi} \int \cos^{2s} \omega d\omega.$$

This integral, taken from $\omega = 0$ to $\omega = 180°$...

This method has the advantage to extend itself to the determination of the middle term of the trinomial $(1+1+1)^s$, and that of the quadrinomial $(1+1+1+1)^{2s}$, and so on. Let us consider the trinomial $(1+1+1)^s$, and let y_s be its middle term; y_s will be equal to the term which is independent of $e^{\omega\sqrt{-1}}$ in the development of the trinomial

$$\left(e^{\omega\sqrt{-1}} + 1 + e^{-\omega\sqrt{-1}}\right)^s;$$

one will therefore have

$$y_s = \frac{1}{\pi} \int \left(2\cos\omega + 1\right)^s d\omega;$$

the integral being taken from $\omega = 0$ to $\omega = \pi$.
(*ibid.*, pp. 270–271)

Laplace's reasoning above is testament to his genius. Let us examine his argument by considering the trinomial $(1+1+1)^s$. Laplace's first insight was that both $(1+1+1)^s$ and $(e^{i\omega} + 1 + e^{-i\omega})^s$, where $i = \sqrt{-1}$, have the same middle term (although their other terms may differ). For example, let $a = b = c = 1$ in the following:

$$\begin{aligned}(a+b+c)^2 &= a^2 + 2ab + \overbrace{2ac + b^2}^{\text{middle terms}} + 2bc + c^2 \\ &= (1)^2 + 2(1)(1) + \overbrace{2(1)(1) + (1)^2}^{\text{middle terms}} + 2(1)(1) + (1)^2.\end{aligned}$$

Also,

$$(e^{i\omega} + 1 + e^{-i\omega})^2 = e^{2i\omega} + 2e^{i\omega} + \overbrace{3}^{\text{middle term}} + 2e^{-i\omega} + e^{-2i\omega}.$$

Therefore the problem of finding the middle term(s) of $(1+1+1)^s$ may be viewed as the problem of finding the middle term of $(e^{i\omega} + 1 + e^{-i\omega})^s$. This was Laplace's first insight.

Laplace's second insight was that the integral of $(e^{i\omega} + 1 + e^{-i\omega})^s$ from $\omega = 0$ to $\omega = 0$ is simply πy_s, where y_s is the middle term of $(e^{i\omega} + 1 + e^{-i\omega})^s$. To see why this is true, note that for each term $\alpha e^{i\beta\omega}$ (where α, β are constants) in the expansion of $(e^{i\omega} + 1 + e^{-i\omega})^s$, there is a corresponding term $\alpha e^{-i\beta\omega}$. Now, by using Euler's formula $e^{i\theta} = \cos\theta + i\sin\theta$, the integral of the sum of these two terms is

$$\begin{aligned}\int_0^\pi \left(\alpha e^{i\beta\omega} + \alpha e^{-i\beta\omega}\right) d\omega &= 2\alpha \int_0^\pi \cos\left(\beta w\right) d\omega \\ &= \frac{2\alpha}{\beta}\left[\sin\left(\beta\omega\right)\right]_0^\pi \\ &= 0.\end{aligned}$$

Hence,

$$\int_0^{\pi}\left(e^{i\omega}+1+e^{-i\omega}\right)^s d\omega=\int_0^{\pi} y_s d\omega=\pi y_s.$$

The above therefore gives Laplace a direct way to obtain y_s and hence $\Pr\{T_s=0\}=y_s/3^s$:

$$y_s=\frac{1}{\pi}\int_0^{\pi}\left(e^{i\omega}+1+e^{-i\omega}\right)^s d\omega=\frac{1}{\pi}\int_0^{\pi}(1+2\cos\omega)^s d\omega. \tag{1.95}$$

Laplace's two insights carry through to any multinomial $(1+1+\cdots+1)^n$.

We now show how Laplace evaluated the integral in (1.95) above. First he wrote

$$y_s=\frac{1}{\pi}\int_0^{2\pi/3}(1+2\cos\omega)^s d\omega+\frac{(-1)^s}{\pi}\int_0^{\pi/3}(2\cos\omega-1)^s d\omega.$$

For the first integral, Laplace made the substitution $(2\cos\omega+1)^s=3^s e^{-t^2}$, which gives

$$3-\omega^2+\frac{\omega^4}{12}-\cdots=3-\frac{3t^2}{s}+\frac{3t^4}{2s^2}-\cdots.$$

By Lagrange's inversion formula,

$$\omega=t\sqrt{\frac{3}{s}}\left(1-\frac{t^2}{8s}+\cdots\right),$$

so that

$$\begin{aligned}\int_0^{2\pi/3}(1+2\cos\omega)^s d\omega&=\frac{3^{s+1/2}}{\sqrt{s}}\int_0^{\infty}e^{-t^2}\left(1-\frac{3t^2}{8s}+\cdots\right)dt\\&=\frac{3^{s+1/2}}{2}\sqrt{\frac{\pi}{s}}\left(1-\frac{3}{16s}+\cdots\right).\end{aligned}$$

Similarly,

$$\int_0^{\pi/3}(2\cos\omega-1)^s d\omega=\frac{1}{2}\sqrt{\frac{\pi}{s}}\left(1-\frac{5}{16s}+\cdots\right)$$

Hence,

$$y_s=\frac{3^{s+1/2}}{2\sqrt{\pi s}}\left(1-\frac{3}{16s}+\cdots\right)+\frac{(-1)^s}{2\sqrt{\pi s}}\left(1-\frac{5\alpha}{16}+\cdots\right)$$

and

$$\frac{y_s}{3^s}\approx\frac{\sqrt{3}}{2\sqrt{\pi s}}$$

(*ibid.*, p. 272).

Laplace also gave formulas for the general case. Consider a $2n$-nomial $(1+1+\cdots+1)^{2s}$, where s is large. Then the middle term is

$$y_s \approx \frac{(2n)^{2s}\sqrt{3}}{\sqrt{2s(2n+1)(n+1)\pi}}$$

so that

$$\frac{y_s}{(2n)^{2s}} \approx \frac{\sqrt{3}}{\sqrt{2s(2n+1)(n+1)\pi}}$$

(*ibid.*, p. 273). Finally, consider the $(2n+1)$-nomial $(1+1+\cdots+1)^s$. Then

$$y_s \approx \frac{(2n+1)^s\sqrt{3}}{\sqrt{2sn(n+1)\pi}}$$

and

$$\frac{y_s}{(2n+1)^s} \approx \frac{\sqrt{3}}{\sqrt{2sn(n+1)\pi}}$$

(*ibidem*).

In modern notation, the CF is defined by

$$\psi_X(t) \equiv \mathsf{E}e^{itX}.$$

If X_j can take the three values -1, 0, and $+1$ each with probability 1/3, then

$$\psi_X(t) = \frac{1}{3}\left(e^{it}+1+e^{-it}\right).$$

Furthermore, if $T_s = X_1 + X_2 + \cdots + X_s$, then its CF is

$$\begin{aligned}\psi_T(t) &= \mathsf{E}e^{itT_s} \\ &= \mathsf{E}e^{it(X_1+X_2+\cdots+X_s)} \\ &= \left(\mathsf{E}e^{itX}\right)^s \\ &= \left\{\frac{1}{3}\left(e^{it}+1+e^{-it}\right)\right\}^s \\ &= \left\{\frac{1}{3}(1+2\cos t)\right\}^s.\end{aligned}$$

Finally, using Laplace's second insight that $\int_0^{\pi}(e^{i\omega}+1+e^{-i\omega})^s d\omega=\pi y_s$, we have

$$\begin{aligned}
\Pr\{T_s=0\} &= \frac{y_s}{3^s} \\
&= \frac{1}{\pi}\int_0^{\pi}\psi_T(t)dt \\
&= \frac{1}{3^s\pi}\int_0^{\pi}(1+2\cos t)^s\,dt \\
&\approx \frac{\sqrt{3}}{2\sqrt{\pi s}}.
\end{aligned}$$

1.2.9 "Mémoire sur les approximations des formules qui sont fonctions de très grands nombres (suite)" (1786): Philosophy of Probability and Universal Determinism, Recognition of Need for Normal Probability Tables

The current memoir is a follow-up to Laplace's previous "Mémoire sur les approximations des formules qui sont fonctions de très grands nombres" (Laplace, 1785), and its numbering continues from the latter.

The first article (XXXII) of the "Suite" contains an important section which characterized Laplace's concepts of probability and determinism throughout his life:

> All events, even those which by their smallness and their irregularity do not seem to belong to the general system of nature, are a consequence as necessary as the revolutions of the Sun. We attribute them to chance, because we ignore the causes which produce them and the laws which link them to the big phenomena of the universe; thus the appearance and movement of comets, which we know today to depend on the same law that brings the seasons, were previously thought of as the effect of chance by those who arranged these stars amongst the meteors. The word *chance* thus expresses only our ignorance on the causes of phenomena which are seen by us to happen and repeat without any apparent order.
>
> Probability is relative in part to this ignorance, in part to our knowledge. (Laplace, 1786a, OC 10, pp. 295–296)

A similar statement was later made by Laplace in *Théorie Analytique des Probabilitiés* (Laplace, 1812, p. 177) and in the *Essai Philosophique sur les Probabilités* (Laplace, 1814, English edition, pp. 2, 6). It expresses Laplace's philosophy of universal determinism (also known as Laplace's Demon): all phenomena are ultimately deterministic in nature, probability is only our relative ignorance of them.

The next two sections of the "Suite" contain a definition of conditional probability and a statement of Bayes' theorem for the discrete case, assuming a uniform prior. There is nothing new here since Laplace had considered these issues in previous memoirs.

However, starting from Article XXXV, Laplace was led to make a noteworthy observation. Let $x\,(0\le x\le 1)$ be the "cause" of an observation $y=y(x)$. Then the probability that x lies between θ and θ', given the observation y, is given by

$$\frac{\int_{\theta}^{\theta'} y dx}{\int_{0}^{1} y dx}$$

(Laplace, 1786a, OC 10, p. 302). The above is the continuous version of Bayes' theorem in the case of a uniform prior for x. By using techniques similar to those of his previous memoir (see Section 1.2.8.1), Laplace was able to show that the posterior probability that $x \leq \theta$ is

$$\frac{\int_T^\infty e^{-t^2} dt}{\sqrt{\pi}} - \frac{e^{-T^2}\left\{\left(\frac{du^2}{dx}\right)_{x=a} + T\frac{1}{1\cdot 2}\left(\frac{d^2u^3}{dx^2}\right)_{x=a} + \cdots\right\}}{2\sqrt{\pi}\left\{(u)_{x=a} + \frac{1}{2}\cdot\frac{1}{1\cdot 2}\left(\frac{d^2u^3}{dx^2}\right)_{x=a} + \cdots\right\}}$$

(*ibid.*, p. 305), where $x=a$ is the point at which y attains its maximum Y, $u=(x-a)/\sqrt{\log(Y/y)}$, and $T^2=\log\{Y/(y)_{x=\theta}\}$. Laplace next made the following pertinent remarks:

> The integral $\int e^{-t^2} dt$ is frequently encountered in this analysis, and, for this reason, it would be very useful to build a Table of its values. (*ibidem*)

Recognizing the ubiquity of the normal curve, Laplace was thus the first to suggest its tabulation.* He also gave the following approximation for $T \geq 3$:

$$\int_T^\infty e^{-t^2} dt = \frac{e^{-T^2}}{2T}\left(1 - \frac{1}{2T^2} + \frac{1\cdot 3}{4T^4} - \frac{1\cdot 3\cdot 5}{8T^6} + \cdots\right)$$

(*ibidem*) which results in a "value alternately more and less than the true value."

1.2.10 "Sur les naissances" (1786): Solution of the Problem of Births by Using Inverse Probability

Laplace devoted the current memoir entirely to a problem he had dealt with both in the "Mémoire sur les probabilités" (Laplace, 1781) and in the "Mémoire sur les approximations des formules qui sont fonctions de très grands nombres (suite)" (Laplace, 1786a), namely, the problem of births. Laplace made a case for inverse probability by pointing that:

> These researches depend on a new and still little known theory, that of the probability of future events given observed events. (Laplace, 1786b, OC 11, p. 38)

*The first modern table of the normal distribution was later built by the French physicist Christian Kramp (1760–1826) in a book on refractions (Kramp, 1799). See also Wolfenden (1942, pp. 160–162), Du Pré (1938), David (2005), and Walker (1929, p. 29).

He then gave the following data relating to the population of France. Existing data showed that the ratio of population size to yearly births was approximately 26 : 1. The average number of births for the period 1781–1782 was 973,054.5. Multiplying this number by 26 gives 25,299,417 for the population of France. Laplace analysis showed that, to be within 500,000 of this estimate with odds 1,000 to 1, one needed to determine the factor of 26 based on 771,469 inhabitants. On the other hand, if a factor of 26.5 was used, the population size would be 25,785,944, and to be within 500,000 of this estimate, the factor 26.5 would need to be determined from 817,219 inhabitants. It follows that up to 1,000,000 or 1,200,000 people needed to be surveyed.

Laplace proceeded to substantiate the above numbers by considering an urn model as follows. Consider an infinite urn from which an initial draw resulted in p white balls and q black balls. Now let a second draw be made, resulting in q' black balls but an unknown number of white balls. A natural estimate of the latter number is pq'/q. Now, for a given ω, what is the probability that the true number of white balls in the second draw will be between $\frac{pq'}{q}(1-\omega)$ and $\frac{pq'}{q}(1+\omega)$?

Let x be the proportion of white balls in the urn. Then the probability of obtaining p' white balls in the second draw is

$$\frac{1\cdot 2\cdot 3\cdots(p'+q')}{1\cdot 2\cdot 3\cdots p'\cdot 1\cdot 2\cdot 3\cdots q'}x^{p'}(1-x)^{q'}.$$

Then Laplace applied Bayes' theorem with a uniform prior on p'. The posterior distribution of p' can then be obtained by dividing the above by

$$\begin{aligned}\sum_{p'=0}^{\infty}\frac{1\cdot 2\cdot 3\cdots(p'+q')}{1\cdot 2\cdot 3\cdots p'\cdot 1\cdot 2\cdot 3\cdots q'}x^{p'}(1-x)^{q'} &= \frac{(1-x)^{q'}}{1\cdot 2\cdot 3\cdots q'}\sum_{p'=0}^{\infty}\frac{1\cdot 2\cdot 3\cdots(p'+q')}{1\cdot 2\cdot 3\cdots p'.}x^{p'}\\ &= (1-x)^{q'}\underbrace{\left\{1+(q'+1)x+\frac{(q'+1)(q'+2)}{1\cdot 2}x^2+\cdots\right\}}_{S}.\end{aligned} \tag{1.96}$$

Now the infinite series S above is $1/(1-x)^{q'+1}$ so that the posterior distribution of p' is

$$\frac{\dfrac{1\cdot 2\cdot 3\cdots(p'+q')}{1\cdot 2\cdot 3\cdots p'\cdot 1\cdot 2\cdot 3\cdots q'}x^{p'}(1-x)^{q'}}{(1-x)^{q'}(1-x)^{-q'-1}} = \frac{1\cdot 2\cdot 3\cdots(p'+q')}{1\cdot 2\cdot 3\cdots p'\cdot 1\cdot 2\cdot 3\cdots q'}x^{p'}(1-x)^{q'+1} \tag{1.97}$$

(*ibid.*, p. 39). Moreover, the posterior distribution of x based on the first draw and based on a uniform prior is (cf. Eq. 1.5)

$$\frac{x^p(1-x)^q}{\int_0^1 x^p(1-x)^q\,dx}. \tag{1.98}$$

Laplace next reasoned that Eq. (1.97) was the posterior distribution of p' for a given x, and Eq. (1.98) was the posterior distribution of x. By multiplying these two equations and integrating with respect to x, the "total" posterior distribution of p' could therefore be obtained as

$$\frac{1\cdot 2\cdot 3\cdots(p'+q')\int_0^1 x^{p+p'}(1-x)^{q+q'+1}\,dx}{1\cdot 2\cdot 3\cdots p'\cdot 1\cdot 2\cdot 3\cdots q'\int_0^1 x^p(1-x)^q\,dx}$$

(*ibid.*, p. 40). Therefore, the posterior probability that p' lies between 0 and s is

$$\frac{1}{\int_0^1 x^p(1-x)^q\,dx}\sum_{p'=0}^{s}\frac{1\cdot 2\cdot 3\cdots(p'+q')\int_0^1 x^{p+p'}(1-x)^{q+q'+1}\,dx}{1\cdot 2\cdot 3\cdots p'\cdot 1\cdot 2\cdot 3\cdots q'}.$$

Using Eq. (1.96), the posterior probability that p' lies between 0 and s becomes

$$\frac{\int_0^1 x^p(1-x)^{q+q'+1}\left\{1+(q'+1)x+\cdots+\frac{(q'+1)(q'+2)\cdots(q'+s)}{1\cdot 2\cdot 3\cdots s}x^s\right\}dx}{\int_0^1 x^p(1-x)^q\,dx}. \tag{1.99}$$

Laplace next referred to his "Mémoire sur les approximations des formules qui sont fonctions de très grand nombres" (Laplace, 1785, OC 10, p. 267) where he had shown that, for large q' and large s,

$$1+(q'+1)x+\cdots+\frac{(q'+1)(q'+2)\cdots(q'+s)}{1\cdot 2\cdot 3\cdots s}x^s=\frac{1}{(1-x)^{q'+1}}\frac{\int_x^1 x'^s(1-x')^{q'}\,dx'}{\int_0^1 x'^s(1-x')^{q'}\,dx'}.$$

Therefore, the posterior probability in Eq. (1.99) that p' lies between 0 and s becomes

$$\frac{\int_0^1\int_x^1 x^p(1-x)^q\,x'^s(1-x')^{q'}\,dx'dx}{\int_0^1\int_0^1 x^p(1-x)^q\,x'^s(1-x')^{q'}\,dx'dx}$$

(Laplace, 1786b, OC 11, p. 40). Referring to the "Mémoire sur les approximations..." again (p. 313), Laplace observed that the above integral could be written, for s slightly smaller than pq'/q, as

$$\frac{\int_T^\infty e^{-t^2}\,dt}{\sqrt{\pi}},$$

and could be written, for s slightly larger than pq'/q, as

$$1-\frac{\int_T^\infty e^{-t^2}\,dt}{\sqrt{\pi}},$$

where

$$T^2=\frac{\left(\dfrac{p}{p+q}-\dfrac{s}{s+q'}\right)^2(p+q)^3(s+q')^3}{2sq'(p+q)^3+2pq(s+q')^3}.$$

Therefore, the posterior probability that p' lies between s and s' is

$$1-\frac{\int_{T'}^\infty e^{-t^2}\,dt}{\sqrt{\pi}}-\frac{\int_T^\infty e^{-t^2}\,dt}{\sqrt{\pi}}, \tag{1.100}$$

where

$$T'^2=\frac{\left(\dfrac{p}{p+q}-\dfrac{s'}{s'+q'}\right)^2(p+q)^3(s'+q')^3}{2s'q'(p+q)^3+2pq(s'+q')^3}.$$

Letting $s=pq'(1-\omega)/q$ and $s'=pq'(1+\omega)/q$, where ω is small (so that ω^2 and higher powers can be neglected), we have

$$T^2=T'^2=\frac{pqq'\omega^2}{2(p+q)(q+q')}=V^2,\quad \text{say.} \tag{1.101}$$

Hence, the posterior probability that p' lies between $pq'(1-\omega)/q$ and $pq'(1+\omega)/q$ is, from Eq. (1.100),

$$P=1-\frac{2\int_V^\infty e^{-t^2}\,dt}{\sqrt{\pi}}$$

(Laplace, 1786b, OC 11, p. 42).

The last formula can now be applied to the population data provided by Laplace at the start of this section. Let the first draw correspond to the sample that is used to estimate the ratio of population size to yearly births and the second draw correspond to the population of France. The sample is made of p inhabitants and q additional yearly births, and the population has p' inhabitants and q' additional yearly births. Let $p=iq$ and $pq'\omega/q=a$. Then $\omega=a/(iq')$ and Eq. (1.101) gives

$$p=\frac{2i^2(i+1)q'^2V^2}{a^2-2i(i+1)q'V^2} \tag{1.102}$$

(*ibid.*, p. 43). In the above, $a=500{,}000$, $q'=973{,}054.5$, and V can be obtained from

$$P=1-\frac{2\int_V^\infty e^{-t^2}dt}{\sqrt{\pi}}=\frac{1000}{1001}\quad\Rightarrow\quad V^2=5415.$$

Regarding i, Laplace took its values successively as 25.5, 26, and 26.5 so that Eq. (1.102) yields

$$p=727{,}510,\quad p=771{,}469,\quad p=817{,}219.$$

The last two numbers are those he gave at the start of the section.

Karl Pearson (1928a) has commented at length on Laplace's above analysis. His two main criticisms are as follows:

> …I can see no justification for Laplace's method of reducing the problem to an urn problem…I see further no ground whatever for considering the first sample and France as a whole as independent samples from an indefinitely large population. (*ibid.*, p. 172)

The second criticism seems to be more justified and is a major weakness of Laplace's analysis.

1.2.11 "Mémoire sur les approximations des formules qui sont fonctions de très grands nombres et sur leur application aux probabilités" (1810): Second Phase of Laplace's Statistical Career, Laplace's First Proof of the Central Limit Theorem

Laplace's "Mémoire sur les approximations des formules qui sont fonctions de très grands nombres et leur application aux probabilités" (Laplace, 1810a) is a landmark paper because it marked the beginning of the second phase in Laplace's statistical career, one in which he used direct, instead of inverse, probability in the solution of various statistical problems (Fig. 1.14). In this paper, Laplace first proved the Central Limit Theorem after de Moivre's initial derivation in 1730 (de Moivre, 1730). Laplace here came back to a

MÉMOIRE

SUR LES

APPROXIMATIONS DES FORMULES

QUI SONT FONCTIONS DE TRÈS GRANDS NOMBRES

ET SUR

LEUR APPLICATION AUX PROBABILITÉS [1].

Mémoires de l'Académie des Sciences, 1re Série, T. X, année 1809; 1810.

L'analyse conduit souvent à des formules dont le calcul numérique, lorsqu'on y substitue de très grands nombres, devient impraticable, à cause de la multiplicité des termes et des facteurs dont elles sont composées. Cet inconvénient a lieu principalement dans la théorie des probabilités, où l'on considère les événements répétés un grand nombre de fois. Il est donc utile alors de pouvoir transformer ces formules en séries d'autant plus convergentes que les nombres substitués sont plus considérables. La première transformation de ce genre est due à Stirling, qui réduisit de la manière la plus heureuse, dans une série semblable, le terme moyen du binôme élevé à une haute puissance; et le théorème auquel il parvint peut être mis au rang des plus belles choses que l'on ait trouvées dans l'analyse. Ce qui frappa surtout les géomètres, et spécialement Moivre, qui s'était occupé longtemps de cet objet, fut l'introduction de la racine carrée de la circonférence dont le rayon est l'unité, dans une recherche qui semblait étrangère à cette transcendante. Stirling y était arrivé au moyen de

[1] Lu le 9 avril 1810.

Œuvres de L. — XII. 38*

FIGURE 1.14 First page of "Mémoire sur les approximations des formules qui sont fonctions de très grands nombres et sur leur application aux probabilités" (Laplace, 1810a)

problem he had considered several times before, namely, that of the inclination of the orbits of comets.* We now describe his derivation† of this fundamental theorem.

Laplace first considered an interval of length h that consisted of $2k$ equal parts. He defined S to be the sum of n random variables each with a discrete uniform distribution on $\{-k,-k+1,\ldots,-1,0,1,\ldots,k-1,k\}$. He next wrote the sum

$$\Psi=\left\{e^{-ik\omega}+e^{-i(k-1)\omega}+\cdots+1+\cdots+e^{i(k-1)\omega}+e^{ik\omega}\right\}^{n}=\left(1+2\cos\omega+\cdots+2\cos k\omega\right)^{n} \quad (1.103)$$

(Laplace, 1810a, OC 12, p. 309), where $i=\sqrt{-1}$. It can be seen that, except for a constant factor, Ψ is the characteristic function of the sum S. Laplace then took the inverse Fourier cosine transform so as to obtain the coefficient of e^{ilw} in Eq. (1.103) as

$$A_l=\frac{1}{\pi}\int_0^{\pi}\Psi\cos(l\omega)\,d\omega. \quad (1.104)$$

Note that the coefficient A_l is also the total number of ways of adding the discrete uniform random variables such that $S=l$. Now, from Eq. (1.103), Laplace wrote

$$\begin{aligned}1+2\cos\omega+\cdots+2\cos(k\omega)&=\frac{\cos(k\omega)-\cos\{(k+1)\omega\}}{1-\cos\omega}\\&=\cos(k\omega)+\frac{\cos(\omega/2)\sin(k\omega)}{\sin(\omega/2)}\\&=\cos t+\frac{\cos\left(\dfrac{t}{2k}\right)\sin t}{\sin\left(\dfrac{t}{2k}\right)},\end{aligned}$$

where $k\omega=t$. The next step is to pass to the continuous limit by letting $k\to\infty$ so that

$$\cos t+\frac{\left\{1-\dfrac{(t/2k)^2}{2!}+\cdots\right\}\sin t}{\dfrac{t}{2k}-\dfrac{(t/2k)^3}{3!}+\cdots}\approx\cos t+2k\frac{\sin t}{t}.$$

Equation (1.104) then becomes

* See Section 1.2.4.

† See also Sheynin (1977, pp. 10–13), Hald (1998, pp. 311–313), Fischer (2010, pp. 21–23), Lubet (2004, pp. 215–221), and Gillispie (2000, Chapter 25).

$$\begin{aligned} A_l &= \frac{1}{\pi}\int_0^{\pi}\left(\cos t + 2k\frac{\sin t}{t}\right)^n \cos(l\omega)d\omega \\ &= \frac{(2k)^n}{\pi}\int_0^{\pi}\left(\frac{\sin t}{t}\right)^n \cos\left(rt\sqrt{n}\right)d\omega \\ &= \frac{(2k)^n}{\pi k}\int_0^{\infty}\left(\frac{\sin t}{t}\right)^n \cos\left(rt\sqrt{n}\right)dt \end{aligned}$$

(*ibid*., p. 310), where $r = l/(k\sqrt{n})$. By doing a series expansion on $\{(\sin t)/t\}^n$, Laplace obtained

$$A_l = \frac{(2k)^n}{\pi k}\int_0^{\infty}\cos\left(rt\sqrt{n}\right)e^{-nt^2/6}\left(1 - \frac{n}{180}t^4 + \cdots\right)dt. \qquad (1.105)$$

By taking each term of $(1 - nt^4/180 + \cdots)$, multiplying by the series expansion of $\cos(rt\sqrt{n})$, and integrating, Laplace obtained

$$A_l = \frac{(2k)^n}{2k\sqrt{\pi}}\sqrt{\frac{6}{n}}e^{-3r^2/2}\left\{1 - \frac{3}{20n}\left(1 - 6r^2 + r^4\right) - \cdots\right\}$$

(*ibid*., p. 314). Now the sum of all coefficients lying in $[-m, m]$ is, in the continuous limit,

$$\sum_{l=-m}^{m} A_l \approx (2k)^n\sqrt{\frac{6}{\pi}}\int_{-m}^{m} e^{-3r^2/2}\left\{1 - \frac{3}{20n}\left(1 - 6r^2 + r^4\right) - \cdots\right\}dr$$

where Laplace had used $l = kr\sqrt{n}$ so that $dl = kdr\sqrt{n} = 1$, that is, $1/k = dr\sqrt{n}$. Recalling that A_l is the total number of ways of adding the discrete uniform random variables such that $S = l$, Laplace obtained

$$\begin{aligned} \Pr\{-m \le S \le m\} &= \frac{\sum_{l=-m}^{m} A_l}{(2k)^n} \\ &\approx \sqrt{\frac{6}{\pi}}\int_{-m}^{m} e^{-3r^2/2}\left\{1 - \frac{3}{20n}\left(1 - 6r^2 + r^4\right) - \cdots\right\}dr \\ &= \sqrt{\frac{6}{\pi}}\left\{\int_{-m}^{m} e^{-3r^2/2}dr - \frac{3r}{20n}\left(1 - r^2\right)e^{-3r^2/2}\right\} \end{aligned}$$

(*ibidem*). For large n, we can "neglect the terms of order $1/n$" so that

$$\Pr\{-m \le S \le m\} \approx \sqrt{\frac{6}{\pi}}\int_{-m}^{m} e^{-3r^2/2}dr \qquad (1.106)$$

(*ibid*., p. 317).* This result implies that the sum of n random variables each with a discrete uniform distribution on $\{-k,-k+1,\ldots,-1,0,1,\ldots,k-1,k\}$ has, for large n, an approximate normal distribution with mean 0 and variance 1/3. The integral in (1.106) also gives the probability that the sample mean S/n lies between $-m/n$ and m/n. By using CF, Laplace was thus able to prove a restricted version of the Central Limit Theorem.

In Article VI of the "Mémoire sur les approximations des formules qui sont fonctions de très grands nombres et leur application aux probabilités" (Laplace, 1810a), Laplace proved a more general version of the Central Limit Theorem† by considering the sum of random variables each of which had an *arbitrarily given distribution*. We now describe Laplace's proof.

Laplace first divided an interval h into the parts $(-j,-j+1)$, $(-j+1,-j+2)$, …, $(-1,0)$, …, $(j'-1, j')$, and denoted the probability of an error s in an observation by

$$\phi\left(\frac{s}{j+j'}\right).$$

In analogy with Eq. (1.103), he then formed the sum

$$\left\{\phi\left(\frac{-j}{j+j'}\right)e^{-ij\omega}+\phi\left(\frac{-j+1}{j+j'}\right)e^{-i(j-1)\omega}+\cdots+1+\cdots+\phi\left(\frac{j'-1}{j+j'}\right)e^{i(j'-1)\omega}+\phi\left(\frac{j'}{j+j'}\right)e^{ij'\omega}\right\}^n$$
$$=\left\{\sum_{r=-j}^{j'}\phi\left(\frac{r}{j+j'}\right)e^{ir\omega}\right\}^n$$

and the coefficient of $e^{ir\omega}$ in it gives the probability that the sum S of n errors is r. This implies that

$$\Pr\{S=r+nq\}=\text{coeff. of } e^{ir\omega} \text{ in}\left\{\sum_{r=-j}^{j'}\phi\left(\frac{r}{j+j'}\right)e^{i(r-q)\omega}\right\}^n \tag{1.107}$$

(*ibid*., p. 323). Like before, Laplace then made a passage to the continuous limit by setting

$$\frac{r}{j+j'}=\frac{x}{h},\quad \frac{q}{j+j'}=\frac{q'}{h},\quad \frac{1}{j+j'}=\frac{dx}{h}.$$

Then the expression containing the power n in Eq. (1.107) becomes

$$\frac{(j+j')^n}{h^n}\left\{\int\phi\left(\frac{x}{h}\right)dx+i(j+j')\omega\int\frac{x-q'}{h}\phi\left(\frac{x}{h}\right)dx-\frac{(j+j')^2}{1\cdot 2}\omega^2\int\frac{(x-q')^2}{h^2}\varphi\left(\frac{x}{h}\right)dx+\cdots\right\}^n, \tag{1.108}$$

*Laplace's formula in Eq. (1.106) seems to be in error and twice what it should really be. What follows in the current text is based on the correct formula.

† See also Sheynin (1977, pp. 13–15) and Hald (1998, pp. 313–314).

where $q' = \int x\phi(x/h)dx$ and all integrals are taken from $x = -jh/(j+j')$ to $x = j'h/(j+j')$. Note that the second integral in (1.108) is zero. By substituting

$$
\begin{aligned}
k &= \int \phi\left(\frac{x}{h}\right)dx, \\
k' &= \int \frac{(x-q')^2}{h^2}\phi\left(\frac{x}{h}\right)dx, \qquad (1.109) \\
&\cdots
\end{aligned}
$$

the expression in (1.108) becomes

$$\frac{(j+j')^n}{h^n}k^n\left\{1-\frac{k'}{2k}(j+j')^2\omega^2+\cdots\right\}^n.$$

In accordance with Eq. (1.104), the probability that $S = nq - l$ or $S = nq + l$ is obtained by multiplying the above by $2\cos(l\omega)$, integrating from $\omega = 0$ to $\omega = \pi$, and dividing by π, that is,

$$\Pr\{S = nq - l \text{ or } S = nq + l\} = \frac{(j+j')^n k^n}{h^n}\frac{2}{\pi}\int_0^\pi \left[\left\{1-\frac{k'}{2k}(j+j')^2\omega^2+\cdots\right\}^n \cos(l\omega)\right]d\omega.$$

Laplace next made the substitution $(j+j')\omega = t$ so that the above becomes

$$
\begin{aligned}
\Pr\{S = nq - l \text{ or } S = nq + l\} &= \frac{(j+j')^n k^n}{h^n}\frac{2}{\pi}\int_0^\pi \left\{\left(1-\frac{k'nt^2}{2nk}\omega^2+\cdots\right)^n \cos(l\omega)\right\}d\omega \\
&\approx \frac{(j+j')^n k^n}{h^n}\frac{2}{\pi}\int_0^\pi e^{-\frac{k'nt^2}{2k}}\cos(l\omega)d\omega \\
&= \frac{(j+j')^n k^n}{h^n}\frac{2}{\pi}\frac{1}{j+j'}\int_0^\infty e^{-\frac{k'nt^2}{2k}}\cos\left(\frac{lt}{j+j'}\right)dt.
\end{aligned}
$$

The next step is to note that $k = h/(j+j')$ and to set $l/(j+j') = r\sqrt{n}$, so that the above becomes

$$\Pr\{S = nq - l \text{ or } S = nq + l\} \approx \frac{2}{\pi(j+j')}\int_0^\infty e^{-\frac{k'nt^2}{2k}}\cos\left(rt\sqrt{n}\right)dt$$

(*ibid.*, p. 324). Note that the integral in the above is similar to that in Eq. (1.105). By doing a series expansion on $\cos\left(rt\sqrt{n}\right)$ and integrating, like in the latter case, we obtain

$$\Pr\{S = nq - l \text{ or } S = nq + l\} \approx \frac{2}{(j+j')\sqrt{\pi}}\sqrt{\frac{k}{2k'n}}e^{-\frac{kr^2}{2k'}}.$$

From the last expression, Laplace wrote

$$\Pr\{nq-l\le S\le nq+l\}\approx\frac{2}{(j+j')\sqrt{\pi}}\sqrt{\frac{k}{2k'n}}\int_0^{(j+j')l\sqrt{n}} e^{-\frac{kr^2}{2k'}}dl^*.$$

Since $dl^*=(j+j')\sqrt{n}dr$, the above becomes

$$\Pr\{nq-l\le S\le nq+l\}\approx\frac{2}{\sqrt{\pi}}\sqrt{\frac{k}{2k'}}\int_0^{l} e^{-\frac{kr^2}{2k'}}dr \quad (1.110)$$

(*ibid.*, p. 325). The above can also be written in terms of the sample mean S/n:

$$\Pr\left\{q-\frac{l}{n}\le\frac{S}{n}\le q+\frac{l}{n}\right\}\approx\frac{2}{\sqrt{\pi}}\sqrt{\frac{k}{2k'}}\int_0^{l} e^{-\frac{kr^2}{2k'}}dr.$$

Laplace thus proved the Central Limit Theorem more generally, and both this version and the earlier (more restricted) one were included in Chapter IV of Book II of his *Théorie Analytique des Probabilités* (Laplace, 1812).*

It is natural to ask how rigorous are Laplace's proofs? They certainly do not stand up to today's standards of rigor. Means and variances (see Eq. 1.109) are assumed to exist and variances, assuming they exist, are also taken to be well behaved. Infinite series expansions are assumed to be made up of continuous terms and to be uniformly convergent so that term-by-term integration is allowed. We are thus told by Gnedenko and Sheynin that:

> Laplace is extremely careless in his reasoning and in carrying out formal transformations. He begins by considering discrete random variables ξ_i, then reasons about variables distributed uniformly on a segment $[-h, h]$ while dealing in the text itself with a variable distributed uniformly on the segment $[0, h]$. He approximates these continuous uniformly distributed random variables by discrete variables by subdividing the segment $[0, h]$ into $2m$ $[2k]$ equal parts of "unit length". But then it turns out that by a "unit length" he means a segment of length h/m and that in fact he subdivides the segment $[-h, h]$. ...The result is generalized for the case of arbitrary distributions (having variances)...This derivation is, however, greatly lacking in rigor. Nevertheless, it is appropriate to emphasize Laplace's exceptional intuition that enabled him to arrive at correct conclusions using non-rigorous and, now and then, simply confused reasoning. (Gnedenko and Sheynin, 1992, p. 224)

Despite the lack of rigor in Laplace's proofs, his effort cannot be regarded as anything less than trailblazing since it paved the way for better and more rigorous later proofs.

* See also Section 1.2.15.1.

1.2.12 "Supplément au Mémoire sur les approximations des formules qui sont fonctions de très grands nombres et sur leur application aux probabilités" (1810): Justification of Least Squares Based on Inverse Probability, The Gauss–Laplace Synthesis

Laplace wrote this Supplement (Laplace, 1810b) right in the wake of Gauss' book, *Theoria Motus Corporum Coelestium* (Gauss, 1809, English edition, 1857). In the latter, Gauss had provided in a somewhat circular way a first probabilistic justification for the method of least squares. Details on these topics are provided in Section 1.5.4.

1.2.13 "Mémoire sur les intégrales définies et leur applications aux probabilités, et spécialement à la recherche du milieu qu'il faut choisir entre les résultats des observations" (1811): Laplace's Justification of Least Squares Based on Direct Probability

Laplace's justification of least squares based on direct probability is described in Section 1.5.5.

1.2.14 *Théorie Analytique des Probabilités (1812)*: The de Moivre–Laplace Theorem

We shall consider Laplace's book, *Théorie Analytique des Probabilités* (Laplace, 1812), in the next section, but for now we shall describe his extension of de Moivre's derivation of the normal distribution as an approximation to the symmetric binomial distribution for a large number of trials. De Moivre's work resulted in the first derivation of the normal distribution and took place in his two memoirs *Miscellanea analytica* (de Moivre, 1730) and *Approximatio ad summam terminorum binomii* $(a+b)^n$ *in seriem expansi* (de Moivre, 1733).* In his opus *Théorie Analytique des Probabilités* (Laplace, 1812, pp. 275–278), Laplace extended de Moivre's result to the case of a general binomial distribution as follows.

Without any mention of de Moivre, Laplace started by considering two events a and b with probabilities p and $1-p$, respectively. The probability that the events a and b occur x and x' times, respectively, is

$$\frac{1\cdot 2\cdot 3\cdots\left(x+x'\right)}{1\cdot 2\cdots x\cdot 1\cdot 2\cdots x'}p^{x}\left(1-p\right)^{x'}$$

(*ibid.*, p. 275), which is the $(x'+1)$th term of the binomial expansion of $\{p+(1-p)\}^{x+x'}$. If the largest term of the latter expansion is denoted by k, then the term preceding the largest term is $\frac{kp}{1-p}\cdot\frac{x'}{x+1}$ and the term succeeding the largest term is $k\frac{1-p}{p}\cdot\frac{x}{x'+1}$. Consequently, Laplace then wrote

$$\frac{p}{1-p}<\frac{x+1}{x'}>\frac{x}{x'+1}. \tag{1.111}$$

*For a description of de Moivre's proof, see Gorroochurn (2012a, Prob. 10).

We note that the last expression can be obtained by first writing

$$\frac{kp}{1-p}\cdot\frac{x'}{x+1}<k>k\cdot\frac{1-p}{p}\cdot\frac{x}{x'+1}.$$

This implies $\frac{x+1}{x'}>\frac{p}{1-p}$ and $\frac{x'+1}{x}>\frac{1-p}{p}$ so that $\frac{x+1}{x'}\frac{x'+1}{x}>1$. Hence, Eq. (1.111) is obtained. This implies that

$$x<(n+1)p>(n+1)p-1,$$

where $x+x'=n$. Therefore, Laplace concluded that the value of x* that corresponded to k was the largest integer contained in $(n+1)p$. Therefore, let

$$x=(n+1)p-s, \quad \text{where} \quad 0\le s<1.$$

This implies

$$p=\frac{x+s}{n+1} \quad \Rightarrow \quad \frac{p}{1-p}=\frac{x+s}{x'+1-s}\approx\frac{x}{x'} \quad \text{for } x \text{ and } x' \text{ both large}. \tag{1.112}$$

Now, the lth term after the largest in the binomial expansion of $\{p+(1-p)\}^{x+x'}$ is

$$\frac{1\cdot2\cdot3\cdots(x+x')}{1\cdot2\cdots(x-l)\cdot1\cdot2\cdots(x'+l)}p^{x-l}(1-p)^{x'+l}. \tag{1.113}$$

Using Stirling's formula $\nu!\sim\sqrt{2\pi\nu}(\nu/e)^{\nu}$ (where ν is large), we have

$$\begin{aligned}\frac{1}{1\cdot2\cdots(x-l)}&=(x-l)^{l-x-1/2}\cdot\frac{e^{x-l}}{\sqrt{2\pi}}\left\{1-\frac{1}{12(x-l)}-\cdots\right\},\\ \frac{1}{1\cdot2\cdots(x'+l)}&=(x'+l)^{-l-x'-1/2}\cdot\frac{e^{x'+l}}{\sqrt{2\pi}}\left\{1-\frac{1}{12(x'+l)}-\cdots\right\}\end{aligned} \tag{1.114}$$

Now,

$$\log\left(1-\frac{l}{x}\right)=-\frac{l}{x}-\frac{l^2}{2x^2}-\frac{l^3}{3x^3}-\cdots.$$

Therefore,

$$\left(l-x-\frac{1}{2}\right)\left\{\log x+\log\left(1-\frac{l}{x}\right)\right\}=\left(l-x-\frac{1}{2}\right)\log x+l+\frac{l}{2x}-\frac{l^2}{2x}-\frac{l^3}{6x^2}-\cdots.$$

* From now on, therefore, x denotes the value for which the binomial probability is largest.

(*ibid.*, p. 277). By taking antilogs in the last equation, the expansion of the term $(x-l)^{l-x-1/2}$ in the first equation of (1.114) can be obtained as

$$(x-l)^{l-x-1/2} = \exp\left(l - \frac{l^2}{2x}\right) x^{l-x-1/2} \left(1 + \frac{l}{2x} - \frac{l^3}{6x^2} - \cdots\right). \tag{1.115}$$

Similarly,

$$(x'+l)^{-l-x'-1/2} = \exp\left(-l - \frac{l^2}{2x'}\right) x'^{-l-x'-1/2} \left(1 - \frac{l}{2x'} + \frac{l^3}{6x'^2} - \cdots\right). \tag{1.116}$$

Laplace's next step was to write $p=(x+s)/(n+1)$ in (1.112) as $p=(x-z)/n$, where $-(n-x)/(x+1) < z \le x/(n+1)$ since $0 \le s < 1$. Therefore, $|z| \le 1$. Now, using Eqs. (1.115), (1.116), and $p=(x-z)/n$, Eq. (1.113) becomes

$$\frac{1\cdot 2\cdot 3\cdots(x+x')}{1\cdot 2\cdots(x-l)\cdot 1\cdot 2\cdots(x'+l)} p^{x-l}(1-p)^{x'+l} = \frac{\sqrt{n}e^{-\frac{nl^2}{2xx'}}}{\sqrt{\pi}\sqrt{2xx'}} \times \left\{1 + \frac{nzl}{xx'} + \frac{l(x'-x)}{2xx'} - \frac{l^3}{6x^2} + \frac{l^3}{6x'^2} - \cdots\right\}$$

(*ibid.*, p. 278). We note that, for large n and x, $p=(x-z)/n$ implies $x \approx np$. Similarly, $x' \approx n(1-p)$. The above expression, which is the lth term after the largest in the binomial expansion of $\{p+(1-p)\}^{x+x'}$, can therefore be approximated by

$$\frac{1}{\sqrt{2\pi np(1-p)}} \exp\left\{-\frac{l^2}{2np(1-p)}\right\}.$$

Thus, the probability that the event a occurs $np+l$ times can be approximated, for large n, by a normal distribution with mean np and variance $np(1-p)$. This is the celebrated de Moivre–Laplace theorem.

1.2.15 Laplace's Probability Books

1.2.15.1 Théorie Analytique des Probabilités (1812) The *Théorie Analytique des Probabilités* (TAP) (Laplace, 1812)* was the apotheosis of Laplace's statistical career. It has been described as "the Mont Blanc of mathematical analysis" (De Morgan, 1837, p. 347): it is majestic and imposing but at the same time quite intimidating and inaccessible. While it is true that no probability book before and none for many years later matched the level of sophistication of TAP, the book itself does not contain much *new* material. This is because a large part of TAP is made up of Laplace's previous memoirs.

*The TAP is also described in Pearson (1978, pp. 704–732), Dale (1999, pp. 224–276), Stigler (2005), Gillispie (2000, Chapter 26), Samueli and Boudenot (2009, pp. 221–243), Molina (1930), Lubet (2004), and Robert (2012).

The book is made up of two major parts: one on generating functions, the other on probabilities per se. The contents of the first edition (1812) are as follows:

- BOOK I: CALCULUS OF GENERATING FUNCTIONS
  - PART I: ON GENERATING FUNCTIONS
    - Chap. I: On generating functions in one variable
    - Chap. II: On generating functions in two variables
  - PART II: THEORY OF ASYMPTOTIC APPROXIMATIONS
    - Chap. I: On the approximate integration of derivatives involving factors raised to high powers
    - Chap. II: On the approximate integration of linear equations in finite and infinitely small differences
    - Chap. III: Application of the preceding methods to the approximation of various functions involving high powers
- BOOK II: GENERAL THEORY OF PROBABILITIES
  - Chap. I: General principles of this theory
  - Chap. II: On the probability of events composed of simple events whose respective possibilities are given
  - Chap. III: On probability laws resulting from the indefinite accumulation of events
  - Chap. IV: On the probability of errors on the means of a large number of observations and on the most advantageous means
  - Chap. V: Application of the calculus of probabilities, on the search for phenomena and their causes
  - Chap. VI: On the probability of causes and future events, based on observed events
  - Chap. VII: On the influence of unknown deviations which might exist in chances which we assume to be perfectly equal
  - Chap. VIII: On the mean duration of life, of marriages and other associations
  - Chap. IX: On benefits depending on the probability of future events
  - Chap. X: On moral expectation

The first edition of TAP also contains a dedication to Napoléon. This was removed in the later editions (second in 1814 and third in 1820). A short description of the contents of the first edition of TAP now follows. In BOOK I, Part I is taken from Laplace's the "Mémoire sur les suites"* (Laplace, 1782). Here Laplace reproduced the method of generating functions and its application to difference equations. Part II is taken from "Mémoire sur les approximations des formules qui sont fonctions de très grands nombres"† (Laplace, 1786a). In this part, Laplace reproduced his method of asymptotic approximations and also the technique of characteristic functions (Laplace, 1812, p. 138).

* See Section 1.2.7.

† See Section 1.2.8.

In BOOK II, Chapter I starts with a statement of Laplace's philosophy of universal determinism, dictated by some vast intelligence (see Section 1.1). This is followed by the (classical) definition of probability and the addition and multiplication laws. Laplace next stated the discrete version of Bayes' theorem (*ibid.*, p. 182), which is given as a *Principe*. Bayes is not mentioned (in fact he is not mentioned at all in the book) and Bayes' theorem is given for the case of a discrete uniform prior. Chapter I also contains a definition of mathematical and moral expectation. This is taken from his memoir "Recherches sur l'intégration des équations différentielles aux différences finis" (Laplace, 1776b).* Chapter II deals with problems of composite events. The first few sections are devoted to various lottery problems, some of which are solved using the techniques of BOOK I. Chapter II also contains a solution to the Problem of Points† extended to n players and solved by the method of generating functions (1812, p. 207) and a solution to de Moivre's Problem‡ applied to the inclination of orbits. In Chapter III, Laplace proved the de Moivre–Laplace theorem (see Section 1.2.14). Chapter IV is taken from Laplace's "Mémoire sur les approximations des formules qui sont fonctions de très grands nombres et sur leur application aux probabilités" (Laplace, 1810a) and proves the Central Limit Theorem (see Section 1.2.11). Chapter IV also contains the results of "Mémoire sur les intégrales définis et leur application aux probabilités" (Laplace, 1811a), where Laplace provided a justification of the principle of least squares based on direct probability and also showed that, for large samples, the method of least squares coincided with that of least absolute error (see Section 1.5.5). Chapter V of TAP considers several applications of the calculus of probabilities, including the diurnal variation of barometers and the rotation of the earth. In this chapter we can also find Buffon's needle problem§ cast in terms of a very thin cylinder thrown between two parallel lines (Laplace, 1812, p. 359). Laplace also considered an extension of the problem to the case of a very thin cylinder thrown on a set of congruent rectangles. Chapter VI is taken from "Mémoire sur les approximations des formules qui sont fonctions de très grands nombres (suite)" (Laplace, 1786b) and contains the continuous version of Bayes' theorem in the case of a uniform prior. The theorem is then applied to the problem of births in Paris, London, and Naples. In Chapter VII, Laplace considered the effect of unequal probabilities, which are assumed to be the same across trials. Here Laplace concluded that, for example, in the game of heads or tails, the odds in favor of similar outcome are still higher than those of different outcomes even if the coin is biased. Chapters VIII and IX of TAP deal with questions of political arithmetic. Finally, Chapter X deals with the issue of moral expectation first introduced by Daniel Bernoulli.

The second edition of TAP was published in 1814 and contained several additions. First an Introduction running 104 pages was added. In the same year, the Introduction was also published as a separate book as *Essai Philosophique sur les Probabilités* (Laplace,

*See Section 1.2.3.2.

†The Problem of Points is also discussed in Edwards (1982), Hald (1990), Todhunter (1865, Chapter II), Smorynski (2012, pp. 4–25), and Gorroochurn (2012a, Prob. 4 and end of Prob. 9; 2014).

‡De Moivre's Problem is also discussed in Hald (1990, pp. 210–211) and Gorroochurn (2012a, Prob. 9).

§Buffon's needle problem is also discussed in Uspensky (1937, pp. 251–257), Solomon (1978, Chapter 1), van Fraassen (1989, pp. 301–302), Aigner and Ziegler (2004, Chapter 21), Gridgeman (1960), Kendall and Moran (1963, pp. 70–77), O'Beirne (1965, pp. 192–197), and Gorroochurn (2012a, Prob. 16).

1814). In addition to the Introduction, the second edition of TAP contained an extra chapter (Chapter XI) on the probability of testimony and three additions involving proofs. The third edition of 1820 was further augmented by three Supplements. A fourth Supplement was added in 1825.

The *Essai Philosophique* was quite popular, but TAP was more revered than read and did not sell well. Reviewing the third edition of TAP, de Morgan admitted that "[of] all masterpieces of analysis, this is perhaps the least known." (De Morgan, 1837, p. 347). De Morgan's review is written in a very lively and sometimes quirky style and is a must-read. It is apparent that de Morgan blamed Laplace's literary style for the low popularity of TAP:

> The genius of Laplace was a perfect sledge hammer in bursting purely mathematical obstacles; but, like that useful instrument, it gave neither finish nor beauty to the results. In truth, in truism if the reader please, Laplace was neither Lagrange nor Euler, as every student is made to feel. The second is power and symmetry, the third power and simplicity; the first is power without either symmetry or simplicity. (*ibid.*, p. 348)

He also reproached Laplace (together with his French countrymen) for failing to acknowledge scholars from whose works he had borrowed in several places of his book. De Morgan later repeated the same charge in his 1839 *Cyclopedia* article (De Morgan, 1839, p. 326). In spite of these misgivings, de Morgan later also admitted that:

> ...Nevertheless, since less had been done to master the difficulties of this subject than in the case of the theory of gravitation, it is here [i.e. in the TAP] that Laplace most shines as a creator of resources. It is not for us to say that, failing such predecessors as he had (Newton only excepted), he would not by his own genius have opened a route for himself. Certainly, if the power of anyone man would have sufficed for the purpose, that man might have been Laplace. (De Morgan, 1837, pp. 849–850)

1.2.15.2 Essai Philosophique sur les Probabilités (1814) Laplace's *Essai Philosophique sur les Probabilités* (Laplace, 1814) was meant to provide an accessible exposition to the concepts and philosophy of probability addressed to a nonmathematical audience. It was based on a lecture given by Laplace in 1795 at the Écoles Normales. The *Essai Philosophique* was published in 1814, both separately as a book and an Introduction to the much more sophisticated *Théorie Analytique des Probabilités*. In all, it went through seven editions, the last one appearing in 1840, well after Laplace's death. An English translation by F.W. Truescott and F.L. Emory was first published as *A Philosophical Essay on Probabilities* (PEP) in 1902 (Laplace, 1902) and was based on the sixth edition. The contents of PEP are as follows:

- PART I. A PHILOSOPHICAL ESSAY ON PROBABILITIES
  - Chap. I: Introduction
  - Chap. II: Concerning Probability
  - Chap. III: General Principles of the Calculus of Probabilities
  - Chap. IV: Concerning Hope
  - Chap. V: Analytical Methods of the Calculus of Probabilities

- PART II. APPLICATION OF THE CALCULUS OF PROBABILITIES
  - Chap. VI: Games of Chance
  - Chap. VII: Concerning the Unknown Inequalities which may Exist among Chances Supposed to be Equal
  - Chap. VIII: Concerning the Laws of Probability which result from the Indefinite Multiplication of Events
  - Chap. IX: Application of the Calculus of Probabilities to Natural Philosophy
  - Chap. X: Application of the Calculus of Probabilities to the Moral Sciences
  - Chap. XI: Concerning the Probability of Testimonies
  - Chap. XII: Concerning the Selections and Decisions of Assemblies
  - Chap. XIII: Concerning the Probability of the Judgments of Tribunals
  - Chap. XIV: Concerning Tables of Mortality, and the Mean Durations of Life Marriage, and Some Associations
  - Chap. XV: Concerning the Benefits of Institutions which Depend upon the Probability of Events
  - Chap. XVI: Concerning Illusions in the Estimation of Probabilities
  - Chap. XVII: Concerning the Various Means of Approaching Certainty
  - Chap. XVIII: Historical Notice of the Calculus of Probabilities to 1816

PEP starts with Laplace's endorsement of a deterministic philosophy of science*:

> ALL events, even those which on account of their insignificance do not seem to follow the great laws of nature, are a result of it just as necessarily as the revolutions of the sun. In ignorance of the ties which unite such events to the entire system of the universe, they have been made to depend upon final causes or upon hazard, according as they occur and are repeated with regularity, or appear without regard to order; but these imaginary causes have gradually receded with the widening bounds of knowledge and disappear entirely before sound philosophy, which sees in them only the expression of our ignorance of the true causes. (*ibid.*, p. 3)

Universal determinism, we are told by Laplace, is due to a "vast intelligence" (which has been dubbed "Laplace demon"):

> We ought then to regard the present state of the universe as the effect of its anterior state and as the cause of the one which is to follow. Given for one instant an intelligence which could comprehend all the forces by which nature is animated and the respective situation of the beings who compose it-an intelligence sufficiently vast to submit these data to analysis-it would embrace in the same formula the movements of the greatest bodies of the universe and those of the lightest atom; for it, nothing would be uncertain and the future, as the past, would be present to its eyes. (*ibid.*, p. 4)

*This is a repeat of his earlier statement in the *Théorie Analytique des Probabilités* (Laplace, 1812, p. 177).

In Chapter III, Laplace gave 10 fundamental principles of probability, all of which are word versions of the same principles given in more sophisticated form in TAP:

> *First Principle* [classical definition of probability]. The first of these principles is the definition itself of probability, which, as has been seen, is the ratio of the number of favorable cases to that of all the cases possible. (*ibid.*, p. 11)
>
> *Second Principle* [calculation of probability when possible cases are not equally likely]. But that supposes the various cases equally possible. If they are not so, we will determine first their respective possibilities, whose exact appreciation is one of the most delicate points of the theory of chance. Then the probability will be the sum of the possibilities of each favorable case. (*ibidem*)
>
> *Third Principle* [product rule for independent events]. One of the most important points of the theory of probabilities and that which lends the most to illusions is the manner in which these probabilities increase or diminish by their mutual combination. If the events are independent of one another, the probability of their combined existence is the product of their respective probabilities. (*ibid.*, p. 12)
>
> *Fourth Principle* [definition of conditional probability]. When two events depend upon each other, the probability of the compound event is the product of the probability of the first event and the probability that, this event having occurred, the second will occur. (*ibid.*, p. 14)
>
> *Fifth Principle* [definition of conditional probability]. If we calculate a priori, the probability of the occurred event and the probability of an event composed of that one and a second one which is expected, the second probability divided by the first will be the probability of the event expected, drawn from the observed event. (*ibid.*, p. 15)
>
> *Sixth Principle* [Bayes' theorem both for a uniform prior and for the general case]. Each of the causes to which an observed event may be attributed is indicated with just as much likelihood as there is probability that the event will take place, supposing the event to be constant. The probability of the existence of anyone of these causes is then a fraction whose numerator is the probability of the event resulting from this cause and whose denominator is the sum of the similar probabilities relative to all the causes; if these various causes, considered *à priori*, are unequally probable, it is necessary, in place of the probability of the event resulting from each cause, to employ the product of this probability by the possibility of the cause itself. This is the fundamental principle of this branch of the analysis of chances, which consists in passing from events to causes. (*ibidem*)
>
> *Seventh Principle* [law of total probability]. The probability of a future event is the sum of the products of the probability of each cause, drawn from the event observed, by the probability that, this cause existing, the future event will occur. (*ibid.*, p. 17)
>
> *Eighth Principle* [definition of mathematical expectation]. When the advantage depends on several events it is obtained by taking the sum of the products of the probability of each event by the benefit attached to its occurrence. (*ibid.*, p. 20)
>
> *Ninth Principle* [definition of mathematical expectation]. In a series of probable events of which the ones produce a benefit and the others a loss, we shall have the advantage which results from it by making a sum of the products of the probability of each favorable event by the benefit which it procures, and subtracting from this sum that of the products of the probability of each unfavorable event by the loss which is attached to it. If the second sum is greater than the first, the benefit becomes a loss and hope is changed to fear. (*ibid.*, p. 21)

> *Tenth Principle* [change in moral value of a sum varies as the increase in sum divided by original sum]. The relative value of an infinitely small sum is equal to its absolute value divided by the total benefit of the person interested. (*ibid.*, p. 23)

We note that, after the seventh principle, Laplace did a famous calculation to illustrate the application of Bayes' theorem and the law of total probability. This application resulted in the *law of succession* (see Eq. 1.6), which Laplace then applied to the rising of the sun. As we shall explain in Section 1.3.3, Laplace received a lot of criticism for this calculation.

In Chapter V, Laplace attempted to give several formulas in words but was much less successful there. Consider the following, for example:

> Let us consider now a new function Z of t, developed like V and T according to the powers of t; let us designate by the character Δ placed before the primitive function the coefficient of the xth power of t in the product of V by Z; this coefficient in the product of V by the nth power of Z will be expressed by the character Δ affected by the exponent n and placed before the primitive function of x. (*ibid.*, 38)

Chapter XVIII gives a brief history of probability up to Laplace's times. Pascal, Fermat, Huygens, James Bernoulli, Montmort, de Moivre, and Bayes are all mentioned. Last, but not least, we mention the following powerful closing sentences of the PEP:

> It is seen in this essay that the theory of probabilities is at bottom only common sense reduced to calculus; it makes us appreciate with exactitude that which exact minds feel by a sort of instinct without being able ofttimes to give a reason for it. It leaves no arbitrariness in the choice of opinions and sides to be taken; and by its use can always be determined the most advantageous choice. Thereby it supplements most happily the ignorance and the weakness of the human mind. If we consider the analytical methods to which this theory has given birth; the truth of the principles which serve as a basis; the fine and delicate logic which their employment in the solution of problems requires; the establishments of public utility which rest upon it; the extension which it has received and which it can still receive by its application to the most important questions of natural philosophy and the moral science; if we consider again that, even in the things which cannot be submitted to calculus, it gives the surest hints which can guide us in our judgments, and that it teaches us to avoid the illusions which ofttimes confuse us, then we shall see that there is no science more worthy of our meditations, and that no more useful one could be incorporated in the system of public instruction. (*ibid.*, p. 196)

1.3. THE PRINCIPLE OF INDIFFERENCE

1.3.1 Introduction

As we mentioned on several occasions in the previous sections, the principle of indifference was an indispensable tool in Laplace's work on probability and statistics. It occurs in several places in his early writings, for example:

> We suppose in the theory that the different ways in which an event can occur are equally probable, or where they are not, that their probabilities are in a given ratio. When we wish then to make use of this theory, we regard two events as equally probable when we see no

* Italics are ours.

> reason that makes one more probable than the other, because if they were unequally possible, since we are ignorant of which side is the greater, this uncertainty makes us regard them as equally probable. (Laplace, 1774a, OC 8, p. 61)
>
> ...
>
> The probability of an event is thus just the ratio of the number of cases favorable to it, to the number of possible cases, *when there is nothing to make us believe that one case should occur rather than any other.** (Laplace, 1776b, OC 8, p. 146)

Not least, the principle appears implicitly in Laplace's use of Bayes' theorem, where the prior is assumed to be uniform, most of the time without any warning or discussion.

The origin of the principle of indifference can be traced to several places in Part IV of James Bernoulli's *Ars Conjectandi*, for example:

> ...all cases are equally possible, or can happen with equal ease. (Bernoulli, 1713, English edition, p. 322)
>
> Moreover these all have equal tendencies to occur; because of the similarity of the faces and the uniform weight of the die, there is no reason why one of the faces should be more prone to fall than another.... (*ibid.*, pp. 326–327)

The principle of indifference had at first been called the "principle of insufficient reason" (possibly as a word play on Leibniz's principle of sufficient reason) by the German physiological psychologist Johannes von Kries (1853–1928) in the *Die Principien der Wahrscheinlichkeitsrechnung* (von Kries, 1886). However, Keynes later stated:

> A rule, adequate to the purpose, introduced by James Bernoulli, who was the real founder of mathematical probability, has been widely adopted, generally under the title of The Principle of Non-Sufficient Reason, down to the present time. This description is clumsy and unsatisfactory, and, if it is justifiable to break away from tradition, I prefer to call it *The Principle of Indifference.* (Keynes, 1921, p. 41)

He went on to provide a definition that is still used today:

> The principle of indifference asserts that if there is no known reason for predicating of our subject one rather than another of several alternatives, then relatively to such knowledge the assertions of each of these alternatives have an equal probability. (*ibid.*, p. 42)

1.3.2 Bayes' Postulate

Many statisticians usually think of Thomas Bayes' posthumous paper, "An Essay towards solving a problem in the doctrine of chances" (Bayes, 1764) (Fig. 1.15), as the first instance where the principle of indifference was explicitly stated. In fact, the use of a uniform prior for a parameter, when there is absolutely no knowledge about it, is often called *Bayes' postulate*. However, as we shall now see, a careful reading of Bayes' paper, especially the scholium to the paper, reveals that the situation is not as straightforward as it seems.*

* See also Edwards (1978), Stigler (1982b), and Good (1988).

[370]
quodque ſolum, certa nitri ſigna præbere, ſed plura concurrere debere, ut de vero nitro producto dubium non relinquatur.

LII. *An Eſſay towards ſolving a Problem in the Doctrine of Chances. By the late Rev. Mr.* Bayes, *F. R. S. communicated by Mr.* Price, *in a Letter to* John Canton, *A. M. F. R. S.*

Dear Sir,

Read Dec. 23, 1763.

I Now ſend you an eſſay which I have found among the papers of our deceaſed friend Mr. Bayes, and which, in my opinion, has great merit, and well deſerves to be preſerved. Experimental philoſophy, you will find, is nearly intereſted in the ſubject of it; and on this account there ſeems to be particular reaſon for thinking that a communication of it to the Royal Society cannot be improper.

He had, you know, the honour of being a member of that illuſtrious Society, and was much eſteemed by many in it as a very able mathematician. In an introduction which he has writ to this Eſſay, he ſays, that his deſign at firſt in thinking on the ſubject of it was, to find out a method by which we might judge concerning the probability that an event has to happen, in given circumſtances, upon ſuppoſition that we know nothing concerning it but that, under the ſame circum-

FIGURE 1.15 First page of Bayes' posthumous paper, "An Essay towards solving a problem in the doctrine of chances" (Bayes, 1764)

It is hard to overemphasize the importance of Bayes' paper. It was the first to explicitly define probability as a subjective measure of belief and calculate how likely it was for the *probability* of an event to lie between specified limits, given the event had occurred a certain number of times. This is the reverse of what had been done by others before, so Bayes' technique came to be known as inverse probability at first, then Bayes' theorem* much later. However, Bayes' theorem as we know it today was never explicitly given by Bayes in his paper, rather it first occurred in Laplace's work.

*As was mentioned on p. 13, *Bayes' theorem* is usually used for the general theorem (see Eq. 1.3) and *inverse probability* is usually used for *Bayes' theorem (or Bayesian methods) together with a uniform prior.* The word "Bayesian" itself started getting used only in the 1950s (Fienberg, 2006).

Bayes wished to solve the following problem:

> *Given* the number of times in which an unknown event has happened and failed: *Required* the chance that the probability of its happening in a single trial lies somewhere between any two degrees of probability that can be named (*ibid.*, 376)

To solve the above problem, Bayes used a level square table on which a ball was thrown. Then a second ball was thrown n times and the number of successes was recorded, depending on whether the second ball was closer than the first ball from a given side of the table. In modern notation, let the first ball land at a distance p from the given side of the table, where $p \sim \text{Unif}\,(0, 1)$. Then the n throws of the second ball corresponds to a binomial experiment with parameters n and p. Suppose the number of successes X is equal to a. It is thus required to find $\Pr\{p_1 < p < p_2 | X = a\}$, where $0 \le p_1 < p_2 \le 1$. In Proposition 8 of his paper, Bayes essentially gave the result*

$$\Pr\{p_1 < p < p_2, X = a\} = \int_{p_1}^{p_2} \frac{n!}{a!(n-a)!} p^a (1-p)^{n-a}\, dp. \tag{1.117}$$

This is obtained by first writing

$$\Pr\{p_1 < p < p_2, X = a\} = \int_{p_1}^{p_2} \Pr\{X = a \mid p\} f(p) dp,$$

and then using the result that the prior $f(p)$ of p is Unif (0, 1). As a corollary to Eq. (1.117), Bayes wrote

$$\Pr\{X = a\} = \int_0^1 \frac{n!}{a!(n-a)!} p^a (1-p)^{n-a}\, dp = \frac{1}{n+1}. \tag{1.118}$$

Dividing Eq. (1.117) by Eq. (1.118), Bayes obtained (in Proposition 9) the solution to the initial problem:

$$\Pr\{p_1 < p < p_2 | X = a\} = (n+1)\int_{p_1}^{p_2} \frac{n!}{a!(n-a)!} p^a (1-p)^{n-a}\, dp = \frac{(n+1)!}{a!(n-a)!}\int_{p_1}^{p_2} p^a (1-p)^{n-a}\, dp.$$

Now, in the Scholium to his paper, Bayes addressed the issue of how to proceed when nothing is known about the parameter p and the device of the initial ball on the table is no longer applicable. Referring to his earlier analysis, he stated:

> …And that the same rule is the proper one to be used in the case of an event concerning the probability of which we absolutely know nothing antecedently to any trials made concerning it, seems to appear from the following consideration; viz. that concerning such an event I have no reason to think that, in a certain number of trials, it should rather happen any one

* Bayes himself did not use integration but areas under curves.

> possible number of times than another. For, on this account, I may justly reason concerning it as if its probability had been at first unfixed, and then determined in such a manner as to give me no reason to think that, in a certain number of trials, it should rather happen any one possible number of times than another... In what follows therefore I shall take for granted that the rule given concerning the event M in prop.9 is also the rule to be used in relation to any event concerning the probability of which nothing at all is known antecedently to any trials made or observed concerning it. (*ibid.*, pp. 392–393)

In the above, Bayes' stated that knowing nothing about the parameter p was equivalent to saying that all possible valued of X were equally likely, that is,

$$\Pr\{X = a\} = \frac{1}{n+1} \quad \text{for} \quad a = 0,1,\ldots,n \tag{1.119}$$

But the above is a corollary to (1.117), in which a uniform prior is assumed. Hence, in situations when nothing is known about the parameter, that is, when Eq. (1.119) can be assumed, then one should proceed as in (1.117) by using a uniform prior for the parameter. It can therefore be seen that Bayes' use of a uniform distribution for the parameter p was not a *postulate*, rather it was an *argument* derived from the assumption of an equiprobable distribution for the random variable X.

1.3.3 Laplace's Rule of Succession. Hume's Problem of Induction

Coming back to Laplace, we have described in Section 1.2.2.2 how he implicitly used the principle of indifference to derive the *rule of succession.** He considered an infinite number of white and black tickets in an unknown ratio x. Suppose that, out of $(p+q)$ tickets drawn, p are white and q black. What is the probability $\tilde{x}$ that the next ticket drawn will be white? Laplace's solution to the problem relied on the assumption that x ~ Unif[0,1], since there is no reason to suppose that x should take one value rather than the other. He was then able to obtain (cf. Eq. 1.6)

$$\tilde{x} = \frac{p+1}{p+q+2}.$$

This is Laplace's rule of succession and has been the subject of much controversy, especially following one particular application Laplace made of it.

But it is necessary for us to go back in time to David Hume's (Fig. 1.16) *An Inquiry Concerning Human Understanding* (Hume, 1748) in order to better understand the controversy around Laplace's application of the rule of succession. In his landmark treatise, Hume distinguished between two types of universal propositions: *relations of ideas* and *matters of fact*. In the former, the truth of the proposition can be inferred by the mere operation of thought, "without dependence on what is anywhere existent in the universe." For example, the truth of the proposition "a triangle has three sides" can be discovered by simple reasoning: three sides is part of the concept of a triangle. On the other hand, we are

*The rule of succession was first so called by Venn (1866, p. 159) and is also discussed in Zabell (1989b), Hald (1998, pp. 256–262), Keynes (1921, p. 376), Sarkar and Pfeifer (2006, p. 47), Hogben (1957, Chapter 6), and Pearson (1892, pp. 168–179).

FIGURE 1.16 David Hume (1711–1776). Wikimedia Commons (Public Domain), http://commons.wikimedia.org/wiki/File:David_Hume.jpg

told by Hume, matters of fact propositions rest on shakier grounds. Their truths can only be gauged empirically because the ideas involved are not logically connected. For example, the statement "Jupiter is the largest planet in the Solar System" involves concepts (a planet named Jupiter, largest, Solar System) that are not logically connected and it would imply no logical contradiction if we believed Mars was the largest planet. Thus, the truth of matters of fact propositions cannot be inferred by reasoning: experience (or the use of the senses) is required.

Hume stressed the fact that, although experience is required for matters of fact inferences, the latter are not based on any further reasoning and therefore have no rational justification. To illustrate, he considered one particular form of matter of fact reasoning, namely, induction (i.e., reasoning that takes the observed as grounds for belief about the unobserved). Hume contended that induction had no rational justification and was based only on impressions formed by experience. This is *Hume's problem of induction** and can be stated as follows:

> *Suppose out of a large number n of occurrences of an event A, an event B occurs m times. Based on these observations, an inductive inference would lead us to believe that approximately m/n of all events of type A is also of type B, i.e. the probability of B given A is approximately m/n. Hume's problem of induction states that such an inference has no rational justification, but arises only as a consequence of impressions formed by experience.*

In Hume's own words:

> That there are no demonstrative arguments in the case [of inductive arguments], seems evident; since it implies no contradiction, that the course of nature may change, and that an object, seemingly like those which we have experienced, may be attended with different or contrary effects. May I not clearly and distinctly conceive, that a body, falling from the clouds, and which, in all other respects, resembles snow, has yet the taste of salt or feeling of fire? Is there any more intelligible proposition than to affirm, that all the trees will flourish in December and January, and decay in May and June? Now whatever is intelligible, and can be distinctly conceived, implies no contradiction, and can never be proved false by any demonstrative argument or abstract reasoning à priori. (Hume, 1748, p. 25)

Earlier in his book, Hume gave the famous "rise-of-the-sun" example, which was meant to illustrate the shaky ground on which inductive reasoning rested:

> Matters of fact, which are the second objects of human reason, are not ascertained in the same manner; nor is our evidence of their truth, however great, of a like nature with the foregoing. The contrary of every matter of fact is still possible; because it can never imply a contradiction, and is conceived by the mind with the same facility and distinctness, as if ever so conformable to reality. That the sun will not rise tomorrow is no less intelligible a proposition, and implies no more contradiction, than the affirmation, that it will rise. We should in vain, therefore, attempt to demonstrate its falsehood. Were it demonstratively false, it would imply a contradiction, and could never be distinctly conceived by the mind (*ibid.*, p. 18)

*Hume's problem of induction is also discussed in Salmon (1967, Chapters I–III), Black (2006, pp. 635–650), BonJour (2009, Chapter 4), Skyrms (2000, Chapter III), and Gorroochurn (2011).

Hume's ideas had a deep impact on philosophers and scientists. His "rise-of-the-sun" example was commented by many, including Buffon, Price (in the appendix to Bayes' Essay*), and later Laplace. In the *Essai Philosophique*, Laplace said:

> Thus we find that an event having occurred successively any number of times, the probability that it will happen again the next time is equal to this number increased by unity divided by the same number, increased by two units. Placing the most ancient epoch of history at five thousand years ago, or at 1826213 days, and the sun having risen constantly in the interval at each revolution of twenty-four hours, it is a bet of 1826214 to one that it will rise again tomorrow. (Laplace, 1814, English edition, p. 19)

Laplace obtained the numbers above from Eq. (1.6) by using $q=0$ and $p=5000\times 365.24$ $26=1{,}826{,}213$:

$$\tilde{x}=\frac{1{,}826{,}214}{1{,}826{,}215}\quad(\approx .9999994).$$

Laplace's calculation was thus meant to counter Hume's criticism of induction, but quite understandably he received a lot of flak for it. Somehow, Laplace must have felt that there was something amiss with his calculations, for his very next sentence read:

> But this number is incomparably greater for him who, recognizing in the totality of phenomena the principal regulator of days and seasons, sees that nothing at the present moment can arrest the course of it. (*ibidem*)

Laplace here seems to warn the reader that his method was correct when based only on the information from the sample, but his statement was too timid. Indeed, applying the rule of succession to the rising of the sun should be viewed with skepticism for several reasons. A major criticism lies in the reliance of the solution on the principle of indifference: the probability of the sun rising is assumed to be equally likely to take any of the values in [0, 1] because there is no reason to favor any particular value for the probability.† To many, this is not a reasonable assumption. Indeed, the kind of ridiculous conclusions that can be reached by recklessly applying the rule of succession is one of the major criticisms of the principle of indifference. As another example of absurdity, consider:

> A boy is 10 years old today. The rule [of succession] says that, having lived ten years, he has probability 11/12 [=.917] of living one more year. On the other hand, his 80-year-old grandfather has probability 81/82 [=.988] of living one more year! (Parzen, 1960, p. 123)

Because of Laplace's authority, at first few questioned the universal applicability of the principle of indifference. According to Keynes (1921, p. 85), the first person to object to the principle of indifference was Robert Leslie Ellis:

> Mere ignorance is no ground for any inference whatever. *Ex nihilo nihil* (Ellis, 1850)

*A recent paper by Stigler (2013) has shown that Bayes was also aware of Hume's problem of induction.

† It is also dubious if the rising of the sun on a given day can be considered a random event at all (see e.g., Schay, 2007, p. 65).

Other major opponents of the principle soon followed, namely, Boole, Venn, and Chrystal. All these mathematicians showed that there were problems with the principle of indifference. These defects seriously undermined the applicability of inverse probability and Bayes' theorem in general.

In particular, John Venn (1834–1923) attacked the rule of succession, saying that it was a mathematician's "duty to reject and denounce it for the future" (Venn, 1866, p. 156). George Chrystal (1851–1911) went even further and attacked the principle of inverse probability itself. Thus, in his book *Algebra*, Chrystal categorically refused to include Bayes' theorem as a legitimate topic:

> All matter of debatable character or of doubtful utility has been excluded. Under this head fall, in our opinion, the theory of *a priori* or inverse probability, and the applications to the theory of evidence. The very meaning of some of the propositions usually stated in parts of these theories seems to us to be doubtful. Notwithstanding the weighty support of La Place, Poisson, De Morgan, and others, we think that many of the criticisms of Mr. Venn on this part of the doctrine of chances are unanswerable. (Chrystal, 1889, p. 576)

Other contemporary books, such as Hall and Knight's *Higher Algebra* (Hall and Knight, 1891), did not follow Chrystal's suit and did include inverse probability.

In a later address delivered before the Actuarial Society of Edinburgh, Chrystal further denounced the "*reduction as absurdum* of the rules of Inverse Probability" (Chrystal, 1891, p. 421). He gave several examples purporting to show the futility of Bayes' theorem.

In retrospect, Chrystal's position was extreme as there are several situations where there is no ambiguity concerning the prior and Bayes' theorem can be put to good use. Whittaker (1921) later pointed this fact in a discussion of Chrystal's paper.

1.3.4 Bertrand's and Other Paradoxes

We now mention some other classic examples that show why the principle of indifference is viewed as deficient.

A first example showing problems with the principle of indifference is through its application in geometric probability in a well-known probability conundrum called *Bertrand's Paradox*. The problem was first adduced by Joseph Bertrand as the fifth problem in his influential book *Calculs des Probabilités* (Bertrand, 1889, p. 4), and is as follows: *A chord is randomly chosen on a circle. What is the probability that the chord is longer than the side of an equilateral triangle inscribed inside the circle?*

Intriguingly, the problem has an infinite number of possible solutions, depending on how we apply the principle of indifference. Three common ones are as follows (Fig. 1.17).

In the first solution (see left circle in Fig. 1.17), to fix ideas let one end of the chord be at A, coinciding with the vertex A of the triangle ABC. Then, by the principle of indifference, the other end of the chord is equally likely to be any other point on the circumference of the circle, such as the ones shown. Now the remaining two vertices (B and C) of the triangle divide the circumference of the circle into three equal arcs (AC, CB, and BA). The chord is longer than the side of the triangle if and only if it intersects the triangle (i.e., the other end of the chord lines on the arc BC). The required probability is thus 1/3.

In the second solution (see middle circle in Fig. 1.17), one of the edges, say AB, of the triangle is bisected at right angles by a radius of the circle. The chord is obtained by choosing its midpoint from a point on the radius such that the chord is parallel to AB.

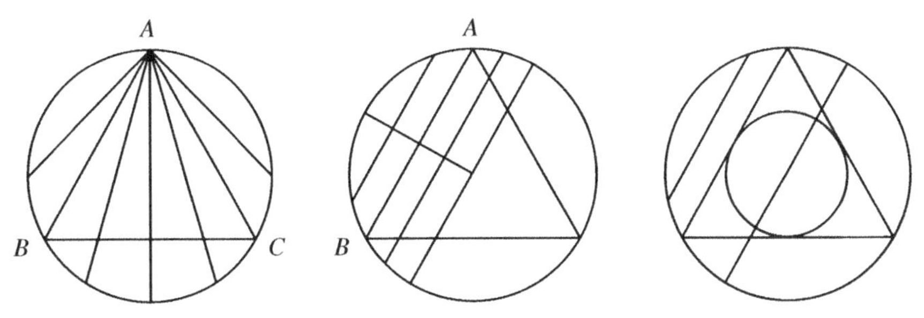

FIGURE 1.17 Three of the ways in which a chord can be randomly chosen on a circle

Now, the chord is longer than the side of the triangle if and only if the selected midpoint lies on the half of the radius that is closer to the center of the circle. Since, by the principle of indifference, the midpoint is equally likely to be anywhere on the radius, the required probability is 1/2.

In the third solution (see right circle in Fig. 1.17), we randomly choose a point inside the circle and obtain the chord so that the chosen point is its midpoint. The chord is longer than the side of the triangle if and only if the chosen point lies inside the inscribed circle of the triangle. By the principle of indifference, the chosen point is equally likely to be anywhere inside the outer circle. Since the inner inscribed circle has radius half that of the outside circle, the required probability is $(1/2)^2 = 1/4$.

Thus, the application of the principle of indifference leads to different results to the same problem.

A second example is provided by von Mises:

> We are given a glass containing a mixture of water and wine. All that is known about the proportions of the liquids is that the mixture contains at least as much water as wine, and at most, twice as much water as wine. The range for our assumptions concerning the ratio of water to wine is thus the interval 1 to 2. Assuming that nothing more is known about the mixture, the indifference or symmetry principle or any other similar form of the classical theory tells us to assume that equal parts of this interval have equal probabilities. The probability of the ratio lying between 1 and 1.5 is thus 50%, and the other 50% corresponds to the probability of the range 1.5 to 2.
>
> But there is an alternative method of treating the same problem. Instead of the ratio water/wine, we consider the inverse ratio, wine/water; this we know lies between 1/2 and 1. We are again told to assume that the two halves of the total interval, i.e., the intervals 1/2 to 3/4 and 3/4 to 1, have equal probabilities (50% each); yet, the wine/water ratio 3/4 is equal to the water/wine ratio 4/3. Thus, according to our second calculation, 50% probability corresponds to the water/wine range 1 to 4/3 and the remaining 50% to the range 4/3 to 2. According to the first calculation, the corresponding intervals were 1 to 3/2 and 3/2 to 2. The two results are obviously incompatible. (von Mises, 1957, p. 77)

A third, more technical, example showing the contradictory nature of the principle of indifference is due to Fisher (1922c, p. 325), one of the key critics of Bayesianism.*

* See also Fisher (1956b, p. 16) for a similar example.

Fisher considered a proportion p about which there is complete ignorance. The principle of indifference would therefore assign it the following distribution:

$$f(p)=1 \quad \text{for} \quad 0<p<1 \tag{1.120}$$

Now suppose $\sin\theta = 2p-1$ where "the quantity, θ, measures the degree of probability, just as well as p, and is even, for some purposes, the more suitable variable." Complete ignorance about p is tantamount to complete ignorance about θ so that, again,

$$f(\theta)=\frac{1}{\pi} \quad \text{for} \quad -\frac{\pi}{2}<\theta<\frac{\pi}{2}.$$

But since $\sin\theta = 2p-1$ we also have

$$f(p)=f(\theta)\left|\frac{d\theta}{dp}\right|=\frac{1}{\pi\sqrt{p(1-p)}} \quad \text{for} \quad 0<p<1,$$

in contradiction to Eq. (1.120).

The above are just a few of the many examples that show why the principle of indifference is usually regarded as deficient.

1.3.5 Invariance

On the other hand, some mathematicians have defended the principle of indifference by using the concept of *invariance*. For example, referring to Bertrand's Paradox, Henry Poincaré (1912, p. 130) (Fig. 1.18) suggested that the arbitrariness in the interpretation of randomness could be removed by associating probability densities that remain invariant under Euclidean transformations. That is to say, the problem could become well posed if we could find probability densities for random variables such that we would get the same answer if the circle and inscribed triangle were to be rotated, translated, or dilated. Poincaré (*ibidem*) and later Jaynes (1973) (Fig. 1.19) concluded that such probability densities exist only for the random radius method. Let us now explain Jaynes' approach and show how "invariance under a transformation group" lead to a probability of 1/2.

Let the circle C have radius R and let the randomly drawn chord have its midpoint at (X, Y), relative to the center O of the circle. Suppose the probability density of (X, Y) is $f(x,y)$, where $(x,y)\in C$. We next draw a smaller concentric circle C' with radius aR, where $0<a\le 1$. Let the density of (X, Y) be $h(x,y)$ for $(x,y)\in C'$. The two densities f and h should be the same except for a constant factor, because their supports are different. For $(x,y)\in C'$, we have

$$h(x,y)=\frac{f(x,y)}{\iint_{C'} f(x,y)dxdy}.$$

FIGURE 1.18 Henri Poincaré (1854–1912). Wikimedia Commons (Public Domain), http://commons.wikimedia.org/wiki/File:Henri_Poincar%C3%A9-2.jpg

FIGURE 1.19 Edwin T. Jaynes (1922–1998). Wikimedia Commons (Public Domain), http://commons.wikimedia.org/wiki/File:ETJaynes1.jpg

Rotational invariance implies that both $f(x, y)$ and $h(x, y)$ depend on (x, y) only through the radial distance $r = \sqrt{x^2 + y^2}$. Therefore, in polar coordinates (r, θ), the previous formula for h becomes

$$\begin{aligned} h(r) &= \frac{f(r)}{\int_0^{2\pi}\int_0^{aR} rf(r)\,dr\,d\theta} \\ &= \frac{f(r)}{2\pi\int_0^{aR} rf(r)\,dr} \\ \therefore f(r) &= 2\pi h(r)\int_0^{aR} f(r)\,r\,dr \end{aligned} \tag{1.121}$$

(*ibid.*, p. 4).

Jaynes next invoked scale invariance. Imagine that the larger circle is shrunk by a factor a, $0 < a \le 1$, to give the smaller circle. Then scale invariance implies

$$\begin{aligned} f(r)\,r\,dr\,d\theta &= h(ar)\cdot ar\cdot d(ar)\,d\theta \\ f(r)\,r\,dr\,d\theta &= a^2 h(ar)\,r\,dr\,d\theta \\ \therefore f(r) &= a^2 h(ar) \end{aligned} \tag{1.122}$$

(*ibid.*, p. 5). From Eqs. (1.121) and (1.122), we have for $0 < a \le 1$ and $0 < r \le R$,

$$a^2 f(ar) = 2\pi f(r)\int_0^{aR} rf(r)\,dr.$$

If we differentiate the above with respect to a, then set $a = 1$, we obtain the differential equation

$$2f(r) + rf'(r) = 2\pi R^2 f(r) f(R)$$

with general solution

$$f(r) = \frac{qr^{q-2}}{2\pi R^q}, \tag{1.123}$$

(*ibidem*) where q is a positive constant.

The final property is that of translational invariance. Jaynes considered the original circle C with radius R and imagined a second circle C' whose center is displaced by an amount b from the first. Then a given chord will have its center at a point $P(r, \theta)$ with respect to C and at a point $P'(r', \theta')$ with respect to C', where (Fig. 1.20)

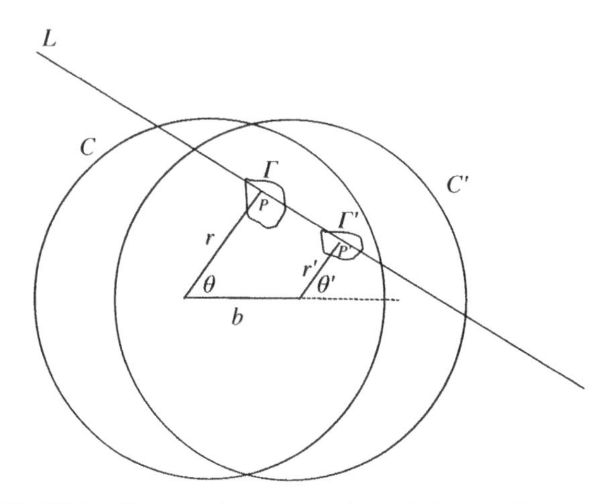

FIGURE 1.20 Translational invariance in Bertrand's Paradox

$$r' = |r - b\cos\theta|,$$
$$\theta' = \begin{cases} \theta, & r > b\cos\theta \\ \theta + \pi, & r < b\cos\theta \end{cases} \tag{1.124}$$

Moreover, the regions Γ and Γ' are such that as P varies over the first region, P' varies over the second, and vice versa. Thus, the probability that the midpoint of a chord lies in Γ is equal to the probability that the midpoint also lies in Γ'. Now the probability that a chord which intersects C has its midpoint in Γ is

$$\int_\Gamma f(r)\,dxdy = \int_\Gamma f(r)\,rdrd\theta = \frac{q}{2\pi R^q}\int_\Gamma r^{q-1}drd\theta. \tag{1.125}$$

Similarly, the probability that a chord which intersects C' has its midpoint in Γ' is

$$\frac{q}{2\pi R^q}\int_{\Gamma'} (r')^{q-1}\,dr'd\theta'.$$

By using the transformations in (1.124), the above probability becomes

$$\frac{q}{2\pi R^q}\int_\Gamma |r - b\cos\theta|^{q-1}\,drd\theta. \tag{1.126}$$

For the two probabilities in (1.125) and (1.126) to be the same, we need $q = 1$ (*ibid.*, p. 6).
Substituting $q = 1$ in Eq. (1.123), Jaynes finally obtained

$$f(r) = \frac{1}{2\pi Rr}, \quad 0 < r \le R, \quad 0 \le \theta \le 2\pi \tag{1.127}$$

(*ibidem*).

Assuming these invariance arguments, the probability that the chord is longer than the side of the inscribed triangle can finally be calculated as

$$\begin{aligned}\int_0^{2\pi}\int_0^{R/2} f(r)\,rdrd\theta &= \int_0^{2\pi}\int_0^{R/2} \frac{1}{2\pi Rr}\,rdrd\theta \\ &= (2\pi)\left(\frac{R}{2}\right)\cdot\frac{1}{2\pi R} \\ &= \frac{1}{2}.\end{aligned}$$

Thus, adherents to the invariance principle favor the answer of half. Moreover, this was the answer obtained when experiments of actually throwing straws into a circle were carried out (*ibid.*, p. 8). However, we hasten to add that not all situations where the principle of indifference breaks down are amenable to the kind of invariance treatment provided above, as Jaynes himself conceded.* Moreoever, the invariance principle itself is not immune to criticism, the main one being the necessity to indentify an appropriate transformation group. For further discussion, see Fine (1973, pp. 170–172).

1.4. FOURIER TRANSFORMS, CHARACTERISTIC FUNCTIONS, AND CENTRAL LIMIT THEOREMS

As we explained in Section 1.2.8.3, Laplace had used only an embryonic form of the characteristic function (CF) in his 1785 memoir (Laplace, 1785). However, he later used it more explicitly in the following manner: if $f_X(x)$ is the density ("loi de facilité") of X then its CF is

$$\psi_X(t) = \mathsf{E}e^{itX} = \int_{-\infty}^{\infty} e^{itx} f_X(x)\,dx, \tag{1.128}$$

where $i = \sqrt{-1}$. Moreover, there is a one-to-one relationship between $\psi_X(t)$ and $f_X(x)$ so that the former can be inverted back to obtain the latter:

$$f_X(x) = \frac{1}{2\pi}\int_{-\infty}^{\infty} e^{-itx}\psi_X(t)\,dt. \tag{1.129}$$

Written in the form (1.128), it turns out that $\psi_X(t)$ is the *Fourier transform* of $f_X(x)$, and Eq. (1.129) gives the *inversion formula* for $f_X(x)$. The CF is especially useful when summing independent and identically distributed (IID) random variables. This is because if $Y = X_1 + X_2 + \ldots + X_n$, where the X_j's are IID, then

* For example, von Mises water and wine example, discussed on p. 107 is not resolved by using invariance.

$$\begin{aligned}\psi_Y(t) &= \mathsf{E}e^{itY} \\ &= \mathsf{E}e^{it(X_1+X_2+\cdots+X_n)} \\ &= \left(\mathsf{E}e^{itX_j}\right)^n \\ &= \left\{\psi_X(t)\right\}^n.\end{aligned}$$

The CF is also related to the moment-generating function (MGF) $M_X(t) = \mathsf{E}e^{tX}$, since

$$\psi_X(t) = M_X(it).$$

However, the CF is often preferred to the latter as a statistical tool because, whereas $M_X(t)$ is undefined for some random variables, $\psi_X(t)$ does not have this disadvantage. That this is so can be seen from

$$\mathsf{E}\left|e^{itX}\right| = \mathsf{E}1 = 1.$$

This implies that $\mathsf{E}e^{itX}$ must always exist, since for a general random variable $Y, \mathsf{E}Y$ exists if and only if $\mathsf{E}|Y|$ exists.

There is no doubt that Laplace's further development of the CF (i.e., the Fourier transform of the density) in 1809 and 1810 was influenced by the recent work of the mathematician Joseph Fourier in 1807. In what follows, we describe the events surrounding Fourier's breakthrough and its aftermath.

1.4.1 The Fourier Transform: From Taylor to Fourier

In 1715, Taylor published his *Methodus Incrementorum* (Taylor, 1715, Nos. 17 and 18) in which he studied a vibrating string. The latter had also been studied in one of his previous papers (Taylor, 1713). Taylor found that the normal acceleration at any point of the string is inversely proportional to the radius of curvature. But Taylor had no notion of partial derivatives and did not explicitly write down the partial differential equation of the vibrating string.

In 1747, d'Alembert considered the problem of the vibrating string. Acknowledging the contribution of Taylor, he obtained the partial differential equation

$$\frac{\partial^2 y}{\partial s^2} = a^2 \frac{\partial^2 y}{\partial t^2} \tag{1.130}$$

(d'Alembert, 1747, p. 214), where $y = y(s,t)$ is the vertical displacement of a point on a vibrating string (of length l) at time t, s is the horizontal distance of the point, and a is a constant. D'Alembert's general solution to Eq. (1.130), with $a^2 = 1$, was

$$y = \psi(t+s) + \Gamma(t-s)$$

(*ibid*, p. 216), where ψ and Γ are arbitrary functions. By using the boundary conditions $y(0,t) = y(l,t) = 0$, d'Alembert wrote the above as

$$y = \psi(t+s) - \psi(t-s). \tag{1.131}$$

Applying the boundary conditions again to Eq. (1.131), d'Alembert obtained $\psi(t+l)=\psi(t-l)$. By substituting $t-l=\tau$, it is seen that ψ must satisfy

$$\psi(\tau)=\psi(\tau+2l).$$

The same problem was considered by Euler in 1748 (Euler, 1748). He reached the same solution as d'Alembert's Eq. (1.131), but with one major difference: only in Euler's solution was ψ allowed to be nondifferentiable. This gave rise to a back and forth between d'Alembert and Euler, the former rejecting the possibility of a nondifferentiable function. Euler's insistence on a nondifferentiable ψ was motivated by the physical possibility that the string could be initially plucked. But d'Alembert also had a point: if ψ was allowed to be nondifferentiable, then its second derivative would surely be undefined and Eq. (1.130) would make no sense.* It is also interesting to note that Euler gave an example of the initial shape of the string in the form

$$y=\alpha\sin\frac{\pi s}{l}+\beta\sin\frac{2\pi s}{l}+\gamma\sin\frac{3\pi s}{l}+\cdots \tag{1.132}$$

(*ibid.*, p. 84), where $\alpha,\beta,\gamma,\ldots$ are arbitrary constants. Euler did not specify whether the above sum was finite or infinite. Since Eq. (1.132) was differentiable, it was not as general as his own solution.

The next mathematician to consider the problem was Daniel Bernoulli (1753). He provided a heuristic argument that *all* initial positions of the string could have the form in (1.132). Euler promptly repudiated Bernoulli's proposal because he believed that it would imply that *every arbitrary curve* should be periodic, which was clearly not the case.†

We now consider Lagrange's 1759 memoir on the topic (Lagrange, 1759). Lagrange first considered a finite number of vibrating particles and obtained Bernoulli's solution. He then passed to the limiting case of an infinite number of vibrating particles (that is, the string itself) by interchanging summations with integrations and bringing the latter outside. Lagrange thus obtained Euler's solution as the most general one. He showed that, for the differential equation in (1.130), the general solution was of the form

$$\begin{aligned}y&=\frac{2}{a}\int_0^a Y(\tilde{x})\sum_{n=1}^{\infty}\left(\sin\frac{n\pi\tilde{x}}{a}\sin\frac{n\pi x}{a}\cos\frac{n\pi ct}{a}\right)d\tilde{x}\\&\quad+\frac{2}{\pi a}\int_0^a V(\tilde{x})\sum_{n=1}^{\infty}\left(\frac{1}{n}\sin\frac{n\pi\tilde{x}}{a}\sin\frac{n\pi x}{a}\sin\frac{n\pi ct}{a}\right)d\tilde{x}\end{aligned} \tag{1.133}$$

*As we shall see, the two points of view would later be reconciled through Fourier's extension of the concept of a function.

†The d'Alembert–Euler–Bernoulli controversy over the vibrating string is also described in Truesdell (1960, Part III), Garber (1999, Chapter II), and Kline (1972, pp. 502–514).

(*ibid.*, p. 100).* In the last formula, Y and V are, respectively, the initial displacement and initial velocity of a string of length a, and c is a constant. As several authors have noted (e.g., Truesdell, 1960, p. 270; Kline, 1972, p. 511), Lagrange's passage to the limit in obtaining Eq. (1.133) was faulty since it introduced the problem of divergent series into the solution. Moreover, it forbade him from recognizing

$$\frac{2}{a}\int_0^a Y(\tilde{x})\sin\frac{n\pi\tilde{x}}{a}\,d\tilde{x}$$

as a Fourier coefficient.

Another attempt which was also very close to the technique that Fourier put to later use was made by Euler in his paper "Methodus facilis inveniendi series per sinus cosinusve" (Euler, 1798). In this paper, Euler considered a function of the form

$$y = A_0 + 2A_1\cos\Phi + 2A_2\cos 2\Phi + 2A_3\cos 3\Phi + \cdots.$$

By multiplying both sides of the above by $\cos n\Phi$, integrating term by term from 0 to π, and using the fact that

$$\int_0^\pi \cos j\Phi \cos k\Phi \, d\Phi = \begin{cases} \pi/2 & \text{for} \quad k = j; j = 1,2,\ldots \\ 0 & \text{for} \quad j \neq k \end{cases}$$

he obtained

$$A_n = \frac{1}{\pi}\int_0^\pi y\cos n\Phi\, d\Phi, \quad n = 0,1,2,\ldots$$

(*ibid.*, p. 94).

We now consider Fourier's (Fig. 1.21) 1808 paper "Mémoire sur la propagation de la chaleur dans les corps solides" (Fourier, 1808) (Fig. 1.22). Previously, Laplace and Lavoisier wrote a paper in 1784 about experiments they had conducted on the specific heat of various substances. In 1804, Biot, who was Laplace's student, published a memoir entitled "Mémoire sur la propagation de la chaleur" (Biot, 1804). Fourier's approach, however, was different. Despite the pathbreaking nature of his paper, it was not enthusiastically received in 1807 when it was submitted to the *Académie des Sciences*. The reasons for this will be explained soon. The committee that reviewed Fourier's paper consisted of Lagrange, Laplace, Lacroix, and Monge. Of these, Lagrange was the one most vehemently opposed to Fourier's work, and the paper did not get published. However, the Academy set a prize problem soon after on the conduction of heat. Fourier reworked his manuscript, submitted again, and won the prize in 1812. This was in spite of the fact that the committee still had reservations about Fourier's paper. Thus, the latter did not get

*Although we have indicated the lower and upper limits on the $\int$ sign itself here, this was not actually done by Lagrange. The notation for the definite integral was first used much later by Fourier in the *Théorie Analytique de la Chaleur* (Fourier, 1822, pp. 200–201). Cajori has further noted that: "But Fourier had used this notation somewhat earlier in the *Mémoires* of the French Academy for the 1819–20, in an article of which the early part of his book of 1822 is a reprint" (Cajori, 1928, p. 250, Vol. II).

FIGURE 1.21 Jean Baptiste Joseph Fourier (1768–1830). Wikimedia Commons (Public Domain), http://commons.wikimedia.org/wiki/Jean_Baptiste_Joseph_Fourier#/media/File:Joseph_Fourier.jpg

MÉMOIRE

SUR LA

PROPAGATION DE LA CHALEUR

DANS LES CORPS SOLIDES,

PAR M. FOURIER (¹).

Présenté le 21 décembre 1807 à l'Institut national.

Nouveau Bulletin des Sciences par la Société philomathique de Paris, t. I, p. 112-116, n° 6; mars 1808. Paris, Bernard.

L'auteur de ce Mémoire s'est proposé de soumettre la Théorie de la chaleur à l'Analyse mathématique et de vérifier, par l'expérience, les résultats du calcul. Pour exposer l'état de la question, supposons une barre de métal, cylindrique et d'une longueur indéterminée, plongée par une de ses extrémités dans un fluide entretenu à une température constante : la chaleur se répandra successivement dans la barre; et, sans la perte qui a lieu à sa surface et à son autre extrémité, elle prendrait dans toute son étendue la température constante du foyer; mais, à cause de cette perte, la chaleur ne s'étendra d'une manière sensible que jusqu'à une distance du foyer dépendante de la grosseur de la barre, de la conductibilité du métal et de son degré de poli, qui influe sur le rayonnement; de sorte que des thermomètres placés dans l'éten-

(¹) Cet Article, que nous avons déjà signalé dans l'Avant-Propos du Tome I, n'est pas de Fourier. Signé de l'initiale P, il a été écrit par Poisson, qui était un des rédacteurs du *Bulletin des Sciences* pour la partie mathématique. A raison de l'intérêt historique qu'il présente comme étant le premier écrit où l'on ait fait connaître la théorie de Fourier, nous avons cru devoir le reproduire intégralement. G. D.

FIGURE 1.22 First page of summary by Poisson of Fourier's "Mémoire sur la propagation de la chaleur dans les corps solides" (Fourier, 1808)

published and almost all of it was later incorporated in his 1822 opus *Théorie Analytique de la Chaleur** (Fourier, 1822).

In order to understand the full impact of this paper, it is important to understand that, in all the works we have outlined previously, the function y originated from a well-defined physical experiment (the vibration of a string) and, as such, its existence was unquestioned. Moreover, a function, as it was understood at that time, was well behaved and differentiable. In Fourier's framework, however, the function was completely arbitrary and given graphically by a curve. The latter was not necessarily differentiable and could be made of separate functions or parts of functions.

In the 1808 memoir, Fourier modeled the diffusion of heat in a lamina. He considered the steady-state differential equation

$$\frac{\partial^2 v}{\partial x^2}+\frac{\partial^2 v}{\partial y^2}=0 \qquad (x\geq 0,-1\leq y\leq 1),$$

(Fourier, 1808, Oeuvres 2, p. 218), where (x, y) is a point on the lamina and $v=v(x,y)$ is the temperature. He obtained the solution

$$v=ae^{-\pi x/2}\cos\left(\frac{\pi y}{2}\right)+a'e^{-3\pi x/2}\cos\left(\frac{3\pi y}{2}\right)+a''e^{-5\pi x/2}\cos\left(\frac{5\pi y}{2}\right)+\cdots, \qquad (1.134)$$

where $a,a',a'',\ldots$ are constants (*ibidem*). To determine the values of the constants $a,a',a'',\ldots$, Fourier used the boundary condition $v(0,y)=\varphi(y)$ so that Eq. (1.134) becomes

$$\varphi(y)=a\cos\left(\frac{\pi y}{2}\right)+a'\cos\left(\frac{3\pi y}{2}\right)+a''\cos\left(\frac{5\pi y}{2}\right)+\cdots.$$

Multiplying both sides of the above by $\cos\{(2k+1)\pi y/2\}$ and integrating from $y=-1$ to $y=1$, the constants were obtained as follows:

$$a_k=\frac{1}{\pi}\int_{-1}^{1}\varphi(y)\cos\left\{(2k+1)\frac{\pi y}{2}\right\}dy.$$

In the above, $\varphi(y)$ is the *Fourier cosine transform* of the coefficients $a,a',a'',\ldots$. A similar formula holds for the *Fourier sine transform*. Moreover, it is often easier to work with e^{iny} $(i=\sqrt{-1})$ instead of sines and cosines, so we define the discrete Fourier complex transform of a function $\varphi(y)$ by $\sum_y e^{iny}\varphi(y)$. This presents no difficulty since e^{iny}, $\cos ny$, and $\sin ny$ are related through Euler's formulas:

$$\cos ny\equiv\frac{e^{iny}+e^{-iny}}{2},$$
$$\sin ny\equiv\frac{e^{iny}-e^{-iny}}{2i}.$$

*This book is examined in detail by Grattan-Guinness (2005).

The continuous version of the Fourier complex transform is given in Eq. (1.128), where the density f_X is replaced by any arbitrary function f.

Continuing with Fourier's work, even more importantly he argued that *every function* $F(x)$ *of "any form whatever"* could be expressed in the form

$$F(x) = a_0 + \sum_{n=1}^{\infty} a_n \cos nx + \sum_{n=1}^{\infty} b_n \sin nx, \tag{1.135}$$

where the coefficients are given by

$$a_0 = \frac{1}{2\pi}\int_{-\pi}^{\pi} F(x)\,dx,$$

$$\left.\begin{aligned} a_n &= \frac{1}{\pi}\int_{-\pi}^{\pi} F(x)\cos nx dx \\ b_n &= \frac{1}{\pi}\int_{-\pi}^{\pi} F(x)\sin nx dx \end{aligned}\right\} \quad \text{for } n = 1,2,\ldots$$

(Fourier, 1822, English translation, p. 204). This was a revolutionary statement that literally changed the concept of the function.

But at least two major questions needed to be answered. First, since in Fourier's framework, $F(x)$ was allowed to be discontinuous, what to make of expressions like $\int F(x)dx$, $\int F(x)\sin nxdx$, and $\int F(x)\cos nxdx$? To this, Fourier urged that integration be interpreted as areas rather than antidifferentiation, the latter being the practice of the time. Second, since there were infinite sums in (1.135), what was the guarantee that these would converge to $F(x)$? This was a thorny issue that was not properly addressed by Fourier, hence the criticism of Fourier's memoir both in 1807 and in 1812 by several mathematicians, and in particular by Lagrange. It was only in 1829 that Dirichlet was able to give a set of sufficient conditions* for the infinite sums in (1.135) to converge to $F(x)$, in a paper entitled "Sur la convergence des séries trigonométriques" (Dirichlet, 1829). Dirichlet's sufficient conditions are:

- $F(x)$ is single valued and bounded.
- $F(x)$ is piecewise continuous (it has only a finite number of discontinuities within one period).
- $F(x)$ is piecewise monotone (it has only a finite number of maxima and minima within one period).

1.4.2 Laplace's Fourier Transforms of 1809

Notwithstanding the problems of rigor in Fourier's 1807 memoir, Laplace had undoubtedly been influenced by it when, in 1809, he performed several important Fourier transforms (Laplace, 1809). For example, to evaluate the integrals

*These were given in sketch form in the concluding sections of Fourier's *Théorie Analytique de la Chaleur* (Fourier, 1822).

$$\int_0^\infty \frac{\cos x}{x}\,dx,$$
$$\int_0^\infty \frac{\sin x}{x}\,dx, \tag{1.136}$$

Laplace started with the Fourier transform

$$\int_0^\infty \frac{e^{ix}}{x^\alpha}\,dx = \int_0^\infty \frac{\cos x}{x^\alpha}\,dx + i\int_0^\infty \frac{\sin x}{x^\alpha}\,dx \tag{1.137}$$

(*ibid.*, p. 245), where $0 < \alpha < 1$ and $i = \sqrt{-1}$. By setting $x = it^{1/(1-\alpha)}$, we have

$$\int_0^\infty \frac{e^{ix}}{x^\alpha}\,dx = \frac{i^{1-\alpha}k}{1-\alpha}, \tag{1.138}$$

where

$$k = \int_0^\infty \exp\left\{-t^{1/(1-\alpha)}\right\}dt.$$

Using Euler's and de Moivre's formulas, for $r \in Z$,

$$\begin{aligned} i^{1-\alpha} &= \left[\cos\left\{(2r+1)\frac{\pi}{2}\right\} + i\sin\left\{(2r+1)\frac{\pi}{2}\right\}\right]^{1-\alpha} \\ &= \cos\left\{(1-\alpha)(2r+1)\frac{\pi}{2}\right\} + i\sin\left\{(1-\alpha)(2r+1)\frac{\pi}{2}\right\}. \end{aligned}$$

Therefore, from Eq. (1.138),

$$\int_0^\infty \frac{e^{ix}}{x^\alpha}\,dx = \frac{k}{1-\alpha}\left[\cos\left\{(1-\alpha)(2r+1)\frac{\pi}{2}\right\} + i\sin\left\{(1-\alpha)(2r+1)\frac{\pi}{2}\right\}\right].$$

Comparing the above with Eq. (1.137) and equating real and imaginary parts,

$$\begin{aligned} \int_0^\infty \frac{\cos x}{x^\alpha}\,dx &= \frac{k}{1-\alpha}\cos\left\{(1-\alpha)(2r+1)\frac{\pi}{2}\right\}, \\ \int_0^\infty \frac{\sin x}{x^\alpha}\,dx &= \frac{k}{1-\alpha}\sin\left\{(1-\alpha)(2r+1)\frac{\pi}{2}\right\}. \end{aligned} \tag{1.139}$$

Laplace next considered the case $\alpha \to 1^-$. Then the second equation in (1.139) becomes

$$\begin{aligned}\lim_{\alpha\to 1^-}\int_0^\infty \frac{\sin x}{x^\alpha}dx = \int_0^\infty \frac{\sin x}{x}dx \\ = \frac{\left(\lim\limits_{\alpha\to 1^-} k\right)}{1-\alpha}\times(1-\alpha)(2r+1)\frac{\pi}{2} \\ = \frac{\pi}{2}(2r+1)\lim_{\alpha\to 1^-} k,\end{aligned} \tag{1.140}$$

where, with $\beta = 1/(1-\alpha)$,

$$\begin{aligned}\lim_{\alpha\to 1^-} k &= \lim_{\beta\to\infty}\int_0^\infty \exp\left(-t^\beta\right)dt \\ &= \lim_{\beta\to\infty}\int_0^1 \exp\left(-t^\beta\right)dt + \lim_{\beta\to\infty}\int_1^\infty \exp\left(-t^\beta\right)dt \\ &= 1+0 \\ &= 1.\end{aligned}$$

Moreover, Laplace reasoned that

$$\int_0^\infty \frac{\sin x}{x}dx < \int_0^\pi \frac{\sin x}{x}dx < \int_0^\pi \frac{x}{x}dx = \pi.$$

Therefore, in (1.140), he took $r=0$ and obtained

$$\int_0^\infty \frac{\sin x}{x}dx = \frac{\pi}{2}$$

(*ibid.*, p. 247). By a similar reasoning, he showed that $\int_0^\infty (\cos x)/x dx = \infty$.

1.4.3 Laplace's Use of the Fourier Transform to Solve a Differential Equation (1810)

In the memoir "Mémoire sur les approximations des formules qui sont fonctions de très grands nombres et sur leur application aux probabilités" (Laplace, 1810a, p. 333), Laplace again used Fourier transforms to solve for the derivative $\Psi'(s)$ in the differential equation

$$\Psi''(s+2)+\Psi''(s) = \{\Psi'(s+2)-\Psi'(s)\}\frac{n-1}{n} - \frac{s+2}{n}\Psi''(s+2)+\frac{s}{n}\Psi''(s). \tag{1.141}$$

He first performed the Fourier cosine transform

$$\Psi'(s) = \int \Pi(t)\cos(st)\,dt, \tag{1.142}$$

where the function $\Pi(t)$ and the integration limits are determined below. By substituting the last expression in Eq. (1.141), he obtained

$$0=\frac{t}{n}\sin t\sin\{(s+1)t\}\Pi(t)+\int\left[\sin\{(s+1)t\}\left\{(t\cos t-\sin t)\Pi(t)-\frac{t\sin t}{n}\Pi'(t)\right\}\right]dt. \quad (1.143)$$

Referring to a method of solution from his previous memoir (Laplace, 1785, p. 249), Laplace next set the expression under the integral sign above to zero, obtaining

$$0=(t\cos t-\sin t)\Pi(t)-\frac{t\sin t}{n}\Pi'(t).$$

Integrating the above,

$$\Pi(t)=A\left(\frac{\sin t}{t}\right)^n,$$

where A is a constant of integration. Then Eq. (1.142) becomes

$$\Psi'(s)=A\int\cos(st)\left(\frac{\sin t}{t}\right)^n dt$$

(*ibid.*, p. 334). To find the integration limits of the above, Laplace solved for the values of t at which the first term of the right side of (1.143) is equal to zero, obtaining $t=0,\infty$. These are respectively the lower and upper limits of the integral above.

1.4.4 Lagrange's 1776 Paper: A Precursor to the Characteristic Function

Together with Laplace, Lagrange was the most distinguished French mathematician of his time. But, unlike his contemporary, he did not make any fundamental contributions to probability and statistics.

Joseph-Louis Lagrange* (Fig. 1.23) was born in Turin, Italy, on January 25, 1736, in a French–Italian family. He was the youngest of 11 children and the only one to survive beyond infancy. His father was a public official but the family lived modestly. Lagrange's father wanted him to be a lawyer. At first, Lagrange did not object but soon realized he was more interested in physics and mathematics. On August 12, 1755, Lagrange sent Euler solutions to some famous problems using a technique that was later to be called the calculus of variations by the latter. Lagrange applied his new method to problems of celestial mechanics. This was so successful that, by the age of 25, he was regarded by many as the greatest living mathematician. In 1776, with Euler's and d'Alembert's backing, Lagrange was invited to Berlin by Frederick the Great. There he succeeded Euler as the Director of Mathematics at the Prussian Academy of Sciences. The next year, he married his cousin Vittoria Conti. In Berlin, Lagrange published one of his most important works, the *Mécanique Analytique* (Lagrange, 1788), which offered the most

*Lagrange's life is also described in Itard (1975a) and Agarwal and Sen (2014, pp. 211–213).

FIGURE 1.23 Joseph-Louis Lagrange (1736–1813). Wikimedia Commons (Public Domain), http://commons.wikimedia.org/wiki/Joseph-Louis_Lagrange#/media/File:Joseph_Louis_Lagrange.jpg

comprehensive treatment of classical mechanics since Newton. The book is also famous for the sentence "one will not find figures in this work, but only algebraic equations" in the preface.* In 1787, Lagrange left Berlin for Paris at the invitation of Louis XVI and was treated with great honor. The years in Paris were spent mainly on writing some of Lagrange's greatest treatises. He married again in 1792, his first wife having died in 1783 after a long illness. Lagrange passed away in Paris on April 10, 1813.

The first use of a technique close to the probabilistic Fourier transform (i.e., the CF) was due to Lagrange.† Our interest here is his first memoir on probability, the "Mémoire sur l'utilité de la méthode de prendre le milieu entre les résultats de plusieurs observations" (Lagrange, 1776) (Fig. 1.24). In Problem X of this memoir, Lagrange wished to determine the probability that the average of n independent errors lies between r and $-s$, where the density ("*facilité*") of each error X is y for $-q \le x \le p$. Lagrange explained the method of solution as follows:

> One shall first start by finding the probability that the average error is z, and this probability being given by a function of z, one would have to only take the integral from $z=r$ to $z=-s$; this will be the required probability.
>
> Now, to obtain the probability that the average error of n observations is z, one must consider the polynomial which is represented by the integral of $ya^x dx$, assuming the integral is taken from $x=p$ to $x=-q$; one will raise this polynomial to the power of n, and one will look for the coefficient of a^z...; this coefficient, which will be a function of z, will express the probability that the mean error will be z. (Lagrange, 1776, Oeuvres 2, p. 227)

Lagrange's algorithm can be translated in modern symbols as follows. Denote the n errors by $X_1, X_2, \ldots, X_n$, where each X_i has density y. The total error is then

$$W = X_1 + X_2 + \cdots + X_n. \tag{1.144}$$

Let the density of W be $f_W(w)$. The required probability is then

$$\int_{-ns}^{nr} f_W(w)\,dw, \tag{1.145}$$

which Lagrange determined as follows. First he used Eq. (1.144) to obtain

$$\int f_W(w) a^w dw = \left(\int ya^x dx\right)^n, \tag{1.146}$$

where a is an arbitrary variable. It can be seen that the above follows by applying the following transform to Eq. (1.144):

$$\begin{aligned} \mathsf{E}a^W &= \mathsf{E}a^{X_1+X_2+\cdots+X_n} \\ &= \mathsf{E}a^{X_1}a^{X_2}\cdots a^{X_n} \\ &= \left(\mathsf{E}a^{X_i}\right)^n. \end{aligned}$$

* "*On ne trouvera point de Figures dans cet Ouvrage. Les methods que j'y expose ne demandent ni constructions, ni raisonnements géométriques ou mécaniques, mais seulement des opérations algébriques...*" (Lagrange, 1788, p. vi).

† Lagrange work is also described in Seal (1949).

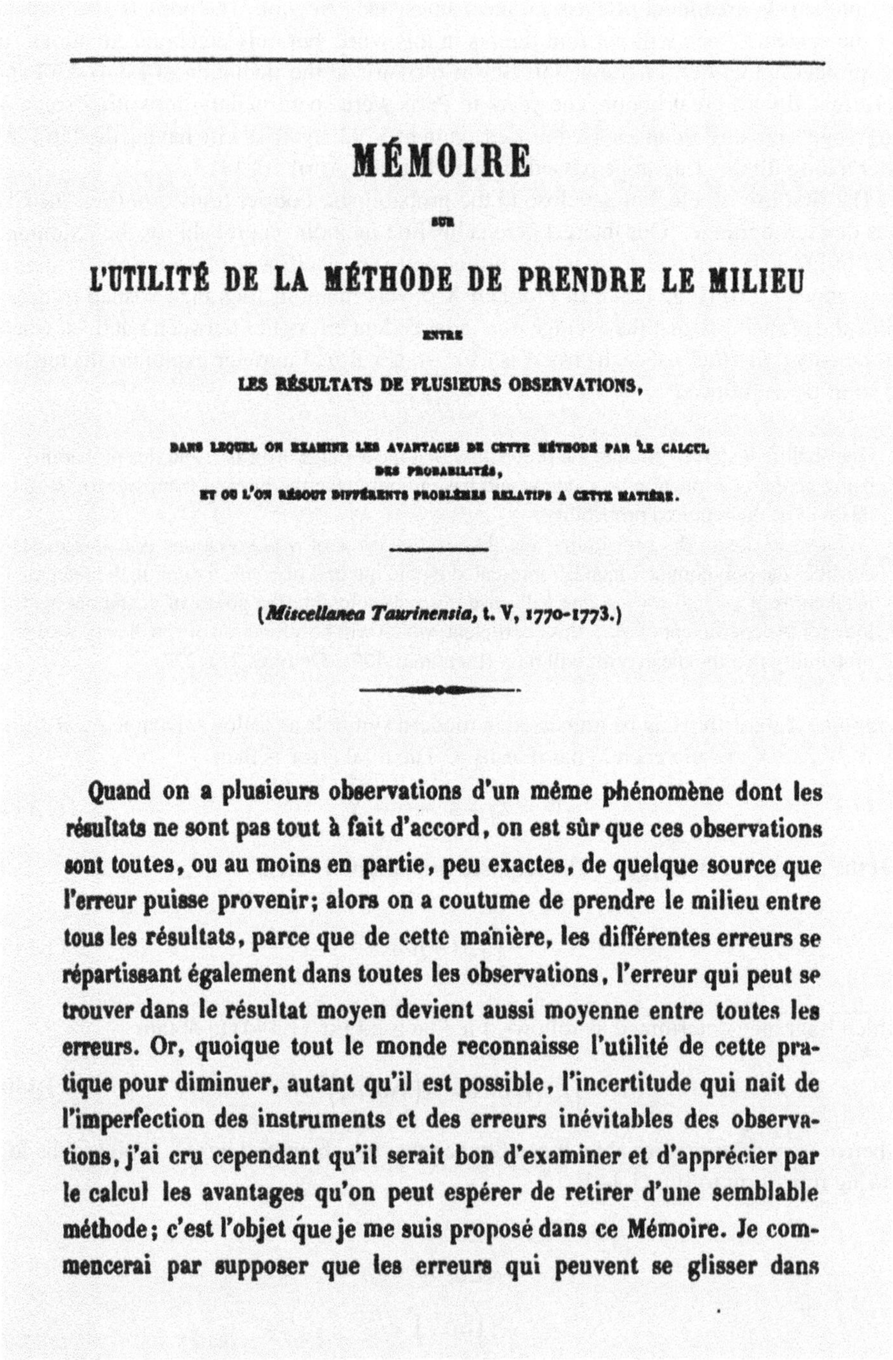

MÉMOIRE

SUR

L'UTILITÉ DE LA MÉTHODE DE PRENDRE LE MILIEU

ENTRE

LES RÉSULTATS DE PLUSIEURS OBSERVATIONS,

DANS LEQUEL ON EXAMINE LES AVANTAGES DE CETTE MÉTHODE PAR LE CALCUL DES PROBABILITÉS,
ET OÙ L'ON RÉSOUT DIFFÉRENTS PROBLÈMES RELATIFS A CETTE MATIÈRE.

(*Miscellanea Taurinensia*, t. V, 1770-1773.)

Quand on a plusieurs observations d'un même phénomène dont les résultats ne sont pas tout à fait d'accord, on est sûr que ces observations sont toutes, ou au moins en partie, peu exactes, de quelque source que l'erreur puisse provenir; alors on a coutume de prendre le milieu entre tous les résultats, parce que de cette manière, les différentes erreurs se répartissant également dans toutes les observations, l'erreur qui peut se trouver dans le résultat moyen devient aussi moyenne entre toutes les erreurs. Or, quoique tout le monde reconnaisse l'utilité de cette pratique pour diminuer, autant qu'il est possible, l'incertitude qui naît de l'imperfection des instruments et des erreurs inévitables des observations, j'ai cru cependant qu'il serait bon d'examiner et d'apprécier par le calcul les avantages qu'on peut espérer de retirer d'une semblable méthode; c'est l'objet que je me suis proposé dans ce Mémoire. Je commencerai par supposer que les erreurs qui peuvent se glisser dans

FIGURE 1.24 First page of Lagrange's "Mémoire sur l'utilité de la méthode de prendre le milieu entre les résultats de plusieurs observations" (Lagrange, 1776)

In the last expression, the equating of the expectation of a product to the product of expectations implies that Lagrange implicitly assumed independence of the errors. Lagrange's next step was to find the coefficient of a^w in $(\int ya^x dx)^n$. This coefficient gives $f_W(w)$ from which the integral in (1.145) can be determined.

As an example, Lagrange considered the case when the density of each X_i is $y = K$, where K is a constant. Then Eq. (1.146) becomes

$$\left(\int_{-q}^{p} Ka^x dx\right)^n = \left\{\frac{K}{\log a}\left[a^x\right]_{-q}^{p}\right\}^n = \left\{\frac{K\left(a^p - a^{-q}\right)}{\log a}\right\}^n.$$

Lagrange wrote the above as

$$\frac{A}{\left(\log a\right)^n}, \tag{1.147}$$

where, with $p + q = t$,

$$A = K^n\left\{a^{pn} - na^{pn-t} + \frac{n(n-1)}{2}a^{pn-2t} - \cdots\right\}.$$

Now, the coefficient of a^{pn-x} in (1.147) is

$$\frac{K^n}{1\cdot 2\cdot 3\cdots(n-1)}\left\{x^{n-1} - n(x-t)^{n-1} + \frac{n(n-1)}{2}(x-2t)^{n-1} - \frac{n(n-1)(n-2)}{2\cdot 3}(x-3t)^{n-1} + \cdots\right\}, \tag{1.148}$$

a result Lagrange had previously shown in his paper (*ibid.*, p. 225). Note that the series above is continued until $(x - mt)$ becomes negative. Putting $pn - x = z$ in the above thus gives the coefficient of a^z in (1.147), the latter coefficient being the density of W evaluated at z. Now to evaluate the probability in (1.145), the above-mentioned coefficient needs to be integrated from $z = -ns$ to $z = nr$, that is, Eq. (1.148) needs to be integrated from $x = pn + ns$ to $x = pn - nr$. Lagrange thus finally obtained the required probability in (1.145) as

$$\begin{aligned}\frac{K^n}{1\cdot 2\cdot 3\cdots n}\Big\{&(pn+s)^n - n(pn+s-t)^n + \frac{n(n-1)}{2}(pn+s-2t)^n - \cdots \\ &-(pn-r)^n + n(pn-r-t)^n - \frac{n(n-1)}{2}(pn-r-2t)^n + \cdots\Big\}\end{aligned} \tag{1.149}$$

(*ibid.*, p. 228).

1.4.5 The Concept of Characteristic Function Introduced: Laplace in 1785

We have already described how Laplace introduced of the concept of CF in his 1785 memoir (Laplace, 1785), in the context of determining the middle term of the series expansion of the multinomial expression $(1+1+\cdots+1)^n$, where n is a large positive integer. Details can be found in Section 1.2.8.3.

1.4.6 Laplace's Use of the Characteristic Function in his First Proof of the Central Limit Theorem (1810)

After the 1785 memoir, Laplace wrote other memoirs where the Fourier transform was used.* The most important instance undoubtedly took place in the "Mémoire sur les approximations des formules qui sont fonctions de très grands nombres et leur application aux probabilités" (Laplace, 1810a). In Article III, Laplace gave his first proof the Central Limit Theorem, in the context of the inclination of the orbits of comets, for the special case of a discrete uniform distribution.

Later, in Article VI of the same memoir, Laplace proved a more general version of the Central Limit Theorem by considering the sum of random variables each of which had an *arbitrarily given distribution*. Here also Laplace used CF. Details of both proofs are provided in Section 1.2.11.

1.4.7 Characteristic Function of the Cauchy Distribution: Laplace in 1811

In the short article "Sur les intégrales définies" (Laplace, 1811b), Laplace was able to obtain the CF of a very important density, which later came to be known as the Cauchy distribution (due to a set of often heated correspondences Cauchy had with Irénée-Jules Bienaymé in 1853). This distribution is important because it is a counterexample to virtually every limit law in the theory of probability. The density function of a Cauchy random variable X is given by

$$f_X(x) = \frac{1}{\pi} \cdot \frac{1}{1+x^2} \quad \text{for} \quad -\infty < x < \infty. \tag{1.150}$$

However, as yet another instance of Stigler's law of eponymy, the Cauchy distribution was not first adduced by Cauchy. Rather, the person to have first discovered this distribution and pointed out its peculiar behavior was Siméon Denis Poisson in 1824. Moreover, Laplace and Poisson were the first to have derived its CF in 1811, although they did not pay much attention to it at that time.

We now explain Laplace's derivation of the CF of the Cauchy distribution. In "Sur les intégrales définies" (Laplace, 1811b),† Laplace wished to determine the double integral

$$\int_0^\infty \int_0^\infty 2y \cos(ax) e^{-y^2(1+x^2)} dx dy \tag{1.151}$$

*Of course, this transform was also fully exploited in the form of the characteristic function in Laplace's *Théorie Analytique des Probabilités* (Laplace, 1812).

†The demonstration that follows was later reproduced in Laplace's "Mémoire sur les intégrales définies et leur application aux probabilités, et spécialement à la recherche du milieu qu'il faut choisir entre les résultats des observations" (Laplace, 1811a).

(*ibid*., p. 262). By integrating Eq. (1.151) first with respect to y, he obtained

$$\int_0^\infty\int_0^\infty 2y\cos(ax)e^{-y^2(1+x^2)}dxdy=\int_0^\infty 2\cos(ax)\left\{\int_0^\infty ye^{-y^2(1+x^2)}dy\right\}dx=\int_0^\infty\frac{\cos(ax)}{1+x^2}dx. \qquad (1.152)$$

Now, by integrating Eq. (1.151) first with respect to x, he obtained

$$\int_0^\infty\int_0^\infty 2y\cos(ax)e^{-y^2(1+x^2)}dxdy=\int_0^\infty 2y\left\{\int_0^\infty\cos(ax)e^{-y^2(1+x^2)}dx\right\}dy. \qquad (1.153)$$

To determine the above integral, let

$$Y=\int_0^\infty\cos(ax)e^{-\gamma^2x^2}dx, \qquad (1.154)$$

where γ is a constant. Then, by differentiating Eq. (1.154) and integrating by parts,

$$\begin{aligned}\frac{dY}{da}&=-\int_0^\infty x\sin(ax)e^{-\gamma^2x^2}dx\\&=\left[\frac{1}{2\gamma^2}\sin(ax)e^{-A^2x^2}\right]_{x=0}^\infty-\frac{a}{2\gamma^2}\int_0^\infty\cos(ax)e^{-\gamma^2x^2}dx\\&=-\frac{a}{2\gamma^2}\int_0^\infty\cos(ax)e^{-\alpha^2x^2}dx\\\therefore\quad\frac{dY}{da}&=-\frac{aY}{2\gamma^2}.\end{aligned}$$

By integrating the above differential equation, we have $Y=Ce^{-r^2/(4a^2)}$, where C is a constant of integration. Thus

$$Y=\int_0^\infty\cos(rx)e^{-A^2x^2}dx=C\exp\left(-\frac{r^2}{4\gamma^2}\right).$$

To determine C, set $r=0$. The left side of the above becomes $\int_0^\infty e^{-\gamma^2x^2}dx=\sqrt{\pi}/(2\gamma)$, and the right side becomes C, that is, $C=\sqrt{\pi}/(2\gamma)$. Hence,

$$Y=\int_0^\infty\cos(rx)e^{-\gamma^2x^2}dx=\frac{\sqrt{\pi}}{2\gamma}\exp\left(-\frac{r^2}{4\gamma^2}\right).$$

We can now use this result and go back to the integral in (1.153):

$$\begin{aligned}
\int_0^\infty\int_0^\infty 2y\cos(ax)e^{-y^2(1+x^2)}dxdy &= \int_0^\infty 2y\left\{\int_0^\infty \cos(ax)e^{-y^2(1+x^2)}dx\right\}dy \\
&= \int_0^\infty 2y\left\{\int_0^\infty \cos(ax)e^{-y^2}e^{-y^2x^2}dx\right\}dy \\
&= \int_0^\infty 2ye^{-y^2}\left\{\int_0^\infty \cos(ax)e^{-y^2x^2}dx\right\}dy \\
&= \int_0^\infty 2ye^{-y^2}\frac{\sqrt{\pi}}{2y}e^{-a^2/(4y^2)}dy \\
&= \sqrt{\pi}\int_0^\infty e^{-y^2-a^2/(4y^2)}dy \\
&= \sqrt{\pi}e^{-a}\int_0^\infty e^{-(2y^2-a)^2/(4y^2)}dy.
\end{aligned} \tag{1.155}$$

Now if $y' = a/(2y)$, then $dy' = -ady/(2y^2)$ and $y' - a/(2y') = a/(2y) - y$ so that

$$\int_0^\infty e^{-\left(y'-\frac{a}{2y'}\right)^2}dy' = -\int_\infty^0 \frac{a}{2y^2}e^{-\left(y-\frac{a}{2y}\right)^2}dy = \int_0^\infty \frac{a}{2y^2}e^{-\left(y-\frac{a}{2y}\right)^2}dy.$$

Letting $z = y - a/(2y)$ so that $dz = dy + ady/(2y^2)$, the above becomes

$$\int_0^\infty e^{-z^2}dy' = \int_0^\infty \frac{a}{2y^2}e^{-z^2}dy = \int_{-\infty}^\infty e^{-z^2}dz - \int_0^\infty e^{-z^2}dy. \tag{1.156}$$

Since

$$\int_0^\infty e^{-z^2}dy' = \int_0^\infty e^{-\left(\frac{a}{2y'}-y'\right)^2}dy' = \int_0^\infty e^{-\left(y-\frac{a}{2y}\right)^2}dy = \int_0^\infty e^{-z^2}dy,$$

Equation (1.156) implies

$$\int_0^\infty e^{-z^2}dy = \frac{1}{2}\int_{-\infty}^\infty e^{-z^2}dz = \frac{1}{2}\sqrt{\pi}.$$

Using this result, Eq. (1.155) thus leads to

$$\begin{aligned}
\int_0^\infty\int_0^\infty 2y\cos(ax)e^{-y^2(1+x^2)}dxdy &= \sqrt{\pi}e^{-a}\int_0^\infty e^{-(2y^2-a)^2/(4y^2)}dy \\
&= \sqrt{\pi}e^{-a}\int_0^\infty e^{-z^2}dy \\
&= \frac{\pi}{2}e^{-a}.
\end{aligned}$$

Using this result and Eq. (1.152), Laplace finally obtained

$$\int_0^\infty \frac{\cos(ax)}{1+x^2}dx = \frac{\pi}{2}e^{-a} \tag{1.157}$$

(*ibid.*, p. 264). He noted that one needed a>0 because otherwise z≠0. With (1.157), Laplace had thus effectively derived the CF of the Cauchy distribution. This is because the above implies

$$\int_{-\infty}^\infty \frac{\cos(ax)}{1+x^2}dx = \pi e^{-a}$$

(the integrand being an even function). Moreover,

$$\int_{-\infty}^\infty \frac{\sin(ax)}{1+x^2}dx = 0$$

(the integrand being an odd function). Hence, for a Cauchy random variable X,

$$\begin{aligned} \mathsf{E}e^{itX} &= \int_{-\infty}^\infty e^{itx} f_X(x)dx \\ &= \int_{-\infty}^\infty \frac{\cos tx}{\pi(1+x^2)}dx + i\int_{-\infty}^\infty \frac{\sin tx}{\pi(1+x^2)}dx \\ &= e^{-t}, t>0. \end{aligned} \tag{1.158}$$

1.4.8 Characteristic Function of the Cauchy Distribution: Poisson in 1811

Siméon Denis Poisson* (Fig. 1.25) was born in Pithiviers, France, on June 21, 1781. He came from a modest family, his father having been a soldier. Poisson was first educated at home and then at the École Centrale of Fontainebleau. He entered the École Polytechnique as the top candidate in 1798 and then joined the academic staff through the support of Laplace, his mentor. Poisson was appointed Associate Professor in 1802, and then Professor in 1806 to replace Fourier. In 1809, he also became Professor of Mechanics at the University of Paris. When Napoléon fell in 1815, Poisson was put in charge of all French scientific institutions. For the next 25 years, he decided on the appointment of Professors and on the curriculum. In spite of these enormous administrative responsibilities, Poisson found time for scientific research. This hardworking mathematician did research across the entire fields of physics and analysis, as they were developed at that time. Among Poisson's most notable statistical work is his refined proof in 1824 of Laplace's Central Limit Theorem. In addition to several books in physics, he also published his opus *Recherches sur la Probabilité des Jugements en Matière Criminelle et en Matière Civile* (Poisson, 1837). When this landmark book came out, it became famous especially

* Poisson's life is also described in Costabel (1978), Seneta (2006), and Féron (1978). A wide-ranging account of his life and work is also provided in the article by Bru (2005).

FIGURE 1.25 Siméon Denis Poisson (1781–1840). Wikimedia Commons (Public Domain), http://commons.wikimedia.org/wiki/File:Simeon_Poisson.jpg

because it contained a proof of Poisson's law of large numbers* (*ibid.*, pp. 138–142). This was an extension of Bernoulli's law of large numbers to the case of unequal success probabilities across trials. The book also introduced the concept of a random variable† (*ibid.*, p. 141). Nowadays, the book is especially famous for containing the so-called Poisson distribution‡ (*ibid.*, pp. 205–207). Poisson passed away in Paris on April 25, 1840.

We shall now examine Poisson's memoir of 1811 (Poisson, 1811). In the same issue where Laplace's "Sur les intégrales définies" (Laplace, 1811b) was published, Poisson gave a more elegant demonstration of Eq. (1.157) as follows.§ He first considered the integral

$$y=\int_0^\infty \frac{\cos(ax)}{1+x^2}dx \tag{1.159}$$

(Poisson, 1811, p. 225), where a is an unknown. Then, by differentiating with respect to a,

$$\begin{aligned}
y-\frac{d^2y}{da^2} &= \int_0^\infty \frac{\cos(ax)}{1+x^2}dx+\int_0^\infty \frac{x^2\cos(ax)}{1+x^2}dx \\
&= \int_0^\infty \frac{(1+x^2)\cos(ax)}{1+x^2}dx \\
&= \int_0^\infty \cos(ax)dx \\
&= 0.
\end{aligned}$$

In the above $\int_0^\infty \cos(ax)dx=0$ follows from $\int_0^\infty e^{-bx}\cos(ax)dx=b/(a^2+b^2)$. Now the general solution to

$$y-\frac{d^2y}{da^2}=0$$

is

$$y=\beta e^{-a}+\gamma e^{a}, \tag{1.160}$$

where β and γ are constants. To determine the latter, note that from Eq. (1.159),

$$(y)_{a=0}=\int_0^\infty \frac{1}{1+x^2}dx=\frac{\pi}{2},$$

so that Eq. (1.160) gives

$$\beta+\gamma=\frac{\pi}{2}. \tag{1.161}$$

* See Sheynin (1978, pp. 270–275) for a detailed description.

† "... one considers a thing A of a certain nature, capable of a number λ of values, known or unknown, which I will represent by $a_1, a_2, \ldots, a_k$, and between which only one will occur in each trial..." (Poisson, 1837, p. 141). The "thing A" is a discrete random variable. On pp. 254–255, Poisson also developed the notion of a continuous random variable.

‡ Poisson's derivation of this distribution was actually based on approximating the distribution function of a negative binomial distribution and has been described by Stigler (1982a).

§ The demonstration that follows is based on that found in Poisson's "Mémoire sur les intégrales définies" (Poisson, 1813).

Moreover, from Eq. (1.159) again,

$$\frac{dy}{da} = -\int_0^\infty \frac{x\sin(ax)}{1+x^2}dx.$$

Substituting $z = ax$ in the above,

$$\frac{dy}{da} = -\int_0^\infty \frac{z\sin z}{a^2+z^2}dz$$

so that

$$\left(\frac{dy}{da}\right)_{a=0} = -\int_0^\infty \frac{\sin z}{z}dx = -\frac{\pi}{2}.$$

Hence Eq. (1.160) gives

$$-\beta+\gamma = -\frac{\pi}{2}. \tag{1.162}$$

Solving Eqs. (1.161) and (1.162) results in $\beta = \pi/2$ and $\gamma = 0$. Hence, Poisson was able to show that

$$y = \int_0^\infty \frac{\cos(ax)}{1+x^2}dx = \frac{\pi}{2}e^{-a}$$

(*ibid.*, p. 233).

1.4.9 Poisson's Use of the Characteristic Function in his First Proof of the Central Limit Theorem (1824)

Following Laplace's demonstration the Central Limit Theorem in 1810,* Poisson paved the way for a more rigorous formulation of this theorem in his memoir "Sur la probabilité des résultats moyens des observations" (Poisson, 1824). Poisson's aim was to make Laplace's demonstration more transparent, making its conditions of validity as clear as possible:

> ...The generality of M. Laplace's analysis, the variety and the importance of the objects to which he has applied it, no doubt leave nothing to be desired; but it seems to me that certain points of this theory could be developed further; and I have thought that the remarks which I made while studying it, would be appropriate to clarify the difficulties, and could also be of some usefulness in practice. (*ibid.*, p. 273)

The main tool in Poisson's proof was again the characteristic function (CF). His demonstration was as follows.†

Poisson considered the sum S when s dice, each of which had faces numbered $-j, -j+1, \ldots, 0, \ldots, j$, are thrown. If the outcome on the lth die is denoted by X_l, then

$$S = X_1 + \cdots + X_s.$$

*See Section 1.2.11.

†Poisson's proof is also described in Hald (1998, pp. 317–321) and Fischer (2010, pp. 36–38).

The quantity S can also be viewed as the sum of individual errors (X_l) each of which has a multinomial distribution. Let $N = \Pr\{X_l = n\}$ and $M = \Pr\{S = m\}$. Then M is the coefficient of t^m in the expansion of $\sum t^m \Pr\{S = m\}$, that is, in the expansion of

$$\left\{\sum_{n=-j}^{j} t^n \Pr\{X_l = n\}\right\}^s = \left\{\sum_{n=-j}^{j} N t^n\right\}^s.$$

Next Poisson substituted $e^{i\theta} = t$ (where $i = \sqrt{-1}$) thus making the above a CF, and then performed an inverse Fourier transform (cf. Eq. 1.129) to obtain M:

$$M = \frac{1}{2\pi}\int_{-\pi}^{\pi}\left\{\sum_{n=-j}^{j} N e^{ni\theta}\right\}^s e^{-im\theta}\,d\theta. \tag{1.163}$$

Poisson next set out to determine

$$p = \Pr\{\mu \le S \le \mu'\} = \sum_{m=\mu}^{\mu'} \Pr\{S = m\}.$$

Using Eq. (1.163),

$$\begin{aligned} p &= \frac{1}{2\pi}\int_{-\pi}^{\pi}\left\{\sum_{n=-j}^{j} N e^{ni\theta}\right\}^s \left(\sum_{m=\mu}^{\mu'} e^{-im\theta}\right) d\theta \\ &= \frac{1}{2\pi}\int_{-\pi}^{\pi}\left\{\sum_{n=-j}^{j} N e^{in\theta}\right\}^s \times \frac{e^{-i\left(\mu-\frac{1}{2}\right)\theta} - e^{-i\left(\mu'+\frac{1}{2}\right)\theta}}{2i\sin(\theta/2)}\,d\theta \\ &= \frac{1}{4\pi i}\int_{-\pi}^{\pi}\left\{\sum_{n=-j}^{j} N e^{in\theta}\right\}^s \left\{\frac{e^{-i\left(\mu-\frac{1}{2}\right)\theta} - e^{-i\left(\mu'+\frac{1}{2}\right)\theta}}{\sin(\theta/2)}\right\} d\theta \end{aligned}$$

(*ibid.*, p. 274). Poisson next considered the continuous case by writing $-j, -j+1, \ldots, 0, \ldots, j$ as $[-a, a]$ and letting

$$\begin{aligned} &\omega = \frac{2a}{2j+1}, \qquad x = n\omega, \qquad \theta = \alpha\omega, \\ &\mu\omega = b - c, \qquad \mu'\omega = b + c, \\ &N = \omega f(x). \end{aligned}$$

Then, from the above, Poisson obtained

$$p = \frac{a}{\pi}\int_{-\pi/\omega}^{\pi/\omega}\left\{\sum_{x=-a}^{a} \omega f(x) e^{ix\alpha}\right\}^s e^{-ib\alpha}\,\frac{\sin\left\{\left(c + \frac{a}{2j+1}\right)\alpha\right\}}{\sin\left(\frac{a\alpha}{2j+1}\right)}\,\frac{d\alpha}{2j+1}.$$

Now let $j \to \infty$ so that $\omega \to 0$. Then $\sin\left(c + \frac{a}{2j+1}\right)\alpha \to \sin(c\alpha)$ and $(2j+1)\sin\left(\frac{a\alpha}{2j+1}\right) \to a\omega$ so that in the continuous limit

$$p = \Pr\{b - c \le S \le b + c\} = \frac{1}{\pi}\int_{-\infty}^{\infty}\left[\left\{\int_{-a}^{a} f(x)e^{ix\alpha}dx\right\}^{s} e^{-ib\alpha}\sin(c\alpha)\right]\frac{d\alpha}{\alpha} \tag{1.164}$$

(*ibid.*, p. 275). The imaginary parts in the above can be eliminated as follows. Let

$$\begin{aligned}
\rho^2 &= \left\{\int_{-a}^{a} f(x)\cos(\alpha x)dx\right\}^2 + \left\{\int_{-a}^{a} f(x)\sin(\alpha x)dx\right\}^2, \\
\cos\phi &= \frac{1}{\rho}\left\{\int_{-a}^{a} f(x)\cos(\alpha x)dx\right\}, \\
\sin\phi &= \frac{1}{\rho}\left\{\int_{-a}^{a} f(x)\sin(\alpha x)dx\right\}.
\end{aligned} \tag{1.165}$$

The quantity ρ can be recognized as the magnitude of the CF $\int_{-a}^{a} e^{i\alpha x} f(x)dx$ of X_r. Now Eq. (1.164) becomes

$$p = \Pr\{b - c \le S \le b + c\} = \frac{2}{\pi}\int_{0}^{\infty}\rho^{s}\cos(s\phi - b\alpha)\left\{\frac{\sin(c\alpha)}{\alpha}\right\}d\alpha \tag{1.166}$$

(*ibid.*, p. 279). The imaginary parts of Eq. (1.164) have thus been eliminated. Now, by writing

$$\rho^2 = \int_{-a}^{a} f(x)\cos(\alpha x)dx\int_{-a}^{a} f(x')\cos(\alpha x')dx' + \int_{-a}^{a} f(x)\sin(\alpha x)dx\int_{-a}^{a} f(x')\sin(\alpha x')dx',$$

Poisson obtained

$$\rho^2 = \int_{-a}^{a}\int_{-a}^{a} f(x)f(x')\cos\{(x - x')\alpha\}dxdx'. \tag{1.167}$$

Note that for $\alpha = 0$ we have $\rho = 1$, and for $\alpha > 0$ we have $\rho < 1$. In the latter case, ρ^s decreases as s increases. Poisson therefore wished to find ρ as a function of the first two terms of the expansion of the right side of Eq. (1.167) in powers of α. First, he assumed the existence of the first two moments of each X_r:

$$\begin{aligned}
k &= \int_{-a}^{a} xf(x)dx, \\
k' &= \int_{-a}^{a} x^2 f(x)dx.
\end{aligned}$$

Then

$$\begin{aligned}
\rho^2 &= \int_{-a}^{a}\int_{-a}^{a} f(x)f(x')\cos\{(x-x')\alpha\}\,dxdx' \\
&\approx \int_{-a}^{a}\int_{-a}^{a} f(x)f(x')\left\{1-\frac{(x-x')^2\alpha^2}{2}\right\}dxdx' \\
&= \int_{-a}^{a}\int_{-a}^{a} f(x)f(x')\left\{1-\frac{x^2\alpha^2}{2}+xx'\alpha^2-\frac{x'^2\alpha^2}{2}\right\}dxdx' \\
&= \int_{-a}^{a}\int_{-a}^{a} f(x)f(x')dxdx' - \alpha^2\int_{-a}^{a}\frac{x^2}{2}f(x)dx + \alpha^2\int_{-a}^{a}xf(x)dx\int_{-a}^{a}x'f(x')dx' - \alpha^2\int_{-a}^{a}\frac{x'^2}{2}f(x')dx' \\
&= 1-\alpha^2\left(k'-k^2\right).
\end{aligned}$$

Therefore,

$$\rho = \left\{1-\alpha^2\left(k'-k^2\right)\right\}^{1/2} \approx 1-\frac{1}{2}\alpha^2\left(k'-k^2\right) < 1 \quad \text{for all} \quad \alpha \neq 0 \tag{1.168}$$

(*ibid.*, p. 280), since $k'-k^2$ (i.e., var X_l) is always positive. In view of the latter, Poisson next wrote

$$\frac{1}{2}\left(k'-k^2\right) = h^2.$$

Substituting both the above and $\alpha = y/\sqrt{s}$ in Eq. (1.168), we have

$$\rho^s = \left(1-\frac{h^2y^2}{s}\right)^s \approx e^{-h^2y^2} \quad \text{for large } s$$

(*ibid.*, p. 281). Also, from Eq. (1.165), $\phi = k\alpha$. Poisson therefore wrote Eq. (1.166) as

$$\begin{aligned}
p &= \Pr\{b-c \le S \le b+c\} \\
&= \frac{2}{\pi}\int_0^\infty e^{-h^2y^2}\cos\left\{(ks-b)\frac{y}{\sqrt{s}}\right\}\sin\left(\frac{cy}{\sqrt{s}}\right)\frac{dy}{y} \\
&= \frac{1}{\pi\sqrt{s}}\int_{-c}^{c}\left[\int_0^\infty e^{-h^2y^2}\cos\left\{(ks-b+z)\frac{y}{\sqrt{s}}\right\}dy\right]dz \\
&= \frac{1}{2h\sqrt{\pi s}}\int_{-c}^{c}\exp\left\{-\frac{(ks-b+z)^2}{4h^2s}\right\}dz.
\end{aligned}$$

Now, in the last expression, the maximum value of p with respect to b occurs at $b = ks$. By making this substitution and taking $c = 2hr\sqrt{s}$, Poisson finally obtained

$$\begin{aligned}
p &= \Pr\left\{ks - 2hr\sqrt{s} \le S \le ks + 2hr\sqrt{s}\right\} \\
&= \frac{1}{2h\sqrt{\pi s}} \int_{-c}^{c} \exp\left(-\frac{z^2}{4h^2 s}\right) dz \\
&= \frac{1}{2h\sqrt{\pi s}} \int_{-c}^{c} \exp\left(-\frac{z^2 r^2}{c^2}\right) dz \\
&= \frac{1}{2h\sqrt{\pi s}} \int_{-r}^{r} e^{-R^2} \cdot \frac{c}{r} dR \\
&= \frac{2}{\sqrt{\pi}} \int_{0}^{r} e^{-R^2} dR
\end{aligned}$$

(*ibid.*, p. 283). In the above, we have used $R = zr / c$, and the final result can be recognized as the integral of the $N(0,1/2)$ distribution. Poisson thus obtained the Central Limit Theorem, which can also be written in terms of the sample mean $\overline{S} = S / s$ as

$$p = \Pr\left\{k - \frac{2hr}{\sqrt{s}} \le \overline{S} \le k + \frac{2hr}{\sqrt{s}}\right\} = \frac{2}{\sqrt{\pi}} \int_{0}^{r} e^{-R^2} dR.$$

1.4.10 Poisson's Identification of the Cauchy Distribution (1824)

Throughout his proof of the Central Limit Theorem in "Sur la probabilité des résultats moyens" (Poisson, 1824), time and again Poisson drew attention to a peculiar distribution that failed to satisfy this fundamental theorem. The distribution in question now goes by the name of the Cauchy distribution (cf. Eq. 1.150), although Cauchy himself worked on it in 1853, much later than Poisson. We now explain this aspect of Poisson's work.*

After Poisson had derived Eq. (1.164), he considered the following example:

$$f(x) = \frac{1}{\pi\left(1 + x^2\right)}, \qquad a = \infty,^{\dagger}$$

(*ibid.*, p. 278). The CF of this distribution is

$$\int_{-\infty}^{\infty} f(x) e^{ix\alpha} dx = \frac{1}{\pi} \int_{-\infty}^{\infty} \frac{e^{ix\alpha}}{1 + x^2} dx = e^{-|\alpha|}$$

* Poisson's identification of the Cauchy distribution is also described in Stigler (1974).

† Consistent with his usage in the previous section, Poisson used $(-a, a)$ to denote the support of a continuous distribution.

(cf. Eq. 1.158). Then Eq. (1.164) becomes

$$
\begin{aligned}
p &= \Pr\{b-c \le S \le b+c\} \\
&= \frac{1}{\pi}\int_{-\infty}^{\infty}\left\{e^{-s\alpha}\cos(b\alpha)\frac{\sin(c\alpha)}{\alpha}\right\}d\alpha \\
&= \frac{2}{\pi}\int_{0}^{\infty}\left\{e^{-s\alpha}\cos(b\alpha)\int_{0}^{c}\cos(c^*\alpha)dc^*\right\}d\alpha \\
&= \frac{2}{\pi}\int_{0}^{c}\left\{\int_{0}^{\infty}e^{-s\alpha}\cos(b\alpha)\cos(c^*\alpha)d\alpha\right\}dc^* \\
&= \frac{1}{\pi}\int_{0}^{c}\left\{\frac{s}{s^2+(b-c^*)^2}+\frac{s}{s^2+(b+c^*)^2}\right\}dc^* \\
&= \frac{1}{\pi}\tan^{-1}\left(\frac{2cs}{s^2+b^2-c^2}\right)
\end{aligned}
$$

(*ibidem*). Next Poisson made the substitution $b' = b/s$ and $c' = c/s$ so that the above becomes

$$
p = \Pr\{b'-c' \le \bar{S} \le b'+c'\} = \frac{1}{\pi}\tan^{-1}\left(\frac{2c'}{1+b'^2-c'^2}\right).
$$

This is a surprising result because:

> By making $b = b's$, $c = c's$, the average error [i.e. $\bar{S}$] will be contained within the limits $b' \pm c'$, the probability p corresponding to this will be independent of the number s of observations; from which it follows that in this particular example, as the this number increases, the average error will not converge towards zero or another fixed term; and, however large the number of observations, there will always be the same probability that the average error will be within given limits. (*ibid.*, pp. 278–279)

Poisson commented that the anomaly arose because, for the Cauchy distribution, the magnitude (ρ) of the CF did not satisfy the condition in Eq. (1.168). Indeed $\rho = e^{-|\alpha|} \approx 1-|\alpha|$ in the Cauchy case, and k (and hence also k') does not even exist.

1.4.11 First Modern Rigorous Proof of the Central Limit Theorem: Lyapunov in 1901

The mid-nineteenth century saw a new era in mathematics with the introduction of rigor in analysis, spearheaded mainly by Cauchy and then followed by many others. Among other things, rigor meant that every mathematical concept used had to be explicitly defined in terms of known concepts and that all assumptions in a mathematical proof had to be made explicit. Rigor also meant that mathematical ideas were based on precise and exact concepts, rather than intuition. For example, the reader will recall that several of Laplace's proofs were based in large part on intuition and thus were far from rigorous.

Although the concept of rigor has slightly evolved through the years, Cauchy effectively set the standards by which mathematics is still done today.*

The probabilistic world was certainly not indifferent to these developments. The necessity for rigor in the proof of the Central Limit Theorem (CLT) was recognized in a large part by the Russian (or St Petersburg) Mathematics School, headed by Pafnuty Chebychev (1821–1894).† For the latter, rigor especially meant that bounds needed to be placed on errors of approximations, a feature that lacked in both Laplace's and Poisson's proof of the CLT. Chebychev strived for a rigorous proof but was not entirely successful. His work was continued by his students of whom Andrei Markov (1856–1922) and Aleksandr Lyapunov (1857–1918) were most prominent. It was the latter who eventually gave the first modern rigorous proof of the Central Limit Theorem in a series of four articles from 1900 to 1901 (Lyapunov, 1900; 1901a; 1901b; 1901c). In the first article, Lyapunov gave an initial proof and in the fourth he gave a more general one. Before describing Lyapunov's proof, we shall say a few words about his life.

Aleksandr Mikhailovich Lyapunov‡ (Fig. 1.26) was born in Yaroslavl, Russia, on June 7, 1857. His father, Mikhail Vasilievich, was then director of the Demidovsk Lycée, a general high school. Lyapunov received his earliest education from his father. Two years after the latter died in 1868, his mother and her three sons moved to Gorki, where Lyapunov received his secondary education. In 1876 he became a student in mathematics at St Petersburg University, and there grew as a scholar under the influence of Chebychev. Lyapunov graduated in 1880 and published two papers on hydrostatics in 1881. His 1884 Master's thesis was entitled *On the Stability of Ellipsoidal Forms of Equilibrium of Rotating Fluids* and was translated in French in 1904. After finishing his masters, Lyapunov joined Kharkov University. He obtained his Ph.D. in 1892 for the thesis *The General Problem of Stability of Motion*, which is now regarded as a classic work (the English version was published by Academic Press in 1966). In 1893, Lyapunov became Professor at Kharkov, which he left in 1902. In 1901, he published a set of proofs on the CLT, culminating in a general form of the theorem. This was the first modern rigorous proof the CLT and is still to be found in textbooks today. In 1902, Lyapunov was elected to the Russian Academy of Sciences in St. Petersburg. Lyapunov enjoyed a close professional relationship with Markov throughout his life. In June 1917, Lyapunov left St Petersburg with his wife (who was also his cousin), since the latter was suffering badly from tuberculosis. In October 31, 1918, Lyapunov's wife died and on that very day Lyapunov shot himself in the head. He died 3 days later after having asked to be buried next to his wife. The inscription on his gravestone reads:

> The creator of the theory of stability of motion, the doctrine of equilibrium figures of rotating liquid, methods of the qualitative theory of differential equations, author of the central limit theorem and other deep investigations in areas of mechanics and mathematical analysis. (Sinai, 2003, p. 2)

* Cauchy thus also gave a proof of the CLT (Cauchy, 1853), which was more rigorous than those of Laplace and Poisson.

† For more on Chebychev and the Russian School, see Maistrov (1974, pp. 188–224) and Seneta (1994).

‡ Lyapunov's life is also described in Smirnov (1992), Grigorian (1974), Sinai (2003, pp. 1–2), and Agarwal and Sen (2014, pp. 310–311).

FIGURE 1.26 Aleksandr Mikhailovich Lyapunov (1857–1918). Wikimedia Commons (Public Domain), http://commons.wikimedia.org/wiki/File:Aleksandr_Lyapunov.jpg

We shall now describe Lyapunov's more general proof of the CLT* as given in his fourth paper (Lyapunov, 1901a) (Fig. 1.27). His general version of the CLT was: *let* $x_1, x_2, x_3, \ldots$ *be an infinite sequence of independent random variables. Let*

$$\mathsf{E}x_i = \alpha_i,$$
$$\mathsf{E}(x_i - \alpha_i)^2 = a_i,$$
$$\mathsf{E}|x_i - \alpha_i|^{2+\delta} = d_i,$$

where δ *is some positive number. If the following condition holds*

$$\lim_{n\to\infty} \frac{(d_1 + d_2 + \cdots + d_n)^2}{(a_1 + a_2 + \cdots + a_n)^{2+\delta}} = 0, \tag{1.169}$$

then

$$\lim_{n\to\infty} \Pr\left\{ z_1 < \frac{x_1 - \alpha_1 + x_2 - \alpha_2 + \cdots + x_n - \alpha_n}{\sqrt{2(a_1 + a_2 + \cdots + a_n)}} < z_2 \right\} = \frac{1}{\sqrt{\pi}} \int_{z_1}^{z_2} e^{-z^2} dz$$

uniformly for all for $z_1 < z_2$ (Lyapunov, 1901a; Adams, 2009, pp. 176–177).

Lyapunov's proof drew heavily on the results of his previous paper (Lyapunov, 1900) and was made up of three stages. In the first stage, he invoked the following lemma: let $x', x'', x''', \ldots$ be a sequence of positive integers and let $f(x)$ be a function that is positive at $x', x'', x''', \ldots$. If l, m, and n are numbers such that $l > m > n \geq 0$, then

$$\left\{\sum f(x) x^m\right\}^{l-n} < \left\{\sum f(x) x^n\right\}^{l-m} \left\{\sum f(x) x^l\right\}^{m-n}. \tag{1.170}$$

Lyapunov noted that in the special case that $\sum f(x) = 1$, the above inequality becomes

$$\left\{\sum f(x) x^m\right\}^{l} < \left\{\sum f(x) x^l\right\}^{m}.$$

This can also be written as

$$\left(\mathsf{E}|x|^m\right)^l < \left(\mathsf{E}|x|^l\right)^m \quad \text{for} \quad l > m > 0$$

and is known as *Lyapunov's inequality*.

With the help of the inequality in (1.170), Lyapunov next wished to prove that if the condition in (1.169) held for a given δ, then it would also hold for smaller values of δ. The importance of this is that, if Lyapunov could show it to be true, then he could take a

*Lyapunov's proof is also described in Fischer (2010, pp. 198–205). Uspensky (1937, pp. 284–292) gives a version that is somewhat close to Lyapunov's proof. For a more general description of Lyapunov's proof and other limit theorems, see Adams (2009), Loève (1950), and Le Cam (1986).

ЗАПИСКИ ИМПЕРАТОРСКОЙ АКАДЕМІИ НАУКЪ.

MÉMOIRES

DE L'ACADÉMIE IMPÉRIALE DES SCIENCES DE ST.-PÉTERSBOURG.

VIIIE SÉRIE.

ПО ФИЗИКО-МАТЕМАТИЧЕСКОМУ ОТДѢЛЕНІЮ. | CLASSE PHYSICO-MATHÉMATIQUE.

Томъ XII. № 5. | Volume XII. № 5.

NOUVELLE FORME DU THÉORÈME

SUR

LA LIMITE DE PROBABILITÉ.

PAR

A. Liapounoff.

(Lu le 25 avril 1901).

С.-ПЕТЕРБУРГЪ. 1901. ST.-PÉTERSBOURG.

Продается у коммиссіонеровъ Императорской Академіи Наукъ:

Н. П. Глазунова, М. Эггерса и Комп. и К. Л. Риккера въ С.-Петербургѣ,
Н. П. Карбасникова въ С.-Петерб., Москвѣ, Варшавѣ и Вильнѣ,
Н. Я. Оглоблина въ С.-Петербургѣ и Кіевѣ,
М. В. Клюкина въ Москвѣ,
Е. П. Распопова въ Одессѣ,
Н. Киммеля въ Ригѣ,
Фоссъ (Г. Гессель) въ Лейпцигѣ,
Люзакъ и Комп. въ Лондонѣ.

Commissionnaires de l'Académie Impériale des Sciences:

J. Glasounof, M. Eggers & Cie. et C. Ricker à St.-Pétersbourg,
N. Karbasnikof à St.-Pétersbourg, Moscou, Varsovie et Vilna,
N. Oglobline à St.-Pétersbourg et Kief,
M. Klukine à Moscou,
E. Raspopoff à Odessa,
N. Kymmel à Riga,
Voss' Sortiment (G. Haessel) à Leipsic,
Luzac & Cie. à Londres.

Цѣна: 80 коп. — Prix: 2 Mark.

FIGURE 1.27 Title page of Lyapunov's 1901 article "Nouvelle forme du théorème sur la limite de probabilité" (Lyapunov, 1901a)

convenient value for δ and he would be assured that the CLT held for all δ-values less than that value. His proof was as follows. Let $\mathsf{E}|x_i - \alpha_i|^{2+\beta} = b_i$ for $\beta < \delta$. Then, (1.170) implies

$$(b_1 + b_2 + \cdots + b_n)^\delta < (a_1 + a_2 + \cdots + a_n)^{\delta-\beta} (d_1 + d_2 + \cdots + d_n)^\beta,$$

whence, by taking the ($\delta\beta/2$)th root on both sides,

$$\left\{ \frac{(b_1 + b_2 + \cdots + b_n)^2}{(a_1 + a_2 + \cdots + a_n)^{2+\beta}} \right\}^{1/\beta} < \left\{ \frac{(d_1 + d_2 + \cdots + d_n)^2}{(a_1 + a_2 + \cdots + a_n)^{2+\delta}} \right\}^{1/\delta}. \tag{1.171}$$

Thus, as $n \to \infty$, if the right side above tends to zero, then so will the left side. This verifies Lyapunov argument.

In the second stage of his proof, Lyapunov set $a_1 + a_2 + \cdots + a_n = A$ and

$$P = \Pr\left\{ z_1\sqrt{2A} < x_1 - \alpha_1 + x_2 - \alpha_2 + \cdots + x_n - \alpha_n < z_2\sqrt{2A} \right\} = \frac{1}{\sqrt{\pi}} \int_{z_1}^{z_2} e^{-z^2} dz + \Delta \tag{1.172}$$

(*ibid.*, p. 178). His aim was to find an upper bound for the magnitude of the error Δ. If he could show that this upper bound converged uniformly to zero as $n \to \infty$, then he would have proved his theorem. To do this, he proceeded through the device of an *auxiliary variable* ξ, which will be defined shortly. But first, he let $f(x_i)$ be the probability that $x = x_i$ and in (1.172) defined

$$\alpha_1 + \alpha_2 + \cdots + \alpha_n + z_1\sqrt{2A} = g - h,$$
$$\alpha_1 + \alpha_2 + \cdots + \alpha_n + z_2\sqrt{2A} = g + h.$$

Now let

$$\Pi = \Pr\{ g - h < x_1 + \cdots + x_n + \xi < g + h \}.$$

In the above, ξ is an auxiliary $N(0, \chi\sqrt{2})$ variable which is independent of the variables $x_1, .., x_n$, where χ is constant which will be dealt with soon. Then

$$\Pi = \frac{1}{2\chi\sqrt{\pi}} \sum f(x_1) \cdots f(x_n) \int_{-h-s}^{h-s} \exp\left(-\frac{x^2}{4\chi^2} \right) dx = \frac{1}{\sqrt{\pi}} \sum f(x_1) \cdots f(x_n) \int_{-(h+s)/(2\chi)}^{(h-s)/(2\chi)} e^{-x^2} dx,$$

where $s = x_1 + x_2 + \cdots + x_n - g$ and the sum in the above is taken over all $x_1, ..., x_n$ such that $g - h - \xi < x_1 + \cdots + x_n < g + h - \xi$. The expression above can be simplified by introducing the Dirichlet discontinuity factor

$$\frac{2}{\pi} \int_0^\infty \frac{\sin(ht)}{t} \cos(st) dt. \tag{1.173}$$

Therefore,

$$\Pi = \frac{2}{\pi}\sum f(x_1)\cdots f(x_n)\int_0^\infty \frac{\sin(ht)}{t}\cos(st)e^{-\chi^2 t^2}\,dt = \frac{2}{\pi}\int_0^\infty \frac{\sin(ht)}{t} q e^{-\chi^2 t^2}\,dt, \tag{1.174}$$

where

$$q = \sum f(x_1)\cdots f(x_n)\cos(st) = \mathrm{Re}\left\{\sum f(x_1)\cdots f(x_n)e^{jst}\right\} \quad \left(j=\sqrt{-1}\right). \tag{1.175}$$

In the above, Re(Z) denotes the real part of Z. Lyapunov next defined two quantities ρ_i, σ_i, which are functions of a parameter t, as follows:

$$\begin{cases} \rho_i \cos\sigma_i = \sum f(x_i)\cos\{(x_i - \alpha_i)t\}, \\ \rho_i \sin\sigma_i = \sum f(x_i)\sin\{(x_i - \alpha_i)t\}. \end{cases}$$

These can be compared to Poisson's (1.165). With the above transformations, Eq. (1.174) becomes

$$\Pi = \frac{2}{\pi}\int_0^\infty \frac{\sin(ht)}{t}\rho_1\rho_2\ldots\rho_n \cos(\sigma_1 + \sigma_2 + \cdots + \sigma_n - gt)e^{-\chi^2 t^2}\,dt$$

(*ibid.*, p. 179). The next step is now to set

$$R = \Pi - \frac{1}{\sqrt{\pi}}\int_{\zeta_1}^{\zeta_2} e^{-z^2}\,dz, \tag{1.176}$$

where

$$\zeta_1 = \frac{g-h}{\sqrt{2A}}, \quad \zeta_2 = \frac{g+h}{\sqrt{2A}}.$$

Before continuing, Lyapunov first modified a previous inequality and showed that, for any positive numbers χ and λ, and for $z_2 > z_1$, Δ in (1.172) satisfies

$$|\Delta| < \underbrace{\frac{8\lambda}{\sqrt{2\pi A}} + \frac{4\lambda^2}{\sqrt{2\pi e A}} + \frac{1}{\sqrt{\pi}}\frac{\chi}{\lambda}e^{-\lambda^2/\chi^2} + 2L}_{\Omega} \tag{1.177}$$

(*ibid.*, p. 180), where L is estimated in Eq. (1.179) below.

Coming back to Eq. (1.176), to evaluate the integral

$$\int_{\zeta_1}^{\zeta_2} e^{-z^2}\,dz,$$

Lyapunov used the Dirichlet discontinuity factor in (1.173) again so that

$$\frac{1}{\sqrt{\pi}}\int_{\zeta_1}^{\zeta_2} e^{-z^2}\,dz = \frac{2}{\pi}\int_0^\infty \frac{\sin(ht)}{t}\cos(gt)e^{-\frac{1}{2}At^2}\,dt.$$

Then Eq. (1.176) becomes

$$R = \frac{2}{\pi}\int_0^\infty \frac{\sin(ht)}{t}Tdt, \tag{1.178}$$

where

$$T = \rho_1\rho_2\ldots\rho_n\cos(\sigma_1+\sigma_2+\cdots+\sigma_n-gt)e^{-\chi^2t^2} - \cos(gt)e^{-\frac{1}{2}At^2}.$$

Lyapunov next wrote $R = R_1 + R_2$, where for $\tau > 0$,

$$R_1 = \frac{2}{\pi}\int_\tau^\infty \frac{\sin(ht)}{t}Tdt, \quad R_2 = \frac{2}{\pi}\int_0^\tau \frac{\sin(ht)}{t}Tdt.$$

He showed that, for $\tau_1 > \tau$,

$$|R_1| < \frac{2}{\pi\tau}\int_\tau^{\tau_1}\rho_1\rho_2\ldots\rho_n dt + \frac{1}{\pi\chi^2\tau_1^2}e^{-\chi^2\tau_1^2} + \frac{2}{\pi A\tau^2}e^{-\frac{1}{2}At^2},$$

$$|R_2| < \frac{2\chi^2}{\pi}\int_0^\tau t\rho_1\rho_2\ldots\rho_n dt + \frac{2}{\pi}\int_0^\tau \frac{Q}{t}dt,$$

where

$$Q = \left|\rho_1\rho_2\cdots\rho_n - e^{-At^2/2}\right| + |\sigma_1+\sigma_2+\cdots+\sigma_n|e^{-\frac{1}{2}At^2}.$$

Lyapunov now demonstrated that $\left|\rho_1\rho_2\ldots\rho_n - e^{-At^2/2}\right| < 2Dt^{2+\delta}e^{-qAt^2/2}$ where $D = d_1 + d_2 + \cdots + d_n$ and $q = 1 - 4D\tau^\delta / A$, and that $|\sigma_1+\sigma_2+\cdots+\sigma_n| < Dt^{2+\delta} < Dt^{2+\delta}e^{2D\tau^\delta t^2}$ for t in $(0, \tau)$. He then took

$$L = \frac{1}{\pi\chi^2\tau_1^2}e^{-\chi^2\tau_1^2} + \frac{3}{\pi}\left(\frac{2}{q}\right)^{\frac{2+\delta}{2}}\frac{D}{\sqrt{A^{2+\delta}}} + \frac{2\chi^2}{\pi qA} + \frac{4}{\pi q_1 A\tau^2}e^{-\frac{1}{2}q_1A\tau^2} \tag{1.179}$$

(*ibid.*, p. 184), where $q_1 = 1 - 4D\tau_1^\delta / A$, so that in Eq. (1.178),

$$|R| < L,$$

provided that $4D\tau_1^\delta / A < 1, \tau < \tau_1$, and $D\tau^{2+\delta} < k^{2+\delta}$, k being a positive root of $k^{2+\delta} = 8(1-k^2)$.

In the third stage of his proof, Lyapunov first made use of his result in stage one (see argument above Eq. 1.171). He chose a value of $\delta \le 1$ such that the condition (1.169) of his theorem held. Next he wrote the square root of the quantity whose limit is being taken in (1.169) as

$$\frac{D}{\sqrt{A^{2+\delta}}} = \varepsilon^{3\delta}$$

Note that $\varepsilon \to 0$ as $n \to \infty$. Now, in view of his stage two result in (1.179), the right side of (1.177) can be written as

$$\Omega = \frac{8\lambda}{\sqrt{2\pi A}} + \frac{6}{\pi}\left(\frac{2}{q}\right)^{\frac{2+\delta}{2}} \varepsilon^{3\delta} + \frac{1}{\sqrt{\pi}}\frac{\chi}{\lambda} e^{-\lambda^2/\chi^2} + \frac{2}{\pi\chi^2\tau_1^2} e^{-\chi^2\tau_1^2} + \Omega' \tag{1.180}$$

(*ibid.*, p. 185), where

$$\Omega' = \frac{4\chi^2}{\pi q A} + \frac{4\lambda^2}{\sqrt{2\pi e A}} + \frac{8}{\pi q_1 A\tau^2} e^{-\frac{1}{2}q_1 A\tau^2}.$$

Thus, it only remained for Lyapunov to now show that $\Omega \to 0$ as $\varepsilon \to 0$. To do this, he set

$$A\tau^2\varepsilon^{2\delta} = \mu^2, \quad \frac{\tau}{\tau_1} = \frac{\mu}{\mu_1}\varepsilon^{3-\delta},$$

where μ and μ_1 are positive numbers independent of n. Then, since $q = 1 - 4D\tau^\delta / A$ and $q_1 = 1 - 4D\tau_1^\delta / A$,

$$\frac{D}{A}\tau_1^\delta = \mu_1^\delta, \quad D\tau^{2+\delta} = \mu^{2+\delta}\varepsilon^{(1-\delta)\delta}.$$

The above also gives

$$\chi^2\tau_1^2 = \mu_1^2 \frac{\chi^2}{A\varepsilon^6}, \qquad A\tau^2 = \frac{\mu^2}{\varepsilon^{2\delta}},$$
$$q = 1 - 4\mu^\delta\varepsilon^{(3-\delta)\delta}, \quad q_1 = 1 - 4\mu_1^\delta.$$

From Eq. (1.180), it is seen that, as $\varepsilon \to 0$ (i.e., as $n \to \infty$), then $\Omega \to 0$ if each of $\lambda / \sqrt{A}$, χ/λ, and $A\varepsilon^6/\chi^2$ also tends to zero (*ibid.*, p. 186). This can be shown by writing

$$\frac{\lambda}{\sqrt{A}} = \frac{3\varepsilon^2}{\mu_1}\log\frac{E}{\varepsilon}, \quad \frac{\chi}{\lambda} = \left(3\log\frac{E}{\varepsilon}\right)^{-1/2},$$

where $E > \varepsilon$. Then

$$\frac{A\varepsilon^6}{\chi^2} = \mu_1^2\left(3\log\frac{E}{\varepsilon}\right)^{-1}$$

(*ibid.*, p. 186). Hence, as $n \to \infty$, each of $\lambda/\sqrt{A}$, χ/λ, and $A\varepsilon^6/\chi^2$ tends to zero, and from Eq. (1.180) Lyapunov was thus able to argue that $\Omega \to 0$. From Eq. (1.177), this implies that $|\Delta| \to 0$ as $n \to \infty$. Hence, the CLT is proved.

Lyapunov's proof was more rigorous than those of Laplace and Poisson. However, unlike these two authors, he did not explicitly use CF. The claim is often made that he did and this presumably stems from Lyapunov's use of Eq. (1.175). Uspensky's demonstration (1937, pp. 284–292) of Lyapunov's proof is a reconstructed version using CF.

1.4.12 Further Extensions: Lindeberg (1922), Lévy (1925), and Feller (1935)

One of the drawbacks of Lyapunov's proof was that it required the existence of moments beyond the second. A simpler and less restrictive proof was later provided by the Finnish mathematician Jarl Lindeberg (1876–1932) (Lindeberg, 1922) and the French mathematician Paul Lévy (1886–1971) (Lévy, 1925): if $X_1, X_2, \ldots, X_n$ are IID* random variables with common mean $\mathsf{E}X_i = \mu$ and common variance $\operatorname{var} X_i = \sigma^2 < \infty$, then

$$\lim_{n\to\infty} \Pr\left\{\frac{X_1 + X_2 + \cdots + X_n - n\mu}{\sigma\sqrt{n}} \le z\right\} = \int_{-\infty}^{z} \frac{1}{\sqrt{2\pi}} \exp\left(-\frac{u^2}{2}\right) du.$$

The advantage of the above result is that, unlike Lyapunov's version, it does not require the existence of moments of higher order than 2.

Sufficient and necessary conditions were provided by Lindeberg (1922) and William Feller (1906–1970) (Feller, 1935),[†] respectively: let $X_1, X_2, \ldots, X_n$ be independent random variables such that $\mathsf{E}X_i = \mu_i$ and $\operatorname{var} X_i = \sigma_i^2 < \infty$. Let $S_n = X_1 + X_2 + \cdots + X_n$, $m_n = \mu_1 + \mu_2 + \cdots + \mu_n$, and $B_n^2 = \sigma_1^2 + \sigma_2^2 + \cdots + \sigma_n^2$. Then

$$\lim_{n\to\infty} \Pr\left\{\frac{S_n - m_n}{B_n} \le z\right\} = \int_{-\infty}^{z} \frac{1}{\sqrt{2\pi}} \exp\left(-\frac{u^2}{2}\right) du$$

and

$$\lim_{n\to\infty} \max_{1\le i\le n} \frac{\sigma_i}{B_n} = 0$$

if and only if, for any $\varepsilon > 0$,

$$\lim_{n\to\infty}\left\{\frac{1}{B_n^2}\sum_{i=1}^{n}\int_{|x-\mu_i|>\varepsilon B_n} (x-\mu_i)^2 f_{X_i}(x)\,dx\right\} = 0.$$

In the above, $f_{X_i}(x)$ is the probability density of each X_i.

* Independent and identically distributed.

[†] See Section 5.2 and Chapter 6 of Fischer's book (Fischer, 2010) for more details on the contributions of Lindeberg, Lévy, and Feller (among several others) to the CLT.

1.5. LEAST SQUARES AND THE NORMAL DISTRIBUTION

1.5.1 First Publication of the Method of Least Squares: Legendre in 1805

Adrien-Marie Legendre[*] (1752–1833) was born in Paris on September 18, 1752, in a well-to-do family. He was educated at the Collège Mazarin, Paris, in science, including physics and mathematics. In 1770, Legendre defended his theses in these subjects at the Collège. Several of these essays were later used in 1774 to write a treatise on mechanics. During the period 1775–1780, Legendre taught mathematics at the École Militaire. In 1782, he won the prize of the Berlin Academy on the topic of exterior ballistics. Legendre's prize-winning essay attracted the attention of Lagrange who asked Laplace about the young author. Meanwhile, Legendre continued research in various fields, including the mutual attractions of planetary spheroids, second degree indeterminate equations, probability, and the rotation of bodies in equilibrium. As a result of his research, Legendre was elected to the *Académie des Sciences* in 1783 as *adjoint mécanicien*, replacing Laplace, who had been promoted to *associé*. After 1783, Legendre published more landmark papers including his "Recherches sur la figure des planètes[†]" (Legendre, 1784) where the famous "Legendre polynomials" first appeared. But his most groundbreaking work took place in 1805 when he published his memoir "Nouvelles méthodes pour la détermination des orbites des comètes[‡]" (Legendre, 1805), where the method of least squares first appeared in print. Prior to that, Laplace's method of situation (i.e., minimizing the sum of absolute errors) had been the method most extensively studied. Legendre's publication of the method of least squares resulted in a controversy between himself and Gauss on the issue of priority on the method, as we shall soon explain (see Section 1.5.3.3). Legendre also did outstanding research in other fields such as number theory and elliptic functions. He died in Paris on January 9, 1833.

As was mentioned, the method of least squares was first published by Legendre in the memoir "Nouvelles méthodes pour la détermination des orbites des comètes" (Legendre, 1805) (Fig. 1.28).[†] In the course of his mathematical investigation of the motion of comets, in Art. LVI Legendre obtained five equations for the latitude errors $E^{I}, E^{II}, E^{III}, E^{IV}$, and E^{V} in terms of the four unknowns z, u, τ, and π:

$$
\begin{aligned}
E^{I} &= -0'500 + z(1'113) + u(3'189) + \tau(26'900) - \pi(0'818),\\
E^{II} &= 0'125 + z(0'194) + u(3'264) + \tau(5'657) - \pi(0'830),\\
E^{III} &= -0'123 + z(4'208) - u(1'761) - \tau(5'529) + \pi(2'154),\\
E^{IV} &= -4'757 - z(0'770) + u(0'136) + \tau(40'693) - \pi(0'670),\\
E^{V} &= 6'350 + z(2'129) + u(2'002) + \tau(9'659) - \pi(7'211)
\end{aligned}
$$

[*] Legendre's life is also described in Itard (1975b), Farebrother (2001), and Agarwal and Sen (2014, pp. 218–220).

[†] Research on the Shape of the Planets.

[‡] New Methods for the Determination of Comet Orbits.

[§] For a survey of the early history of the method of least squares, see Farebrother (1999), Merriman (1877), Eisenhart (1964), and Wolfenden (1942, pp. 170–175).

APPENDICE.

Sur la Méthode des moindres quarrés.

Dans la plupart des questions où il s'agit de tirer des mesures données par l'observation, les résultats les plus exacts qu'elles peuvent offrir, on est presque toujours conduit à un système d'équations de la forme

$$E = a + bx + cy + fz + \&c.$$

dans lesquelles a, b, c, f, &c. sont des coëfficiens connus, qui varient d'une équation à l'autre, et x, y, z, &c. sont des inconnues qu'il faut déterminer par la condition que la valeur de E se réduise, pour chaque équation, à une quantité ou nulle ou très-petite.

Si l'on a autant d'équations que d'inconnues x, y, z, &c., il n'y a aucune difficulté pour la détermination de ces inconnues, et on peut rendre les erreurs E absolument nulles. Mais le plus souvent, le nombre des équations est supérieur à celui des inconnues, et il est impossible d'anéantir toutes les erreurs.

Dans cette circonstance, qui est celle de la plupart des problêmes physiques et astronomiques, où l'on cherche à déterminer quelques élémens importans, il entre nécessairement de l'arbitraire dans la distribution des erreurs, et on ne doit pas s'attendre que toutes les hypothèses conduiront exactement aux mêmes résultats; mais il faut sur-tout faire en sorte que les erreurs extrêmes, sans avoir égard à leurs signes, soient renfermées dans les limites les plus étroites qu'il est possible.

De tous les principes qu'on peut proposer pour cet objet, je pense qu'il n'en est pas de plus général, de plus exact, ni d'une application plus facile que celui dont nous avons fait usage dans les recherches précédentes, et qui consiste à rendre

FIGURE 1.28 Legendre's method of least squares, taken from the *Nouvelles Méthodes* (Legendre, 1805)

(*ibid.*, p. 63). How should the last equations be solved? Legendre noted that if all five errors were set to zero, then we would have five equations in four unknowns, and a solution for z, u, τ, and π would not in general be possible. It is thus more reasonable to reduce the errors. As a possible solution, Legendre set $E^{IV} = E^{V} = 0$. Then, by substituting the expressions for τ and π in the first three equations, and setting $z = u = 0$, he observed that $E^{II} \approx 0$ and that both E^{I} and E^{III} are approximately $2'$. Now, these "errors [are] very acceptable in the theory of comets" but:

> ...it is possible to reduce them more by finding the *minimum* of the sum of the squares of the quantities E^{I}, E^{II}, E^{III} (*ibid.*, p. 64)

This is the first published statement of the principle of least squares. It should be noted, however, that it is not a general statement of the principle, since Legendre did not refer to the minimization of the sum of squares of *all* five errors. This shortcoming was addressed in the appendix of the paper, where Legendre enunciated the principle of least squares in its full generality.

In the appendix (*ibid.*, pp. 72–80), entitled "On the method of least squares," Legendre introduced the method in a purely algebraic way. He stated that in almost all cases where observations are used for measurement of certain quantities, one is led to a system of equations of the form

$$E = a + bx + cy + fz + \cdots,$$

where $a, b, \ldots$ are known coefficients and $x, y, \ldots$ are unknowns to be determined under the condition that the value of the error E for each equation is either zero or very small.

The other equations in the system are

$$\begin{aligned} E^{I} &= a' + b'x + c'y + f'z + \cdots \\ E^{II} &= a'' + b''x + c''y + f''z + \cdots \\ &\cdots \end{aligned}$$

where $a', b', \ldots, a'', b'', \ldots$ are known. Next, Legendre observed that if there are as many equations are there are unknowns, then the above system can be solved exactly by setting all errors to zero. It should be noted that this statement is not correct since it requires the additional condition that the equations need to be independent of each other. Legendre further observed that in most cases the number of equations is larger than the number of unknowns. This is the case he wished to consider. In order to determine the unknowns in such a situation, he enunciated the principle of least squares in its generality as follows:

> Of all the principles which can be proposed for that objective, I think there is none more general, more exact, and more easily applicable, than that which we have used in the preceding researches, and which consists of making the sum of the squares of the errors a *minimum*. By this means, there is established between the errors a kind of equilibrium which, while preventing the extremes to prevail, is very well suited to reveal that state of the system which is closest to the truth. (*ibid.*, pp. 72–73)

Now, the sum of squares of the errors is

$$(a + bx + cy + fz + \cdots)^2 + (a' + b'x + c'y + f'z + \ldots)^2 + (a'' + b''x + c''y + f''z + \cdots)^2 + \cdots$$

For this sum to be a minimum, one needs to differentiate it with respect to $x, y, z, \ldots$ and set to zero each time:

$$\begin{aligned} 0 &= \sum ab + x\sum b^2 + y\sum bc + z\sum bf + \cdots \\ 0 &= \sum ac + x\sum bc + y\sum c^2 + z\sum cf + \cdots \\ 0 &= \sum af + x\sum bf + y\sum cf + z\sum f^2 + \cdots, \\ &\ldots. \end{aligned} \tag{1.181}$$

Note that the above equations, known as the *normal equations*, contain as many equations as there are unknowns, which can then be solved for by the usual methods. Legendre observed that a shortcut for obtaining the equations in (1.181) was to take each equation of the system, multiply by the coefficient of the unknown in that equation, and then add.

Legendre next showed that his method of least squares leads to the arithmetic mean when several observations $a', a'', a''', \ldots$ are made on an unknown quantity x. For then, the sum of squares of the errors is

$$(a' - x)^2 + (a'' - x)^2 + (a''' - x)^2 + \cdots$$

(*ibid.*, p. 75). This is a minimum when

$$0 = (a' - x) + (a'' - x) + (a''' - x) + \cdots \quad \Rightarrow \quad x = \frac{a' + a'' + a''' + \cdots}{n},$$

where n is the number of observations.

As pathbreaking as Legendre's paper was, it lacked an essential element. What was the mathematical basis for the method of least squares? Legendre did not give a reason why the *sum of squares* of the observations should be minimized. Surely, the fact that it leads to the arithmetic mean cannot count as a fundamental justification of the method.

1.5.2 Adrain's Research Concerning the Probabilities of Errors (1808): Two Proofs of the Normal Law

Robert Adrain* (Fig. 1.29) was a mathematician of exceptional brilliance. He was born in Carrickfergus, Ireland, on September 30, 1775. His father was a schoolmaster and maker of mathematical instruments. When his father died, Adrain assumed the latter's vacant position as a teacher. With the Irish uprising in 1798, Adrain joined the insurgent forces and was wounded in the back. However, he managed to escape to the United States with his wife and daughter. He first arrived in New York, but because of the cholera epidemic he moved to Princeton, New Jersey. Adrain worked for a brief period as Mathematics Master at Princeton Academy, then moved to York, Pennsylvania, in 1800 to serve as the headmaster of the York Academy. Adrain began contributing to the *Correspondent*, and even became its editor, but the journal soon went out of circulation. In 1808, Adrain

*The life of Robert Adrain is also described in Struik (1970b), Swetz (2008), Coolidge (1926), and Gehman (1955).

FIGURE 1.29 Robert Adrain (1775–1843). Wikimedia Commons (Public Domain), http://commons.wikimedia.org/wiki/Category:Robert_Adrain#/media/File:Robert_Adrain.jpg

began his own journal *The Analyst or Mathematical Museum*. It was in that same journal and year that Adrain famously gave an early demonstration of the normal law of error. This was one year before Gauss' 1809 demonstration but Adrain's paper went largely unnoticed until it was reprinted in 1871. In 1809, Adrain was appointed the first Professor of Mathematics at Queens College. He joined Columbia University in 1813. With the death of *The Analyst* in 1814, Adrain started yet another journal, *The Mathematical Diary*, in 1825. That journal too ceased in 1832. Adrain died in his family home at New Brunswick, New Jersey, in 1843.

In the 1808 paper entitled "Research concerning the probabilities of the errors which happen in making observations" (Adrain, 1808) (Fig. 1.30), Adrain was the first to study the simultaneous occurrence of at least two independent errors in position of a point. The 1808 paper is also historically important because it contains an early analytic deduction of the normal law (the first being due to de Moivre in 1733 (cf. Gorroochurn, 2012a, p. 95)). Adrain gave two deductions of the law.*

In his first proof, Adrain assumed that two unknown values are respectively observed as a and b with errors x and y. To maximize the probability of the two errors, Adrain postulated that the following condition should hold:

$$\frac{x}{y}=\frac{a}{b} \tag{1.182}$$

(Adrain, 1808, p. 93), that is, the errors ought to be proportional to their measured values. The next step was to denote the probabilities of the errors x and y by X and Y. Adrain now wished to maximize the probability XY under the condition $x+y=E$.

The logarithm of the probability can be written as

$$\log(XY)=\log X+\log Y=X'+Y', \tag{1.183}$$

where $X'=\log X$ and $Y'=\log Y$. Using Newton's fluxion notation (which writes the derivative df/dt or "fluxion" of a function f as $\dot{f}$), Adrain differentiated the above to obtain

$$\frac{d}{dt}X'+\frac{d}{dt}Y'=0 \quad \Rightarrow \quad \frac{dX'}{dx}\dot{x}+\frac{dY'}{dy}\dot{y}=0.$$

He wrote the last equation as

$$X''\dot{x}+Y''\dot{y}=0. \tag{1.184}$$

Moreover, since $x+y=E$, we also have

$$\dot{x}+\dot{y}=0. \tag{1.185}$$

*Adrain's derivation is also described in Dutka (1990), Coolidge (1926), Glaisher (1872), Merriman (1877, pp. 163–165), and Farebrother (1999, pp. 85–86).

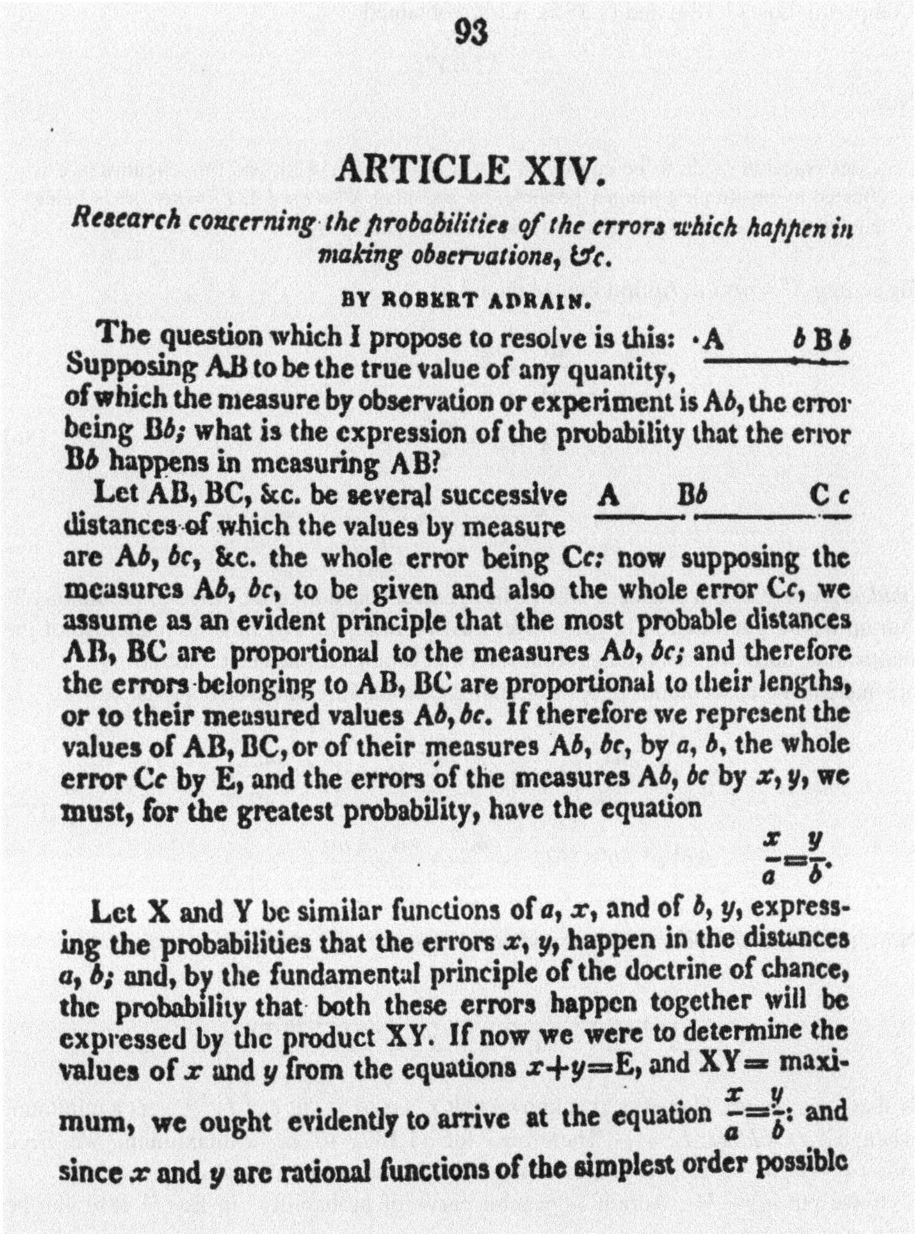

93

ARTICLE XIV.

Research concerning the probabilities of the errors which happen in making observations, &c.

BY ROBERT ADRAIN.

The question which I propose to resolve is this: Supposing AB to be the true value of any quantity, of which the measure by observation or experiment is A*b*, the error being B*b*; what is the expression of the probability that the error B*b* happens in measuring AB?

A *b* B *b*

Let AB, BC, &c. be several successive distances of which the values by measure are A*b*, *bc*, &c. the whole error being C*c*: now supposing the measures A*b*, *bc*, to be given and also the whole error C*c*, we assume as an evident principle that the most probable distances AB, BC are proportional to the measures A*b*, *bc*; and therefore the errors belonging to AB, BC are proportional to their lengths, or to their measured values A*b*, *bc*. If therefore we represent the values of AB, BC, or of their measures A*b*, *bc*, by a, b, the whole error C*c* by E, and the errors of the measures A*b*, *bc* by x, y, we must, for the greatest probability, have the equation

A B*b* C *c*

$$\frac{x}{a}=\frac{y}{b}.$$

Let X and Y be similar functions of a, x, and of b, y, expressing the probabilities that the errors x, y, happen in the distances a, b; and, by the fundamental principle of the doctrine of chance, the probability that both these errors happen together will be expressed by the product XY. If now we were to determine the values of x and y from the equations $x+y=\text{E}$, and XY= maximum, we ought evidently to arrive at the equation $\frac{x}{a}=\frac{y}{b}$: and since x and y are rational functions of the simplest order possible

FIGURE 1.30 First page of Robert Adrain's 1808 paper "Research concerning the probabilities of the errors which happen in making observations." (Adrain, 1808)

Comparing Eqs. (1.184) and (1.185), Adrain obtained

$$X'' = Y''.$$

Now,

> ...this equation ought to be equivalent $x/a = y/b$ [Eq. (1.182)]; and this circumstance is effected in the simplest manner possible, by assuming $X'' = mx/a$, $Y'' = my/b$; m being any fixed number which the question may require. (*ibid.*, p. 94)

By setting $X'' = mx/a$, Adrain thus obtained

$$\begin{aligned} \frac{dX'}{dx} &= \frac{mx}{a} \\ X' &= a' + \frac{mx^2}{2a} \\ X &= \exp\left(a' + \frac{mx^2}{2a}\right) \end{aligned} \tag{1.186}$$

(*ibidem*), where a' is a constant. This is the "general equation of the *curve of probability*."* Adrain next argued that m in the above must be negative. For this, he made use of the multivariate normal distribution, as follows. First Adrain reasoned that if the errors $x, y, z, \ldots$ are made in the observations $a, b, c, \ldots$, then the probability of the errors must be

$$\begin{aligned} &\exp\left(a' + \frac{mx^2}{2a}\right) \times \exp\left(b' + \frac{my^2}{2b}\right) \times \exp\left(c' + \frac{mz^2}{2c}\right) \times \cdots \\ &= \exp\left(a' + b' + c' + \cdots + \frac{mx^2}{2a} + \frac{my^2}{2b} + \frac{mz^2}{2c} + \cdots\right). \end{aligned} \tag{1.187}$$

Now, the above will be a maximum when

$$\frac{mx^2}{2a} + \frac{my^2}{2b} + \frac{mz^2}{2c} + \cdots = \frac{m}{2}\left(\frac{x^2}{a} + \frac{y^2}{b} + \frac{z^2}{c} + \cdots\right)$$

is also a maximum. However, it is known that $x^2/a^2 + y^2/b^2 + z^2/c^2 + \cdots$ is a minimum when $x/a = y/b = z/c = \cdots$ Therefore, for (1.187) to be a maximum, we need $m(x^2/a^2 + y^2/b^2 + z^2/c^2 + \cdots)/2$ to be negative, that is, we need $m<0$.

If we put $m = -M^2$, Adrain's "general curve of probability" in Eq. (1.186) can be written as a normal curve:

$$X = A\exp\left(-\frac{M^2x^2}{2a}\right),$$

where $A = e^{a'}$.

* Italics are Adrain's.

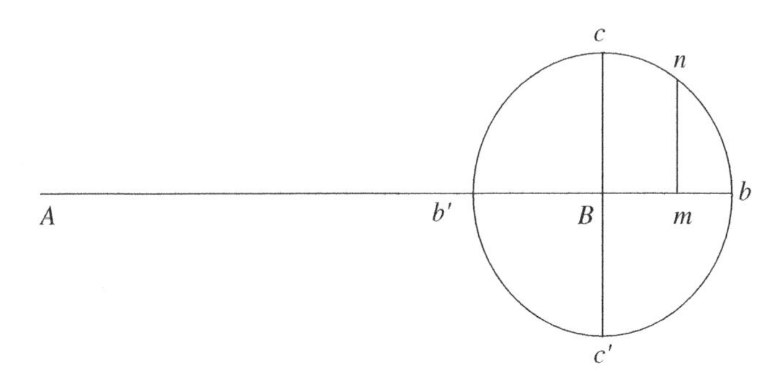

FIGURE 1.31 Adrain's second derivation of the normal law

Some comments about Adrain's proof are in order. First, X and Y are not really the probabilities of errors x and y, since these probabilities are always zero. Rather they are the probability densities of the errors, that is, if dx is infinitesimally small then Xdx is the probability that the error lies in the interval $(x, x+dx)$. Second, Adrain's assumption in (1.182) that the errors are proportional to the measured quantities is arbitrary and questionable. Third, the reason why Adrain sought for the simplest form for X'' in particular, and not for any other function, is not explained.

Having given his proof, Adrain now wished to "confirm what has been said by a different method of investigation." In his second proof (*ibid.*, p. 97), Adrain considered a chord AB (Fig. 1.31), which was subject to the equal errors Bb and Bb' in its length and to the equal errors Bc and Bc' in its bearing (angle from A). Now assume that the vertical and horizontal errors are the same. Then Adrain argued that the curve passing through the points b, c, b', and c' and equidistant from B must be the circumference of a circle with the latter point as center.

The aim now is to find the joint probability of the errors $Bm=x$ and $mn=y$. Assume the probabilities of the errors x and y are X and Y, respectively, and let $X' = \log X$ and $Y' = \log Y$. Next Adrain wrote XY=constant so that $X'+Y'$ = constant and

$$\dot{X}' + \dot{Y}' = 0,$$

where, as before, $\dot{f} \equiv df/dt$ for any function f. Thus, Adrain wished to maximize XY, as he did in his first proof.

From the last equation, we obtain the same relations as (1.184), namely,

$$X''\dot{x} + Y''\dot{y} = 0 \quad \Rightarrow \quad X''\dot{x} = -Y''\dot{y}. \tag{1.188}$$

Now, from the argument used at the start of the second proof, $x^2 + y^2 = Bb^2 = r^2$ so that by differentiating with respect to t,

$$x^2 + y^2 = Bb^2 = r^2 \quad \Rightarrow \quad x\dot{x} = -y\dot{y}. \tag{1.189}$$

Dividing Eq. (1.188) by (1.189), Adrain obtained

$$\frac{X''}{x} = \frac{Y''}{y}$$

whence he deduced that each of the ratios must be a constant quantity, say n. Solving $X''/x = n$, we obtain

$$X' = C + \frac{nx^2}{2},$$

where C is a constant. Similarly, $Y' = C + ny^2/2$. Hence, $XY = e^{C+nx^2/2}e^{C+ny^2/2}$, where n must be negative since the joint probability XY decreases with increasing x. Substituting $C = a'$ and $n = m/a$, Adrain also wrote

$$u = X = \exp\left(a' + \frac{m}{2a}x^2\right)$$

(*ibid*., p. 98), which is the same as his previous Eq. (1.186).

Essentially the same proof as above was later given verbally by Herschel (1850, pp. 19–20).

Merriman has criticized Adrain's second proof on two grounds:

> ...I regard it as defective in taking $X \times Y = \text{constant}$ or in considering the probabilities of the x and y deviations as independent. (Merriman, 1877, p. 164)

However, Merriman is wrong in his first criticism because Adrain wished to maximize the product XY, as he did in his first proof, and at the point of maximum we have $XY = \text{constant}$. Merriman's second criticism is weak since it is not unreasonable that the two deviations should be independent. Although neither proof is entirely satisfactory, Adrain's second proof is an improvement over his first because the former does not make as strong an assumption as (1.182). This opinion, however, is not shared by some writers (e.g., Coolidge, 1926, p. 69).

In any case, after his two derivations of the normal law, Adrain considered the following problem:

> *Supposing a, b, c, d, & e to be the observed measures of any quantity x, the most probable value of x is required**. (Adrain, 1808, p. 98)

To solve the above Adrain noted that the errors were $x-a, x-b, x-c, \ldots$. From Eq. (1.187), for the joint probability of the errors to be maximized, we need

$$(x-a)^2 + (x-b)^2 + (x-c)^2 + \cdots = \min$$

(*ibidem*). In the above, Adrain effectively stated the principle of least squares. By differentiating the last expression, he obtained the most probable value of x as

$$x = \frac{a+b+c+\cdots}{n},$$

where n is the number of observations.

* Italics are Adrain's.

Adrain thus not only gave two derivations of the normal law but also stated the principle of least squares in his 1808 paper. It is natural to ask if he was aware of Legendre's 1805 work. Although Adrain made no mention of the latter, he had a copy of Legendre's memoir in his library and could have consulted it while writing his paper.

1.5.3 Gauss' First Justification of the Principle of Least Squares (1809)

1.5.3.1 Gauss' Life Gauss (Fig. 1.32) was the uncontested Prince of Mathematics. So much has been written about him that it is hard to avoid being repetitive. Carl Friedrich Gauss* was born in Brunswick, Germany, on April 30, 1777, in a family with modest means. As a child he attended a squalid school, but his talent for mathematics showed at a very early age. The story goes that he discovered the formula $1+2+\cdots+n=n(n+1)/2$ in elementary school when his teacher asked a question to the class. With the financial support of the Duke of Brunswick, he graduated from Caroline College, Brunswick, in 1796, then from the University of Göttingen in 1798. Thereupon, Gauss returned to Brunswick and obtained his doctorate from Helmstedt University in 1799. As part of his dissertation, Gauss gave a first proof of the so-called fundamental theorem of algebra in 1797 using advanced techniques from outside algebra. The next year, he completed his masterpiece *Disquisitiones Arithmeticae*† (Gauss, 1801), which marked the true beginning of the theory of numbers. In 1807, Gauss was appointed Professor of Astronomy and first director of the new observatory at Göttingen. Two years later, Gauss published another masterpiece *Theoria Motus Corporum Coelestium*‡ (Gauss, 1809). The book was not only the definite treatment of astronomy at the time but it also contained the first probabilistic justification of the method of least squares together with one of the first derivations of the normal law. In the 1820s, Gauss was interested in geodesy and made further breakthroughs in the field. In the 1830s, he diverted more into physics, but there too he enriched the field. Gauss died in Göttingen on February 23, 1855, but the influence of his wide-ranging scientific work is still felt today.

1.5.3.2 Derivation of the Normal Law. Postulate of the Arithmetic Mean In the *Theoria Motus* (Gauss, 1809, English edition, p. 253),§ Gauss considered μ observations $M, M', M'', \ldots$, respectively, on $V, V', V'', \ldots$, the latter being functions of the ν unknown quantities p, q, r, s, etc. Let the probability density of each error $\Delta = V - M$ be $\phi(\Delta)$. Now, "since we are authorized to regard all the observations as events independent of each other," the joint probability (density) of all the observations is

$$\Omega = \phi(M-V)\phi(M'-V')\phi(M''-V'')\cdots \tag{1.190}$$

* Of the several biographies of Gauss, we mention the ones by Bühler (1987), May (1972), and Dunnington (1955).

† Arithmetical Observations.

‡ Theory of the Motion of the Heavenly Bodies

§ Gauss' first justification of least squares is also described in Sheynin (1979, pp. 29–32), Goldstine (1977, pp. 213–215), Whittaker and Robinson (1944, pp. 218–222), Sprott (1978), Waterhouse (1990), and Chabert (1989).

FIGURE 1.32 Carl Friedrich Gauss (1777–1855). Wikimedia Commons (Public Domain), http://commons.wikimedia.org/wiki/File:Carl_Friedrich_Gauss.jpg

(*ibid.*, p. 255). Then the posterior density of the quantities p, q, r, s, … given the observations is obtained from Bayes' theorem as

$$f\left(p,q,r,s,\ldots|M,M',M'',M''',\ldots\right)=\frac{f\left(M,M',M'',M''',\ldots|p,q,r,s,\ldots\right)f\left(p,q,r,s,\ldots\right)}{\int f\left(M,M',M'',M''',\ldots|p,q,r,s,\ldots\right)f\left(p,q,r,s,\ldots\right)dpdqdrds\ldots}$$
$$=\frac{\Omega}{\int\Omega dpdqdrds\ldots} \tag{1.191}$$

(*ibid.*, p. 257). In the above, Gauss used a uniform prior for $p,q,r,s,...$, that is, $f(p,q,r,s,\cdots)=1$, based on the assumption that "all systems of values of these unknown quantities $[p,q,r,s,...]$ were equally probable previous to the observations." From Eq. (1.191), $f(p,q,r,s,\ldots\mid M,M',M'',M''',\ldots)$ is maximized when Ω, hence $\log\Omega$, attains its maximum value. By writing $V-M=v, V'-M'=v', V''-M''=v'',\cdots$, this occurs when

$$\frac{\partial}{\partial p}\left\{\log\phi(v)+\log\phi(v')+\log\phi(v'')+\cdots\right\}=0,$$
$$\frac{\partial}{\partial q}\left\{\log\phi(v)+\log\phi(v')+\log\phi(v'')+\cdots\right\}=0,$$
$$\frac{\partial}{\partial r}\left\{\log\phi(v)+\log\phi(v')+\log\phi(v'')+\cdots\right\}=0,$$
$$\cdots\cdot$$

The above can be written as

$$\frac{1}{\phi(v)}\frac{d\phi(v)}{dv}\frac{\partial v}{\partial p}+\frac{1}{\phi(v')}\frac{d\phi(v')}{dv'}\frac{\partial v'}{\partial p}+\frac{1}{\phi(v'')}\frac{d\phi(v'')}{dv''}\frac{\partial v''}{\partial p}+\cdots=0,$$
$$\frac{1}{\phi(v)}\frac{d\phi(v)}{dv}\frac{\partial v}{\partial q}+\frac{1}{\phi(v')}\frac{d\phi(v')}{dv'}\frac{\partial v'}{\partial q}+\frac{1}{\phi(v'')}\frac{d\phi(v'')}{dv''}\frac{\partial v''}{\partial q}+\cdots=0,$$
$$\frac{1}{\phi(v)}\frac{d\phi(v)}{dv}\frac{\partial v}{\partial r}+\frac{1}{\phi(v')}\frac{d\phi(v')}{dv'}\frac{\partial v'}{\partial r}+\frac{1}{\phi(v'')}\frac{d\phi(v'')}{dv''}\frac{\partial v''}{\partial r}+\cdots=0,$$
$$\cdots\cdot$$

By making the substitution $\frac{1}{\phi(\Delta)}\cdot\frac{d\phi(\Delta)}{d\Delta}=\phi^*(\Delta)$, the above system becomes

$$\phi^*(v)\frac{\partial v}{\partial p}+\phi^*(v')\frac{\partial v'}{\partial p}+\phi^*(v'')\frac{\partial v''}{\partial p}+\cdots=0,$$
$$\phi^*(v)\frac{\partial v}{\partial q}+\phi^*(v')\frac{\partial v'}{\partial q}+\phi^*(v'')\frac{\partial v''}{\partial q}+\cdots=0, \tag{1.192}$$
$$\phi^*(v)\frac{\partial v}{\partial r}+\phi^*(v')\frac{\partial v'}{\partial r}+\phi^*(v'')\frac{\partial v''}{\partial r}+\cdots=0,$$
$$\cdots$$

(*ibidem*). The last equations can be solved once ϕ^* is known. But "[s]ince this cannot be defined *a priori*, we will be approaching the subject from another point of view, inquire about what function…assumed as a base…the common principle…depends." Thereupon, Gauss made the debatable assumption that the arithmetic mean was the most probable value of the observations:

> …It has been customary certainly to regard as an axiom the hypothesis that if any quantity has been determined by several direct observations, made under the same circumstances and with equal care, the arithmetical mean of the observed values affords the most probable value, if not rigorously, yet very nearly at least, so that it is always most safe to adhere to it. (*ibid.*, p. 258)

Now, assuming that the observations $M, M', M'', \ldots$ were made on the same function $V = V' = V'' \cdots = p$, and since $V - M = v$, Eq. (1.192) becomes

$$\phi^*(M-p)+\phi^*(M'-p)+\phi^*(M''-p)+\cdots=0. \tag{1.193}$$

Gauss' *postulate of the arithmetic mean* implies that $p = (M + M' + M'' + \cdots)/\mu$. Moreover, consider a particular set of observations for which $M' = M'' = \cdots = M - \mu N$, where N is a real number. Therefore,

$$p = \frac{M + M' + M'' + \cdots}{\mu} = \frac{M + (M - \mu N) + (M - \mu N) + \cdots}{\mu} = M - (\mu - 1)N.$$

Hence, Eq. (1.193) becomes

$$\phi^*((\mu-1)N)+(\mu-1)\phi^*(-N)=0$$

(*ibidem*). From this, it follows that

$$\frac{\phi^*((\mu-1)N)}{(\mu-1)N} = \frac{\phi^*(-N)}{-N},$$

implying that

$$\frac{\phi^*(\Delta)}{\Delta} = \frac{\phi'(\Delta)}{\Delta\phi(\Delta)} = k,$$

where k is a constant. Solving the above differential equation, Gauss obtained

$$\phi(\Delta) = \chi \exp\left(\frac{1}{2}k\Delta^2\right),$$

where χ is a constant (*ibidem*). Noting that k must be negative for Ω to become a maximum, Gauss next set $k = -2h^2$, where h is a constant, so that the above equation becomes

$$\phi(\Delta) = \chi \exp(-h^2\Delta^2).$$

Finally, by using "the elegant theorem first discovered by Laplace," Gauss noted that $\int_{-\infty}^{\infty} \exp(-h^2\Delta^2)d\Delta = \sqrt{\pi}/h$ so that the formula for $\phi(\Delta)$ becomes

$$\phi(\Delta) = \frac{h}{\sqrt{\pi}} \exp(-h^2\Delta^2)$$

(*ibid.*, p. 259). Having derived the normal density above, it was then easy for Gauss to argue that the joint density in (1.190), namely,

$$\Omega = \phi(v)\phi(v')\phi(v'')\cdots = \left(\frac{h}{\sqrt{\pi}}\right)^{\mu} \exp\left\{-h^2\left(v^2 + v'^2 + v''^2 + \cdots\right)\right\},$$

is maximized when the sum of squares of the errors, that is, $v^2 + v'^2 + v''^2 + \cdots$, is minimized:

> ...It is evident, in order that the product
>
> $$\Omega = h^{\mu}\pi^{-\mu/2}e^{-hh(vv+v'v'+v''v''+\cdots)}$$
>
> may become a maximum, that the sum
>
> $$vv + v'v' + v''v'' + etc.,$$
>
> must become a minimum. *Therefore, that will be the most probable system of values of the unknown quantities p, q, r, s, etc., in which the sum of the squares of the differences between the observed and computed values of V, V′, V″, etc. is a minimum....** (*ibid.*, p. 260)

Gauss was thus able to probabilistically justify the method of least squares. However, his reasoning contains an inherent circularity because the normal distribution emerges as a consequence of the postulate of the arithmetic mean, which is in fact a consequence of the normality assumption!

1.5.3.3 Priority Dispute with Legendre After having provided his probabilistic justification, Gauss went on to make the following comments regarding the principle of least squares:

> Our principle, which we have made use of since the year 1795, has lately been published by LEGENDRE in the work *Nouvelles méthodes pour la détermination des orbites des comètes*, Paris, 1806, where several other properties of this principle have been explained, which, for the sake of brevity, we here omit. (*ibid.*, p. 270)

Extracts from Gauss' *Theoria Motus* were passed to Legendre by Sophie Germain (1776–1831).† A dispute thereby erupted between Legendre and Gauss, and has been comprehensively documented by Plackett (1972). It is noteworthy that the two men had already been at odds before. In 1801, Gauss claimed priority on the law of reciprocity of quadratic residues although Legendre had already stated that law in

* Italics are Gauss'.

† Sophie Germain knew Legendre personally and exchanged several letters with Gauss under the pen name Mr. Leblanc. Gauss had the highest esteem for the brilliance of Germain's mathematical work.

1785. Now, the issue was that of least squares, which we describe in what follows. All quotations that follow are from Plackett's paper.*

Legendre took umbrage at Gauss' claim, especially his use of the term "our principle" (*principium nostrum*). In a letter dated May 9, 1809, to Gauss, after initially praising parts of Gauss' work, Legendre wrote:

> I will therefore not conceal from you, Sir, that I felt some regret to see that in citing my memoir p. 221 you say *principium nostrum quo jam inde ab anno 1795 usi sumus etc*. There is no discovery that one cannot claim for oneself by saying that one had found the same thing some years previously; but if one does not supply the evidence by citing the place where one has published it, this assertion becomes pointless and serves only to do a disservice to the true author of the discovery. In Mathematics it often happens that one discovers the same things that have been discovered by others and which are well known; this has happened to me a number of times, but I have never mentioned it and I have never called *principium nostrum* a principle which someone else had published before me. (*ibid.*, p. 243)

Gauss did not have any published work to back his claim of priority and so did not directly reply to Legendre's letter. However, he sought the support of his friends to vindicate him. Thus, writing to Heinrich Olbers (1758–1840) on October 4, 1809, Gauss pleaded:

> ...Do you still remember, dearest friend, that on my first visit to Bremen in 1803 I talked with you about the principle which I used to represent observations most exactly, namely that the sum of squares of the differences must be minimized when the observations have equal weights? That we discussed the matter in Rehburg in 1804 of that I still clearly recollect all the circumstances. It is important to me to know this. The reason for the question can wait.... (*ibidem*)

Laplace also got involved in the controversy and wrote a letter to Gauss on November 15, 1811:

> M. Gauss says in his work on elliptical movement that he was conversant with it before M. Le Gendre had published it; I would greatly like to know whether before this publication anything was printed in Germany concerning this method and I request M. Gauss to have the kindness to inform me about it. (*ibidem*)

Gauss replied to Laplace on January 30, 1812, and gave more details on his prior use of least squares:

> I have used the method of least squares since the year 1795 and I find in my papers, that the month of June 1798 is the time when I reconciled it with the principles of the calculus of probabilities: a note about this is contained in a diary which I kept about my mathematical work since the year 1796, and which I showed at that time to Mr. De Lindenau.
>
> However my frequent applications of this method only date from the year 1802, since then I use it as you might say every day in my astronomical calculations on the new planets. As I had intended since then to assemble all the methods which I have used in one extensive

*The priority dispute between Legendre and Gauss is also discussed in Stigler (1977; 1981) and Dutka (1996).

> work (which I began in 1805 and of which the manuscript originally in German, was completed in 1806, but which at the request of Mr Perthes I afterwards translated into Latin: printing began in 1807 and was finished only in 1809), I am in no hurry to publish an isolated fragment, therefore Mr. Legendre has preceded me. Nevertheless I had already communicated this same method, well before the publication of Mr Legendre's work, to several people among other to Mr. Olbers in 1803 who must certainly remember it. Therefore, in my theory of the motions of planets, I was able to discuss the method of least squares, which I have applied thousands of times during the last 7 years, and for which I had developed the theory, in section 3 of book II of this work, in German at least, well before having seen Mr Legendre's work—I say, could I have discussed this principle, which I had made known to several of my friends already in 1803 as being likely to form part of a work which I was preparing,—as a method derived from Mr. Legendre? I had no idea that Mr. Legendre would have been capable of attaching so much value to an idea so simple that, rather than being astonished that it had not been thought of a hundred years ago, he should feel annoyed at my saying that I had used it before he did. In fact, it would be very easy to prove it to everyone by evidence which could not be refuted, if it were worth the trouble. But I thought that all those who know me would believe it on my word alone, just as I would have believed it with all my heart if Mr Legendre had stated that he had already been conversant with the method before 1795. I have many things in my papers, which I may perhaps lose the chance of being first to publish; but so be it, I prefer to let things ripen. (*ibid.*, p. 245)

Two points need to be made from the above. First, it appears that Gauss did *not* include the section on least squares in his book only after seeing Legendre's 1805 article. Second, Gauss regarded least squares as an elementary principle that should have been discovered a hundred years before his own time!

On March 10, 1812, Olbers wrote to Gauss and backed the latter's claim of priority on least squares:

> ...I have now received the November issue of M.C., and with it your beautiful elimination method for *moindres carrés*. I can attest publicly at the first opportunity, and will do so with pleasure, that you had already told me the basic principle in 1803. I remember this quite well as if it had happened today. There must also be something concerning it written down among my papers. For I noted it then, together with your interpolation formula, which you communicated to me at that time .... (*ibidem*)

In spite of this, the controversy persisted for a long time after. In 1827, writing to a young Jacobi, Legendre continued to express his anger toward Gauss. Legendre died in 1833, but years after that Gauss continued to stake his claim on least squares.

In retrospect, there is little doubt that Gauss did use least squares or some form of it around 1795, well before Legendre's publication. He most likely also used the method in 1801 when he correctly predicted the position of the planet Ceres (which had disappeared after its initial discovery on January 1, 1801, by Piazza). But, in the absence of any publication, it is difficult to gauge to what extent he understood or developed the method. There must surely be many mathematical discoveries some part or the whole of which might have been conjectured by others at some prior times but never put on paper. If one were to rely on such *a priori* conjectures to establish priority, the resulting situation would be one of chaos. It is only fair, therefore, that the claim of priority on the principle of least squares should go to Legendre.

There is another lingering question. If Gauss had known about least squares almost a decade before Legendre's paper, why did he withhold publication? Gauss, as he admitted, might have regarded it as too elementary. But it is also possible that he did not publish because he did not believe he had a valid justification for the method. Gauss might have decided to come out with the method in 1809 only after having obtained some kind of justification for it. However, as we have seen, even his 1809 justification was far from perfect. It was only in 1823 that Gauss was able to give a definite justification of the least squares principle (see Section 1.5.6).

1.5.4 Laplace in 1810: Justification of Least Squares Based on Inverse Probability, the Gauss–Laplace Synthesis

Laplace must have read Gauss' work (see Section 1.5.3) and immediately made a connection between the Central Limit Theorem he proved in the "Mémoire sur les approximations des formules qui sont fonctions de très grands nombres et sur leur application aux probabilités" (Laplace, 1810a) and the method of least squares. He quickly published a sequel entitled "Supplément au mémoire sur les approximations des formules qui sont fonctions de très grands nombres et sur leur application aux probabilités" (Laplace, 1810b) where he gave a probabilistic justification of the method of least squares without invoking the postulate of the arithmetic mean. We now describe this aspect of Laplace's work.

After reproducing his previous formula (see Eq. 1.110) for the Central Limit Theorem, Laplace considered a set of n observations. Each observation has the same probability distribution ("*loi de facilité*"), and the mean of the n observations is denoted by A. Additional sets of $n', n'', \ldots$ observations are similarly assumed to have means $A+q, A+q', \ldots$, with each set having a different probability distribution. The aim is to estimate the mean of *all* the observations.

Assuming the true mean of all observations is $A+x$, the error in the mean result for the set of n observations is $-x$. Laplace wrote the posterior density, assuming a uniform prior, of this error as

$$\frac{1}{\sqrt{\pi}}\sqrt{\frac{k}{2k'}}\frac{dr}{dx}e^{-\frac{k}{2k'}r^2} \tag{1.194}$$

(Laplace, 1810b, OC 12, p. 350), where as before (see Eq. 1.109),

$$k = \int \phi\left(\frac{x}{h}\right)dx,$$

$$k' = \int \frac{(x-q')^2}{h^2}\phi\left(\frac{x}{h}\right)dx,$$

$\phi(x/h)$ is the probability of an error x, and $x = rh/\sqrt{n}$. Equation (1.194) can be obtained as follows. Let x be the error in the mean of the set of n observations. Then using Eq. (1.110), the posterior distribution function of x is

$$\Pr\left\{x \le \frac{rh}{\sqrt{n}}\right\} = \frac{1}{2} + \frac{1}{\sqrt{\pi}}\sqrt{\frac{k}{2k'}}\int_0^{rh/\sqrt{n}} e^{-\frac{kr^2}{2k'}}dr.$$

The posterior density of x is therefore

$$f\left(\frac{rh}{\sqrt{n}}\right)=\frac{1}{\sqrt{\pi}}\sqrt{\frac{k}{2k'}}e^{-\frac{kr^2}{2k'}},$$

from which Eq. (1.194) follows.

By letting $a=\frac{1}{h}\sqrt{\frac{k}{2k'}}$ and using $x=rh/\sqrt{n}$, Eq. (1.194) becomes

$$\frac{1}{\sqrt{\pi}}a\sqrt{n}e^{-na^2x^2}.$$

Similarly, the error in the mean of the set of n' observations is $q-x$, and its posterior density is

$$\frac{1}{\sqrt{\pi}}a'\sqrt{n'}e^{-n'a'^2(q-x)^2},$$

with a' "expressing relative to these observations what a expresses relative to the n observations." Similar probability densities can also be obtained for the other sets of observations. Writing the successive posterior densities as $\Psi(-x),\Psi'(q-x),\Psi''(q'-x),\cdots$, Laplace obtained the joint posterior density of the errors by the following *courbe des probabilités* (probability curve):

$$y=\Psi(-x)\Psi'(q-x)\Psi''(q'-x)\cdots. \qquad (1.195)$$

It is seen that Laplace's approach here is similar to the one in his "Mémoire sur les Probabilités" (Laplace, 1781). In particular, the *courbe des probabilités* he used previously (see Eq. 1.73) is similar to the present one. Proceeding as before (see Eq. 1.75), Laplace chose the value of the mean l as one which satisfied

$$\int_0^l ydz=\int_l^\infty ydz,$$

where z is the abscissa corresponding to the ordinate y. This choice of l corresponds to the median and minimizes the expected value of the absolute error (see Eq. 1.74).

Thereupon, Laplace criticized Daniel Bernoulli, Euler, and Gauss on their preference for the method of maximum likelihood. He claimed that his method, on the other hand, was based on the theory of probability. However, Laplace was not completely correct in his criticism. Euler did not endorse maximum likelihood (Sheynin, 1977), and Gauss used the maximum posterior density instead of maximum likelihood (see Sec. 1.5.3).

Continuing with his analysis, Laplace next substituted $x=X+z$ so that Eq. (1.195) becomes

$$y=pp'\,p''\cdots\exp\left\{-p^2\pi(X+z)^2-p'^2\pi(q-X-z)^2-p''^2\pi(q'-X-z)^2-\cdots\right\} \qquad (1.196)$$

(Laplace, 1810b, OC 12, p. 352). In the last expression, $p = a\sqrt{n}/\sqrt{\pi}$, which is the mode of the density for the n observations. Similarly, $p', p'', \ldots$ are the modes for the other sets of observations. Now, by setting the coefficient of z in the exponent of (1.196) to zero, Laplace obtained a value of X, which corresponded to the median sought and which also maximized the corresponding posterior density:

$$X = \frac{p'^2 q + p''^2 q' + \cdots}{p^2 + p'^2 + \cdots}. \tag{1.197}$$

The maximum value of y is then of the form

$$y = pp' \, p'' \cdots e^{-M - Nz^2},$$

and the required mean to be taken from the observed means $A, A+q, A+q', \cdots$ is $A+X$. Laplace further noted that the value of X in Eq. (1.197) also minimized

$$(pX)^2 + \{p'(q-X)\}^2 + \{p''(q'-X)\}^2 + \cdots \tag{1.198}$$

(*ibid.*, p. 353). It is seen that, for the case $p = p' = p'' = \cdots$, the expression in (1.198) is the least squares criterion. Laplace concluded:

> Thus this property [that the sum of squares is minimized] which is only hypothetical when considering results from a single observation or from a small number of observations, becomes necessary when the results between which one has to take a mean are each given by a large number of observations, whatever the law of facility of the errors in these observations. It is a reason for using it in all cases. (*ibidem*)

Laplace had thus achieved an important synthesis with Gauss' results on least squares. In particular, he had shown that there was no need to a priori *assume* a normal distribution for least squares to be valid. This is because if the errors involved were the sum of large number of components, Laplace's Central Limit Theorem would *a fortiori* imply that the errors would be normally distributed. The least squares property thus followed as a natural consequence of the Central Limit Theorem. Moreover, Laplace also showed that, for a large number of observations, his criterion of *least absolute error* coincided with that of *least squared error*.

Laplace's current justification of the method of least squares can be compared with that of Gauss in 1809 (see Section 1.5.3). Both authors used inverse probability. In addition, Gauss appealed to the postulate of arithmetic mean, although his method had the advantage of being valid for all sample sizes. On the other hand, Laplace used large sample methods but avoided the postulate of the arithmetic mean. One year later, he used direct, instead of inverse, probability to prove the method of least squares for large samples (see next section). Almost a decade later, Gauss gave the definite justification of the method of least squares, which avoided both inverse probability and the postulate of the arithmetic mean, and which also was valid for all sample sizes (see Section 1.5.6).

1.5.5 Laplace's Justification of Least Squares Based on Direct Probability (1811)

In "Mémoire sur les intégrales définis et leur application aux probabilités" (Laplace, 1811a), Laplace provided a justification of the method of least squares for large samples using direct probability. He first (Laplace, 1811a, OC 12, p. 387) considered the case of a single "element" (parameter). His proof was as follows.*

Laplace first considered an unknown "element" (parameter) whose approximate value is known and to which a correction z needs to be applied. Suppose an observation ζ, which is a function of the element, is made. By doing a series expansion of ζ in powers of z, and ignoring z^2 and higher powers, one obtains

$$\zeta = m + pz.$$

Assuming an error ε is made in ζ, the above becomes

$$\zeta + \varepsilon = m + pz.$$

Letting $\zeta - m = \phi$, we have

$$\varepsilon = pz - \phi.$$

If several observations are made resulting in errors $\varepsilon, \varepsilon^{(1)}, \varepsilon^{(2)}, \ldots$, then

$$\varepsilon^{(i)} = p^{(i)}z - \phi^{(i)}, \quad i = 0, 1, \cdots, s-1. \tag{1.199}$$

By summing over all values of i,

$$\sum_i \varepsilon^{(i)} = z\sum_i p^{(i)} - \sum_i \phi^{(i)} = 0 \quad \Rightarrow \quad z = \frac{\sum_i \phi^{(i)}}{\sum_i p^{(i)}}.$$

Laplace called z above "the mean result of the observations." He next considered the linear combination

$$q\varepsilon + q^{(1)}\varepsilon^{(1)} + q^{(2)}\varepsilon^{(2)} + \cdots + q^{(s-1)}\varepsilon^{(s-1)}, \tag{1.200}$$

which is equal to zero and where $q, q^{(1)}, q^{(2)}, \ldots$ are integers. By substituting Eq. (1.199) into the above,

$$z = \frac{\sum_i q^{(i)}\phi^{(i)}}{\sum_i q^{(i)}p^{(i)}} \tag{1.201}$$

* See also Sheynin (1977, pp. 18–20) and Farebrother (1999, pp. 107–111).

(*ibid.*, p. 389). If an error z' is made in z, then

$$z = \frac{\sum_i q^{(i)}\phi^{(i)}}{\sum_i q^{(i)}p^{(i)}} + z', \tag{1.202}$$

so that (1.200) becomes

$$z'\sum_i p^{(i)}q^{(i)}.$$

Laplace now wished to calculate the probability of the error z'. For this, he first considered

$$\sum_x \Psi\left(\frac{x}{a}\right)e^{jqx\omega} \times \sum_x \Psi\left(\frac{x}{a}\right)e^{jq^{(1)}x\omega} \times \cdots \times \sum_x \Psi\left(\frac{x}{a}\right)e^{jq^{(s-1)}x\omega},$$

where $j = \sqrt{-1}$ and $\Psi(x/a)$ is the probability of the error x. Now, the coefficient of $e^{jl\omega}$ in the above is the probability that the linear combination in (1.200) will be equal to l. This probability is therefore also the term independent of ω in the product $e^{-jl\omega}$ and the above expression, that is, the term independent of ω in

$$e^{-jl\omega} \times 2\sum_x \Psi\left(\frac{x}{a}\right)\cos(qx\ \omega) \times 2\sum_x \Psi\left(\frac{x}{a}\right)\cos\left(q^{(1)}x\ \omega\right) \times \cdots \times 2\sum_x \Psi\left(\frac{x}{a}\right)\cos\left(q^{(s-1)}x\ \omega\right),$$

with $\Psi(x/a)$ being assumed to be even. Now, the probability that the linear combination in (1.200) will be equal to $-l$ will be the term independent of ω in the expression above with l replaced by $-l$. Therefore, the probability that the linear combination in (1.200) will be equal to $\pm l$ is obtained by addition of the two expressions and is the term independent of ω in

$$2\cos(l\omega) \times 2\sum_x \Psi\left(\frac{x}{a}\right)\cos(qx\ \omega) \times 2\sum_x \Psi\left(\frac{x}{a}\right)\cos\left(q^{(1)}x\ \omega\right) \times \cdots$$
$$\times 2\sum_x \Psi\left(\frac{x}{a}\right)\cos\left(q^{(s-1)}x\ \omega\right).$$

By using a reasoning similar to the one that led to Eq. (1.95), the probability that the linear combination in (1.200) will be equal to $\pm l$ is therefore

$$\frac{2}{\pi}\int_0^\pi \cos(l\omega) \times 2\sum_x \Psi\left(\frac{x}{a}\right)\cos(qx\omega) \times 2\sum_x \Psi\left(\frac{x}{a}\right)\cos\left(q^{(1)}x\omega\right) \times \cdots$$
$$\times 2\sum_x \Psi\left(\frac{x}{a}\right)\cos\left(q^{(s-1)}x\omega\right)d\omega \tag{1.203}$$

(*ibid.*, p. 390). The cosine terms can be expanded in powers of x:

$$\sum_x \Psi\left(\frac{x}{a}\right)\cos(qx\omega) = \sum_x \Psi\left(\frac{x}{a}\right) - \frac{1}{2}q^2a^2\omega^2\sum_x\left(\frac{x^2}{a^2}\right)\Psi\left(\frac{x}{a}\right) + \cdots.$$

Laplace next made a passage to the continuous limit. He set $x/a = x'$ so that, assuming the variation in x is unity, we have $1/a = dx'$. Now, let

$$k = 2\int_0^1 \Psi(x')dx', \quad k' = \int_0^1 x'^2 \Psi(x')dx'.$$

Therefore, by using the last two equations,

$$\begin{aligned}
2\sum_x \Psi\left(\frac{x}{a}\right)\cos(qx\omega) &= 2\sum_x \Psi\left(\frac{x}{a}\right) - q^2a^2\omega^2\sum_x\left(\frac{x^2}{a^2}\right)\Psi\left(\frac{x}{a}\right)+\cdots \\
&= 2a\int_0^1 \Psi(x')dx' - q^2a^3\omega^2\int_0^1 x'^2\Psi(x')dx' + \cdots \\
&= ak\left(1-\frac{k'}{k}q^2a^2\omega^2+\cdots\right).
\end{aligned}$$

Using $ak = 1$, the logarithm of the above becomes

$$-\frac{k'}{k}q^2a^2\omega^2 - \cdots.$$

Therefore,

$$\begin{aligned}
&\log\left\{2\sum_x \Psi\left(\frac{x}{a}\right)\cos(qx\omega)\times 2\sum_x \Psi\left(\frac{x}{a}\right)\cos\left(q^{(1)}x\omega\right)\times\cdots\times 2\sum_x \Psi\left(\frac{x}{a}\right)\cos\left(q^{(s-1)}x\omega\right)\right\} \\
&= -\frac{k'}{k}\sum_i \left(q^{(i)}\right)^2 a^2\omega^2 - \cdots.
\end{aligned}$$

By taking antilogarithms of the above, the probability in Eq. (1.203) becomes

$$\begin{aligned}
\Pr\left\{\sum_i q^{(i)}\varepsilon^{(i)} = \pm l\right\} &= \frac{2}{\pi a}\int_0^{\pi} a\cos(l\omega)\exp\left\{-\frac{k'}{k}a^2\omega^2\sum_i q^{(i)2}\right\}d\omega \\
&= \frac{2}{\pi a}\int_0^{a\pi} a\cos\left(\frac{l}{a}t\right)\exp\left\{-\frac{k'}{k}t^2\sum_i\left(q^{(i)}\right)^2\right\}dt \\
&= \frac{1}{a\sqrt{\pi}}\frac{1}{\sqrt{\frac{k'}{k}\sum_i\left(q^{(i)}\right)^2}}\exp\left\{-\frac{kl^2}{4k'a^2\sum_i\left(q^{(i)}\right)^2}\right\},
\end{aligned}$$

where $a\omega = t$. Therefore, by substituting $l = ar$ so that $adr = 1$,

$$\Pr\left\{-ar \le \sum_i q^{(i)}\varepsilon^{(i)} \le ar\right\} = \frac{1}{\sqrt{\frac{k'\pi}{k}\sum_i\left(q^{(i)}\right)^2}}\int \exp\left\{-\frac{kr^2}{4k'\sum_i\left(q^{(i)}\right)^2}\right\}dr \quad (1.204)$$

(*ibid.*, p. 392). One can recognize the above as a form of the Central Limit Theorem.

With the result (1.204) in hand, Laplace went back to the problem he considered at the start of the section. In Eq. (1.202), an error z' is made in z, and (1.200) becomes

$$z'\sum_i p^{(i)}q^{(i)}.$$

By substituting $ar = z'\sum_i p^{(i)}q^{(i)}$, Laplace obtained

$$\frac{\exp\left\{-\dfrac{kr^2}{4k'\sum_i\left(q^{(i)}\right)^2}\right\}dr}{2\sqrt{\dfrac{k'\pi}{k}\sum_i\left(q^{(i)}\right)^2}} = \underbrace{\frac{\sum_i p^{(i)}q^{(i)}}{2a\sqrt{\pi}\sqrt{\dfrac{k'}{k}\sum_i\left(q^{(i)}\right)^2}}\exp\left[-\frac{kz'^2\left\{\sum_i p^{(i)}q^{(i)}\right\}^2}{4k'a^2\sum_i\left(q^{(i)}\right)^2}\right]}_{H}dz' \quad (1.205)$$

(*ibid.*, p. 393). He observed that H in the above was the density ("*courbe des probabilités*") of the error z'. Laplace then noted:

> ...any error, whether positive, whether negative, must be considered a disadvantage or a loss in some game; now, by the known principles of the Calculus of probabilities, one evaluates the disadvantage by taking the sum of the products of each disadvantage by its probability; the mean value of the error is thus the sum of the products of each error, ignoring its sign, by its probability.... (*ibidem*)

Laplace's therefore wished to calculate the *mean absolute error*. The latter quantity is, from Eq. (1.205),

$$\int_0^\infty \frac{|z'|\sum_i p^{(i)}q^{(i)}}{2a\sqrt{\pi}\sqrt{\dfrac{k'}{k}\sum_i\left(q^{(i)}\right)^2}}\exp\left[-\frac{kz'^2\left\{\sum_i p^{(i)}q^{(i)}\right\}^2}{4k'a^2\sum_i\left(q^{(i)}\right)^2}\right]dz' = 2a\sqrt{\frac{k'}{k\pi}}\frac{\sqrt{\sum_i q^{(i)2}}}{\sum_i p^{(i)}q^{(i)}}. \quad (1.206)$$

Now the values of $p, p^{(1)}, p^{(2)}, \ldots$ are known, but those of $q, q^{(1)}, q^{(2)}, \ldots$ are unknown and need to be chosen so as minimize the mean absolute error. By differentiating the latter with respect to $q^{(i)}$,

$$\frac{q^{(i)}}{\sum_i q^{(i)}} = \frac{p^{(i)}}{\sum_i p^{(i)}q^{(i)}} \quad\Rightarrow\quad q^{(i)} = \frac{\sum_i q^{(i)}}{\sum_i p^{(i)}q^{(i)}}p^{(i)} \equiv \mu p^{(i)},$$

where μ is a constant independent of i such that $q^{(i)}$ are whole numbers. The minimum absolute error is then

$$2a\sqrt{\frac{k'}{k\pi}}\frac{1}{\sqrt{\sum_i \left(p^{(i)}\right)^2}}.$$

Now, the above minimum corresponds to the "mean result of observations" given in Eq. (1.201), namely,

$$z=\frac{\sum_i q^{(i)}\phi^{(i)}}{\sum_i q^{(i)}p^{(i)}}.$$

But this mean result is in fact the one obtained by the method of least squares, for which the sum of squares is

$$\left(pz-\phi\right)^2+\left(p^{(1)}z-\phi^{(1)}\right)^2+\cdots+\left(p^{(s-1)}z-\phi^{(s-1)}\right)^2.$$

When the above is minimized with respect to z, the result is given by z above. Hence, Laplace was able to justify the method of least squares based on the principle of minimum absolute error.

A variant of the first principle (least squared error) was later called the *most advantageous method* by Laplace. On the other hand, the second principle (least absolute error) was one that Laplace had preferred since 1793 (Laplace, 1793) and had adopted from the works of Roger Boscovich (1711–1787). Laplace later called the principle of least absolute error the *method of situation.*

Laplace next (*ibid.*, p. 401)* considered two unknown "elements" (parameters) η and η' whose approximate values $\hat{\eta}$ and $\hat{\eta}'$ are known and to which two unknown corrections z and z' need to be applied. Next, suppose an observation $\zeta=\zeta(\eta,\eta')$ is made. Then, by doing a series expansion on ζ and ignoring nonlinear terms in z and z', Laplace obtained

$$\zeta=\zeta\left(\hat{\eta}+z,\hat{\eta}'+z'\right)=A+pz+qz'.$$

Now, suppose several observations are made. By writing $\zeta-A=\alpha$ and denoting the error in the ith observation by ε_i, Laplace obtained the following system of equations:

$$\varepsilon_i=p_iz+q_iz'-\alpha_i,\quad i=1,2,\cdots,s, \tag{1.207}$$

(*ibid.*, p. 402). In the above, p_i, q_i, and α_i are known but the errors ε_i are unknown. By multiplying the ith equation in (1.207) by a constant m_i and adding,

$$\sum_i m_i\varepsilon_i=z\sum_i m_ip_i+z'\sum_i m_iq_i-\sum_i m_i\alpha_i.$$

* See also Sheynin (1977, pp. 22–24) and Farebrother (1999, pp. 111–117).

By multiplying the same equation by another constant n_i and adding,

$$\sum_i n_i\varepsilon_i = z\sum_i n_i p_i + z'\sum_i n_i q_i - \sum_i n_i\alpha_i.$$

Next Laplace reasoned as follows. If the sums $\sum_i m_i\varepsilon_i$ and $\sum_i n_i\varepsilon_i$ are each set to zero, then the unknowns z and z' can be determined from the following two equations:

$$\left.\begin{aligned} 0 &= z\sum_i m_i p_i + z'\sum_i m_i q_i - \sum_i m_i\alpha_i \\ 0 &= z\sum_i n_i p_i + z'\sum_i n_i q_i - \sum_i n_i\alpha_i \end{aligned}\right\} \tag{1.208}$$

But z and z' may themselves be subjected to errors u and u', respectively, in which case the sums $\sum_i m_i\varepsilon_i$ and $\sum_i n_i\varepsilon_i$ will take values l and l', each different from zero. That is,

$$\left.\begin{aligned} l &= (z+u)\sum_i m_i p_i + (z'+u')\sum_i m_i q_i - \sum_i m_i\alpha_i \\ l' &= (z+u)\sum_i n_i p_i + (z'+u')\sum_i n_i q_i - \sum_i n_i\alpha_i \end{aligned}\right\}. \tag{1.209}$$

By subtracting each of the equations in (1.208) from those in (1.209),

$$\left.\begin{aligned} l &= u\sum_i m_i p_i + u'\sum_i m_i q_i \\ l' &= u\sum_i n_i p_i + u'\sum_i n_i q_i \end{aligned}\right\} \tag{1.210}$$

(*ibid.*, p. 403). Laplace now wished to determine the constants $m_1, \ldots, m_s, n_1, \ldots, n_s$ such that "the mean error is a minimum." What is implied here is that the mean error *on each of the elements* is a minimum, as Laplace had previously clarified in the memoir* (*ibid.*, p. 362). To achieve the required minimization, Laplace considered the product

$$\begin{aligned} &\sum_{x=-a}^{a}\phi\left(\frac{x}{a}\right)\exp\{-jx(m_1\omega+n_1\omega')\}\sum_{x=-a}^{a}\phi\left(\frac{x}{a}\right)\exp\{-jx(m_2\omega+n_2\omega')\}\cdots \\ &\qquad\qquad \sum_{x=-a}^{a}\phi\left(\frac{x}{a}\right)\exp\{-jx(m_s\omega+n_s\omega')\}, \end{aligned} \tag{1.211}$$

where $j=\sqrt{-1}$, x is the error made on an observation and lies between $\mp a$, and $\phi(x/a)$ is the probability of the error x and is assumed to be even. Equation (1.211) then becomes

$$\begin{aligned} &2\sum_{x=0}^{a}\phi\left(\frac{x}{a}\right)\cos(m_1\omega x+n_1\omega' x)\times 2\sum_{x=0}^{a}\phi\left(\frac{x}{a}\right)\cos(m_2\omega x+n_2\omega' x) \\ &\times\cdots\times 2\sum_{x=0}^{a}\phi\left(\frac{x}{a}\right)\cos(m_s\omega x+n_s\omega' x). \end{aligned} \tag{1.212}$$

* See also Sheynin (1977, p. 22).

Next Laplace observed that the joint probability $\Pr\left\{\sum_i m_i\varepsilon_i = l, \sum_i n_i\varepsilon_i = l'\right\}$ is equal to the coefficient of $\exp\{-j(l\omega + l'\omega')\}$ in Eq. (1.211), that is, to the term independent of ω and ω' in the product of Eq. (1.212) and $\exp\{-j(l\omega + l'\omega')\}$. Therefore, assuming $m_1, \ldots, m_s, n_1, \ldots, n_s$ are integers and using a reasoning similar to the one that led to Eq. (1.95),

$$\Pr\left\{\sum_i m_i\varepsilon_i = l, \sum_i n_i\varepsilon_i = l'\right\} = \frac{1}{4\pi^2}\int_{-\pi}^{\pi}\int_{-\pi}^{\pi} \exp\left\{-j\left(\omega l + \omega' l'\right)\right\} \times \underbrace{2\sum_{x=0}^{a}\phi\left(\frac{x}{a}\right)\cos\left(m_1\omega x + n_1\omega' x\right) \times \cdots \times 2\sum_{x=0}^{a}\phi\left(\frac{x}{a}\right)\cos\left(m_s\omega x + n_s\omega' x\right)}_{P} d\omega d\omega' \tag{1.213}$$

(*ibid.*, p. 404). Laplace next made a passage to the continuous limit through the following substitutions:

$$\frac{x}{a} = x', \quad \frac{1}{a} = dx',$$

$$K = 2\int_0^1 \phi(x')dx', \quad K'' = \int_0^1 x'^2\phi(x')dx', \quad K^{IV} = \int_0^1 x'^4\phi(x')dx', \ldots .$$

By doing series expansions of cosines, integrating, and then using $aK = 1$, he thus obtained

$$2\sum_{x=0}^{a}\phi\left(\frac{x}{a}\right)\cos\left(m_i\omega x + n_i\omega' x\right) = 1 - \frac{K''a^2}{K}\left(m_i\omega + n_i\omega'\right)^2 + \frac{K^{IV}}{12K}\left(m_i\omega + n_i\omega'\right)^4 + \cdots .$$

Next, by taking the logarithm of the right side of the last equation for each i ($i = 1, 2, \ldots, s$) and adding, Laplace was able to simplify $\log P$ (where P is defined in Eq. 1.213). By then exponentiating $\log P$, he wrote Eq. (1.213) as

$$\Pr\left\{\sum_i m_i\varepsilon_i = l, \sum_i n_i\varepsilon_i = l'\right\} = \frac{1}{4\pi^2}\int_{-\pi}^{\pi}\int_{-\pi}^{\pi}\left\{1 + \frac{KK^{IV} - 6K''^2}{12K^2}a^4\left(\omega^4\sum_i m_i^4 + \cdots\right)\right\}$$
$$\times \exp\left\{-j\omega l - j\omega' l' - \frac{K''}{K}a^2\left(\omega^2\sum_i m_i^2 + 2\omega\omega'\sum_i m_i n_i + \omega'^2\sum_i n_i^2\right)\right\} d\omega d\omega'.$$

The next steps were to make the transformations $\omega\sqrt{s} = t$ and $a\omega'\sqrt{s} = t'$ and integrate with respect to t and t', each from $-\infty$ to ∞. Laplace finally obtained the double integral of the bivariate normal density

$$\Pr\left\{\sum_i m_i\varepsilon_i = l, \sum_i n_i\varepsilon_i = l'\right\}$$
$$= \int_{-\infty}^{\infty}\int_{-\infty}^{\infty} \frac{K}{4K''\pi a^2}\frac{I}{\sqrt{E}} \exp\left\{-\frac{K\left(Fu^2 + 2Guu' + Hu'^2\right)}{4K''a^2E}\right\} du du' \tag{1.214}$$

(*ibid.*, p. 407),* where l and l' are given in Eq. (1.210), and

$$E=\sum_i m_i^2\sum_i n_i^2-\left(\sum_i m_in_i\right)^2,$$
$$F=\sum_i n_i^2\left(\sum_i m_ip_i\right)^2-2\sum_i m_in_i\sum_i m_ip_i\sum_i n_ip_i+\sum_i m_i^2\left(\sum_i n_ip_i\right)^2,$$
$$G=\sum_i n_i^2\sum_i m_ip_i\sum_i m_iq_i+\sum_i m_i^2\sum_i n_ip_i\sum_i n_iq_i-\sum_i m_in_i\left(\sum_i n_ip_i\sum_i m_iq_i+\sum_i m_ip_i\sum_i n_iq_i\right),$$
$$H=\sum_i n_i^2\left(\sum_i m_iq_i\right)^2-2\sum_i m_in_i\sum_i m_iq_i\sum_i n_iq_i+\sum_i m_i^2\left(\sum_i n_iq_i\right)^2,$$
$$I=\sum_i m_ip_i\sum_i n_iq_i-\sum_i n_ip_i\sum_i m_iq_i=\frac{FH-G^2}{E}.$$

Next, in Eq. (1.214), the integration with respect to u' is performed by first setting

$$t=\frac{\sqrt{\dfrac{KH}{4K''}}\left(u'+\dfrac{Gu}{H}\right)}{a\sqrt{E}}$$

and then integrating from $t=-\infty$ to $t=\infty$. The result is

$$\int_{-\infty}^{\infty}\sqrt{\frac{K}{4K''\pi}}\frac{I}{a\sqrt{H}}\exp\left(-\frac{KI^2u^2}{4K''a^2H}\right)du.$$

From the above, Laplace wrote the "the curve of probability of errors u," that is, the density of u, as

$$\sqrt{\frac{K}{4K''\pi}}\frac{I}{a\sqrt{H}}\exp\left(-\frac{KI^2u^2}{4K''a^2H}\right).$$

Taking the expectation of the above density, Laplace obtained the mean absolute error in the first element:

$$\pm\sqrt{\frac{K}{4K''\pi}}\frac{I}{a\sqrt{H}}\int_0^{\infty}u\exp\left(-\frac{KI^2u^2}{4K''a^2H}\right)du=\pm\sqrt{\frac{K''}{K\pi}}\frac{a\sqrt{H}}{I}$$

*The formula as given in the Oeuvres Completes 12 (Laplace, 1811a, p. 407) erroneously has an a, instead of an a^2, in the denominator of the exponential function.

(*ibid.*, p. 408). Similarly, the mean absolute error in the second element is

$$\pm\sqrt{\frac{K''}{K\pi}}\frac{a\sqrt{F}}{I}$$

(*ibidem*). For the mean absolute error in each of the elements to be a minimum, Laplace differentiated each of $\log(\sqrt{H}/I)$ and $\log(\sqrt{F}/I)$ with respect to m_i and n_i. The result is that the mean absolute errors are minimized when

$$m_i = \mu p_i, \quad n_i = \mu q_i$$

(*ibid.*, p. 409) where μ is a constant independent of i such that m_i and n_i are whole numbers. Using the above forms of m_i and n_i in Eq. (1.208), the corrections z and z' are

$$z = \frac{\sum_i q_i^2 \sum_i p_i\alpha_i - \sum_i p_i q_i \sum_i q_i\alpha_i}{\sum_i p_i^2 \sum_i q_i^2 - \left(\sum_i p_i q_i\right)^2},$$

$$z' = \frac{\sum_i p_i^2 \sum_i q_i\alpha_i - \sum_i p_i q_i \sum_i q_i\alpha_i}{\sum_i p_i^2 \sum_i q_i^2 - \left(\sum_i p_i q_i\right)^2}$$

(*ibidem*). From the above formulas, Laplace concluded:

> These corrections are those that are obtained by the method of least squares on the errors of observations, or the condition of the minimum of the function
>
> $$\sum_i (p_i z + q_i z - \alpha_i)^2.$$
>
> (*ibidem*)

1.5.6 Gauss' Second Justification of the Principle of Least Squares in 1823: The Gauss–Markov Theorem

In Section 1.5.3 we described how Gauss gave a first justification of the principle of least squares. We also mentioned that Gauss' derivation was circular because it was based on the postulate of the arithmetic mean. In Art. 20 of his book, *Theoria Combinationis Observationum Erroribus Minimis Obnoxia** (Gauss, 1823), he gave a more definite justification of the principle.[†] However, let us first describe the mathematical terminology used by Gauss.

* "Theory of the Combination of Observations Least Subject to Errors."

[†] Gauss' second justification of least squares is also described in Goldstine (1977, pp. 216–220), Farebrother (1999, pp. 134–140), Sprott (1978), and Chabert (1989).

Gauss considered a probability density $\phi(x)$ with mean zero and "mean error" m, where

$$m^2 = \int_{-\infty}^{\infty} x^2 \phi(x)\,dx$$

(Gauss, 1823, English edition, p. 11). It should be noted that when the mean is zero, then the mean error is the population standard deviation. For classes of observations, or quantities derived from them that have different precisions, Gauss defined their *relative weights* to be inversely proportional to m^2. The *precisions* of the observations are inversely proportional to m. Finally, to be able to use the weights numerically, one of the classes of observations is usually assigned a weight of unity.

Furthermore, suppose U is a function of the unknown quantities $V, V', V'', \ldots$. When the latter are observed, let the respective errors be $e, e', e'', \ldots$, each with mean zero. Then the error in U is

$$E = \frac{\partial U}{\partial V}e + \frac{\partial U'}{\partial V'}e' + \frac{\partial U''}{\partial V''}e'' + \cdots = \lambda e + \lambda' e' + \lambda'' e'' + \cdots$$

(*ibid.*, p. 32). The mean error in U is M, which is the square root of the mean of

$$\begin{aligned} E^2 &= \left(\lambda e + \lambda' e' + \lambda'' e'' + \cdots\right)^2 \\ &= \lambda^2 e^2 + \lambda'^2 e'^2 + \lambda''^2 e''^2 + \cdots + 2\lambda\lambda' ee' + 2\lambda\lambda'' ee'' + \cdots + 2\lambda'\lambda'' e'e'' + \cdots \end{aligned}$$

Since the mean of terms of the type $2\lambda\lambda' ee', 2\lambda\lambda'' ee'', \ldots$ are zero ($e, e', e'', \ldots$ being mutually independent, each with mean zero), the mean error of U becomes

$$M = \sqrt{\lambda^2 m^2 + \lambda'^2 m'^2 + \lambda''^2 m''^2 + \cdots},$$

where $m, m', m'', \ldots$ are the mean errors of $e, e', e'', \ldots$, respectively. The above relationship can also be written in term of weights, rather than mean errors. The weight of U is

$$P = \frac{1}{\dfrac{\lambda^2}{p} + \dfrac{\lambda'^2}{p'} + \dfrac{\lambda''^2}{p''} + \cdots} \tag{1.215}$$

(*ibid.*, p. 35), where $p, p', p'', \ldots$ are the weights of $e, e', e'', \ldots$, respectively.

Finally, Gauss assumed $V, V', V'', \ldots$ are π functions of the ρ unknowns $x, y, z, \ldots$, where $\rho \le \pi$. The unknowns $x, y, z, \ldots$ can be obtained by solving linear equations relating them to the observed values of $V, V', V'', \ldots$. Denoting these observed values by $L, L', L'', \ldots$, respectively, Gauss defined*

$$v = (V - L)\sqrt{p}, v' = (V' - L')\sqrt{p'}, v'' = (V'' - L'')\sqrt{p''}, \ldots \tag{1.216}$$

*Gauss' version of these equations incorrectly has $\sqrt{p}, \sqrt{p'}, \sqrt{p''}, \ldots$ in the denominator, rather than the numerator.

Thus, $\nu, \nu', \nu'', \ldots$ are the "standardized" observations. They have the same mean error $m\sqrt{p} = m'\sqrt{p'} = m''\sqrt{p''} = \cdots$ and the same weight of unity. Moreover, the equations to be solved become

$$v = 0, \quad v' = 0, \quad v'' = 0, \cdots. \tag{1.217}$$

We are now ready to explain Gauss' second justification of the principle of least squares. He first considered the following problem:

PROBLEM. *Let* v, v', v'', etc. *denote the following functions of the unknowns x, y, z, etc*:

$$\begin{aligned} v &= ax + by + cz + \text{etc.} + l \\ v' &= a'x + b'y + c'z + \text{etc.} + l' \\ v'' &= a''x + b''y + c''z + \text{etc.} + l'' \text{etc.} \end{aligned} \tag{1.218}$$

Of all systems of coefficients $\kappa, \kappa', \kappa''$, *etc., for which the general solution*

$$\kappa v + \kappa' v' + \kappa'' v'' + \text{etc.} = x - k \tag{1.219}$$

holds for some constant k independent of x, y, z, etc. find the one for which $\kappa\kappa + \kappa'\kappa' + \kappa''\kappa'' + \text{etc.}$ *is a minimum.** (*ibid*., p. 39)

To solve the above problem, Gauss first set

$$\left.\begin{aligned} av + a'v' + a''v'' + \cdots &= \xi \\ bv + b'v' + b''v'' + \cdots &= \eta \\ cv + c'v' + c''v'' + \cdots &= \zeta \\ &\cdots \end{aligned}\right\}. \tag{1.220}$$

By substituting Eq. (1.218) into Eq. (1.220),

$$\begin{aligned} \xi &= x\sum a^2 + y\sum ab + z\sum ac + \cdots + \sum al \\ \eta &= x\sum ab + y\sum b^2 + z\sum bc + \cdots + \sum bl \\ \zeta &= x\sum ac + y\sum bc + z\sum c^2 + \cdots + \sum cl, \\ &\cdots \end{aligned} \tag{1.221}$$

where $\sum a^2 = a^2 + a'^2 + \cdots$, $\sum ab = ab + a'b' + \cdots$, and so on. Noting that the number of unknowns in the above is equal to the number of equations, Gauss wrote the solution obtained by elimination as

$$\begin{aligned} x &= A + [\alpha\alpha]\xi + [\alpha\beta]\eta + [\alpha\gamma]\zeta + \cdots \\ &\cdots \end{aligned} \tag{1.222}$$

* Italics are Gauss'.

where the quantities $[\alpha\alpha]$, $[\alpha\beta]$, $[\alpha\gamma]$,* … are obtained later. The next step was to substitute (1.220) into the last equations and let

$$\begin{aligned} a[\alpha\alpha]+b[\alpha\beta]+c[\alpha\gamma]+\cdots&=\alpha, \\ a'[\alpha\alpha]+b'[\alpha\beta]+c'[\alpha\gamma]+\cdots&=\alpha', \\ a''[\alpha\alpha]+b''[\alpha\beta]+c''[\alpha\gamma]+\cdots&=\alpha'', \\ &\cdots. \end{aligned} \tag{1.223}$$

The equations in (1.222) then become

$$\begin{aligned} &x-A=\alpha v+\alpha' v'+\alpha'' v''+\cdots \\ &\cdots \end{aligned} \tag{1.224}$$

so that one set of coefficients in (1.219) is given by $\kappa=\alpha,\kappa'=\alpha',\kappa''=\alpha'',\cdots$ (*ibid.*, p. 41).

To show that $\kappa=\alpha,\kappa'=\alpha',\kappa''=\alpha'',\cdots$ minimizes the quantity $\kappa^2+\kappa'^2+\kappa''^2+\cdots$, Gauss used Eqs. (1.219) and (1.224) to write

$$(\kappa-\alpha)v+(\kappa'-\alpha')v'+(\kappa''-\alpha'')v''+\cdots=A-k.$$

By substituting the Eq. (1.218) in the above and equating coefficients of x, y, z, …,

$$\begin{aligned} &(\kappa-\alpha)a+(\kappa'-\alpha')a'+(\kappa''-\alpha'')a''+\cdots=0 \\ &(\kappa-\alpha)b+(\kappa'-\alpha')b'+(\kappa''-\alpha'')b''+\cdots=0 \\ &(\kappa-\alpha)c+(\kappa'-\alpha')c'+(\kappa''-\alpha'')c''+\cdots=0. \\ &\cdots \end{aligned}$$

By multiplying the above equations in turn by $[\alpha\alpha]$, $[\alpha\beta]$, $[\alpha\gamma]$, … and adding, and then using Eq. (1.223),

$$(\kappa-\alpha)\alpha+(\kappa'-\alpha')\alpha'+(\kappa''-\alpha'')\alpha''+\cdots=0.$$

This implies that

$$\begin{aligned} \kappa^2+\kappa'^2+\kappa''^2+\cdots&=(\kappa-\alpha+\alpha)^2+(\kappa'-\alpha'+\alpha')^2+(\kappa''-\alpha''+\alpha'')^2+\cdots \\ &=\alpha^2+\alpha'^2+\alpha''^2+\cdots+(\kappa-\alpha)^2+(\kappa'-\alpha')^2+(\kappa''-\alpha'')^2+\cdots- \\ &\quad 2(\kappa-\alpha)\alpha+2(\kappa'-\alpha')\alpha'+2(\kappa''-\alpha'')\alpha''+\cdots \\ &=\alpha^2+\alpha'^2+\alpha''^2+\cdots+(\kappa-\alpha)^2+(\kappa'-\alpha')^2+(\kappa''-\alpha'')^2+\cdots \end{aligned}$$

(*ibidem*). From the above, Gauss concluded that $\kappa^2+\kappa'^2+\kappa''^2+\cdots$ is a minimum when $\kappa=\alpha,\kappa'=\alpha',\kappa''=\alpha'',\cdots$, which is what he wished to prove.

*The reader may think of these quantities as P, Q, R, … so as not to be confused with what follows.

Now, how should $\alpha, \alpha', \alpha'', \ldots$, and hence the minimum, be determined? To do this, Eq. (1.218) can be substituted into (1.224) and the coefficient of x equated on both sides of the latter equation. This yields

$$\begin{aligned}\alpha a + \alpha' a' + \alpha'' a'' + \cdots &= 1\\ \alpha b + \alpha' b' + \alpha'' b'' + \cdots &= 0\\ \alpha c + \alpha' c' + \alpha'' c'' + \cdots &= 0\\ &\cdots,\end{aligned}$$

from which $\alpha, \alpha', \alpha'', \ldots$ can be determined. Moreover, by multiplying the above equations in turn by $[\alpha\alpha]$, $[\alpha\beta]$, $[\alpha\gamma]$, …, adding, and then using Eq. (1.223), Gauss obtained

$$\alpha^2 + \alpha'^2 + \alpha''^2 + \cdots = [\alpha\alpha]$$

(*ibid.*, p. 43). Now, from Eq. (1.217), we have $v = v' = v'' = \cdots = 0$. Therefore, from Eq. (1.224), we obtain

$$\alpha v + \alpha' v' + \alpha'' v'' + \cdots = 0 = x - A.$$

The above gives the estimate $x = A$ with weight which is given by Eq. (1.215) as

$$\frac{1}{\alpha^2/1 + \alpha'^2/1 + \alpha''^2/1 + \cdots} = \frac{1}{\alpha^2 + \alpha'^2 + \alpha''^2 + \cdots} = \frac{1}{[\alpha\alpha]}.$$

Since $[\alpha\alpha]$ is the minimum value of $\kappa^2 + \kappa'^2 + \kappa''^2 + \cdots$, the weight given above is a maximum.

Gauss next noted that the estimate $x = A$ could also have been obtained by substituting $\xi = 0, \eta = 0, \zeta = 0, \cdots$ in Eq. (1.222). By writing

$$\begin{aligned}\Omega &= v^2 + v'^2 + v''^2 + \cdots\\ &= (ax + by + cz + \cdots)^2 + (a'x + b'y + c'z + \cdots)^2 + (a''x + b''y + c''z + \cdots)^2 + \cdots\end{aligned}$$

one sees from (1.221) that

$$\frac{\partial\Omega}{\partial x} = 2\xi, \quad \frac{\partial\Omega}{\partial y} = 2\eta, \quad \frac{\partial\Omega}{\partial z} = 2\zeta, \cdots$$

(*ibid.*, p. 45). But $\xi = 0, \eta = 0, \zeta = 0, \ldots$ occur when Ω is a minimum. From Eq. (1.216), the sum of squares of the errors are then also minimized (assuming the errors all have the same weights of one). Hence, Gauss was able to show that the coefficients $\kappa = \alpha, \kappa' = \alpha', \kappa'' = \alpha'', \cdots$ in the equation $\kappa v + \kappa' v' + \kappa'' v'' + \cdots = x - k$ result in an estimated x with maximum weight (minimum variance), and these coefficients also result from the principle of least squares.

With the above demonstration, Gauss achieved an important milestone in so far as a statistically sound justification of the method of least squares was concerned. The simplicity of his method should be contrasted with the complexity of Laplace's similar attempts (see Sections 1.5.4 and 1.5.5).

Gauss also had a definite preference for the current justification of the least squares principles, compared to his 1809 proof (see Section 1.5.3.2), as he himself conceded to Friedrich Bessel (1784–1846) in a February 28, 1839, letter* (Fig. 1.33):

> That the metaphysic employed in my Theoria Motus Corporum Coelestium to justify the method of least squares has been subsequently allowed by me to drop (*Dass ich...habe fallen lassen*) has occurred chiefly for a reason that I have myself not mentioned publicly. The fact is, I cannot but think it in every way less important to ascertain that value of an unknown magnitude the probability of which is the greatest—which probability is nevertheless infinitely small—rather than that value by employing which we render the Expectation of detriment a minimum (*an welchen sich haltend man das am wenigsten nachteilige Spiel hat*). (Gauss and Bessel, 1880, p. 523)

Thus, Gauss viewed the minimization of error (sum of squares) as more fundamental than the maximization of the joint probability density. Another point which he did not mention is that Gauss' 1809 proof was based on inverse probability (his joint probability density was a posterior density) whereas his 1823 proof used direct probability.

Gauss' 1823 result is known as the *Gauss–Markov theorem* because Markov dealt with the same problem in his book *Ischislenie Veroiatnostei*† (Markov, 1900). Markov's own contribution was minimal but was highlighted by Neyman (1934) who thought that Markov's contribution had been overlooked. It therefore seems reasonable that the theorem should only be called Gauss' theorem, though the name "Gauss–Markov" has stuck. See also Plackett (1949) and Sheynin (1989).

1.5.7 Hagen's Hypothesis of Elementary Errors (1837)

Gotthilf Heinrich Ludwig Hagen (1797–1884) was a German physicist who studied mathematics, architecture, and civil engineering at the University of Konïgsberg. In 1837, he wrote a book on probability entitled *Grundzüge der Wahrscheinlichkeits-Rechnung*‡ (Hagen, 1837), where he put forward the hypothesis of elementary errors. According to this hypothesis:

> ...the error in the result of any measurement is the algebraic sum of an infinitely large number of elementary errors, which are all equally large, and of which each single one can be just as positive as negative. (*ibid.*, p. 34)

The hypothesis of elementary errors was to prove to be highly influential. Hagen first considered a total of $2n$ errors, each of which can take the values 1/2 and $-1/2$ with equal probability. Then the number of ways of obtaining a sum of zero is given by

$$\frac{(2n)(2n-1)(2n-2)(2n-3)\cdots(n+3)(n+2)(n+1)}{1\cdot 2\cdot 3\cdot 4\cdots(n-2)(n-1)n}$$

*The translation that follows is taken from Edgeworth's paper "On the probable errors of frequency-constants" (Edgeworth, 1908, p. 386).

† "Calculus of Probabilities."

‡ "Fundamentals of the Calculus of Probability."

523

mehr, da ich fürchte, dass man ihn in Berlin, aus gewöhnlichen Gründen, zu unterdrücken sucht. Er ist gewiss ein ausgezeichneter Mann, dessen auf einander folgende Arbeiten auch ein schnelles Wachsen seiner Reife darthun. Leider halten ihn die Geschäfte, welche die Bekanntmachung der Früchte seiner Reise mit sich führt, noch gänzlich von neuen Arbeiten zurück.

Nr. 176. **Gauss an Bessel.** [67

Göttingen 28. Februar 1839.

Hochverehrter Freund.

Ich habe Ihnen noch meinen verbindlichsten Dank abzustatten für das gütige Geschenk, welches Sie mir mit dem Werke über Ihre Gradmessung und mit dem neuesten Bande der Königsberger Beobachtungen gemacht haben.

Ihren Aufsatz in den Astronomischen Nachrichten über die Annäherung des Gesetzes für die Wahrscheinlichkeit aus zusammengesetzten Quellen entspringender Beobachtungsfehler an die Formel $e^{-\frac{xx}{hh}}$ habe ich mit grossem Interesse gelesen; doch bezog sich, wenn ich aufrichtig sprechen soll, diess Interesse weniger auf die Sache selbst als auf Ihre Darstellung. Denn jene ist mir seit vielen Jahren familiär, während ich selbst niemals dazu gekommen bin, die Entwickelung vollständig auszuführen.

Dass ich übrigens die in der Theoria Motus Corporum Coelestium angewandte Metaphysik für die Methode der kleinsten Quadrate späterhin habe fallen lassen, ist vorzugsweise auch aus einem Grunde geschehen, den ich selbst öffentlich nicht erwähnt habe. Ich muss es nämlich in alle Wege für weniger wichtig halten, denjenigen Werth einer unbekannten Grösse auszumitteln, dessen Wahrscheinlichkeit die grösste ist, die ja doch immer nur unendlich klein bleibt, als vielmehr denjenigen, an welchen sich haltend man das am wenigsten nachtheilige Spiel hat; oder wenn fa die Wahrscheinlichkeit des Werthes a für die Unbekannte x bezeichnet, so ist weniger daran gelegen, dass fa ein Maximum werde, als daran, dass $\int fx - F(x-a) \cdot dx$ ausgedehnt durch alle möglichen Werthe des x ein Minimum werde,

FIGURE 1.33 Gauss' February 28, 1839, letter to Bessel (Gauss and Bessel, 1880)

(*ibid.*, p. 42), which one can recognize as the binomial coefficient $\binom{2n}{n}$. Denote the previous expression by y', and let the number of ways of obtaining a sum of x be y. Then the number of ways of obtaining a sum of one is $\binom{2n}{n-1}$, so Hagen wrote

$$\text{for} \quad x=1, \quad y=y'\frac{n}{n+1}.$$

Similarly,

$$\text{for} \quad x=2, \quad y=y'\frac{n}{n+1}\cdot\frac{n-1}{n+2},$$

$$\text{for} \quad x=3, \quad y=y'\frac{n}{n+1}\cdot\frac{n-1}{n+2}\cdot\frac{n-2}{n+3}.$$

Hence,

$$\text{for} \quad x=m-1, \quad y=y'\frac{n}{n+1}\cdot\frac{n-1}{n+2}\cdots\frac{n-m}{n+m-1}=Y,$$

$$\text{for} \quad x=m, \quad y=y'\frac{n}{n+1}\cdot\frac{n-1}{n+2}\cdots\frac{n-m}{n+m-1}\cdot\frac{n-m-1}{n+m}=Y'.$$

Therefore,

$$Y'=\frac{n-m-1}{n+m}Y \quad \Rightarrow \quad Y'-Y=-Y\frac{2m-1}{n+m}$$

Hagen now let n grow infinitely large so that, with $m=x$, the above becomes

$$\frac{dy}{dx}=-y\frac{2x}{n}$$

(*ibid.*, p. 44). This differential equation has solution

$$\log y=-\frac{1}{n}x^2+C,$$

where C is a constant. At $x=0$, $y=y'$ so $C=\log y'$, and hence

$$y=y'e^{-x^2/n}$$

(*ibidem*). Since each of the two errors is equally likely, the number of ways of obtaining a particular sum is proportional to its corresponding probability. Hence, Hagen had effectively obtained the normal law for the sum of a large number of elementary errors, albeit in a nonrigorous way.

PART TWO: FROM GALTON TO FISHER

2

GALTON, REGRESSION, AND CORRELATION

2.1 FRANCIS GALTON (1822–1911)

Francis Galton (Fig. 2.1) was a gentleman scholar and the father of the eugenics movement. Indeed, questions in heredity were primarily responsible for much of his work in statistics. Galton* was born in Birmingham, England, on February 16, 1822, and was a child prodigy. He came from an intellectual family. His mother was the aunt of Charles Darwin (the latter was Galton's first cousin) and his grandfather was a fellow of the Royal Society. Galton's father ran a bank. Wishing to pursue a medical career, Galton began his medical training at the Birmingham General Hospital in 1838 and continued it at King's College, London, in 1839. He stopped his training to study mathematics at Trinity College the next year but got only a pass degree. When his father died in 1844, Galton was left a fortune and spent the next 5 years in amusements. During the period 1850–1852, he traveled extensively in the territory of the Ottoman Empire and documented his trips. Due to his explorations, Galton was awarded a Gold Medal from the Geographical Society and then became a fellow of the Royal Society in 1860. Some of his early contributions were in the field of meteorology, where he first recognized and named the anticyclone. From the age 43, genetics and statistics became his primary focus. In his book *Hereditary Genius* (Galton, 1869), Galton argued that descendants of eminent individuals were more likely to be eminent than others but to a lesser degree than their ancestors. This later gave rise to the concept of regression to the mean. Galton's regression ideas were subsequently written in his major book *Natural Inheritance* (Galton, 1889). Galton strongly believed in the inheritance of talent and in the development of programs to foster talent and eliminate illness and sickliness. Galton coined such programs as "eugenics" in 1883. Although Galton was a staunch believer in Darwin's theory

* The definitive biography of Galton can be found in Karl Pearson's 4-volume *The Life, Letters and Labours of Francis Galton* (Pearson, 1930). Two other excellent biographies are Bulmer (2003) and Gridgeman (1972b).

Classic Topics on the History of Modern Mathematical Statistics: From Laplace to More Recent Times, First Edition. Prakash Gorroochurn.

FIGURE 2.1 Francis Galton (1822–1911). Wikimedia Commons (Public Domain), http://commons.wikimedia.org/wiki/File:Francis_Galton_1850s.jpg

of evolution by natural selection, he had disagreements with his cousin on the source of variation upon which natural selection acted. Unlike Darwin, Galton believed in discontinuous variations (or sports).* Galton's other noteworthy contribution includes his establishment of fingerprinting as a tool for human identification. Galton was knighted in 1909 but never held any academic or professional position. He passed away in Haslemere, England, on January 17, 1911.

A distinctive characteristic of Galton's scientific approach was his predilection for quantitative methods. Although his mathematics was somewhat limited, Galton was a statistical pioneer in his own right, being responsible for such terms as "quartile," "decile," "percentile," "ogive," "regression," and "correlation."

Much of Galton's statistical work took place in the context of his continued investigation of the normal distribution (at that time known as *the law of error*). Galton had been introduced to this distribution though his friend William Spottiswoode, who was a geologist. Once he started learning about the normal curve, Galton was greatly inspired by the Belgian mathematician Adolphe Quetelet (1796–1874) who had pioneered the use of the curve in the social sciences. In particular, using data from French and Scottish soldiers, Quetelet had shown that many physical measurements were distributed according to the normal curve. Galton's work on the normal curve, however, broke away from the Gaussian tradition in which error was a quantity that needed to be either eliminated or at least minimized. For Galton, error (or, rather, deviation from the mean, as he called it) was an important quantity that could be used to categorize subjects, depending on the amount of deviation.

In one of Galton's first major works, *Heredity Genius* (Galton, 1869), Galton made much use of the normal curve in the context of mental heredity. Following Quetelet, he argued that if data could be proved to follow a normal curve, then it could be inferred that the data came from a homogeneous population with respect to the attribute being measured. In his own words:

> The law may, therefore, be used as a most trustworthy criterion, whether or not the events of which an average has been taken, are due to the same or to dissimilar classes of conditions. (*ibid.*, p. 29)

Even more importantly, Galton argued that, in analogy with physical measurements, mental characteristics would also likely be normally distributed. He used tables of the cumulative normal distribution to classify and rank men into various categories of mental ability, measured in probable errors† above or below the mean. In several paragraphs that are very indicative of the prevailing prejudices of the time, Galton stated:

> Let us, then, compare the negro race with the Anglo-Saxon, with respect to those qualities alone which are capable of producing judges, statesmen, commanders, men of literature and science, poets, artists, and divines…

*The disagreement between Galton and Darwin, and Galton's views on evolution are described in Provine (2001, pp. 11–24).

†The probable error of a random variable X is simply its semi-interquartile range, that is, for a symmetric distribution the value e_p such that $\Pr\{|X-m|<e_p\}=.5$, where m is the mean. At one time, it was a popular measure of variability but was replaced by the standard deviation (SD), a term coined by Pearson (1894). It can be shown that when X is normally distributed, $e_p=.6745$ SD. For more details, see Cowles and Davis (1982) and David (1995).

> First, the negro race has occasionally, but very rarely, produced such men as Toussaint l'Ouverture,* who are of our class F [well above average]...
>
> Secondly, the negro race is by no means wholly deficient in men capable of becoming good factors, thriving merchants, and otherwise considerably raised above the average of white...the average intellectual standard of the negro race is some two grades below our own. (*ibid.*, p. 338)

In his 1875 paper, "Statistics by intercomparison, with remarks on the law of frequency of error" (Galton, 1875), Galton further developed the above ideas but then made a highly debatable extrapolation, which we now discuss. Galton first made use of his newly discovered "ogive" (or cumulative frequency distribution) to graph some easily measurable normally distributed characteristic (e.g., height). But then he argued that such an ogive could be used to rank individuals with respect to more elusive characteristics such as intelligence by using the ogive in reverse fashion:

> There is another method, which I have already advocated and adopted, for gaining an insight into the absolute efficacies of qualities, on which there remains more to say. Whenever we have grounds for believing the law of frequency of error to apply, we may *work backwards*, and, from the relative frequency of occurrence of various magnitudes, derive a knowledge of the true relative values of those magnitudes, expressed in units of probable error. The law of frequency of error says that "magnitudes differing from the mean value by such and such multiples of the probable error, will occur with such and such degrees of frequency." My proposal is to reverse the process and to say, "since such and such magnitudes occur with such and such degrees of frequency, therefore the differences between them and the mean value are so and so, as expressed in units of probable error." (*ibid.*, pp. 37–38)

Galton's argument can be explained through an example, as follows. Suppose the height of a large sample of individuals from a population is measured and found to fit a normal curve. Then an ogive of these data is constructed with the horizontal axis scaled in terms of probable errors from the mean (or median) and the vertical axis in terms of percentiles. Now, suppose we wish to "measure the intelligence" of another individual from the same population. Then by measuring the height of the individual and using the ogive for height, we can obtain the percentile for the individual. From this percentile, we can use the ogive backward and claim that the intelligence of the individual is at a certain number of probable errors from the mean. Of course, Galton's argument is erroneous in general and would hold only in the case of a strong correlation between the two attributes.

2.2 GENESIS OF REGRESSION AND CORRELATION

2.2.1 Galton's 1877 Paper, "Typical Laws of Heredity": Reversion

Regression, as we know it today, was born from Galton's investigations of the laws of heredity.† His groundbreaking 1877 paper, "Typical laws of heredity" (Galton, 1877) (Fig. 2.2), dealt with a problem that had preoccupied Galton for a while, indeed since

*Toussaint Louverture (1743–1803) was a (black) military genius and a man of great political acumen who led the Haitian Revolution and the establishment of the independent state of Haiti.

†A general description of Galton's work is also provided by Cowan (1972), Stigler (1997; 2010), Porter (1986, pp. 270–296), and Denis (2001).

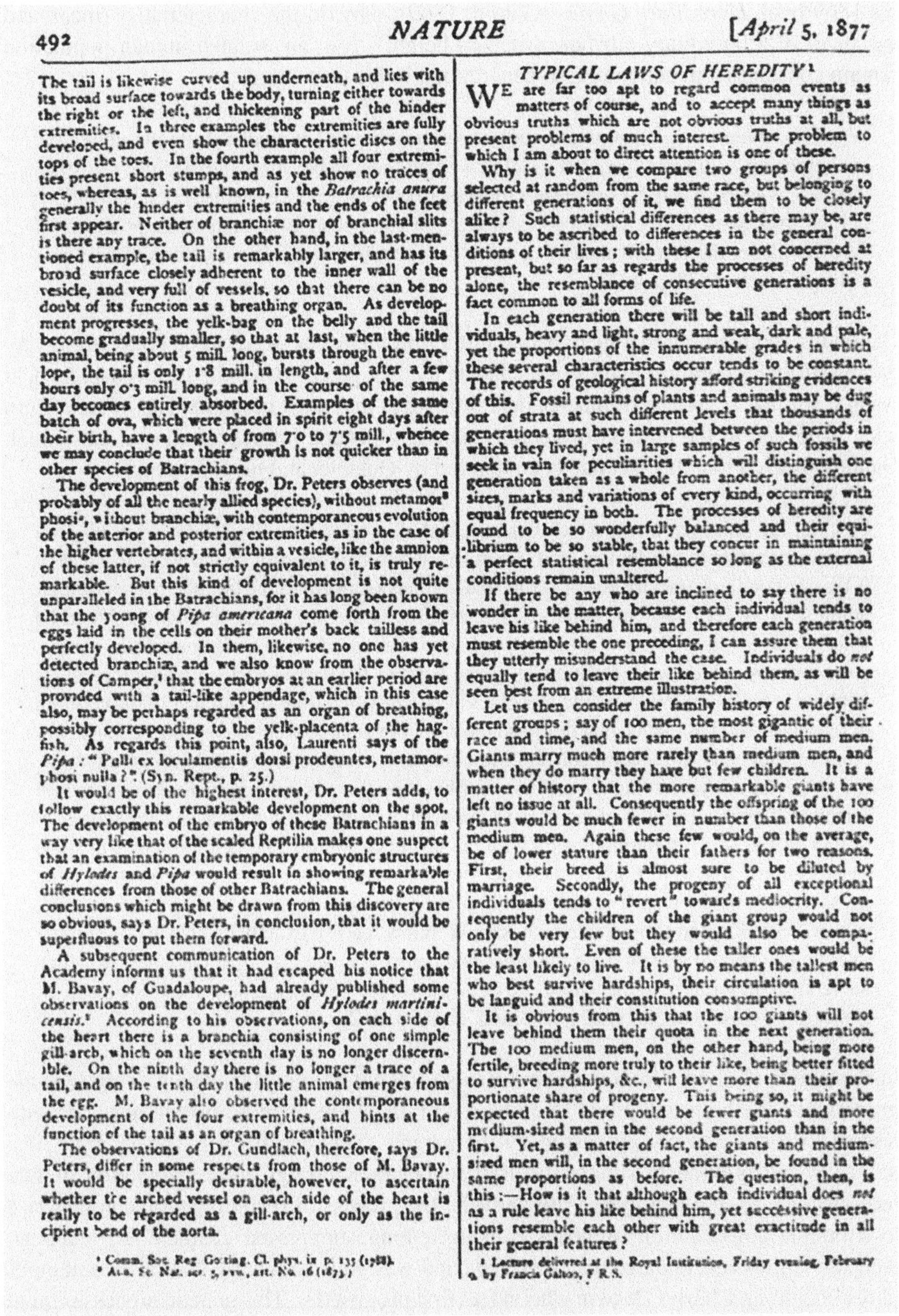

492 NATURE [April 5, 1877

The tail is likewise curved up underneath, and lies with its broad surface towards the body, turning either towards the right or the left, and thickening part of the hinder extremities. In three examples the extremities are fully developed, and even show the characteristic discs on the tops of the toes. In the fourth example all four extremities present short stumps, and as yet show no traces of toes, whereas, as is well known, in the *Batrachia anura* generally the hinder extremities and the ends of the feet first appear. Neither of branchiæ nor of branchial slits is there any trace. On the other hand, in the last-mentioned example, the tail is remarkably larger, and has its broad surface closely adherent to the inner wall of the vesicle, and very full of vessels, so that there can be no doubt of its function as a breathing organ. As development progresses, the yelk-bag on the belly and the tail become gradually smaller, so that at last, when the little animal, being about 5 mill. long, bursts through the envelope, the tail is only 1·8 mill. in length, and after a few hours only 0·3 mill. long, and in the course of the same day becomes entirely absorbed. Examples of the same batch of ova, which were placed in spirit eight days after their birth, have a length of from 7·0 to 7·5 mill., whence we may conclude that their growth is not quicker than in other species of Batrachians.

The development of this frog, Dr. Peters observes (and probably of all the nearly allied species), without metamorphosis, without branchiæ, with contemporaneous evolution of the anterior and posterior extremities, as in the case of the higher vertebrates, and within a vesicle, like the amnion of these latter, if not strictly equivalent to it, is truly remarkable. But this kind of development is not quite unparalleled in the Batrachians, for it has long been known that the young of *Pipa americana* come forth from the eggs laid in the cells on their mother's back tailless and perfectly developed. In them, likewise, no one has yet detected branchiæ, and we also know from the observations of Camper,[1] that the embryos at an earlier period are provided with a tail-like appendage, which in this case also, may be perhaps regarded as an organ of breathing, possibly corresponding to the yelk-placenta of the hag-fish. As regards this point, also, Laurenti says of the *Pipa*: "Pulli ex loculamentis dorsi prodeuntes, metamorphosi nulla?" (Syn. Rept., p. 25.)

It would be of the highest interest, Dr. Peters adds, to follow exactly this remarkable development on the spot. The development of the embryo of these Batrachians in a way very like that of the scaled Reptilia makes one suspect that an examination of the temporary embryonic structures of *Hylodes* and *Pipa* would result in showing remarkable differences from those of other Batrachians. The general conclusions which might be drawn from this discovery are so obvious, says Dr. Peters, in conclusion, that it would be superfluous to put them forward.

A subsequent communication of Dr. Peters to the Academy informs us that it had escaped his notice that M. Bavay, of Guadaloupe, had already published some observations on the development of *Hylodes martinicensis*.[2] According to his observations, on each side of the heart there is a branchia consisting of one simple gill-arch, which on the seventh day is no longer discernible. On the ninth day there is no longer a trace of a tail, and on the tenth day the little animal emerges from the egg. M. Bavay also observed the contemporaneous development of the four extremities, and hints at the function of the tail as an organ of breathing.

The observations of Dr. Gundlach, therefore, says Dr. Peters, differ in some respects from those of M. Bavay. It would be specially desirable, however, to ascertain whether the arched vessel on each side of the heart is really to be regarded as a gill-arch, or only as the incipient bend of the aorta.

[1] Comm. Soc. Reg. Goting. Cl. phys. ix p. 135 (1788).
[2] Ann. Sc. Nat. ser. 5, xvii, art. No. 16 (1873).

TYPICAL LAWS OF HEREDITY[1]

WE are far too apt to regard common events as matters of course, and to accept many things as obvious truths which are not obvious truths at all, but present problems of much interest. The problem to which I am about to direct attention is one of these.

Why is it when we compare two groups of persons selected at random from the same race, but belonging to different generations of it, we find them to be closely alike? Such statistical differences as there may be, are always to be ascribed to differences in the general conditions of their lives; with these I am not concerned at present, but so far as regards the processes of heredity alone, the resemblance of consecutive generations is a fact common to all forms of life.

In each generation there will be tall and short individuals, heavy and light, strong and weak, dark and pale, yet the proportions of the innumerable grades in which these several characteristics occur tends to be constant. The records of geological history afford striking evidences of this. Fossil remains of plants and animals may be dug out of strata at such different levels that thousands of generations must have intervened between the periods in which they lived, yet in large samples of such fossils we seek in vain for peculiarities which will distinguish one generation taken as a whole from another, the different sizes, marks and variations of every kind, occurring with equal frequency in both. The processes of heredity are found to be so wonderfully balanced and their equilibrium to be so stable, that they concur in maintaining a perfect statistical resemblance so long as the external conditions remain unaltered.

If there be any who are inclined to say there is no wonder in the matter, because each individual tends to leave his like behind him, and therefore each generation must resemble the one preceding, I can assure them that they utterly misunderstand the case. Individuals do *not* equally tend to leave their like behind them, as will be seen best from an extreme illustration.

Let us then consider the family history of widely different groups; say of 100 men, the most gigantic of their race and time, and the same number of medium men. Giants marry much more rarely than medium men, and when they do marry they have but few children. It is a matter of history that the more remarkable giants have left no issue at all. Consequently the offspring of the 100 giants would be much fewer in number than those of the medium men. Again these few would, on the average, be of lower stature than their fathers for two reasons. First, their breed is almost sure to be diluted by marriage. Secondly, the progeny of all exceptional individuals tends to "revert" towards mediocrity. Consequently the children of the giant group would not only be very few but they would also be comparatively short. Even of these the taller ones would be the least likely to live. It is by no means the tallest men who best survive hardships, their circulation is apt to be languid and their constitution consumptive.

It is obvious from this that the 100 giants will not leave behind them their quota in the next generation. The 100 medium men, on the other hand, being more fertile, breeding more truly to their like, being better fitted to survive hardships, &c., will leave more than their proportionate share of progeny. This being so, it might be expected that there would be fewer giants and more medium-sized men in the second generation than in the first. Yet, as a matter of fact, the giants and medium-sized men will, in the second generation, be found in the same proportions as before. The question, then, is this:—How is it that although each individual does *not* as a rule leave his like behind him, yet successive generations resemble each other with great exactitude in all their general features?

[1] Lecture delivered at the Royal Institution, Friday evening, February 9, by Francis Galton, F.R.S.

FIGURE 2.2 First page of Galton's "Typical laws of heredity" (Galton, 1877)

his 1869 book *Hereditary Genius* (Galton, 1869): why do the characteristics (mean and variance) of a hereditary attribute (such as height) from an isolated human population remain constant from generation to generation? In his own words:

> Why is it, when we compare two groups of persons selected at random from the same race, but belonging to different generations of it, we find them to be closely alike? Such statistical differences as there may be, are always to be ascribed to differences in the general conditions of their lives; with these I am not concerned at present; but so far as regards the processes of heredity alone, the resemblance of consecutive generations is a fact common to all forms of life. (Galton, 1877, p. 492)

Galton also pointed out that there is one further problem that needs to be considered: why did the attributes conform to the normal law in the first place? After all, the law usually held when it was the aggregate of independent influences that must be *small*. This was far from what happened in the process of heredity whereby genetic influences can be substantial. Galton observed that this fact had been ignored by Quetelet and his followers:

> First, let me point out a fact which Quetelet and all writers who have followed in his path have unaccountably overlooked, and which has an intimate bearing on our work to-night. It is that, although characteristics of plants and animals conform to the law, the reason of their doing so is as yet totally unexplained. The essence of the law is the differences should be wholly due to the collective actions of a host of independent *petty* influences in various combinations, which were represented by the teeth of the harrow, among which the pellets tumbled in various ways. Now the processes of heredity that limit the number of the children of one class, such as giants, that diminish their resemblance to their fathers, and kill many of them, are not petty influences, but very important ones. Any selective tendency is ruin to the law of deviation, yet among the processes of heredity there is the large influence of natural selection. The conclusion is of the greatest importance to our problem. It is, that the processes of heredity must work harmoniously with the law of deviation, and be themselves in some sense conformable to it. Each of the processes must show this conformity separately, quite irrespectively of the rest. It is not an admissible hypothesis that any two or more of them, such as reversion and natural selection, should follow laws so exactly inverse to one another that the one should reform what the other had deformed; because characteristics, in which the relative importance of the various processes is very different, are none the less capable of conforming closely to the typical condition. (*ibid*., p. 512)

In the above, Galton invoked *reversion*, also known as atavism, which is the *genetic process* by which an individual resembled a grandparent or more distant ancestor with respect to some trait not possessed by the parents. This can happen, for example, if a recessive and previously suppressed trait reappeared through the combination of two recessive alleles in a genotype. Alternatively, the process of recombination can give rise to a unique constellation of genes causing a long suppressed character to reappear. Atavism is thus *reversion to ancestral type* and was well known by Galton's contemporaries, including Charles Darwin who in fact first proposed it. This genetic process is quite different from the purely statistical phenomenon that Galton soon discovered and at first identified with reversion.

Continuing with the discussion of the 1877 paper, Galton explained that he resorted to experiments with sweet peas in order to answer his questions. He sorted a large number of sweet pea seeds into seven equally spaced size (weight) classes and sent each of his friends seven packets, each containing 10 seeds of a given class size. The seeds from the offspring were then collected and sent back to Galton. Before discussing his results, Galton made the following clarification:

> The only processes concerned in simple descent that can affect the characteristics of a sample of a population are those of Family Variability and Reversion. It is well to define these words clearly. By family variability is meant the departure of the children of the same or similarly descended families, from the ideal mean type of all of them. Reversion is the tendency of that ideal mean filial type to depart from the parent type, "reverting" towards what may be roughly and perhaps fairly described as the average ancestral type. If family variability had been the only process in simple descent that affected the characteristics of sample, the dispersion of the race from its mean ideal type would indefinitely increase with the number of the generations; but reversion checks this increase, and brings it to a standstill, under conditions which will now be explained. (*ibid.*, p. 513)

In the above, Galton asserted that the filial generation is the product of two opposing forces, namely, family variability and reversion. Family variability is kept in check (i.e., remains constant) due to reversion.

However, Galton seems to use reversion in the sense of the genetic process described previously. This is similar to his earlier statement of 1865:

> Lastly, though the talent and character of both of the parents might, in any particular case, be of a remarkably noble order, and thoroughly congenial, yet they would necessarily have such mongrel antecedents that it would be absurd to expect their children to invariably equal them in their natural endowments. The law of atavism prevents it. (Galton, 1865, p. 319, Part II)

In the 1877 paper, Galton next considered his sweet pea results and made the following two key conclusions:

> On weighing and sorting large samples of the produce of each of the seven different classes of the peas, I found in every case the law of deviation to prevail, and in every case the value of 1° of deviation to be the same...The next great fact was that reversion followed the simplest possible law; the proportion being constant between the deviation of the mean weight of the produce generally and the deviation of the parent seed, reckoning in every case from one standard point. (Galton, 1877, p. 513)

The last two observations, respectively, mean that:

1. For a given parental class size, the size of the filial seeds was normally distributed, with the same probable error e_p within each class (i.e., the family variability within in each class was the same).
2. Reversion was a simple linear function of the deviation from the grand mean (M). Thus, if the parent size in a given class has mean $M + ke_p (k = 0, \pm 1, \pm 2, \pm 3)$, then the corresponding filial size will have mean $M + k\rho e_p$, where $0 < \rho < 1$.

In an appendix to the 1877 paper, Galton gave the algebraic conditions necessary for stability in population variability in terms of the modulus c of a distribution.* Galton first

* The modulus of a normal distribution was historically used to represent $\sigma\sqrt{2}$.

considered the case of two parents each of whom could be productive but later turned to the case of simple descent. It is the latter case we shall describe here. First, Galton wrote the distribution of the "amount of deviation" x in a present population as

$$y = \frac{1}{c\sqrt{\pi}} \exp\left(-\frac{x^2}{c^2}\right).$$

Second, "reversion is expressed by a simple fractional coefficient of the deviation," which we shall denote by ρ* $(0 < \rho < 1)$. Then, in the "reverted parentages,"

$$y = \frac{1}{\rho c\sqrt{\pi}} \exp\left(-\frac{x^2}{\rho^2 c^2}\right).$$

Galton then denoted the modulus of the present population by c_1 and that of the "reverted parentages" by c_2, so that

$$c_2 = \rho c_1. \tag{2.1}$$

The next step for Galton was to consider the variation of the number of progeny for a given parental class, that is, the family variability:

> Family variability was shown by experiment to follow the law of deviation, its modulus, which we will write v, being the same for all classes. Therefore the amount of deviation of anyone of the offspring from the mean of his race is due to the combination of two influences—the deviation of his "reverted" parentage and his own family variability; both of which follow the law of deviation. This is obviously an instance of the well-known law of the "sum of two fallible measures."† (*ibid*., p. 533)

Denoting the modulus of the family variability by v, Galton wrote the above law as

$$c_4^2 = c_2^2 + v^2, \tag{2.2}$$

where c_4 is the overall modulus in the progeny. Combining Eqs. (2.1) and (2.2),

$$c_4^2 = \rho^2 c_1^2 + v^2.$$

Now, for variability to remain constant across generations, we need $c_4 = c_1$, which implies

$$c_1^2 = \frac{v^2}{1-\rho^2} \tag{2.3}$$

(*ibidem*). It should be noted that Eq. (2.3) expresses the fact that the variability across generations remains constant due to the balancing forces of family variability (which tends to increase spread) and reversion (which tends to reduce spread).

A modern version of Galton's analysis can be written as follows. Let X and Y be the present and parental population class sizes such that (X, Y) is bivariate normal. Then each

*Galton himself used the symbol r, but we shall reserve the latter for the *sample* correlation coefficient.

†Airy, "Theory of Errors," Section 43 (*Galton's footnote*).

of X and Y is normally distributed. Let their variances be σ_X^2, σ_Y^2, respectively. The variance of the present population generation for a given parental class is $\sigma_{X|Y}^2 = \text{Var}(Y \mid X)$, where

$$\sigma_{X|Y}^2 = \sigma_X^2 \left(1 - \rho^2\right)^*$$

and $\rho = \text{Corr}(X,Y)$. Stability in population variance means that $\sigma_X^2 = \sigma_Y^2$, so that we finally obtain

$$\sigma_{X|Y}^2 = \sigma_Y^2 \left(1 - \rho^2\right).$$

In sum, we can view Galton's 1877 paper as groundbreaking but still not fully satisfactory. The paper provided no justification for why reversion should be proportional to the deviation from the grand mean. More importantly, Galton's use of reversion to signify "the tendency of that ideal mean filial type to depart from the parent type, reverting towards what may be roughly and perhaps fairly described as the average ancestral type" showed that, at this stage, he probably thought of the process as being both *unidirectional* and *genetic*. To his amazement, the contrary turned out to be true, and this was to be shown in his 1885 presidential address to the *Anthropology Section of the British Association* (Galton, 1885a). This is also where the term "regression" was first used by Galton. We shall soon turn our attention to this lecture.

2.2.2 Galton's Quincunx (1873)

Galton's ideas were greatly influenced by the ingenious quincunx[†] that he had built in 1873 (e.g., see Stigler, 1986b; 1989; 1997) (cf. Fig. 2.3, left).

The quincunx has a funnel at the top, a succession of rows of pins below it, and a series of vertical compartments at the bottom. Lead shot are dropped from the top and scamper

> ... deviously down through the pins in a curious and interesting way; each of them darting a step to the right or left, as the case may be, every time it strikes a pin. The pins are disposed in a quincunx fashion, so that every descending shot strikes against a pin in each successive row ... and, at length, every shot finds itself caught in a compartment immediately after freeing itself from the last row of pins. The outline of the columns of shot that accumulate in the successive compartments approximates to the [Normal] Curve of Frequency, and is closely of the same shape however often the experiment is repeated. The outline of the columns would become more nearly identical with the Normal Curve of Frequency, if the rows of pins were much more numerous, the shot smaller, and the compartments narrower; also if a larger quantity of shot were used. (Galton, 1889, p. 64)

Galton's original quincunx thus approximated a symmetric (equal probability of each lead shot being deflected to the right or left) binomial distribution with a large number of

* See Section 2.2.5.1 for why this is true.

[†] The quincunx can be seen today at the Galton Laboratory at University College London. The name "quincunx" follows from the pattern in the construction consisting of *five* points arranged in a cross.

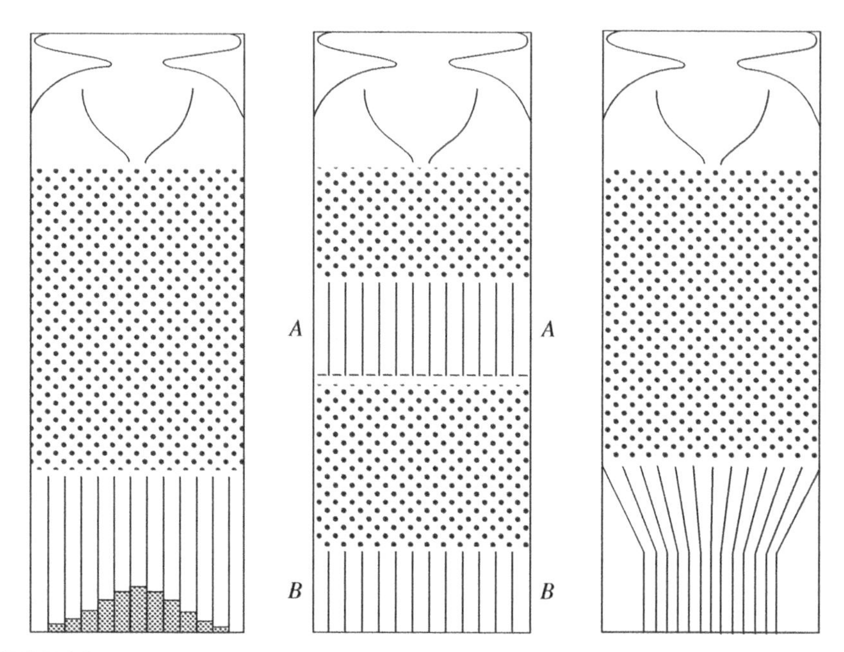

FIGURE 2.3 The quincunx (left), the double quincunx (middle), and the convergent quincunx (right) (From Galton (1889)). Only the quincunx was actually constructed

trials (the number of rows of pins) by a normal distribution. Later in 1877, Galton proposed a second version of the quincunx, the double quincunx, which had two sets of compartments, AA and BB (cf. Fig. 2.3, middle). Here, the AA compartments are all closed when the lead shot are thrown from the top. Then, one of the compartments is opened, so that its lead shot "cascade downwards and disperse themselves among the BB compartments on either side of the perpendicular lines from its starting point." Then all of the AA compartments are opened in turn and the final results "must be to reproduce the identically same system in the BB compartments" that was obtained before. The double quincunx thus illustrated that a normal mixture of normal distributions was itself normal. To Galton, this illustrated how a normal distribution (the distribution of the size of all offspring pea seeds) could be aggregated from a set of smaller normal distributions (the distribution of the size of offspring pea seeds for a given parent class size). However, it was Galton's third version of the quincunx, the convergent quincunx (cf. Fig. 2.3, right), that was most crucial in fixing his ideas on the stability of population variability. One issue with the double quincunx was that there the variability at the bottom level was higher than that at the top level, which was contrary to Galton's own observations of stability of variability in natural populations. Therefore, in his convergent quincunx, Galton had included inclined chutes at the bottom. As the lead shot fell from the top, there was an increase in distribution variability mimicking the effect of parental variability. Now, as they passed through the inclined chutes, the lead shot still produced a normal distribution but the latter's variability was reduced so as to equal the original variability. The inclined chutes thus mimicked the effect of reversion.

2.2.3 Galton's 1885 Presidential Lecture and Subsequent Related Papers: Regression, Discovery of the Bivariate Normal Surface

In 1885, Galton was to make another major breakthrough following his initial 1877 paper on reversion. In his 1885 Presidential Lecture to the Anthropology Section of the British Association (Galton, 1885a) (Fig. 2.4), Galton recalled his initial observation of the regression effect (which he had called reversion at that time):

> It appeared from these experiments that the offspring did *not* tend to resemble their parent seeds in size, but to be always more mediocre than they were—to be smaller than the parents, if the parents were large; to be larger than the parents, if the parents were small. (*ibid*., p. 507)

Galton also admitted:

> ...I was then [in 1877] blind to what I now perceive to be the simple explanation of the phenomenon.... (*ibidem*)

The presidential lecture was a major breakthrough because Galton was able to not only show that reversion (which he now called *regression*) was symmetric in nature but also to give the correct mechanism for it.* We now consider these topics.

Galton explained that he had taken pains to collect the heights of parents and adult children from 205 families. He then multiplied the height of each female parent by 1.08 so as to make the male and female parents comparable and "no objection grounded on the sexual difference of stature need be raised." He then took the average for each pair of parents to obtain a "midparent" height. The data are shown in Figure 2.5, which appeared in the accompanying paper "Regression towards mediocrity in hereditary stature" (Galton, 1885b). From these data, Galton calculated the median for both midparents (X) and offspring (Y) as 68.25 in. and the probable deviations as $e_{\mathrm{p},X} = 1.2$ and $e_{\mathrm{p},Y} = 1.7$, respectively. Next, for each row in Figure 2.5, Galton calculated the median height of the adult children and plotted the latter median heights against the midparental heights. The line is fitted by inspection and Galton obtained a slope of 2/3. Galton was now ready to define his law of regression for these data:

> It is that the height deviate of the offspring is, on the average, two-thirds of the height deviate of its midparentage. (Galton, 1885a, p. 508)

In modern notation, we can write this as

$$\begin{aligned} \mathsf{E}(Y \mid X) - M &= \frac{2}{3}(X - M) \\ \therefore\quad \mathsf{E}(Y \mid X) &= \frac{2}{3}X + \frac{1}{3}M \end{aligned} \tag{2.4}$$

*Galton was correct in explaining regression from one generation to the other. However, Galton also made an argument for perpetual regression which turned out to be erroneous. For more details, see Bulmer (2003, p. 285).

Sept. 24, 1885] *NATURE* 507

Natural History memoir, made from new observations during the same journey. In addition the Committee have received from Mr. Guy Le Strange, and published, observations and notes made by him during a recent journey east of Jordan. The results of the survey, so far as it has been completed, will appear in a map reduced to a scale of about three miles to an inch, showing the country on both sides of the river Jordan, instead of on the western side only. This portion of the work is under the direction of Col. Sir Charles Wilson, K.C.M.G., F.R.S. The Society has also issued during the last year a popular account, by Prof. Hule, of his recent journey, called "Mount Seir," and reprints of Capt. Conder's popular books, "Tent Work in Palestine" and "Heth and Moab." Finally, the Committee have completed the issue of their great work, the "Survey of Western Palestine," with the last volumes of "Jerusalem," the "Flora and Fauna," and a portfolio of plates showing the excavations and their results.

SECTION H

ANTHROPOLOGY

OPENING ADDRESS BY FRANCIS GALTON, F.R.S., ETC., PRESIDENT OF THE ANTHROPOLOGICAL INSTITUTE, PRESIDENT OF THE SECTION

THE object of the Anthropologist is plain. He seeks to learn what mankind really are in body and mind, how they came to be what they are, and whither their races are tending; but the methods by which this definite inquiry has to be pursued are extremely diverse. Those of the geologist, the antiquarian, the jurist, the historian, the philologist, the traveller, the artist, and the statistician, are all employed, and the Science of Man progresses through the help of specialists. Under these circumstances, I think it best to follow an example occasionally set by presidents of sections, by giving a lecture rather than an address, selecting for my subject one that has long been my favourite pursuit, on which I have been working with fresh data during many recent months, and about which I have something new to say.

My data were the Family Records entrusted to me by persons living in all parts of the country, and I am now glad to think that the publication of some first-fruits of their analysis will show to many careful and intelligent correspondents that their painstaking has not been thrown away. I shall refer to only a part of the work already completed, which in due time will be published, and must be satisfied if, when I have finished this address, some few ideas that lie at the root of heredity shall have been clearly apprehended, and their wide bearings more or less distinctly perceived. I am the more desirous of speaking on heredity, because, judging from private conversations and inquiries that are often put to me, the popular views of what may be expected from inheritance seem neither clear nor just.

The subject of my remarks will be "Types and their Inheritance." I shall discuss the conditions of the stability and instability of types, and hope in doing so to place beyond doubt the existence of a simple and far-reaching law that governs hereditary transmission, and to which I once before ventured to draw attention, on far more slender evidence than I now possess.

It is some years since I made an extensive series of experiments on the produce of seeds of different size but of the same species. They yielded results that seemed very noteworthy, and I used them as the basis of a lecture before the Royal Institution on February 9, 1877. It appeared from these experiments that the offspring did *not* tend to resemble their parent seeds in size, but to be always more mediocre than they—to be smaller than the parents, if the parents were large; to be larger than the parents, if the parents were very small. The point of convergence was considerably below the average size of the seeds contained in the large bagful I bought at a nursery-garden, out of which I selected those that were sown.

The experiments showed further that the mean filial regression towards mediocrity was directly proportional to the parental deviation from it. This curious result was based on so many plantings, conducted for me by friends living in various parts of the country, from Nairn in the north to Cornwall in the south, during one, two, or even three generations of the plants, that I could entertain no doubt of the truth of my conclusions. The exact ratio of regression remained a little doubtful, owing to variable influences; therefore I did not attempt to define it. After the lecture had been published, it occurred to me that the grounds of my misgivings might be urged as objections to the general conclusions. I did not think them of moment, but as the inquiry had been surrounded with many small difficulties and matters of detail, it would be scarcely possible to give a brief and yet a full and adequate answer to such objections. Also, I was then blind to what I now perceive to be the simple explanation of the phenomenon, so I thought it better to say no more upon the subject until I should obtain independent evidence. It was anthropological evidence that I desired, caring only for the seeds as means of throwing light on heredity in man. I tried in vain for a long and weary time to obtain it in sufficient abundance, and my failure was a cogent motive, together with others, in inducing me to make an offer of prizes for family records, which was largely responded to, and furnished me last year with what I wanted. I especially guarded myself against making any allusion to this particular inquiry in my prospectus, lest a bias should be given to the returns. I now can securely contemplate the possibility of the records of height having been frequently drawn up in a careless fashion, because no amount of unbiassed inaccuracy can account for the results, contrasted in their values but concurrent in their significance, that are derived from comparisons between different groups of the returns.

An analysis of the records fully confirms and goes far beyond the conclusions I obtained from the seeds. It gives the numerical value of the regression towards mediocrity as from 1 to $\frac{2}{3}$ with unexpected coherence and precision, and it supplies me with the class of facts I wanted to investigate—the degrees of family likeness in different degrees of kinship, and the steps through which special family peculiarities become merged into the typical characteristics of the race at large.

The subject of the inquiry on which I am about to speak was Hereditary Stature. My data consisted of the heights of 930 adult children and of their respective parentages, 205 in number. In every case I transmuted the female statures to their corresponding male equivalents and used them in their transmuted form, so that no objection grounded on the sexual difference of stature need be raised when I speak of averages. The factor I used was 1·08, which is equivalent to adding a little less than one-twelfth to each female height. It differs a very little from the factors employed by other anthropologists, who, moreover, differ a trifle between themselves; anyhow it suits my data better than 1·07 or 1·09. The final result is not of a kind to be affected by these minute details, for it happened that, owing to a mistaken direction, the computer to whom I first entrusted the figures used a somewhat different factor, yet the result came out closely the same.

I shall explain with fulness why I chose stature for the subject of inquiry, because the peculiarities and points to be attended to in the investigation will manifest themselves best by doing so. Many of its advantages are obvious enough, such as the ease and frequency with which its measurement is made, its practical constancy during thirty-five years of middle life, its small dependence on differences of bringing up, and its inconsiderable influence on the rate of mortality. Other advantages which are not equally obvious are no less great. One of these lies in the fact that stature is not a simple element, but a sum of the accumulated lengths or thicknesses of more than a hundred bodily parts, each so distinct from the rest as to have earned a name by which it can be specified. The list of them includes about fifty separate bones, situated in the skull, the spine, the pelvis, the two legs, and the two ankles and feet. The bones in both the lower limbs are counted, because it is the average length of these two limbs that contributes to the general stature. The cartilages interposed between the bones, two at each joint, are rather more numerous than the bones themselves. The fleshy parts of the scalp of the head and of the soles of the feet conclude the list. Account should also be taken of the shape and set of many of the bones which conduce to a more or less arched instep, straight back, or high head. I noticed in the skeleton of O'Brien, the Irish giant, at the College of Surgeons, which is, I believe, the tallest skeleton in any museum, that his extraordinary stature of about 7 feet 7 inches would have been a trifle increased if the faces of his dorsal vertebræ had been more parallel and his back consequently straighter.

The beautiful regularity in the statures of a population, whenever they are statistically marshalled in the order of their heights,

FIGURE 2.4 First page of Galton's 1885 Presidential Lecture to the Anthropology Section of the British Association (Galton, 1885a)

TABLE I.

NUMBER OF ADULT CHILDREN OF VARIOUS STATURES BORN OF 205 MID-PARENTS OF VARIOUS STATURES.
(All Female heights have been multiplied by 1·08).

Heights of the Mid-parents in inches.	Heights of the Adult Children.														Total Number of		Medians.
	Below	62·2	63·2	64·2	65·2	66·2	67·2	68·2	69·2	70·2	71·2	72·2	73·2	Above	Adult Children.	Mid-parents.	
Above ..	..	..	..	..	..	..	..	..	..	..	..	1	3	..	4	5	..
72·5	..	..	..	..	..	..	..	1	2	1	2	7	2	4	19	6	72·2
71·5	..	..	..	..	1	3	4	3	5	10	4	9	2	2	43	11	69·9
70·5	1	..	1	..	1	1	3	12	18	14	7	4	3	3	68	22	69·5
69·5	..	..	1	16	4	17	27	20	33	25	20	11	4	5	183	41	68·9
68·5	1	..	7	11	16	25	31	34	48	21	18	4	3	..	219	49	68·2
67·5	..	3	5	14	15	36	38	28	38	19	11	4	..	..	211	33	67·6
66·5	..	3	3	5	2	17	17	14	13	4	..	..	..	..	78	20	67·2
65·5	1	..	9	5	7	11	11	7	7	5	2	1	..	..	66	12	66·7
64·5	1	1	4	4	1	5	5	..	2	..	..	..	..	..	23	5	65·8
Below ..	1	..	2	4	1	2	2	1	1	..	..	..	..	..	14	1	..
Totals ..	5	7	32	59	48	117	138	120	167	99	64	41	17	14	928	205	..
Medians ..	..	..	66·3	67·8	67·9	67·7	67·9	68·3	68·5	69·0	69·0	70·0	..	..	..	..	..

NOTE.—In calculating the Medians, the entries have been taken as referring to the middle of the squares in which they stand. The reason why the headings run 62·2, 63·2, &c., instead of 62·5, 63·5, &c., is that the observations are unequally distributed between 62 and 63, 63 and 64, &c., there being a strong bias in favour of integral inches. After careful consideration, I concluded that the headings, as adopted, best satisfied the conditions. This inequality was not apparent in the case of the Mid-parents.

FIGURE 2.5 Galton's table on stature (Galton, 1885b, p. 248)

Galton now gave the correct explanation for the regression effect:

> The explanation of it is as follows. The child inherits partly from his parents, partly from his ancestry. Speaking generally, the further his genealogy goes back, the more numerous and varied will his ancestry become, until they cease to differ from any equally numerous sample taken at haphazard from the race at large. Their mean stature will then be the same as that of the race; in other words, it will be mediocre. Or, to put the same fact into another form, the most probable value of the mid-ancestral deviates in any remote generation is zero.
>
> For the moment let us confine our attention to the remote ancestry and to the mid-parentages, and ignore the intermediate generations. The combination of the zero of the ancestry with the deviate of the mid-parentage, is that of nothing with something, and the result resembles that of pouring a uniform proportion of pure water into a vessel of wine. It dilutes the wine to a constant fraction of its original alcoholic strength, whatever that strength may have been. (*ibidem*)

Indeed, the above explanation is one of the most intuitive ways of understanding the regression effect: in general, suppose a first measurement X is made on a given subject, followed by a second measurement Y. Assume X is exceptionally high. As long as X and Y are imperfectly correlated, X can be thought to be made up of two components:

1. A first component that is unusually extreme and is expected to remain extreme
2. A second component that is not extreme and is expected to remain near the center of the distribution

The first measurement X is extreme because both of the components are high. However, for the second measurement Y, the first component is expected to remain high but the

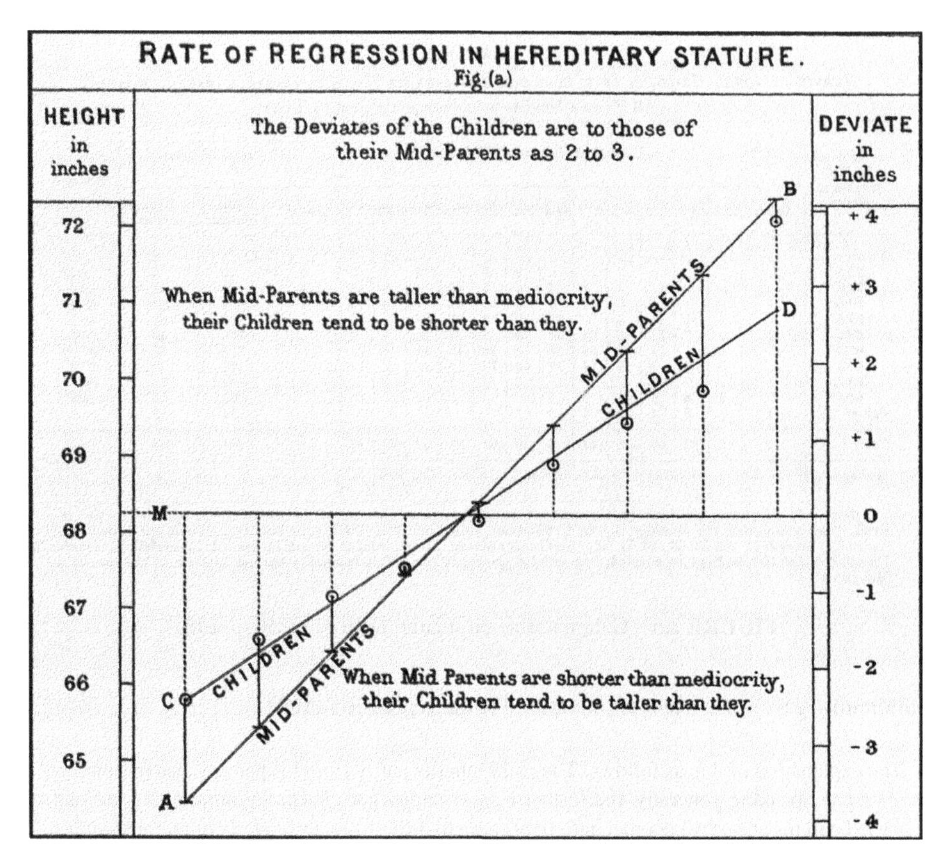

FIGURE 2.6 Galton's plot of offspring median heights (Y) against the midparental heights (X) (Galton, 1885b)

second component is expected to be near the center. Hence, the average value of Y will be less extreme than X and closer to the center of the distribution.

Notice that the previous explanation does not require any biological, economic, or other force to be present for regression to occur. The regression effect (or regression to the mean) is a purely mathematical artifact arising from the imperfect correlation* between X and Y. As such, it is also symmetric, in the sense that the same reasoning as above can made to argue that, for a given extreme Y, the value of X is expected to be less extreme and closer to the center.

The last regression, that of X on Y, is the issue that Galton next came to in his 1885 lecture and was probably the reason for his realization of the correct, nongenetic mechanism of regression. Having plotted offspring median heights (Y) against the midparental heights (X), Galton also plotted the median midparental heights against the offspring heights and obtained a slope of 1/3 (cf. Fig. 2.6).

*Of course, the concept of correlation was not known to Galton when he first discovered the regression effect.

In modern notation,

$$\begin{aligned} \mathrm{E}(X \mid Y) - M &= \frac{1}{3}(Y - M) \\ \therefore \quad \mathrm{E}(X \mid Y) &= \frac{1}{3}Y + \frac{2}{3}M \end{aligned} \tag{2.5}$$

Galton correctly warned:

> Because the most probable deviate of the son is only two-thirds that of his mid-parentage, it does not in the least follow that the most probable deviate of the mid-parentage is $\frac{3}{2}$, or $1\frac{1}{2}$. (*ibid.*, p. 509)

As we now show, if X and Y had the same variance, the two most probable deviates (or regression coefficients) would have been the same, namely, the correlation coefficient between X and Y. In general, using modern notation, the regression coefficient of Y on X is

$$\beta_{Y|X} = \rho \frac{\sigma_Y}{\sigma_X}, \tag{2.6}$$

where $\sigma_X = \sqrt{\text{var } X}$, $\sigma_Y = \sqrt{\text{var } Y}$, and ρ is the correlation between X and Y. Similarly, the regression coefficient of X on Y is

$$\beta_{X|Y} = \rho \frac{\sigma_X}{\sigma_Y}. \tag{2.7}$$

If $\sigma_X = \sigma_Y$, then $\beta_{Y|X} = \beta_{X|Y} = \rho$.

On the other hand, in Galton's case, we *do not* have $\sigma_X = \sigma_Y$ because X is the average of the male parent height (X_1) and adjusted female height (X_2), that is, $X = (X_1 + X_2)/2$. Thus, if we assume Y, X_1, and X_2 all have the same variance σ^2, then

$$\begin{aligned} \text{var } Y &= \sigma^2 \\ \text{var } X &= \text{var}\left(\frac{X_1 + X_2}{2}\right) = \frac{1}{4}\left(\sigma^2 + \sigma^2\right) = \frac{\sigma^2}{2}, \end{aligned}$$

so that $\beta_{Y|X} = \rho\sqrt{2}$, $\beta_{X|Y} = \rho/\sqrt{2}$, and $\beta_{Y|X}/\beta_{X|Y} = 2$.*

However, back in 1885, the fact that the two regression coefficients were not the same troubled Galton because "some suspicion may remain of a paradox lurking in these

* The above variances are also in line with the probable errors reported by Galton on p. 19: $e_{p,X} = 1.2$, $e_{p,Y} = 1.7$ so that $e_{p,X}/e_{p,Y} = .71 = \sqrt{\text{var } X/\text{var } Y}$. Finally, note that, from Eqs. (2.6) and (2.7), $\rho^2 = \beta_{Y|X}\beta_{X|Y}$, which gives $\rho = \sqrt{\left(\frac{1}{3}\right)\left(\frac{2}{3}\right)} = .471$.

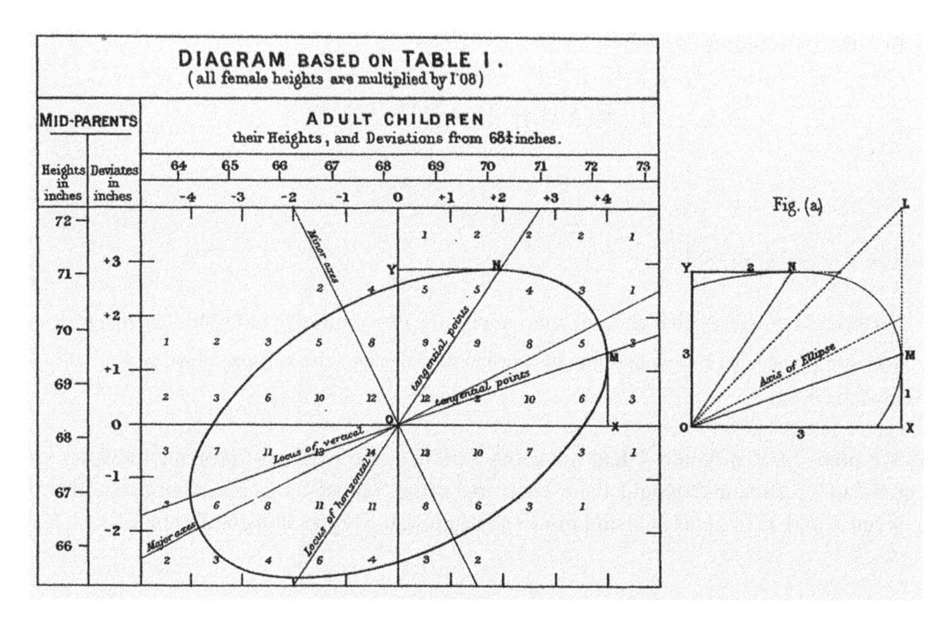

FIGURE 2.7 Galton's discovery of the bivariate normal distribution (Galton, 1885b)

strongly contrasted results." Therefore, Galton set out to graph the joint distribution of X and Y and, in the process, discovered the concept of the bivariate normal distribution. We now describe this aspect of Galton's work.

Galton used a sheet of squared paper and entered the frequencies on it (Fig. 2.7). He described the procedure in more detail in the 1885 paper, "Regression towards mediocrity in hereditary stature," as follows:

> It is deduced from a large sheet on which I entered every child's height, opposite to its mid-parental height, and in every case each was entered to the nearest tenth of an inch. Then I counted the number of entries in each square inch, and copied them out as they appear in the table. The meaning of the table is best understood by examples. Thus, out of a total of 928 children who were born to the 205 mid-parents on my list, there were 18 of the height of 69.2 inches (counting to the nearest inch), who were born to mid-parents of the height of 70.5 inches (also counting to the nearest inch). So again there were 25 children of 70.2 inches born to mid-parents of 69.5 inches. I found it hard at first to catch the full significance of the entries in the table, which had curious relations that were very interesting to investigate. They came out distinctly when I "smoothed" the entries by writing at each intersection of a horizontal column with a vertical one, the sum of the entries in the four adjacent squares, and using these to work upon. I then noticed (see [Fig 5]) that lines drawn through entries of the same value formed a series of concentric and similar ellipses. (Galton, 1885b, pp. 253–254)

Galton then reported the observations he made:

> Their common center lay at the intersection of the vertical and horizontal lines, that corresponded to 68.25 inches. Their axes were similarly inclined. The points where each ellipse in succession was touched by a horizontal tangent, lay in a straight line inclined to the

> vertical in the ratio of 2/3; those where they were touched by a vertical tangent lay in a straight line inclined to the horizontal in the ratio of 1/3. These ratios confirm the values of average regression already obtained by a different method, of 2/3 from mid-parent to offspring, and of 1/3 from offspring to mid-parent, because it will be obvious on studying [Figure 2.7] that the point where each horizontal line in succession is touched by an ellipse, the greatest value in that line must appear at the point of contact. The same is true in respect to the vertical lines. These and other relations were evidently a subject for mathematical analysis and verification. (*ibid.*, p. 255)

The remaining step was to mathematically prove these results, a task that Galton thought was beyond his analytical skills. Therefore, in the paper "Family likeness in stature" (Galton, 1886), he solicited the help of the able mathematician Hamilton Dickson. In modern mathematical language,* Dickson was provided with the information that $Y \sim N\left(0,\sigma_Y^2\right)$ and $X \mid Y \sim N\left(\beta_{X|Y}y,\sigma_{X|Y}^2\right)$ and asked the following questions:

a) What is the joint density of (X, Y), and why is the shape of the contours of equal probability density elliptical?
b) How can the regression coefficient $\beta_{Y|X}$ be calculated?
c) What is the density of Y given X?
d) What is the relationship between $\beta_{Y|X}$ and $\beta_{X|Y}$?

Dickson answered each of the above questions without much trouble, and the solution was published as an appendix (Fig. 2.8) to Galton's paper. In modern notation, the joint density of X and Y is

$$f_{XY}(x,y)=f_Y(y)f_{X|Y}(x\mid y)\propto \exp\left[-\left\{\frac{y^2}{2\sigma_Y^2}+\frac{\left(x-\beta_{X|Y}y\right)^2}{2\sigma_{X|Y}^2}\right\}\right]^{\dagger}. \tag{2.8}$$

To obtain the contours of equal probability, Dickson set the expression in the exponent above to a constant (say, K):

$$\frac{y^2}{\sigma_Y^2}+\frac{\left(x-\beta_{X|Y}y\right)^2}{\sigma_{X|Y}^2}=K, \tag{2.9}$$

which by varying K gives the equation of a set of ellipses.

To obtain the required regression coefficient $\beta_{Y|X}$, first Eq. (2.9) is differentiated:

$$\frac{y\,dy/dx}{\sigma_Y^2}+\frac{\left(x-\beta_{Y|X}y\right)\left(1-\beta_{X|Y}dy/dx\right)}{\sigma_{X|Y}^2}=0,$$

*This is exactly how the problem was phrased: "A point P is capable of moving along a straight line P′OP, making an angle $\tan^{-1} 2/3$ with the axis of y, which is drawn through O the mean position of P; the probable error of the projection of P on Oy is 1.22 inch: another point p, whose mean position at any time is P, is capable of moving from P parallel to the axis of x (rectangular co-ordinates) with a probable error of 1.50 inch. To discuss the 'surface of frequency' of p" (Galton, 1886).

†Pearson has expressed his puzzle why Galton did not himself derive the joint density of X and Y since he already knew both $f_Y(y)$ and $f_{Y|X}(x|y)$ (Pearson, 1920, p. 37).

sible problems are evidently very various and complicated, I do not propose to speak further about them now. It is some consolation to know that in the commoner questions of hereditary interest, the genealogy is fully known for two generations, and that the average influence of the preceding ones is small.

In conclusion, it must be borne in mind that I have spoken throughout of heredity in respect to a quality that blends freely in inheritance. I reserve for a future inquiry (as yet incomplete) the inheritance of a quality that refuses to blend freely, namely, the colour of the eyes. These may be looked upon as extreme cases, between which all ordinary phenomena of heredity lie.

Appendix. By J. D. Hamilton Dickson.

Problem 1.

A point P is capable of moving along a straight line P′OP, making an angle $\tan^{-1}\frac{2}{3}$ with the axis of y, which is drawn through O the mean position of P; the probable error of the projection of P on Oy is 1·22 inch: another point p, whose mean position at any time is P, is capable of moving from P parallel to the axis of x (rectangular co-ordinates) with a probable error of 1·50 inch. To discuss the "surface of frequency" of p.

1. Expressing the "surface of frequency" by an equation in x, y, z, the exponent, with its sign changed, of the exponential which appears in the value of z in the equation of the surface is, save as to a factor,

$$\frac{y^2}{(1{\cdot}22)^2}+\frac{(3x-2y)^2}{9(1{\cdot}50)^2} \quad . \quad . \quad . \quad . \quad . \quad . \quad . \qquad (1)$$

hence all sections of the "surface of frequency" by planes parallel to the plane of xy are ellipses, whose equations may be written in the form,

$$\frac{y^2}{(1{\cdot}22)^2}+\frac{(3x-2y)^2}{9(1{\cdot}50)^2}=\text{C, a constant} \quad . \quad . \quad . \quad . \qquad (2)$$

2. Tangents to these ellipses parallel to the axis of y are found, by differentiating (2) and putting the coefficient of dy equal to zero, to meet the ellipses on the line,

$$\left.\begin{aligned} &\frac{y}{(1{\cdot}22)^2}-2\frac{3x-2y}{9(1{\cdot}50)^2}=0, \\ \text{that is} \qquad &\frac{y}{x}=\frac{\dfrac{6}{9(1{\cdot}50)^2}}{\dfrac{1}{(1{\cdot}22)^2}+\dfrac{4}{9(1{\cdot}50)^2}}=\frac{6}{17{\cdot}6} \end{aligned}\right\} \quad . \quad . \quad . \quad . \quad . \qquad (3)$$

or, approximately, on the line $y=\frac{1}{3}x$. Let this be the line OM.

FIGURE 2.8 First page of Dickson's analysis (Galton, 1886, p. 63)

so that

$$\frac{y}{\sigma_Y^2}dy + \frac{(x - \beta_{Y|X}y)(dx - \beta_{X|Y}dy)}{\sigma_{X|Y}^2} = 0.$$

By setting the coefficient of dy to zero, tangents to the ellipse in (2.9) parallel to the y-axis can be obtained and these intersect the ellipse at points lying on the line OM (cf. Fig. 2.7) with equation

$$\frac{y}{\sigma_Y^2} - \frac{(x - \beta_{X|Y}y)\beta_{X|Y}}{\sigma_{X|Y}^2} = 0,$$

or

$$y = \frac{\beta_{X|Y}\sigma_Y^2}{\sigma_{X|Y}^2 + \beta_{X|Y}^2\sigma_Y^2}x.$$

Thus,

$$\beta_{Y|X} = \frac{\beta_{X|Y}\sigma_Y^2}{\sigma_{X|Y}^2 + \beta_{X|Y}^2\sigma_Y^2}. \tag{2.10}$$

By substituting $\beta_{X|Y}=2/3$, $\sigma_{X|Y}=1.50/.675$, and $\sigma_Y=1.22/.675$ (cf. footnote on p. 201), Dickson obtained $\beta_{Y|X} \approx 1/3$ (*ibid.*, p. 63). He could actually have simplified the above result: since $X = \beta_{X|Y}Y + e$, he knew that $\beta_{X|Y}^2\sigma_Y^2 + \sigma_{X|Y}^2 = \sigma_X^2$ so that Eq. (2.10) becomes

$$\beta_{Y|X} = \beta_{X|Y}\left(\frac{\sigma_Y^2}{\sigma_X^2}\right). \tag{2.11}$$

To obtain the conditional density $f_{Y|X}(y \mid x)$, with the aid of Eq. (2.10), the exponent in Eq. (2.8) can be written as

$$\begin{aligned}\frac{y^2}{2\sigma_Y^2} + \frac{(x - \beta_{X|Y}y)^2}{2\sigma_{X|Y}^2} &= \frac{x^2 + \left(\frac{\beta_{X|Y}^2\sigma_Y^2 + \sigma_{X|Y}^2}{\sigma_Y^2}\right)\left(y^2 - \frac{2x\beta_{X|Y}}{\beta_{X|Y}^2 + \sigma_{X|Y}^2/\sigma_Y^2}y\right)}{2\sigma_{X|Y}^2} \\ &= \frac{x^2}{2\sigma_{X|Y}^2/(1 - \beta_{X|Y}\beta_{Y|X})} + \frac{(y - \beta_{Y|X}x)^2}{2\sigma_{X|Y}^2\beta_{Y|X}/\beta_{X|Y}},\end{aligned}$$

which gives the two results:

$$\begin{aligned} X &\sim N\left(0, \frac{\sigma_{X|Y}^2}{1 - \beta_{X|Y}\beta_{Y|X}}\right), \\ Y \mid X &\sim N\left(\beta_{Y|X}x, \sigma_{X|Y}^2\frac{\beta_{Y|X}}{\beta_{X|Y}}\right).^* \end{aligned} \tag{2.12}$$

* We now know that $\beta_{Y|X}\beta_{X|Y} = \rho^2$, the square of the correlation coefficient between X and Y. See also Eqs. (2.6) and (2.7).

The relationship between $\beta_{Y|X}$ and $\beta_{X|Y}$ is now not difficult to deduce. It is already given in Eq. (2.11). But Dickson did not know this; instead, he wrote the relationship as in Eq. (2.10)

Galton was clearly elated when he received Dickson's analysis, since it confirmed all his empirical results:

> I may be permitted to say that I never felt such a glow of loyalty and respect towards the sovereignty and magnificent sway of mathematical analysis as when his [Dickson's] answer reached me, confirming, by purely mathematical reasoning, my various and laborious statistical conclusions with far more minuteness than I had dared to hope, for the original data ran somewhat roughly, and I had to smooth them with tender caution. (Galton, 1885a, p. 509)

As Galton had suspected, with only the three pieces of information $\sigma_Y^2, \beta_{X|Y}$ and $\sigma_{X|Y}^2$ (together with the normal distribution assumption), the elliptical contours he had observed could be constructed. Moreover, given that there was linear regression of X on Y, Dickson had not only shown that there was also linear regression of Y on X but also had been able to deduce the coefficient for the regression ($\beta_{Y|X}$). These were powerful results which confirmed Galton's hypothesis that regression was both symmetrical and intrinsically statistical.*

It is also very likely that Galton's appreciation of the consequence of symmetry was to be further consolidated, in addition to Dickson's analysis, through his later examination of the data on brothers (Galton, 1889, p. 210; Stigler, 1986b, p. 290). The data were obtained from returns of 295 families and consisted of 783 brothers in all. From these, Galton constructed a table showing the distributions of heights of brothers of men with a given height. The resulting table turned out to be symmetric, reflecting the symmetric nature of regression.†

All of Galton's previous works on regression (including Dickson's work) were further discussed and elaborated in Galton's 1889 opus *Natural Inheritance* (Galton, 1889).

2.2.4 First Appearance of Correlation (1888)

"Correlation," or rather "co-relation," first appeared in Galton's 1888 "Co-relations and their Measurements" (Fig. 2.9). This is how Galton defined it:

> Two variable organs are said to be co-related when the variation of the one is accompanied on the average by more or less variation on the other, and in the same direction. (Galton, 1888)‡

* See also Denis (2001) and Maraun et al. (2011).

† Although Stigler points out that the degree of symmetry had been exaggerated due to Galton erroneously counting each pair of brothers twice (Stigler, 1986b, p. 293).

‡ Galton was, of course, referring to *positive* correlation. Apart from this issue, his definition is quite accurate. Note that a positive correlation between X and Y does *not* mean that when X increases, Y must also increase: rather Y would increase *on average*. On any *one* occasion, though, Y could remain unchanged, increase, or even decrease. Similarly, a negative correlation does not mean that when X increases Y should decrease, rather that Y would decrease *on average*.

1888.] *Co-relations and their Measurement.* 135

Observations, &c. (*continued*).
Order. 1884. 4to. *London* 1888; Daily Weather Reports. 1888. January to June. 4to. *London.* The Office.
Nautical Almanac Office. The Nautical Almanac for 1892. 8vo. *London* 1888. The Office.
Madrid:—Observatorio. Observaciones Meteorológicas. 1883–85. 8vo. *Madrid* 1887–88. The Observatory.

December 20, 1888.

Professor G. G. STOKES, D.C.L., President, in the Chair.

The Presents received were laid on the table, and thanks ordered for them.

The following Papers were read:—

I. "Co-relations and their Measurement, chiefly from Anthropometric Data." By FRANCIS GALTON, F.R.S. Received December 5, 1888.

"Co-relation or correlation of structure" is a phrase much used in biology, and not least in that branch of it which refers to heredity, and the idea is even more frequently present than the phrase; but I am not aware of any previous attempt to define it clearly, to trace its mode of action in detail, or to show how to measure its degree.

Two variable organs are said to be co-related when the variation of the one is accompanied on the average by more or less variation of the other, and in the same direction. Thus the length of the arm is said to be co-related with that of the leg, because a person with a long arm has usually a long leg, and conversely. If the co-relation be close, then a person with a very long arm would usually have a very long leg; if it be moderately close, then the length of his leg would usually be only long, not very long; and if there were no co-relation at all then the length of his leg would on the average be mediocre. It is easy to see that co-relation must be the consequence of the variations of the two organs being partly due to common causes. If they were wholly due to common causes, the co-relation would be perfect, as is approximately the case with the symmetrically disposed parts of the body. If they were in no respect due to common causes, the co-relation would be *nil*. Between these two extremes are an endless number of intermediate cases, and it will be shown how the

FIGURE 2.9 First page of Galton's 1888 paper "Co-relations and their measurements" (Galton, 1888)

In a later 1890 paper, Galton explained his reason for the 1888 paper:

> In a book of mine called "Natural Inheritance," published about a year ago, I showed that the problems of family likeness fell entirely within the scope of the higher laws of chance....
>
> After the proofs of my book had been finally revised and had passed out of my hands, it happened that there was a delay of a few months before its actual publication. In the interim I was busily at work upon a new inquiry that had been suggested to me by two concurrent circumstances. One was a renewed discussion among anthropologists as to the information that the length of a particular bone-say a solitary thigh- bone dug out of an ancient grave-might afford concerning the stature of the unknown man to whom it belonged... The other circumstance arose out of the interest excited by M. Alphonse Bertillon, who proved that it was feasible to identify old criminals by an anthropometric process....
>
> These two [above] problems...are clearly identical....
>
> Reflection soon made it clear to me that not only were the two new problems identical in principle with the old one of kinship which I had already solved, but that all three of them were no more than special cases of a much more general problem-namely, that of Correlation.
>
> Fearing that this idea, which had become so evident to myself, would strike many others as soon as "Natural Inheritance" was published, and that I should be justly reproached for having overlooked it, I made all haste to prepare a paper for the Royal Society with the title of "Correlation." (Galton, 1890, p. 81)

Galton had thus realized that his previous investigations of regression were not limited to heredity problems but they could be applied to many other fields. Indeed, all these problems were the manifestations of one underlying phenomenon, namely, correlation.

Continuing with the 1888 paper, Galton gave the fundamental insight through an example:

> The relation between the cubit* and the stature will be shown to be such that for every inch, centimetre, or other unit of absolute length that the cubit deviates from the mean length of cubits, the stature will on the average deviate from the mean length of statures to the amount of 2.5 units, and. in the same direction. Conversely, for each unit of deviation of stature, the average deviation of the cubit will be 0.26 unit. These relations are not numerically reciprocal, but the exactness of the co-relation becomes established when we have transmuted the inches or other measurement of the cubit and of the stature into units dependent on their respective scales of variability... After this has been done, we shall find the deviation of the cubit as compared to the mean of the corresponding deviations of the stature, to be as 1 to 0.8. Conversely, the deviation of the stature as compared to the mean of the corresponding deviations of the cubit will also be as 1 to 0.8. Thus the existence of the co-relation is established, and its measure is found to be 0.8. (Galton, 1888, p. 136)

Galton here made the fundamental observation that the regression coefficients for the two regressions, that of Y on X and that of X on Y, are generally different. However, when both X and Y are measured in terms of their respective variabilities (σ_X and σ_Y), the

* Derived from the Latin *cubitus*, the lower arm.

two regression coefficients become exactly the same, namely, the correlation. In modern notation, Galton had just stated the relationships in Eqs. (2.6) and (2.7), which we now write as

$$\beta_{Y|X}\frac{\sigma_X}{\sigma_Y}=\beta_{X|Y}\frac{\sigma_Y}{\sigma_X}=\rho. \tag{2.13}$$

Galton also mentioned that the number ρ* measures "the closeness of co-relation in any particular case."

* 2.2.5 Some Results on Regression Based on the Bivariate Normal Distribution: Regression to the Mean Mathematically Explained

2.2.5.1 Basic Results Based on the Bivariate Normal Distribution We provide a few basic results on simple regression derived from the bivariate normal distribution, using modern notation. The reader can thus hopefully better appreciate how much was achieved by Galton.

In general, suppose (X, Y) is bivariate normal with means μ_X, μ_Y, variances σ_X^2, σ_Y^2, and correlation ρ, that is, $(X, Y) \sim \text{BivNormal}\left(\mu_X, \mu_Y; \sigma_X^2, \sigma_Y^2; \rho\right)$. Then the joint density of (X, Y) is

$$f_{XY}(x,y)=\frac{1}{2\pi\sigma_X\sigma_Y\sqrt{1-\rho^2}}\exp\left[-\frac{1}{2\left(1-\rho^2\right)}\left\{\left(\frac{x-\mu_X}{\sigma_X}\right)^2-2\rho\left(\frac{x-\mu_X}{\sigma_X}\right)\left(\frac{y-\mu_Y}{\sigma_Y}\right)+\left(\frac{y-\mu_Y}{\sigma_Y}\right)^2\right\}\right]. \tag{2.14}$$

Integrating the right side of the above with respect to y yields the marginal density of X:

$$f_X(x)=\frac{1}{\sigma_X\sqrt{2\pi}}\exp\left\{-\frac{\left(x-\mu_X\right)^2}{2\sigma_X^2}\right\}.$$

Similarly, the marginal density of Y is

$$f_Y(y)=\frac{1}{\sigma_Y\sqrt{2\pi}}\exp\left\{-\frac{\left(y-\mu_Y\right)^2}{2\sigma_Y^2}\right\}.$$

The above forms imply that

$$\begin{aligned} X &\sim N\left(\mu_X,\sigma_X^2\right),\\ Y &\sim N\left(\mu_Y,\sigma_Y^2\right). \end{aligned} \tag{2.15}$$

Having obtained the marginal distributions of X and Y, we can obtain the conditional density

* Galton actually used the symbol r, but we shall reserve the latter for the *sample* correlation coefficient.

$$f_{Y|X}(x) = \frac{f_{XY}(x,y)}{f_X(x)}$$

$$= \frac{1}{\sigma_Y \sqrt{1-\rho^2}\sqrt{2\pi}} \exp\left[-\frac{\left\{ y - \mu_Y - \rho \frac{\sigma_Y}{\sigma_X}(x - \mu_X) \right\}^2}{2\sigma_Y^2 (1-\rho^2)} \right].$$

The above is a normal distribution with mean

$$\mathsf{E}(Y \mid X) = \mu_Y + \rho \frac{\sigma_Y}{\sigma_X}(X - \mu_X), \tag{2.16}$$

and variance

$$\text{var}(Y \mid X) = \sigma_Y^2 (1 - \rho^2). \tag{2.17}$$

By symmetry, the conditional density of X given Y is normal with

$$\mathsf{E}(X \mid Y) = \mu_X + \rho \frac{\sigma_X}{\sigma_Y}(Y - \mu_Y), \tag{2.18}$$

and

$$\text{var}(X \mid Y) = \sigma_X^2 (1 - \rho^2). \tag{2.19}$$

Equations (2.16) and (2.18) are the true regression lines (or functions) of Y on X and of X on Y, respectively, with regression coefficients

$$\beta_{Y|X} = \rho \frac{\sigma_Y}{\sigma_X},$$

$$\beta_{X|Y} = \rho \frac{\sigma_X}{\sigma_Y}.$$

These are Eqs. (2.6) and (2.7) we gave previously. Multiplying these two equations gives the relationship

$$\beta_{Y|X}\beta_{X|Y} = \rho^2.$$

This implies that the two regression coefficients are inversely proportional to each other and further that their product is always between 0 and 1.

Equations (2.17) and (2.19) also enable us to calculate from variances

$$\rho = \sqrt{1 - \frac{\sigma_{Y|X}^2}{\sigma_Y^2}} = \sqrt{1 - \frac{\sigma_{X|Y}^2}{\sigma_X^2}}.$$

2.2.5.2 Regression to the Mean Mathematically Explained Using Eq. (2.16), we can also explain regression to the mean. Suppose two measurements, X and Y, are taken on a population, such that (X, Y) is bivariate normal $\text{BivNormal}\left(\mu_X, \mu_Y; \sigma_X^2, \sigma_Y^2; \rho\right)$. Assume a stable population with $\mu_Y = \mu_X = \mu$ and $\sigma_Y^2 = \sigma_X^2 = \sigma^2$. Equation (2.16) becomes

$$\mathsf{E}(Y \mid X) = \mu + \rho(X - \mu)$$

so that

$$\mathsf{E}(Y \mid X) - X = -(1-\rho)(X - \mu). \tag{2.20}$$

Equation (2.20) shows that the amount of regression (or "reversion") is proportional to the deviation of X from the mean μ, that is,

$$\left\{\mathsf{E}(Y \mid X) - X\right\} \propto -(X - \mu).$$

Suppose now, without loss of generality, that $\rho > 0$ and an observation is made on X that is larger than the mean, that is, $X > \mu$. From Eq. (2.20), we obtain

$$\mathsf{E}(Y \mid X) - X < 0.$$

Hence, the value of Y given a large X (such that $X > \mu$) will, on average, be less than X and thus closer than X to the mean μ. Similarly, again assuming $\rho > 0$, the value of Y given a small X (such that $X < \mu$) will, on average, be greater than X and thus closer than X to the mean μ. These were the revolutionary observations made by Galton.

2.3 FURTHER DEVELOPMENTS AFTER GALTON

2.3.1 Weldon (1890; 1892; 1893)

Walter Frank Raphael Weldon* (Fig. 2.10) was born in Highgate, London, on March 15, 1860, and was the second child in the family. His father, Walter Weldon, was a journalist and industrial chemist. Weldon was sent to a boarding school at Caversham in 1873. He joined University College London in 1876 with the intention of pursuing a medical career. At University College, Weldon learned mathematics from Olaus Henrici and biology from the zoologist E. Ray Lankester. In the next year, he was transferred to King's College, London, then to St. John's College, Cambridge, in 1878. At Cambridge, Weldon studied under the zoologist Francis Maitland Balfour and decided to give up his medical career for zoology. In 1881, he obtained a first-class degree in the Natural Sciences Tripos.

But soon a series of misfortunes befell Weldon, with the death of his mother first, then of Balfour, and finally that of his father. Thereupon, he left for Naples to work at the Zoological Station and published his first paper entitled "Note on the early development of *Lacerta muralis*" (Weldon, 1883). In September 1882, Weldon returned to Cambridge as demonstrator for Adam Sedgwick, who had succeeded Balfour. Weldon wrote his second paper from the Sedgwick laboratory. On March 14, 1883, the anniversary of his parents' marriage, he married Miss Florence Tebb. In 1884, Weldon was elected to a Fellowship at St John's College, and then shortly after, he was appointed as University Lecturer in Invertebrate

*Weldon's life is also described in Pearson (1908) and Magnello (2001b, pp. 261–264).

FIGURE 2.10 Walter Frank Raphael Weldon (1860–1906). Wikimedia Commons (Public Domain), http://commons.wikimedia.org/wiki/Category:Walter_Frank_Raphael_Weldon#/media/File:Weldon_Walter_F_R.jpg

Morphology. Five years later, Weldon succeeded E. Ray Lankester in the Jodrell Chair of Zoology at University College London. At around that time, Weldon read Galton's *Natural Inheritance* (Galton, 1889). Now Weldon had for a while been enthusiastic about Darwin's theory of evolution. But many of the hypotheses made by the latter needed to be tested. A key task had been how to represent the distribution of variations of organic character. Galton's book provided Weldon with the means to proceed with these investigations.

Thus, Weldon was the first biometrician to apply Galton's regression method. Between 1890 and 1892, Weldon published two major papers (Weldon, 1890; 1892) applying regression to the study of variation in the shrimp *Crangon vulgaris*. In the first paper, he showed that body measurements for large populations of shrimp were normally distributed and thus provided the first empirical evidence of normality in a wild population under natural selection* (Cowan, 1981, p. 251). This corroborated a hypothesis Galton had previously made. The second paper was the first instance of a correlation coefficient calculated for a wild population. Weldon here showed that the correlation between pairs of organ lengths in shrimps were high ($r \approx .8$). Moreover, he reported a negative correlation with due appreciation of its meaning for the first time. In an 1893 paper, Weldon called Galton's correlation coefficient "Galton function" (Weldon, 1893).

Although Weldon was not strong in statistics or mathematics, he did recognize the need for quantitative methods in order to provide empirical evidence for Darwin's theory. Many of the biological questions Weldon asked were responsible for Pearson's contributions to modern statistics. The collaboration between the two scientists was so close that, with the support of Galton, they established their own journal, *Biometrika*, in 1901. This happened in the wake of the negative reactions Pearson got from the Royal Society to biological papers of a mathematical nature. Weldon and Pearson remained close until the former's death in Oxford on April 13, 1906.

2.3.2 Edgeworth in 1892: First Systematic Study of the Multivariate Normal Distribution

Francis Ysidro Edgeworth† (Fig. 2.11) was born in Edgeworthstown, County Longford, Ireland, on February 8, 1845. His grandfather, Richard Lovell Edgeworth, was interested in the applications of science to industry and had won gold and silver medals from the Society of Encouragement of Arts and also a Fellowship of the Royal Society in 1761. Edgeworth's father, Francis Beaufort, died about 8 months after Edgeworth was born. During his childhood, Edgeworth was privately tutored at home. In 1862, he entered Trinity College, Dublin, but took no degree there. Five years later, Edgeworth moved to Oxford and was admitted to Balliol the next year. In 1869, he was awarded first-class honors in Literis Humanioribus but took the B.A. degree only in 1873. His M.A. followed in 1877, the same year he was admitted to the bar, but Edgeworth does not seem to have practiced. During the period 1867–1877, Edgeworth seems to have been interested in philosophy, ethics, and economics, but not mathematics. He published his first major book *New and Old Methods of Ethics* (Edgeworth, 1877) in 1877, after which he held various minor lecturing jobs in London. In 1881, Edgeworth published his most important work in economics entitled

*Galton's pea and height studies had been natural selection free.

†Edgeworth's life is also described in Bowley (1934) and Dale (2001). His contributions to mathematical statistics are described in Bowley (1928).

FIGURE 2.11 Francis Ysidro Edgeworth (1845–1926). Wikimedia Commons (Public Domain), http://commons.wikimedia.org/wiki/Category:Francis_Ysidro_Edgeworth#/media/File:Edgeworth.jpeg

Mathematical Psychics: An Essay on the Application of Mathematics to the Moral Sciences (Edgeworth, 1881). After 1881, he became more interested in probability and statistics, publishing his first statistics paper in 1883. Edgeworth became Professor of Political Economy at King's College in 1888. In 1891, he became the first editor of the *Economic Journal*, which occupied many of his later years. In much of his statistical work, Edgeworth used inverse probability. Although he published only about 40 statistical papers during 1893–1926, he seems to have anticipated, albeit often in crude form, many of the developments Fisher was to later make in the 1920s. Thus, Edgeworth had essentially derived the large-sample formula for the variance of the maximum likelihood estimator as early as 1909 (see Eq. 5.145). When Fisher later presented the formula (Fisher, 1935d) after having initially derived it in 1922 (Fisher, 1922c),* he was chastised by Arthur Bowley (1869–1957) for not acknowledging Edgeworth's priority on the formula. However, to be fair to Fisher, it must be admitted that Edgeworth's writings on mathematical statistics were often meandering and confusing, and it is very likely that Fisher had not seen Edgeworth's derivation. Bowley explained Edgeworth's mathematical style as follows:

> ...[Edgeworth's] mathematical writing indicates a want of systematic training. Though he shows great insight into the principles of mathematics, there is a want of facility and neatness in his handling of problems. Familiar as he was with the work of Laplace, Todhunter, and Clerk Maxwell, he had difficulty in elementary applications. His line of thought is often a little obscure; sometimes he labors the obvious, and at others is so brief as to be difficult to follow. He was always the victim of numerical mistakes and errors in writing and printing. (Bowley, 1934, p. 114)

From 1912 to 1914, Edgeworth was President of the Royal Statistical Society. He never married and passed away in London on February 13, 1926.

We are here interested in Edgeworth's 1892 paper,† "Correlated averages" (Edgeworth, 1892) (Fig. 2.12), where the multivariate normal distribution was first systematically studied. The paper was also (a) the first serious attempt to extend Galton's bivariate correlation to three or more variables, (b) where the term "coefficient of correlation" was introduced, although not for the usual estimate of ρ. However, "Correlated averages" also contained an unusual number of errors, many of which were not typographical.

Right at the start of his paper, Edgeworth stated that the "correlation"‡ between the members of a system such as the limbs or other measurable attributes of an organism could be expressed by the general formula

$$\Pi = Je^{-R} \tag{2.21}$$

where

$$R = p_1(\mathbf{x}_1 - x_1)^2 + p_2(\mathbf{x}_2 - x_2)^2 + \cdots + 2q_{12}(\mathbf{x}_1 - x_1)(\mathbf{x}_2 - x_2) + 2q_{13}(\mathbf{x}_1 - x_1)(\mathbf{x}_3 - x_3) + \cdots;$$

$\mathbf{x}_1, \mathbf{x}_2, \mathbf{x}_3, \ldots$ are the mean values of the respective organs; $x_1, x_2, x_3, \ldots$ are particular values; $p_1, p_2, p_3, \ldots, q_{12}, q_{13}, \ldots$ are constants; and J is a normalizing constant. Equation (2.21) thus expresses the joint density of $x_1, x_2, x_3, \ldots$. For Edgeworth, the joint density was useful because

* See also Eq. (5.26).

† Edgeworth's paper is also described in Pearson (1920) and Plackett (1983).

‡ Meaning the "correlation surface."

[190]

XXII. *Correlated Averages.*
By Professor F. Y. EDGEWORTH, *M.A., D.C.L.**

THE "correlation"† between the members of a system such as the limbs or other measurable attributes of an organism may in general be expressed by the formula

$$\Pi = Je^{-R}dx_1, dx_2, dx_3, \&c.;$$

where

$$R = p_1(\mathrm{x}_1 - x_1)^2 + p_2(\mathrm{x}_2 - x_2)^2 + \&c.,$$
$$+ 2q_{12}(\mathrm{x}_1 - x_1)(\mathrm{x}_2 - x_2) + 2q_{13}(\mathrm{x}_1 - x_1)(\mathrm{x}_3 - x_3) + \&c.;$$

x_1, x_2, x_3 &c. are the average values of the respective organs; x_1, x_2, &c. are particular values of the same; $p_1, p_2 \ldots q_{12}, q_{13}$ are constants to be obtained from observation; J is a constant deduced from the condition that the integral of Π between extreme limits should be unity. The expression Π represents the probability that any particular values of x_1, x_2, &c. should concur. It enables us to answer the questions: What is the *most probable* value of one deviation x_r corresponding to *assigned* values x_1', x_2' &c. of the other variables? and What is the *dispersion* of the values of x_r about its mean (the other variables being assigned)?

This general formula for the concurrence of particular values of several organs is deducible from the proposition, proved by theory and observation, that each organ considered by itself assumes different values according to the exponential law of error. In a subsequent paper I hope to justify this principle; at present, assuming the propriety of the above-written formula, I propose to show how the constants $p_1, p_2 \ldots q_{12}, q_{13} \ldots$ are calculated. This problem has been solved by Mr. Galton for the case of two variables. The happy device of measuring each deviation by the corresponding quartile taken as unit enables him to express the sought quadratic in terms of a single parameter; as thus:—

$$R = \frac{x_1^2}{1-\rho^2} - \frac{2\rho x_1 x_2}{1-\rho^2} + \frac{x_2^2}{1-\rho^2};$$

where our ρ is Mr. Galton's r, and the x_1, x_2 of our general formula are zero. The parameter is found by observing the

* Communicated by the Author.
† See Galton, Proc. Roy. Soc. 1888, "Co-relations and their Measurement;" and Weldon, Proc. Roy. Soc. 1892, "Certain Correlated Variations in *Crangon vulgaris*."

FIGURE 2.12 First page of Edgeworth's "Correlated averages" (Edgeworth, 1892)

> ...it [Π] enables us to answer the questions: What is the *most probable* value of one deviation x_r corresponding to *assigned* values x_1', x_2' &c of the other variables? and What is the *dispersion* of the values of x_r about its mean (and the other variables being assigned)? (*ibid.*, p. 190)

Edgeworth next explained that his assumption of the law in (2.21) rested on the supposition ("proved by theory and observation") that each organ considered alone was normally distributed. However, he was not strictly correct because for the multivariate normal law to hold, not only must each organ be normally distributed but *every* linear combination of the organs must also be normally distributed.* Edgeworth also stated he would provide a formal justification of the multivariate normal law in a later paper. But as we shall see, he was never able to give a general demonstration, although he did consider two particular cases (trivariate and 4-variate).

Edgeworth first (*ibid.*, p. 194) set out to consider the trivariate normal case. His strategy was to determine the type of trivariate normal distribution, which, when one variable was integrated out, would result in Galton's bivariate normal distribution. Thus, let $\rho_{12}, \rho_{13}, \rho_{23}$ be the correlations between the pairs (x_1, x_2), (x_1, x_3), and (x_2, x_3), respectively. Suppose x_1, x_2, and x_3 are measured in units of their respective standard deviations, and all means are assumed to be zero. Then the bivariate density of (x_1, x_2) is Ke^{-S}, where

$$S = \frac{x_1^2}{1-\rho_{12}^2} - \frac{2\rho_{12}x_1x_2}{1-\rho_{12}^2} + \frac{x_2^2}{1-\rho_{12}^2}. \tag{2.22}$$

Now, let the trivariate density of (x_1, x_2, x_3) be Je^{-R} where, from (2.21),

$$R = p_1x_1^2 + p_2x_2^2 + p_3x_3^2 + 2q_{12}x_1x_2 + 2q_{13}x_1x_3 + 2q_{23}x_2x_3. \tag{2.23}$$

The aim is to determine the p's and q's. Suppose Je^{-R} is integrated with respect to x_3 from $-\infty$ to ∞. If, in the exponent of the result obtained, the coefficients of x_1^2, x_1x_2, x_2^2 are equated with those in Eq. (2.22), then

$$\frac{1}{1-\rho_{12}^2} = \frac{p_2p_3 - q_{13}^2}{p_3},$$

$$\frac{-2\rho_{12}}{1-\rho_{12}^2} = 2\frac{q_{12}p_3 - q_{13}q_{23}}{p_3},$$

$$\frac{1}{1-\rho_{12}^2} = \frac{p_1p_3 - q_{23}^2}{p_3}.$$

To obtain the value of ρ_{12} from the above, Edgeworth reasoned as follows: for a given value x_2' of x_2 in Eq. (2.22), what is the most frequent value of x_1? Call this value ξ_1. Then

$$\left(\frac{\partial S}{\partial x_1}\right)_{x_1=\xi_1, x_2=x_2'} = 0 \Rightarrow \rho_{12} = \frac{\xi_1}{x_2'}.^{\dagger} \tag{2.24}$$

* See, for example, Stuart and Ord (1994, p. 512).

† Edgeworth explained that this formula could be used by taking the sum or mean of all or only the positive (or negative) values on one variable and the corresponding quantity on the other variable. Application of the formula then yields the "coefficient of correlation."

The same technique can be applied to R in Eq. (2.23). For $x_2 = x_2'$, let the most frequent values of x_1, x_3 be ξ_1, ξ_3, respectively. Then differentiating R with respect to x_1, x_3,

$$2p_1\xi_1 + 2q_{12}x_2' + 2q_{13}\xi_3 = 0,$$
$$2p_3\xi_3 + 2q_{13}\xi_1 + 2q_{23}x_2' = 0.$$

Equating the expressions for ξ_3 in the above two equations,

$$-\frac{p_1\xi_1 + q_{12}x_2'}{q_{13}} = -\frac{q_{13}\xi_1 + q_{23}x_2'}{p_3} \Rightarrow \frac{\xi_1}{x_2'} = \frac{q_{13}q_{23} - q_{12}p_3}{p_1p_3 - q_{13}^2}.$$

Therefore, from Eq. (2.24),

$$\rho_{12} = \frac{q_{13}q_{23} - q_{12}p_3}{p_1p_3 - q_{13}^2} = \frac{Q_{12}}{P_2}^{*}$$

(*ibid*., p. 195), where Q_{12} and P_2 are minors, corresponding, respectively, to the (1, 2)th and (2, 2)th entries, of the determinant

$$\Delta = \begin{vmatrix} p_1 & q_{12} & q_{13} \\ q_{12} & p_2 & q_{23} \\ q_{13} & q_{23} & p_3 \end{vmatrix}.$$

The next step for Edgeworth was to integrate x_2 out of the bivariate density Ke^{-S} (whose exponent is given in Eq. (2.22)). This gave the marginal density of x_1 as

$$L\exp\left(-x_1^2 \frac{\Delta}{P_2}\right).$$

But since x_1 is measured in units of its standard deviation, its modulus $\left(=\sqrt{2}\,\text{SD}\right)$ should be unity so that $\Delta = P_2$.

By symmetry, the following equations also hold:

$$\begin{aligned} \Delta &= P_3 = P_1, \\ \rho_{13} &= \frac{Q_{13}}{P_3} = \frac{Q_{13}}{P_1}^{\dagger}, \\ \rho_{23} &= -\frac{Q_{23}}{P_3} = -\frac{Q_{23}}{P_2}. \end{aligned} \tag{2.25}$$

* In Edgeworth's paper, this equation is wrongly given as $\rho_{12} = \frac{q_{13}q_{23} - q_{12}p_3}{p_2p_3 - q_{13}^2} = \frac{Q_{12}}{P_1}$.

† Edgeworth's version of these equations is erroneous. Similarly for the two equations that follow.

Therefore,

$$Q_{12} = -\Delta\rho_{12}; \quad Q_{13} = \Delta\rho_{13}; \quad Q_{23} = -\Delta\rho_{23}.$$

Now the reciprocal of a determinant is the determinant whose members (constituents) are the minors of the given one (Salmon, 1866, p. 24), that is, the reciprocal Δ' of the determinant Δ is

$$\Delta' = \begin{vmatrix} P_1 & Q_{12} & Q_{13} \\ Q_{12} & P_2 & Q_{23} \\ Q_{13} & Q_{23} & P_3 \end{vmatrix} = \begin{vmatrix} \Delta & -\Delta\rho_{12} & \Delta\rho_{13} \\ -\Delta\rho_{12} & \Delta & -\Delta\rho_{23} \\ \Delta\rho_{13} & -\Delta\rho_{23} & \Delta \end{vmatrix}. \tag{2.26}$$

Now if a_{ij} is a minor of Δ' and a_{ij} is the corresponding minor of Δ, then $\mathsf{a}_{ij} = \Delta^{n-2}a_{ij} = \Delta a_{ij}$ (*ibidem*), since $n=3$. Thus, for the minors P_1 and Q_{12} of Δ',

$$\Delta^2 - \Delta^2\rho_{23}^2 = \Delta p_1,$$
$$\Delta^2\rho_{23}\rho_{13} - \Delta^2\rho_{12} = \Delta q_{12}.$$

Hence,

$$p_1 = \Delta\left(1-\rho_{23}^2\right),$$
$$q_{12} = \Delta\left(\rho_{23}\rho_{13} - \rho_{12}\right),$$

and similarly the other p's and q's in Eq. (2.23) can be found.

Edgeworth's final task was to determine Δ. For this, he used $\Delta = P_1$ from Eq. (2.25). Therefore,

$$P_1 = \begin{vmatrix} p_2 & q_{23} \\ q_{23} & p_3 \end{vmatrix} = \begin{vmatrix} \Delta\left(1-\rho_{13}^2\right) & \Delta\left(\rho_{12}\rho_{23} - \rho_{13}\right) \\ \Delta\left(\rho_{12}\rho_{23} - \rho_{13}\right) & \Delta\left(1-\rho_{12}^2\right) \end{vmatrix} = \Delta,$$

so that

$$\Delta^2\left(1-\rho_{13}^2\right)\left(1-\rho_{12}^2\right) - \Delta^2\left(\rho_{12}\rho_{23} - \rho_{13}\right)^2 = \Delta.$$

The above implies

$$\Delta = \frac{1}{\left(1-\rho_{13}^2\right)\left(1-\rho_{12}^2\right) - \left(\rho_{12}\rho_{23} - \rho_{13}\right)^2}$$

(Edgeworth, 1892, p. 196). Hence, Edgeworth was able to completely specify the trivariate normal distribution whose exponent is given in Eq. (2.23).

Having derived the trivariate normal distribution, Edgeworth then stated that his

> …reasoning is quite general; and accordingly, replacing the symbols x_1, x_2, x_3, we may extend to four and higher number of variables the solution which has been given above for the case of three variables. (*ibid.*, p. 201)

For example, in the 4-variate case, in analogy with (2.26), the reciprocal Δ' of the determinant Δ is

$$\Delta' = \begin{vmatrix} \Delta & -\Delta\rho_{12} & \Delta\rho_{13} & -\Delta\rho_{14} \\ -\Delta\rho_{12} & \Delta & -\Delta\rho_{23} & \Delta\rho_{24} \\ \Delta\rho_{13} & -\Delta\rho_{23} & \Delta & -\Delta\rho_{34} \\ -\Delta\rho_{14} & \Delta\rho_{24} & -\Delta\rho_{34} & \Delta \end{vmatrix}$$

(*ibidem**), and the minors of a_{ij} of Δ' are related to the minors a_{ij} of Δ through the equation $\mathsf{a}_{ij} = \Delta^2 a_{ij}$.

2.3.3 Origin of Pearson's *r* (Pearson et al., 1896)

Pearson's well-known product-moment formula for the sample correlation coefficient r made its first appearance in the 1896 fundamental paper, "Regression, heredity and panmixia" (Pearson et al., 1896) (Fig. 2.13). Pearson stated:

> The question now arises as to what is *practically* the best method of determining $r[\rho]$. I do not feel satisfied that the method used by Mr. GALTON and Professor WELDON will give the best results. (*ibid.*, p. 264)

Pearson assumed that $(X, Y) \sim \text{BivNormal}\left(0, 0; \sigma_1^2, \sigma_2^2; \rho\right)$. Then, the joint density of a random sample $(X_1, Y_1), (X_2, Y_2), \ldots, (X_n, Y_n)$, for a given value of ρ, is proportional to†

$$\frac{1}{\left(1-\rho^2\right)^{n/2}} \exp\left[-\frac{1}{2}\left\{\frac{x_1^2}{\sigma_1^2\left(1-\rho^2\right)} - \frac{2\rho x_1 y_1}{\sigma_1\sigma_2\left(1-\rho^2\right)} + \frac{y_1^2}{\sigma_2^2\left(1-\rho^2\right)}\right\}\right] \times \exp\left[-\frac{1}{2}\left\{\frac{x_2^2}{\sigma_1^2\left(1-\rho^2\right)} - \frac{2\rho x_2 y_2}{\sigma_1\sigma_2\left(1-\rho^2\right)} + \frac{y_2^2}{\sigma_2^2\left(1-\rho^2\right)}\right\}\right] \times \exp\left[-\frac{1}{2}\left\{\frac{x_3^2}{\sigma_1^2\left(1-\rho^2\right)} - \frac{2\rho x_3 y_3}{\sigma_1\sigma_2\left(1-\rho^2\right)} + \frac{y_3^2}{\sigma_2^2\left(1-\rho^2\right)}\right\}\right] \times \cdots \tag{2.27}$$

(*ibid.*, p. 265). Next, Pearson set $\sigma_1^2 = \sum x_i^2/n$, $\sigma_2^2 = \sum y_i^2/n$, and $\lambda = \sum x_i y_i/(n\sigma_1\sigma_2)$, so that (2.27) becomes

* Edgeworth's own Δ' did not contain any negative signs, but he did realize they were necessary and introduced them later in his formulas for the $\Delta\rho$'s. However, he failed to realize that his treatment of the trivariate case contained many errors.

† See Eq. (2.14).

[253]

VII. *Mathematical Contributions to the Theory of Evolution.—III. Regression, Heredity, and Panmixia.*

By KARL PEARSON, *University College, London.*

Communicated by Professor HENRICI, *F.R.S.*

Received September 28,—Read November 28, 1895.
Revised November 29, 1895.

CONTENTS.

7.5.96

FIGURE 2.13 First page of Pearson's "Regression, heredity, and panmixia" (Pearson et al., 1896)

$$\frac{1}{\left(1-\rho^2\right)^{n/2}}\exp\left\{\frac{-n(1-\lambda\rho)}{1-\rho^2}\right\}.$$

Pearson now wrote

$$\begin{aligned}u(\rho)&=\frac{1}{\left(1-\rho^2\right)^{n/2}}\exp\left\{\frac{-n(1-\lambda\rho)}{1-\rho^2}\right\}\\&=\exp\left[n\left\{-\frac{1}{2}\log\left(1-\rho^2\right)-\frac{1-\lambda\rho}{1-\rho^2}\right\}\right].\\\therefore\quad \frac{1}{n}\log u(\rho)&=-\frac{1}{2}\log\left(1-\rho^2\right)-\frac{1-\lambda\rho}{1-\rho^2}.\end{aligned}$$

By using Taylor's theorem, he obtained

$$\frac{1}{n}\log u(\rho+\delta\rho)=\frac{1}{n}\log u(\rho)+\frac{\left(1+\rho^2\right)(\lambda-\rho)}{\left(1-\rho^2\right)^2}\delta\rho+\frac{1}{2}\frac{\lambda\left(2\rho^3+6\rho\right)-1-6\rho^2-\rho^4}{\left(1-\rho^2\right)^3}(\delta\rho)^2+\cdots. \tag{2.28}$$

Thus, $\log u(\rho)$, and hence $u(\rho)$, is a maximum when the coefficient of δp is set to zero, that is, when

$$\frac{\left(1+\hat{\rho}^2\right)(\lambda-\hat{\rho})}{\left(1-\hat{\rho}^2\right)^2}=0,$$

$$\text{i.e.,}\quad \hat{\rho}=\lambda=\frac{\sum_{i=1}^{n}x_iy_i}{n\sigma_1\sigma_2}=\frac{\sum_{i=1}^{n}x_iy_i}{\sqrt{\sum_{i=1}^{n}x_i^2\sum_{i=1}^{n}y_i^2}} \tag{2.29}$$

(*ibidem*).*

Pearson's result can be generalized as follows: if $(X,Y)\sim \text{BivNormal}\left(\mu_1,\mu_2;\sigma_1^2,\sigma_2^2;\rho\right)$, then ρ can be estimated by the sample correlation coefficient

$$r=\frac{\sum_{i=1}^{n}(x_i-\bar{x})(y_i-\bar{y})}{\sqrt{\sum_{i=1}^{n}(x_i-\bar{x})^2\sum_{i=1}^{n}(y_i-\bar{y})^2}},$$

where $\bar{x}$ and $\bar{y}$ are sample means of the X and Y values, respectively.

* $u(\rho)$ is indeed a *maximum* when $\hat{\rho}=\lambda$ because the coefficient of $(\delta\rho)^2$ in the Taylor expansion above, that is, the second derivative of $\frac{1}{n}\log u(p)$, is then *negative*.

Although Pearson's final result in (2.29) is correct, his derivation suffers from a major drawback. The parameters σ_1^2 and σ_2^2 are treated as known constants, which is rarely the case in practice, and are set to $\sigma_1^2 = \sum x_i^2 / n$ and $\sigma_2^2 = \sum y_i^2 / n$.*

A correct version of Pearson's analysis proceeds through the method of likelihood and is as follows. As before, we assume a random sample $(x_1, y_1), \ldots, (x_n, y_n)$ from $(X, Y) \sim \text{BivNormal}\left(0, 0; \sigma_1^2, \sigma_2^2; \rho\right)$. Then

$$f_{XY}\left(x_i, y_i \mid \sigma_1, \sigma_2, \rho\right) = \frac{1}{2\pi\sigma_1\sigma_2\sqrt{1-\rho^2}} \exp\left\{-\frac{1}{2\left(1-\rho^2\right)}\left(\frac{x_i^2}{\sigma_1^2} - \frac{2\rho x_i y_i}{\sigma_1\sigma_2} + \frac{y_i^2}{\sigma_2^2}\right)\right\}.$$

The likelihood function is

$$\begin{aligned} L\left(\sigma_1, \sigma_2, \rho \mid (\mathbf{x}, \mathbf{y})\right) &= \prod_{i=1}^{n} f_{XY}\left(x_i, y_i \mid \sigma_1, \sigma_2, \rho\right) \\ &= \left(\frac{1}{2\pi\sigma_1\sigma_2\sqrt{1-\rho^2}}\right)^n \exp\left\{-\frac{1}{2\left(1-\rho^2\right)}\sum_{i=1}^{n}\left(\frac{x_i^2}{\sigma_1^2} - \frac{2\rho x_i y_i}{\sigma_1\sigma_2} + \frac{y_i^2}{\sigma_2^2}\right)\right\}, \end{aligned}$$

so that the log-likelihood function is

$$l = \log L = -n \log\left(2\pi\sigma_1\sigma_2\sqrt{1-\rho^2}\right) - \frac{1}{2\left(1-\rho^2\right)}\sum_{i=1}^{n}\left(\frac{x_i^2}{\sigma_1^2} - \frac{2\rho x_i y_i}{\sigma_1\sigma_2} + \frac{y_i^2}{\sigma_2^2}\right).$$

Differentiating l with respect to σ_1, σ_2 and ρ and setting to zero each time, we obtain the (maximum likelihood) estimates of these parameters

$$\hat{\sigma}_1 = \sqrt{\frac{1}{n}\sum_{i=1}^{n} x_i^2},$$

$$\hat{\sigma}_2 = \sqrt{\frac{1}{n}\sum_{i=1}^{n} y_i^2},$$

$$\hat{\rho} = \frac{\sum_{i=1}^{n} x_i y_i}{\sqrt{\sum_{i=1}^{n} x_i^2 \sum_{i=1}^{n} y_i^2}}.$$

*Note that if no substitutions are made for σ_1^2 and σ_2^2, then the resulting estimate $\hat{\rho}$ is *not* the product-moment correlation coefficient.

2.3.4 Standard Error of r (Pearson et al., 1896; Pearson and Filon, 1898; Student, 1908; Soper, 1913)

In the same 1896 paper, "Regression, heredity and panmixia" (Pearson et al., 1896), Pearson made a first attempt at deriving the standard deviation of the sample correlation coefficient r. Substituting $\lambda = \rho$ in Eq. (2.28) and remembering that the coefficient of $\delta\rho$ is zero, Pearson obtained

$$\frac{1}{n}\log u(\rho+\delta\rho) \approx \frac{1}{n}\log u(\rho) - \frac{(1+\rho^2)}{(1-\rho^2)^2}\frac{(\delta\rho)^2}{2} - \frac{2\rho(3+\rho^2)}{(1-\rho^2)^3}\frac{(\delta\rho)^3}{3} - \cdots$$

$$u(\rho+\delta\rho) \approx u(\rho)\exp\left\{-\frac{n(1+\rho^2)}{(1-\rho^2)^2}\frac{(\delta\rho)^2}{2} - \frac{2\rho n(3+\rho^2)}{(1-\rho^2)^3}\frac{(\delta\rho)^3}{3} - \cdots\right\}$$

(*ibid*., p. 265). By substituting $\rho+\delta\rho = r$ and neglecting terms that are $o(\delta\rho)^2$, the above becomes

$$u(r) \approx u(\rho)\exp\left\{-\frac{n(1+\rho^2)}{2(1-\rho^2)^2}(r-\rho)^2\right\}.$$

The function $u(r)$ can be recognized as the normal density function of r with mean ρ and variance

$$\text{var } r \approx \frac{(1-\rho^2)^2}{n(1+\rho^2)}. \tag{2.30}$$

Instead of $(1-\rho^2)^2/\{n(1+\rho^2)\}$, Pearson actually used $(1-r^2)^2/\{n(1+r^2)\}$ (the estimated variance of r) and stated that the formula had "sufficient accuracy for most cases."

However, Pearson was soon to realize that Eq. (2.30) underestimated the true variance of r because it is based on the assumption that both σ_1^2 and σ_2^2 are known, as we explained in the last section. The appropriate formula when both σ_1^2 and σ_2^2 are treated as unknown was provided in a subsequent paper by Pearson and Filon (1898). We now describe their proof.*

First, let $\eta_1, \eta_2, \ldots, \eta_p$ be p unknown parameters and let $(X_1, X_2, \ldots, X_m)$ be a multivariate random variable with density $z = f(x_1,\ldots,x_m,\eta_1,\eta_1,\ldots,\eta_p)$. If n independent observations are made on the multivariate random variable, then their joint density is

$$P_0 = \prod f(x_1,\ldots,x_m;\eta_1,\ldots,\eta_p).$$

* See also Plummer (1940, pp. 259–263).

Now, suppose P_δ is the joint density at $x_1+\delta h_1,\ldots,x_m+\delta h_m$ and $\eta_1+\delta\eta_1,\ldots,\eta_p+\delta\eta_p$. Then

$$\frac{P_\delta}{P_0}=\frac{\Pi f\left(x_1+\delta h_1,\ldots,x_m+\delta h_m;\eta_1+\delta\eta_1,\ldots,\eta_p+\delta\eta_p\right)}{\Pi f\left(x_1,\ldots,x_m;\eta_1,\ldots,\eta_p\right)}$$

(*ibid*., p. 231), and

$$\begin{aligned}\log\frac{P_\delta}{P_0}&=\Sigma f\left(x_1+\delta h_1,\ldots,x_m+\delta h_m;\eta_1+\delta\eta_1,\ldots,\eta_p+\delta\eta_p\right)-\Sigma f\left(x_1,\ldots,x_m;\eta_1,\ldots,\eta_p\right)\\&=\sum_t\left\{\delta h_t\sum_j\frac{\partial}{\partial x_j}(\log f)\right\}+\frac{1}{2}\sum_t\left\{(\delta h_t)^2\sum_j\frac{\partial^2}{\partial x_j^2}(\log f)\right\}\\&\quad+\sum_{t\neq t'}\left\{\delta h_t\ \delta h_{t'}\sum_{j,k}\frac{\partial^2}{\partial x_j\partial x_k}(\log f)\right\}+\sum_s\left\{\delta\eta_s\sum_j\frac{\partial}{\partial\eta_j}(\log f)\right\}\\&\quad+\frac{1}{2}\sum_s\left\{(\delta\eta_s)^2\sum_j\frac{\partial^2}{\partial\eta_j^2}(\log f)\right\}+\sum_{s\neq s'}\left\{\delta\eta_s\ \delta\eta_{s'}\sum_{j,k}\frac{\partial^2}{\partial\eta_j\partial\eta_k}(\log f)\right\}\\&\quad+\sum_{t,s}\left\{\delta h_t\ \delta\eta_s\sum_{j,k}\frac{\partial^2}{\partial x_j\partial\eta_k}(\log f)\right\}+\cdots\end{aligned}\tag{2.31}$$

Now, if the last equation is written as

$$\begin{aligned}\log\frac{P_\delta}{P_0}&=\sum_t A_t\delta h_t-\frac{1}{2}\sum_t B_t(\delta h_t)^2+\sum_{t\neq t'}C_{tt'}\delta h_t\delta h_{t'}+\sum_s D_s\delta\eta_s-\frac{1}{2}\sum_s E_s(\delta\eta_s)^2\\&\quad+\sum_{s\neq s'}F_{ss'}\delta\eta_s\delta\eta_{s'}+\sum_{t,s}G_{ts}\delta h_r\delta\eta_s+\cdots,\end{aligned}\tag{2.32}$$

and if the inside summations in Eq. (2.31) are replaced by integrals, then the constants $A_t, B_t, C_{tt'},\ldots$ can be found as follows. First,

$$\begin{aligned}A_t&=\iiint\cdots\int f\frac{\partial\log f}{\partial x_t}dx_1dx_2\ldots dx_m\\&=\iiint\cdots\int\frac{\partial f}{\partial x_t}dx_1dx_2\ldots dx_m\\&=\iiint\cdots\int[f]dx_1dx_2\ldots dx_{r-1}dx_{r+1}\ldots dx_m\end{aligned}$$

(*ibid*., p. 232). In the above, [*f*] is the difference in the values of *f* taken between the largest and smallest values of x_r. Since "in most cases of frequency the frequencies for extreme values of any organ are zero," we have $[f]=0$ so that $[A_r]=0$. Next,

$$B_t = -\iiint \cdots \int f \frac{\partial^2 \log f}{\partial x_t^2} dx_1 dx_2 \ldots dx_m,$$

$$C_{t,t'} = \iiint \cdots \int f \frac{\partial^2 \log f}{\partial x_t \partial x_{t'}} dx_1 dx_2 \ldots dx_m,$$

$$\begin{aligned}
D_s &= \iiint \cdots \int f \frac{\partial \log f}{\partial \eta_s} dx_1 dx_2 \ldots dx_m \\
&= \iiint \cdots \int \frac{\partial f}{\partial \eta_s} dx_1 dx_2 \ldots dx_m \\
&= \frac{\partial}{\partial \eta_s} \iiint \cdots \int f dx_1 dx_2 \ldots dx_m \\
&= \frac{\partial}{\partial \eta_s}(1) \\
&= 0.
\end{aligned}$$

Finally,

$$E_s = -\iiint \cdots \int f \frac{\partial^2 \log f}{\partial \eta_s^2} dx_1 dx_2 \ldots dx_m,$$

$$F_{ss'} = \iiint \cdots \int f \frac{\partial^2 \log f}{\partial \eta_s \partial \eta_{s'}} dx_1 dx_2 \ldots dx_m,$$

$$G_{rs} = \iiint \cdots \int f \frac{\partial^2 \log f}{\partial x_r \partial \eta_s} dx_1 dx_2 \ldots dx_m$$

(*ibid.*, p. 233).

Now, consider a random sample of n bivariate observations following a BivNormal $(\mu_1, \mu_2; \sigma_1^2, \sigma_2^2; \rho)$ distribution. The density of the deviations (x_1, x_2) from their respective means of any one pair of observations is

$$f = \frac{1}{2\pi\sigma_1\sigma_2\sqrt{1-\rho^2}} \exp\left[-\frac{1}{2}\left\{\frac{x_1^2}{\sigma_1^2(1-\rho^2)} - \frac{2\rho x_1 x_2}{\sigma_1\sigma_2(1-\rho^2)} + \frac{x_2^2}{\sigma_2^2(1-\rho^2)}\right\}\right]^* \tag{2.33}$$

(*ibid.*, p. 237). Since interest is in the standard errors of estimators of parameters (in particular, the standard error of the sample correlation coefficient), derivatives with respect to parameters only will be considered:

* Pearson's version of this density contains an extra n in the numerator since his density integrates to n.

$$\frac{\partial^2(\log f)}{\partial\sigma_1^2}=\frac{1}{\sigma_1^2}\left\{1-\frac{3x_1^2}{\sigma_1^2(1-\rho^2)}+\frac{2\rho x_1x_2}{\sigma_1\sigma_2(1-\rho^2)}\right\},$$

$$\frac{\partial^2(\log f)}{\partial\sigma_2^2}=\frac{1}{\sigma_2^2}\left\{1-\frac{3x_2^2}{\sigma_2^2(1-\rho^2)}+\frac{2\rho x_1x_2}{\sigma_1\sigma_2(1-\rho^2)}\right\},$$

$$\frac{\partial^2(\log f)}{\partial\sigma_1\partial\sigma_2}=\frac{\rho x_1x_2}{\sigma_1^2\sigma_2^2(1-\rho^2)},$$

$$\frac{\partial^2(\log f)}{\partial\sigma_1\partial\rho}=\frac{1}{\sigma_1}\left\{\frac{2\rho x_1^2}{\sigma_1^2(1-\rho^2)^2}-\frac{x_1x_2(1+\rho^2)}{\sigma_1\sigma_2(1-\rho^2)^2}\right\},$$

$$\frac{\partial^2(\log f)}{\partial\sigma_2\partial\rho}=\frac{1}{\sigma_2}\left\{\frac{2\rho x_2^2}{\sigma_2^2(1-\rho^2)^2}-\frac{x_1x_2(1+\rho^2)}{\sigma_1\sigma_2(1-\rho^2)^2}\right\},$$

$$\frac{\partial^2(\log f)}{\partial\rho^2}=\frac{1+\rho^2}{(1-\rho)^2}-\frac{1+3\rho^2}{(1-\rho^2)^3}\left(\frac{x_1^2}{\sigma_1^2}+\frac{x_2^2}{\sigma_2^2}\right)+\frac{6\rho+2\rho^3}{(1-\rho^2)^3}\frac{x_1x_2}{\sigma_1\sigma_2}.$$

Using the integral forms for the constants $E_s, F_{ss'}$ and the above derivatives, Pearson and Filon obtained

$$E_1=\frac{n}{\sigma_1^2}\frac{2-\rho^2}{1-\rho^2}=a_{11},\qquad E_2=\frac{n}{\sigma_2^2}\frac{2-\rho^2}{1-\rho^2}=a_{22}$$

$$F_{12}=\frac{n\rho^2}{\sigma_1\sigma_2(1-\rho^2)}=a_{12},\quad F_{13}=\frac{n\rho}{\sigma_1(1-\rho^2)}=a_{13}$$

$$F_{23}=\frac{n\rho}{\sigma_2(1-\rho^2)}=a_{23},\qquad E_3=\frac{n(1+\rho^2)}{(1-\rho^2)}=a_{33}$$

(*ibid*., p. 239). The final step was to appeal to a result in his previous paper, "Regression, heredity and panmixia" (Pearson et al., 1896, p. 301), where he had shown that the standard errors of various statistics could be obtained by using the a_{ij}'s above. In particular, the standard error of the correlation coefficient r is

$$\Sigma_r=\sqrt{\frac{A_{33}}{\Delta}}$$

where the Δ is the determinant of the symmetric matrix

$$\begin{bmatrix} a_{11} & -a_{12} & -a_{13} \\ -a_{12} & a_{22} & -a_{23} \\ -a_{13} & -a_{23} & a_{33} \end{bmatrix}$$

and A_{33} is the minor of the (3, 3)rd entry of the matrix. By substituting for the a_{ij}'s, Pearson and Filon finally obtained the standard error of the correlation coefficient r as

$$\Sigma_r = \sqrt{\frac{4n^2/\{\sigma_1^2\sigma_2^2(1-\rho^2)\}}{4n^3/\{\sigma_1^2\sigma_2^2(1-\rho^2)^3\}}} = \frac{1-\rho^2}{\sqrt{n}} \tag{2.34}$$

(Pearson and Filon, 1898, p. 242, Eq. (xvi.)). Pearson and Filon's formula underestimates the true variance when r is high and the kurtosis is small, and overestimates the true variance when both r and the kurtosis are high. It also applies only to large samples.

We make three comments on Pearson and Filon's derivation. First, as Fisher has correctly pointed out:

> The probable errors obtained in this way are those appropriate to the method of maximum likelihood, but not in other cases to statistics obtained by the method of moments, by which method the examples given were fitted. (Fisher, 1922c)

Second, although Pearson and Filon seem to be using the method of maximum likelihood in their derivation, this turns out not to be true (e.g., see Edwards, 1974; Stigler, 2007). Pearson and Filon did not spell it out, but their (2.33) is a posterior density based on a uniform prior rather than a direct probability.* Third, Pearson and Filon's method essentially boiled down to the following important large-sample formula for the variance of a *maximum likelihood* estimator $\hat{\theta}$:

$$\text{var } \hat{\theta} = \frac{1}{n\mathsf{E}\left(\dfrac{\partial \log f}{\partial \theta}\right)^2}. \tag{2.35}$$

In the above, f is the density of each observation and is assumed to satisfy certain regularity conditions (most notably the support of f should not depend on θ). The expression $\mathsf{E}(\partial \log f / \partial\theta)^2$ in the above was later (cf. 5.41) termed by Fisher the intrinsic accuracy (Fisher, 1922c, p. 339) or the (expected) information (I) provided by a single observation (Fisher, 1925c, p. 709),† that is,

$$I = \mathsf{E}\left(\frac{\partial \log f}{\partial \theta}\right)^2 = \mathsf{E}\left(\frac{\partial f/\partial\theta}{f}\right)^2 = \int_{-\infty}^{\infty}\left\{\frac{(\partial f/\partial\theta)^2}{f}\right\}d\theta. \tag{2.36}$$

Note that since $\int f dx = 1$, we have

*Edwards further adds: "I conclude from this phrase ['the frequency distribution for errors in the values of the frequency constants'] that the authors were thinking in terms of inverse probability..." (Edwards, 1974).

†More precisely, it is the expected information of a single observation regarding the parameter θ. The expected information of the whole sample is nI.

$$0=\int\frac{\partial f}{\partial\theta}dx=\int\frac{\partial\log f}{\partial\theta}fdx$$
$$\text{i.e., }\mathsf{E}\frac{\partial\log f}{\partial\theta}=0. \tag{2.37}$$

Differentiating Eq. (2.37) again with respect to θ,

$$\begin{aligned}0&=\int\left(\frac{\partial^2\log f}{\partial\theta^2}f+\frac{\partial\log f}{\partial\theta}\frac{\partial f}{\partial\theta}\right)dx\\&=\int\left(\frac{\partial^2\log f}{\partial\theta^2}+\frac{\partial\log f}{\partial\theta}\frac{\partial\log f}{\partial\theta}\right)fdx\\&=\mathsf{E}\frac{\partial^2\log f}{\partial\theta^2}+\mathsf{E}\left(\frac{\partial\log f}{\partial\theta}\right)^2.\end{aligned}$$

The expected information in Eq. (2.36) can thus be written as

$$I=\mathsf{E}\left(\frac{\partial\log f}{\partial\theta}\right)^2=-\mathsf{E}\frac{\partial^2\log f}{\partial\theta^2}.$$

The formula in (2.35) was also later (see Eq. 5.145) suggested by Edgeworth (1908), first explicitly (see Eq. 5.26) described by Fisher (1922c), and first rigorously proved by Hotelling (1930). Using it, we can derive both Eqs. (2.30) and (2.34) under the assumption that, for given population means, the population variances are known and unknown, respectively.

The first small-sample results concerning r and its standard error were obtained by Student (1908a). Through his admirable sense of intuition and the use of Pearson type II curves*, Student was able to correctly give the distribution of r when $\rho=0$ as

$$f(r)\propto\left(1-r^2\right)^{(n-4)/2}. \tag{2.38}$$

From the above, he was able to deduce that, again when $\rho=0$,

$$\text{var } r=\frac{1}{n-1}.$$

A generalization of Student's variance was obtained by Soper (1913), by giving the variance of r to a second approximation for any ρ:

$$\text{var } r\approx\frac{\left(1-\rho^2\right)^2}{n}\left(1+\frac{11\rho^2}{2n}\right). \tag{2.39}$$

* See Table 5.1 (p. 392)

The last formula also avoids the restrictions inherent in Pearson and Heron's formula (cf. Eq. 2.34). Soper also improved Pearson's formula for the expectation of r by showing that

$$\mathsf{E}r \approx \rho\left(1 - \frac{1-\rho^2}{2n}\right).$$

2.3.5 Development of Multiple Regression, Galton's Law of Ancestral Heredity, First Explicit Derivation of the Multivariate Normal Distribution (Pearson et al., 1896)

2.3.5.1 Development of Multiple Regression. Galton's Law of Ancestral Heredity

After having developed the bivariate normal surface in the landmark paper "Regression, heredity and panmixia" (Pearson et al., 1896), as we described in the two previous sections, Pearson next (*ibid*., p. 286) considered the trivariate case. He wished to study the (correlated) measurements made on three organs of an animal.

Pearson started by writing the trivariate normal (or correlation) surface in the form

$$P = C\exp\left\{-\frac{1}{2}\left(\lambda_1\frac{x^2}{\sigma_1^2} + \lambda_2\frac{y^2}{\sigma_2^2} + \lambda_3\frac{z^2}{\sigma_3^2} - \frac{2yz}{\sigma_2\sigma_3}\nu_1 - \frac{2xz}{\sigma_1\sigma_3}\nu_2 - \frac{2xy}{\sigma_1\sigma_2}\nu_3\right)\right\}, \quad (2.40)$$

where x, y, z are deviations from the means of the three organs, $\sigma_1, \sigma_2, \sigma_3$ are their respective standard deviations, and $C, \lambda_1, \lambda_2, \lambda_3, \nu_1, \nu_2, \nu_3$ are constants. Next, Pearson rewrote the above in two equivalent forms:

$$P = C\exp\left\{-\frac{1}{2}\lambda_1\left(\frac{x}{\sigma_1} - \frac{\nu_3 y}{\lambda_1\sigma_2} - \frac{\nu_2 z}{\lambda_1\sigma_3}\right)^2\right\} \times \exp\left\{-\frac{\lambda_2\lambda_1 - \nu_3^2}{2\lambda_1}\left(\frac{y}{\sigma_2} - \frac{z}{\sigma_3}\frac{\nu_1\lambda_1 + \nu_2\nu_3}{\lambda_2\lambda_1 - \nu_3^2}\right)^2\right\}$$
$$\times\exp\left\{-\frac{\lambda_1\lambda_2\lambda_3 - 2\nu_1\nu_2\nu_3 - \lambda_1\nu_1^2 - \lambda_2\nu_2^2 - \lambda_3\nu_3^2}{2(\lambda_1\lambda_2 - \nu_3^2)}\left(\frac{z}{\sigma_3}\right)^2\right\} \quad (2.41)$$

$$P = C\exp\left\{-\frac{1}{2}\lambda_1\left(\frac{x}{\sigma_1} - \frac{\nu_3 y}{\lambda_1\sigma_2} - \frac{\nu_2 z}{\lambda_1\sigma_3}\right)^2\right\}$$
$$\times\exp\left[-\frac{1}{2}\left\{\left(\frac{y}{\sigma_2}\right)^2\frac{\lambda_1\lambda_2 - \nu_3^2}{\lambda_1} + \left(\frac{z}{\sigma_3}\right)^2\frac{\lambda_1\lambda_3 - \nu_2^2}{\lambda_1} - \frac{2yz}{\sigma_2\sigma_3}\frac{\nu_1\lambda_1 + \nu_2\nu_3}{\lambda_1}\right\}\right] \quad (2.42)$$

By integrating the probability density in (2.41) with respect to x, y, and z, we obtain

$$C = \frac{n\sqrt{\chi}}{(2\pi)^{3/2}\sigma_1\sigma_2\sigma_3},^*$$

where

$$\chi = \lambda_1\lambda_2\lambda_3 - 2\nu_1\nu_2\nu_3 - \lambda_1\nu_1^2 - \lambda_2\nu_2^2 - \lambda_3\nu_3^2.$$

Furthermore, by integrating Eq. (2.42) with respect to x only and comparing with the bivariate normal surface,

$$\begin{aligned} &\frac{\lambda_1}{\lambda_1\lambda_2 - \nu_3^2} = \frac{\lambda_1}{\lambda_1\lambda_3 - \nu_2^2} = 1 - r_1^2 \\ &\frac{\nu_1\lambda_1 + \nu_2\nu_3}{\lambda_1\lambda_2 - \nu_3^2} = r_1, \end{aligned} \tag{2.43}$$

where r_1 is the correlation coefficient for the pair yz. Next, by integrating Eq. (2.41) with respect to x and y only and comparing to the univariate normal curve,

$$\lambda_1\lambda_2 - \nu_3^2 = \chi.$$

Therefore, using symmetry, the equations in Eq. (2.43) can be written as

$$\begin{aligned} &\lambda_1 = \chi(1-r_1^2), \quad \lambda_2 = \chi(1-r_2^2), \quad \lambda_3 = \chi(1-r_3^2), \\ &\nu_1\lambda_1 + \nu_2\nu_3 = \chi r_1, \quad \nu_2\lambda_2 + \nu_3\nu_1 = \chi r_2, \quad \nu_3\lambda_3 + \nu_1\nu_2 = \chi r_3, \end{aligned}$$

where r_2, r_3 are the correlation coefficients for the pairs xz and xy, respectively. Solving for the unknowns, Pearson finally obtained

$$\begin{aligned} &\lambda_1 = \chi(1-r_1^2), \quad \lambda_2 = \chi(1-r_2^2), \quad \lambda_3 = \chi(1-r_3^2), \\ &\nu_1 = \chi(r_1 - r_2 r_3), \quad \nu_2 = \chi(r_2 - r_3 r_1), \quad \nu_3 = \chi(r_3 - r_1 r_2), \\ &\chi = \frac{1}{1 - r_1^2 - r_2^2 - r_3^2 + 2r_1 r_2 r_3} \end{aligned}$$

(*ibid.*, p. 287). Hence, the trivariate correlation surface in Eq. (2.40) can be written as

$$\begin{aligned} P = \frac{n\sqrt{\chi}}{(2\pi)^{3/2}\sigma_1\sigma_2\sigma_3}\exp\Bigg[-\frac{1}{2}\chi\Bigg\{&\frac{x^2}{\sigma_1^2}(1-r_1^2) + \frac{y^2}{\sigma_2^2}(1-r_2^2) + \frac{z^2}{\sigma_3^2}(1-r_3^2) - 2(r_1 - r_2 r_3)\frac{yz}{\sigma_2\sigma_3} \\ &-2(r_2 - r_3 r_1)\frac{zx}{\sigma_3\sigma_1} - 2(r_3 - r_1 r_2)\frac{xy}{\sigma_1\sigma_2}\Bigg\}\Bigg]. \end{aligned}$$

* The n in the numerator of C implies that P here integrates to n (and not to unity).

By dividing the last trivariate surface with the bivariate surface for y and z, the conditional distribution of x given y and z can at once be found. Since the conditional distribution is normal, its mean and standard deviation can be directly read off. Pearson thus wrote the conditional standard deviation of x given y and z as

$$\Sigma_1 = \frac{\sigma_1}{\sqrt{\chi\left(1-r_1^2\right)}} = \sigma_1\sqrt{\frac{1-r_1^2-r_2^2-r_3^2+2r_1r_2r_3}{1-r_1^2}}$$

and the conditional mean of x given y and z as

$$h_1 = \frac{r_3 - r_1r_2}{1-r_1^2}\frac{\sigma_1}{\sigma_2}h_2 + \frac{r_2 - r_1r_3}{1-r_1^2}\frac{\sigma_1}{\sigma_3}h_3, \tag{2.44}$$

where h_2 and h_3 are the given values of y and z, respectively. With Eq. (2.44), Pearson had effectively derived the first multiple regression model in terms of the two predictors h_2 and h_3. He further added:

> Expressions of the form $\frac{r_3 - r_1r_2}{1-r_1^2}$ will be spoken of as coefficients of double correlation, and expressions of the form $\frac{r_3 - r_1r_2}{1-r_1^2}\frac{\sigma_1}{\sigma_2}$ as coefficients of double regression. (*ibidem*)

Pearson's "coefficients of double regression" are nowadays called *coefficients of partial regression.*

Equipped with the formula in Eq. (2.44), Pearson was to make an important use of it, namely, the verification of the so-called Galton's law of Ancestral Heredity. In his 1885 paper "Regression towards mediocrity in heredity stature," Galton had stated:

> The influence, pure and simple, of the midparent may be taken as 1/2, of the mid-grand-parent 1/4, of the mid-greatgrandparent 1/8, and so on. That of the individual parent would therefore be 1/4, of the individual grandparent 1/16, of an individual in the next generation 1/64, and so on. (Galton, 1885b, p. 261)

In the same 1885 paper, Galton gave a derivation of the law, which amounts to the formula

$$\mathsf{E}\left(y_0 \mid y_1, y_2, y_3, \ldots\right) = \frac{1}{2}y_1 + \frac{1}{4}y_2 + \frac{1}{8}y_3 + \cdots \tag{2.45}$$

(see Bulmer, 2003, p. 240), where y_0 is the offspring value of some measurement, y_1 is the midparent value, y_2 is the mid-grandparent value, and so on. Galton incorrectly assumed that the form (2.45) of his law was equivalent to the one he gave previously (in the quotation above).

In the 1896 paper (Pearson et al., 1896), Pearson investigated Galton's law in (2.45) which at this stage Pearson called "The Theory of the Midparent." In Eq. (2.44), let h_1, h_2, h_3 be the deviations of an offspring, a male parent, and a female parent from its respective mean. Now, the correlation r_1 for the same or different organs in two parents is usually small, so that Eq. (2.44) becomes

$$h_1 = r_3 \frac{\sigma_1}{\sigma_2} h_2 + r_2 \frac{\sigma_1}{\sigma_3} h_3$$

(*ibid.*, p. 288). But Galton's law in (2.45) implies $h_1 = (h_2 + h_3)/4 + \cdots$. Therefore, Pearson concluded that Galton's "mid-parent theory must be looked upon as only an approximation of a rough kind."

Following the 1896 paper, Pearson later did a more thorough investigation of the subject matter in 1897 (Pearson, 1897a). This resulted in the formulation of his own version of Galton's law of Ancestral Heredity as an extension of Eq. (2.44).

2.3.5.2 First Explicit Derivation of the Multivariate Normal Distribution The 1896 fundamental paper "Regression, heredity and panmixia" (Pearson et al., 1896) also contained the first explicit derivation of the multivariate normal distribution in its general (and modern) form. In Section 10 of this fundamental paper, Pearson wished to study measurements made on several organs of an animal. He distinguished between two sets of organs: one affected by natural selection (i.e., "selected") and the other unselected. His first question was:

> Given $p+1$ normally correlated organs, p out of these organs are selected in the following manner: each organ is selected normally *round a given mean*, and the p selected organs, pair and pair, are correlated in any arbitrary manner. What will be the nature of the distribution of the remaining $(p+1)$th organ? (*ibid.*, p. 298)

Pearson started by considering the unselected organs. Let $\beta_1, \beta_2, \beta_3, \ldots$ be the estimated regression coefficients of the $(p+1)$th organ on the p organs. If the values of these p organs are $h_1, h_2, h_3, \ldots$, then the $(p+1)$th organ will be normally distributed with mean $\beta_1 h_1 + \beta_2 h_2 + \beta_3 h_3 + \cdots$ and standard deviation σ "given by the general theory of correlation." Now, if the values of these p organs are $h_1 + x_1, h_2 + x_2, h_3 + x_3, \ldots$, then the $(p+1)$th organ will be normally distributed with standard deviation σ and mean

$$\beta_1 (h_1 + x_1) + \beta_2 (h_2 + x_2) + \beta_3 (h_3 + x_3) + \cdots = \zeta + \sum_i \beta_i x_i,$$

where $\zeta = \sum_i \beta_i h_i$. Thus, the density of the $(p+1)$th organ evaluated at v is proportional to

$$\exp\left\{ -\frac{(v - \zeta - \Sigma \beta_i x_i)^2}{2\sigma^2} \right\}. \tag{2.46}$$

Pearson next wrote the correlation surface for the unselected organs centering round the values $h_1, h_2, h_3, \ldots$ in the form of a multivariate normal distribution as

$$z = \text{constant} \times \exp\left\{ -\frac{1}{2}\left(a_{11} x_1^2 + a_{22} x_2^2 + \cdots + 2a_{12} x_1 x_2 + \cdots \right) \right\}. \tag{2.47}$$

Since (2.46) is the conditional density of v given $x_1, x_2, x_3, \ldots$ and (2.47) is the joint density of $x_1, x_2, x_3, \ldots$, the "total" density of v is

$$z = \text{constant} \times \int_{-\infty}^{\infty}\int_{-\infty}^{\infty}\int_{-\infty}^{\infty} \cdots \exp\left\{-\frac{(v-\zeta-\Sigma\beta_i x_i)^2}{2\sigma^2} - \frac{1}{2}\left(\Sigma a_{ii}x_i^2 + 2\Sigma a_{ij}x_i x_j\right)\right\} dx_1 dx_2 dx_3 \ldots \tag{2.48}$$

(*ibid*., p. 299). To perform the integration above with respect to $x_1, x_2, x_3, \ldots$, Pearson centered the exponent as follows. He first let $v - \zeta = u$ and denoted the coordinates of the center as $x'_1, x'_2, x'_3, \ldots$, where the centering equations* are

$$\begin{aligned}
\beta_1\left(u - \textstyle\sum_i \beta_i x'_i\right)/\sigma^2 &= a_{11}x'_1 + a_{12}x'_2 + a_{13}x'_3 + \cdots, \\
\beta_2\left(u - \textstyle\sum_i \beta_i x'_i\right)/\sigma^2 &= a_{21}x'_1 + a_{22}x'_2 + a_{23}x'_3 + \cdots, \\
\beta_3\left(u - \textstyle\sum_i \beta_i x'_i\right)/\sigma^2 &= a_{31}x'_1 + a_{32}x'_2 + a_{33}x'_3 + \cdots, \\
&\cdots
\end{aligned}$$

Then,

$$\begin{aligned}
\Delta x'_1 &= \left(\beta_1 A_{11} + \beta_2 A_{12} + \beta_3 A_{13} + \cdots\right)\left(u - \textstyle\sum \beta_i x'_i\right)/\sigma^2, \\
\Delta x'_2 &= \left(\beta_1 A_{21} + \beta_2 A_{22} + \beta_3 A_{23} + \cdots\right)\left(u - \textstyle\sum \beta_i x'_i\right)/\sigma^2, \\
\Delta x'_3 &= \left(\beta_1 A_{31} + \beta_2 A_{32} + \beta_3 A_{33} + \cdots\right)\left(u - \textstyle\sum \beta_i x'_i\right)/\sigma^2, \\
&\cdots
\end{aligned} \tag{2.49}$$

where Δ and the A_{ij}'s are, respectively, the determinant and minors of the matrix $\{a_{ij}\}$. Multiplying the above equations, respectively, by $\beta_1, \beta_2, \beta_3, \ldots$ and adding,

$$\begin{aligned}
\sigma^2\Delta\textstyle\sum \beta_i x'_i &= \left(\beta_1^2 A_{11} + \beta_2^2 A_{22} + \beta_3^2 A_{33} + \cdots + 2A_{12}\beta_1\beta_2 + 2A_{13}\beta_1\beta_3 + \cdots\right)\left(u - \textstyle\sum \beta_i x'_i\right) \\
&= \left(\textstyle\sum \beta_i^2 A_{ii} + 2\textstyle\sum \beta_i\beta_j A_{ij}\right)\left(u - \textstyle\sum \beta_i x'_i\right).
\end{aligned}$$

Hence,

$$\Sigma\beta_i x'_i = \frac{u\chi}{\sigma^2\Delta + \chi},$$

where

*To understand how the centering equations are obtained, consider the univariate case: $(u-\beta_1 x_1)^2/(2\sigma^2) - a_{11}x_1^2/2$ needs to be written in the form $k_1(x_1 - x'_1)^2 + k_2$, where x_1 is the center and k_1, k_2 are constants. To obtain the value of k_1, differentiate both expressions with respect to x_1: $(-\beta_1)(u-\beta_1 x_1)/\sigma^2 - a_{11}x_1 \equiv 2k_1(x_1 - x'_1)$; then set $x_1 = x'_1$ resulting in $(-\beta_1)(u-\beta_1 x'_1)/\sigma^2 - a_{11}x'_1 = 0$. This is the centering equation. In this simple case, $x'_1 = \beta_1 u/(a_{11}\sigma^2 + \beta_1^2)$.

$$\chi = \Sigma\beta_i^2 A_{ii} + 2\Sigma A_{ij}\beta_i\beta_j.$$

By substituting the formula just found for $\Sigma\beta_i x_i'$ in Eq. (2.49), the centers $x_1', x_2', x_3', \ldots$ can now be determined. Thus, the density of v in (2.48) can be written as

$$z = \text{constant} \times \int_{-\infty}^{\infty}\int_{-\infty}^{\infty}\int_{-\infty}^{\infty}\cdots$$
$$\exp\left[-\frac{u^2}{2}\left(\frac{1}{\sigma^2} - \frac{\chi}{\sigma^2\Delta + \chi}\right) - \frac{1}{2}\left\{\Sigma x_i^2\left(a_{ii} + \frac{\beta_i^2}{\sigma^2}\right) + 2\Sigma x_i x_j\left(a_{ij} - \frac{\beta_i\beta_j}{\sigma^2}\right)\right\}\right] dx_1 dx_2 dx_3 \ldots,$$

where $u = v - \zeta$ and $x_1, x_2, x_3, \ldots$ are now the coordinates with the centers $x_1', x_2', x_3', \ldots$ as origin. By performing the multiple integral above, Pearson finally obtained the univariate normal density

$$z = \text{constant} \times \exp\left\{-\left(\frac{u^2/2}{\sigma^2 + \chi/\Delta}\right)\right\}$$

(*ibid*., p. 300). From the above form, Pearson concluded that the $(p+1)$th organ also followed a normal distribution, and its mean and variance were $\beta_1 h_1 + \beta_2 h_2 + \beta_3 h_3 + \cdots$ and $\sigma^2 + \chi/\Delta$, respectively.

Pearson next considered a selected correlation surface of the form

$$z = \text{constant} \times \exp\left\{-\frac{1}{2}\left(a_{11}x_1^2 + a_{22}x_2^2 + \cdots + 2a_{12}x_1x_2 + \cdots\right)\right\}, \tag{2.50}$$

where the first two variables x_1 and x_2 are assigned the values η_1 and η_2. Then, to center the remaining variables $x_3, x_4, \ldots$ in Eq. (2.50), the required equations are

$$\begin{aligned} 0 &= \eta_1 a_{31} + \eta_2 a_{32} + a_{33}x_3' + a_{34}x_4' + \cdots \\ 0 &= \eta_1 a_{41} + \eta_2 a_{42} + a_{43}x_3' + a_{44}x_4' + \cdots \\ &\cdots. \end{aligned}$$

For mathematical convenience, Pearson introduced two extra identities before the equations above:

$$\begin{aligned} f + \eta_1 a_{11} + \eta_2 a_{12} &= \eta_1 a_{11} + \eta_2 a_{12} + a_{13}x_3' + a_{14}x_4' + \cdots \\ g + \eta_1 a_{21} + \eta_2 a_{22} &= \eta_1 a_{21} + \eta_2 a_{22} + a_{23}x_3' + a_{24}x_4' + \cdots \\ 0 &= \eta_1 a_{31} + \eta_2 a_{32} + a_{33}x_3' + a_{34}x_4' + \cdots \\ 0 &= \eta_1 a_{41} + \eta_2 a_{42} + a_{43}x_3' + a_{44}x_4' + \cdots \\ &\cdots \end{aligned} \tag{2.51}$$

(*ibid*., p. 301), where $f = a_{13}x_3' + a_{14}x_4' + \cdots$ and $g = a_{23}x_3' + a_{24}x_4' + \cdots$. From the last expressions,

$$f = \frac{\eta_1 A_{22} - \eta_2 A_{12}}{A_{11}A_{22} - A_{12}^2}\Delta - \eta_1 a_{11} - \eta_2 a_{12},$$

$$g = \frac{\eta_1 A_{11} - \eta_1 A_{12}}{A_{11}A_{22} - A_{12}^2}\Delta - \eta_1 a_{12} - \eta_2 a_{22},$$

where Δ and the A_{ij}'s are, respectively, the determinant and minors of the a_{ij}'s in (2.50). Then the correlation surface in (2.50) becomes

$$z = \text{constant} \times \exp\left\{-\frac{1}{2}\left(a_{11}\eta_1^2 + 2a_{12}\eta_1\eta_2 + a_{22}\eta_2^2\right)\right\}$$
$$\times \exp\left\{-\frac{1}{2}\left(a_{33}x_3^2 + a_{44}x_4^2 + \cdots + 2a_{34}x_3x_4 + \cdots\right)\right\}.$$

Integrating $x_3, x_4, \ldots$ out from the above, Pearson obtained the correlation surface for η_1, η_2 as

$$z' = \text{constant} \times \exp\left\{-\frac{1}{2} \cdot \frac{\Delta}{1 - \left(\dfrac{A_{12}^2}{A_{11}A_{22}}\right)}\left(\frac{\eta_1^2}{A_{11}} + \frac{\eta_2^2}{A_{22}} - 2\eta_1\eta_2\frac{A_{12}}{A_{11}A_{22}}\right)\right\}.$$

The above represents the joint density of η_1, η_2 (viewed as random variables) and is the bivariate normal distribution. If x_1, x_2 have correlation ρ_{12} and variances s_1^2, s_2^2, then

$$\rho_{12}^2 = \frac{A_{12}^2}{A_{11}A_{22}}, \quad s_1^2 = \frac{A_{11}}{\Delta}, \quad s_2^2 = \frac{A_{22}}{\Delta}. \tag{2.52}$$

The next section of Pearson's paper (*ibid*., p. 301) is even more important because it is where he gave an explicit expression for the multivariate normal distribution (which he called "Edgeworth's theorem"). Pearson stated that he wished to use the results in (2.52) to deduce Edgeworth's theorem.

Referring to the theory of minors found in Salmon's *Higher Algebra* (Salmon, 1866, p. 24), Pearson first wrote

$$\Delta^{p-1} = \begin{vmatrix} A_{11} & A_{12} & A_{13} & \cdots \\ A_{21} & A_{22} & A_{23} & \cdots \\ A_{31} & A_{32} & A_{33} & \cdots \\ \cdots & \cdots & \cdots & \cdots \end{vmatrix}.$$

Using the results in (2.52) and similar results for other pairs of variables, the above becomes

$$\Delta^{p-1} = \Delta^p s_1^2 s_2^2 s_3^2 \cdots \begin{vmatrix} 1 & \rho_{12} & \rho_{13} & \cdots \\ \rho_{21} & 1 & \rho_{23} & \cdots \\ \rho_{31} & \rho_{32} & 1 & \cdots \\ \cdots & \cdots & \cdots & \cdots \end{vmatrix}.$$

Writing the determinant of the correlation coefficients above as R, Pearson obtained

$$\frac{1}{\Delta} = s_1^2 s_2^2 s_3^2 \ldots R.$$

Now, let $B_{11}, B_{22}, \ldots, B_{12}, \ldots$ be the minors of the A-determinant and let $R_{11}, R_{22}, \ldots, R_{12}, \ldots$ be the minors of the R-determinant. Then using the above result and Salmon's *Higher Algebra* again,

$$a_{11} = \frac{B_{11}}{\Delta^{p-2}} = \frac{\Delta R_{11} s_1^2 s_2^2 s_3^2 \ldots}{s_1^2} = \frac{R_{11}}{R s_1^2},$$

$$a_{22} = \frac{B_{22}}{\Delta^{p-2}} = \frac{\Delta R_{22} s_1^2 s_2^2 s_3^2 \ldots}{s_2^2} = \frac{R_{22}}{R s_2^2},$$

$$a_{12} = \frac{B_{12}}{\Delta^{p-2}} = \frac{\Delta R_{12} s_1^2 s_2^2 s_3^2 \ldots}{s_1 s_2} = \frac{R_{12}}{R s_1 s_2},$$

$$\ldots.$$

Using the above values of the a_{ij}'s and a suitable constant, Pearson wrote the correlation surface (for the selected organs) in (2.50) as

$$z = \frac{\mu n}{(2\pi)^{p/2} s_1 s_2 s_3 \ldots \sqrt{R}} \exp\left\{ -\frac{1}{2R}\left(R_{11}\frac{x_1^2}{s_1^2} + R_{22}\frac{x_2^2}{s_2^2} + \cdots + 2R_{12}\frac{x_1 x_2}{s_1 s_2} + \cdots \right) \right\}^{*} \quad (2.53)$$

(Pearson et al., 1896, p. 302), where n is the sample size ("the total number of sets of p organs") and μ is a "numerical factor denoting the number of $(p+1)$th organs corresponding to each set…which is assumed to be practically constant…over the selected portion of it" (i.e., of the correlation surface). With Eq. (2.53), Pearson had effectively derived the general form of the multivariate normal distribution.

2.3.6 Marriage of Regression with Least Squares (Yule, 1897)

Both Galton's and Pearson's developments of regression and correlation, as detailed in the preceding sections, were explicitly based on the assumption of normality for the underlying variables. The question remained on how to proceed when that assumption was no longer tenable. Was it then reasonable to talk of regression of one variable on the

* The n in the numerator of z implies that the latter here integrates to n (and not to unity).

other, and if so how should the regression line be obtained? This was essentially the question that Yule addressed in his landmark 1897 paper "On the significance of Bravais' formulae for regression, &c., in the case of skew correlation" (Yule, 1897). But before describing Yule's work, we shall say a few words about his life.

George Udny Yule* (Fig. 2.14) was born at Morham, near Haddington, Scotland, on February 18, 1871. His father, who had been knighted, was an administrator in the Indian Civil Service and belonged to an old established family with a history of some government, military, and literary distinction. In 1875, the Yule family moved to London. Yule received his early education at Winchester College and then at University College London when he was 16 years old. There he studied civil engineering in 1887–1890 but did not take a degree since there was none in the subject at that time. After graduating, Yule spent one year in a small engineering works (George Wailes) and then another at Arc Works, Chelmsford. In 1892, he realized his interest was not in engineering and spent one year under Heinrich Hertz in Bonn. There he wrote his first paper on the passage of electric waves through electrolytes. When Yule returned to London in 1893, he was offered a position by Karl Pearson as demonstrator at University College. In 1896, Yule was promoted to Assistant Professor of Applied Mathematics. Three years later he married May Winifred. After his marriage, Yule felt obliged to earn a higher salary, so he resigned his position at University College to take a dreary administrative post at the City and Guilds of London Institute. However, owing to his continued interest in statistics, he also kept a position as Newmarch Lecturer on Statistics at University College during the period 1902–1909. His evening lectures during that period lead to the publication of the important book *An Introduction to the Theory of Statistics* (Yule, 1911) (Fig. 2.15). Apart from this, Yule also published some landmark papers, including his 1897 and 1907 memoirs on correlation (Yule, 1897; 1907) and his 1900 memoir on association (Yule, 1900). After the publication of his book, Yule was offered a position as Lecturer of Statistics in the School of Agriculture at the University of Cambridge. During World War I, he was statistician to the director of army contracts, later to the Ministry of Food. In 1922, Yule became a fellow of the Royal Society. He resigned his university position in 1931 and passed away in Cambridge, England, on June 26, 1951.

We now come back to Yule's 1897 paper (Yule, 1897). Right at the start, Yule made clear that he wished to develop a theory of regression without the assumption of normality:

> The only theory of correlation at present available for practical use is based on the normal law of frequency, but, unfortunately, this law is not valid in great many cases which are both common and important. It does not hold good, to take examples from biology, for statistics of fertility in man, for measurements on flowers, or for weight measurements even on adults. In economics statistics, on the other hand, normal distributions appear to be highly exceptional: variation of wages, prices, valuations, pauperism, and so forth, are always skew. In cases like these we have at present no means of measuring the correlation by one or more "correlation coefficients" such as afforded by the normal theory. (*ibid.*, p. 477)

In the above, Yule also wished to extend the application of regression and correlations to socioeconomic studies. To proceed, Yule considered the Ox and Oy axes (cf. Fig. 2.16)

*Yule's life is also described in Yates (1952) and Edwards (2001).

FIGURE 2.14 George Udny Yule (1871–1951). Wikimedia, http://en.wikipedia.org/wiki/Udny_Yule

AN INTRODUCTION TO THE

THEORY OF STATISTICS

BY

G. UDNY YULE,

HONORARY SECRETARY OF THE ROYAL STATISTICAL SOCIETY OF LONDON;
MEMBER OF THE INTERNATIONAL STATISTICAL INSTITUTE;
FELLOW OF THE ROYAL ANTHROPOLOGICAL INSTITUTE.

With 53 Figures and Diagrams.

LONDON:
CHARLES GRIFFIN AND COMPANY, LIMITED,
EXETER STREET, STRAND.
1911.
[*All Rights Reserved.*]

FIGURE 2.15 Title page of first edition of Yule's *Introduction to the Theory of Statistics* (Yule, 1911)

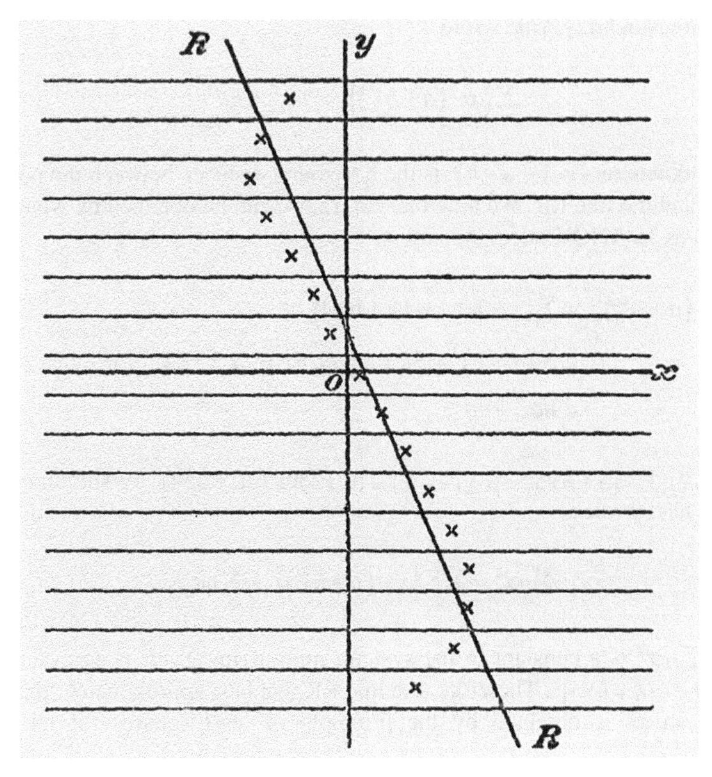

FIGURE 2.16 Plot of mean x-values (denoted by ×) at given y-values for a three-dimensional frequency surface. From Yule (1897, p. 478)

from an arbitrary three-dimensional frequency (density) surface. The points denoted by × represent the means of successive x-arrays (where each array is made up of several individual points) for given values of y and trace a curve which Yule called “the curve of regression of x on y.” The line RR is then drawn “by subjecting the distances of the means from the line to some minimal condition.” Yule further noted that, in the case of a bivariate normal frequency surface, the curve of regression would coincide with RR, that is, the curve of regression of x on y would be a straight line (as had been found by Galton and Pearson).

Now, let n be the number of observations in any x-array and let d be the horizontal distance between the corresponding × point and the line RR. Then came the key sentence:

> I propose to subject the line to the condition that the sum of all quantities like nd^2 shall be a minimum, i.e., I shall use the condition of least squares. (*ibid.*, p. 478)

Let the individual point taken from a given array have coordinates (x, y), the standard deviation of the points in that array be σ, and the line RR have equation

$$X = a + bY.$$

Then, for the given array, Yule wrote

$$\Sigma\{x-(a+bY)\}^2 = n\sigma^2 + nd^2 \tag{2.54}$$

(*ibid*., p. 479), where $d = \bar{x} - a - bY$ is the horizontal distance between the point X in the given array and the line RR. We note that Eq. (2.54) can be obtained by what later came to be known as an ANOVA decomposition:

$$\begin{aligned}\Sigma\{x-(a+bY)\}^2 &= \Sigma\{x-\bar{x}+\bar{x}-(a+bY)\}^2\\ &= \Sigma(x-\bar{x})^2 + \Sigma\{\bar{x}-(a+bY)\}^2 + 2\Sigma(x-\bar{x})\{\bar{x}-(a+bY)\}\\ &= n\sigma^2 + nd^2,\end{aligned}$$

since $\Sigma(x-\bar{x})\{\bar{x}-(a+bY)\} = d\Sigma(x-\bar{x}) = 0$. From Eq. (2.54), by summing across all arrays, Yule next wrote

$$\Sigma nd^2 = \Sigma\Sigma\{x-(a+bY)\}^2 - \Sigma n\sigma^2.$$

Now since $\Sigma n\sigma^2$ is a constant in the system, minimizing Σnd^2 is equivalent to minimizing $\Sigma\Sigma\{x-(a+bY)\}^2$. Therefore, the line RR that best approximates "the regression curve of x on y" is obtained by the principle of least squares (of the horizontal deviations):

> ...we may regard our method in another light. We may say that we form a single-valued relation
>
> $$x = a + by$$
>
> between a pair of associated deviations, such that the sum of the squares of our errors in estimating anyone x from its y by the relation is a minimum. This single-valued relation, which we may call the characteristic relation, is simply the equation to the line of regression RR. There will be two such equations to be formed corresponding to the two lines of regression. (*ibidem*)

Yule thus unified regression (and correlation) with least squares. Before Yule's paper, least squares had mainly been used for the adjustment of observations subject to errors (see Section 1.5.1). By extending Galton's regression to the case of a nonnormal bivariate surface, Yule obtained a nonlinear regression curve. By then minimizing the sum of squares of the deviations between the regression curve and a straight line, Yule was able to obtain the line that best approximated the regression curve, that is, the "line of best fit" (which he called the "characteristic relation").

The extension of Yule's method to the case of multiple variables $x_1, x_2, x_3, \ldots$ was natural. To perform the regression of x_1 on $x_2, x_3, x_4, \ldots$, one used the "characteristic" equation

$$x_1 = a_{12}x_2 + a_{13}x_3 + a_{14}x_4 + \cdots.$$

The regression coefficients $a_{12}, a_{13}, a_{14}, \ldots$ were then obtained by minimizing the sum of squares:

$$\Sigma\left\{x_1 - \left(a_{12}x_2 + a_{13}x_3 + a_{14}x_4 + \cdots\right)\right\}^2$$

across all observations.

Having done that much, there was one important additional task for Yule. How did his method compare with the so-called Bravais' (or Galton's) formula? Recall that in the latter, under the assumption of a bivariate normal distribution, the regression line of x on y was given by

$$x = \frac{rs_x}{s_y}y \tag{2.55}$$

and of y on x was given by

$$y = \frac{rs_y}{s_x}x \tag{2.56}$$

(cf. Eq. 2.13). In the above, x, y are, respectively, deviations from the means, s_x, s_y are standard deviations of the x- and y-values, and r is the product-moment correlation coefficient. Yule's approach was to first write the "characteristic" lines (obtained by least squares) for the two regressions as

$$\begin{aligned} x &= a_1 + b_1 y, \\ y &= a_2 + b_2 x. \end{aligned} \tag{2.57}$$

Using N to denote the number of (x, y) pairs, the normal equations for the first line are

$$\begin{aligned} \Sigma x &= Na_1 + b_1 \Sigma y, \\ \Sigma xy &= a_1 \Sigma y + b_1 \Sigma y^2 \end{aligned}$$

(*ibid.*, p. 480). Solving these simultaneously gives

$$a_1 = 0, \quad b_1 = \frac{\Sigma xy}{\Sigma y^2}.$$

Writing $(\Sigma x^2)/N = \sigma_1^2$, $(\Sigma y^2)/N = \sigma_2^2$, and $(\Sigma xy)/(N\sigma_1\sigma_2) = r$, the above can be expressed as

$$a_1 = 0, \quad b_1 = r\frac{\sigma_1}{\sigma_2}. \tag{2.58}$$

Similarly, for the second characteristic line in (2.57),

$$a_2 = 0, \quad b_2 = r\frac{\sigma_2}{\sigma_1}. \tag{2.59}$$

Now, the least squares intercepts and slopes in (2.58) and (2.59) are exactly those given by the "Bravais" lines in (2.55) and (2.56), respectively. In Yule's own words:

> *In any case, then, where the regression appears to be linear, Bravais' formulae may be used at once without troubling to investigate the normality of the distribution. The exponential character of the surface appears to have nothing whatever to do with the result.** (*ibid.*, 481)

In other words, the method of least squares, which makes no distributional assumptions, gives exactly the same characteristic lines as Bravais' formula, which is based on the assumption of normality. Therefore, we are told by Yule, we can simply apply the method of least squares to obtain the characteristic lines without troubling ourselves with normality.

A key critic of Yule's method was Karl Pearson. Even before the publication of the 1895 paper, in his communication with Yule, Pearson had expressed reservations about the use of least squares in Yule's treatment of the problem. Writing much later in 1920, Pearson left no doubts about how he felt on the matter:

> As early as 1897 Mr G. U. Yule, then my assistant, made an attempt in this direction. He fitted a line or plane by the method of least squares to a swarm of points, and this has been extended later to n-variates and is one of the best ways of reaching the multiple regression equations and the coefficient of multiple correlation. Now while these methods are convenient or utile, we may gravely doubt whether they are more accurate theoretically than the assumption of a normal distribution. Are we not making a fetish of the method of least squares as others made a fetish of the normal distribution? For how shall we determine that we are getting a "best fit" to our system by the method of least squares? (Pearson, 1920)

In spite of Pearson's criticism, Yule's paper was a major breakthrough and represents the way in which regression is still taught today, namely, through the method of least squares. However, in retrospect, there was some legitimacy in Pearson's tempering of the overoptimistic statements made by Yule concerning the ubiquity of the method of least squares. Although, as Yule claimed, it is true that the method of least squares is not based on any distributional assumption, inferential procedures on regression parameters can be undermined if normality of the dependent variable does not hold, especially when sample sizes are small.

2.3.7 Correlation Coefficient for a 2×2 Table (Yule, 1900). Feud Between Pearson and Yule

Pearson's disagreement with Yule's application of least squares to the problem of regression was nothing compared to the bitter dispute[†] that opposed them in the early 1900s over the 2×2 contingency table. The fight was over the correct measure of association in the 2×2 table and has often been described in the literature as the "politics of the contingency table." Pearson and Yule's aim here was not to test for the *existence* of

* Italics are Yule's.

[†] The Yule–Pearson dispute is also described in MacKenzie (1978, Chapter 7) and Cowles (2001, pp. 148–153).

TABLE 2.1 Table of Attributes for Yule's Coefficient of Association

	B	β
A	(AB)	$(A\beta)$
α	(αB)	$(\alpha\beta)$

The attribute in the first column has two levels A and α, and the attribute in the first row has two levels B and β. The frequencies of joint attributes are denoted with brackets, for example, (AB).

association, a task that Pearson had already performed with his X^2-test (cf. Chapter 3), but rather to measure the *amount* or *strength* of association.

The story really started with Yule's 1900 paper "On the association of attributes in statistics: with illustrations from the material of the childhood society, &c" (Yule, 1900). Since Pearson's 1896 paper (Pearson et al., 1896) had already developed a measure of association, namely the correlation coefficient r, for continuous variables, Yule wished to find a corresponding measure for two categorical variables or "attributes":

> Now it seems to me that one of the chief needs in handling statistics of the kind we are considering is some sort of "coefficient of association," which should take the place of the "*coefficient of correlation**" for continuous variables, and be a measure of the approach of association towards complete independence on the one hand and complete association on the other. (Yule, 1900, p. 271)

But such a measure should satisfy certain desirable conditions. Referring to Table 2.1:

> (1) Be zero when the variables or attributes A, B, are independent, and only when they are independent.
> (2) It should be +1 when, and only when, A and B are completely associated, i.e. when either
>
> $$\left.\begin{array}{r}\text{all } A\text{'s are } B\\ \text{all } \beta\text{'s are } \alpha\end{array}\right\}$$
>
> or
>
> $$\left.\begin{array}{r}\text{all } B\text{'s are } A\\ \text{all } \alpha\text{'s are } \beta\end{array}\right\}$$
>
> or when both of these statements are true together, which can only be when
>
> $$(A)=(B), \quad (\alpha)=(\beta).^{\dagger}$$

* Italics are Yule's.

† *Author's note:* (X) denotes the number of subjects with attribute X, and (XY) denotes the number of subjects with both attributes X and Y.

(3) It should be −1 when, and only when, A and β or B and α are completely associated, i.e., when either

$$\left.\begin{array}{l}\text{all } A\text{'s are } \beta \\ \text{all } B\text{'s are } \alpha\end{array}\right\}$$
$$\left.\begin{array}{l}\text{all } \beta\text{'s are } A \\ \text{all } \alpha\text{'s are } B\end{array}\right\}$$

or when both of these statements are true together, which can only be when

$$(A)=(\alpha), \quad (B)=(\beta).$$

(*ibid.*, p. 272)

Based on these three postulates, which he called "theorems," Yule proposed the following quantity as a measure of association for two attributes:

$$Q=\frac{(AB)(\alpha\beta)-(A\beta)(\alpha B)}{(AB)(\alpha\beta)+(A\beta)(\alpha B)} \tag{2.60}$$

(*ibidem*). Yule used the symbol Q in acknowledgment of Adolphe Quetelet* who had previously worked on "the association of some one attribute A with a series of others, B, C, D, &c." Yule also admitted that Q in Eq. (2.60) was not the only function that satisfied his three postulates for a reasonable measure of association:

> It is perfectly possible that other simple functions of the frequencies might be devised which should have the same properties, but Q at any rate will serve. (*ibidem*)

He further acknowledged that Pearson had pointed to him a property, which Yule had found quite remarkable, of the numerator of Q. Denoting by $(AB)_0$ the value (AB) would have if Q were zero, and so on, Pearson had pointed out that

> …The excesses of (AB) and $(\alpha\beta)$ above $(AB)_0$ and $(\alpha\beta)_0$, and of $(A\beta)_0$ and $(\alpha B)_0$ above $(A\beta)$ and (αB), are all equal, and equal to the ratio of the difference of the cross-products [in the numerator of Eq. (2.60)] to the number of observations. (*ibid.*, p. 273)

However, Yule was unable to find a

> …similar relation for the sum of the cross-products [in the denominator of Eq. (2.60)] so as to give a complete physical meaning to Q. (*ibidem*)

Shortly after Yule's paper, Pearson came back with his own measure of association for attributes in a 2×2 table (Pearson, 1900a), which we now describe. Pearson first assumed that a 2×2 table is derived from a double dichotomy of a BivNormal$(0, 0; \sigma_1^2, \sigma_2^2; \rho)$ distribution (Fig. 2.17):

*Yule explicitly acknowledged this in a later paper (Yule, 1912, p. 586).

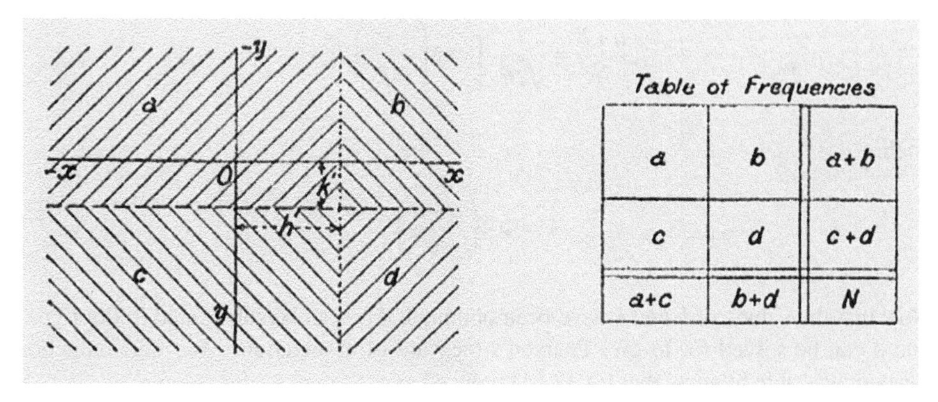

a	b	a+b
c	d	c+d
a+c	b+d	N

FIGURE 2.17 Plane and table representation for Pearson's measure of association for attributes (Pearson, 1900a, p. 2)

$$z = \frac{1}{2\pi\sigma_1\sigma_2\sqrt{1-\rho^2}} \exp\left\{ -\frac{1}{2(1-\rho^2)} \left(\frac{x^2}{\sigma_1^2} - \frac{2\rho xy}{\sigma_1\sigma_2} + \frac{y^2}{\sigma_2^2} \right) \right\}.$$

Thus, the levels of the first attribute can be written as $B = \{x : x \leq h'\}$ and $\beta = \{x : x > h'\}$, where h' is a constant. Similarly, the levels of the second attribute can be written as $A = \{y : y \leq k'\}$ and $\alpha = \{y : y > k'\}$.

Pearson first equated the observed frequency d to its theoretical counterpart:

$$\begin{aligned} \frac{d}{N} &= \frac{1}{2\pi\sqrt{1-\rho^2}} \int_{h'}^{\infty}\int_{k'}^{\infty} \exp\left\{ -\frac{1}{2(1-\rho^2)} \left(\frac{x^2}{\sigma_1^2} - \frac{2\rho xy}{\sigma_1\sigma_2} + \frac{y^2}{\sigma_2^2} \right) \right\} dydx \\ &= \frac{1}{2\pi\sqrt{1-\rho^2}} \int_{h}^{\infty}\int_{k}^{\infty} \exp\left\{ -\frac{1}{2(1-\rho^2)} \left(x^2 - 2\rho xy + y^2 \right) \right\} dydx, \end{aligned} \tag{2.61}$$

where $h = h'/\sigma_1$ and $k = k'/\sigma_2$ (*ibid.*, p. 2). The next step was to equate the observed marginal frequencies in the 2×2 table to their theoretical counterparts. For the first attribute,

$$\frac{a+c}{N} = \frac{1}{\sigma_1\sqrt{2\pi}} \int_{-\infty}^{h'} \exp\left(-\frac{x^2}{2\sigma_1^2} \right) dx,$$

so that h can be determined from standard normal tables:

$$h = \Phi^{-1}\left(\frac{a+c}{N} \right),$$

where Φ is the cumulative distribution function of a N (0, 1) variable. Similarly, for the second attribute,

$$\frac{a+b}{N} = \frac{1}{\sqrt{2\pi}} \int_{-\infty}^{k} \exp\left(-\frac{y^2}{2}\right) dx,$$

so that

$$k = \Phi^{-1}\left(\frac{a+b}{N}\right).$$

Now that the values of h and k have been obtained, they can be substituted in Eq. (2.61) and ρ can be solved for to give Pearson's measure of association. After much algebra, Pearson was able to show that Eq. (2.61) reduced to

$$\frac{d}{N} = \frac{(b+d)(c+d)}{N^2} + \sum_{n=1}^{\infty} \left(\frac{\rho^n}{n!} HK \overline{v}_{n-1} \overline{w}_{n-1} \right) \tag{2.62}$$

(*ibid.*, p. 6), where

$$H = \frac{1}{\sqrt{2\pi}} e^{-h^2/2}, \qquad K = \frac{1}{\sqrt{2\pi}} e^{-k^2/2},$$
$$\overline{v}_n = h\overline{v}_{n-1} - (n-1)\overline{v}_{n-2}, \qquad \overline{w}_n = k\overline{w}_{n-1} - (n-1)\overline{w}_{n-2},$$
$$\overline{v}_0 = 1, \overline{v}_1 = h, \qquad \overline{w}_0 = 1, \overline{w}_1 = k.$$

Finally, from Eq. (2.62), Pearson was able to obtain

$$\begin{aligned} \frac{2\pi(ad-bc)}{N^2} \exp\left(\frac{h^2+k^2}{2}\right) = \rho &+ \frac{\rho^2}{2!} hk + \frac{\rho^3}{3!}(h^2-1)(k^2-1) + \frac{\rho^4}{4!} hk(h^2-3)(k^3-3) \\ &+ \frac{\rho^5}{5!}(h^4-6h^2+3)(k^4-6k^2+3) \\ &+ \frac{\rho^6}{6!} hk(h^4-10h^2+15)(k^4-10k^2+15) \\ &+ \frac{\rho^7}{7!}(h^6-15h^4+45h^2-15)(k^6-15k^4+45k^2-15) \\ &+ \frac{\rho^8}{8!} hk(h^6-21h^4+105h^2-105)(k^6-21k^4+105k^2-105) + \cdots \end{aligned}$$

(*ibid.*, p. 6, Eq. (xix.)). The estimated value of ρ from this equation gives Pearson's measure of association for attributes in a 2×2 table. This measure is now called a *tetrachoric correlation* (r_T). According to Walker (1929, p. 189), the term tetrachoric in this connection was first used by Everitt (1910) while referring to the functions $H\overline{v}_{n-1}/\sqrt{n!}$ and $K\overline{w}_{n-1}/\sqrt{n!}$ in Eq. (2.62).

Three comments are in order. First, as the reader can see, there is no closed-form expression for r_T, a fact that limits its usefulness. Pearson was well aware of this and went to great length in the 1900 paper to give numerical procedures to facilitate the evaluation of r_T. For example, when X and Y are each split at the median (so that $h=k=0$), Pearson showed that

$$r_T \approx \cos\left(\pi \frac{b}{a+b}\right)$$

(Pearson, 1900a, p. 7). Second, it should be noted that r_T is *not* a product-moment correlation; rather it is an estimate of the parameter ρ from a bivariate normal distribution. Finally, Pearson's assumption that the level of an attribute is derived by choosing a suitable cutoff from a normal distribution is reasonable in some cases (e.g., tall vs. short, hot vs. cold, etc.). However, such an assumption is hardly tenable in many other cases, such as when the attribute clearly does not arise from a continually varying quantity (e.g., dead vs. alive, black vs. white, etc.).

Concerning Yule's Q, Pearson treated it rather dismissively. He proposed four other measures (Q_3, Q_4, Q_5 and κ), which satisfied Yule's three conditions (postulates), and together with Yule's Q (which Pearson denoted by Q_2) compared them to r_T, which Pearson viewed as the gold standard:

> Now an examination of this table shows that notwithstanding the extreme elegance and simplicity of Mr. YULE's coefficient of association Q_2, the coefficients Q_3, Q_4, and Q_5, which satisfy also his requirements, are much nearer to the values assumed by the correlation [r_T]. I take this to be such great gain that it more than counterbalances the somewhat greater labour of calculation. (*ibid*., p. 7)

A final important aspect of Pearson's 1900 paper concerned his proposal of the following measure:

$$r_{hk} = \frac{ad - bc}{\sqrt{(a+b)(c+d)(a+c)(b+d)}} \tag{2.63}$$

(*ibid*., p. 12). Pearson stated that r_{hk} "expresses the correlation between errors in the position of the means of the two characters under consideration." In other words, Pearson defined r_{hk} as the correlation between the errors (from their mean values) in h and k, *not* as the correlation between X and Y. Note, however, that r_{hk} can be shown to be identical to the product-moment correlation between the two attributes when each attribute is assigned a 0 or 1 depending on the level it assumes. As we shall soon see, r_{hk} had a considerable role in the Yule–Pearson dispute.

Cognizant that his previous measures of association were dependent on the *order* in which the levels of an attribute were arranged, Pearson sought yet another measure of association that would circumvent this limitation. In 1904, Pearson was able to use his X^2-test statistic (cf. Section 3.2.2) to achieve this (Pearson, 1904). He proposed:

$$\phi^2 = \frac{X^2}{N}, \tag{2.64}$$

and

$$C = \sqrt{\frac{\phi^2}{1+\phi^2}} = \sqrt{\frac{X^2}{X^2+N}}, \tag{2.65}$$

where N is the total frequency. Pearson called ϕ^2 the mean square contingency (X^2 itself being the square contingency) and C the first coefficient of contingency (nowadays known simply as the *contingency coefficient*). Several points should be noted about these measures:

- For a 2×2 table, $X^2 = \frac{N(ad-bc)^2}{(a+b)(c+d)(a+c)(b+d)}$ so that $\phi \equiv r_{hk}$.
- As Pearson noted in the 1904 paper, as the number of cells in a contingency table approaches infinity, $C \rightarrow \rho$.
- Since $0 \le X^2 \le N(\zeta - 1)$, where $\zeta = \min(R, C)$ for an $R\times C$ contingency table, we have $0 \le \phi^2 \le \zeta - 1$ and $0 \le C \le \sqrt{(\zeta-1)/\zeta}$. The latter two facts should be considered as limitations of both ϕ^2 and C as measures of the strength of association, since they are dependent on the size of the table.

It was now Yule's turn to respond. In a 1906 paper (Yule, 1906a), Yule criticized Pearson's 1900 paper on the tetrachoric correlation. He stated that if the underlying continuous variables, from which the levels of the attributes were derived, were normal, then the property of *isotropy* should be satisfied, that is, for a contingency table of arbitrary size, the sign of $ad-bc$ for any four *adjacent* cells should be the same. Yule then tested for isotropy in Pearson's contingency tables and found some of his results "most chaotic." For Yule, this raised serious doubts on the validity of the tetrachoric correlation, dependent as it was on the assumption of normality for isotropy to hold. Another paper published in the year (Yule, 1906b) claimed that the deviation from normality might be due to "subjective influences." In 1907, still trying to maintain civil tone, Pearson retaliated by stating:

> I cannot therefore accept Mr. Yule's test as very likely to be helpful in measuring deviation from Gaussian distribution. (Pearson, 1907, p. 472)

The Yule–Pearson dispute took an altogether different turn after the 1911 publication of Yule's influential book *Introduction to the Theory of Statistics* (1911). In that book, Chapter III, Yule discussed his Q-coefficient in (2.60) at great length. He also introduced the "correlation coefficient for a two-$\times$two-fold table":

$$Q' = \frac{N\{a-(a+c)(a+b)/N\}}{\sqrt{(a+b)(c+d)(a+c)(b+d)}} \tag{2.66}$$

(*ibid*., p. 213). However, Pearson's r_{T} was completely snubbed and the reaction came very soon. In a highly charged article entitled "The danger of certain formulae suggested as substitutes for the correlation coefficient" (Heron, 1911), Pearson's associate David Heron was furious:

> STATISTICAL Theory has suffered much in the past from the illegitimate application of processes which, when applied to appropriate data, are perfectly sound; but the introduction, without a single word of warning, of methods which in no circumstances can give correct results is much more dangerous. Especially is this the case when the methods claim to shorten the labour of the calculation of statistical constants, since they are invariably adopted by those who, unable or unwilling to examine critically their claims to validity, are dependent on any formula that is offered to them.
>
> A Text-Book of Statistical Theory should above all be free from such blunders and it is therefore much to be regretted that in Mr. G. Udny Yule's recent textbook*, greater care has not been taken to ensure that the processes described there have a sound theoretical foundation.
>
> On the present occasion, attention will only be directed to a single point, the methods suggested by Mr. Yule for the measurement of the degree of association between characters which are classified alternatively. (*ibid.*, p. 109)

Heron then cited an example, which results in $Q = .91$ and $Q' = .02$ at the same time. This allegedly put a serious dent in Yule's measures. Heron then pointed out that Yule's Q' in Eq. (2.66) was simply Pearson's 1900 r_{hk} in Eq. (2.63), a fact that Yule had actually overlooked because of the different notation he had used to define his Q'. Heron also took issue with Yule's implication that Q' was a correlation coefficient between the attributes, since Pearson had previously shown that r_{hk} was the correlation between the errors in the cutoffs, not between the attributes.

Heron continued his paper by further attacking the validity of Q and concluded as follows:

> ...much harm will arise to statistical science, owing to the postulating of false conclusions resulting from the use of inadequate formulae. (Heron, 1911, p. 122)

Heron's hostile paper could not go without any reply from Yule. In his 1912 paper, Yule retaliated but the tone was far from conciliatory:

> The normal [i.e. tetrachoric] coefficient has derived its repute solely from the belief that it gave the true correlation between the variables which the classification was supposed to represent; it is usually, indeed, termed "the correlation" without qualification, and the method spoken of as "Professor Pearson's fourfold-table method of determining the correlation," or some equivalent phrase. It is true that the author of the method gave several warnings as to its untrustworthiness in his original memoir, but these seem to have been almost immediately forgotten, even by himself. (Yule, 1912, pp. 629–630)

Yule continued, saying that he had been aware that Q and Q' could be different since they possessed essentially different properties:

> I was well aware they gave different values; I point out in the *Introduction* (p. 213) that the two coefficients possess essentially different properties, and I am unable to agree that Dr. Heron's "test of validity" is, in fact, any test. (*ibid.*, p. 631)

More importantly, Yule charged that Pearson and his colleagues had erroneously identified his Q' with Pearson's r_{hk}. Yule stated that, in fact, Q' was simply the product–sum

* *An Introduction to the Theory of Statistics*, Griffin and Co. Ltd., 1911 (*Yule's footnote*).

correlation coefficient (when each attribute is assigned a 0 or 1 depending on its level, as we explained on p. 249). He further added that "the comedy of errors has, it is hoped, ended."

Aside from the criticism of Pearson and his associates, Yule's 1912 paper also introduced an alternative measure of association, which he called the "coefficient of colligation":

$$w_{\text{coll}} = \frac{1-\sqrt{(bc)/(ad)}}{1+\sqrt{(bc)/(ad)}}$$

(*ibid.*, p. 592).

Yule was probably under no illusion that his 1912 rebuke would provoke no reaction from the Pearson camp. However, little could he have anticipated the *magnitude* of the reaction. The latter came in the form of a 157–page (!) scathing paper, "On theories of association," by Pearson and Heron (1913). Some of the sections of this paper had the revealing titles: "On the Idleness of Mr. Yule's Coefficient of Association…," "The Fallacy of Mr. Yule's Coefficient of Association…," "General Protest against the Use of Mr. Yule's Coefficient of Association," and "Further Criticisms of Mr. Yule's Methods of Controversy."

Pearson and Heron devoted several pages of their paper further pointing out the deficiencies in Yule's Q:

> …we have in fact no idea at all of what Mr. Yule's coefficients of association and colligation really measure…divisions taken at very slight distances apart may give hopelessly divergent values of Q, of which differences of values Mr. Yule has given no intelligible interpretation. (*ibid.*, p. 201)

They also reiterated the major point that Q lacked a statistical basis:

> He [Yule] then guesses a formula $\dfrac{ad-bc}{ad+bc}$ or another $\dfrac{\sqrt{ad}-\sqrt{bc}}{\sqrt{ad}+\sqrt{bc}}$ out of many thousands which can be invented. (*ibid.*, p. 177)

We shall avoid further elaborating on the highly hostile tone of the paper, but suffice it to say that many were surprised at just how acrimonious Pearson could become when seriously challenged.

In the end, the reader might ask which of the measures proposed by Yule and Pearson are to be preferred. There is no clear-cut answer. They are all useful in their own way and they all have their own share of disadvantages. The truth of the matter seems to lie in Yule's own words, spelled in the last paper on the matter:

> This brief review does not, of course, cover all the possible coefficients that have been suggested, but will suffice to confirm the argument of this section. Our condemnation of these coefficients, for the present purpose, may at least claim to be impartial, inasmuch as two of them were originally proposed by one of us.
>
> The results of this section are disappointing in so far as they fail to provide a simple answer to important practical questions. On the other hand, we venture to hope that they will be of value to subsequent inquirers. (Greenwood and Yule, 1915, p. 189)

2.3.8 Intraclass Correlation (Pearson, 1901; Harris, 1913; Fisher, 1921; 1925)

It is not very difficult to calculate Pearson's r when trying to determine the sample correlation coefficient between two characteristics (which might be the same or different) for pairs of members belonging to *different* class types. Such a correlation is said to be an *interclass* correlation. Examples include the correlation between the heights of brothers and sisters, or the correlation between the heights of brothers and weights of sisters. A difficulty arises, though, when the formula is applied on the same characteristic for pairs belonging to the *same* class type (so that each member of a given pair is *interchangeable* with the other), as in the correlation between the heights of brothers. The problem is that of choosing the *order* of entry in the correlation table (i.e., which one in a pair is the X variable and which one is the Y variable?).

Pearson was the first to explicitly address the issue in the ninth of a long series of papers entitled "Mathematical contributions to the theory of evolution" (Pearson, 1901a). Pearson proposed to deal with the problem by putting a double entry in the correlation table for each pair in a class and then calculating the usual r. In case each class consisted of multiple members, an entry was made for each pair of members. Thus, in his discussion of the correlation between the number of veins in leaves from a tree, Pearson stated:

> ...One hundred trees fairly of the same age and belonging to the same district, were selected, and twenty-six leaves specified by the letters of the alphabet were gathered from each of these. The leaves were gathered so far as possible all round the outside of the tree, roughly about the same height from the ground, and scattered over different parts of the individual boughs. Thus each tree was supposed to be individualized by twenty-six leaves. The veins on these leaves were then counted, and varied for beech-leaves in general between ten and twenty-two. All the possible pairs were now taken, i.e., $\frac{1}{2}(26 \times 25) = 325$ in number, and entered on a correlation table in the usual manner, the two variables being the number of veins in the first leaf and the number of veins in the second leaf. But as either member of the pair might be a "first" leaf, the table so formed was rendered symmetrical by starting with either leaf in the pair as first or second. Thus a single tree led to 650 entries in the correlation table, or with 100 trees there were 65,000 entries. (*ibid.*, pp. 292–293)

Pearson called the correlation coefficient thus calculated the "fraternal correlation" (R). Pearson's suggestion is sensible: if there are two brothers with heights a and b in a given class, two entries ($x = a$, $y = b$) and ($x = b$, $y = a$) should be made in the correlation table. On the other hand, if there are three brothers with heights a, b, and c, then six entries ($x = a$, $y = b$), ($x = b$, $y = a$), ($x = a$, $y = c$), ($x = c$, $y = a$), ($x = b$, $y = c$), and ($x = c$, $y = b$) should be made. The correlation table thus becomes symmetrical.

Pearson's method had been anticipated to some extent by Galton. In his book *Natural Inheritance*, while discussing the relationship between brothers, Galton had stated:

> ...These Tables were constructed by registering the differences between each possible pair of brothers in each family: thus if there were three brothers, A, B, and C, in a particular family, I entered the differences of stature between A and B, A and C, and B and C, four brothers gave rise to 6 entries, and five brothers to 10 entries. (Galton, 1889, p. 92)

One difficulty with Pearson's symmetric table method is its cumbersomeness, especially when the number of members in a given class becomes large. J. Arthur Harris* (1880–1930) clearly explained the issue in his 1913 paper "On the calculation of intra-class and inter-class coefficients of correlation from class moments when the number of possible combinations is large":

> When the number of observations is as large as desirable, the formation and verification of the ordinary correlation table (where each individual...is entered only once) is an irksome task, but when each measurement is compared with a number of others, the purely clerical labour involved in tabulation becomes onerous in the extreme. If n is the number of individuals in any class...the number of combinations for each class is $n(n-1)$ which gives for the m classes constituting the population $S\,[n\,(n-1)]$† or, where n is constant, $m\,[n\,(n-1)]$ entries...With n as low as 20 and m only 250, this gives 95,000 combinations. (Harris, 1913, p. 447)

Thereupon, Harris proposed a shortcut formula based on the *original table* to calculate the fraternal correlation, which he called the "intraclass" correlation (ICC).‡ Harris' key insight was that the symmetrical table was augmented by *permutations* of the entries within classes in the original table. Therefore, the ICC could be calculated by computing means and variances within a class in the original table and then averaging across classes. Harris' shortcut expression for the ICC was based on the well-known product-moment formula for the correlation coefficient between X and Y:

$$r_{xy} = \frac{\Sigma(xy)/N - m_x m_y}{S_x S_y} \tag{2.67}$$

(*ibid*., p. 448), where N is the *total number of measurements*, $m_x = (\Sigma x)/N$, and $S_x = \sqrt{(\Sigma x^2)/N - m_x^2}$ (and similarly for m_y and S_y). In a later section of the 1913 paper (*ibid*., p. 450), Harris showed how the formula in (2.67) could be used in the calculation of the ICC. We now describe his procedure.

Let m be the number of classes and n be the number of subjects within one class, so that $N = mn$. Let x_{ik} be the observation on the ith ($i = 1, \ldots, n$) subject in class k ($k = 1, \ldots, m$).

*James Arthur Harris was born in Plantsville, Ohio, on September 29, 1880. When he was a child, his family moved to Kansas. At the age of 21, he obtained his A.B. from Kansas University and already had four publications to his credit. Harris had a keen interest in biology and obtained his first appointment as assistant botanist at the Missouri Botanical Gardens during 1901–1903. In 1903, he obtained his Ph.D. from Washington University. Harris was a great admirer of Karl Pearson, whom he regarded as "the Giant of our time in Biological Science." He got the opportunity to work as an assistant to Pearson in 1908–1909. Harris was a botanist by profession and had little college training in mathematics. Nevertheless, he was quick to grasp the philosophy of mathematical thought, as is evidenced by the superlative nature of his statistics papers. In 1921, Harris received the Weldon Medal and Weldon Memorial Prize from the University of Oxford. In 1924, he joined the University of Minnesota where he soon became head of the Department of Botany. Harris died on April 24, 1930, after a brief illness. For more information on Harris, see Rosendahl et al. (1936), A.E.T (1930), Davenport (1930), and Chapman et al. (1930).

†The symbol S was historically used to represent summation.

‡A history of the intraclass correlation is also provided in the book by Haggard (1958, Chapter I).

Within class k, the first and second (uncentered) moments of x_{ik} are $\left(\sum_i x_{ik}\right)/n$ and $\left(\sum_i x_{ik}^2\right)/n$, and the number of pairs is $n(n-1)$. Therefore, the product moment for class k is

$$\frac{\sum_{i\neq j} x_{ik}x_{jk}}{n(n-1)}=\frac{\left(\sum_{i=1}^{n} x_{ik}\right)^2-\sum_{i=1}^{n} x_{ik}^2}{n(n-1)}.$$

By averaging the moments and product moment across classes, Eq. (2.67) gives the ICC:

$$r_{\text{IC}}=\frac{\dfrac{\sum_{k=1}^{m}\left\{\left(\sum_{i=1}^{n} x_{ik}\right)^2-\sum_{i=1}^{n} x_{ik}^2\right\}}{mn(n-1)}-\overline{x}.\overline{x}}{\sqrt{\dfrac{\sum_{k=1}^{m}\sum_{i=1}^{n} x_{ik}^2}{mn}-\overline{x}^2}\sqrt{\dfrac{\sum_{k=1}^{m}\sum_{i=1}^{n} x_{ik}^2}{mn}-\overline{x}^2}}=\frac{\sum_{k=1}^{m}\left\{\left(\sum_{i=1}^{n} x_{ik}\right)^2-\sum_{i=1}^{n} x_{ik}^2\right\}-N(n-1)\overline{x}^2}{(n-1)\left(\sum_{k=1}^{m}\sum_{i=1}^{n} x_{ik}^2-N\overline{x}^2\right)}, \tag{2.68}$$

where $\overline{x}=\frac{1}{N}\sum_{k=1}^{m}\sum_{i=1}^{n} x_{ik}$ is the grand mean. Note that Eq. (2.68) leads to an important relationship:

$$\begin{aligned}
r_{\text{IC}}&=\frac{\sum_{k=1}^{m}\left\{\left(\sum_{i=1}^{n} x_{ik}\right)^2-\sum_{i=1}^{n} x_{ik}^2\right\}-N(n-1)\overline{x}^2}{(n-1)\left(\sum_{k=1}^{m}\sum_{i=1}^{n} x_{ik}^2-N\overline{x}^2\right)}\\
&=\frac{\sum_{k=1}^{m} n^2\overline{x}_{.k}^2-\sum_{k=1}^{m}\sum_{i=1}^{n} x_{ik}^2-N(n-1)\overline{x}^2}{(n-1)\left(\sum_{k=1}^{m}\sum_{i=1}^{n} x_{ik}^2-N\overline{x}^2\right)}\\
&=\frac{\left(\sum_{k=1}^{m} n^2\overline{x}_{.k}^2-Nn\overline{x}^2\right)-\left(\sum_{k=1}^{m}\sum_{i=1}^{n} x_{ik}^2-N\overline{x}^2\right)}{(n-1)\left(\sum_{k=1}^{m}\sum_{i=1}^{n} x_{ik}^2-N\overline{x}^2\right)}\\
&=\frac{Nn\left(\sum_{k=1}^{m}\overline{x}_{.k}^2/m-\overline{x}^2\right)-Ns^2}{(n-1)Ns^2}\\
&=\frac{n\left(s_{\text{class}}^2/s^2\right)-1}{n-1}.
\end{aligned} \tag{2.69}$$

Here, $\bar{x}_{.k} = \left(\sum_i x_{ik}\right)/n$ is the mean of class k, $s^2 = \left(\sum_k \sum_i x_{ik}^2 - N\bar{x}^2\right)/N$ is the sample variance of all observations, and $s_{\text{class}}^2 = \sum_k \bar{x}_{.k}^2 / m - \bar{x}^2$ is the sample variance between classes. Equation (2.69) is known as *Harris' formula*.

It can be seen that, since $0 \le s_{\text{class}}^2 / s^2 \le 1$ in Eq. (2.69), we have

$$-\frac{1}{n-1} \le r_{\text{IC}} \le 1,$$

the lower bound occurring when $s_{\text{class}}^2 = 0$, that is, all class means are the same, and the upper bound occurring when $s_{\text{class}}^2 = s^2$, that is, all observations within a given class are the same. Thus, r_{IC} measures the relative amount of internal (within-class) homogeneity. More specifically, it is the proportion of the total variance that is between classes.

Following Harris' work, the intraclass correlation coefficient became even more popular after R.A. Fisher derived its exact distribution in 1921 (Fisher, 1921a) and also chose to include it as a whole chapter in his widely influential book *Statistical Methods for Research Workers* (Fisher, 1925b, Chapter 7, pp. 176–210). In the 1921 paper, Fisher used his 1915 technique for the distribution of the ordinary correlation coefficient* (Fisher, 1915) to obtain the distribution of the ICC. For n classes each with k subjects,† the density of r_{IC} (evaluated at r) is proportional to

$$\frac{(1-r)^{\frac{(k-1)n-2}{2}}\left(\frac{1}{k-1}+r\right)^{\frac{n-3}{2}}}{\{1+(k-2)\rho-(k-1)\rho r\}^{\frac{kn-1}{2}}}$$

(Fisher, 1921a, p. 7), where ρ is the population ICC.

In *Statistical Methods for Research Workers*, Fisher acknowledged Harris' formula in (2.69) (Fisher, 1925b, p. 180) and gave the following transformation to normality for r_{IC}:

$$z = \frac{1}{2}\log\frac{1+(k-1)r_{\text{IC}}}{1-r_{\text{IC}}}$$

(*ibid.*, p. 185), where z is approximately normal with mean $\frac{1}{2}\log\frac{1+(k-1)\rho}{1-\rho}$ and variance $\frac{k}{2(k-1)(n-2)}$, for reasonably large n.

Further in his book, Fisher made the following important remark:

> A very great simplification is introduced into questions involving intraclass correlation when we recognise that in such cases the correlation merely measures the relative importance of two groups of factors causing variation. (*ibid.*, p. 188)

* See Section 2.3.9.

† Note that Fisher's notation throughout is different from that of Harris.

Fisher thus observed that the intraclass correlation could be given an interpretation in terms of analysis of variance (ANOVA)*. The basis for this observation rested on the fact that, as Fisher pointed out, the formula for r_{IC} in Eq. (2.69) required the two quantities s^2 and s^2_{class}. Now the latter could be calculated by suitably dividing each of the following, respectively:

$$\sum_{i=1}^{n}\sum_{j=1}^{k}\left(x_{ij}-\bar{x}\right)^2, \quad k\sum_{i=1}^{n}\left(\bar{x}_{i.}-\bar{x}\right)^2,$$

where

$$\sum_{i=1}^{n}\sum_{j=1}^{k}\left(x_{ij}-\bar{x}\right)^2 = k\sum_{i=1}^{n}\left(\bar{x}_{i.}-\bar{x}\right)^2 + \sum_{i=1}^{n}\sum_{j=1}^{k}\left(x_{ij}-\bar{x}_{i.}\right)^2 \tag{2.70}$$

(*ibid*., p. 188). In the above, x_{ij} is the jth ($j=1, \ldots, k$) observation in class i ($i=1, \ldots, n$), $\bar{x}_{i.}$ is the mean of class i, and $\bar{x}$ is the grand mean. Note that the relationship in (2.70) is the fundamental equation of ANOVA. It expresses the fact that the total sum of squares is equal to the between-class sum of squares and the within-class sum of squares.

Having provided the motivation for an ANOVA interpretation of the ICC, Fisher next considered a variance components model in which a quantity was made up of two parts, each normally and independently distributed. Let the variances of the first and second components be A and B, respectively, so that the variance of the quantity is $A+B$. Next, Fisher imagined a two-stage sampling in which a random sample of n values (classes) was first chosen, followed by a sample of k values for each class. Then, "in the infinite population from which these are drawn the correlation between the pairs of numbers of the same class [i.e. the population ICC] will be"

$$\rho = \frac{A}{A+B} \tag{2.71}$$

(*ibid*., p. 190). The variances A and B are estimated through an ANOVA:

> …we may make estimates of the values A and B, or in other words we may analyse the variance into the portions contributed by the two causes; the intraclass correlation will be merely the fraction of the total variance due to that cause which observations in the same class have in common. (*ibidem*)

First, the quantity B, which is the within-class variance, can be estimated through

$$\sum_{i=1}^{n}\sum_{j=1}^{k}\left(x_{ij}-\bar{x}_{i.}\right)^2 = n(k-1)\hat{B}. \tag{2.72}$$

Second, we note that A is the between-class variance. To estimate A, Fisher observed that the mean of observations in a given class is made up of two parts: a first part with

* See Sec. 5.4 for the origin and development of ANOVA

variance $\hat{A}$ and a second part that is the mean of k values and therefore has variance $\hat{B}/k$. Therefore, by considering the observed variation of class means,

$$\frac{\sum_{i=1}^{n}\left(\bar{x}_{i.}-\bar{x}\right)^2}{n-1}=\hat{A}+\frac{\hat{B}}{k} \tag{2.73}$$

(*ibid.*, p. 191).

If we now define the within-class and between-class variances* by

$$s_w^2 \equiv \frac{\sum_{i=1}^{n}\sum_{j=1}^{k}\left(x_{ij}-\bar{x}_{i.}\right)^2}{n(k-1)},$$

$$s_b^2 \equiv \frac{k\sum_{i=1}^{n}\left(\bar{x}_{i.}-\bar{x}\right)^2}{n-1}.$$

Equations (2.72) and (2.73) lead to

$$\left.\begin{array}{r}\hat{B}=s_w^2\\ \hat{A}+\dfrac{\hat{B}}{k}=\dfrac{s_b^2}{k}\end{array}\right\}\Rightarrow\left\{\begin{array}{l}\hat{A}=\dfrac{s_b^2-s_w^2}{k}\\ \hat{B}=s_w^2\end{array}\right..$$

By using these estimates of A and B, ρ_{IC} in (2.71) can be estimated by

$$r_{\text{IC}}=\frac{\hat{A}}{\hat{A}+\hat{B}}=\frac{\left(s_b^2-s_w^2\right)/k}{\left(s_b^2-s_w^2\right)/k+s_w^2}=\frac{s_b^2-s_w^2}{s_b^2+(k-1)s_w^2}. \tag{2.74}$$

2.3.9 First Derivation of the Exact Distribution of *r* (Fisher, 1915)

Fisher's formal entrance on the correlation scene in 1915 (Fisher, 1915) was prompted by Soper's 1913 paper, to which we referred at the end of Section 2.3.4. At the start of his 1915 paper, Fisher explained:

> My attention was drawn to the problem of the frequency distribution of the correlation coefficient by an article published by Mr. H. E. Soper in 1913. Seeing that the problem might be attacked by means of geometrical ideas, which I had previously found helpful in the consideration of samples.... (Fisher, 1915, p. 507)

Recall that (cf. Eq. 2.38), in 1908, Student had been able to correctly guess the exact distribution of r when $\rho=0$ as

$$f(r)\propto\left(1-r^2\right)^{(n-4)/2}.$$

*The between-class variance was previously denoted by s_{class}^2.

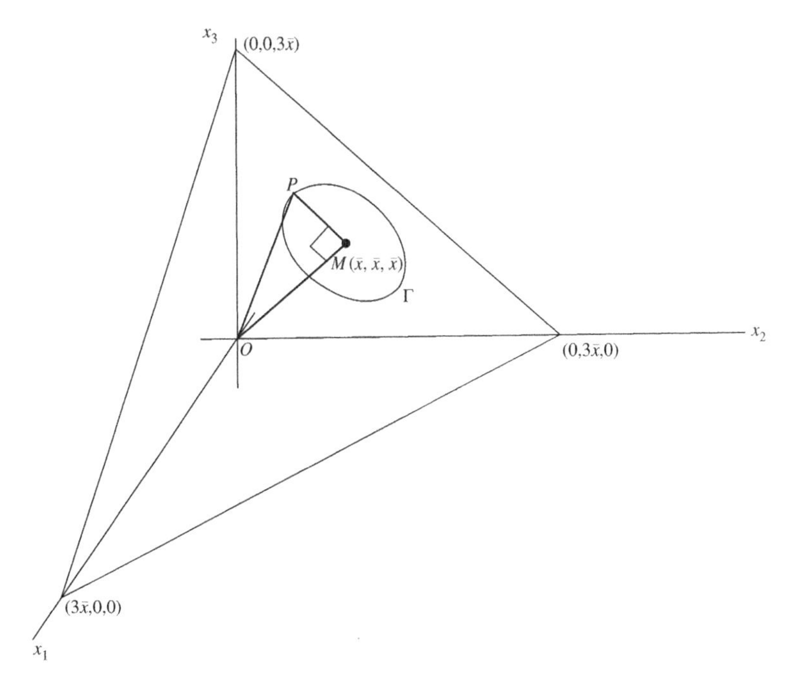

FIGURE 2.18 Fisher's geometric derivation of the exact distribution of the correlation coefficient

However, a rigorous derivation of the distribution of r for any ρ was still lacking. This was exactly what Fisher set out to do in his 1915 paper.

Fisher used his powerful geometrical insights* to obtain the actual sampling distribution of r. In his paper, he stated:

> The problem of the frequency distribution of the correlation coefficient r, derived from a sample of n pairs, taken at random from an infinite population, may be solved, when that population can be represented by a normal surface, with the aid of certain very general conceptions derived from the geometry of n dimensional space. (Fisher, 1915, p. 508)

Let $(x_1, y_1), (x_2, y_2), \ldots, (x_n, y_n)$ be values from a BivNormal $\left(0, 0; \sigma_X^2, \sigma_Y^2, \rho\right)$ distribution. Fisher first considered the X-space. He represented the x-coordinates of the n values by the point $P(x_1, x_2, \ldots, x_n)$ (Fig. 2.18). Let OM be the line

$$x_1 = x_2 = \cdots = x_n$$

*Fisher had a talent for geometric proofs that was unmatched by most of his peers. Edwards (2005, p. 857) explains: "[Fisher] distinguished himself in mathematics despite being handicapped by poor eyesight that prevented him working by artificial light. His teachers used to instruct him by ear, and Fisher developed a remarkable capacity for pursuing complex mathematical arguments in his head. This manifested itself later in life in an ability to reach a conclusion whilst forgetting the argument, to handle complex geometrical trains of thought, and to develop and report essentially mathematical arguments in English (only for students to have to reconstruct the mathematics later)." Fisher's geometrical derivation of the distribution of r is also described in Das Gupta (1980).

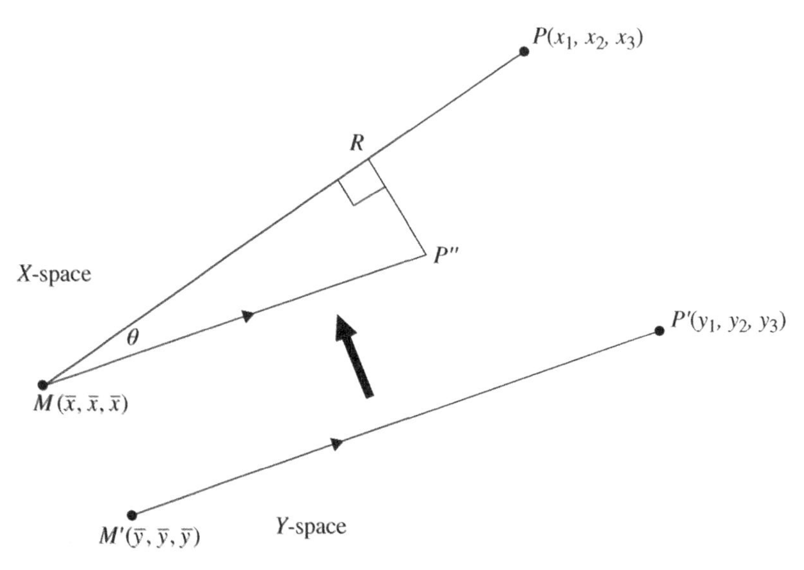

FIGURE 2.19 Projection of Y-space onto X-space

and M be the point $(\bar{x}, \bar{x}, \ldots, \bar{x})$. Then the plane through MP and perpendicular to OM has equation

$$\begin{aligned}(x_1, x_2, \ldots, x_n)\cdot(x_1, x_1, \ldots, x_1) &= (\bar{x}, \bar{x}, \ldots, \bar{x})\cdot(x_1, x_1, \ldots, x_1)\\ &\Rightarrow x_1 + x_2 + \cdots + x_n = n\bar{x}.\end{aligned}$$

Furthermore, the locus of points that are at a fixed distance PM from M lies on the n-dimensional sphere

$$(x_1 - \bar{x})^2 + (x_2 - \bar{x})^2 + \cdots + (x_n - \bar{x})^2 = ns_X^2.$$

Now the above-mentioned plane and hypersphere intersect in an $(n-1)$-dimensional hypersphere with center $M(\bar{x}, \bar{x}, \ldots, \bar{x})$ and radius $s_X\sqrt{n}$. Its "area" is proportional to $\left(s_X\sqrt{n}\right)^{n-1}$ and the element of volume is

$$dx_1 dx_2 \ldots dx_n \propto d\bar{x}d\left(s_X\sqrt{n}\right)^{n-1} \propto s_X^{n-2}d\bar{x}ds_X$$

(*ibid*., p. 509). Similarly, considering the y-values, the corresponding $(n-1)$-dimensional hypersphere has center $M'(\bar{y}, \bar{y}, \ldots, \bar{y})$ and radius $s_Y\sqrt{n}$. Given $P(x_1, x_2, \ldots, x_n)$ and $Q(y_1, y_2, \ldots, y_n)$, the sample correlation coefficient is $r = \cos\theta$, where θ is the angle between MP and $M'P'$. We now project the line $M'P'$ in the Y-space to the line MP'' in the X-space so that $MP'' = s_Y\sqrt{n}$ and $RP'' = MP''\sin\theta = s_Y\sqrt{n}\sin\theta$ (Fig. 2.19). In the Y-space, each of $\bar{y}, s_Y, r$ imposes one restriction. If we make the infinitesimal changes $d\bar{y}, ds_Y$, and $d\theta$, the element of volume generated in the Y-space is

$$\begin{aligned} dy_1 dy_2 \ldots dy_n &\propto d\bar{y} ds_Y \left(RP''\right)^{n-3}\left(MP'' d\theta\right) \\ &= d\bar{y} ds_Y \left(s_Y \sqrt{n}\sin\theta\right)^{n-3}\left(s_Y \sqrt{n} d\theta\right) \\ &\propto s_Y^{n-2}\left(1-r^2\right)^{(n-4)/2} dr d\bar{y} ds_Y \end{aligned}$$

(*ibid*., p. 510). Since the element of volume in the X-space is $dx_1 dx_2 dx_3 \propto s_X^{n-2} d\bar{x} ds_X$, the total element of volume is

$$dx_1 dx_2 \ldots dx_n dy_1 dy_2 \ldots dy_n \propto s_X^{n-2} s_Y^{n-2}\left(1-r^2\right)^{(n-4)/2} d\bar{x} d\bar{y} ds_X ds_Y dr.$$

Since

$$f\left(\bar{x}, \bar{y}, s_X, s_Y, r\right) d\bar{x} ds_X d\bar{y} ds_Y dr = \prod_{i=1}^{n} f\left(x_i, y_i\right) dx_i dy_i,$$

the joint density $f(\bar{x}, \bar{y}, s_X, s_Y, r)$ is given by

$$\begin{aligned} f\left(\bar{x}, \bar{y}, s_X, s_Y, r\right) d\bar{x} ds_X d\bar{y} ds_Y dr &\propto \exp\left[-\frac{1}{2\left(1-\rho^2\right)}\sum_{i=1}^{n}\left\{\left(\frac{x_i}{\sigma_X}\right)^2 - 2\rho\left(\frac{x_i y_i}{\sigma_X \sigma_Y}\right) + \left(\frac{y_i}{\sigma_Y}\right)^2\right\}\right] \\ &\quad \times s_X^{n-2} s_Y^{n-2}\left(1-r^2\right)^{(n-4)/2} d\bar{x} ds_X d\bar{y} ds_Y dr, \\ f\left(\bar{x}, \bar{y}, s_X, s_Y, r\right) &\propto \exp\left[-\frac{1}{2\left(1-\rho^2\right)}\sum_{i=1}^{n}\left\{\left(\frac{x_i}{\sigma_X}\right)^2 - 2\rho\left(\frac{x_i y_i}{\sigma_X \sigma_Y}\right) + \left(\frac{y_i}{\sigma_Y}\right)^2\right\}\right] \\ &\quad \times s_X^{n-2} s_Y^{n-2}\left(1-r^2\right)^{(n-4)/2}. \end{aligned}$$

Because $\bar{x}$ and $\bar{y}$ are independent of s_X, s_Y, and r, they can be integrated out of the above. Fisher then obtained

$$f\left(s_X, s_Y, r\right) \propto s_X^{n-2} s_Y^{n-2}\left(1-r^2\right)^{(n-4)/2} \exp\left\{-\frac{n}{2\left(1-\rho^2\right)}\left(\frac{s_X^2}{\sigma_X^2} - \frac{2r s_X s_Y}{\sigma_X \sigma_Y} + \frac{s_Y^2}{\sigma_Y^2}\right)\right\}.$$

He next made the following transformations:

$$\begin{aligned} \zeta &= \left(s_X / \sigma_X\right)\left(s_Y / \sigma_Y\right), \\ z &= \log\left(\frac{s_X / \sigma_X}{s_Y / \sigma_Y}\right), \\ r &= r, \end{aligned}$$

with Jacobian

$$J\left(\frac{\zeta, z, r}{s_X, s_Y, r}\right) = -\frac{2}{\sigma_X \sigma_Y}.$$

Therefore,

$$f(\zeta, z, r) = f(s_X, s_Y, r)\left|J\left(\frac{\zeta, z, r}{s_X, s_Y, r}\right)\right|^{-1}$$
$$\propto \exp\left\{-\frac{n}{2(1-\rho^2)}\zeta(\cosh z - \rho r)\right\}\zeta^{n-2}(1-r^2)^{(n-4)/2}.$$

By integrating out ζ,

$$f(z, r) \propto \frac{(1-r^2)^{(n-4)/2}}{(\cosh z - \rho r)^{n-1}}.$$

The density of r can then be written as

$$f(r) \propto \int_0^\infty \frac{(1-r^2)^{(n-4)/2}}{(\cosh z - \rho r)^{n-1}} dz. \tag{2.75}$$

By substituting $-\rho r = \cos\lambda$, we have

$$\int_0^\infty \frac{dz}{\cosh z + \cos\lambda} = \frac{\lambda}{\sin\lambda},$$

so that

$$\int_0^\infty \frac{dz}{(\cosh z + \cos\lambda)^{n-1}} = \frac{1}{(n-2)!}\left(\frac{d}{\sin\lambda d\lambda}\right)^{n-2}\frac{\lambda}{\sin\lambda}.$$

From Eq. (2.75), Fisher finally obtained

$$f(r) \propto (1-r^2)^{(n-4)/2}\left(\frac{d}{\sin\lambda d\lambda}\right)^{n-2}\frac{\lambda}{\sin\lambda} \propto (1-r^2)^{(n-4)/2}\frac{d^{n-2}}{dr^{n-2}}\left(\frac{\lambda}{\sin\lambda}\right) \tag{2.76}$$

(*ibid.*, p. 511). Normalizing the above, we can also write

$$f(r) = \frac{(1-\rho^2)^{(n-1)/2}}{\pi\Gamma(n-2)}\frac{(1-r^2)^{(n-4)/2}}{\rho^{n-2}}\frac{d^{n-2}}{dr^{n-2}}\left(\frac{\lambda}{\sin\lambda}\right)$$
$$= \frac{(1-\rho^2)^{(n-1)/2}}{\pi\Gamma(n-2)}(1-r^2)^{(n-4)/2}\frac{d^{n-2}}{d(\rho r)^{n-2}}\left\{\frac{\cos^{-1}(-\rho r)}{\sqrt{1-\rho^2 r^2}}\right\}.$$

Note that when $\rho = 0$, the above reduces to

$$f(r) = \frac{1}{\pi\Gamma(n-2)}(1-r^2)^{(n-4)/2}.$$

Apart from the normalizing constant, the right side of the last formula is the same density that Student had heuristically obtained (cf. Eq. 2.38).

Having obtained the exact distribution of r for arbitrary ρ, Fisher however noted:

> The use of the correlation coefficient r as independent variable of these frequency curves is in some respects highly unsatisfactory. For high values of r the curve becomes extremely distorted and cramped, and although this very cramping forces the mean r to approach p, the difference compared with $1-r$ becomes inordinately great. Even for high values of n, the distortion in this region becomes extreme, and since at the same time the curve rapidly changes its shape, the values of the mean and standard deviation cease to have any very useful meaning. It would appear essential in order to draw just conclusions from an observed high value of the correlation coefficient, say .99, that the frequency curves should be reasonably constant in form. (*ibid.*, p. 517)

Fisher thus explained that the strong dependency of the mean and variance of r on the population correlation ρ made these two measures ineffective. In an attempt to counter the erratic behavior of the distribution of r, especially when the latter is large, Fisher proposed the following transformation:

$$t = \frac{r}{\sqrt{1-r^2}}, \qquad (2.77)$$

$$\tau = \frac{\rho}{\sqrt{1-\rho^2}}$$

(*ibidem*). He correctly noted that when $\rho = 0$, "the frequency…curves [of t] are identical with those found by 'Student' for z [with n replaced by $n-1$]."* However, Fisher also noted that the t-transformation had the same cramping effect as with the original distribution of r, only in the opposite direction. Thus,

> …by using t, we spread out the region of high values, producing asymmetry in the opposite sense. (*ibid.*, p. 520)

In the very last paragraph of the 1915 paper, Fisher proposed a second transformation, namely,

$$\tanh^{-1} \rho = \frac{1}{2}\log\left(\frac{1+\rho}{1-\rho}\right) \qquad (2.78)$$

(*ibid.*, p. 521), and claimed that this transformation "does not tend to simplify the analysis, and approaches relative constancy at the expense of the constancy proportionate to the variable…." As it turns out, the transformations in (2.77) for $\rho = 0$ and in (2.78) for $\rho \neq 0$

*That is, $r/\sqrt{1-r^2}$ has a distribution proportional to Student's t-distribution with $(n-2)$ degrees of freedom. Note: (a) The concept of degrees of freedom was not yet known in 1915. (b) More precisely, $\sqrt{n-2}(r/\sqrt{1-r^2}) \sim t_{n-2}$.

are the ones that are preferred today. In a 1921 paper, published in the journal *Metron*, Fisher further elaborated on the transformation in (2.78) and showed that its distribution approaches normality fairly rapidly (Fisher, 1921a). However, before embarking on that issue, we make a brief pause to explain how a specific portion of Fisher's 1915 paper was the starting point in an escalating feud with Karl Pearson.

2.3.10 Controversy between Pearson and Fisher on the Latter's Alleged Use of Inverse Probability (Soper et al., 1917; Fisher, 1921)

As we explained in the last section, with his 1915 paper (Fisher, 1915), Fisher had achieved the major feat of deriving the exact distribution of the sample correlation r. In that paper, Fisher also tackled the issue of determining "the most likely value of ρ." Now this was a relatively minor task, which Fisher performed by simply differentiating the density in Eq. (2.76) with respect to ρ and setting to zero:

$$\frac{\partial}{\partial\rho}\left\{\left(1-\rho^2\right)^{(n-1)/2}\left(\frac{d}{\sin\lambda d\lambda}\right)^{n-1}\left(\frac{\lambda^2}{2}\right)\right\}=0.$$

After some algebra, Fisher was able to show that, as a first approximation,

$$\hat{\rho}=\frac{r}{\left(1+\dfrac{1-r^2}{2n}\right)} \tag{2.79}$$

(*ibid.*, p. 521), so that "the most likely value of the correlation will in general be less than the observed." All of this is fine, except that Fisher used the following words before deriving the result in Eq. (2.79):

> I have given elsewhere* a criterion, independent of scaling, suitable for obtaining the relation between an observed correlation of a sample and the most probable value of the correlation of the whole population. (*ibid.*, p. 520)

Now Fisher had referred to inverse probability in his first paper (Fisher, 1912). Pearson and his associates (Soper et al., 1917) pounced on this and accused Fisher of having made "indiscriminate use of Bayes' Theorem" in his derivation of the "most likely value of ρ" in the 1915 paper. In particular, they criticized Fisher for having used a uniform prior for ρ while experience shows that the latter is not distributed in this way. All of this took place in spite of the fact that the latter paper had been accepted by Pearson himself to be published in *Biometrika*.† But let us examine Fisher's 1912 paper more carefully to see if he was really referring to Bayes' theorem.

Fisher started the 1912 paper‡ by pointing out defects in both the methods of least squares and of moments. Concerning the first method, Fisher considered a function

*R.A. Fisher, "On an absolute criterion for fitting frequency curves," *Messenger of Mathematics*, February, 1912 (*Fisher's footnote*).

†This journal was founded in 1901 by Pearson, Weldon, and Galton as a result of the Royal Society refusing to publish biological papers of a mathematical nature.

‡Fisher's paper is also described in Aldrich (1997) and Edwards (1974).

$f(x;\theta_1,\ldots,\theta_r)$ for which the observed value is y. Then the method of least squares consisted in minimizing $\int_{-\infty}^{\infty}(f-y)^2\,dy$. Fisher then explained that if x is rescaled to $\xi=\xi(x)$, then minimizing $\int_{-\infty}^{\infty}(f-y)^2\,d\xi$ would result in a different set of $(\theta_1,\ldots,\theta_r)$. Concerning the method of moments, Fisher explained that choosing the first r equations in $\int x^q f dx=\sum x^q/n$, $q=1,2,\ldots$ was arbitrary at best. Next was his "criterion":

> But we may solve the real problem directly.
> If f is an ordinate of the theoretical curve of unit area, then $p=f\,dx$ is the chance of an observation falling within the range dx; and if
>
> $$\log P'=\sum_1^n \log p$$
>
> then P' is proportional to the chance of a given set of observations occurring. The factors dx are independent of the theoretical curve, so the probability of any particular set of θ's is proportional to P, where
>
> $$\log P=\sum_1^n \log f$$
>
> The most probable set of values for the θ's will make P a maximum. (Fisher, 1912, pp. 156–157)

The reader should see in the above the first statement of what is now called the "principle of maximum likelihood," but the word "likelihood" itself is not used. Although the use of the phrase "the probability of any particular set of θ's" might give the impression that Fisher was using Bayes' theorem, this turns out not to be case. Thus, while discussing the estimation of the mean (m) and the precision (h) in a normal distribution, Fisher clearly warned:

> We shall see... that the integration with respect to m is illegitimate and has no definite meaning with respect to inverse probability. (*ibid.*, p. 159)

But what "inverse probability" was Fisher referring to in the above? As he later conceded (Fisher, 1922c, p. 326), he had used inverse probability incorrectly in 1912, having referred to inverse probability when what Fisher really meant was subsequently to be called likelihood. But the harm was already done in 1912, when for instance Fisher said:

> Corresponding to any pair of values, m and h, we can find the value of P, and the inverse probability system may be represented by the surface traced out by a point at a height P above the point on a plane, of which m and h are the coordinates (Fisher, 1912, p. 158).

Edwards has thus commented that

> It turned out to be one of the most influential errors of terminology in statistics, for it led directly to his first quarrel with Pearson, who did not look beyond the phrase to Fisher's account and subsequent use of the method, which was non-Bayesian. (Edwards, 1974)

Thus, in the 1915 paper, when Fisher referred to his 1912 criterion, "stained" as the latter was by inverse probability, Fisher had opened himself to the cooperative attack by Pearson and colleagues (Soper et al., 1917). The latter were, as expected, quite harsh:

> It will thus be evident that in problems like the present the indiscriminate use of Bayes' Theorem is to be deprecated. It has unfortunately been made into a fetish by certain purely mathematical writers on the theory of probability, who have not adequately appreciated the limits of Edgeworth's justification of the theorem by appeal to general experience. (*ibid.*, p. 359)

In 1920, Fisher drafted a response to what he believed was undue criticism of his 1915 paper and submitted to *Biometrika*, whose editor was none other than Pearson. The latter, however, would have none of it:

> I should be unlikely to publish it [Fisher's response] in its present form, or without a reply to your criticisms which would involve also a criticism of your work of 1912-I would prefer you published elsewhere. Under present printing and financial conditions, I am regretfully compelled to exclude all that I think erroneous on my own judgment, because I cannot afford controversy. (Pearson, 1968, p. 453)

This was the third time that Pearson had rejected Fisher's paper, having previously done so in 1916 and in 1918. Fisher thereafter never submitted to *Biometrika*.

His 1920 response having been rejected by Pearson, Fisher published the paper in the journal *Metron* (Fisher, 1921a). The last section of the paper is entitled "Note on the confusion between BAYES' Rule and my method of the evaluation of the optimum" and is the place where Fisher clarified and defended his 1912 criterion:

> My treatment of this problem differs radically from that of BAYES. BAYES (1763) attempted to find, by observing a sample, the actual *probability* that the population value lay in any given range. In the present instance the complete solution of this problem would be to find the probability integral of the distribution of r. Such a problem is indeterminate without knowing the statistical mechanism under which different values of r come into existence; it cannot be solved from the data supplied by a sample, or any number of samples, of the population. What we can find from a sample is the likelihood of any particular value of r, if we define the likelihood as a quantity proportional to the probability that, from a population having that particular value of r, a sample having the observed value r, should be obtained.
>
> So defined, probability and likelihood are quantities of an entirely different nature. (*ibid.*, p. 24)

In the above, Fisher had not also succinctly described the fundamental problem of Bayesian inference, but he had also used and explicitly defined *likelihood* for the first time, drawing a sharp contrast between it and probability. Thus, if $f(x \mid \theta)$ is the density of X depending on some parameter θ, then the *probability of* X lying in $(x, x+\delta x)$ is $f(x \mid \theta)\delta x$. On the other hand, the *likelihood of the parameter* θ is $L(\theta \mid x)$, where

$$L(\theta \mid x) \equiv f(x \mid \theta).$$

Moreover:

> Probability is transformable as a differential element; thus if the probability that r falls in the range dr
>
> $$y\, dr$$

and we use the transformation

$$r = \tanh z$$

then the probability that z falls into the range dz, is

$$y \operatorname{sech}^2 z \, dz$$

The likelihood of a particular value of r on the other hand, is equal to the likelihood of the corresponding value of z, being unchanged by any transformation. It is not a differential element, and is incapable of integration. (*ibidem*)

Fisher came back to his 1912 criterion one more time, in his groundbreaking article "On the mathematical foundations of theoretical statistics" in 1922, admitting to his poor choice of words (Fisher, 1922c, p. 326).

This part of the controversy between Pearson and Fisher thus ended. However, an even more vitriolic battle was brewing between the two. This happened when Fisher attacked Pearson's inappropriate use of chi-squared distribution in the case of parameters estimated from data. This topic is treated in Section 3.3.

2.3.11 The Logarithmic (or *Z*-) Transformation (Fisher, 1915; 1921)

Fisher's 1921 paper (Fisher, 1921a), published in *Metron*, contains a mathematical treatment of the distribution of the logarithmic transformation of the correlation coefficient. Fisher had already alluded to this transformation (cf. Eq. 2.78) in his 1915 paper (Fisher, 1915) on the exact distribution of the correlation coefficient. We now examine the logarithmic transformation of r.

Recall from Eq. (2.76) that the density of r can be written, upon normalization, as

$$f(r) = \frac{\left(1-\rho^2\right)^{(n-1)/2}}{\pi\Gamma(n-2)} \frac{\left(1-r^2\right)^{(n-4)/2}}{\rho^{n-2}} \frac{d^{n-2}}{dr^{n-2}}\left(\frac{\lambda}{\sin\lambda}\right) \tag{2.80}$$

(Fisher, 1915, p. 511), where $-\rho r = \cos\lambda$. This distribution becomes increasingly skewed to the left as r increases in magnitude. From Eq. (2.75), the density of r can also be written as

$$\begin{aligned} f(r) &= \frac{n-2}{\pi}\left(1-\rho^2\right)^{(n-1)/2}\left(1-r^2\right)^{(n-4)/2}\int_0^\infty \frac{1}{\left(\cosh u + \cos\lambda\right)^{n-1}}\,du \\ &= \frac{n-2}{\pi}\left(1-\rho^2\right)^{(n-1)/2}\left(1-r^2\right)^{(n-4)/2}\int_0^\infty \frac{1}{\left(\cosh u - \rho r\right)^{n-1}}\,du \end{aligned}$$

(Fisher, 1921a, p. 12). He next used the transformations*

$$\begin{aligned} z &= \tanh^{-1} r = \frac{1}{2}\log\left(\frac{1+r}{1-r}\right), \\ \zeta &= \tanh^{-1}\rho = \frac{1}{2}\log\left(\frac{1+\rho}{1-\rho}\right) \end{aligned} \tag{2.81}$$

*We shall explain how Fisher could have obtained these transformations in Section 2.3.12.

(*ibid*., p. 7). Therefore,

$$f(z)=\left\{\frac{n-2}{\pi}\left(1-\rho^2\right)^{(n-1)/2}\left(1-r^2\right)^{(n-4)/2}\int_0^\infty\frac{1}{\left(\cosh u-\rho r\right)^{n-1}}du\right\}\left|\frac{dr}{dz}\right|$$

$$=\frac{n-2}{\pi}\operatorname{sech}^{n-1}\zeta\operatorname{sech}^{n-2}z\int_0^\infty\frac{1}{\left(\cosh u-\rho r\right)^{n-1}}du.$$

By writing $z-\zeta=x$, Fisher then expanded the above in powers of x and inverse powers of $(n-1)$:

$$f(x)=\frac{n-2}{\sqrt{2\pi(n-1)}}e^{-(n-1)x^2/2}\left[1+\frac{1}{2}\rho x+\left\{\frac{2+\rho}{8(n-1)}+\frac{4-\rho^2}{8}x^2+\frac{n-1}{12}x^4\right\}\right.$$

$$+\rho x\left\{\frac{4-\rho^2}{16(n-1)}+\frac{4+3\rho^2}{48}x^2+\frac{n-1}{24}x^4\right\}+\left\{\frac{4+12\rho^2+9\rho^4}{128(n-1)^2}\right.$$

$$\left.\left.+\frac{8-2\rho^2+3\rho^4}{64(n-1)}x^2+\frac{8+4\rho^2-5\rho^4}{128}x^4+\frac{28-15\rho^2}{1440}x^6(n-1)+\frac{(n-1)^2}{288}x^8\right\}+\cdots\right]$$

(*ibid*., p. 13). By taking expectations, Fisher was then able to show that*

$$\mu_1'=\mathsf{E}X=\frac{\rho}{2(n-1)}\left\{1+\frac{5+\rho^2}{4(n-1)}+\frac{11+2\rho^2+3\rho^4}{8(n-1)^2}+\cdots\right\},$$

$$\mu_2=\operatorname{var}\ X=\frac{1}{n-1}\left\{1+\frac{4-\rho^2}{2(n-1)}+\frac{22-6\rho^2-3\rho^4}{6(n-1)^2}+\cdots\right\},$$

$$\mu_3=\mathsf{E}\left(X-\mu_X\right)^3=\frac{\rho^3}{(n-1)^3}+\cdots,$$

$$\mu_4=\mathsf{E}\left(X-\mu_X\right)^4=\frac{1}{(n-1)^2}\left\{3+\frac{14-3\rho^2}{n-1}+\frac{184-48\rho^2-21\rho^4}{4(n-1)^2}+\cdots\right\}.$$

The mean and variance of Z are, respectively,

$$\mathsf{E}Z\approx\zeta+\frac{\rho}{2(n-1)}\left\{1+\frac{5+\rho^2}{4(n-1)}\right\},$$

$$\operatorname{var}\ Z\approx\frac{1}{n-1}\left\{1+\frac{4-\rho^2}{2(n-1)}\right\}.$$

*Fisher's own equations were incorrect and were corrected by A.H. Pollard (1945, unpublished) and by Gayen (1951). The correct equations are therefore given in what follows.

Compared to the variance of r (cf. Eq. 2.39), namely,

$$\text{var } r \approx \frac{\left(1-\rho^2\right)^2}{n}\left(1+\frac{11\rho^2}{2n}\right),$$

it is seen that, for moderately large n, var Z is much less dependent on ρ than var r is. Furthermore, for small ρ, the variance of Z is

$$\text{var } Z \approx \frac{1}{n-1}+\frac{2}{\left(n-1\right)^2} \approx \frac{1}{n-3}.$$

Now the skewness and kurtosis of Z are, respectively,

$$\gamma_1 = \frac{\mathsf{E}\left(X-\mu_X\right)^3}{\sqrt{\text{var}^3 X}} \approx \frac{\rho^3}{\left(n-1\right)^{3/2}},$$

$$\gamma_2 = \frac{\mathsf{E}\left(X-\mu_X\right)^4}{\sqrt{\text{var}^4 X}} - 3 \approx \frac{2}{n-1}+\frac{4+2\rho^2-3\rho^4}{\left(n-1\right)^2}.$$

Comparing these with the skewness and kurtosis of r, namely,

$$\frac{\mathsf{E}\left(r-\mu_r\right)^3}{\sqrt{\text{var}^3 r}} \approx \frac{-6\rho}{n^{1/2}},$$

$$\frac{\mathsf{E}\left(r-\mu_r\right)^4}{\sqrt{\text{var}^4 r}} - 3 \approx \frac{6\left(12\rho^2-1\right)}{n},$$

it is seen that, for moderately large n, the skewness of Z is much less than that of r in magnitude. In view of the nearly constant variance of Z and its small skewness, Fisher proposed the following approximation for moderately large n:

$$\frac{1}{2}\log\left(\frac{1+r}{1-r}\right) \stackrel{\cdot}{\sim} N\left[\frac{1}{2}\log\left(\frac{1+\rho}{1-\rho}\right), \frac{1}{n-3}\right]$$

(*ibid.*, p. 14). The above result is an invaluable tool for testing a null statistical hypothesis of the type $H_0 : \rho = \rho_0 (\neq 0)$. For $H_0 : \rho = 0$, we use the result (cf. Eq. 2.77)

$$\frac{r}{\sqrt{1-r^2}}\sqrt{n-2} \sim t_{n-2}.$$

* 2.3.12 Derivation of the Logarithmic Transformation

Although Fisher did not explain how he arrived at the logarithmic transformations in Eq. (2.81), it is not difficult to see how he might have obtained them. Recall Pearson and Filon's result in Eq. (2.34) that

$$\sqrt{n}\left(r-\rho\right)\xrightarrow{D}N\left[0,\left(1-\rho^2\right)^2\right].$$

Fisher sought a transformation $g(r)$ such that var $g(r)$ does not involve ρ (i.e., a variance-stabilization transformation). Now, from Cramer's theorem*,

$$\begin{aligned}&\sqrt{n}\ (r-\rho)\xrightarrow{D}\mathrm{N}\left[0,\left(1-\rho^2\right)^2\right]\\ \Rightarrow&\sqrt{n}\ \{g(r)-g(\rho)\}\xrightarrow{D}\mathrm{N}\left[0,\{g'(\rho)\}^2\left(1-\rho^2\right)^2\right]\end{aligned}$$

To obtain the simplest $g(\rho)$, we can therefore set

$$g'(\rho)\left(1-\rho^2\right)=1.$$

Hence,

$$\begin{aligned}g(\rho)&=\int\frac{1}{(1-\rho)(1+\rho)}d\rho\\ &=\frac{1}{2}\int\left\{\frac{1}{(1-\rho)}+\frac{1}{(1+\rho)}\right\}d\rho\\ &=\frac{1}{2}\log\frac{1+\rho}{1-\rho},\end{aligned}$$

where we have omitted the constant of integration so as to obtain the simplest $g(\rho)$.

2.4 WORK ON CORRELATION AND THE BIVARIATE (AND MULTIVARIATE) NORMAL DISTRIBUTION BEFORE GALTON

There is no question that Galton was the first to have discovered the concept of correlation as an inherent or, to use Person's word (Pearson, 1920), "organic" relationship between variables. However, correlation was mathematically implied, although never explicitly stated, in a few of the theoretical works preceding Galton's own 1877 paper,

* See, for example, Ferguson (1996, pp. 45 and 54).

"Typical laws of heredity" (Galton, 1877). Before describing some of these, let us recall that if $(X,Y) \sim$ BivNormal $(0,0;\sigma_X^2,\sigma_Y^2;\rho)$, then

$$f_{XY}(x,y)=\frac{1}{2\pi\sigma_X\sigma_Y\sqrt{1-\rho^2}}\exp\left\{-\frac{1}{2(1-\rho^2)}\left(\frac{x^2}{\sigma_X^2}-\frac{2\rho xy}{\sigma_X\sigma_Y}+\frac{y^2}{\sigma_Y^2}\right)\right\}. \tag{2.82}$$

The "normal correlation surface" is then

$$\frac{x^2}{\sigma_X^2}-\frac{2\rho xy}{\sigma_X\sigma_Y}+\frac{y^2}{\sigma_Y^2}=C, \tag{2.83}$$

where C is a constant and ρ is the population correlation coefficient. That is, when a plane parallel to the xy-plane is moved up along the z-axis, it intersects the joint density surface of $f_{XY}(x, y)$ in ellipses given by Eq. (2.83). Those ellipses have axes that are parallel to the x- and y-axes only when $\rho = 0$, in which case X and Y are also independent.

2.4.1 Lagrange's Derivation of the Multivariate Normal Distribution from the Multinomial Distribution (1776)

Lagrange's derivation of the multivariate normal distribution took place in the "Mémoire sur l'utilité de la méthode de prendre le milieu entre les résultats de plusieurs observations*" (Lagrange, 1776).† In Problem VI (see also Section 5.7.2), he considered several errors $p,q,r,\ldots$ that have been observed $\alpha,\beta,\gamma,\ldots$ times in n repeated measurements. The "number of cases" that give the respective errors are $a,b,c,\ldots$. Lagrange then showed that the likelihood function in (5.136), that is,

$$\frac{1\cdot2\cdot3\cdots n}{1\cdot2\cdot3\cdots\alpha\cdot1\cdot2\cdot3\cdots\beta\cdot1\cdot2\cdot3\cdots\gamma}\frac{a^\alpha b^\beta c^\gamma\cdots}{s^n}, \tag{2.84}$$

was maximized when

$$a=\frac{s\alpha}{n},\; b=\frac{s\beta}{n},\; c=\frac{s\gamma}{n},\ldots \tag{2.85}$$

(Lagrange, 1776, Oeuvres 2, p. 202), where $s = a+b+c+\cdots$, that is, the probabilities of the various errors are estimated by their observed relative frequencies.

Lagrange next wished to determine

> ...the probability that these values [of *a, b, c, ...*] will not deviate from the true value by some quantity $\pm\frac{rs}{n}$.... (*ibidem*)

* Memoir on the usefulness of the method of taking the mean of the results of many observations.

† Lagrange's derivation is also described in Hald (1998, pp. 43–46).

To obtain this probability, Lagrange first substituted the values of a, b, c, ... in (2.85) in expression (2.84) and denoted the resulting quantity by P:

$$P = \frac{1\cdot 2\cdot 3\cdots n}{1\cdot 2\cdot 3\cdots \alpha\cdot 1\cdot 2\cdot 3\cdots \beta\cdot 1\cdot 2\cdot 3\cdots \gamma}\frac{(s\alpha/n)^{\alpha}(s\beta/n)^{\beta}(s\gamma/n)^{\gamma}\cdots}{s^n}$$
$$= \frac{1\cdot 2\cdot 3\cdots n}{n^n}\frac{\alpha^{\alpha}}{1\cdot 2\cdot 3\cdots \alpha}\frac{\beta^{\beta}}{1\cdot 2\cdot 3\cdots \beta}\cdots. \tag{2.86}$$

By substituting $a = s(\alpha + x)/n, b = s(\beta + y)/n, c = s(\gamma + z)/n, \ldots$ in (2.84) and denoting the resulting quantity by Q,

$$Q = \frac{1\cdot 2\cdot 3\cdots n}{1\cdot 2\cdot 3\cdots \alpha\cdot 1\cdot 2\cdot 3\cdots \beta\cdot 1\cdot 2\cdot 3\cdots \gamma}\frac{\{s(\alpha+x)/n\}^{\alpha}\{s(\beta+y)/n\}^{\beta}\{s(\gamma+z)/n\}^{\gamma}\cdots}{s^n}$$
$$= P\left(1+\frac{x}{\alpha}\right)^{\alpha}\left(1+\frac{y}{\beta}\right)^{\beta}\left(1+\frac{z}{\gamma}\right)^{\gamma}\cdots. \tag{2.87}$$

Then the required probability that the observed relative frequencies do not deviate from the true probabilities by more than rs/n is the sum of Q for $x, y, z, \ldots$ equal to $\pm 1, \pm 2, \ldots, \pm r$ while holding $x + y + z + \cdots = 0$. This is the same as saying that the required probability is $P\int V$ where

$$V = \frac{Q}{P} = \left(1+\frac{x}{\alpha}\right)^{\alpha}\left(1+\frac{y}{\beta}\right)^{\beta}\left(1+\frac{z}{\gamma}\right)^{\gamma}\cdots \tag{2.88}$$

and $\int V$ denotes the sum of V for all x, y, z from 0 to r such that $x + y + z + \cdots = 0$.

Lagrange noted that it is not easy to determine the integral $\int V$, especially when more than two variables are involved, so that recourse to an approximation is needed. One such approach consists of multiplying the mean value of V by the number of distinct values of V that enter the integral $\int V$. In view of the fact the number of distinct values will be such that $x + y + z + \cdots = 0$ in (2.88), one sees that this number is also the constant term in the series expansion of V. Lagrange wrote the constant term as

$$T = \frac{(mr+1)(mr+2)(mr+3)\cdots(mr+m-1)}{1\cdot 2\cdot 3\cdots(m-1)}$$
$$-m\frac{\{(m-2)r\}\{(m-2)r+1\}\{(m-2)r+2\}\cdots\{(m-2)r+m-2\}}{1\cdot 2\cdot 3\cdots(m-1)}$$
$$+\frac{m(m-1)}{2}\frac{\{(m-4)r-1\}\{(m-4)r\}\{(m-4)r+1\}\cdots\{(m-4)r+m-3\}}{1\cdot 2\cdot 3\cdots(m-1)}-\cdots$$

(*ibid.*, p. 203), where m is the number of variables $\alpha, \beta, \gamma, \ldots$ and the series above is continued until one of the terms $mr+1$, $(m-2)r$, $(m-4)r-1$, ... becomes negative. Thus, if the mean value of V is W, then the required probability $P\int V$ can be approximated by PTW.

Lagrange next proceeded to find an approximation for the quantity Q in (2.88), where $Q = PV$. By assuming to be large and by applying Stirling's formula $1\cdot 2\cdot 3\cdots u = \sqrt{2\pi u}\,(u/e)^u$,* Eq. (2.86) becomes

$$P = \sqrt{\frac{\pi n}{\pi a \times \pi b \times \pi c \cdots}},$$

where $n = \alpha + \beta + \gamma + \cdots$. Moreover, from Eq. (2.88),

$$\begin{aligned}\log V &= \alpha \log\left(1+\frac{x}{\alpha}\right) + \beta \log\left(1+\frac{y}{\beta}\right) + \gamma \log\left(1+\frac{z}{\gamma}\right) + \cdots \\ &= \alpha\left(\frac{x}{\alpha} - \frac{x^2}{2\alpha^2} + \cdots\right) + \beta\left(\frac{\gamma}{\beta} - \frac{\gamma^2}{2\beta^2} + \cdots\right) + \gamma\left(\frac{z}{\gamma} - \frac{z^2}{2\gamma^2} + \cdots\right) + \cdots \\ &= -\frac{1}{2}\left(\frac{x^2}{\alpha^2} + \frac{y^2}{\beta^2} + \frac{z^2}{\gamma^2} + \cdots\right),\end{aligned} \tag{2.89}$$

the terms linear in $x, y, z, \ldots$ above disappearing since $x + y + z + \cdots = 0$ by assumption. From the above,

$$V = e^{-\frac{1}{2}\left(\frac{\xi^2}{A} + \frac{\psi^2}{B} + \frac{\zeta^2}{C} + \cdots\right)}.$$

Further simplifications can be made by substituting $x = \xi\sqrt{n}$, $y = \psi\sqrt{n}$, $z = \zeta\sqrt{n}, \ldots$ and $\alpha/n = A$, $\beta/n = B$, $\gamma/n = C$, …. Then $\xi + \psi + \zeta + \cdots = 0$, $A + B + C + \cdots = 1$, and

$$P = \frac{1}{(2\pi n)^{\frac{m-1}{2}}\sqrt{ABC\ldots}},$$

$$V = e^{-\frac{1}{2}\left(\frac{\xi^2}{A} + \frac{\psi^2}{B} + \frac{\zeta^2}{C} + \cdots\right)}$$

(*ibid.*, p. 205). The next step in Lagrange's analysis was to reason that since the increments in the quantities $x, y, z, \ldots$ are unity, so that the increments in $\xi, \psi, \zeta, \ldots$ will be $1/\sqrt{n}$, a very small quantity. Denoting the latter by $d\theta$, P in the above formula becomes

$$P = \frac{(d\theta)^{m-1}}{\sqrt{(2\pi)^{m-1} ABC\ldots}}$$

so that

$$PV = Q = \frac{e^{-\frac{1}{2}\left(\frac{\xi^2}{A} + \frac{\psi^2}{B} + \frac{\zeta^2}{C} + \cdots\right)}(d\theta)^{m-1}}{\sqrt{(2\pi)^{m-1} ABC\ldots}}$$

*Lagrange own formula does not have the "2" because he used π to denote the ratio of the circumference to the radius of a circle!

(*ibidem*). The last expression is an $(m-1)$-dimensional multivariate normal distribution and approximates the probability that the observed frequencies $(\alpha,\beta,\gamma,\ldots)$ deviate from the true ones $(na/s, nb/s, nc/s,\ldots)$ by x, y, z, ….

Karl Pearson (1978) has spoken very highly of Lagrange's derivation, even calling Todhunter (1865, p. 305) "myopic"* for failing to appreciate its full scope. By writing the last equation as

$$Q=\frac{1}{(2\pi)^{\frac{m-1}{2}}}\cdot\frac{n^{\frac{1}{2}}}{(\alpha\beta\gamma\ldots)^{\frac{1}{2}}}e^{-\frac{1}{2}\left(\frac{x^2}{\alpha}+\frac{y^2}{\beta}+\frac{z^2}{\gamma}+\cdots\right)}$$

(Pearson, 1978, p. 600), Pearson pointed that Lagrange had effectively shown that the probability of the deviations x, y, z, … is proportional to

$$\frac{x^2}{\alpha}+\frac{y^2}{\beta}+\frac{z^2}{\gamma}+\cdots. \tag{2.90}$$

Note that this is *nearly* the same sum that appears in Pearson's X^2 (see Eq. 3.9). The latter is of the form

$$\sum\frac{(\text{expected}-\text{observed})^2}{\text{expected}},$$

while Lagrange's (2.90) is of the form

$$\sum\frac{(\text{expected}-\text{observed})^2}{\text{observed}}.$$

Furthermore, the latter can be written as

$$\begin{aligned}
X'^2 &= \frac{x^2}{\alpha}+\frac{y^2}{\beta}+\cdots\\
&= \frac{x^2}{an/s}\cdot\frac{an/s}{\alpha}+\frac{y^2}{bn/s}\cdot\frac{bn/s}{\beta}+\cdots\\
&= \frac{x^2}{an/s}\cdot\frac{an/s}{an/s-x}+\frac{y^2}{bn/s}\cdot\frac{bn/s}{bn/s-y}+\cdots\\
&= \frac{x^2}{an/s}\left(1-\frac{x}{an/s}\right)^{-1}+\frac{y^2}{bn/s}\left(1-\frac{y}{bn/s}\right)^{-1}+\cdots\\
&= \frac{x^2}{an/s}\left(1+\frac{x}{an/s}+\cdots\right)+\frac{y^2}{bn/s}\left(1+\frac{y}{bn/s}+\cdots\right)+\cdots\\
&= \underbrace{\left(\frac{x^2}{an/s}+\frac{y^2}{bn/s}+\cdots\right)}_{X^2}+\left\{\frac{x^3}{(an/s)^2}+\frac{y^3}{(bn/s)^3}+\cdots\right\}.
\end{aligned}$$

*A description of Todhunter not shared by the current author.

Hence, the difference between Lagrange's X'^2 and Pearson's X^2 is

$$\frac{x^3}{(an/s)^2}+\frac{y^3}{(bn/s)^2}+\cdots,$$

which is precisely the order of terms neglected by Lagrange in (2.89). Moreover, both sums have asymptotically the same distribution, although, unlike Pearson, Lagrange did not investigate the latter distribution.

2.4.2 Adrain's Use of the Multivariate Normal Distribution (1808)

Adrain's use of the multivariate normal distribution took place in his 1808 paper (Adrain, 1808) while deriving the normal distribution. This is described in Section 1.5.2.

2.4.3 Gauss' Use of the Multivariate Normal Distribution in the *Theoria Motus* (1809)

Right after his derivation (see Section 1.5.3) of the normal distribution in the book *Theoria Motus Corporum Coelestium** (Gauss, 1809), Gauss proceeded to consider the multivariate version of this distribution, as we now explain.

As before, Gauss considered μ observations $M, M', M'', \ldots$ on the respective functions $V, V', V'', \ldots$, where

$$\begin{aligned} V &= V(p,q,r,\ldots), \\ V' &= V'(p,q,r,\ldots), \\ V'' &= V''(p,q,r,\ldots), \\ &\cdots \end{aligned}$$

and $p, q, r, \ldots$ are ν unknown quantities. Then he noted that the joint multivariate density

$$\Omega = \frac{h^\mu}{\pi^{\mu/2}}\exp\left\{-h^2\left(v^2+v'^2+v''^2+\cdots\right)\right\}$$

(*ibid*., p. 260) is a maximum when the sum $(v^2+v'^2+v''^2+\cdots)$ is a minimum. In the above, h is the *precision* of each observation and $v = V - M$, $v' = V' - M'$, $v'' = V'' - M''$,....

Gauss later (*ibid*., p. 263) considered the case of only one unknown p whose estimation is based on the functions $ap+n$, $a'p+n'$, $a''p+n''+\cdots$, the latter having been, respectively, observed as $M, M', M'', \ldots$. Then the most probable value of p is

$$A = \frac{am + a'm' + a''m'' + \cdots}{a^2 + a'^2 + a''^2 + \cdots},$$

where $m = M - n$, $m' = M' - n'$, $m'' = M'' - n''$,.... Note that the value of A above can be obtained by minimizing the sum of errors $(M-ap-n)^2+(M'-a'p-n')^2+\cdots$. Gauss

* "Theory of the Motion of Heavenly Bodies."

now wished to determine the "degree of accuracy to be attributed to this value." He proceeded by writing the probability density of an error Δ in the observations as

$$\frac{h}{\sqrt{\pi}}\exp\left(-h^2\Delta^2\right).$$

Therefore, the joint density of the observations $M, M', M'', \ldots$ is the multivariate normal distribution, which is proportional to

$$\exp\left[-h^2\left\{(ap-m)^2+(a'p-m')^2+(a''p-m'')^2+\cdots\right\}\right].$$

Hence, the posterior density of p (assuming a uniform prior) is also proportional to the function above. Next, Gauss noted that the exponent of this function can be written in the form

$$-h^2\left(a^2+a'^2+a''^2+\cdots\right)\left(p^2-2pA+B\right),$$

where B is independent of p. Therefore, Gauss concluded that the accuracy of A is higher than that of a single observation by a factor

$$\sqrt{a^2+a'^2+a''^2+\cdots}$$

(*ibidem*). In modern statistical parlance, this is the same as saying that if the variance of each of $m, m', m'', \ldots$ is $\sigma^2 = [=1/(2h^2)]$, then the variance of A is

$$\text{var }A=\frac{\left(a^2+a'^2+a''^2+\cdots\right)\sigma^2}{\left(a^2+a'^2+a''^2+\cdots\right)^2}=\frac{\sigma^2}{\left(a^2+a'^2+a''^2+\cdots\right)}$$

and its standard deviation is

$$\text{SDA}=\frac{\sigma}{\sqrt{a^2+a'^2+a''^2+\cdots}}.$$

2.4.4 Laplace's Derivation of the Joint Distribution of Linear Combinations of Two Errors (1811)

We next mention Laplace's "Mémoire sur les intégrales définis et leur application aux probabilités" (Laplace, 1811a). In this memoir, Laplace explicitly derived the bivariate normal distribution while justifying the method of least squares in the case of two unknown parameters. Details are provided in the second part of Section 1.5.5 (see Eq. 1.214 in particular).

2.4.5 Plana on the Joint Distribution of Two Linear Combinations of Random Variables (1813)

Giovanni Antonio Amedeo Plana* (Fig. 2.20) was one of the greatest Italian scientists of his time. He was born in Voghera, Italy, on November 6, 1781. He was sent in 1796 by his father to Grenoble in France where he became well known for his mathematical

* Plana's life is also described in Tricomi (1981).

FIGURE 2.20 Giovanni Antonio Amedeo Plana (1781–1864). Wikimedia Commons (Public Domain), http://commons.wikimedia.org/wiki/File:Giovanni_Plana.jpg

ability. In 1800, he joined the École Polytechnique in Paris where one of his teachers was Joseph-Louis Lagrange (1736–1813). The latter had tremendous influence on his life and future career. In 1803, Plana returned to Italy. Joseph Fourier (1768–1830) was very impressed by him and tried to get him a position as Professor of Mathematics at the artillery school of Grenoble. Although that did not succeed, Fourier managed to get a similar position for Plana at the artillery school of Piedmont. In 1811, Plana was named Professor of Astronomy at the University of Turin in 1811 on Lagrange's recommendation. It is reported that Plana's teaching was on the highest quality, of the same level as the *grandes écoles* of Paris. He had a wide range of interests, including mathematical analysis, mathematical physics, geodesy, and astronomy. Plana passed away in Turin, Italy, on January 20, 1864.

In 1813, Plana published a paper entitled "Mémoire sur divers problèmes de probabilité" (Plana, 1813).* In it, Plana first considered the problem of throwing a die with N (where N was first set to $2n$, then to $2n+1$) faces a total of p times. By using the technique of characteristic functions, he obtained the probability of obtaining a sum of zero.

The next problem Plana solved was a generalization of the first. Now, p dice each with $2n$ faces are thrown giving the results $\varepsilon_1, \varepsilon_2, \ldots, \varepsilon_p$, and the problem is to find the probability that the sum $q_1\varepsilon_1 + q_2\varepsilon_2 + \cdots + q_p\varepsilon_p$, where $q_1, q_2, \ldots, q_p$ are given integers, is equal to some quantity q.

It is the third problem, though, that interests us here. Generalizing the second problem, Plana wished to determine the joint probability mass function of Q and Q', where

$$\begin{aligned} Q &= q_1\varepsilon_1 + q_2\varepsilon_2 + \cdots + q_p\varepsilon_p, \\ Q' &= q_1'\varepsilon_1 + q_2'\varepsilon_2 + \cdots + q_p'\varepsilon_p \end{aligned} \tag{2.91}$$

(*ibid*., p. 376). In the above, q_i and q_i' are known integers, and ε_i is assumed to be discrete uniform on $\{-n, -n+1, -1, 1, \ldots, n-1, n\}$. First, Plana defined the polynomials

$$\begin{aligned} X' &= \left(x^{-nq_1}y^{-nq_1'} + x^{-(n-1)q_1}y^{-(n-1)q_1'} + \cdots + x^{-2q_1}y^{-2q_1'} + x^{-q_1}y^{-q_1'}\right) \\ &\quad + \left(x^{q_1}y^{q_1'} + x^{2q_1}y^{2q_1'} + \cdots + x^{(n-1)q_1}y^{(n-1)q_1'} + x^{nq_1}y^{nq_1'}\right), \\ X'' &= \left(x^{-nq_2}y^{-nq_2'} + x^{-(n-1)q_2}y^{-(n-1)q_2'} + \cdots + x^{-2q_2}y^{-2q_2'} + x^{-q_2}y^{-q_2'}\right) \\ &\quad + \left(x^{q_2}y^{q_2'} + x^{2q_2}y^{2q_2'} + \cdots + x^{(n-1)q_2}y^{(n-1)q_2'} + x^{nq_2}y^{nq_2'}\right). \\ &\ldots \end{aligned}$$

Then the probability z that the two sums on the right side of Eq. (2.91) are equal to Q and Q' is the coefficient of $x^Q y^{Q'}$ in the expansion of the product $X'X'' \ldots X^{(p)}$. By defining $x = e^{j\omega}\left(j = \sqrt{-1}\right)$ and $y = e^{j\omega'}$, the polynomials above become

$$\begin{aligned} X' &= 2\sum_{i=1}^{n}\cos\{i(q_1\omega + q_1'\omega')\}, \\ X'' &= 2\sum_{i=1}^{n}\cos\{i(q_2\omega + q_2'\omega')\}, \\ &\ldots \end{aligned}$$

*Plana's work is also described in Walker (1928) and Pearson (1928b).

and the probability z is the term independent of ω and ω' in the expansion of the product

$$X'X''...X^{(p)}\cos(Q\omega+Q'\omega').$$

Now, Plana noted that, in the above product, all terms with the exception of z were of the form

$$A\cos(\alpha\omega+\beta\omega'),\ B\cos(M\omega),\ C\cos(N\omega').$$

Since $\int_{-\pi}^{\pi}A\cos(\alpha\omega+\beta\omega')d\omega=\int_{-\pi}^{\pi}B\cos(M\omega)d\omega=\int_{-\pi}^{\pi}C\cos(N\omega')d\omega'=0$, he obtained

$$z=\frac{1}{4\pi^2}\int_{-\pi}^{\pi}\int_{-\pi}^{\pi}X'X''...X^{(P)}\cos(Q\omega+Q'\omega')d\omega d\omega' \tag{2.92}$$

(*ibid.*, p. 377). His next step was to simplify the above by using Laplace's method of asymptotic expansion.* First, he set

$$X'X''...X^{(p)}=(2n)^p e^{-a\Pi}e^{\frac{(2b^2-a^2)\Pi'}{2}},$$

where

$$\begin{aligned}
a&=\frac{1}{2!n}\sum_{i=1}^{n}i^2=\frac{1}{12}(n+1)(2n+1),\\
b&=\frac{1}{4!n}\sum_{i=1}^{n}i^4=\frac{1}{720}(n+1)(2n+1)(3n^2+3n-1),\\
\Pi&=(q_1\omega+q_1'\omega')^2+(q_2\omega+q_2'\omega')^2+\cdots+(q_p\omega+q_p'\omega')^2,\\
\Pi'&=(q_1\omega+q_1'\omega')^4+(q_2\omega+q_2'\omega')^4+\cdots+(q_p\omega+q_p'\omega')^4.
\end{aligned} \tag{2.93}$$

Then by defining

$$\begin{aligned}
A&=q_1^2+q_2^2+\cdots+q_p^2,\\
B&=q_1q_1'+q_2q_2'+\cdots+q_pq_p',\\
C&=q_1'^2+q_2'^2+\cdots+q_p'^2,
\end{aligned}$$

Equation (2.92) becomes

$$z=\frac{(2n)^p}{4\pi^2}\int_{-\pi}^{\pi}\int_{-\pi}^{\pi}e^{-(aA\omega^2+2aB\omega\omega'+aC\omega'^2)}\cos(Q\omega+Q'\omega')d\omega d\omega',$$

where the factor $(2b^2-a^2)\Pi'/2$ has been omitted because "it produces very small terms in the integration." A change of variables in the above integral to $x=\omega\sqrt{ap}$ and $x'=\omega'\sqrt{ap}$ results in

$$z=\frac{(2n)^p}{4a\pi^2 p}\int_{-\infty}^{\infty}\int_{-\infty}^{\infty}e^{-\frac{Ax^2}{p}-\frac{2Bxx'}{p}-\frac{Cx'^2}{p}}\cos\left(\frac{Qx}{\sqrt{ap}}+\frac{Q'x'}{\sqrt{ap}}\right)dxdx'.$$

* See Section 1.2.2.3.

In the last formula, the limits are taken between $-\infty$ and ∞ because p is assumed to be very large. To be able to perform the integral, Plana used $\cos u = (e^{ju} + e^{-ju})/2$ and then made the substitutions $\alpha = A/p$, $\beta = C/p$, $\gamma = 2B/p$, $\alpha = A/p$, $\delta = jQ/\sqrt{ap}$, and $\Xi = jQ'/\sqrt{ap}$. The result is

$$z = \frac{(2n)^p}{8a\pi^2 p}\int_{-\infty}^{\infty}\int_{-\infty}^{\infty}\exp\left\{-\left(\alpha x^2 + \beta x'^2 + \gamma xx' + \delta x + \Xi x'\right)\right\}dxdx'$$
$$+\frac{(2n)^p}{8a\pi^2 p}\int_{-\infty}^{\infty}\int_{-\infty}^{\infty}\exp\left\{-\left(\alpha x^2 + \beta x'^2 + \gamma xx' - \delta x - \Xi x'\right)\right\}dxdx'.$$

The expression $Y = \alpha x^2 + \beta x'^2 + \gamma xx' + \delta x + \Xi x'$ in the above needs to be expressed as the sum of two squares. By making the substitutions $x = u - (\gamma u')/(2a) + f$, $x' = u' - h$, $f = (\gamma\Xi - 2\beta\delta)/(4\alpha\beta - \gamma^2)$, and $h = (2\alpha\Xi - \gamma\delta)/(4\alpha\beta - \gamma^2)$, Plana obtained

$$z = \frac{(2n)^p}{8a\pi^2 p}\left(\int_{-\infty}^{\infty}\int_{-\infty}^{\infty} e^{-\frac{Au^2}{p}} e^{-\frac{Eu'^2}{Ap}} e^{-H} dxdx'\right)\times 2$$
$$= \frac{(2n)^p}{4a\pi^2 p} e^{-H}\frac{\pi p}{\sqrt{E}}$$
$$= \frac{(2n)^p}{4a\pi\sqrt{E}}\exp\left\{-\frac{1}{4aE}\left(CQ^2 - 2BQQ' + AQ'^2\right)\right\},$$

where

$$H = \frac{CQ^2 - 2BQQ' + AQ'^2}{4a\left(AC - B^2\right)} \quad \text{and} \quad E = AC - B^2.$$

By substituting for a from Eq. (2.93) in the above, Plana finally obtained the (unscaled) bivariate normal distribution

$$z = \frac{3(2n)^p}{\pi(n+1)\,(2n+1)\sqrt{E}}\exp\left\{\frac{-3\left(CQ^2 - 2BQQ' + AQ'^2\right)}{E(n+1)\,(2n+1)}\right\} \tag{2.94}$$

(*ibid*., p. 381) for the joint probability that the two sums on the right side of Eq. (2.91) are equal to Q and Q'.

When Eq. (2.94) is divided by $(2n)^p$, we obtain the joint density of Q and Q':

$$f(Q_1, Q_2) = \frac{3}{\pi(n+1)\,(2n+1)\sqrt{E}}\exp\left\{\frac{-3\left(CQ^2 - 2BQQ' + AQ'^2\right)}{E(n+1)\,(2n+1)}\right\}.$$

Plana's analysis is remarkably accurate (though not completely rigorous) because $f(Q_1, Q_2)$ corresponds exactly to the bivariate normal density in Eq. (2.82). Indeed, the former equation can be recovered from the latter by making the following substitutions*:

$$\begin{aligned} x &= Q_1, \quad y = Q_2 \\ \sigma_X^2 &= \text{var } Q \\ &= \text{var}\left(q_1\varepsilon_1 + q_2\varepsilon_2 + \cdots + q_p\varepsilon_p\right) \\ &= \frac{1}{6}(n+1)\,(2n+1)\sum_{i=1}^{p} q_i^2 \\ &= \frac{A}{6}(n+1)\,(2n+1), \\ \sigma_Y^2 &= \text{var } Q' = \frac{C}{6}(n+1)\,(2n+1) \end{aligned}$$

and

$$\begin{aligned} \rho &= \frac{\text{cov}(Q_1, Q_2)}{\sqrt{\text{var } Q_1}\sqrt{\text{var } Q_2}} \\ &= \frac{\frac{1}{6}(n+1)\,(2n+1)\sum_{i=1}^{p} q_i q_i'}{\sqrt{\frac{A}{6}(n+1)\,(2n+1)}\sqrt{\frac{C}{6}(n+1)\,(2n+1)}} \\ &= \frac{B}{\sqrt{AC}}. \end{aligned}$$

2.4.6 Bravais' Determination of Errors in Coordinates (1846)

Of all the mathematicians we have considered in this section, none came as close to discovering the concept of correlation than the scholar we shall now consider, namely, Bravais. Before examining Bravais' work, we shall say a few words about his life.

Auguste Bravais† (Fig. 2.21) was born in Annonay, France, on August 23, 1811, and was the ninth of ten children. His father, François-Victor Bravais, was a physician. Bravais first studied at the Collège Stanislas, Paris, in 1827. In 1828, he went to Nîmes to take the polytechnical examination, but failed. However, Bravais was not discouraged. At the end of the year, he won first prize in mathematics in the general examination and was then accepted at the École Polytechnique, ranking second. Toward the end of his studies, Bravais became a naval officer. In 1832, he sailed the Mediterranean and was assigned to

*Remembering that $\mathsf{E}\varepsilon_i = 0$ and var $\varepsilon_i = (n+1)(2n+1)/6$, since ε_i is discrete uniform on $\{-n, -n+1, -1, 1, \ldots, n-1, n\}$.

†The life of Auguste Bravais is also described in De Beaumont (1865) and Birembaut (1970).

FIGURE 2.21 Auguste Bravais (1811–1863). Wikimedia Commons (Public Domain), http://commons.wikimedia.org/wiki/Auguste_Bravais#/media/File:Bravais2.gif

map the coast of Algeria. During the period 1833–1835, Bravais studied plant organography with his elder brother Louis and his friend Charles Martins. Bravais' doctoral thesis in 1837 was on the subject of nautical surveying and the stability of ships. Soon after, Bravais and Martins became part of the Commission Scientifique du Nord. In 1841, the navy gave Bravais permission to teach astronomy as the Faculté des Sciences at Lyons. Together with Pierre Lortet and Joseph Fournet, Bravais founded the Hydrometric Society of Lyons. In 1844, he was elected to the *Académie Royale des Sciences, Belles-Lettres et Arts de Lyon*. Bravais became Professor of Physics at the École Polytechnique in 1845. In 1848, Bravais made a comprehensive study of lattices and derived the 14 possible arrangements of points in space. In his book *Études Cristallographiques* (Bravais, 1866), Bravais made an exhaustive study of the geometry of molecular polyhedra. The book turned out to be very influential for future scientists. However, starting from 1853, Bravais experienced a string of misfortunes. Both his father and only son passed away. In 1856, Bravais became seriously ill from a brain disease and resigned from the École Polytechnique. He passed away on May 1863.

In 1846, Bravais published a memoir* on the theory of errors entitled "Analyse mathématique sur les probabilités des erreurs de situation d'un point" (Bravais, 1846) (Fig. 2.22). His aim was to determine the probability distribution of the errors in the coordinates x, y, z of a point when the coordinates were not directly measurable but rather were functions of the observed quantities a, b, c, ...:

$$\begin{aligned} x &= \varphi(a,b,c,\ldots), \\ y &= \psi(a,b,c,\ldots), \\ z &= \chi(a,b,c,\ldots) \end{aligned}$$

(*ibid.*, p. 255). Bravais now took differentials and wrote

$$\begin{aligned} \delta x &= \frac{\partial\varphi}{\partial a}\delta a + \frac{\partial\varphi}{\partial b}\delta b + \frac{\partial\varphi}{\partial c}\delta c + \cdots = A\delta a + B\delta b + C\delta c + \cdots, \\ \delta y &= \frac{\partial\psi}{\partial a}\delta a + \frac{\partial\psi}{\partial b}\delta b + \frac{\partial\psi}{\partial c}\delta c + \cdots = A'\delta a + B'\delta b + C'\delta c + \cdots, \\ \delta z &= \frac{\partial\chi}{\partial a}\delta a + \frac{\partial\chi}{\partial b}\delta b + \frac{\partial\chi}{\partial c}\delta c + \cdots = A''\delta a + B''\delta b + C''\delta c + \cdots. \end{aligned} \tag{2.95}$$

Bravais next considered the first of the equations in (2.95). He wrote x, m, n, p, ... for $\delta x, \delta a, \delta b, \delta c, \ldots$, obtaining

$$x = Am + Bn + Cp + \cdots. \tag{2.96}$$

Bravais called m, n, p, ... the independent variables while x the dependent variable. He then stated that Laplace had shown that the probability that x lies between x and $x + dx$ was

$$\sqrt{\frac{h_x}{\pi}}\exp\left(-h_x x^2\right)dx, \tag{2.97}$$

*This memoir had been submitted in 1838. Bravais' derivation is also described in Pearson (1920), Denis (2001), Walker (1928), Plackett (1983), and Lancaster (1972).

DES ERREURS DE SITUATION D'UN POINT. 273

Le rayon vecteur, correspondant à ces valeurs de x et de y, sera le demi-diamètre conjugué de l'axe des x, et si T représente l'angle de ces deux diamètres, $\frac{y}{\sin.T} = \frac{1}{\sin.T}\sqrt{\frac{aD}{ab-e^2}}$ sera la grandeur OM de ce rayon vecteur (voyez figure ci-dessous).

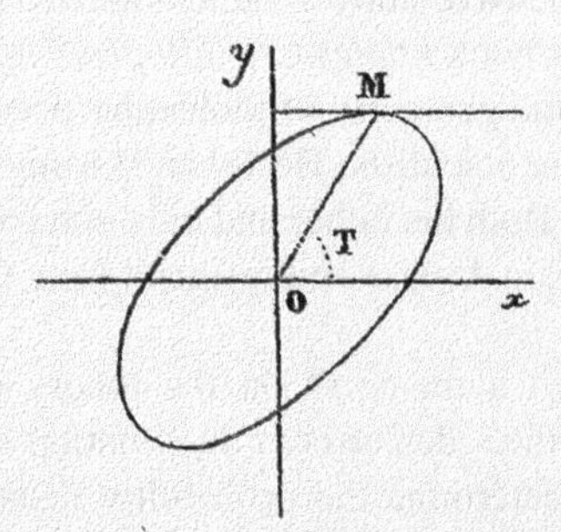

L'autre demi-diamètre s'obtient en faisant $y = 0$ dans l'équation de l'ellipse et a pour valeur $\sqrt{\frac{D}{a}}$. Le produit de ces deux demi-diamètres, multiplié par π sin. T, donne, comme on sait, l'aire de l'ellipse, que nous représenterons par s : on aura ainsi

$$s = \pi \frac{D}{\sqrt{ab-e^2}} \qquad D = \frac{s\sqrt{ab-e^2}}{\pi}. \qquad (31)$$

Ainsi l'on peut, dans les formules (18) et (29), remplacer l'exposant de e par cette valeur de D, et l'on trouve, en observant que $\sqrt{ab-e^2} = K$,

$$\frac{d^2\varpi}{dx\,dy} = \frac{K}{\pi} e^{-\frac{K}{\pi}s}, \qquad (32)$$

ce qui montre clairement que la probabilité différentielle reste la même tout le long de l'arc d'une des ellipses données par l'équation (30), et qu'elle varie en passant d'une ellipse à l'autre, en raison inverse de l'exponentielle de la surface.

Si nous intégrons dans toute l'étendue de l'espace annulaire compris entre deux ellipses infiniment voisines, s restera cons-

9. 35

FIGURE 2.22 Extract from Bravais' 1846 memoir (Bravais, 1846, p. 273)

where

$$\frac{1}{h_x} = \frac{A^2}{h_m} + \frac{B^2}{h_n} + \frac{C^2}{h_p} + \cdots$$

(*ibid*., p. 261). Here, the h_i's are the precisions of the respective distributions defined by $h = 1/\left(\sigma\sqrt{2}\right)$ in modern notation. Bravais then considered the first two equations in (2.95):

$$\begin{aligned} x &= Am + Bn + Cp + \cdots, \\ y &= A'm + B'n + C'p + \cdots \end{aligned} \tag{2.98}$$

(*ibid*., p. 263) with respective precisions

$$\frac{1}{h_x} = \frac{A^2}{h_m} + \frac{B^2}{h_n} + \frac{C^2}{h_p} + \cdots = \alpha_0,$$

$$\frac{1}{h_y} = \frac{A'^2}{h_m} + \frac{B'^2}{h_n} + \frac{C'^2}{h_p} + \cdots = \alpha_1.$$

He next made the following important observation regarding (2.98):

> The co-existence of the same variables *m*, *n*, *p*, … in the simultaneous equations for *x* and *y*, produces a correlation such that, the moduli h_x, h_y cease to represent the probability of the simultaneous values of (*x*, *y*) under the true viewpoint of the question. (*ibidem*)

Bravais wrote the densities of the independent variables *m*, *n*, *p*, … in (2.98), respectively, as

$$\sqrt{\frac{h_m}{\pi}}\exp\left(-h_m m^2\right),$$

$$\sqrt{\frac{h_n}{\pi}}\exp\left(-h_n n^2\right),$$

$$\sqrt{\frac{h_p}{\pi}}\exp\left(-h_p p^2\right),$$

$$\cdots.$$

By assuming independence, the joint density of *m*, *n*, *p*, … is

$$\sqrt{\frac{h_m}{\pi}\cdot\frac{h_n}{\pi}\cdot\frac{h_p}{\pi}\cdots}\exp\left\{-\left(h_m m^2 + h_n n^2 + h_p p^2 + \cdots\right)\right\}.$$

The task was now to write the joint density of *x*, *y*, *z*, … in terms of that of *m*, *n*, *p*, …. The first density is in fact the product of the second and the magnitude of the Jacobian of the transformation. Bravais gave the formula for the Jacobian in some simple cases. For example, if $x = \phi_0(m, n, p, \ldots)$ and $y = \phi_1(m, n, p, \ldots)$, then

$$dxdy = \left(\frac{\partial\phi_0}{\partial m}\cdot\frac{\partial\phi_1}{\partial n} - \frac{\partial\phi_0}{\partial n}\cdot\frac{\partial\phi_1}{\partial m}\right)dmdn = \left(AB'\right)dmdn$$

(*ibid.*, p. 265), where (AB') denotes the Jacobian* of the transformation from (x,y) to (m,n). In general, the joint density of x, y, z, … is

$$\frac{1}{(AB'C''\ldots)}\sqrt{\frac{h_m}{\pi}\cdot\frac{h_n}{\pi}\cdot\frac{h_p}{\pi}\cdots}\exp\left\{-\left(h_m m^2+h_n n^2+h_p p^2+\cdots\right)\right\},$$

where $(AB'C''\ldots)$ is the Jacobian of the transformation from $(x,y,z,\ldots)$ to $(m,n,p,\ldots)$. By assuming the equations in (2.98) can be inverted to give

$$\begin{aligned} m&=\alpha x+\beta y+\gamma z+\cdots,\\ n&=\alpha' x+\beta' y+\gamma' z+\cdots,\\ p&=\alpha'' x+\beta'' y+\gamma'' z+\cdots, \end{aligned}$$

the joint density of x, y, z, … is

$$\frac{1}{(AB'C''\ldots)}\sqrt{\frac{h_m}{\pi}\cdot\frac{h_n}{\pi}\cdot\frac{h_p}{\pi}\cdots}\exp\Big[-\Big\{h_m(\alpha x+\beta y+\gamma z+\cdots)^2+h_n(\alpha' x+\beta' y+y' z+\cdots)^2$$
$$+h_p(\alpha'' x+\beta'' y+\gamma'' z+\cdots)^2\Big\}\Big]$$

(*ibid.*, p. 267). Bravais next observed that if z, w, … are integrated out of the above, the joint density of x and y is obtained and is of the form

$$\frac{K}{\pi}\exp\left\{-\left(ax^2+2exy+by^2\right)\right\}. \tag{2.99}$$

In the above, K, a, b, and e are constants which Bravais determined as follows. First integrating Eq. (2.99) with respect to y, he obtained

$$\frac{K}{\sqrt{\pi}}b^{-1/2}\exp\left\{-\frac{\left(ab-e^2\right)}{b}x^2\right\}.$$

Equating this with the density in (2.97),

$$\frac{1}{h_x}=\frac{b}{ab-e^2}=\frac{b}{K^2}.$$

Similarly, by integrating Eq. (2.99) with respect to x and comparing to the density of y,

$$\frac{1}{h_y}=\frac{a}{ab-e^2}=\frac{a}{K^2}.$$

The last two expressions give the three equations

$$\begin{aligned} a&=K^2/h_y=K^2\alpha_1,\\ b&=K^2/h_x=K^2\alpha_0,\\ K^2&=ab-e^2 \end{aligned}$$

* In fact, the *magnitude* of the Jacobian should be used to prevent the occurrence of negative densities.

(*ibid.*, p. 270). Since there are four constants (K, a, b, and e) to be determined, one more equation is required. The latter is obtained by rotating the x–y coordinate system and is

$$\frac{1}{K^2}=\left(\frac{A^2}{h_m}+\frac{B^2}{h_n}+\frac{C^2}{h_p}+\cdots\right)\left(\frac{A'^2}{h_m}+\frac{B'^2}{h_n}+\frac{C'^2}{h_p}+\cdots\right)-\left(\frac{AA'}{h_m}+\frac{BB'}{h_n}+\frac{CC'}{h_p}+\cdots\right)^2$$
$$=\frac{(AB'-A'B)^2}{h_m h_n}+\frac{(AC'-A'C)^2}{h_m h_p}+\cdots$$
$$=\sum\frac{(AB'-A'B)^2}{h_m h_n}.$$

Solving for K, a, b, and e in the preceding four equations, Bravais finally wrote the joint density of x and y in Eq. (2.99) as

$$\frac{1}{\pi\sqrt{\sum\frac{(AB'-A'B)^2}{h_m h_n}}}\exp\left[-\frac{\left\{x^2\sum\left(\frac{A'^2}{h_m}\right)-2xy\sum\left(\frac{AA'}{h_m}\right)+y^2\sum\left(\frac{A^2}{h_m}\right)\right\}}{\sum\frac{(AB'-A'B)^2}{h_m h_n}}\right] \tag{2.100}$$

(*ibid.*, p. 272). Equation (2.100) is of the same form given in Eq. (2.82). Bravais wrote it as

$$\frac{K}{\pi}\exp\left\{-K^2\left(\alpha_1 x^2-2\beta_0 xy+\alpha_0 y^2\right)\right\} \tag{2.101}$$

and made some key conclusions. He noted that by setting the exponent in Eq. (2.101) to a constant, the equation

$$\alpha_1 x^2-2\beta_0 xy+\alpha_0 y^2=\text{const.} \tag{2.102}$$

could be obtained. He then wrote:

> This equation represents an ellipse whose center is at the origin, whose principal axes are in general different from the coordinate axes, and, by varying the constant of the second member, it furnishes an infinite series of similar ellipses, whose areas increases proportionally to the constant. (*ibidem*)

By rewriting the ellipse in Eq. (2.102) as

$$ax^2+2exy+by^2=D,$$

Bravais then took differentials and set $dy=0$, thus obtaining the coordinates of the points with zero slope, namely,

$$x=-\frac{e}{a}y, \tag{2.103}$$
$$y^2=\frac{aD}{ab-e^2}.$$

With Eq. (2.103), Bravais had in effect obtained the equation of the regression line of x on y (the line OM in Fig. 2.22). He did not realize this because he was concerned with the areas of ellipses, not with the relationship between the x and y variables. However, it is evident he was a few inches away from the concepts of correlation and regression. That he was extremely close to these concepts is further evidenced when he set $\Sigma(AA'/h_m)=0$ in Eq. (2.100) obtaining the joint density of x and y as

$$\frac{1}{\pi}\sqrt{h_x h_y}\exp\left\{-\left(h_x x^2 + h_y y^2\right)\right\}$$

(*ibid*., p. 279). He then made these pertinent remarks:

> Thus, in this case, the probability of the simultaneous values $x=x$, $y=y$, is exactly the same as when the variables x and y were completely independent of each other. This case occurs when the equations (9) [Eq. 2.98] is of the form
>
> $$x = Am + Bn + Cp + \cdots$$
> $$y = A'm_1 + B'n_1 + C'p_1 + \cdots,$$
>
> the variables $m, n, p, \ldots$ in the equation for x being essentially distinct of the variables $m_1, n_1, p_1, \ldots$ in the equation for y. (*ibidem*)

From the above, some might think that Bravais actually discovered correlation before Galton. However, this is not the case. At no instant did Bravais wish to quantity the relationship between his x and y variables. In the words of Karl Pearson:

> He [Bravais] is merely seeking to express the variability of x and y in terms of the directly determined constants and certain differential coefficients. This is one of the fundamental problems of the Method of Least Squares and had already been solved by Gauss. Bravais adds so far nothing whatever to Gauss' solution of 20 years earlier. If Bravais discovered correlation, then Gauss had done so previously. (Pearson, 1920, p. 31)

2.4.7 Bullet Shots on a Target: Bertrand's Derivation of the Bivariate Normal Distribution (1888)

Bertrand's work did not exactly take place before Galton. In the same year as the latter published "Co-relations and their Measurements" (Galton, 1888), Bertrand derived the bivariate normal distribution in the second (Bertrand, 1888) of a set of articles relating to bullet shots on a target. Before examining Bertrand's derivation, we shall say a few words about his life.

Joseph Louis François Bertrand* (Fig. 2.23) was born in Paris on March 11, 1822. His father, Alexandre Bertrand, was a physician who graduated at the École Polytechnique but died when Bertrand was 9 years old. Thereupon, the latter was raised by his uncle Jean Marie Constant Duhamel, a famous mathematician. Bertrand was a precocious child and was allowed to follow lectures at the École Polytechnique at the age of 11. He then received his B.A. and B.Sc. at the age of 16. The next year, in 1839, Bertrand obtained his Ph.D. for a thesis in thermomechanics. He was then of age to enter the École Polytechnique,

* Bertrand's life is also described in Struik (1970a) and Bru and Jongmans (2001).

FIGURE 2.23 Joseph Louis François Bertrand (1822–1900). Wikimedia Commons (Public Domain), http://commons.wikimedia.org/wiki/File:Bertrand.jpg

having already obtained his Ph.D.! During 1841–1848, Bertrand was Professor of Elementary Mathematics at the Collège Saint-Louis. In 1844, he married Mlle. Aclocque and then became tutor in analysis (*répétiteur d'analyse*) at the École Polytechnique. Bertrand was made interim Professor of Mathematical Physics at the Collège de France. During the revolution in 1848, he was a captain in the national guard and managed to publish substantially in mathematical physics, analysis, and differential geometry. In 1856, Bertrand replaced Jacques Charles Francois Sturm as Professor of Analysis at the École Polytechnique. In the same year, he was also elected member of the *Académie des Sciences*. He became permanent secretary for the mathematical sciences from 1874 till his death. Apart from research papers, Bertrand published several textbooks, of which his *Calculs des Probabilités* (Bertrand, 1889) deserves special mention. The book was quite popular and contains the famous *Bertrand Chord Problem** (*ibid*., p. 4), which is still used in probability textbooks today. Bertrand passed away in Paris on April 5, 1900.

We now examine Bertrand's 1888 article (Bertrand, 1888) where he derived the bivariate normal distribution. Referring to the latter as Bravais' theorem, Bertrand stated that his derivation rested on an assumption very similar to Gauss' assumption about the sample mean in his derivation of the normal distribution (see Section 1.5.3.2). Bertrand's assumption was that

> *If we know the points where the target has been hit, then the most probable position of the point aimed at is the center of gravity of the system of points obtained.*[†] (*ibid*., p. 389)

Bertrand credited the above postulate to the English mathematician Roger Cotes (1682–1716) in 1722.[‡] His derivation was as follows.[§]

Suppose the targeted point has coordinates (X, Y) in the xy-plane and the points actually hit have coordinates $(x_1,y_1),(x_2,y_2),\ldots,(x_n,y_n)$. Moreover, let $F(X-x,Y-y)dxdy$ be the probability that the bullet hits an element of area $dxdy$ at the point (x, y), that is, F is the bivariate density of the deviations from (X, Y) of the points hit. Then, by the assumed postulate, the product

$$F(X-x_1,Y-y_1)F(X-x_2,Y-y_2)\cdots F(X-x_n,Y-y_n) \tag{2.104}$$

is a maximum when

$$\begin{aligned} X &= \frac{x_1+x_2+\cdots+x_n}{n}, \\ Y &= \frac{y_1+y_2+\cdots+y_n}{n}. \end{aligned} \tag{2.105}$$

From thereon, Bertrand thus assumed that X and Y were assigned the values of their respective sample means. Next, he let

$$\frac{\partial \log F(x,y)}{\partial x} = \phi(x,y), \quad \frac{\partial \log F(x,y)}{\partial y} = \omega(x,y). \tag{2.106}$$

* See p. 106.

† Italics are Bertrand's.

‡ In his book *Calculs des Probabilités* (Bertrand, 1889, p. 228), Bertrand revised the date to 1709.

§ Bertrand's derivation is also described in Whittaker and Robinson (1944, pp. 321–323) and Plackett (1983).

Because (2.104), and hence its logarithm, is a maximum when (2.105) holds, we have

$$\phi(X-x_1,Y-y_1)+\phi(X-x_2,Y-y_2)+\cdots+\phi(X-x_n,Y-y_n)=0,$$
$$\omega(X-x_1,Y-y_1)+\omega(X-x_2,Y-y_2)+\cdots+\omega(X-x_n,Y-y_n)=0.$$

With $\alpha_i = X - x_i$ and $\beta_i = Y - y_i$, Bertrand wrote

$$\begin{aligned}\phi(\alpha_1,\beta_1)+\phi(\alpha_2,\beta_2)+\cdots+\phi(\alpha_n,\beta_n)=0,\\ \omega(\alpha_1,\beta_1)+\omega(\alpha_2,\beta_2)+\cdots+\omega(\alpha_n,\beta_n)=0.\end{aligned} \tag{2.107}$$

He also stated that the above were the necessary consequences ("*conséquences nécessaires*") of

$$\begin{aligned}\alpha_1+\alpha_2+\cdots+\alpha_n=0,\\ \beta_1+\beta_2+\cdots+\beta_n=0.\end{aligned} \tag{2.108}$$

The first equation in (2.107) then becomes

$$\phi(-\alpha_2-\alpha_3-\cdots-\alpha_n,-\beta_2-\beta_3-\cdots-\beta_n)+\phi(\alpha_2,\beta_2)+\cdots+\phi(\alpha_n,\beta_n)=0.$$

By differentiating the above with respect to α_2 and letting $\partial\phi(x,y)/\partial x=\phi_1$,

$$-\phi_1(-\alpha_2-\alpha_3-\cdots-\alpha_n,-\beta_2-\beta_3-\cdots-\beta_n)+\phi_1(\alpha_2,\beta_2)=0,$$

whence it can be concluded that $\phi_1(x,y)$ is functionally independent of x and y. Similarly, it can be shown that $\phi_2=\partial\phi(x,y)/\partial y$ is functionally independent of x and y. This implies that ϕ must be in the form $\phi(x,y)=ax+by+c$, where a, b, and c are constants. Substituting this in Eq. (2.107) and applying the conditions in (2.108) show that $c=0$. Hence,

$$\phi(x,y)=ax+by$$

and similarly

$$\omega(x,y)=a'x+b'y,$$

where a' and b' are constant. Therefore, integrating the left equation in (2.106) with respect to x and the right equation with respect to y,

$$\log F(x,y)=\frac{ax^2}{2}+bxy+g_1(y),$$

$$\log F(x,y)=a'xy+\frac{b'y^2}{2}+g_2(y),$$

so that

$$\log F(x,y)=\frac{ax^2}{2}+bxy+\frac{b'y^2}{2}+C,$$

where C is a constant. Using the above, Bertrand finally expressed $F(x,y)$ as

$$Ge^{-k^2x^2-2\lambda xy-k'^2y^2} \tag{2.109}$$

(*ibid.*, p. 390), where G is a constant. Eq. (2.109) is a bivariate normal density and Bertrand called it "Bravais' theorem." By double integrating the last equation and setting to unity,

$$G = \frac{\sqrt{k^2 k'^2 - \lambda^2}}{\pi}.$$

He also noted that the curves of equal probability are the ellipses with equation

$$k^2 x^2 + 2\lambda xy + k'^2 y^2 = H.$$

The next task for Bertrand was to determine the three constants k, k', and λ in (2.109). First, he equated the mean value of x^2 obtained from (2.109) to the observed average (denoted by A), that is,

$$\frac{x_1^2 + x_2^2 + \cdots + x_n^2}{n} = \frac{k'^2}{2\left(k^2 k'^2 - \lambda^2\right)} = A.$$

Second, equating the mean value of y^2 to the observed average (denoted by B),

$$\frac{y_1^2 + y_2^2 + \cdots + y_n^2}{n} = \frac{k^2}{2\left(k^2 k'^2 - \lambda^2\right)} = B.$$

Finally, equating the mean value of xy to the observed average (denoted by C),

$$\frac{x_1 y_1 + x_2 y_2 + \cdots + x_n y_n}{n} = \frac{\lambda}{2\left(k^2 k'^2 - \lambda^2\right)} = C.$$

By solving the last three equations for the three unknowns k, k', and λ, the latter can be estimated:

$$k^2 = \frac{B}{2\left(AB - C^2\right)},$$

$$k'^2 = \frac{A}{2\left(AB - C^2\right)},$$

$$\lambda = \frac{-C}{2\left(AB - C^2\right)}.$$

3

KARL PEARSON'S CHI-SQUARED GOODNESS-OF-FIT TEST

3.1 KARL PEARSON (1857–1936)

Karl Pearson will remain one of the towering geniuses of mathematical statistics. For how else could we describe the man who, in 1892 at the age of 35, did not know much about statistics, yet within the next 10 years or so had introduced much of the statistical terminology that we still use today? Terms such as "standard deviation," "contingency table," "goodness of fit," "correlation coefficient," "skewness," "kurtosis," "moment," and many more were first introduced by none other than Pearson.

Karl Pearson* (Fig. 3.1) was born in London, England, on March 27, 1857. He was the second of three children, and his father was a barrister of the Inner Temple. Pearson was educated at home until the age of nine, when he was sent to University College School, London, for 7 years. During the final year Pearson withdrew because of health reasons and spent the next year with a private tutor. In 1875, he obtained a scholarship at King's College, Cambridge, where he was taught mathematics by James Clerk Maxwell, Arthur Cayley, and William Burnside. Pearson obtained his B.A. in 1879 and soon after went to Germany. In Heidelberg, he read Berkeley, Fichte, Locke, Kant, and Spinoza but afterward abandoned philosophy. Pearson then studied physics under G.H. Quincke and metaphysics under Kuno Fischer. In November 1880, Pearson went back to England, took up rooms in the Inner Temple, and read law at Lincoln's Inn. He was called to the Bar the next year and practiced law of a very short time. During the period 1880–1881, Pearson also lectured on socialism, Marx and Lassalle. Provine has noted that Pearson was a most stimulating teacher but also that:

> Pearson was an intelligent young man. He knew it, and was quick to criticize the incompetence of others. An example was his attack on an exhibition in 1883 at the British Museum

* Karl Pearson was originally called Carl Pearson. While he was at the University of Heidelberg, Germany, his first name was misspelled. Pearson decided to keep the name Karl owing to his love of Germany and possibly as a tribute to Karl Marx, whose writings Pearson admired. To his surrounding, Pearson was known simply as K.P. Pearson's life is also described in Porter (2010), Eisenhart (1974), and Magnello (2001a, pp. 248–256).

Classic Topics on the History of Modern Mathematical Statistics: From Laplace to More Recent Times, First Edition. Prakash Gorroochurn.

FIGURE 3.1 Karl Pearson (1857–1936). Wikimedia Commons (Public Domain), http://commons.wikimedia.org/wiki/File:Karl_Pearson.jpg

> celebrating the three hundredth anniversary of Martin Luther's birth. When others reacted to his criticism, he engaged in a number of literary duels, brandishing sharp-edged rhetoric.... (Provine, 2001, p. 27)

The above aspect of Pearson's character was to have heavy consequences with his interactions with several statisticians, most notably Yule and Fisher.

In 1884, Pearson was appointed Goldsmid Professor of Applied Mathematics and Mechanics at University College London. In June 1890, he married Marie Sharpe. From 1891 to 1894, he took up the position of Gresham Chair of Geometry while retaining his other position at University College. At Gresham College, Pearson delivered 12 lectures a year, free to the public. The first eight lectures formed the basis of his book *The Grammar of Science* (Pearson, 1892), which was subsequently translated into several languages. In the lecture of November 18, 1891, one can find Pearson's earliest teachings in probability and statistics, namely, probability applied to actuarial methods and graphical statistics. The lecturers were influenced by the works of Edgeworth, Jevons, and Venn. At around the same time, Galton's *Natural Inheritance* (Galton, 1889) had come out, and Pearson was exposed to the ideas of regression and correlation. The book had a profound effect on his future career. But a major impetus for this was to come from the zoologist W.F.R. Weldon, who took up the position of Jodrell Professor of Zoology at University College in 1891. Weldon was interested in corroborating Darwin's theory of natural selection and believed that statistical studies of variation and correlation, as detailed in Galton's book, would be the means of achieving this. Weldon interacted with Pearson by asking the biological questions that resulted in many of Pearson's seminal contributions to statistics.

An important incident occurred in 1900, the same year that Pearson published his landmark chi-squared paper (Pearson, 1900b), when Pearson submitted his important "homotyposis paper" (Pearson, 1901a) to the Royal Society. The paper did get accepted but so also did a severe criticism of it by Pearson's rival William Bateson* (Bateson, 1901). Pearson protested in vain that only original work ought to be published by the Royal Society, not criticisms of other works, and that if Bateson would publish his criticism then Pearson should have the right to reply. The whole incident distressed both Pearson and his ally Weldon, and they decided to launch their own journal *Biometrika* with the financial support of Galton. *Biometrika* was thus founded in 1901.

After receiving a grant in 1903, Pearson set up the Draper's Biometric Laboratory, followed by an Astronomical Laboratory one year later. In 1911, he gave up the Goldsmid Chair to become the first Galton Professor of Eugenics. The Draper's Biometric and Galton Eugenics laboratories then became incorporated into the Department of Applied Statistics.

In the early 1920s, Pearson became increasingly embroiled in a bitter fight with Fisher, who successfully attacked an implementation of his greatest work, the chi-squared test.

* Bateson's controversy with Pearson stemmed from the former's endorsing a theory of evolution that proceeded in a discontinuous manner (the Mendelian viewpoint), as opposed to Pearson's view that evolution proceeded continuously (the Biometrician viewpoint).

In 1928, Pearson's wife passed away, and one year later Pearson married Margaret Victoria Child, a co-worker in the Biometric Laboratory. He was made Emeritus Professor in 1933. In 1935, he was offered a knighthood but refused. A similar fate had previously occurred to the Order of the British Empire, offered to him in 1920. Pearson passed away in Surrey, England, on April 27, 1936.

Apart from his contributions to statistics, Pearson also wrote on the philosophy of science.* His contributions in this field can be found in his book *The Grammar of Science* (Pearson, 1892) and the collection of essays and lectures entitled *The Ethic of Freethought* (Pearson, 1901b). Pearson advocated that modern science had several benefits for human society through its educational value:

> *Modern Science, as training the mind to an exact and impartial analysis of facts, is an education specially fitted to promote sound citizenship.*[†] (Pearson, 1892, p. 11)

Pearson was against a utilitarian view of science: the value of science does not reside in only how useful it can be to society. In his own words:

> I want the reader to appreciate clearly that science justifies itself in its methods, quite apart from any serviceable knowledge it may convey…That form of popular science which merely recites the results of investigations, which merely communicates useful knowledge, is from this standpoint bad science, or no science at all. (*ibid.*, pp. 11–12)

Apart for the idealism above, in other respects Pearson embraced the sensationalist philosophy of Ernst Mach (1838–1916). According to the latter, science rests on sense experience, and all branches of sciences are united in that they are concerned with the study of sensations.[‡] Moreover, science does not *explain* natural phenomena, rather it tries to *describe* them:

> One of the greatest of German physicists, Kirchhoff, thus commences his classical treatise on mechanics:
>
> "Mechanics is the science of motion; we define as its object the complete *description* in the *simplest* possible manner of such motions as occur in nature."
>
> In this definition of Kirchhoff's lies, I venture to think, the only consistent view of mechanism and the true conception of scientific law…When, therefore, we say that we have reached a "mechanical explanation" of any phenomenon, we only mean that we have described in the concise language of mechanics a certain routine of perceptions. We are neither able to explain why sense-impressions have a definite sequence, nor to assert that there is really an element of necessity in the phenomenon. (*ibid.*, pp. 138–139)

Science, we are thus told by Pearson, is about "how," not about "why." Pearson also rejected the notion that science can establish causation:

> …That a certain sequence has occurred and recurred in the past is a matter of experience to which we give expression in the concept *causation*; that it will continue to recur in the future is a matter of belief to which we give expression in the concept *probability*. Science in no case can demonstrate any inherent necessity in a sequence, nor prove with absolute certainty

*See also Alexander (2006, pp. 159–161).

[†] Italics are Pearson's

[‡] Mach's philosophy was a successor to Auguste Comte's positivism.

> that it must be repeated. Science for the past is a description, for the future a belief; it is not, and has never been, an explanation, if by this word is meant that science shows the necessity of any sequence of perceptions. (*ibid.*, p. 136)

We see that one of the reasons Pearson latter embraced Galton's concept of correlation was that the latter was a tool of description rather than one of causation.

3.2 ORIGIN OF PEARSON'S CHI-SQUARED

3.2.1 Pearson's Work on Goodness of Fit Before 1900

Pearson's first encounter with the issue of goodness of fit seems to have occurred in 1892 (Plackett, 1983). Pearson tossed a coin 2400 times, which was supplemented with 8178 more tosses performed by his former student, Mr C.L.T. Griffith. Pearson needed more data. These were obtained from records published in *Le Monaco* and were related to the outcomes of Monte Carlo roulette tables. To analyze these experiments, Pearson compared the observed standard deviations to the theoretical ones, as we shall soon explain.

In 1895, Pearson proposed the "mean percentage error in the ordinates" of a frequency polygon as a measure of goodness of fit (Pearson et al., 1895). Hald has pointed out that, at that time, Pearson was most likely thinking in terms of sums of absolute errors, rather than sums of squares of errors (Hald, 1998, p. 652).

Additional insight can be gleaned from Pearson's 1897 book *The Chances of Death* (Pearson, 1897b). The first two chapters of this book show that Pearson had already started to seriously consider measuring goodness of fit, prior to his 1900 groundbreaking paper. We now briefly review the relevant sections in these two chapters. In Chapter I, Pearson showed the observed and theoretical frequencies in four distribution curves (Fig. 3.2): (1) the occurrence of red counters in a set of 10 drawings from a bag containing 25 counters each of red, black, yellow, and green (9148 drawings); (2) the occurrence of hearts in 10 cards drawn once from a full pack (18,600 drawings); (3) the occurrence of fives or sixes in 26,306 throws of 12 dice; and (4) the occurrence of heads in 2048 tosses of 10 shillings. All four cases show remarkable agreement between the observed and theoretical frequencies. Pearson remarked:

> It is not theory, but actual statistical experience, which forces us to the conclusion that, however little we know of what will happen in the individual instance, yet the frequency of a large number of instances is distributed round the mode in a manner more and more smooth and uniform the greater the number of individual instances. When this distribution round the mode does not take place... then we assert than some cause other than chance is at work. (*ibid.*, p. 15)

In Chapter II, Pearson examined the frequency of runs (red or black) in 4274 throws of the roulette ball at Monte Carlo and 4191 sets of tossing a penny (due to his pupil, Mr Griffith) (cf. Table 3.1).

Pearson declared:

> ...whenever the actual deviation reaches three to four times the standard deviation, we are approaching the very improbable. In the case of the tossing the actual deviation is slightly

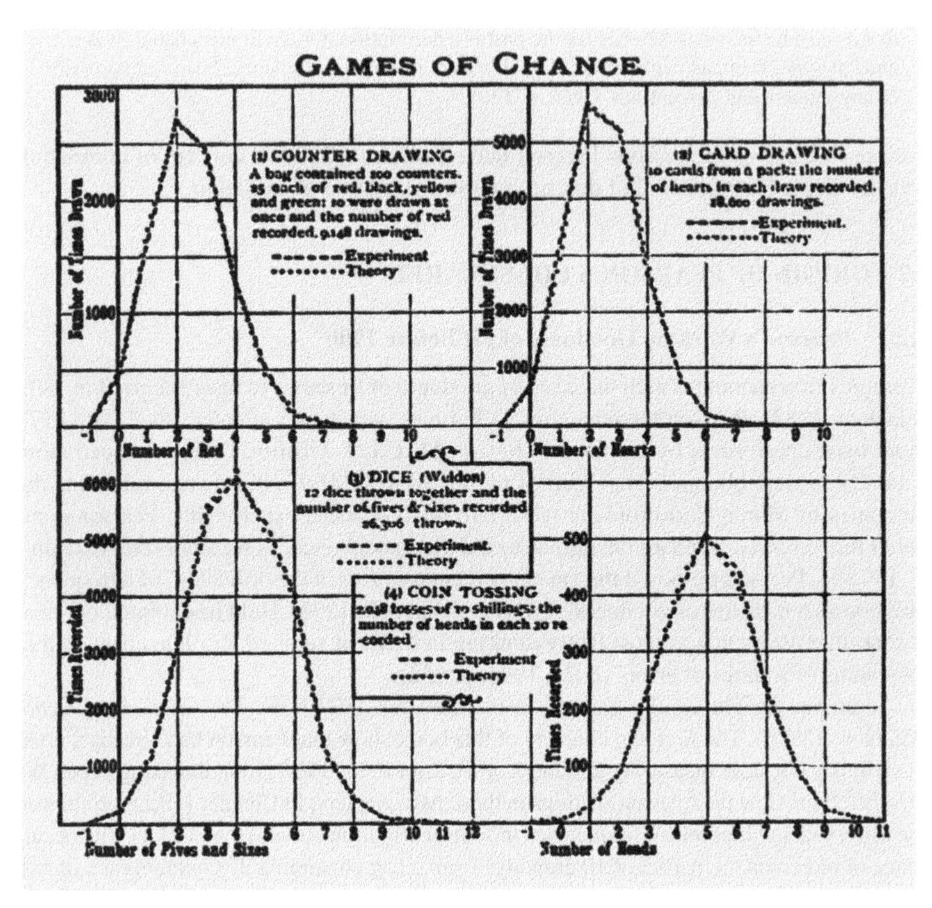

FIGURE 3.2 From Pearson's *The Chances of Death* (Pearson, 1897b, p. 13)

TABLE 3.1 Roulette and Coin-Tossing Results

Runs	1	2	3	4	5	6	7	8	9	10	11	12
Theoretical SD	33	28	22	16	12	8	6	4	3	2	1.5	1
Actual deviation (roulette)	325	123	201	47	1	14	10	13	4	3	3	0
Actual deviation (coin tossing)	68	8	45	22	11	3	7	1	4	1	2	0

twice the standard on two occasions; in the case of the roulette on one occasion the actual deviation is nearly *ten* times the standard, on another occasion *nine* times the standard, on a third occasion *four* times, and twice it is *three* times. The odds are thousand millions to one against such a deviation as nine or ten times the standard. (*ibid.*, p. 54)

We can write Pearson's rule of thumb for a deviation to be due to "some cause other than chance" as

$$|X - \mathsf{E}X| > 3\sigma. \tag{3.1}$$

Here $\mathsf{E}X$ is the expected value of some random variable X and σ is the (theoretical) standard deviation. For the cases considered by Pearson, X can be approximated by a normal distribution, and the probability of the event in (3.1) is

$$\Pr\{|X-\mathsf{E}X|>3\sigma\}\approx\Pr\{|Z|>3\}=.0027$$

where $Z\sim N(0,1)$. However, (3.1) is a *point-by-point* rule of thumb, not an *aggregate* formal goodness-of-fit test. The latter was precisely what Pearson was able to obtain in 1900, which we describe next.

3.2.2 Pearson's 1900 Paper

Prior to Pearson, several mathematicians had derived the bivariate and multivariate normal distribution, as we described in Section 2.4. In 1892, Edgeworth (1892) had obtained the multivariate normal distribution, albeit in an inelegant form.* Pearson was the first (Pearson et al., 1896) to have explicitly stated the multivariate normal distribution in its general (and modern) form.† This fact played a key role in Pearson's formulation of his goodness-of-fit test in the groundbreaking 1900 paper entitled "On the criterion that a given system of deviations from the probable in the case of a correlated system of variables is such that it can be reasonably supposed to have arisen from random sampling" (Pearson, 1900b) (Fig. 3.3). We now describe this paper.‡

First, Pearson considered a system $x_1,x_2,\ldots,x_n$ of normal deviations from the means of n variables with standard deviations $\sigma_1,\sigma_2,\ldots,\sigma_n$ and with correlations $\rho_{12},\rho_{13},\ldots,\rho_{n-1,n}$.§ The joint density of $x_1,x_2,\ldots,x_n$ can be written as

$$Z=Z_0\exp\left\{-\frac{1}{2}\left(\sum_{i=1}^{n}\frac{R_{ii}}{R}\frac{x_i^2}{\sigma_i^2}+2\sum_{i<j}^{n}\frac{R_{ij}}{R}\frac{x_ix_j}{\sigma_i\sigma_j}\right)\right\} \tag{3.2}$$

(*ibid.*, p. 157), where R is the determinant of the correlation matrix

$$\begin{bmatrix} 1 & \rho_{12} & \rho_{13} & \cdots & \rho_{1n} \\ \rho_{21} & 1 & \rho_{23} & \cdots & \rho_{2n} \\ \rho_{31} & \rho_{32} & 1 & \cdots & \rho_{3n} \\ \cdots & \cdots & \cdots & & \cdots \\ \rho_{n1} & \rho_{n2} & \rho_{n3} & \cdots & 1 \end{bmatrix},$$

* See Section 2.3.2.

† See Section 2.3.5.2.

‡ Pearson's paper is also described in Barnard (1992) and Magnello (2005). In particular, the derivation of the distribution of X^2 is also described in Plummer (1940, pp. 246–250), Uspensky (1937, pp. 326–327), Cochran (1952), and Plackett (1983).

§ Pearson used the symbol r for the population correlation, but we shall stick to ρ, in accordance with common usage.

[157]

X. *On the Criterion that a given System of Deviations from the Probable in the Case of a Correlated System of Variables is such that it can be reasonably supposed to have arisen from Random Sampling.* By KARL PEARSON, *F.R.S., University College, London**.

THE object of this paper is to investigate a criterion of the probability on any theory of an observed system of errors, and to apply it to the determination of goodness of fit in the case of frequency curves.

(1) *Preliminary Proposition.* Let $x_1, x_2 \ldots x_n$ be a system of deviations from the means of n variables with standard deviations $\sigma_1, \sigma_2 \ldots \sigma_n$ and with correlations $r_{12}, r_{13}, r_{23} \ldots r_{n-1,n}$.

Then the frequency surface is given by

$$Z = Z_0 e^{-\frac{1}{2}\left\{S_1\left(\frac{R_{pp}}{R}\frac{x_p^2}{\sigma_p^2}\right)+2S_2\left(\frac{R_{pq}}{R}\frac{x_p}{\sigma_p}\frac{x_q}{\sigma_q}\right)\right\}}, \quad \ldots \quad \text{(i.)}$$

where R is the determinant

$$\begin{vmatrix} 1 & r_{12} & r_{13} \ldots r_{1n} \\ r_{21} & 1 & r_{23} \ldots r_{2n} \\ r_{31} & r_{32} & 1 \ldots r_{3n} \\ \cdot & \cdot & \cdot \\ \cdot & \cdot & \cdot \\ r_{n1} & r_{n2} & r_{n3} \ldots 1 \end{vmatrix}$$

and R_{pp}, R_{pq} the minors obtained by striking out the pth row and pth column, and the pth row and qth column. S_1 is the sum for every value of p, and S_2 for every pair of values of p and q.

Now let

$$\chi^2 = S_1\left(\frac{R_{pp}}{R}\frac{x_p^2}{\sigma_p^2}\right)+2S_2\left(\frac{R_{pq}}{R}\frac{x_p x_q}{\sigma_p \sigma_q}\right) \quad \ldots \quad \text{(ii.)}$$

Then: χ^2 = constant, is the equation to a generalized "ellipsoid," all over the surface of which the frequency of the system of errors or deviations $x_1, x_2 \ldots x_n$ is constant. The values which χ must be given to cover the whole of space are from 0 to ∞. Now suppose the "ellipsoid" referred to its principal axes, and then by squeezing reduced to a sphere, $X_1, X_2, \ldots X$ being now the coordinates; then the chances of a system of errors with as great or greater frequency than

* Communicated by the Author.

FIGURE 3.3 First page of Pearson's 1900 paper "On the criterion that a given system of deviations from the probable in the case of a correlated system of variables is such that it can be reasonably supposed to have arisen from random sampling" (Pearson, 1900b)

R_{ij} is the (i,j)th cofactor of the matrix, and Z_0 is a normalizing constant. From the exponent of Z, Pearson next defined

$$X^2 = \sum_{i=1}^{n} \frac{R_{ii}}{R} \frac{x_i^2}{\sigma_i^2} + 2\sum_{i<j}^{n} \frac{R_{ij}}{R} \frac{x_i x_j}{\sigma_i \sigma_j} \tag{3.3}$$

and described X^2 = constant as the "equation to a generalized 'ellipsoid,' all over the surface of which the frequency of the system of errors or deviations was constant." Pearson was interested in calculating the probability $P = \Pr\{X > X_0\}$, which could be obtained by integrating Eq. (3.2) over the region $\{X > X_0\}$. However, he noted that the calculation can be simplified by turning the ellipsoid X^2 = constant into a sphere through a linear transformation, so that

$$P = \frac{\int \cdots \int_{X>X_0} e^{-X^2/2} dt_1 dt_2 \ldots dt_n}{\int \cdots \int_{X>0} e^{-X^2/2} dt_1 dt_2 \ldots dt_n} \tag{3.4}$$

(*ibid.*, p. 158), where $t_1, t_2, \ldots, t_n$ are the new coordinates. Next Pearson stated that a transformation of the above to generalized polar coordinates would give the simplified result:

$$P = \frac{\int_{X_0}^{\infty} e^{-X^2/2} X^{n-1} dX}{\int_{0}^{\infty} e^{-X^2/2} X^{n-1} dX}. \tag{3.5}$$

We shall prove this result soon, but for now it should be noted that, with Eq. (3.5), Pearson had effectively shown that X^2 as defined in (3.3) has an χ^2-distribution with, as we call it today, n degrees of freedom. This is because the denominator of the right side of (3.5) is

$$\int_{0}^{\infty} e^{-X^2/2} X^{n-1} dX = 2^{(n/2)-1} \Gamma(n/2),$$

(where $\Gamma(\cdot)$ is the gamma function) so that the density of X is, from Eq. (3.5),

$$f_X(x) = \frac{1}{2^{(n/2)-1}\Gamma(n/2)} e^{-x^2/2} x^{n-1}, \quad x > 0;\ n = 1, 2, \ldots$$

and the density of $U = X^2$ (where $X > 0$) is

$$\begin{aligned} f_U(u) &= f_X(x) \left| \frac{dx}{du} \right| \\ &= \frac{1}{2^{(n/2)-1}\Gamma(n/2)} e^{-u/2} u^{(n-1)/2} \cdot \frac{1}{2u^{1/2}} \\ &= \frac{1}{2^{n/2}\Gamma(n/2)} e^{-u/2} u^{(n/2)-1}, \quad u > 0;\ n = 1, 2, \ldots \end{aligned} \tag{3.6}$$

The latter is the standard χ_n^2-distribution. Moreover, Pearson's criterion in (3.5) for goodness of fit was

$$P = \Pr\{X > X_0\} = \int_{X_0^2}^{\infty} f_U(u)\,du,$$

which today is simply the p-value, a concept that did not yet exist when Pearson wrote his 1900 paper.

Although these are seldom used nowadays, Pearson also gave approximate formulas for the numerical calculation of P. By using integration by parts on the numerator of the right side of Eq. (3.5), he showed that

$$P = \begin{cases} \sqrt{\frac{2}{\pi}}\int_{X_0}^{\infty} e^{-X^2/2}dX + \sqrt{\frac{2}{\pi}}e^{-X_0^2/2}\left\{X_0 + \frac{X_0^3}{1\cdot 3} + \frac{X_0^5}{1\cdot 3\cdot 5} + \cdots + \frac{X_0^{n-2}}{1\cdot 3\cdot 5\cdots(n-2)}\right\} & \text{for } n \text{ odd} \\ e^{-X_0^2/2}\left\{1 + \frac{X_0^2}{2} + \frac{X_0^4}{2\cdot 4} + \cdots + \frac{X_0^{n-2}}{2\cdot 4\cdot 6\cdots(n-2)}\right\} & \text{for } n \text{ even} \end{cases} \tag{3.7}$$

(*ibid*., p. 159). For $n < 13$, Pearson recommended the use of tables (for different values of $n' = n + 1$) he had included at the end of his paper.

Although Pearson did not give the details, this is how Eq. (3.5) can be obtained. Starting from Eq. (3.3), we apply the following transformations to the variables $X, \phi_1, \phi_2, \ldots, \phi_{n-1}$:

$$\begin{aligned} t_1 &= X\cos\phi_1\cos\phi_2\ldots\cos\phi_{n-1} \\ t_2 &= X\cos\phi_1\cos\phi_2\ldots\cos\phi_{n-2}\sin\phi_{n-1} \\ &\ldots \\ t_j &= X\cos\phi_1\cos\phi_2\ldots\cos\phi_{n-j}\sin\phi_{n-j+1} \\ &\ldots \\ t_n &= X\sin\phi_1 \end{aligned}$$

The Jacobian of the transformation is

$$J\left(\frac{t_1, t_2, \ldots, t_n}{X, \phi_1, \phi_2, \ldots, \phi_{n-1}}\right) = X^{n-1}|D|, \tag{3.8}$$

where $|D|$ is the absolute value of the determinant

$$D = \begin{vmatrix} \cos\phi_1\cos\phi_2\ldots\cos\phi_{n-1} & -\sin\phi_1\cos\phi_2\ldots\cos\phi_{n-1} & \cdots & -\cos\phi_1\cos\phi_2\ldots\sin\phi_{n-1} \\ \cos\phi_1\cos\phi_2\ldots\cos\phi_{n-2}\sin\phi_{n-1} & -\sin\phi_1\cos\phi_2\ldots\cos\phi_{n-2}\sin\phi_{n-1} & \cdots & \cos\phi_1\cos\phi_2\ldots\cos\phi_{n-1} \\ \vdots & \vdots & \cdots & \vdots \\ \sin\phi_1 & \cos\phi_1 & \cdots & 0 \end{vmatrix}.$$

Equation (3.4) then becomes

$$\Pr\{X > X_0\} = \frac{\underset{X>X_0}{\int\cdots\int} e^{-X^2/2} X^{n-1} |D| dX d\phi_1 d\phi_2 \ldots d\phi_{n-1}}{\underset{X>0}{\int\cdots\int} e^{-X^2/2} X^{n-1} |D| dX d\phi_1 d\phi_2 \ldots d\phi_{n-1}},$$

$$= \frac{\int_{X_0}^{\infty} e^{-X^2/2} X^{n-1} dX \times \int\cdots\int |D| d\phi_1 d\phi_2 \ldots d\phi_{n-1}}{\int_{0}^{\infty} e^{-X^2/2} X^{n-1} dX \times \int\cdots\int |D| d\phi_1 d\phi_2 \ldots d\phi_{n-1}}$$

$$= \frac{\int_{X_0}^{\infty} e^{-X^2/2} X^{n-1} dX}{\int_{0}^{\infty} e^{-X^2/2} X^{n-1} dX},$$

and we have proved Eq. (3.5). In the above, we note that since D is a function of $\phi_1, \phi_2, \ldots, \phi_{n-1}$ only, it does not need to be evaluated as it gets cancelled out.

Having done that much of theory, Pearson was now ready for its application. Right at the start of Section 3 of his paper, he announced:

> Now let us apply the above results to the problem of the fit of an observed to a theoretical frequency distribution. Let there be an $(n + 1)$-fold grouping and let the observed frequencies of the groups be
>
> $$m'_1, m'_2, m'_3, \ldots, m'_n, m'_{n+1},$$
>
> and the theoretical frequencies supposed known a priori be
>
> $$m_1, m_2, m_3, \ldots, m_n, m_{n+1}.$$
>
> (*ibid*., p. 160)

If N is the total frequency then $N = \sum_{i=1}^{n+1} m'_i = \sum_{i=1}^{n+1} m_i$. Moreover, the ith error is

$$e_i = m'_i - m_i, \quad i = 1, 2, \ldots, n, n+1,$$

where $\sum_{i=1}^{n+1} e_i = 0$. Next, without mentioning it, Pearson assumed a multinomial distribution for $(m'_1, m'_2, m'_3, \ldots, m'_n, m'_{n+1})$ with respective known probabilities $(p_1, p_2, p_3, \ldots, p_n, p_{n+1})$, where $\sum_{i=1}^{n+1} p_i = 1$. This enabled him to write the standard deviation of e_i and the covariance between e_i and e_j as

$$\sigma_i = \sqrt{N\left(1 - \frac{m_i}{N}\right)\frac{m_i}{N}},$$

$$\sigma_i \sigma_j \rho_{ij} = -\frac{m_i m_j}{N}$$

(*ibid*., p. 161).

The aim of Pearson now was to find the appropriate form for X^2 in (3.3) in the case of the multinomial distribution. However, Pearson noted that only n variables needed to be treated in (3.3) since $e_{n+1} = -\sum_{i=1}^{n} e_i$. Pearson made a trigonometric transformation:

$$\frac{m_i}{N} = \sin^2 \beta_i,$$

where β_i is an "auxiliary angle easily found." This implies

$$\sigma_i = \sqrt{N} \sin \beta_i \cos \beta_i,$$
$$\rho_{ij} = -\tan \beta_i \tan \beta_j$$

so that in (3.3) the determinant R becomes

$$R = \begin{vmatrix} 1 & -\tan \beta_2 \tan \beta_1 & -\tan \beta_3 \tan \beta_1 & \cdots & -\tan \beta_n \tan \beta_1 \\ -\tan \beta_1 \tan \beta_2 & 1 & -\tan \beta_3 \tan \beta_2 & \cdots & -\tan \beta_n \tan \beta_2 \\ \vdots & \vdots & \vdots & & \vdots \\ -\tan \beta_1 \tan \beta_n & -\tan \beta_2 \tan \beta_n & -\tan \beta_3 \tan \beta_n & \cdots & 1 \end{vmatrix}.$$

After some algebra, Pearson obtained

$$\frac{R_{ii}}{R} \frac{1}{\sigma_i^2} = \frac{1}{m_i} + \frac{1}{m_{n+1}},$$
$$\frac{R_{ij}}{R} \frac{1}{\sigma_i \sigma_j} = \frac{1}{m_{n+1}}$$

(*ibid*., p. 163), so that Eq. (3.3) becomes (with x_i replaced by e_i)

$$\begin{aligned} X^2 &= \sum_{i=1}^{n} \left(\frac{1}{m_i} + \frac{1}{m_{n+1}} \right) e_i^2 + 2 \sum_{i<j}^{n} \frac{e_i e_j}{m_{n+1}} \\ &= \sum_{i=1}^{n} \frac{e_i^2}{m_i} + \frac{1}{m_{n+1}} \left(\sum_{i=1}^{n} e_i \right)^2 \\ &= \sum_{i=1}^{n} \frac{e_i^2}{m_i} + \frac{1}{m_{n+1}} \left(-e_{n+1} \right)^2 \\ &= \sum_{i=1}^{n+1} \frac{e_i^2}{m_i}. \end{aligned}$$

Hence, from Eq. (3.6), the following asymptotic result is obtained:

$$X^2 = \sum_{i=1}^{n+1} \frac{\left(m_i - m_i' \right)^2}{m_i} \sim \chi_n^2 \qquad (3.9)$$

(*ibid*., p. 163), where m_i' and m_i are, respectively, the observed frequency and known theoretical frequency in group i ($i = 1, 2, \ldots, n+1$).

With Eq. (3.9) Pearson had thus obtained one of the most popular and easily recognizable formulas in the whole of statistics. X^2 is seen to be a measure of goodness of fit, being the sum of squares of the differences between theoretical and observed frequencies, relative to the theoretical frequencies. Moreover, once it is calculated, the probability associated with it could be computed from a χ_n^2-distribution so as to give a sense of how improbable the deviations were, assuming chance only. Indeed, Pearson's X^2 was the first *formal* test of significance in the history of statistics.

Pearson also considered the situation when the theoretical frequencies $m_1, m_2, m_3, \ldots, m_n, m_{n+1}$ were unknown. This was a very important case, one that resulted in him making a slip, and will be treated in detail in Section 3.3.1.

In Section 6 of his paper, Pearson considered some examples in the case of known theoretical frequencies. The very first example concerns dice data obtained from Weldon (see Table 3.2).

Based on the data in the table above, Pearson asked:

> The results show a bias from the theoretical results, 5 and 6 points occurring more frequently than they should do. Are the deviations such as to forbid us to suppose the results due to random selection? Is there in apparently true dice a real bias towards those faces with the maximum number of points appearing uppermost? (*ibid.*, p. 167)

Thereupon, Pearson used Eq. (3.9) to calculate $X^2 = 43.87241$ with $n' = n + 1 = 13$. Using the second formula in (3.7), he obtained $P = .000016$. Pearson therefore concluded:

> ...the odds are 62,499 to 1 against such a system of deviations on a random selection. With such odds it would be reasonable to conclude that dice exhibit bias towards the higher points. (*ibid.*, p. 168)

TABLE 3.2 Weldon's Data On Frequencies of Dice Showing 5 or 6 Points When 12 Dice Are Cast 26,306 Times

No. of Dice in Cast with 5 or 6 Points	Observed Frequency (m')	Theoretical Frequency (m)	Deviation (e)
0	185	203	−18
1	1,149	1,217	−68
2	3,265	3,345	−80
3	5,475	5,576	−101
4	6,114	6,273	−159
5	5,194	5,018	+176
6	3,067	2,927	+140
7	1,331	1,254	+77
8	403	392	+11
9	105	87	+18
10	14	13	+1
11	4	1	+3
12	0	0	0
	26,306	26,306	

Pearson considered several other examples where similar calculations to the previous one were used. In Section 5 of his paper, however, he considered examples for the case where the theoretical frequencies were unknown. As we mentioned before, he made a fundamental mistake in the latter case. We take up this issue in the next section.

3.3 PEARSON'S ERROR AND CLASH WITH FISHER

3.3.1 Error by Pearson on the Chi-Squared When Parameters Are Estimated (1900)

In the previous section, Pearson had considered the case of observed frequencies $(m'_1, m'_2, m'_3, \ldots, m'_n, m'_{n+1})$ with respective *known* probabilities $(p_1, p_2, p_3, \ldots, p_n, p_{n+1})$. Having dealt with this case, he was quick to point out that the p_i's are rarely known in practice and have to be estimated from the available data. In Section 5 of the 1900 paper, Pearson stated:

> Hitherto we have been considering cases in which the theoretical probability is known *à priori*. But in a great many cases this is not the fact; the theoretical distribution has to be judged from the sample itself. The question we wish to determine is whether the sample may be reasonably considered to represent a random system of deviations from the theoretical frequency distribution of the general population, but this distribution has to be inferred from the sample itself. (Pearson, 1900b, p. 164)

The critical question is, will X^2 in (3.9) still have a χ_n^2-distribution when the p_i's, hence also the theoretical frequencies (i.e., the m_i's), are unknown and m_i will have to be estimated as $m_i^{(s)}$ using the data? Unfortunately, Pearson thought so and this was one of his greatest errors. Let us now examine his argument.*

Pearson first set $\mu_i = m_i - m_i^{(s)}$ and let the value of X^2 when m_i is estimated by $m_i^{(s)}$ be

$$X_{(s)}^2 = \sum_{i=1}^{n+1} \frac{\left(m_i^{(s)} - m'_i\right)^2}{m_i^{(s)}}. \tag{3.10}$$

Also, Eq. (3.9) can be expanded as a Taylor series about $m_i = m_i^{(s)}$ to obtain

$$X^2 = \sum_{i=1}^{n+1} \left\{ \frac{\left(m_i^{(s)} - m'_i\right)^2}{m_i^{(s)}} + \left(m_i - m_i^{(s)}\right) \left[\frac{d}{dm_i} \left\{ \frac{(m_i - m'_i)^2}{m_i} \right\} \right]_{m_i = m_i^{(s)}} + \right.$$

$$\left. \frac{\left(m_i - m_i^{(s)}\right)^2}{2!} \left[\frac{d^2}{dm_i^2} \left\{ \frac{(m_i - m'_i)^2}{m_i} \right\} \right]_{m_i = m_i^{(s)}} + \cdots \right\}$$

$$= \sum_{i=1}^{n+1} \frac{\left(m_i^{(s)} - m'_i\right)^2}{m_i^{(s)}} - \sum_{i=1}^{n+1} \frac{\mu_i}{m_i^{(s)}} \cdot \frac{m_i'^2 - \left(m_i^{(s)}\right)^2}{m_i^{(s)}} + \sum_{i=1}^{n+1} \left(\frac{\mu_i}{m_i^{(s)}} \right)^2 \cdot \frac{m_i'^2}{m_i^{(s)}} + \cdots.$$

*See also Stigler (2008) and Baird (1983).

Dropping out terms of the order $(\mu_i/m_i^{(s)})^3$ and using (3.10), Pearson obtained

$$X^2 - X_{(s)}^2 \approx \underbrace{-\sum_{i=1}^{n+1} \frac{\mu_i}{m_i^{(s)}} \cdot \frac{m_i'^2 - \left(m_i^{(s)}\right)^2}{m_i^{(s)}}}_{T_1} + \underbrace{\sum_{i=1}^{n+1} \left(\frac{\mu_i}{m_i^{(s)}}\right)^2 \cdot \frac{m_i'^2}{m_i^{(s)}}}_{T_2} \tag{3.11}$$

(*ibid.*, p. 165). Pearson correctly noted that $X^2 > X_{(s)}^2$ because

> ...otherwise the general population distribution or curve would give a better fit than the distribution or curve actually fitted to the sample. But we are supposed to fit a distribution or curve to the sample so as to get the "best" values of constants. Hence the right-hand of the above equation [Eq. (3.11)] must be positive. (*ibid.*, p. 165)

Next Pearson tried to argue that the right side of Eq. (3.11) was small:

- For T_1, Pearson argued that this term must either be negative and cancel out part of T_2, or it must be very small. Pearson reasoned that T_1 cannot be positive because this would imply that $\mu_i = m_i - m_i^{(s)}$ and $m_i'^2 - \left(m_i^{(s)}\right)^2$ are negatively correlated. In turn, a negative correlation means that $m_i > m_i^{(s)}$ co-occurs with $m_i' < m_i^{(s)}$ more frequently than by chance, that is, $m'_i < m_i^{(s)} < m_i$. Pearson contended that then "the general population and the observed population would always tend to fall on opposite sides of the sample theoretical population" and that "this seems impossible." While the argument put forward by Pearson is debatable (Stigler, 2008), his conclusion is correct in general.
- For T_2, Pearson contended that "the ratio of $\mu[= m - m^{(s)}]$ to $m^{(s)}$ will, as a rule be small" so that T_2 could be neglected. This is the source of Pearson's major error because, in actual fact, $\mu/m^{(s)}$ can be non-negligible. The upshot is that when the theoretical frequencies m_i are unknown and are estimated by $m_i^{(s)}$, then $X_{(s)}^2$ as calculated in (3.10) no longer has a χ_n^2-distribution.

In retrospect, it can be argued that with a little bit more care, Pearson could have seen the error in his reasoning. In the biography of his father, *Karl Pearson: Some Aspects of His Life and Work* (Pearson, 1938, p. 30), Egon Pearson gave a simple example that clearly shows that Karl Pearson's argument that $T_2 \approx 0$ cannot be correct. Egon considered the extreme case of a multinomial distribution with $n+1$ cells and n parameters to be estimated from the data. Then there is perfect fit so that each estimated frequency $(m_i^{(s)})$ is equal to the corresponding observed one (m_i'). In Eq. (3.11), we therefore have

$$\begin{aligned} \text{LHS} &= X^2 - X_{(s)}^2 \\ &= X^2 - \sum_{i=1}^{n+1} \frac{\left(m_i^{(s)} - m_i'\right)^2}{m_i^{(s)}} \\ &= X^2 - 0 \\ &= X^2, \end{aligned}$$

$$\begin{aligned}\text{RHS} &= -\sum_{i=1}^{n+1} \frac{\mu_i}{m_i^{(s)}} \cdot \frac{m_i'^2 - \left(m_i^{(s)}\right)^2}{m_i^{(s)}} + \sum_{i=1}^{n+1} \left(\frac{\mu_i}{m_i^{(s)}}\right)^2 \cdot \frac{m_i'^2}{m_i^{(s)}} \\ &= -(0) + T_2 \\ &= T_2.\end{aligned}$$

This means that in the case of a perfect fit for a multinomial distribution, T_2 is equivalent to X^2 but certainly not negligibly small, as Karl Pearson had thought.

3.3.2 Greenwood and Yule's Observation (1915)

In 1915, Greenwood and Yule (1915) (Fig. 3.4) published a famous paper on the efficacy of cholera and typhoid vaccination in the prevention of attack. Greenwood and Yule represented their data in a 2×2 table (Table 3.3).

However, the probabilities of vaccination (α) and of attack (β) are unknown. In order to apply Pearson's chi-squared goodness-of-fit test, these probabilities have to be estimated by

$$\begin{aligned}\hat{\alpha} &= \frac{a+b}{N}, \\ \hat{\beta} &= \frac{a+c}{N}.\end{aligned} \tag{3.12}$$

Once this is done, it is then possible to compute the expected cell frequencies $a^{(s)}, b^{(s)}, c^{(s)}, d^{(s)}$ under the hypothesis of independence. If we denote the true (unknown) probability of vaccination and attack by P_{VA},

$$\begin{aligned}a^{(s)} &= N \cdot \hat{P}_{\text{VA}} \\ &= N\hat{\alpha}\hat{\beta} \\ &= N \cdot \frac{a+b}{N} \cdot \frac{a+c}{N} \\ &= \frac{(a+b)(a+c)}{N}.\end{aligned}$$

Similarly,

$$b^{(s)} = \frac{(a+b)(b+d)}{N}, \quad c^{(s)} = \frac{(c+d)(a+c)}{N}, \quad d^{(s)} = \frac{(c+d)(b+d)}{N}.$$

Note that by estimating the two probabilities α and β, we have effectively imposed two constraints on the observed frequencies, namely,

$$\begin{aligned}a+b &= \frac{(a+b)(a+c+b+d)}{N} \\ &= \frac{(a+b)(a+c)}{N} + \frac{(a+b)(b+d)}{N} \\ &= a^{(s)} + b^{(s)},\end{aligned}$$

and similarly

FIGURE 3.4 Major Greenwood (1880–1949). Courtesy of the London School of Hygiene and Tropical Medicine

TABLE 3.3 2×2 Table in Greenwood and Yule's Paper

	Attacked	Not Attacked	Total
Vaccinated	a	b	$a+b$
Not vaccinated	c	d	$c+d$
Total	$a+c$	$b+d$	$N=a+b+c+d$

From Greenwood and Yule (1915).

$$c+d=c^{(s)}+d^{(s)}.$$

Although Greenwood and Yule were aware of this fact, they had not yet fully appreciated the consequence of these restrictions on their X^2-test. So they proceeded as Pearson had prescribed, namely, by using the estimated frequencies instead of the theoretical frequencies in the computation of X^2 (i.e., by using $X^2_{(s)}$ instead) and then assuming that the latter has a χ^2_3-distribution (since $n+1=4$ here). However, they made these most interesting remarks:

> A further point is worthy of remark. In our subsequent discussion we shall frequently compare the two ratios $\frac{\text{inoculated attacked}}{\text{all inoculated}}$ and $\frac{\text{uninoculated attacked}}{\text{all uninoculated}}$ which we may denote by p_1 and p_2. It might all therefore be asked, why we should not adopt as our criterion of significance the ratio of p_1–p_2 to its standard error, counting as significant all differences greater than some assigned multiple of the standard error. It will be found that if this plan is adopted deviations which, judged by the χ^2 test, as not improbable are much less likely to occur as the result of random sampling. This divergence between the results of the two tests is at first sight rather surprising and is not due to neglect of the correlation in errors between the subgroup frequencies. If the standard errors of p_1 and p_2 are worked out from first principles it will be found that the ordinary binomial form results, and that there is no correlation in errors between p_1 and p_2.
>
> The explanation is, we think, as follows: The total number of distributions into, for example, four "cells" of n things which differ from the expected distribution by more than a certain margin is greater than the number of those which fulfil the further condition that the difference between the ratios $\frac{b}{a+b}$ and $\frac{d}{c+d}$ (p_1 and p_2 of our previous remarks, a being the number of inoculated who recovered, b the number of inoculated who died, d and c corresponding frequencies of inoculated) shall exceed a certain magnitude. The result will be that the probability of any arrangement having arisen from random sampling, or of any less probable arrangement, will be greater when estimated by the χ^2 method than when the other test is applied. It can, of course, be urged that we are really only concerned with the probability that chance might give rise to those arrangements which exhibit a difference $\frac{b}{a+b}$ between $\frac{d}{c+d}$ and we are not convinced that the objection is invalid. We think, however, that the point is not free from difficulty and merits further consideration from the theoretical side, to which we have no space in the present paper to devote. (Greenwood and Yule, 1915, p. 118)

Greenwood and Yule's point in the above was that when the test of independence was performed by dividing the difference in proportions by its standard error, the associated probability (which is what we now call the p-value) obtained was different (in particular, it is less) from that obtained by Pearson's X^2-test. This in itself was an ominous sign that there might be a problem in the way the X^2-test was being used. Nonetheless, in all their examples, Greenwood and Yule went on to use X^2 because they wanted to "err on the side of caution.*"

3.3.3 Fisher's 1922 Proof of the Chi-Squared Distribution: Origin of Degrees of Freedom

Although Greenwood and Yule had, in 1915, already noticed serious problems with Pearson's 1900 paper, they did not pursue the matter any further because "we have no space in the present paper to devote." It was left to R.A. Fisher to fully and decisively refute Pearson's approximation of X^2 by $X^2_{(s)}$ while keeping the same χ^2_n-distribution. This first happened in a 1922 paper, which also marked Fisher's introduction of the concept of *degrees of freedom (d.f.)* for the first time (Fisher, 1922b). In 1924, Fisher came back to revisit the issue with a more complete deconstruction of Pearson's arguments. We now describe these topics.

Fisher's relationship with Karl Pearson (a man 33 years his senior) had at first been reasonably cordial, when Fisher started to communicate with Pearson (and William Gosset) while still an undergraduate at Cambridge University in 1912. In 1915, Fisher published a remarkable paper on the exact distribution of the correlation coefficient, a concept whose development owed so much to Pearson. The paper was accepted by Pearson to be published in his journal *Biometrika*, yet two years later a "cooperative study" headed by Pearson appeared criticizing Fisher's 1915 work (Soper et al., 1917).† In 1916, Fisher sent a note to *Biometrika* criticizing the method of minimum chi-squared advocated by Kirstine Smith, a Danish pupil of Thiele, who was studying under Karl Pearson. The latter rejected the note, chiding Fisher for maximizing inverse probability, which Pearson described as the Gaussian method‡:

> If you will write me a defence of the Gaussian method, I will certainly consider it for publication, but if I were to publish your note, it would have to be followed by another note saying that it missed the point.... (Pearson, 1968, p. 455)

Two subsequent rejections of his papers by Pearson led to worsening relationships between Fisher and Pearson until a boiling point was reached in 1922. In that year, Fisher mounted his first major attack on Pearson's use of the X^2 statistic, in a paper entitled "On the interpretation of χ^2 from contingency tables and the calculation of P" (Fisher, 1922b).

In the 1922 paper, Fisher referred to the discrepancies noted by Yule and Greenwood (cf. Section 3.3.2). He also introduced the concept of *degrees of freedom* right in the beginning section of the paper without much explanation, as follows:

* Had the authors gone further and used the fact that the square of the proportions test is identical to Pearson's X^2, they would have realized the latter should have a $Z^2 = \chi^2_1$ distribution, where $Z \sim N(0, 1)$, not Pearson's prescribed χ^2_3 distribution.

† What is even more disconcerting is that Soper et al.'s paper was meant to be an appendix to Fisher's (and Student's) paper! See Sec. 2.3.10 for more details.

‡ See Section 1.5.3.

> ...the value of n' with which the table should be entered* is not now equal to the number of cells, but to *one more than the number of degrees of freedom in the distribution*. Thus for a contingency table of r rows and c columns we should take $n'=(c-1)(r-1)+1$ instead of $n'=cr$. This modification often makes a very great difference to the probability (P) that a given value of $\chi^2[=X^2]$ should have been obtained by chance. (*ibid*., p. 88)

Fisher thus stated that the associated probability of the Pearson X^2 for an $r\times c$ contingency table should be calculated from a χ_ν^2-distribution, where $\nu(=n'-1)$ is *not* one less the number of cells in the table, as prescribed by Pearson, but rather *is* the number of degrees of freedom $(r-1)$ $(c-1)$. The index ν of the χ_ν^2-distribution should therefore equal the number of degrees of freedom.

Fisher did not formally prove the above correct claim, but rather provided a geometric argument. This ran as follows. Consider the set of deviations $e_i = m'_i - m_i$ $(i=1,2,\ldots,n')$, where m'_i and m_i are the observed and theoretical (unknown) frequencies in class i, such that $\text{SD}(e_i)=\sigma_i^2$. Further, assume that there are no linear constraints on the m'_i's so that $N=\sum_i m'_i$ is a random variable. Then the variance of N is

$$\sigma_N^2 = \text{var}\sum_{i=1}^{n'} m'_i = \sum_{i=1}^{n'}\sigma_i^2. \tag{3.13}$$

(*ibidem*). Fisher noted that the variance of e_i can be decomposed into two independent causes, namely, the variance of N and the variance of the proportion that falls into any one class:

$$\sigma_i^2 = p_i^2\sigma_N^2 + p_i(1-p_i)\mu_N, \quad i=1,2,\ldots,n', \tag{3.14}$$

where p_i is the (unknown) probability of falling in class i and μ_N is the mean of N. By summing the above across classes,

$$\sum_{i=1}^{n'}\sigma_i^2 = \sigma_N^2\sum_{i=1}^{n'}p_i^2 + \mu_N\left(\sum_{i=1}^{n'}p_i - \sum_{i=1}^{n'}p_i^2\right).$$

Using Eq. (3.13) and the fact that $\sum_{i=1}^{n'}p_i = 1$, we have

$$\sigma_N^2 = \mu_N.$$

Substituting $\sigma_N^2 = \mu_N$ in Eq. (3.14), Fisher obtained

$$\begin{aligned}\sigma_i^2 &= p_i^2\mu_N + p_i(1-p_i)\mu_N \\ &= \mu_N p_i \\ &= m_i.\end{aligned}$$

Therefore, if we consider m'_i as a Poisson variable with mean (and variance) m_i, we obtain

$$\sum_{i=1}^{n'}\frac{(m'_i-m_i)^2}{m_i} = \sum_{i=1}^{n'}\left(\frac{m'_i-m_i}{\sigma_i}\right)^2,$$

* Following Pearson's X^2 paper, William P. Elderton built tables for testing goodness of fit (Elderton, 1902). The tables were constructed so that $n'=n+1$ stood for the number of groupings. For various values of n' and X^2, the table gave the values of the right-tail probability P in Eq. (3.7). In the 1920s, as we next explain in the text, Fisher realized that it was the number of degrees of freedom that mattered, not n', and gave his own tables for different values of n, which now stood for the number of degrees of freedom.

a sum of squares of (n') standard normal random variables. Thus, under the assumption of no linear constraints on the observed frequencies,

$$X^2 = \sum_{i=1}^{n'} \frac{\left(m'_i - m_i\right)^2}{m_i} \sim \chi^2_{n'}.$$

We note that the ellipsoid of constant density, $X^2 = \text{constant}$, can be linearly transformed to a hypersphere H of constant density in n' dimensions.* Suppose now there are l linear constraints in the observed frequencies (m_i), corresponding to l hyperplanes. The first hyperplane cuts H in a hypersphere of constant density but with one dimension less (i.e., in $n' - 1$ dimensions). Similarly, each of the other hyperplanes further reduces the dimensions of the hypersphere by unity. The net contribution of all l hyperplanes is to reduce the observed frequencies (m'_i) to lie in a hypersphere in $n' - l$ dimensions, corresponding to $n' - l$ degrees of freedom for X^2. Thus, *under the assumption of l linear constraints on the observed frequencies,*

$$X^2 = \sum_{i=1}^{n'} \frac{\left(m'_i - m_i\right)^2}{m_i} \sim \chi^2_{n'-l}.$$

In the case of an $r \times c$ contingency table, the linear constraints are that the sum of observed values for each row and for each column must equal to the sum of expected values for the corresponding row and column. This gives a total $r + c - 1$ constraints so that the number of degrees of freedom is

$$\begin{aligned} v &= n' - (r + c - 1) \\ &= rc - r - c + 1 \\ &= (r-1)(c-1). \end{aligned}$$

This completes Fisher's geometric argument.

*3.3.4 Further Details on Degrees of Freedom

It can be seen that the concept of degrees of freedom was an offshoot of Fisher's geometric intuition and, as such, it is an extremely powerful idea in statistics. Therefore, we now spend some time to explain it further. Let us start with the simple example of three observations x_1, x_2, x_3 *with no restrictions on them*. Since we know nothing about these observations (apart from their values), the point $P(x_1, x_2, x_3)$ lies anywhere in a 3-dimensional space, that is, x_1, x_2, x_3 have *three* degrees of freedom. Now suppose we *know* the sample mean of x_1, x_2, x_3 is $\bar{x}$ $[= (x_1 + x_2 + x_3)/3]$. The point $P(x_1, x_2, x_3)$ is no longer free to lie anywhere in a 3-dimensional space but has to lie on the plane $x_1 + x_2 + x_3 = 3\bar{x}$. It is seen that $P(x_1, x_2, x_3)$ lies in a 2-dimensional space, that is, x_1, x_2, x_3 now have *two* degrees of freedom. Another way to look at this is to realize that the equation $x_1 + x_2 + x_3 = 3\bar{x}$ consists of four quantities, one of which (viz., $\bar{x}$) is known. How many of the remaining quantities can freely vary while still preserving the validity of the equation? The answer is two. Therefore, knowing $\bar{x}$, the observations x_1, x_2, x_3 have *two* degrees of freedom. In general, the number of degrees of freedom associated with a random sample $X_1, X_2, \ldots, X_n$

* See p. 301.

(or with a statistic calculated from it) is the number of the X_i's that can freely vary, when subject to linear constraints. Thus, for a normal sample, while the statistic $\sum_{i=1}^{n}\{(X_i-\mu_0)/\sigma_0\}^2$ has n degrees of freedom, the statistic $\sum_{i=1}^{n}\{(X_i-\bar{X})/\sigma_0\}^2$ has only $n-1$ (where μ_0 and σ_0^2 are the *known* mean and variance, respectively, of each X_i).

3.3.5 Reaction to Fisher's 1922 Paper: Yule (1922), Bowley and Connor (1923), Brownlee (1924), and Pearson (1922)

Let us now briefly describe the immediate aftermath to Fisher's 1922 paper. In that year, Yule published a paper, "On the application of the χ^2 method to association and contingency tables, with experimental illustrations" (Yule, 1922), in which he corroborated Fisher's result both theoretically and experimentally, and admitted to Yule and Greenwood's use of the incorrect χ^2-distribution in 1915 (Greenwood and Yule, 1915). On the face of it, Yule thus showed wisdom: when presented with the evidence, he was willing to admit error. However, this admission was also a major blow to Pearson, given the fact that Yule and Pearson had been previously engaged in a bitter feud during which Pearson had been particularly acrimonious (cf. Section 2.3.7).

However, although Yule was a major help in spreading the new concept of "degrees of freedom," not everybody was sold on Fisher's correction to Pearson X^2 statistic. Thus, in 1923, Bowley and Connor published a paper "to ascertain whether the actual grouping to some assigned classification of statistical observations is consistent with a particular hypothesis of the nature of the grouping" (Bowley and Connor, 1923). However, when fitting a distribution that involved the estimation of parameters from data, Bowley and Connor concluded that Fisher's 1922 result was "not applicable to the problem of determining whether observations are consistent with a law of assigned form." Fisher was quite distressed by this:

> It is the more surprising that Bowley should have reverted to the Pearsonian mode of testing fourfold tables since the actual distribution of χ^2 in this case has been determined experimentally by Yule. (Fisher, 1923b, p. 144)

Furthermore, in 1924, John Brownlee did a set of experiments all of which except one confirmed Fisher's 1922 result on the appropriate degrees of freedom when population parameters are estimated from the data (Brownlee, 1924). Concerning that one case, Brownlee complained that he "probably, therefore, misunderstood Mr. Fisher upon this point." Fisher later showed the discrepant case arose from a false null hypothesis (Fisher, 1924b).

However, the major reaction to Fisher's 1922 paper of course came from Pearson himself. In that same year, Pearson retaliated swiftly and harshly in the *Biometrika* paper "On the χ^2 test of goodness of fit":

> The above re-description of what seems to me very elementary considerations would be unnecessary had not a recent writer in the *Journal of the Royal Statistical Society* appeared to have wholly ignored them. He considers that I have made serious blunders in not linking my degrees of freedom by the number of moments I have taken; ... I hold that such a view is entirely erroneous and that the writer has done no service to the science of statistics by giving it broad-cast circulation in the pages of the *Journal of the Royal Statistical Society*. (Pearson, 1922, p. 187)

> ...I trust my critic [Fisher] will pardon me for comparing him with Don Quixote tilting at the windmill; he must either destroy himself, or the whole theory of probable errors.... (*ibid.*, p. 191)

Pearson's comments can appear overly harsh,* especially that now we know he was wrong. However, Fisher had made an attack on his greatest invention, the X^2 goodness-of-fit test. Moreover, although Pearson had not been thinking in terms of *degree of freedom*, he was quite aware that the index p of the χ_p^2-distribution needed to be changed in some situations. Thus, in 1911, Pearson extended his X^2-test in a paper entitled "On the probability that two independent distributions of frequency are really samples from the same population" (Pearson, 1911). Here Pearson considered a $2\times c$ contingency table in which the row totals had been fixed. In the illustrative examples of his paper, Pearson correctly entered the tables for testing goodness of fit at $n'=c$ (i.e., the number of degrees of freedom is $c-1$), not at $n'=2c$. In 1916, Pearson gave a new derivation of the X^2-test, at the suggestion of Soper, based on the Poisson asymptotic limit of the class frequencies of a multinomial distribution (Pearson, 1916). Pearson here again correctly pointed out that if q linear constraints are imposed on the n' cell frequencies in addition to the usual $\sum_i m'_i = N$ (where m'_i is the observed frequency in class i), then one must enter the tables at $n'-q$ (i.e., the number of degrees of freedom is $n'-q-1$). In spite of all this, Pearson *never* accepted the validity of Fisher's 1922 modification of the values of n' with which the tables had to be entered in the original 1900 problem. Pearson's error arose from his belief that the value of n' (and of the number of degrees of freedom) should depend on whether the margins of a contingency table are fixed or not. In fact, it does not.

3.3.6 Fisher's 1924 Argument: "*Coup de Grâce*" in 1926

3.3.6.1 The 1924 Argument Let us now describe Fisher's 1924 definite refutation of Pearson's 1900 approximation of X^2 by $X_{(s)}^2$ while keeping the same χ^2-distribution (cf. Section 3.3.1). Fisher's paper, entitled "The conditions under which χ^2 measures the discrepancy between observation and hypothesis" (Fisher, 1924b), considered the goodness-of-fit formula (cf. Eq. 3.9)

$$
\begin{aligned}
X^2 &= \sum_{i=1}^{n+1} \frac{(m'_i - m_i)^2}{m_i} \\
&= \sum_{i=1}^{n+1} \frac{m_i'^2}{m_i} - \sum_{i=1}^{n+1}(2m'_i - m_i) \\
&= \sum_{i=1}^{n+1} \frac{m_i'^2}{m_i} - N.
\end{aligned}
$$

Similarly (cf. Eq. 3.10),

$$
X_{(s)}^2 = \sum_{i=1}^{n+1} \frac{m_i'^2}{m_i^{(s)}} - N, \tag{3.15}
$$

*Fisher immediately sent a response to the Royal Statistical Society but the latter rejected it to avoid further controversy. Thereupon, Fisher resigned from the Society, but his animosity toward Pearson was far from over.

so that

$$X^2 - X_{(s)}^2 = \sum_{i=1}^{n+1} m_i'^2 \left(\frac{1}{m_i} - \frac{1}{m_i^{(s)}} \right) \tag{3.16}$$

(*ibid*., p. 447). Fisher now wrote $m_i \equiv m_i(\theta)$ and $m_i^{(s)} \equiv m_i(\theta^{(s)})$, where θ is a parameter and $\theta^{(s)}$ is its value estimated from the sample. If we define $f(\theta) = 1/m_i(\theta) - 1/m_i(\theta^{(s)})$, a Taylor series approximation about $\theta = \theta^{(s)}$ gives

$$\begin{aligned} f(\theta) &= f\left(\theta^{(s)}\right) + \left(\theta - \theta^{(s)}\right) f'\left(\theta^{(s)}\right) + \frac{\left(\theta - \theta^{(s)}\right)^2}{2!} f''\left(\theta^{(s)}\right) + \cdots \\ &= 0 + \left(\theta - \theta^{(s)}\right) \cdot \frac{-1}{\left(m_i^{(s)}\right)^2} \cdot \frac{\partial m_i^{(s)}}{\partial \theta} + \frac{\left(\theta - \theta^{(s)}\right)^2}{2!} \cdot \left\{ \frac{2}{\left(m_i^{(s)}\right)^3} \left(\frac{\partial m_i^{(s)}}{\partial \theta} \right)^2 - \frac{1}{\left(m_i^{(s)}\right)^2} \frac{\partial^2 m_i^{(s)}}{\partial \theta^2} \right\} + \cdots . \end{aligned}$$

Multiplying the above by $m_i'^2$, summing over i, and using Eq. (3.16), we obtain

$$X^2 - X_{(s)}^2 = -\left(\theta - \theta^{(s)}\right) \sum_{i=1}^{n+1} \left\{ \frac{m_i'^2}{\left(m_i^{(s)}\right)^2} \cdot \frac{\partial m_i^{(s)}}{\partial \theta} \right\} + \frac{\left(\theta - \theta^{(s)}\right)^2}{2} \sum_{i=1}^{n+1} \left\{ \frac{2 m_i'^2}{\left(m_i^{(s)}\right)^3} \left(\frac{\partial m_i^{(s)}}{\partial \theta} \right)^2 - \frac{m_i'^2}{\left(m_i^{(s)}\right)^2} \frac{\partial^2 m_i^{(s)}}{\partial \theta^2} \right\} + \cdots \tag{3.17}$$

Now, the method of estimation chooses a value for θ so that $X_{(s)}^2$ is a minimum. From Eq. (3.15), this means that

$$\sum_{i=1}^{n+1} \left(\frac{m'_i}{m_i^{(s)}} \right)^2 \left(\frac{\partial m_i^{(s)}}{\partial \theta} \right) = 0$$

and the first term on the right side of Eq. (3.17) vanishes. Fisher made use of two additional facts, namely,

$$\frac{m'_i}{m_i^{(s)}} \to 1, \quad \text{as } n \to \infty,$$

$$\sum_{i=1}^{n+1} \frac{\partial^2 m_i^{(s)}}{\partial \theta^2} = \frac{\partial^2}{\partial \theta^2} \sum_{i=1}^{n+1} m_i^{(s)} = \frac{\partial^2 N}{\partial \theta^2} = 0.$$

Equation (3.17) thus becomes

$$X^2 - X_{(s)}^2 \approx \left(\theta^{(s)} - \theta\right)^2 \sum_{i=1}^{n+1} \left\{ \frac{1}{m_i^{(s)}} \left(\frac{\partial m_i^{(s)}}{\partial \theta} \right)^2 \right\} \equiv \frac{\left(\theta^{(s)} - \theta\right)^2}{V} \tag{3.18}$$

(*ibid.*, p. 448). Now, based on his 1922 paper (Fisher, 1922c), if $\theta^{(s)}$ is regarded as a maximum likelihood estimate of θ, then

$$\frac{\theta^{(s)}-\theta}{\sqrt{V}} \xrightarrow{D} N(0,1),^*$$

with V defined in Eq. (3.18). From the latter equation, Fisher obtained the final result:

$$X^2 - X^2_{(s)} \xrightarrow{D} \chi^2_1.$$

Thus, $X^2_{(s)}$ could not just be approximated by X^2 in the 1900 goodness-of-fit paper, as Pearson had vainly tried to argue.

3.3.6.2 'Coup de Grâce' in 1926 Following Fisher's 1924 demonstration, Pearson commented on the issue of degrees of freedom on several occasions, but he never acknowledged there was any problem with his own formulation (e.g., see Pearson, 1932). However, Fisher did deliver what Box (1978, p. 88) has aptly called a "*coup de grâce*" in 1926, with the unintentional help of none other than Karl Pearson's son, Egon Pearson. The latter had joined his father at the Galton laboratory in 1923. Two years later, E. Pearson published a paper in *Biometrika* (Pearson, 1925) to test Bayes' theorem. In that paper, he examined 12,448 different events and observed the frequency of occurrence of each in two samples of 20 and 15. Thus, 12,448 fourfold tables could be generated under approximately random sampling. In 1926, Fisher divided these tables according to the total number of successes in the two samples combined and calculated the average value of X^2 in each (Fisher, 1926a). Fisher explained:

> Professor [Karl] Pearson originally took the view, which more recently he has stoutly defended, that the distribution was such that the average value must be 3, while the writer in 1922 (J.R.S.S., LXXFV, pp. 87–94) put forward what he regarded as a corrected distribution, having an average of only 1. Although the writer considers that the point can be settled rigorously by purely mathematical methods, and although Mr. Yule has carried out extensive experiments, which completely confirm his results, yet as the point is still disputed, it is worth while [*sic*] to see what the actual average value of X^2 is in the large group of fourfold tables here presented. (*ibid.*, pp. 32–33)

Fisher was then able to show that "in every case the average value (of X^2) is near to 1, in no case is it near to 3" (*ibid.*, p. 33). This was a very important result in that it corroborated Fisher's 1922 and 1924 arguments that the chi-squared statistic in a fourfold table should have one degree of freedom.[†] Fisher triumphantly concluded:

> It is hoped that this example will remove all doubts as to the correct treatment of the fourfold table, and of other applications of the X^2 test.[‡] (*ibidem*)

* D denotes convergence in distribution.

[†] Recall that if $X^2 \sim \chi^2_\nu$, then $\mathsf{E}X^2 = \nu$. If $\nu = 1$, then $\mathsf{E}X^2 = 1$.

[‡] Fisher also found that "[t]he mean value from E. S. Pearson's sample is thus lower than expectation by a very small but statistically significant amount; showing that the conditions of independent random sampling, though very nearly, were not exactly realised" (Fisher, 1926a).

The fact that K. Pearson never came back to dispute Fisher's findings speaks for itself.

In retrospect, there can be no doubt that Fisher's animosity with Pearson was both at a professional and personal level, the issue of degrees of freedom in the contingency table being one of the highlights of this contentious relationship. Fisher's dissention on this and several other issues with Pearson, and the ensuing retaliation from the latter, invariably left an indelible mark on Fisher. It is therefore not surprising, albeit regrettable, that he still felt some animosity toward Pearson long after the latter's demise. Thus, we can read in Fisher's first edition of *Statistical Methods and Scientific Inference*, published in 1956, 20 years after Pearson's death:

> The terrible weakness of his [Pearson's] mathematical and scientific work flowed from his incapacity in self-criticism, and his unwillingness to admit the possibility that he had anything to learn from others, even in biology, of which he knew very little. His mathematics, consequently, though always vigorous, were usually clumsy, and often misleading. In controversy, to which he was much addicted, he constantly showed himself to be without a sense of justice. In his dispute with Bateson on the validity of Mendelian inheritance he was the bull to a skillful matador. His immense personal output of writings, his great enterprise in publication, and the excellence of production characteristic of the Royal Society and the Cambridge Press, left an impressive literature. The biological world, for the most part, ignored it, for it was indeed both pretentious and erratic. (Fisher, 1956b, p. 3)

Although a few of the above statements made by Fisher may have some elements of truth, one that is definitely not true is the charge that Pearson's mathematical and scientific work was weak and that his mathematics was clumsy. On the contrary, Pearson was mathematically quite sophisticated, although he did make a few errors (just as Fisher did).

3.4 THE CHI-SQUARED DISTRIBUTION BEFORE PEARSON

Although the X^2 goodness-of-fit test was originally due to Pearson (1900b), the χ^2-distribution itself had been derived well before him. According to Sheynin (1966) and Pfanzagl and Sheynin (1996), the original derivation of the χ^2-distribution was due to the German physicist Ernst Abbe (1863). However, Lancaster (1966) reported that the French mathematician Irenée-Jules Bienaymé (1796–1878) had obtained the same result in 1852. We now outline the work on the χ^2-distribution prior to Pearson.

3.4.1 Bienaymé's Derivation of Simultaneous Confidence Regions (1852)

Irenée-Jules Bienaymé* (Fig. 3.5) was a distinguished French mathematician and statistician. He was born in Paris on August 28, 1796, two years after the fall of Robespierre and three years before Napoléon Bonaparte came to power. Bienaymé received his secondary education in Bruges and later at the *Lycée Louis-le-Grand* in Paris. He joined the École Polytechnique in 1815. That same year Napoléon fell and the École Polytechnique was

*Irenée-Jules Bienaymé's life is also described in Heyde and Seneta (1977, Chapter 1), Heyde (1976), and Dugué (1978).

FIGURE 3.5 Irenée-Jules Bienaymé (1796–1878). Wikimedia Commons (Public Domain), http://commons.wikimedia.org/wiki/File:Ir%C3%A9n%C3%A9e-Jules_Bienaym%C3%A9.jpeg

dissolved by Louis XVIII when students persisted in their sympathy toward the ex-emperor. In 1818, Bienaymé became Professor of Mathematics at the Military Academy of Saint-Cyr, but two years later he joined the civil service as a general inspector of finance. Soon Bienaymé started studying actuarial science, probability, and statistics. In 1848, Paris workers led an uprising, and Louis Napoléon was elected President of the Second Republic. Bienaymé was removed from his position because of lack of republican spirit. In 1852, he was elected to the *Académie des Sciences* and acted as a referee for 23 years for the *Prix Montyon*, which was the highest French award for achievement in statistics. Bienaymé published only 23 papers during his career, but his 1853 paper contains an important result, which later came to be known as *Chebychev's inequality*. This was 16 years before *Chebychev's* own derivation of the same result. Bienaymé was also no stranger to controversies. He got embroiled in a heated debate with Cauchy regarding the relative merits of the method of least squares and Cauchy's own interpolation procedure. Bienaymé also (unjustly) criticized Poisson's extension of Bernoulli's weak law of large numbers to the case of unequal success probabilities from trial to trial. Throughout his life, Bienaymé remained a faithful disciple of Laplace, defending the latter whenever he felt Laplace's work was being diminished. Bienaymé died in Paris in 1878.

In 1838, Bienaymé obtained an expression that was almost identical to Pearson's 1900 goodness-of-fit statistic (Bienaymé, 1838). However, the χ^2-distribution itself appeared incidentally in 1852, when Bienaymé was working on the joint distribution of errors (Bienaymé, 1852) (Fig. 3.6).

Bienaymé's 1852 paper was enthusiastic in style and started by acknowledging the ubiquity of the method of least squares. But Bienaymé also claimed that there had been a major oversight in the application of the method:

> ...There would be perhaps more than one defect to point out in the application of the method of least-squares: here it will be question of that which is most conspicuous to the eyes...it is thus a defect to assign a probability on the error of an unknown which is part of a system to be determined, that which it would have if it were alone, instead of giving the rules for calculating the probability of the whole of the errors of the system which cannot, in fact, be isolated from each other. (*ibid.*, pp. 34–35)

In the above, Bienaymé thus made a case for joint (instead of marginal) probability statements about errors in a system.

In Section II of his paper, Bienaymé stated that the method of least squares assumed that a large number of observations $\omega_1, \omega_2, \ldots, \omega_n$ are made, and are each a linear combination of several unknown "elements" $x_1, x_2, \ldots, x_m$. For the hth observation, he wrote

$$\omega_h = a_{1,h}x_1 + a_{2,h}x_2 + \cdots + a_{m,h}x_m, \quad h = 1, 2, \ldots, n, \tag{3.19}$$

where the $a_{i,h}$'s are known coefficients. To solve for $x_1, x_2, \ldots, x_n$, one only needed to consider m linearly independent equations from the system in (3.19). Now, as errors are always associated with the observations made, Bienaymé stated it was better to consider all n equations in order to achieve better precision. To proceed in this way, one needed to multiply (3.19) by some factor $K_{i,h}$ (to be determined later) subject to the conditions

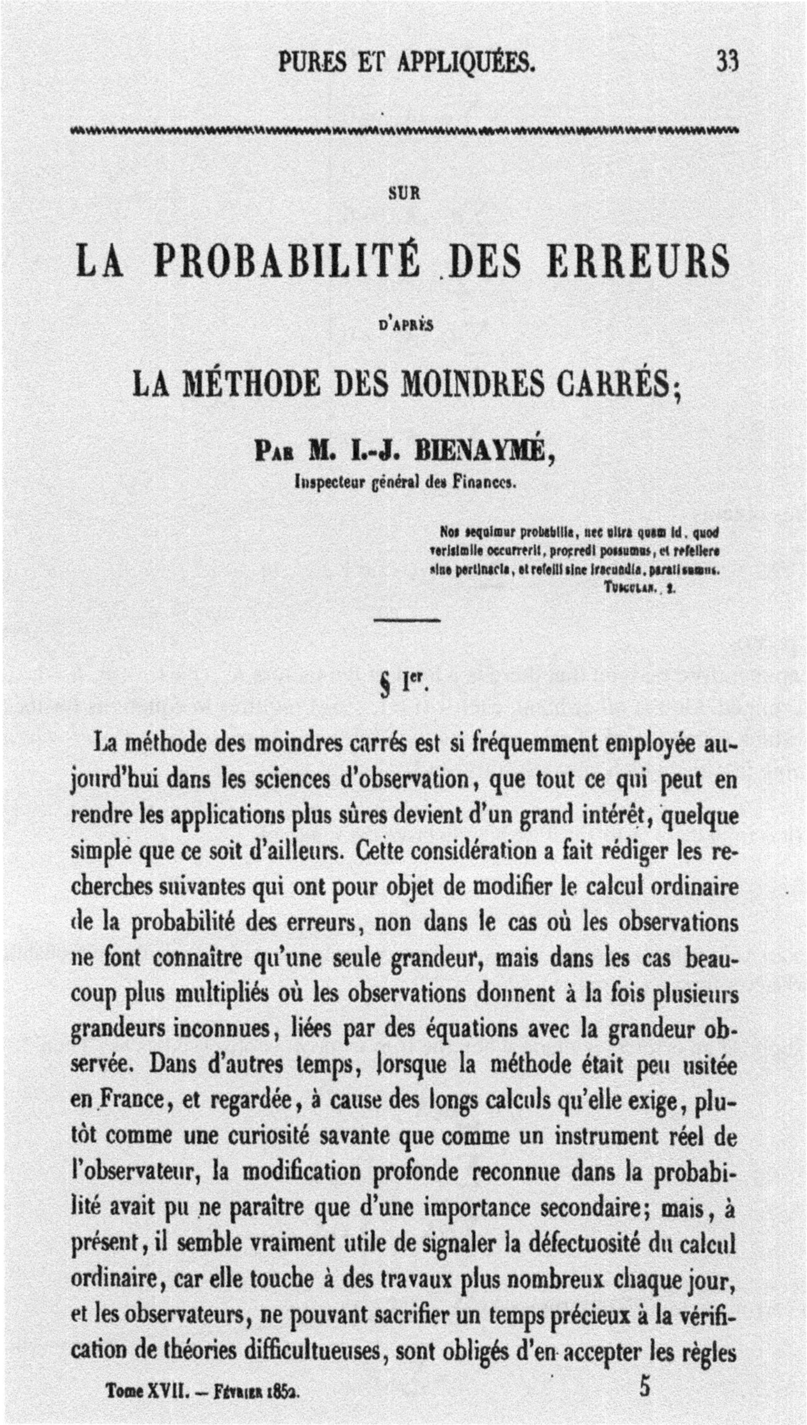

PURES ET APPLIQUÉES. 33

SUR

LA PROBABILITÉ DES ERREURS

D'APRÈS

LA MÉTHODE DES MOINDRES CARRÉS;

Par M. I.-J. BIENAYMÉ,
Inspecteur général des Finances.

Nos sequimur probabilia, nec ultra quam id, quod verisimile occurrerit, progredi possumus, et refellere sine pertinacia, et refelli sine iracundia, parati sumus.
TUSCULAN., 2.

§ Ier.

La méthode des moindres carrés est si fréquemment employée aujourd'hui dans les sciences d'observation, que tout ce qui peut en rendre les applications plus sûres devient d'un grand intérêt, quelque simple que ce soit d'ailleurs. Cette considération a fait rédiger les recherches suivantes qui ont pour objet de modifier le calcul ordinaire de la probabilité des erreurs, non dans le cas où les observations ne font connaître qu'une seule grandeur, mais dans les cas beaucoup plus multipliés où les observations donnent à la fois plusieurs grandeurs inconnues, liées par des équations avec la grandeur observée. Dans d'autres temps, lorsque la méthode était peu usitée en France, et regardée, à cause des longs calculs qu'elle exige, plutôt comme une curiosité savante que comme un instrument réel de l'observateur, la modification profonde reconnue dans la probabilité avait pu ne paraître que d'une importance secondaire; mais, à présent, il semble vraiment utile de signaler la défectuosité du calcul ordinaire, car elle touche à des travaux plus nombreux chaque jour, et les observateurs, ne pouvant sacrifier un temps précieux à la vérification de théories difficultueuses, sont obligés d'en accepter les règles

Tome XVII. — Février 1852. 5

FIGURE 3.6 First page of Bienaymé's article "Sur la probabilité des erreurs d'aprés la méthode des moindres carrés" (Bienaymé, 1852)

$$\left.\begin{aligned}\sum_h a_{1,h}K_{i,h}&=0,\\ \sum_h a_{2,h}K_{i,h}&=0,\\ &\dots\\ \sum_h a_{i-1,h}K_{i,h}&=0,\\ \sum_h a_{i,h}K_{i,h}&=1,\\ \sum_h a_{i+1,h}K_{i,h}&=0,\\ &\dots\\ \sum_h a_{m,h}K_{i,h}&=0.\end{aligned}\right\} \tag{3.20}$$

One thus obtains

$$x_i=\sum_h \omega_h K_{i,h}, \quad i=1,2,\ldots,m \tag{3.21}$$

(*ibid*., p. 39).

It can therefore be seen that there is a total of mn factors $K_{i,h}(i=1,\ldots,m;\ h=1,\ldots,n)$ to be determined. On the other hand, each $x_i(i=1,\ldots,m)$ requires m equations for its evaluation, which gives a total of m^2 equations (conditions). There is thus $mn-m^2=m(n-m)$ remaining factors, which stay arbitrary, and

> …[i]t is from these factors that we have to make the values of
>
> $$x_i=\sum_h \omega_h K_{i,h}$$
>
> as exact as possible; and this is why we must necessarily resort to the calculus of probability, in whatever form it is disguised. (*ibid*., p. 40)

Now, suppose an error ε_h is made when the true value ω is observed as ω_h. Then

$$\begin{aligned}x_i&=\sum_h \omega_h K_{i,h}\\ &=\sum_h(\omega+\varepsilon_h)K_{i,h}\\ &=\sum_h \omega K_{i,h}+\sum_h \varepsilon_h K_{i,h},\end{aligned}$$

and the error made in estimating x_i by $\sum_h w_h K_{i,h}$ is

$$r_i=\sum_h \varepsilon_h K_{i,h}$$

(*ibidem*). Bienaymé now proceeded to determine the joint probability distribution of the (simultaneous) errors $r_1, r_2, \ldots, r_m$, as we next describe. First, he denoted the density of the error ε_h by $\phi(\varepsilon_h)$. Then, while acknowledging the works of Laplace, Fourier, and more recently of Dirichlet, he wrote the following Fourier transform:

$$\begin{aligned} P &= \int\cdots\int \phi(\varepsilon_1)\cdots\phi(\varepsilon_n)\exp(j\alpha_1 r_1)\cdots\exp(j\alpha_m r_m)\,d\varepsilon_1\ldots d\varepsilon_n \\ &= \int\cdots\int \phi(\varepsilon_1)\cdots\phi(\varepsilon_n)\exp\left\{j\alpha_1\left(\varepsilon_1 K_{1,1}+\varepsilon_2 K_{1,2}+\cdots+\varepsilon_n K_{1,n}\right)\right\}\cdots\exp \\ &\quad \left\{j\alpha_m\left(\varepsilon_1 K_{m,1}+\varepsilon_2 K_{m,2}+\cdots+\varepsilon_n K_{m,n}\right)\right\}d\varepsilon_1\ldots d\varepsilon_n, \end{aligned} \tag{3.22}$$

where $j=\sqrt{-1}$. Bienaymé next noted that P above can be written as

$$P=\int\cdots\int \Phi(r_1,\ldots,r_m)\exp\left\{j\left(r_1\alpha_1+\cdots+r_m\alpha_m\right)\right\}dr_1\ldots dr_m$$

(*ibid*., p. 43), where $\Phi(r_1,\ldots,r_m)$ is the joint density of $r_1, r_2, \ldots, r_m$. By Fourier inversion of the above,

$$Q=\Phi(r_1,\ldots,r_m)=\frac{1}{(2\pi)^m}\int\cdots\int P\exp\left\{-j\left(r_1\alpha_1+\cdots+r_m\alpha_m\right)\right\}d\alpha_1\ldots d\alpha_m. \tag{3.23}$$

Now, from Eq. (3.22),

$$\begin{aligned} P &= \int\cdots\int \phi(\varepsilon_1)\exp\left\{j\varepsilon_1\left(\alpha_1 K_{1,1}+\alpha_2 K_{2,1}+\cdots+\alpha_m K_{m,1}\right)\right\}\cdots\phi(\varepsilon_n)\exp \\ &\quad \left\{j\varepsilon_n\left(\alpha_1 K_{1,n}+\alpha_2 K_{2,n}+\cdots+\alpha_m K_{m,n}\right)\right\}d\varepsilon_1\ldots d\varepsilon_n. \end{aligned} \tag{3.24}$$

Bienaymé wrote one of the integrals in the above as

$$\begin{aligned} \int\phi(\varepsilon)\exp\left(j\varepsilon_h\sum_i\alpha_i K_{i,h}\right)d\varepsilon &= \int\phi(\varepsilon)\exp\left(j\varepsilon_h S_h\right)d\varepsilon \\ &= \int\phi(\varepsilon)\left(1+j\varepsilon S_h-\frac{\varepsilon^2 S_h^2}{2}-j\frac{\varepsilon^3 S_h^3}{6}+\frac{\varepsilon^4 S_h^4}{24}-\cdots\right)d\varepsilon \\ &= 1+\mu_1 jS_h-\frac{\mu_2 S_h^2}{2}-\frac{\mu_3 jS_h^3}{6}+\frac{\mu_4 S_h^4}{24}-\cdots \\ &= \exp\left\{\mu_1 jS_h-\frac{S_h^2}{2}\left(\mu_2-\mu_1^2\right)-\frac{jS_h^3}{6}\left(\mu_3-3\mu_2\mu_1+2\mu_1^3\right)+\cdots\right\}, \end{aligned}$$

where $S_h=\sum_i \alpha_i K_{i,h}$ and $\mu_i=\mathsf{E}\varepsilon^i$. Therefore, Eq. (3.24) becomes

$$\begin{aligned} P &= \exp\left\{\mu_1 j\left(S_1+S_2+\cdots+S_n\right)-\frac{\left(\mu_2-\mu_1^2\right)}{2}\left(S_1^2+S_2^2+\cdots+S_n^2\right)+\cdots\right\} \\ &= \exp\left(jT_1-\frac{1}{2}T_2+\cdots\right) \end{aligned}$$

(*ibid.*, p. 45), where

$$\begin{aligned} T_1 &= \mu_1 \sum_h S_h \\ &= \mu_1 \sum_h \sum_i \alpha_i K_{i,h} \\ &= \mu_1 \sum_i \left(\alpha_i \sum_h K_{i,h} \right) \end{aligned}$$

and

$$T_2 = \left(\mu_2 - \mu_1^2\right) \sum_h S_h^2.$$

Having obtained a convenient form for P, the joint density Q of $r_1, r_2, \ldots, r_m$ in Eq. (3.23) becomes

$$Q = \frac{1}{(2\pi)^m} \int \cdots \int \exp\left\{ -j \sum_i r_i \alpha_i + j\mu_1 \sum_i \left(\alpha_i \sum_h K_{i,h} \right) - \frac{T_2}{2} + \cdots \right\} d\alpha_1 \ldots d\alpha_m$$

(*ibid.*, p. 48), with $\sum_h S_h^3$ and higher powers being assumed to be negligible. Now, the probability that $(r_1, r_2, \ldots, r_m)$ lie in a region R_m is

$$p = \int_{R_m} \cdots \int Q\, dr_1 dr_2 \ldots dr_m.$$

The next step is to make the substitutions

$$\begin{aligned} r_i &= \mu_1 \sum_h K_{i,h} + \rho_i \sqrt{2\left(\mu_2 - \mu_1^2\right)}, \\ \alpha_i &= \frac{z_i}{\sqrt{\frac{1}{2}\left(\mu_2 - \mu_1^2\right)}}, \end{aligned} \tag{3.25}$$

the ρ_i's being the transformed errors. The probability p then becomes

$$p = \int_{R_m} \cdots \int \left\{ \int \cdots \int \exp\left(-2j \sum_i \rho_i z_i - \cdots \right) dz_1 dz_2 \ldots dz_m \right\} d\rho_1 d\rho_2 \ldots d\rho_m. \tag{3.26}$$

The following substitutions are now further made: $\sum_h K_{i,h} K_{i',h} = b_{i,i'} = b_{i',i}$,

$$\begin{aligned} \zeta_1 &= z_1 h_{1,1} + z_2 h_{1,2} + z_3 h_{1,3} + \cdots + z_m h_{1,m} + jt_1 \\ \zeta_2 &= z_2 h_{2,2} + z_3 h_{2,3} + \cdots + z_m h_{2,m} + jt_2 \\ &\cdots \\ \zeta_m &= z_m h_{m,m} + jt_m \end{aligned}$$

and

$$\begin{aligned} \rho_1 &= t_1 h_{1,1}, \\ \rho_2 &= t_1 h_{1,2} + t_2 h_{2,2}, \\ &\cdots \\ \rho_m &= t_1 h_{1,m} + t_2 h_{2,m} + \cdots + t_m h_{m,m}. \end{aligned} \tag{3.27}$$

Note that these substitutions imply that

$$b_{i,i'} = \sum_h K_{i,h} K_{i',h} = h_{1,i} h_{1,i'} + h_{2,i} h_{2,i'} + \cdots + h_{i',i} h_{i',i'}. \tag{3.28}$$

Equation (3.26) then becomes

$$\begin{aligned} p &= \frac{1}{\pi^m} \int \underset{R_m}{\cdots} \int \left\{ \int \cdots \int e^{-\zeta_1^2 - \zeta_2^2 - \cdots - \zeta_m^2} dz_1 dz_2 \ldots dz_m \right\} e^{-t_1^2 - t_2^2 - \cdots - t_m^2} dt_1 dt_2 \ldots dt_m \\ &= \frac{1}{\pi^{m/2}} \int \underset{R_m}{\cdots} \int e^{-t_1^2 - t_2^2 - \cdots - t_m^2} dt_1 dt_2 \ldots dt_m \end{aligned} \tag{3.29}$$

since the integral inside the brackets above is approximately $\pi^{m/2}$. The limits of the integral in (3.29) are determined by the equations in (3.27). Also, instead of negative and similar positive values, the limits are each taken from zero to positive values, and p is multiplied by 2^m.

Bienaymé next stated that the integral in (3.29) is generally very difficult to determine but that

> …[h]owever there exists some error combinations whose probability can, on the contrary, be expressed without much difficulty.
>
> These are those for which the power of e [in Eq. (3.29)] can only take values less than a certain constant γ^2. One sees that p must be integrated under this hypothesis for all values of the t_i which satisfy the condition
>
> $$t_1^2 + t_2^2 + \cdots + t_i^2 + \cdots + t_m^2 < \gamma^2.$$
>
> (*ibid*., p. 55)

Bienaymé thus chose the region of integration

$$R_m = \left\{ (t_1, t_2, \ldots, t_m) : t_1^2 + t_2^2 + \cdots + t_i^2 + \cdots + t_m^2 < \gamma^2 \right\}$$

and the probability he wished to determine was

$$p = \Pr\left\{ t_1^2 + t_2^2 + \cdots + t_i^2 + \cdots + t_m^2 < \gamma^2 \right\}.$$

To perform the integral in (3.29), he substituted $t_1^2 + t_2^2 + \cdots + t_i^2 + \cdots + t_m^2 = u$ so that

$$t_m dt_m = u du \quad \Rightarrow \quad dt_m = \frac{u du}{\sqrt{u^2 - t_1^2 - t_2^2 - \cdots - t_{m-1}^2}}.$$

Equation (3.29) then becomes

$$\begin{aligned} p &= \left(\frac{2}{\sqrt{\pi}} \right)^m \int_0^\gamma \int \underset{R_{m-1}}{\cdots} \int \frac{u e^{-u^2} dt_1 dt_2 \ldots dt_{m-1} du}{\sqrt{u^2 - t_1^2 - \cdots - t_{m-1}^2}} \\ &= \left(\frac{2}{\sqrt{\pi}} \right)^m \int_0^\gamma u e^{-u^2} \left\{ \int \underset{R_{m-1}}{\cdots} \int \frac{dt_1 dt_2 \ldots dt_{m-1}}{\sqrt{u^2 - t_1^2 - \cdots - t_{m-1}^2}} \right\} du \end{aligned}$$

$$=\left(\frac{2}{\sqrt{\pi}}\right)^m \int_0^\gamma ue^{-u^2}\left\{\underbrace{\int_0^u \int_0^{\sqrt{u^2-t_1^2}} \cdots \int_0^{\sqrt{u^2-t_1^2-t_2^2-\cdots-t_{m-3}^2}} \int_0^{\sqrt{u^2-t_1^2-t_2^2-\cdots-t_{m-2}^2}} \frac{1}{\sqrt{u^2-t_1^2-\cdots-t_{m-1}^2}} dt_{m-1}dt_{m-2}\ldots dt_2 dt_1}_{I}\right\}du$$

(*ibid.*, p. 57), where R_{m-1} is the region $t_1^2+t_2^2+\cdots+t_{m-1}^2<u^2$. Note that each of the integrals in I above is in the form $\int_0^{\sqrt{a}}(a-t^2)^{\delta/2}dt$, where

$$\int_0^{\sqrt{a}}\left(a-t^2\right)^{\frac{\delta}{2}}dt=\frac{\sqrt{\pi}}{2}a^{\frac{\delta+1}{2}}\frac{\Gamma\left(\frac{\delta}{2}+1\right)}{\Gamma\left(\frac{\delta+1}{2}+1\right)}.$$

The integral I therefore becomes

$$\begin{aligned}I&=u^{m-2}\left(\frac{\sqrt{\pi}}{2}\right)^{m-1}\times\frac{\Gamma\left(\frac{1}{2}\right)}{\Gamma(1)}\times\frac{\Gamma(1)}{\Gamma\left(\frac{3}{2}\right)}\times\frac{\Gamma\left(\frac{3}{2}\right)}{\Gamma(2)}\times\cdots\times\frac{\Gamma\left(\frac{m-1}{2}\right)}{\Gamma\left(\frac{m}{2}\right)}\\&=u^{m-2}\left(\frac{\sqrt{\pi}}{2}\right)^m\frac{2}{\Gamma\left(\frac{m}{2}\right)}.\end{aligned}$$

Hence, Bienaymé was able to obtain the important formula:

$$p=\Pr\left\{t_1^2+t_2^2+\cdots+t_i^2+\cdots+t_m^2<\gamma^2\right\}=\frac{2}{\Gamma\left(\frac{m}{2}\right)}\int_0^\gamma u^{m-1}e^{-u^2}du \tag{3.30}$$

(*ibid.*, p. 58). We note that this is simply $\Pr\{X^2<2\gamma^2\}$, where $X^2\sim\chi_m^2$.

Continuing with the analysis, Bienaymé then used integration by parts to show that

$$\int_0^\gamma u^{m-1}e^{-u^2}du=\frac{-e^{-\gamma^2}}{2}\left\{\begin{aligned}&\gamma^{m-2}+\frac{(m-2)}{2}\gamma^{m-4}+\frac{(m-2)(m-4)}{2^2}\gamma^{m-6}\\&+\cdots+\frac{(m-2)(m-4)\cdots(m-2i)}{2^i}\gamma^{m-2i-2}\end{aligned}\right\}$$

$$+\frac{(m-2)(m-4)\cdots(m-2i-2)}{2^{i+1}}\int_0^\gamma u^{m-2i-3}e^{-u^2}du.$$

Therefore, Eq. (3.30) becomes, respectively, for even and odd m,

$$p_{2g} = 1 - e^{-\gamma^2}\left\{\frac{\gamma^{2g-2}}{\Gamma\left(\frac{2g}{2}\right)} + \frac{\gamma^{2g-4}}{\Gamma\left(\frac{2g-2}{2}\right)} + \cdots + \frac{\gamma^2}{\Gamma\left(\frac{4}{2}\right)} + 1\right\},$$

$$p_{2g-1} = \frac{2}{\sqrt{\pi}}\int_0^{\gamma} e^{-u^2}du - e^{-\gamma^2}\left\{\frac{\gamma^{2g-3}}{\Gamma\left(\frac{2g-1}{2}\right)} + \frac{\gamma^{2g-5}}{\Gamma\left(\frac{2g-3}{2}\right)} + \cdots + \frac{\gamma^3}{\Gamma\left(\frac{5}{2}\right)} + \frac{\gamma}{\Gamma\left(\frac{3}{2}\right)}\right\}$$

(*ibid*., p. 59).

Bienaymé's next task was to determine bounds, subject to the condition that $t_1^2 + t_2^2 + \cdots + t_i^2 + \cdots + t_m^2 < \gamma^2$, for the maximum and minimum values of the transformed error ρ_i, where from Eq. (3.27),

$$\rho_i = t_1 h_{1,i} + t_2 h_{2,i} + \cdots + t_i h_{i,i}. \tag{3.31}$$

To obtain the maximum value of ρ_i, he let $t_1^2 + t_2^2 + \cdots + t_m^2 = u$ so that

$$t_1^2 + t_2^2 + \cdots + t_i^2 = u - t_{i+1}^2 - t_{i+2}^2 - \cdots - t_m^2 = \nu, \tag{3.32}$$

where ν is a positive quantity. By setting $d\rho_i / dt_{i'} = 0$ in (3.31),

$$h_{i',i} + h_{i,i}\frac{dt_i}{dt_{i'}} = 0.$$

Equation (3.32) also gives $t_{i'}dt_{i'} + t_i dt_i = 0$. Therefore

$$h_{i',i} + h_{i,i}\left(-\frac{t_{i'}}{t_i}\right) = 0 \;\Rightarrow\; \frac{t_{i'}}{t_i} = \frac{h_{i',i}}{h_{i,i}}$$

(*ibid*., p. 62). The last equation implies that $t_{i'}$ is proportional to $h_{i',i}$, that is,

$$t_{i'} = f h_{i',i},$$

where f is a constant of proportionality. Equation (3.32) then yields

$$f^2\left(h_{1,i}^2 + h_{2,i}^2 + \cdots + h_{2,i}^2\right) = \nu.$$

Substituting the expression inside the brackets in the equation above by $b_{i,i}$ (see Eq. 3.28), Bienaymé obtained $f = \sqrt{\nu / b_{i,i}}$. The maximum value of ρ_i is then (3.31) evaluated at $t_{i'} = f h_{i',i}$, where $f = \sqrt{\nu / b_{i,i}}$, and is

$$\begin{aligned}\rho_i &= t_1 h_{1,i} + t_2 h_{2,i} + \cdots + t_i h_{i,i} \\ &= f\left(h_{1,i}^2 + h_{2,i}^2 + \cdots + h_{i,i}^2\right) \\ &= f b_{i,i} \\ &= b_{i,i}\sqrt{\frac{\nu}{b_{i,i}}} \\ &= \sqrt{\nu b_{i,i}}.\end{aligned}$$

Bienaymé also argued that the minimum value of ρ_i occurred when f was the negative of its value above, that is, when $p_i = -\sqrt{\nu b_{i,i}}$. Now, from Eq. (3.32), it is seen that the maximum value of ν is u, where $u < \gamma^2$. Hence, he was able to write the maximum and minimum values of ρ_i as

$$\rho_i = \pm\gamma\sqrt{b_{i,i}}$$

(*ibid*., p. 64).

Now, in view of Eq. (3.28), the above can also be written as

$$\rho_i = \pm\gamma\sqrt{\sum_h K_{i,h}^2}. \tag{3.33}$$

To minimize the length of the interval $\left(-\gamma\sqrt{\sum_h K_{i,h}^2}, \gamma\sqrt{\sum_h K_{i,h}^2}\right)$ for a given γ, one needs to keep $\sum_h K_{i,h}^2$ a minimum. Bienaymé observed that this happens when the unknowns in Eq. (3.19) are determined by the method of least squares. He went on to prove this last statement by the method used previously by Gauss (see Section 1.5.6), whom he also cited. Because this proof is in fact the Gauss–Markov theorem for the multivariate case, we shall now describe it.

Bienaymé's proof started by first denoting by $A_{i,h}$ the values of $K_{i,h}$, which result when the unknowns $x_1, x_2, \ldots, x_m$ in (3.19) are solved by the method of least squares. The normal equations will then be

$$\begin{aligned}x_1\sum_h a_{1,h}a_{1,h} + x_2\sum_h a_{1,h}a_{2,h} + \cdots + x_i\sum_h a_{1,h}a_{i,h} + \cdots &= \sum_h a_{1,h}\omega_h, \\ x_1\sum_h a_{2,h}a_{1,h} + x_2\sum_h a_{2,h}a_{2,h} + \cdots + x_i\sum_h a_{2,h}a_{i,h} + \cdots &= \sum_h a_{2,h}\omega_h, \\ &\cdots \\ x_1\sum_h a_{i,h}a_{1,h} + x_2\sum_h a_{i,h}a_{2,h} + \cdots + x_i\sum_h a_{i,h}a_{i,h} + \cdots &= \sum_h a_{i,h}\omega_h, \\ &\cdots \\ x_1\sum_h a_{m,h}a_{1,h} + x_2\sum_h a_{m,h}a_{2,h} + \cdots + x_i\sum_h a_{m,h}a_{i,h} + \cdots &= \sum_h a_{m,h}\omega_h\end{aligned}$$

(*ibid*., p. 65). The system of equations can be solved by the process of elimination: the equations are successively multiplied by the factors $B_1, B_2, \ldots, B_m$, they are then added, and by imposing suitable conditions on the factors, a typical x_i can be solved for as

$$x_i' = B_1 \sum_h a_{1,h}\omega_h + B_2 \sum_h a_{2,h}\omega_h + \cdots + B_m \sum_h a_{m,h}\omega_h,$$

where x'_i is the least squares estimate of x_i. The above can also be written in terms of a linear combination of the observations $\omega_1, \omega_2, \ldots, \omega_n$:

$$x_i' = \omega_1 \sum_{i'} B_{i'} a_{i',1} + \omega_2 \sum_{i'} B_{i'} a_{i',2} + \cdots + \omega_n \sum_{i'} B_{i'} a_{i',n}.$$

But from Eq. (3.21), we also have

$$x_i' = \sum_h \omega_h A_{i',h}, \quad i' = 1,2,\ldots,m.$$

Comparing the last two equations,

$$A_{i,h} = \sum_i B_i a_{i,h}. \tag{3.34}$$

The next step for Bienaymé was to note that similar to the equations in (3.20), corresponding ones can be obtained by replacing the value of $K_{i,h}$ by $A_{i,h}$. Then, by subtracting each corresponding pair of equations, he obtained

$$\begin{aligned}
\sum_h a_{1,h}\left(K_{i,h} - A_{i,h}\right) &= 0,\\
\sum_h a_{2,h}\left(K_{i,h} - A_{i,h}\right) &= 0,\\
&\cdots\\
\sum_h a_{i-1,h}\left(K_{i,h} - A_{i,h}\right) &= 0,\\
\sum_h a_{i,h}\left(K_{i,h} - A_{i,h}\right) &= 0,\\
\sum_h a_{i+1,h}\left(K_{i,h} - A_{i,h}\right) &= 0,\\
&\cdots\\
\sum_h a_{m,h}\left(K_{i,h} - A_{i,h}\right) &= 0.
\end{aligned}$$

Multiplying each of the above equations by B_i and adding:

$$\sum_i B_i \sum_h a_{i,h}\left(K_{i,h} - A_{i,h}\right) = \sum_h \left(\sum_i B_i a_{i,h}\right)\left(K_{i,h} - A_{i,h}\right) = 0. \tag{3.35}$$

Using Eq. (3.34), the above becomes

$$\sum_h A_{i,h}\left(K_{i,h} - A_{i,h}\right) = 0 \;\Rightarrow\; \sum_h A_{i,h} K_{i,h} = \sum_h A_{i,h}^2$$

so that

$$\sum_h K_{i,h}^2 - \sum_h A_{i,h}^2 = \sum_h K_{i,h}^2 - 2\sum_h A_{i,h}^2 + \sum_h A_{i,h}^2 = \sum_h K_{i,h}^2 - 2\sum_h A_{i,h} K_{i,h} + \sum_h A_{i,h}^2.$$

Hence, Bienaymé obtained

$$\sum_h K_{i,h}^2 = \sum_h A_{i,h}^2 + \sum_h \left(K_{i,h} - A_{i,h}\right)^2$$

(*ibid.*, p. 66), from which he concluded that for $\sum_h K_{i,h}^2$ to be a minimum, we need $K_{i,h} = A_{i,h}$, that is, the unknowns in Eq. (3.19) are determined by the method of least squares.

Moving on, in Section IV of the paper, Bienaymé gave some numerical examples on his result in (3.33). First, he considered the case of a single unknown x in Eq. (3.19). The most useful situation for him was when p in Eq. (3.30) was assigned the value of 1/2. Using the value provided by Bessel of $\gamma = .4769$ (to 4 d.p.), then

$$p_1 = \frac{2}{\sqrt{\pi}} \int_0^{\gamma} e^{-u^2} du = \frac{1}{2}.$$

There is thus a probability of 1/2 that the transformed error ρ is contained within the limits (cf. Eq. 3.33)

$$\left(-.4769\sqrt{\sum_h K_{1,h}^2},\ .4769\sqrt{\sum_h K_{1,h}^2}\right).$$

From Eq. (3.25), the 50% confidence limits for the error r in x is therefore

$$\left[\begin{array}{l} \mu_1 \sum_h K_{1,h} - .4769\sqrt{\sum_h K_{1,h}^2}\sqrt{2\left(\mu_2 - \mu_1^2\right)}, \\ \mu_1 \sum_h K_{1,h} + .4769\sqrt{\sum_h K_{1,h}^2}\sqrt{2\left(\mu_2 - \mu_1^2\right)} \end{array}\right]$$

(*ibid.*, p. 70).

For the case of two unknowns, again setting $p=½$ in Eq. (3.30),

$$2\int_0^{\gamma} ue^{-u^2} du = \frac{1}{2} \quad \Rightarrow \quad \gamma = \sqrt{\log 2} = .8326.$$

There is thus a probability of ½ that simultaneously the errors $r_i\ (i = 1,2)$ will be contained within the limits

$$\left[\begin{array}{l} \mu_1 \sum_h K_{i,h} - .8326\sqrt{\sum_h K_{i,h}^2}\sqrt{2\left(\mu_2 - \mu_1^2\right)}, \\ \mu_1 \sum_h K_{i,h} + .8326\sqrt{\sum_h K_{i,h}^2}\sqrt{2\left(\mu_2 - \mu_1^2\right)} \end{array}\right] \tag{3.36}$$

(*ibid.*, p. 71). Bienaymé noted that the intervals for two unknowns are almost twice in length compared to those for a single unknown. He also claimed that, allowing the errors to vary, there was a probability of ½ for the following system of equations to be satisfied:

$$\begin{cases} r_1 - \mu_1 \sum_h K_{1,h} = t_1 \sqrt{2\left(\mu_2 - \mu_1^2\right)\sum_h K_{1,h}^2}, \\ r_2 - \mu_1 \sum_h K_{2,h} = t_1 \sqrt{2\left(\mu_2 - \mu_1^2\right)\frac{\left(\sum_h K_{1,h}K_{2,h}\right)^2}{\sum_h K_{1,h}^2}} \\ \qquad\qquad + t_2 \sqrt{2\left(\mu_2 - \mu_1^2\right)\left\{\sum_h K_{2,h}^2 - \frac{\left(\sum_h K_{1,h}K_{2,h}\right)^2}{\sum_h K_{1,h}^2}\right\}}, \\ t_1^2 + t_2^2 < .8326^2 = .693. \end{cases} \tag{3.37}$$

The quantities t_1 and t_2 can be obtained from the first two equations in Eq. (3.37) and substituted into the third, resulting in an ellipse for the confidence region. On the other hand, the confidence intervals given in Eq. (3.36) represent a square. Thus, the two confidence regions given by Bienaymé in (3.36) and (3.37) are incompatible. In fact, the ellipse is inscribed inside the square (Hald, 1998, p. 508).

Notwithstanding the above oversight, Bienaymé's paper was quite well received and responsible in a major way for his election to the *Académie des Sciences*. In his summary of the paper, Liouville wrote:

> …The new point of view adopted by Bienaymé has naturally led him to do calculations which are new as well as more complicated. The author has accomplished this with great skill; one sees that he is aware of all the progress, even in detail, made in the mathematical sciences lately. (Liouville, 1852, p. 32)

3.4.2 Abbe on the Distribution of Errors in a Series of Observations (1863)

Ernst Karl Abbe* (Fig. 3.7) was an illustrious German physicist, optical scientist, and social reformer. Abbe was born in Eisenach, Germany, on January 23, 1840. His parents were of humble origins, but thanks to his father's employer Abbe was able to attend secondary school. After graduating from the Eisenach Gymnasium in 1857, Abbe studied physics at the Universities of Jena and Gottïngen. He gave private tuition to improve his income and obtained his Ph.D. in Gottïngen in March of 1861. At Gottïngen, Abbe was greatly influenced by the famous mathematician Bernhard Riemann (1826–1866) and the physicist Wilhelm Weber (1804–1891), assistant to Carl Gauss (1777–1855). In August 1863, at the age of 23, Abbe was appointed lecturer at Jena University in

* Ernst Abbe's life is also described in Günther (1970) and Glatzer (1913).

FIGURE 3.7 Ernst Karl Abbe (1840–1905). Wikimedia Commons (Public Domain), http://commons.wikimedia.org/wiki/Ernst_Abbe#/media/File:Abbe.jpg

mathematics, physics, and astronomy. In 1866, Abbe became a research director at the Zeiss Optical Works and in 1869 invented the apochromatic lens (a microscope lens that eliminates both the primary and secondary color distortion). It was only the next year, in 1870, that he became financially stable, when he was promoted to Associate Professor. In 1871, Abbe married Elise Snell, the daughter of Karl Snell, head of the physics department at Jena. Meanwhile, Abbe designed the first refractometer and created the Abbe number and Abbe criterion. In 1876 he was offered a partnership with Carl Zeiss, and this resulted in a rise of Abbe's personal fortune. Zeiss Optical Works also reached worldwide fame with their superlative microscopes and telescopes. When the founder of the company died in 1888, Abbe became the sole proprietor of Zeiss Optical. Instead of enjoying his enormous wealth, the Professor Extraordinaire turned into a social reformer of equal stature. Abbe created the "Carl Zeiss Foundation" to which he bequeathed his personal fortune. The foundation was responsible for providing for the workers of Zeiss Optical, for opening up opportunities for education for the working class, and for sponsoring Jena University. Abbe died in Jena in 1905, a model of deep erudition and equally profound humanism.

Ernst Abbe's important statistical contributions had all but been forgotten until they were pointed out in 1966 by Oscar Sheynin in a *Nature* article (Sheynin, 1966). Abbe's work on the distribution of errors took place when he was 23 and wrote a dissertation entitled *Über die Gesetzmässigkeit in der Vertheilung der Fehler bei Beobachtungsreihen** (Abbe, 1863) (Fig. 3.8) so as to obtain a lectureship at Jena University. In his work, Abbe derived the χ^2-distribution, as we now explain.†

First, Abbe gave the formula for a normal distribution with mean 0 and precision constant h $[=1/(\text{SD}\sqrt{2})]$ as

$$w = \frac{h}{\sqrt{\pi}} e^{-h^2 x^2}.$$

He next defined the two quantities

$$\Delta = \sum x^2 = x_1^2 + x_2^2 + \cdots + x_n^2$$

(*ibid.*, p. 5) and

$$\theta = (x_1 - x_2)^2 + (x_2 - x_3)^2 + \cdots + (x_{n-1} - x_n)^2 + (x_n - x_1)^2$$

(*ibid.*, p. 6). Abbe's aim was to determine "for any number of observations, the probability of occurrence of a system of errors for which the function Δ has a certain value (or a value lying between certain limits) while the function θ also has a certain other value (or a value lying between certain limits)?" We shall here describe Abbe's derivation of the distribution function of Δ because this is where the χ^2-distribution occurs.

* "On the Law of Distribution of Errors in Observation Series."

† See also Kendall (1971).

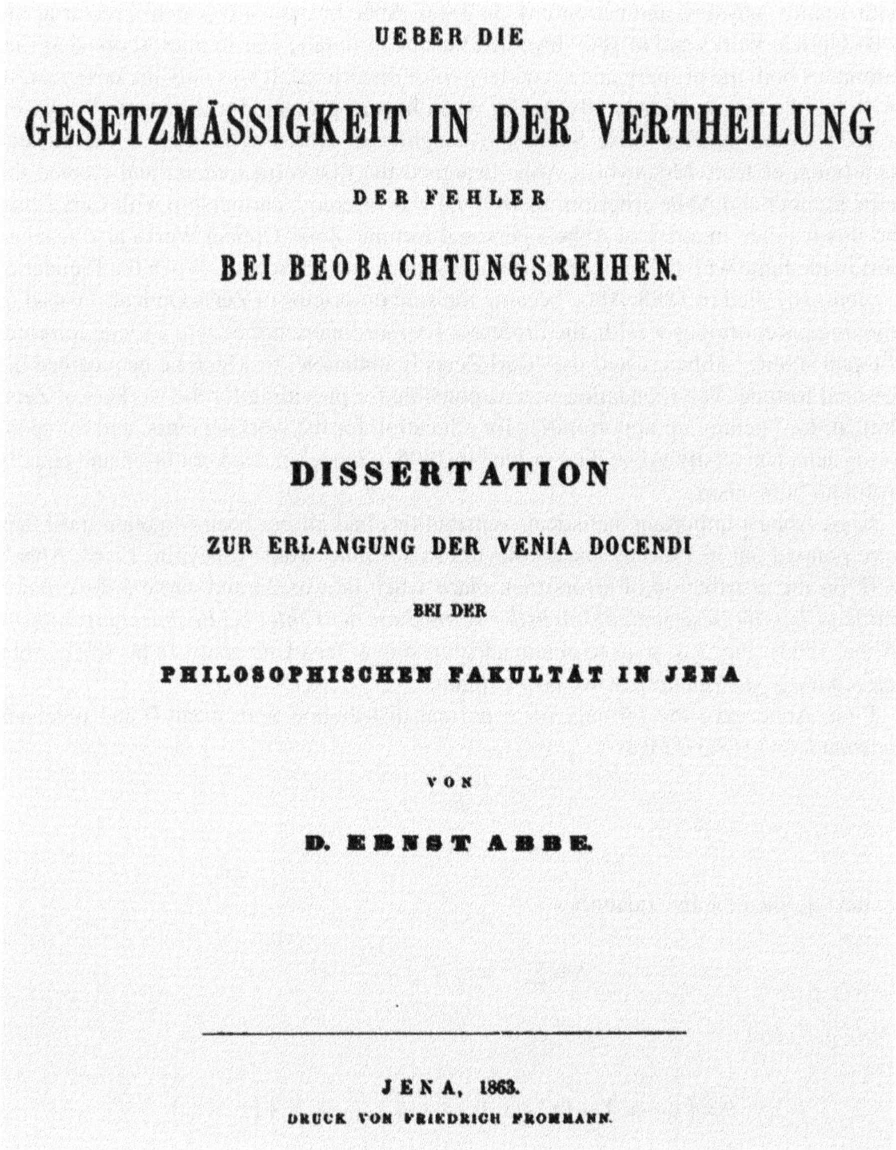

UEBER DIE

GESETZMÄSSIGKEIT IN DER VERTHEILUNG

DER FEHLER

BEI BEOBACHTUNGSREIHEN.

DISSERTATION

ZUR ERLANGUNG DER VENIA DOCENDI

BEI DER

PHILOSOPHISCHEN FAKULTÄT IN JENA

VON

D. ERNST ABBE.

JENA, 1863.

DRUCK VON FRIEDRICH FROMMANN.

FIGURE 3.8 Title page of Abbe's dissertation (Abbe, 1863)

The probability of a system of errors with a sum of squares lying between 0 and Δ is

$$\Phi(\Delta)=\frac{h^n}{\pi^{n/2}}\int\cdots\int e^{-h^2\left(x_1^2+x_2^2+\cdots+x_n^2\right)}dx_1\ldots dx_n, \tag{3.38}$$

the region of integration being given by

$$0<x_1^2+x_2^2+\cdots+x_n^2<\Delta.$$

To perform the latter integral, Abbe introduced the Dirichlet discontinuity factor*

$$\frac{1}{2\pi}\int_{-\infty}^{\infty}\frac{e^{\sigma(a+i\phi)}}{a+i\phi}d\phi \quad \left(i^2=-1\right), \tag{3.39}$$

which equals 1 for $\sigma > 0$ and equals 0 for $\sigma < 0$. Equation (3.38) then becomes, with $\sigma = \Delta - (x_1^2 + x_2^2 + \cdots + x_n^2)$,

$$\Phi(\Delta)=\frac{1}{2\pi}\cdot\frac{h^n}{\pi^{n/2}}\int_{-\infty}^{\infty}\frac{e^{\Delta(a+i\phi)}}{a+i\phi}\left\{\int_{-\infty}^{\infty}e^{-\left(h^2+a+i\phi\right)x_1^2}dx_1\ldots\int_{-\infty}^{\infty}e^{-\left(h^2+a+i\phi\right)x_n^2}dx_n\right\}d\phi$$

(*ibid.*, p. 8). The last equation has a simpler region of integration than Eq. (3.38). Now since

$$\int_{-\infty}^{\infty}e^{-\left(h^2+a+i\phi\right)x_k^2}dx_k=\frac{\sqrt{\pi}}{\sqrt{h^2+a+i\phi}},$$

$\Phi(\Delta)$ becomes

$$\Phi(\Delta)=\frac{h^n}{2\pi}\int_{-\infty}^{\infty}\left\{\frac{e^{\Delta(a+i\phi)}}{(a+i\phi)\left(h^2+a+i\phi\right)^{n/2}}\right\}d\phi.$$

Abbe next made the substitution

$$\frac{1}{\left(h^2+a+i\phi\right)^{n/2}}=\frac{1}{\Gamma\left(\frac{n}{2}\right)}\int_{-\infty}^{\infty}e^{-\left(h^2+a+i\phi\right)y}y^{\frac{n}{2}-1}dy$$

and further simplified $\Phi(\Delta)$ to

$$\begin{aligned}\Phi(\Delta)&=\frac{h^n}{2\pi\Gamma\left(\frac{n}{2}\right)}\int_{-\infty}^{\infty}\left\{\frac{e^{\Delta(a+i\phi)}}{(a+i\phi)}\int_{-\infty}^{\infty}e^{-\left(h^2+a+i\phi\right)y}y^{\frac{n}{2}-1}dy\right\}d\phi\\&=\frac{h^n}{2\pi\Gamma\left(\frac{n}{2}\right)}\int_{-\infty}^{\infty}e^{-h^2y}y^{\frac{n}{2}-1}\left[\int_{-\infty}^{\infty}\left\{\frac{e^{(\Delta-y)(a+i\phi)}}{(a+i\phi)}\right\}d\phi\right]dy\\&=\frac{h^n}{\Gamma\left(\frac{n}{2}\right)}\int_{0}^{\Delta}e^{-h^2y}y^{\frac{n}{2}-1}dy\end{aligned}$$

*The discontinuous factor technique had been introduced by Dirichlet to transform a complicated region of integration into an easier one. See the book by Fischer for more details (Fischer, 2010, p. 46).

(*ibidem*), the last line following from Eq. (3.39), which makes the integral inside the square brackets in the penultimate line equal to 2π for $y < \Delta$. Now, if we make the transformation $v = 2h^2 y$, we obtain

$$\Phi(\Delta) = \int_0^{2h^2\Delta} \frac{v^{\frac{n}{2}-1} e^{-\frac{v}{2}}}{2^{\frac{n}{2}} \Gamma\left(\frac{n}{2}\right)} dv.$$

The integrand above is the χ_n^2-distribution.

3.4.3 Helmert on the Distribution of the Sum of Squares of Residuals (1876): The Helmert Transformations

Friedrich Robert Helmert* (Fig. 3.9) was a famous German geodesist who also made important contributions to the theory of errors. Helmert was born in Freiberg, Kingdom of Saxony, on July 31, 1843. His father was the treasurer of the Johannishospitalgut of Freiberg and of the former Christiana Friederika Linke. Helmert attended secondary school in Freiberg and then entered the Annen-Realschule in Dresden. In 1859, he studied engineering at the Polytechnische Schule in Dresden. Since he was keen in geodesy, Helmert was hired by his teacher, August Nagel, to work on the triangulation of the coal-field of the Erzgebirge and the drafting of the trigonometric network for Saxony. In 1863, Helmert became Nagel's assistant on the Central European Arc Measurement. Based on his work with Nagel, Helmert obtained his doctorate in 1867 from the University of Leipzig for his work on the theory of the "ellipse of error." In 1872, he was appointed Professor of Geodesy at the technical school of Aachen. Shortly after, Helmert wrote the masterpiece *Die mathematischen und physikalischen Theorien der höheren Geodäsie*,† which became a standard text. From 1887, Helmert was Professor of Advanced Geodesy at the University of Berlin and director of the Prussian Geodetic Institute. He was also member of the Prussian Academy of Sciences in Berlin and the Royal Swedish Academy of Sciences. He received about 25 German and foreign decorations and published more than 130 papers. In 1916, Helmert had a stroke and died the next year in Potsdam.

In 1876, Helmert derived the chi-squared distribution in the paper "Über die Wahrscheinlichkeit der Potenzsummen der Beobachtungsfehler und über einige damit im Zusammenhange stehende Fragen"‡ (Helmert, 1876b), after having previously given the result without proof in a paper one year earlier (Helmert, 1875, p. 303). In another 1876 paper (Helmert, 1876a), Helmert proposed a classic transformation of variables that is still used today.

Until the 1960s, the chi-squared distribution had been attributed to Helmert, until the prior works of Bienaymé and Abbe were rediscovered. We now describe the relevant work of Helmert.§

* Helmert's life is also described in Fischer (1973).

† "Mathematical and Physical Theories of Higher Geodesy."

‡ "On the probability of sums of powers of errors of observations and some relevant remaining issues."

§ See also Kruskal (1946).

FIGURE 3.9 Friedrich Robert Helmert (1843–1917). Wikimedia Commons (Public Domain), http://commons.wikimedia.org/wiki/Category:Friedrich_Robert_Helmert#/media/File:F-R_Helmert_1.jpg

In the 1875 paper (Helmert, 1875), Helmert considered a sequence of experimental errors $\varepsilon_1, \ldots, \varepsilon_n$, where each ε_i is normally distributed with mean 0 and precision h. He defined the sum of squares of the ε_i's by $[\varepsilon^2]$ and set

$$\left[\varepsilon^2\right] = n\sigma_2.$$

Helmert then wrote (without proof) the probability that $[\varepsilon^2]$ lies between $n(\sigma_2 - \delta/2)$ and $n(\sigma_2 + \delta/2)$ as

$$w = \frac{h^n}{\Gamma\left(\frac{n}{2}\right)}(n\delta)(n\sigma_2)^{\frac{n}{2}-1} e^{-h^2 n\sigma_2}$$

(*ibid.*, p. 303), where δ is infinitesimally small. This implies that the density of $[\varepsilon^2]$ evaluated at $n\sigma_2$ is

$$\frac{w}{n\delta} = \frac{h^n}{\Gamma\left(\frac{n}{2}\right)}(n\sigma_2)^{\frac{n}{2}-1} e^{-h^2 n\sigma_2}. \tag{3.40}$$

If we write $[\varepsilon^2] = n\sigma_2 = u$, then the density of u is

$$\frac{h^n}{\Gamma\left(\frac{n}{2}\right)} u^{\frac{n}{2}-1} e^{-h^2 u}.$$

Therefore the density of $v = 2h^2 u$ is

$$\frac{v^{\frac{n}{2}-1} e^{-\frac{v}{2}}}{2^{\frac{n}{2}}\Gamma\left(\frac{n}{2}\right)},$$

which is the χ_n^2-distribution.

In the paper "Über die Wahrscheinlichkeit der Potenzsummen der Beobachtungsfehler und über einige damit im Zusammenhange stehende Fragen" (Helmert, 1876b), Helmert provided a proof of the χ^2-distribution in (3.40). We now describe his derivation.

Helmert started his paper by considering n observation errors with *magnitudes* $\varepsilon_1, \varepsilon_2, \ldots, \varepsilon_n$ and then defining the mth raw moment:

$$S_m = \int_0^a \varepsilon^m \psi(\varepsilon)\, d\varepsilon.$$

In the above, the density $\psi(\varepsilon) = \phi(\varepsilon) + \phi(-\varepsilon)$ is the sum of the densities for the positive and negative errors such that $\varepsilon_i < a$. His aim was to calculate

> ...the probability that the sum of the mth powers of n observation errors will lie between the limits

$$n\left(\sigma_m - \frac{\delta_m}{2}\right) \quad \text{and} \quad n\left(\sigma_m + \frac{\delta_m}{2}\right).$$

(*ibid*., p. 192)

Consistent with his previous paper, Helmert used the notation $\sigma_m = [\varepsilon^m]/n = \left(\sum_{i=1}^{n} \varepsilon_i^m\right)/n$. Moreover, the density of $[\varepsilon^m]/n$ evaluated at σ_m is denoted by $\phi(\sigma_m)_n$.

As a simple example, he considered the case $n=1$. Then the probability that ε_1^m lies in the interval $\sigma_m \mp \delta_m/2$ is

$$\phi(\sigma_m)_1 \delta_m = \int_{\sqrt[m]{\sigma_m - \delta_m/2}}^{\sqrt[m]{\sigma_m + \delta_m/2}} \psi(\varepsilon) d\varepsilon.$$

By doing a series expansion on the limits of the above integral and neglecting δ_m^2 and higher powers, the density $\phi(\sigma_m)_1$ becomes

$$\begin{aligned}
\phi(\sigma_m)_1 &= \frac{\left(\sqrt[m]{\sigma_m + \frac{\delta_m}{2}} - \sqrt[m]{\sigma_m - \frac{\delta_m}{2}}\right)}{\delta_m} \psi\left(\sqrt[m]{\sigma_m}\right) \\
&= \frac{\left\{\sigma_m^{1/m} + \frac{1}{m}\sigma_m^{(1/m)-1}\frac{\delta_m}{2}\right\} - \left\{\sigma_m^{1/m} - \frac{1}{m}\sigma_m^{(1/m)-1}\frac{\delta_m}{2}\right\}}{\delta_m} \psi\left(\sqrt[m]{\sigma_m}\right) \\
&= \frac{\psi\left(\sqrt[m]{\sigma_m}\right)}{m\sigma_m^{1-(1/m)}}
\end{aligned}$$

(*ibid*., p. 193). By using double integrals, Helmert next gave a formula for $\phi(\sigma_m)_2$, and the general case $\phi(\sigma_m)_n$ is then treated by inserting a Dirichlet discontinuity factor inside a multiple integral.

In Section 2 of his paper, Helmert considered the case when ψ is a uniform distribution. But it is Section 3 that is of interest here since the normal distribution is then considered:

$$\psi(\varepsilon) = \frac{2h}{\sqrt{\pi}} e^{-h^2\varepsilon^2}.$$

Since ψ is the density of the magnitude of each error, this means that each error is assumed to be normal with mean 0 and precision h.

For each of $n=1, 2$, Helmert gave formulas for $\phi(\sigma_m)_n$ when $m=1, 2$, and 3. For $n=1$ and $m=2$, we have $\sigma_m = [\varepsilon^m]/n = \varepsilon_1^2$ with density

$$\begin{aligned}
\phi(\sigma_2)_1 &= \psi\left(\sqrt{\sigma_2}\right) \cdot \frac{1}{2\sqrt{\sigma_2}} \\
&= \frac{2h}{\sqrt{\pi}} e^{-h^2\sigma_2} \cdot \frac{1}{2\sqrt{\sigma_2}} \\
&= \frac{h}{\sqrt{\pi}} e^{-h^2\sigma_2} \sigma_2^{-1/2}
\end{aligned}$$

(*ibid*., p. 199). For $n=2$ and $m=2$, we have $\sigma_m = [\varepsilon^m]/n = (\varepsilon_1^2 + \varepsilon_2^2)/2$ with density

$$
\begin{aligned}
\phi(\sigma_2)_2 &= 2\int_0^{2\sigma_2} \phi(2\sigma_2 - v)_1 \phi(v)_1 \, dv \\
&= 2\int_0^{2\sigma_2} \frac{h}{\sqrt{\pi}} e^{-h^2(2\sigma_2 - v)} (2\sigma_2 - v)^{-1/2} \cdot \frac{h}{\sqrt{\pi}} e^{-h^2 v} v^{-1/2} dv \\
&= 2\frac{h^2}{\pi} e^{-2h^2\sigma_2} \int_0^{2\sigma_2} (2\sigma_2 - v)^{-1/2} v^{-1/2} dv \\
&= 2h^2 e^{-2h^2\sigma_2}
\end{aligned}
$$

(*ibid*., p. 200).

In Section 4 of his paper, Helmert provided a proof by induction of (3.40). Having given $\phi(\sigma_2)_1$ and $\phi(\sigma_2)_2$ in Section 3, he stated the density of $[\varepsilon^2]/n = \left(\sum_{i=1}^n \varepsilon_i^2\right)/n$ evaluated at σ_2 as

$$
\phi(\sigma_2)_n = \frac{nh^n}{\Gamma\left(\frac{n}{2}\right)} (n\sigma_2)^{\frac{n}{2}-1} e^{-h^2 n\sigma_2} \tag{3.41}
$$

(*ibid*., p. 203). To prove this, he set $x = \sum_{i=1}^n \varepsilon_i^2$ and $y = \varepsilon_{n+1}^2 + \varepsilon_{n+2}^2$. Then, the density of $z = \sum_{i=1}^{n+2} \varepsilon_i^2$ is that of $x+y$ and, by the induction hypothesis, is

$$
\begin{aligned}
\int_0^z \phi(z-u)_n \phi(u)_2 \, du &= \int_0^z \frac{h^n}{\Gamma\left(\frac{n}{2}\right)} (z-u)^{\frac{n}{2}-1} e^{-h^2(z-u)} \cdot h^2 e^{-h^2 u} du \\
&= \frac{h^{n+2} e^{-h^2 z}}{\Gamma\left(\frac{n}{2}\right)} \int_0^z (z-u)^{\frac{n}{2}-1} du \\
&= \frac{2h^{n+2} z^{n/2} e^{-h^2 z}}{n\Gamma\left(\frac{n}{2}\right)}.
\end{aligned}
$$

Hence, the density of $\sigma_2 = \left(\sum_{i=1}^{n+2} \varepsilon_i^2\right)/(n+2)$ is

$$
\frac{h^{n+2}\{(n+2)\sigma_2\}^{n/2} e^{-h^2(n+2)\sigma_2}(n+2)}{\frac{n}{2}\Gamma\left(\frac{n}{2}\right)} = \frac{h^{n+2}\{(n+2)\sigma_2\}^{n/2} e^{-h^2(n+2)\sigma_2}(n+2)}{\Gamma\left(\frac{n+2}{2}\right)}
$$

(*ibidem*). Hence, Eq. (3.41) (or Eq. 3.40) is proved by induction.

In "Die Genauigkeit der Formel von Peters zur Berechnung des wahrscheinlichen Beobachtungsfehler direkter Beobachtungen gleicher Genauigkeit" (Helmert, 1876a), Helmert again considered a sequence of experimental errors $\varepsilon_1, \ldots, \varepsilon_n$, where each ε_i is normally distributed with mean 0 and precision h. He then decomposed the errors as follows:

$$\begin{aligned}
\varepsilon_1 &= \lambda_1 + \sigma, \\
\varepsilon_2 &= \lambda_2 + \sigma, \\
&\cdots \\
\varepsilon_{n-1} &= \lambda_{n-1} + \sigma, \\
\varepsilon_n &= -\lambda_1 - \lambda_2 - \cdots - \lambda_{n-1} + \sigma
\end{aligned} \tag{3.42}$$

(*ibid.*, p. 121), where $\sigma = [\varepsilon]/n = \left(\sum_{i=1}^n \varepsilon_i\right)/n$. The λ_i's are thus deviations of the errors from their average value, so that $[\lambda] = \sum_{i=1}^n \lambda_i = 0$. The usefulness of these deviations stems from the fact they are directly observable, unlike the errors. To see this, an observation y_i with expectation μ is written as $y_i = \mu + \varepsilon_i$. Then $\bar{y} = \mu + \sigma$. By subtracting the last two equations, we obtain $y_i - \bar{y} = \varepsilon_i - \sigma = \lambda_i$, which shows that the λ_i's are directly observable.

Helmert's aim was to obtain the distribution of $[\lambda\lambda] = \sum_{i=1}^n \lambda_i^2$. First, he determined the joint density of $\lambda_1, \ldots, \lambda_{n-1}, \sigma$. This is simply the product of the joint density of $\varepsilon_1, \ldots, \varepsilon_n$ and the absolute value of the Jacobian of the transformation. The joint density of $\varepsilon_1, \ldots, \varepsilon_n$ is

$$\left(\frac{h}{\sqrt{\pi}}\right)^n \exp\left(-h^2 \sum_{i=1}^n \varepsilon_i^2\right)$$

and the Jacobian ("*Functionaldeterminante*") is given by

$$\begin{aligned}
J\left(\frac{\varepsilon_1, \varepsilon_2, \ldots, \varepsilon_n}{\lambda_1, \lambda_2, \ldots, \lambda_{n-1}, \sigma}\right) &= \begin{vmatrix}
\partial\varepsilon_1/\partial\lambda_1 & \partial\varepsilon_1/\partial\lambda_2 & \ldots & \partial\varepsilon_1/\partial\lambda_{n-1} & \partial\varepsilon_1/\partial\sigma \\
\partial\varepsilon_2/\partial\lambda_1 & \partial\varepsilon_2/\partial\lambda_2 & \ldots & \partial\varepsilon_2/\partial\lambda_{n-1} & \partial\varepsilon_2/\partial\sigma \\
\vdots & \vdots & & \vdots & \vdots \\
\partial\varepsilon_{n-1}/\partial\lambda_1 & \partial\varepsilon_{n-1}/\partial\lambda_2 & \ldots & \partial\varepsilon_{n-1}/\partial\lambda_{n-1} & \partial\varepsilon_{n-1}/\partial\sigma \\
\partial\varepsilon_n/\partial\lambda_1 & \partial\varepsilon_n/\partial\lambda_2 & \ldots & \partial\varepsilon_n/\partial\lambda_{n-1} & \partial\varepsilon_n/\partial\sigma
\end{vmatrix} \\
&= \begin{vmatrix}
1 & \ldots & \ldots & \ldots & 1 \\
\ldots & 1 & \ldots & \ldots & 1 \\
\vdots & \vdots & & \vdots & \vdots \\
\ldots & \ldots & \ldots & 1 & 1 \\
-1 & -1 & \ldots & -1 & 1
\end{vmatrix} \\
&= n.
\end{aligned}$$

121 Nr. 2096 122

§ 1. **Wahrscheinlichkeit eines Systems Verbesserungen $\lambda_1 \ldots \lambda_n$ auf's arithmetische Mittel.** In dem Ausdrucke (2) führen wir die Variablen $\lambda_1 \ldots \lambda_{n-1}$ und σ anstatt der ε nach den Gleichungen ein:

$$\varepsilon_1 = \lambda_1 + \sigma$$
$$\varepsilon_2 = \lambda_2 + \sigma$$
$$\cdot \quad \cdot \quad \cdot$$
$$\varepsilon_{n-1} = \lambda_{n-1} + \sigma$$
$$\varepsilon_n = -\lambda_1 - \lambda_2 \ldots - \lambda_{n-1} + \sigma.$$

Hierdurch werden die bekannten Beziehungen zwischen wahren Fehlern ε und Verbesserungen λ erfüllt, denn die Addition der Gleichungen giebt $n\sigma = [\varepsilon]$ und zugleich ist der Bedingung $[\lambda] =$ Null genügt. Die Functionaldeterminante der Substitution, nämlich die Determinante n-Grades

$$\begin{vmatrix} 1 & . & . & . & 1 \\ . & 1 & . & . & 1 \\ . & . & 1 & . & 1 \\ . & . & . & 1 & 1 \\ -1 & -1 & -1 & -1 & 1 \end{vmatrix} = n,$$

es geht daher der Ausdruck (2) über in

$$n\left(\frac{h}{\sqrt{\pi}}\right)^n e^{-h^2[\lambda\lambda] + h^2 n\sigma^2} d\lambda_1 . d\lambda_2 \ldots d\lambda_{n-1}\, d\sigma,$$

wobei $[\lambda\lambda] = \lambda_1^2 + \lambda_2^2 + \ldots + \lambda_n^2$; $\lambda_n = -\lambda_1 - \lambda_2 \ldots - \lambda_{n-1}$ gesetzt ist. Integriren wir nun über alle möglichen σ, so folgt weiter als **Wahrscheinlichkeit eines Systems** $\lambda_1 \ldots \lambda_n$ der Ausdruck

$$(3) \qquad \sqrt{n}\left(\frac{h}{\sqrt{\pi}}\right)^{n-1} e^{-h^2[\lambda\lambda]} d\lambda_1 \ldots d\lambda_{n-1}.$$

Zur Verification kann man denselben über alle möglichen $\lambda_1 \ldots \lambda_{n-1}$ integriren und wird dann gerade eins erhalten, wie es sein muss. Die Integration ist besonders bequem, wenn man zunächst $\lambda_1 + \lambda_2 + \ldots + \lambda_{n-1}$ zwischen s und $s + ds$, also constant annimmt. Mit Benutzung der Formel (9) auf Seite 116 erhält man dann, $k = n - 1$ gesetzt,

$$\sqrt{n}\, e^{-h^2 s^2} . \frac{h}{\sqrt{\pi(n-1)}} ds\, e^{-\frac{h^2 s^2}{n-1}},$$

welcher Ausdruck nach s zwischen $-\infty$ und $+\infty$ integrirt, eins ergiebt.

§ 2. **Günstigste Hypothese über h bei gegebenen Verbesserungen λ.** Wir setzen die Wahrscheinlichkeit einer Hypothese über h proportional dem Ausdruck (3), wenn ein System λ gegeben ist und die gewöhnliche Schlussweise ergiebt weiter als günstigste Annahme über h diejenige, für welche (3) ein Maximum wird. Die Differentiation zeigt an, dass dies eintritt für

$$\frac{1}{2h^2} = \frac{[\lambda\lambda]}{n-1}$$

und hiermit ist denn der erste Theil der Formel (1) für μ bewiesen *).

§ 3. **Wahrscheinlichkeit einer Summe $[\lambda\lambda]$ der Quadrate der Verbesserungen λ.** Die Wahrscheinlichkeit, dass $[\lambda\lambda]$ zwischen σ und $\sigma + d\sigma$ liegen werde, ist mit Benutzung von (3) gleich

$$\sqrt{n}\left(\frac{h}{\sqrt{\pi}}\right)^{n-1} \int d\lambda_1 \ldots \int d\lambda_{n-1}\, e^{-h^2[\lambda\lambda]}$$

integrirt über alle $\lambda_1 \ldots \lambda_{n-1}$, in der Weise, dass der Bedingung Genüge geschieht:

$$\sigma \leq [\lambda\lambda] \leq \sigma + d\sigma.$$

Wir führen nun $n - 1$ neue Variable t mittelst der Gleichungen ein:

$$t_1 = \sqrt{2}\,(\lambda_1 + \tfrac{1}{2}\lambda_2 + \tfrac{1}{2}\lambda_3 + \tfrac{1}{2}\lambda_4 \ldots + \tfrac{1}{2}\lambda_{n-1})$$
$$t_2 = \sqrt{\tfrac{3}{2}}\,(\lambda_2 + \tfrac{1}{3}\lambda_3 + \tfrac{1}{3}\lambda_4 \ldots + \tfrac{1}{3}\lambda_{n-1})$$
$$(4) \qquad t_3 = \sqrt{\tfrac{4}{3}}\,(\lambda_3 + \tfrac{1}{4}\lambda_4 \ldots + \tfrac{1}{4}\lambda_{n-1})$$
$$\cdot \quad \cdot \quad \cdot$$
$$t_{n-1} = \sqrt{\frac{n}{n-1}}\,\lambda_{n-1}$$

zu denen die Functionaldeterminante $\sqrt{n}$ gehört. Damit geht jener Ausdruck über in

$$\left(\frac{h}{\sqrt{\pi}}\right)^{n-1} \int dt_1 \ldots \int dt_{n-1}\, e^{-h^2[tt]},$$

wofür sich die Integrationsgrenzen bestimmen nach der Bedingung

$$\sigma \leq [tt] \leq \sigma + d\sigma.$$

Jetzt erkennt man aber, dass die **Wahrscheinlichkeit für die Quadratsumme der n Verbesserungen λ, $[\lambda\lambda] = \sigma$ gerade so gross ist, wie die Wahrscheinlichkeit, dass eine Quadratsumme [tt] von $n - 1$ wahren Fehlerquadraten t ist gleich σ.** Letztere habe ich in Schlömilch's Zeitschr. 1875 S. 303 angegeben und hat man darnach

$$(5) \qquad \frac{h^{n-1}}{\Gamma\left(\frac{n-1}{2}\right)} \sigma^{\frac{n-3}{2}} e^{-h^2\sigma} d\sigma$$

als **Wahrscheinlichkeit, dass die Quadratsumme $[\lambda\lambda]$ der n Verbesserungen λ von n directen gleichgenauen Beobachtungen auf ihr arithmetisches Mittel zwischen σ und $\sigma + d\sigma$ liege.** Zwischen $\sigma = 0$ und ∞ integrirt, ergiebt sich gerade eins.

§ 4. **Der mittlere Fehler der Formel $\mu = \sqrt{[\lambda\lambda] : (n-1)}$.** Da es schwer ist, eine allgemeine gültige Formel für den wahrscheinlichen Fehler dieser

*) Auf dieselbe Art kann man auch die Formel für μ^2 bei n vermittelnden Beobachtungen mit m Unbekannten streng nach der Wahrscheinlichkeitsrechnung ableiten, wie sich Verf. überzeugt hat und an anderer Stelle mittheilen wird.

FIGURE 3.10 The transformations (p. 122) as first used by Helmert in his paper "Die Genauigkeit der Formel von Peters zur Berechnung des wahrscheinlichen Beobachtungsfehler direkter Beobachtungen gleicher Genauigkeit" (Helmert, 1876a) to obtain the distribution of $[\lambda\ \lambda]$

Therefore, the joint density of $\lambda_1,\ldots,\lambda_{n-1},\sigma$ is

$$\begin{aligned} n\left(\frac{h}{\sqrt{\pi}}\right)^n \exp\left(-h^2\sum_{i=1}^{n}\varepsilon_i^2\right) &= n\left(\frac{h}{\sqrt{\pi}}\right)^n \exp\left\{-h^2\sum_{i=1}^{n}(\lambda_i+\sigma)^2\right\} \\ &= n\left(\frac{h}{\sqrt{\pi}}\right)^n \exp\left\{-h^2\left(\sum_{i=1}^{n}\lambda_i^2+2\sigma\sum_{i=1}^{n}\lambda_i+n\sigma^2\right)\right\} \\ &= n\left(\frac{h}{\sqrt{\pi}}\right)^n e^{-h^2[\lambda\lambda]-h^2n\sigma^2} \end{aligned} \tag{3.43}$$

(*ibidem*), since $\sum_i \lambda_i = [\lambda] = 0$. From Eq. (3.43), we note that the joint density can be written as $g_1(\lambda_1,\lambda_2,\ldots,\lambda_{n-1})g_2(\sigma)$ (where g_1 is a function of $\lambda_1,\lambda_2,\ldots,\lambda_{n-1}$ only and g_2 is a function of σ only) across a cross-product support. Under the assumption of normality, the deviations λ_i and the sample mean σ are therefore independent of each other. Since the standard deviation is a function of the residuals, Helmert had thus actually proved, without realizing it, that the standard deviation is independent of the sample mean when sampling from a normal distribution.

Continuing with the analysis, Helmert next determined the joint density of $\lambda_1,\lambda_2,\ldots,\lambda_{n-1}$. This can be done by integrating σ out of Eq. (3.43), resulting in

$$\begin{aligned} \int_{-\infty}^{\infty} n\left(\frac{h}{\sqrt{\pi}}\right)^n e^{-h^2[\lambda\lambda]-h^2n\sigma^2}d\sigma &= n\left(\frac{h}{\sqrt{\pi}}\right)^n e^{-h^2[\lambda\lambda]}\int_{-\infty}^{\infty} e^{-h^2n\sigma^2}d\sigma \\ &= n\left(\frac{h}{\sqrt{\pi}}\right)^n e^{-h^2[\lambda\lambda]}\cdot\frac{\sqrt{\pi}}{h\sqrt{n}} \\ &= \sqrt{n}\left(\frac{h}{\sqrt{\pi}}\right)^{n-1} e^{-h^2[\lambda\lambda]} \end{aligned} \tag{3.44}$$

(*ibidem*). Helmert's final major step in his aim to determine the distribution of $[\lambda\lambda]$ was to make the following classic transformations (Fig. 3.10)*:

$$\begin{aligned} t_1 &= \sqrt{2}\left(\lambda_1+\frac{1}{2}\lambda_2+\frac{1}{2}\lambda_3+\frac{1}{2}\lambda_4+\cdots+\frac{1}{2}\lambda_{n-1}\right) \\ t_2 &= \sqrt{\frac{3}{2}}\left(\lambda_2+\frac{1}{3}\lambda_3+\frac{1}{3}\lambda_4+\cdots+\frac{1}{3}\lambda_{n-1}\right), \\ t_3 &= \sqrt{\frac{4}{3}}\left(\lambda_3+\frac{1}{4}\lambda_4+\cdots+\frac{1}{4}\lambda_{n-1}\right), \\ &\cdots \\ t_i &= \sqrt{\frac{i+1}{i}}\left(\lambda_i+\frac{1}{i+1}\lambda_{i+1}+\cdots+\frac{1}{i+1}\lambda_{n-1}\right), \\ &\cdots \\ t_{n-1} &= \sqrt{\frac{n}{n-1}}\lambda_{n-1} \end{aligned} \tag{3.45}$$

* The rationale for these transformations is given in Sec. 3.4.4.

(*ibid*., p. 122). Now the joint density of $t_1, t_2, \ldots, t_{n-1}$ is obtained by dividing the joint density in Eq. (3.44) by the absolute value of the Jacobian $J\left(\frac{t_1, t_2, \ldots, t_{n-1}}{\lambda_1, \lambda_2, \ldots, \lambda_{n-1}}\right) = \sqrt{n}$, which gives

$$\left(\frac{h}{\sqrt{\pi}}\right)^{n-1} e^{-h^2[\lambda\lambda]}.$$

Helmert noted that $[\lambda\lambda] = \sum_{i=1}^{n} \lambda_i^2 = \sum_{i=1}^{n-1} t_i^2 = [tt]$, so that the above joint density of $t_1, t_2, \ldots, t_{n-1}$ becomes

$$\left(\frac{h}{\sqrt{\pi}}\right)^{n-1} e^{-h^2[tt]}.$$

Hence, the probability that $\sigma \le [tt] \le \sigma + d\sigma$ is given by

$$\left(\frac{h}{\sqrt{\pi}}\right)^{n-1} \int \cdots \int_G e^{-h^2[tt]} dt_1 dt_2 \ldots dt_{n-1}, \tag{3.46}$$

where G is the region $\{(t_1, t_2, \ldots, t_{n-1}) : \sigma \le [tt] \le \sigma + d\sigma\}$. Now, the probability in (3.46) is also the probability that $\sigma \le \sum_{i=1}^{n-1} \varepsilon_i^2 \le \sigma + d\sigma$, where ε_i is normally distributed with mean 0 and precision h. This probability can be obtained from Helmert's 1875 and 1876 papers (Helmert, 1875; 1876b) (see Eqs. 3.40 and 3.41) by suitable scaling and replacing n by $n-1$ and is

$$\frac{h^{n-1}}{\Gamma\left(\frac{n-1}{2}\right)} \sigma^{\frac{n-3}{2}} e^{-h^2\sigma}$$

(*ibidem*). Writing the above in terms of $v = 2h^2\sigma$ gives the χ_{n-1}^2-distribution.

Although Helmert was not the first to have derived the distribution of the variance and associated statistics, as we have shown, the distribution of the standard deviation is often called the Helmert distribution (e.g., Karl Pearson, 1931; Deming, 1950, p. 511). Others associate Helmert's name with the chi-squared distribution (e.g., Epstein, 2009, p. 29).

*3.4.4 Derivation of the Transformations Used by Helmert

We now come back to an issue that was previously skipped: why did we choose the particular transformations in Eq. (3.45)? A primary motivation is that from the normally distributed λ_i's ($i = 1, 2, \ldots, n$) in Eq. (3.42), we wish to construct the following: (i) new normally distributed random variables t_i's ($i = 1, 2, \ldots, n-1$) (ii) such that $\sum_{i=1}^{n} \lambda_i^2 = \sum_{i=1}^{n-1} t_i^2$. These are the two facts that enabled Helmert to recognize (3.46) as the distribution of the sum of squares of $n-1$ normal random variables (with zero mean). Let us see how the transformations can be obtained.* We first need to get rid of λ_n^2 in $\sum_{i=1}^{n}\lambda_i^2$. Remembering that $\sum_{i=1}^{n}\lambda_i=0$, we have

$$\lambda_n = -(\lambda_1+\lambda_2+\cdots+\lambda_{n-1})$$

so that

$$\begin{aligned}\lambda_n^2 &= (\lambda_1+\lambda_2+\cdots+\lambda_{n-1})^2\\ &= \lambda_1^2 + 2(\lambda_1\lambda_2+\lambda_1\lambda_3+\cdots+\lambda_1\lambda_{n-1})\\ &\quad + \lambda_2^2 + 2(\lambda_2\lambda_3+\lambda_2\lambda_4+\cdots+\lambda_2\lambda_{n-1})\\ &\quad + \cdots \qquad\qquad\qquad +\\ &\qquad\qquad\qquad\qquad\quad \vdots\\ &\qquad\qquad\qquad\qquad\quad +\lambda_{n-1}^2\end{aligned}$$

Therefore,

$$\begin{aligned}\sum_{i=1}^{n}\lambda_i^2 &= \sum_{i=1}^{n-1}\lambda_i^2 + \lambda_n^2\\ &= \sum_{i=1}^{n-1}\lambda_i^2 + \lambda_1^2 + 2(\lambda_1\lambda_2+\lambda_1\lambda_3+\cdots+\lambda_1\lambda_{n-1})\\ &\quad + \lambda_2^2 + 2(\lambda_2\lambda_3+\lambda_2\lambda_4+\cdots+\lambda_2\lambda_{n-1})\\ &\quad + \cdots \qquad\qquad\qquad +\\ &\qquad\qquad\qquad\qquad\quad \vdots\\ &\qquad\qquad\qquad\qquad\quad +\lambda_{n-1}^2\\ &= 2\big(\lambda_1^2+\lambda_1\lambda_2+\lambda_1\lambda_3+\cdots+\lambda_1\lambda_{n-1}\\ &\quad + \lambda_2^2+\lambda_2\lambda_3+\lambda_2\lambda_4+\cdots+\lambda_2\lambda_{n-1}\\ &\quad + \cdots \qquad\qquad\qquad +\\ &\qquad\qquad\qquad\qquad\quad \vdots\\ &\qquad\qquad\qquad\qquad\quad +\lambda_{n-1}^2\big)\end{aligned} \tag{3.47}$$

* See also Hald (1952, pp. 267–269).

The first row in the last equation above involves λ_1 and can be written as*

$$2\left(\lambda_1^2+\lambda_1\lambda_2+\lambda_1\lambda_3+\cdots+\lambda_1\lambda_{n-1}\right)=\left\{\sqrt{2}\left(\lambda_1+\frac{\lambda_2+\lambda_3+\cdots+\lambda_{n-1}}{2}\right)\right\}^2$$
$$-\frac{1}{2}\lambda_2^2-\lambda_2\lambda_3-\cdots-\lambda_2\lambda_{n-1}$$
$$-\frac{1}{2}\lambda_3^2-\lambda_3\lambda_4-\cdots-\lambda_3\lambda_{n-1} \tag{3.48}$$
$$\vdots$$
$$-\frac{1}{2}\lambda_{n-1}^2.$$

Eq. (3.47) then becomes

$$\sum_{i=1}^{n}\lambda_i^2=\left\{\sqrt{2}\left(\lambda_1+\frac{\lambda_2+\lambda_3+\cdots+\lambda_{n-1}}{2}\right)\right\}^2$$
$$+\frac{3}{2}\lambda_2^2+\lambda_2\lambda_3+\lambda_2\lambda_4+\cdots+\lambda_2\lambda_{n-1}$$
$$+\cdots \qquad +$$
$$\vdots$$
$$+\frac{3}{2}\lambda_{n-1}^2.$$

If we similarly "complete the squares" for λ_2 in the second row of the above, we can obtain an expression like we did in (3.48). The square part of that expression is

$$\left\{\sqrt{\frac{3}{2}}\left(\lambda_2+\frac{\lambda_3+\lambda_4+\cdots+\lambda_{n-1}}{3}\right)\right\}^2.$$

If we keep on completing the squares in this way, we eventually reach

$$\sum_{i=1}^{n}\lambda_i^2=t_1^2+t_2^2+\cdots+t_{n-1}^2,$$

* Compare what follows to the "completion of squares": $2(x^2+ax)\equiv\left\{\sqrt{2}(x+a/2)\right\}^2-a^2/2$.

where

$$t_1 = \sqrt{2}\left(\lambda_1 + \frac{\lambda_2 + \lambda_3 + \cdots + \lambda_{n-1}}{2}\right),$$

$$t_2 = \sqrt{\frac{3}{2}}\left(\lambda_2 + \frac{\lambda_3 + \lambda_4 + \cdots + \lambda_{n-1}}{3}\right),$$

$$\cdots$$

$$t_i = \sqrt{\frac{i+1}{i}}\left(\lambda_i + \frac{\lambda_{i+1} + \lambda_{i+2} + \cdots + \lambda_{n-1}}{i+1}\right),$$

$$\cdots$$

$$t_{n-1} = \sqrt{\frac{n}{n-1}}\lambda_{n-1}.$$

These are exactly the same transformations Helmert used in (3.45).

4

STUDENT'S *t*

4.1 WILLIAM SEALY GOSSET (1876–1937)

Very few achievements in statistics have been as momentous as Student's discovery of the *t*-distribution. Student's real name was William Sealy Gosset (Fig. 4.1). Gosset* was born in Canterbury, England, on June 13, 1876, and was the eldest of five children. His father was a colonel in the Royal Engineers, but the family was of quite modest means. However, Gosset was able to secure scholarships at Winchester College in 1889–1895 and later at New College, Oxford. He obtained a first in mathematical moderations in 1897 and a first-class degree in chemistry in 1899. In the same year, Gosset joined the brewers Arthur Guinness and Sons in Dublin as an Oxford-trained chemist. There, he became interested in the application of statistics, especially those of small samples, to brewing problems. During his work at Guinness, Gosset submitted a report entitled "The application of the 'Law of Errors' to the work of the brewery."

Gosset married Marjory Surtees Phillpotts in Tunbridge Wells on January 16, 1906. In that same year, while still an employee of Guinness, he was sent to work under Karl Pearson at University College London. Soon after, Gosset made his most memorable contribution in the 1908 paper "The Probable Error of a Mean" (Student, 1908b) (Fig. 4.2). The 1908 paper, as most of Gosset's statistical work, was the result of practical small-sample problems arising in the brewery. In his papers, Gosset used the pseudonym "Student." Two possible reasons have been advanced for this. A common one is that Guinness wanted to keep it a secret that one of their workers was writing scientific papers. According to a less common explanation, Karl Pearson suggested the pseudonym because he did not want it to be known that a paper in his journal *Biometrika* had been written by a brewer.

*Gosset's statistical biography has been written by Pearson et al. (1990). Other biographies include Hacking (1972), Fienberg and Lazar (2001, pp. 312–317), and Pearson (1939).

Classic Topics on the History of Modern Mathematical Statistics: From Laplace to More Recent Times, First Edition. Prakash Gorroochurn.

FIGURE 4.1 William Sealy Gosset ("Student") (1876–1937). Wikimedia Commons (Public Domain), http://commons.wikimedia.org/wiki/File:William_Sealy_Gosset.jpg

VOLUME VI MARCH, 1908 No. 1

BIOMETRIKA.

THE PROBABLE ERROR OF A MEAN.

BY STUDENT.

Introduction.

ANY experiment may be regarded as forming an individual of a "population" of experiments which might be performed under the same conditions. A series of experiments is a sample drawn from this population.

Now any series of experiments is only of value in so far as it enables us to form a judgment as to the statistical constants of the population to which the experiments belong. In a great number of cases the question finally turns on the value of a mean, either directly, or as the mean difference between the two quantities.

If the number of experiments be very large, we may have precise information as to the value of the mean, but if our sample be small, we have two sources of uncertainty:—(1) owing to the "error of random sampling" the mean of our series of experiments deviates more or less widely from the mean of the population, and (2) the sample is not sufficiently large to determine what is the law of distribution of individuals. It is usual, however, to assume a normal distribution, because, in a very large number of cases, this gives an approximation so close that a small sample will give no real information as to the manner in which the population deviates from normality: since some law of distribution must be assumed it is better to work with a curve whose area and ordinates are tabled, and whose properties are well known. This assumption is accordingly made in the present paper, so that its conclusions are not strictly applicable to populations known not to be normally distributed; yet it appears probable that the deviation from normality must be very extreme to lead to serious error. We are concerned here solely with the first of these two sources of uncertainty.

The usual method of determining the probability that the mean of the population lies within a given distance of the mean of the sample, is to assume a normal distribution about the mean of the sample with a standard deviation equal to $s/\sqrt{n}$, where s is the standard deviation of the sample, and to use the tables of the probability integral.

FIGURE 4.2 First page of Student's 1908 paper. First page in Student (1908b). Courtesy of Oxford University Press

Over the next years, Student interacted with the leading statisticians of the time, including Karl Pearson, R.A. Fisher, Egon Pearson, and Jerzy Neyman. While the latter were at various stages embroiled in several controversies, Student made sure to keep a cordial relationship with all of them. In particular, Student had a formidable ally in Fisher, who would always point out the groundbreaking nature of his work on small samples. However, that did not prevent Student to clash with Fisher in the final years of Student's life. The issue was over the merits of systematic experimental plans, which Student favored, over randomization. Another key fact in Student's life, one which is not often mentioned, is that he played an important part in the early formulation of the Neyman–Pearson theory of hypothesis testing.

In 1935, Student left Dublin to take up the position of head brewer at the new Guinness brewery at Park Royal in northwestern London. He passed away in Beaconsfield, England, on October 16, 1937, from a heart attack.

4.2 ORIGIN OF STUDENT'S TEST: THE 1908 PAPER

In the landmark 1908 paper "The Probable Error of a Mean",* Student announced his aim as follows:

> The usual method of determining the probability that the mean of the population lies within a given distance of the mean of the sample, is to assume a normal distribution about the mean of the sample with a standard deviation equal to $s/\sqrt{n}$, where s is the standard deviation of the sample, and to use the tables of the [normal] probability integral.
>
> But, as we decrease the number of experiments, the value of the standard deviation as found from the sample of experiments becomes itself subject to an increasing error, until judgments reached in this way may become altogether misleading.
>
> ...
>
> The aim of the present paper is to determine the point at which we may use the tables of the probability integral in judging of the significance of the mean of a series of experiments, and to furnish alternative tables for use when the number of experiments is too few. (Student, 1908b, pp. 1–2)

Student's point was that the customary practice of calculating the standard error of the mean through the formula $\text{SE} = s/\sqrt{n}$ and then assigning a normal distribution to $(\overline{x} - \mu)/\text{SE}$ (where $\overline{x}$ and μ are the sample and population means, respectively) was not accurate when dealing with small sample sizes. This is because when n is small, there is extra uncertainty introduced by having to estimate the population standard deviation σ through the standard deviation s, and this is not being accounted for through the conventional formulas. Student therefore set out to determine the "exact" distribution of

$$z = \frac{\overline{x}}{s}, \tag{4.1}$$

* Student's paper is also described in Lehmann (1992b) and Welch (1958).

where

$$s=\sqrt{\frac{\sum_{i=1}^{n}\left(x_i-\bar{x}\right)^2}{n}}, \tag{4.2}$$

and $x_1, x_2, \ldots, x_n$ are assumed to be $N(0,\sigma^2)$.*

Student's first step was the derivation of the distribution of s, but as we shall see, his derivation was heuristic at best. Using $\mathsf{E}s^{2i}=M_i'$ and $\mathsf{E}x^i=\mu_i$, he obtained the first four raw moments of s^2:

$$\begin{aligned}
M_1' &= \mu_2\frac{(n-1)}{n},\\
M_2' &= \mu_2^2\frac{(n-1)(n+1)}{n^2},\\
M_3' &= \mu_2^3\frac{(n-1)(n+1)(n+3)}{n^3},\\
M_4' &= \mu_2^4\frac{(n-1)(n+1)(n+3)(n+5)}{n^4}.
\end{aligned} \tag{4.3}$$

The centered moments of s^2 are given by $\mathsf{E}\left(s^2-\mathsf{E}s^2\right)^i=M_i$, where

$$\begin{aligned}
M_1 &= 0,\\
M_2 &= 2\mu_2^2\frac{(n-1)}{n^2},\\
M_3 &= 8\mu_2^3\frac{(n-1)}{n^3},\\
M_4 &= \frac{12\mu_2^4(n-1)(n+3)}{n^4}.
\end{aligned}$$

From the above, Student calculated Pearson's coefficients of skewness and kurtosis (for the distribution of s^2) respectively as

$$\beta_1=\frac{M_3^2}{M_2^3}=\frac{8}{n-1},\quad \beta_2=\frac{M_4}{M_2^2}=\frac{3(n+3)}{n-1},$$

so that

$$2\beta_2-3\beta_1-6=0.$$

But this is exactly the condition satisfied by a Pearson type III curve (see Table 5.1) and

> …[c]onsequently a curve of Professor Pearson Type III. may be expected to fit the distribution of s^2. (*ibid.*, p. 4)

* Note that the symbol z is different from what we use today, but more on this in Sec. 4.3.2.

Thereupon, Student set the density of s^2 evaluated at x in the form

$$y = Cx^p \exp(-\gamma x),$$

where

$$\gamma = 2\frac{M_2}{M_3} = \frac{n}{2\mu_2}, \quad \text{and} \quad p = \frac{4}{\beta_1} - 1 = \frac{n-3}{2}.$$

Therefore, the density of s^2 evaluated at x becomes

$$y = Cx^{(n-3)/2} \exp\left(-\frac{nx}{2\mu_2}\right) \tag{4.4}$$

(*ibidem*). Student fully acknowledged that he had no actual proof of the above distribution:

> Hence it is probable that the curve found [Eq. (4.4)] represents the theoretical distribution of s^2; so that although we have no actual proof we shall assume it to do so in what follows. (*ibid.*, p. 5)

From Eq. (4.4), he also obtained the density of $s\left[=\sqrt{x}\right]$ as

$$\begin{aligned} y_2 &= y\frac{dx}{ds} \\ &= Cx^{(n-3)/2} \exp\left(-\frac{nx}{2\mu_2}\right) \cdot 2s \\ &= 2Cs^{n-2} \exp\left(-\frac{ns^2}{2\mu_2}\right). \end{aligned}$$

By evaluating the constant C through integration, Student was finally able to write the density of s evaluated at x in the form

$$y = \begin{cases} \dfrac{1}{(n-3)(n-5)\cdots(3)(1)}\sqrt{\dfrac{2}{\pi}}\left(\dfrac{n}{\sigma^2}\right)^{(n-1)/2} x^{n-2} \exp\left(-\dfrac{nx^2}{2\sigma^2}\right) & (n \text{ even}) \\ \dfrac{1}{(n-3)(n-5)\cdots(4)(2)}\left(\dfrac{n}{\sigma^2}\right)^{(n-1)/2} x^{n-2} \exp\left(-\dfrac{nx^2}{2\sigma^2}\right) & (n \text{ odd}) \end{cases} \tag{4.5}$$

(*ibid.*, p. 6), where $\sigma^2 = \mu_2$.

In the second step of his derivation of the exact distribution of z in (4.1), Student announced in Sec. II of his paper that he wished

> ...[t]o show that there is no correlation between (a) the distance of the mean of a sample from the mean of the population and (b) the standard deviation of a sample with normal distribution. (*ibidem*)

Student noted that it was obvious that there would be no correlation between $|\bar{x}|$ and s (since "positive and negative positions of the mean of the sample are equally likely") but that there *could* be a nonzero correlation between $\bar{x}^2$ and s^2. To verify this, he set

$$u^2 = \left(\frac{\sum x_i}{n}\right)^2,$$

$$s^2 = \frac{\sum x_i^2}{n} - \left(\frac{\sum x_i}{n}\right)^2,$$

u and s^2 being the sample mean and sample variance, respectively. By using $M_1' = \mathsf{E}s^2$ and $m_1' = \mathsf{E}u^2$, he wrote from Eq. (4.3)

$$M_1' = \mu_2 \frac{(n-1)}{n}$$

and also, since $\mathsf{E}x_i$ is assumed to be zero,

$$\begin{aligned} m'_1 &= \mathsf{E}\bar{x}^2 \\ &= \operatorname{var} \bar{x} \\ &= \frac{\mu_2}{n}. \end{aligned}$$

Now,

$$\begin{aligned} u^2 s^2 &= \left(\frac{\sum x_i}{n}\right)^2 \left\{\frac{\sum x_i^2}{n} - \left(\frac{\sum x_i}{n}\right)^2\right\} \\ &= \left(\frac{\sum x_i}{n}\right)^2 \left(\frac{\sum x_i^2}{n}\right) - \left(\frac{\sum x_i}{n}\right)^4 \\ &= \frac{\sum x_i^2 + 2\sum x_i x_j}{n^2}\left(\frac{\sum x_i^2}{n}\right) - \frac{\sum x_i^4 + 6\sum x_i^2 x_j^2}{n^4} + O \\ &= \frac{\left(\sum x_i^2\right)^2}{n^3} + 2\frac{\sum x_i x_j \sum x_i^2}{n^3} - \frac{\sum x_i^4}{n^4} - \frac{6\sum x_i^2 x_j^2}{n^4} + O. \end{aligned}$$

In the above, O denotes "other terms of odd order which will vanish on summation (i.e. on taking expectations)." Taking expectations on both sides of the above,

$$\mathsf{E}u^2 s^2 = \frac{\mathsf{E}\left(\sum x_i^2\right)^2}{n^3} + 2\frac{\mathsf{E}\sum x_i x_j \sum x_i^2}{n^3} - \frac{\mathsf{E}\sum x_i^4}{n^4} - \frac{6\mathsf{E}\sum x_i^2 x_j^2}{n^4} + \mathsf{E}O$$

$$\operatorname{cov}\left(u^2, s^2\right) + \mathsf{E}u^2 \mathsf{E}s^2 = \frac{n\mathsf{E}x_i^4 + n(n-1)\left(\mathsf{E}x_i^2\right)^2}{n^3} + 2(0) - \frac{n\mathsf{E}x_i^4}{n^4} - \frac{6\cdot\frac{1}{2}n(n-1)\mathsf{E}\left(x_i^2\right)^2}{n^4} + 0$$

$$R_{u^2s^2}\sqrt{\operatorname{var}u^2}\sqrt{\operatorname{var}s^2}+\frac{\mu_2}{n}\cdot\mu_2\frac{(n-1)}{n}=\frac{\mu_4}{n^2}+\mu_2^2\frac{(n-1)}{n^2}-\frac{\mu_4}{n^3}-3\mu_2^2\frac{(n-1)}{n^3}$$

$$=\frac{3\mu_2^2}{n^3}(n-1)+\frac{\mu_2^2(n-1)}{n^3}(n-3)$$

$$=\frac{\mu_2^2(n-1)}{n^2}$$

$$\therefore\quad R_{u^2s^2}=0$$

(*ibid*., p. 7). In the above, $R_{u^2s^2}$ denotes the correlation between u^2 and s^2, and we have made use of the fact that, for a normal distribution, $\mu_4=3\mu_2^2$.

Having shown that $R_{u^2s^2}=0$, Student took it to imply that s^2 is independent of $\bar{x}^2$ and hence also of $\bar{x}$. The implication fortunately turns out to be true in this case (of a normal distribution for each x_i) but is false in general.*

In the third and final step of Student's derivation of the exact distribution of $z=\bar{x}/s$ in (4.1), in Sec. III of his paper, he first recalled that each x_i ($i=1, 2, \ldots, n$) had a $N(0,\sigma^2)$ distribution. Student then quoted G.B. Airy's book *Theory of Errors of Observations* (Airy, 1861) and wrote the density function of $\bar{x}$ evaluated at x as

$$\frac{\sqrt{n}}{\sigma\sqrt{2\pi}}\exp\left(-\frac{nx^2}{2\sigma^2}\right)^{\dagger}$$

(*ibid*., p. 7). Because of the independence between $\bar{x}$ and s^2,

$$f_{z|s}(z\mid s)=f_{\bar{X}}(\bar{x})\left|\frac{\partial\bar{x}}{\partial z}\right|=f_{\bar{X}}(zs)s$$

so that the conditional density of z for a given s is

$$\frac{s\sqrt{n}}{\sigma\sqrt{2\pi}}\exp\left(-\frac{ns^2z^2}{2\sigma^2}\right).$$

Now, Student wrote the density of s in (4.5) as

$$\frac{\dfrac{C}{\sigma^{n-1}}s^{n-2}\exp\left(-\dfrac{ns^2}{2\sigma^2}\right)}{\displaystyle\int_0^\infty\frac{C}{\sigma^{n-1}}s^{n-2}\exp\left(-\frac{ns^2}{2\sigma^2}\right)ds}.$$

*Neyman reported that even Karl Pearson did not understand the difference between zero correlation and independence in as late as 1925 (Reid, 1998, p. 57). As an example, if $Z\sim N(0,1)$, then $R_{Z,Z^2}=0$ but Z and Z^2 are obviously not independent.

† The reader needs to be cautious here because Student used the same letter y for different distributions.

Therefore, the joint density of z and s can be obtained by multiplying the last two distributions:

$$\frac{\dfrac{C}{\sigma_n}\sqrt{\dfrac{n}{2\pi}}s^{n-1}\exp\left\{-\dfrac{ns^2(1+z^2)}{2\sigma^2}\right\}}{\int_0^\infty \dfrac{C}{\sigma^{n-1}}s^{n-2}\exp\left(-\dfrac{ns^2}{2\sigma^2}\right)ds}=\frac{\sqrt{\dfrac{n}{2\pi}}s^{n-1}\exp\left\{-\dfrac{ns^2(1+z^2)}{2\sigma^2}\right\}}{\sigma\int_0^\infty s^{n-2}\exp\left(-\dfrac{ns^2}{2\sigma^2}\right)ds}.$$

The density of z can now be obtained by integrating s out, resulting in

$$\frac{\sqrt{\dfrac{n}{2\pi}}}{\sigma}\frac{\int_0^\infty s^{n-1}\exp\left\{-\dfrac{ns^2(1+z^2)}{2\sigma^2}\right\}ds}{\int_0^\infty s^{n-2}\exp\left(-\dfrac{ns^2}{2\sigma^2}\right)ds}.$$

By using previous calculations (*ibid*., pp. 5–6), Student was finally able to write the density of z as

$$y=\begin{cases}\dfrac{1}{2}\dfrac{n-2}{n-3}\cdot\dfrac{n-4}{n-5}\cdots\dfrac{5}{4}\cdot\dfrac{3}{4}(1+z^2)^{-n/2} & (n\text{ odd})\\[2ex] \dfrac{1}{\pi}\dfrac{n-2}{n-3}\cdot\dfrac{n-4}{n-5}\cdots\dfrac{4}{3}\cdot\dfrac{1}{2}(1+z^2)^{-n/2} & (n\text{ even})\end{cases} \tag{4.6}$$

(*ibid*., p. 8). Nowadays, this is written as

$$f(z)=\frac{\Gamma\left(\dfrac{n}{2}\right)}{\Gamma\left(\dfrac{n-1}{2}\right)\sqrt{\pi}}(1+z^2)^{-n/2},\quad -\infty<z<\infty, n=2,3,\ldots. \tag{4.7}$$

Student then rightly noted:

> Since this equation is independent of s it will give the distribution of the distance of the mean of a sample from the mean of the population expressed in terms of the standard deviation of the sample for any normal population. (*ibidem*)

Now that the distribution of z had been obtained, Student could further integrate it to obtain, for a given n, the areas to the left of different values of z. In his 1908 paper (*ibid*., p. 19), Student gave table values of these areas for z=.1 (.1) 3.0 and n=4 (1) 10. These could now be used in assessing how extreme an observed value of z (for a given n) was.

This was precisely what Student set out to do in the later part of his paper, through three illustrations. We shall examine the first illustration because it was the one that was later used by R.A. Fisher (who will soon be discussed) in his landmark book *Statistical*

TABLE 4.1 Hours of Sleep Gained by the Use of Hyoscyamine Hydrobromide

Patient	Dextro-	Laevo-	Difference (Dextro- Minus Laevo-)
1	+.7	+1.9	+1.2
2	−1.6	+.8	+2.4
3	−.2	+1.1	+1.3
4	−1.2	+.1	+1.3
5	−1	−.1	0
6	+3.4	+4.4	+1.0
7	+3.7	+5.5	+1.8
8	+.8	+1.6	+.8
9	0	+4.6	+4.6
10	+2.0	+3.4	+1.4
	Mean = +.75	Mean = +2.33	Mean = +1.58
	SD = 1.70	SD = 1.90	SD = 1.17

From Student (1908b, p. 20).

Methods for Research Workers (Fisher, 1925b, p. 108). Student used data from Cushny and Peebles (1905) relating to the different effects of the optical isomers of hyoscyamine in producing sleep. The sleep of 10 patients was measured, first without hypnotic, then after treatment with (i) D. hyoscyamine hydrobromide (Dextro-), (ii) L. hyoscyamine hydrobromide (Laevo-). The number of hours of sleep gained by each patient is shown in Table 4.1.

Student then performed three statistical tests:

1. Student first tested whether the mean hours of sleep gained by "Dextro-" was nil. The observed $z = .75/1.70 = .44$ and $\Pr\{z > .44\} = .113$. The latter probability was obtained from his table of the Z-distribution "for ten experiments."* Student concluded: "It is then very likely that 1 ['Dextro-'] gives an increase of sleep, but would occasion no surprise if the results were reversed by further experiments." In modern language, Student could not reject the null hypotheses of no effect.
2. Student then tested whether the mean hours of sleep gained by "Laevo-" was nil. The observed $z = 2.33/1.90 = 1.23$ and $\Pr\{z > 1.23\} = .0026$. Student now concluded: "… the odds are 400 to 1 that such is the case ['Laevo-' is actually a soporific]."
3. Student finally tested the real hypothesis of the experiment, namely, the difference between "Dextro-" and "Laevo-" was nil. For this, Student used the fourth column of Table 2.1. Here Student obtained an observed $z = 1.58/1.17 = 1.35$ and $\Pr\{z > 1.35\} = .0015$. Student concluded: "Of course odds of this kind make it

*The reader can see that Student used the incorrect degrees of freedom (the correct degrees of freedom should have been $n-1=9$) but the latter concept had not yet been developed in 1908 and had to wait for Fisher in 1922 (Fisher, 1922b). Furthermore, Student used the divisor $n = 10$ (instead of $n-1=9$) in calculating s. Both of these shortcoming were later corrected by Fisher in his book *Statistical Methods for Research Workers* (Fisher, 1925b, p. 108).

almost certain that 2 ['Laevo-'] is the better soporific, and in practical life such a high probability is in most matters considered as a certainty."*

From Student's interpretations above, we note that he was reasoning in terms of probabilities of hypotheses and thus his derivation was actually based on inverse probability.

In any case, Student had ushered in a new era of statistics, namely, that of small-sample or exact statistics (i.e., statistics that did not rely on asymptotic arguments). Although his work initially met with little interest, he was soon to find a formidable ally, whom we shall describe next.

4.3 FURTHER DEVELOPMENTS

4.3.1 Fisher's Geometrical Derivation of 1923

Student's pathbreaking work found a champion in none other than R.A. Fisher. As early as 1912, while still an undergraduate student, Fisher sent a geometrical proof for the distribution of z to Student. The latter distribution was given in Fisher's 1915 paper on the distribution of the correlation coefficient (Fisher, 1915, p. 518)† where Fisher noted that he had obtained "curves [that] are identical with those found by 'Student' for z." In 1923, Fisher published the geometrical proof of the distribution of z (Fisher, 1923a). The following sentence, taken from the 1923 paper, shows the high regard Fisher had for Student's pioneering contributions:

> Student's work is so fundamental from the theoretical standpoint, and has so direct a bearing on the practical conclusions to be drawn from small samples, that it deserves to be far more widely known than it is at present. (Fisher, 1923a, p. 655)

However, Fisher was also keenly aware of the gaps in Student's derivation. In a paper written much later, he remarked:

> As it was, he [Student] satisfied himself with showing somewhat laboriously that the distribution of s^2 was uncorrelated both with $\bar{x}$ and with $\bar{x}^2$. This was the most striking gap in his argument, for, in truth it was not merely the distribution of s..., but the exact simultaneous distribution of s and $\bar{x}$ that "Student" needed to develop his test. (Fisher, 1939a, p. 3)

Fisher's first task was to determine the joint distribution of $\bar{x}$ and s for a sample $x_1, x_2, \ldots, x_n$ from a $N(m, \sigma^2)$ distribution, where

$$s = \sqrt{\frac{\sum_{i=1}^{n}(x_i - \bar{x})^2}{n}}.$$

*Not too long ago, Senn and Richardson (1994) stirred a little controversy by pointing out that Student had incorrectly labeled the data from Cushny and Peebles (1905) and that Student's (and later Fisher's) interpretation of the experiment was wrong. However, the *statistical principles* that Student wanted to illustrate in his 1908 paper remain valid (except for his incorrect degrees of freedom and incorrect divisor for s).

† See Section 2.3.9.

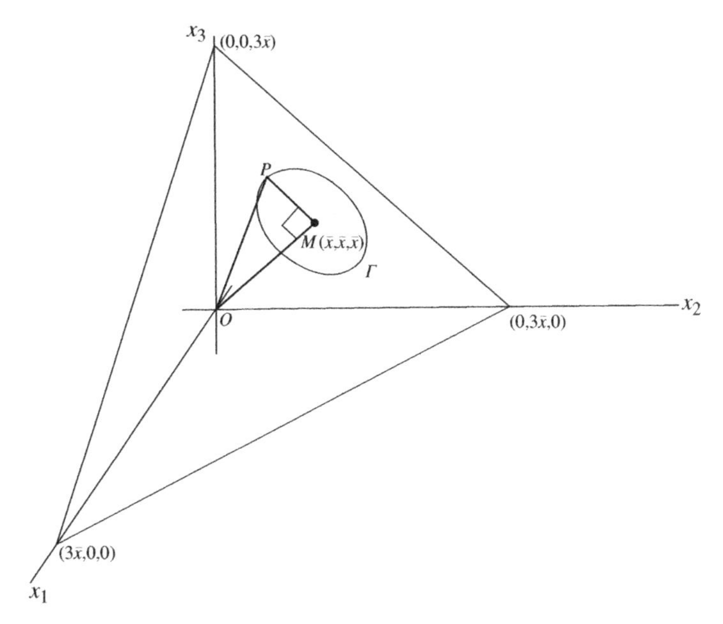

FIGURE 4.3 Fisher's geometric derivation of the joint distribution of $\bar{x}$ and s

To achieve this, he used geometric arguments, as we now explain.*

Fisher considered the hyperplane $x_1 + x_2 + \cdots + x_n = n\bar{x}$ that intersects the n-dimensional hypersphere $(x_1 - \bar{x})^2 + (x_2 - \bar{x})^2 + \cdots + (x_n - \bar{x})^2 = ns^2$ in an $(n-1)$-dimensional hypersphere Γ with center $M(\bar{x}, \bar{x}, ..., \bar{x})$ and radius $s\sqrt{n}$ (Fig. 4.3).

The "area" of Γ is proportional to $\left(s\sqrt{n}\right)^{n-1}$ and the element of volume is

$$\begin{aligned} dx_1 dx_2 \ldots dx_n &\propto d\bar{x} d\left(s\sqrt{n}\right)^{n-1} \\ &\propto s^{n-2} d\bar{x} ds. \end{aligned}$$

The joint density of $\bar{x}$ and s is $f(\bar{x}, s)$, where

$$\begin{aligned} f(\bar{x},s)\, d\bar{x} ds &= f(x_1, x_2, \ldots, x_n)\, dx_1 dx_2 \ldots dx_n \\ &\propto \exp\left\{-\frac{\Sigma(x_i - m)^2}{2\sigma^2}\right\} s^{n-2} d\bar{x} ds \\ &= \exp\left[-\frac{\Sigma\{(x_i - \bar{x}) + (\bar{x} - m)\}^2}{2\sigma^2}\right] s^{n-2} d\bar{x} ds \qquad (4.8) \\ &= \exp\left\{-\frac{ns^2 + n(\bar{x} - m)^2}{2\sigma^2}\right\} s^{n-2} d\bar{x}\, ds \end{aligned}$$

* The reader should also consult Section 2.3.9, where Fisher's geometric arguments are first explained in the context of finding the exact distribution of r.

(Fisher, 1923a, p. 656). Since this joint density is factorizable into functions of $\bar{x}$ and s across a cross-product support, Fisher deduced that two random variables $\bar{x}$ and s are independent of each other and wrote their respective densities as

$$\frac{\sqrt{n}}{\sigma\sqrt{2\pi}}\exp\left\{-\frac{n}{2\sigma^2}(\bar{x}-m)^2\right\}$$

and

$$\frac{n^{(n-1)/2}}{2^{(n-3)/2}\Gamma\left(\dfrac{n-1}{2}\right)}\frac{s^{n-2}}{\sigma^{n-1}}\exp\left(-\frac{ns^2}{2\sigma^2}\right).$$

Fisher's next task was to determine the density of the ratio

$$z=\frac{\bar{x}-m}{s}.$$

First, he found the joint density of z and s. By defining the transformations $z=(\bar{x}-m)/s$ and $s=s$, it is seen that the latter joint density is the product of the joint density of $\bar{x}$ and s and the absolute value of the Jacobian ($=s$) of the transformations. Thus, the joint density of z and s is

$$f(\bar{x},s)s=f(zs+m,s)s=\frac{n^{n/2}}{2^{(n-1)/2}\sqrt{\pi}\Gamma\left(\dfrac{n-1}{2}\right)}\cdot\frac{s^{n-1}}{\sigma^n}\exp\left\{-\frac{ns^2}{2\sigma^2}(1+z^2)\right\}.$$

By integrating s out of the above, Fisher finally obtained the density of z as

$$\frac{\Gamma\left(\dfrac{n}{2}\right)}{\Gamma\left(\dfrac{n-1}{2}\right)\sqrt{\pi}}\cdot\frac{1}{(1+z^2)^{n/2}}$$

(*ibid*., p. 657), which is the same as Student's (4.6).

4.3.2 From Student's z to Student's t

The reader will have noticed that Student's use of z and s in Eqs. (4.1) and (4.2) is different from standard usage. Nowadays, we would write

$$t=\frac{\bar{x}-\mu}{s'/\sqrt{n}},\tag{4.9}$$

where

$$s'=\sqrt{\frac{\sum_{i=1}^{n}(x_i-\bar{x})^2}{n-1}}.\tag{4.10}$$

Student himself in 1908 used*

$$z = \frac{\bar{x} - \mu}{s},$$

where

$$s = \sqrt{\frac{\sum_{i=1}^{n}(x_i - \bar{x})^2}{n}}.$$

Comparing Student's z with t, it is seen that

$$t = z\sqrt{n-1}. \tag{4.11}$$

An interesting question is, "How was the transition from Student's early z to today's t made?" Eisenhart (1979) and Box (1978, Chapter 5) have both provided details of this transition, as we now describe. Following Student's pioneering work (Student, 1908a; 1908b) and Soper's extension in 1913 (Soper, 1913) (cf. end of Section 2.3.4), Fisher was able to derive the exact distribution of the sample correlation r in 1915 (Fisher, 1915) (cf. Section 2.3.9) and show that, for $\rho = 0$, the transformation $r/\sqrt{1-r^2}$ had a distribution "identical with those found by 'Student' for z [with n replaced by $n-1$]."†
Fisher was aware that more theoretical work was needed, for he stated in his 1922 foundations paper that there exists an

> absence of investigation of other important statistics, such as the regression coefficients, multiple correlations, and the correlation ratio. (Fisher, 1922c, p. 315)

The necessary prodding for Fisher to undertake this investigation himself came from none other than Student. In two letters written on April 1922, Student stated:

> But seriously I want to know the frequency distribution of $r(s_x/s_y)$ for small samples, in my work I want that more than the r distribution now happily solved. (Gosset, 1970, Letter 5)

> I forgot to put up another problem to you in my last letter, that of the prob. error of partial $\left\{\begin{matrix}\text{correlation}\\ \text{regression}\end{matrix}\right\}$ coefficients for small samples. (*ibid.*, Letter 6)

* See p. 351

† That is, $r/\sqrt{1-r^2}$ has a distribution proportional to Student's t-distribution with $(n-2)$ degrees of freedom. Note: (i) The concept of degrees of freedom was not yet known in 1915. (ii) More precisely, $\sqrt{n-2}\left(r/\sqrt{1-r^2}\right) \sim t_{n-2}$.

Fisher responded within a few days and showed that the distribution of the sample regression coefficients could be expressed in terms of Student's z-distribution.* These results were promptly published by Fisher (1922d). In his review, Eisenhart has added that:

> It is clear that Fisher had already reached his unified treatment of tests of significance of a mean, of the difference between two means, and of simple and partial coefficients of correlation and regression in terms of "Student's" distribution; and may have already decided that it would be advantageous to change from z to
>
> $$t = z\sqrt{\nu}$$
>
> where ν is the "number of degrees of freedom" .... (Eisenhart, 1979)

Having introduced the pathbreaking concept of degrees of freedom in 1922 itself (cf. Section 3.3.3), Fisher had thus realized it is the available degrees of freedom, not the sample size, that determined where Student's tables had to be entered. Now the latter tables were based on the sample size n and went only up to $n=10$, since they relied on calculations that became increasingly cumbersome with n. Although these tables were extended for larger n by Student in 1917 (Student, 1917), they still suffered from computational difficulties. Therefore, Fisher proposed a new tabulation based on a simpler computational algorithm. In her review, Box has further explained that:

> Two changes were introduced in this tabulation. First, the number of observations, n', for which the table was entered was changed to the number of degrees of freedom, $n=n'-1$, which was the more appropriate number. Second, the quantity t now tabulated was applicable directly to the ratio of a normally distributed quantity to its estimated standard error. It was, therefore, a more natural quantity to use than that originally tabulated by Student (that is, $t/\sqrt{n}$); moreover, it lent itself much more readily to tabulation. (Box, 1978, p. 115)

These new tables were published by Student in 1925 (Student, 1925) and correspond to the ones that are used today.

Although in 1922 Fisher found it more convenient to use the t-form of Student's statistic, he apparently used the symbol x. The symbol t was first used by Student in 1925, in a letter to Fisher:

> I have been working a little at the Type VII.... In the course of that study I calculated all the values for $t=1$ from $n=2$ to $n=30$ to seven places (accurate to 6 places). (Gosset, 1970, Letter 13)

Therefore, in summary:

> ...the decision to shift from the z- to the t-form originated with Fisher, but the choice of the letter "t" to denote the new form was due to "Student." (Eisenhart, 1979)

As for the first "public appearance" of Student's t, this took place in Fisher's widely influential book *Statistical Methods for Research Workers* (Fisher, 1925b, p. 22). A unified

*Concerning the second question, Fisher showed that the distribution of a partial correlation coefficient is the same as that of the corresponding total correlation coefficient for a sample size equal to n minus the number of conditioning variables (Fisher, 1924c).

treatment of the t-distribution was provided by Fisher also in 1925, in the *Metron* paper "Applications of 'Student's' distribution" (Fisher, 1925a). Note that the distribution of t with $n-1$ degrees of freedom can be obtained from Eqs. (4.7) and (4.11) as

$$f(t)=f(z)\left|\frac{dz}{dt}\right|=\frac{\Gamma\left(\frac{n}{2}\right)}{\Gamma\left(\frac{n-1}{2}\right)\sqrt{\pi(n-1)}}\left(1+\frac{t^2}{n-1}\right)^{-n/2}, \quad -\infty<t<\infty, n=2,3,\ldots. \quad (4.12)$$

4.4 STUDENT ANTICIPATED

We now outline some of the studies that anticipated Student's work to some extent.

4.4.1 Helmert on the Independence of the Sample Mean and Sample Variance in a Normal Distribution (1876)

The first major figure is the German geodesist Friedrich Robert Helmert (1843–1917). Helmert's contributions (Helmert, 1875; 1876a; 1876b), however, did not involve deriving Student's distribution. Rather, among several other pioneering results, he proved the independence of the sample mean and sample variance and effectively derived the chi-squared distribution. Recall that the issue of independence was one of the key components in Student's 1908 "derivation" of the Z-distribution in Eq. (4.6).* The relevant work of Helmert is explained in Section 3.4.3 (see in particular Eq. 3.43).

4.4.2 Lüroth and the First Derivation of the t-Distribution (1876)

We now turn to Lüroth (or Lueroth), who was the first to have actually obtained the t-distribution† in an 1876 paper entitled "Vergleichung von zwei Werten des wahrscheinlichen Fehlers"‡ (Lüroth, 1876). Before discussing Lüroth's work, we shall say a few words about his life.

Jacob Lüroth§ (Fig. 4.4) was born in Mannheim, Germany, on February 18, 1844. He was an only child and his father was a member of the Mannheim's municipal parliament. Right at the start, Lüroth was interested in astronomy and in 1862 he published calculations of the orbits of two minor planets. But his poor eyesight made him switch to mathematics. He attended the University of Heidelberg, Berlin, and wrote his Ph.D. dissertation in 1865 on the Pascal configuration. In 1869, still 25 years old, Lüroth became Professor Ordinaries at the Technische Hochschule in Karlsruhe. He worked in many fields including mathematical logic, theory of invariants, various branches of geometry, mechanics, geodesy, and the theory of errors. Lüroth published 70 papers and two books in all. He passed away in 1910 at the University of Freiburg, where he served as rector in 1899–1900.

* Helmert's pioneering work had been unknown to Student and was in fact popularized much later by Pearson (1931).

† This fact was pointed out by Pfanzagl and Sheynin (1996).

‡ "Comparison of two values of the probable error"

§ Lüroth's life is also described in Pfanzagl and Sheynin (2006).

FIGURE 4.4 Jacob Lüroth (1844–1910). Wikimedia Commons (Public Domain), http://commons.wikimedia.org/wiki/File:Jacob_Lueroth.jpeg

Astronomische Nachrichten.

Expedition auf der Königlichen Sternwarte bei Kiel.
Herausgeber: Prof. Dr. C. A. F. Peters.

Bd. 87. **Nr. 2078.** 14.

Vergleichung von zwei Werthen des wahrscheinlichen Fehlers.

Die Bestimmung der wahrscheinlichsten Werthe von n Unbekannten $x_1, x_2 \ldots x_n$ aus einem System von m ($>$ n) linearen Gleichungen, mit Hilfe der Methode der kleinsten Quadrate, gelingt bekanntlich, ohne dass man das Präcisionsmaass h der Gewichtseinheit, welches in den Ausdruck der Wahrscheinlichkeit eines Beobachtungsfehlers eingeht, zu kennen braucht. Dagegen ist die Bestimmung des wahrscheinlichen Fehlers einer der Unbekannten von h abhängig und man kann entweder, wie dies gewöhnlich geschieht, für h den wahrscheinlichsten Werth setzen, der ihm nach den Beobachtungen zukömmt, oder aber man muss die Wahrscheinlichkeit dafür suchen, dass eine Unbekannte zwischen bestimmten Grenzen liegt, während die übrigen und das Präcisionsmaass alle möglichen Werthe haben können. Die auf beide Arten entstehenden Werthe des wahrscheinlichen Fehlers sollen hier verglichen werden.

Sei x_1 die Unbekannte, um deren wahrscheinlichen Fehler es sich handelt, so ist dieser nach der ersten Methode

$$(1) \qquad r_1 = \rho \sqrt{\frac{2}{p}} \sqrt{\frac{a}{a_1}}$$

Hierbei bezeichnet

$\rho = 0.47694\ldots$ die Wurzel der Gleichung

$$\frac{\sqrt{\pi}}{4} = \int_0^{\rho} dx\, e^{-x^2}$$

a die Summe der Quadrate der Fehler, welche übrig bleiben, wenn man die Beobachtungen mit den wahrscheinlichsten Werthen der Unbekannten berechnet,

a_1 den Nenner des Ausdrucks, der sich für x_1 ergiebt, wenn man bei der Elimination aus den Normalgleichungen nach dem Gauss'schen Verfahren x_1 zur letzten Unbekannten macht; endlich ist

$p = m - n$ gesetzt *).

*) Vergl. Gauss, Th. M. C. C., § 182; Encke, Anhang zum Berliner Jahrbuch für 1835, Seite 287.

87. Bd.

Um nach dem zweiten Verfahren den wahrscheinlichen Fehler zu erhalten, muss man die Wahrscheinlichkeit aufstellen dafür, dass x_1 zwischen zwei Grenzen a' und a'' liege, gleichgültig, welches die Werthe von $x_2 \ldots x_n$ und h seien.

Nun ist bekanntlich die Wahrscheinlichkeit, dass die Unbekannten zwischen den Grenzen x_1 und $x_1 + dx_1$, x_2 und $x_2 + dx_2, \ldots x_n$ und $x_n + dx_n$ und das Präcisionsmaass zwischen den Grenzen h und $h + dh$ liegen, proportional der Wahrscheinlichkeit, dass die diesen Werthen entsprechenden Fehler stattfinden werden, also, wenn man mit g das Gewicht einer Beobachtung bezeichnet, bei der der Fehler v übrig bleibt,

$$= C\, dh\, dx_1\, dx_2 \ldots dx_n\, h^m e^{-(gvv)}$$

unter C eine Constante verstanden. Die Wahrscheinlichkeit, dass einige der Grössen $x_1\; x_2 \ldots x_n$ h zwischen weiteren Grenzen liegen, findet sich hieraus, indem man nach diesen zwischen den betreffenden Grenzen integrirt. Folglich ist die Wahrscheinlichkeit, dass x_1 zwischen a' und a'' $x_2\, x_3 \ldots x_n$ zwischen $-\infty$ und $+\infty$, und h zwischen 0 und $+\infty$ liege

$$= C \int_0^{+\infty} dh \int_{a'}^{a''} dx_1 \int_{-\infty}^{+\infty} dx_2 \ldots \int_{-\infty}^{+\infty} dx_n\, h^m e^{-h^2 (gvv)}.$$

Die Wahrscheinlichkeit muss zur Gewissheit also $= 1$ werden, wenn $a' = -\infty$ $a'' = +\infty$ ist. Hieraus ergiebt sich C und damit folgt die gesuchte Wahrscheinlichkeit endlich

$$(2) \qquad W_1 = \frac{\int_0^{+\infty} dh \int_{a'}^{a''} dx_1 \int_{-\infty}^{+\infty} dx_2 \ldots \int_{-\infty}^{+\infty} dx_n . h^m e^{-h^2 (gvv)}}{\int_0^{+\infty} dh \int_{-\infty}^{+\infty} dx_1 \int_{-\infty}^{+\infty} dx_2 \ldots \int_{-\infty}^{+\infty} dx_n . h^m e^{-h^2 (gvv)}}$$

Um die n-fachen nach $x_1, x_2 \ldots x_n$ genommenen Integrale auszuführen, führt man, wie bekannt (vergl. Encke l. c.), diejenigen Functionen als neue Variabeln ein, auf welche man bei der Gauss'schen Elimination

14

FIGURE 4.5 First page of Lüroth's paper (Lüroth, 1876)

In 1876, Lüroth's derived the *t*-distribution by using inverse probability (Lüroth, 1876) (Fig. 4.5). Because Lüroth's paper was also an extension of Gauss' previous work in the book *Theoria Motus Corporum Coelestium** (Gauss, 1809, English edition, 1857), we shall review the relevant portions of the latter before moving on to Lüroth's derivation.

In his book, Gauss considered μ respective observations $M, M', M'', \ldots$ on $V, V', V'', \ldots$, the latter being functions of the ν unknown quantities p, q, r, s, etc. (*ibid*., p. 253). He then defined the errors of observations by

*"Theory of the Motion of Heavenly Bodies"

$$v = V - M,$$
$$v' = V' - M',$$
$$v'' = V'' - M'',$$
$$\cdots.$$

In Section 182 of his book, Gauss defined the error sum of squares by

$$W = v^2 + v'^2 + v''^2 + \cdots$$

so that

$$W = \{V(p,q,r,\ldots) - M\}^2 + \{V'(p,q,r,\ldots) - M'\}^2 + \{V''(p,q,r,\ldots) - M''\}^2 + \cdots$$
$$\frac{dW}{dp} = 2(V - M)\frac{dV}{dp} + 2(V' - M')\frac{dV'}{dp} + 2(V'' - M'')\frac{dV''}{dp} + \cdots.$$

He then set

$$\frac{1}{2}\frac{dW}{dp} = p' = \lambda + \alpha p + \beta q + \gamma r + \sigma s + \cdots$$

and

$$W - \frac{p'^2}{\alpha} = W'. \tag{4.13}$$

Now, since

$$\frac{dW'}{dp} = \frac{dW}{dp} - \frac{2p'}{\alpha}\frac{dp'}{dp}$$
$$= 2p' - \left(\frac{2p'}{\alpha}\right)\alpha$$
$$= 0,$$

W' is independent of p. Similarly, Gauss defined

$$\frac{1}{2}\frac{dW'}{dp} = q' = \lambda' + \beta' q + \gamma' r + \sigma' s + \cdots$$

and

$$W' - \frac{q'^2}{\beta'} = W'' \tag{4.14}$$

so that W'' is independent of both p and q. By continuing in this way and using Eqs. (4.13), (4.14), and other such equations, Gauss was able to reach the following decomposition for the error sum of squares:

$$W = \frac{p'^2}{\alpha} + \frac{q'^2}{\beta'} + \cdots + a \tag{4.15}$$

(*ibid*., p. 265), where a is a (positive) constant representing the residual sum of squares and $\alpha, \beta', \ldots$ are positive quantities. We note that the process of obtaining the set of equations

$$\begin{cases} p' = \lambda + \alpha p + \beta q + \gamma r + \sigma s + \cdots \\ q' = \lambda' + \beta' q + \gamma' r + \sigma' s + \cdots \\ \cdots \end{cases}$$

is now known as *Gaussian elimination*: each succeeding equation has one less unknown than the previous one. Thus, by starting with the last equation (involving only one unknown), one can solve for the unknowns p, q, r, s, ... by back substitution. We also note that Eq. (4.15) represents what is now called the orthogonal decomposition of the (total) error sum of squares (W) into the regression sum of squares ($p'^2/\alpha + q'^2/\beta' + \cdots$) and the residual sum of squares (a).

In Lüroth's paper (1876), the ν unknowns $p, q, r, \ldots$ are written as $x_1, x_2, \ldots, x_n$. The quantities $p', q', r', \ldots$ are written in reverse index notation as $\xi_n, \xi_{n-1}, \ldots, \xi_1$, where

$$\begin{aligned} \xi_n &= a_n x_n + \cdots, \\ \xi_{n-1} &= a_{n-1} x_{n-1} + \cdots, \\ &\cdots \\ \xi_1 &= a_1 x_1 + \alpha_1 \end{aligned} \tag{4.16}$$

(*ibid*., p. 211). In the above, ξ_1 is a function of x_1 only, ξ_2 is a function of x_1 and x_2 only, and so on. Assuming a uniform prior for all parameters and a $N\left(0, \dfrac{1}{2h^2}\right)$ distribution (where h is the precision) for each of the m observations $M, M', M'', \ldots$, Lüroth wrote the posterior distribution of $x_1, x_2, \ldots, x_n, h$ as

$$Ch^m e^{-(gvv)},$$

where C is a constant and (gvv) is the error sum of squares W in Eq. (4.15), that is,

$$(gvv) = \frac{\xi_1^2}{a_1} + \frac{\xi_2^2}{a_2} + \cdots + \frac{\xi_n^2}{a_n} + a.$$

The posterior probability that x_1 lies in the interval (a', a'') is then

$$W_1 = C \int_0^\infty \int_{a'}^{a''} \cdots \int_{-\infty}^{\infty} h^m e^{-h^2 (gvv)} dx_n \ldots dx_1 dh.$$

Since C is a normalizing constant, Lüroth wrote the above as

$$W_1 = \frac{\int_0^\infty \int_{a'}^{a''} \cdots \int_{-\infty}^{\infty} h^m e^{-h^2 (gvv)} dx_n \ldots dx_1 dh}{\int_0^\infty \int_{-\infty}^{\infty} \cdots \int_{-\infty}^{\infty} h^m e^{-h^2 (gvv)} dx_n \ldots dx_1 dh}$$

(*ibid*., p. 210). By changing variables according to (4.16),

$$W_1 = \frac{\int_0^{\infty}\int_{a_1a'+\alpha_1}^{a_1a''+\alpha_1}\cdots\int_{-\infty}^{\infty} h^m \exp\left(-h^2 a\right)\exp\left(-h^2\frac{\xi_1^2}{a_1}\right)\cdots\exp\left(-h^2\frac{\xi_n^2}{a_n}\right)d\xi_n\ldots d\xi_1 dh}{\int_0^{\infty}\int_{-\infty}^{\infty}\cdots\int_{-\infty}^{\infty} h^m \exp\left(-h^2 a\right)\exp\left(-h^2\frac{\xi_1^2}{a_1}\right)\cdots\exp\left(-h^2\frac{\xi_n^2}{a_n}\right)d\xi_n\ldots d\xi_1 dh}$$

$$= \frac{1}{\sqrt{\pi a_1}}\frac{\int_0^{\infty}\int_{a_1a'+\alpha_1}^{a_1a''+\alpha_1} h^{p+1}\exp\left\{-h^2\left(\frac{\xi_1^2}{a_1}+a\right)\right\}d\xi_1 dh}{\int_0^{\infty} h^p \exp\left(-h^2 a\right)dh},$$

where $p = m - n$. Now, by using

$$\int_0^{\infty} h^p \exp\left(-h^2 a\right)dh = \begin{cases} \dfrac{\sqrt{\pi}}{2}\cdot\dfrac{(p-1)(p-3)\cdots 3\cdot 1}{2^{p/2}a^{(p+1)/2}} & (p \text{ even}), \\ \dfrac{(p-1)(p-3)\cdots 4\cdot 2}{2^{(p+1)/2}a^{(p+1)/2}} & (p \text{ odd}), \end{cases}$$

and by making the substitution $x = \xi_1/\sqrt{aa_1}$, Lüroth finally obtained

$$W_1 = \Pr\left\{-R_1\sqrt{\frac{a_1}{a}} < x < R_1\sqrt{\frac{a_1}{a}}\right\} = \begin{cases} \dfrac{1}{\pi}\cdot\dfrac{p(p-2)\cdots(4)(2)}{(p-1)(p-3)\cdots(3)(1)}\displaystyle\int_{-R_1\sqrt{\frac{a_1}{a}}}^{R_1\sqrt{\frac{a_1}{a}}}\frac{dx}{\left(1+x^2\right)^{(p/2)+1}} & (p \text{ even}) \\ \dfrac{1}{2}\cdot\dfrac{p(p-2)\cdots(3)(1)}{(p-1)(p-3)\cdots(4)(2)}\displaystyle\int_{-R_1\sqrt{\frac{a_1}{a}}}^{R_1\sqrt{\frac{a_1}{a}}}\frac{dx}{\left(1+x^2\right)^{(p/2)+1}} & (p \text{ odd}) \end{cases}$$

(*ibid*., pp. 211–212), with $a' = -\alpha_1/a_1 - R_1$ and $a'' = -\alpha_1/a_1 + R_1$. By comparing with Eq. (4.12), it is seen that

$$u = x\sqrt{p+1} = \frac{a_1 x_1 + \alpha_1}{\sqrt{aa_1}}\sqrt{p+1}$$

has a Student's t-distribution with $p+1$ d.f.:

$$\frac{\Gamma\left(\frac{p+2}{2}\right)}{\Gamma\left(\frac{p+1}{2}\right)\sqrt{\pi(p+1)}}\cdot\frac{1}{\left(1+\frac{u^2}{p+1}\right)^{(p/2)+1}}.$$

4.4.3 Edgeworth's Derivation of the *t*-Distribution Based on Inverse Probability (1883)

The third and final noteworthy figure we shall consider who anticipated Student to some extent is the Irish mathematician and philosopher Francis Ysidro Edgeworth (1845–1926). Edgeworth's life has been described in Section 2.3.2. Like Lüroth, Edgeworth's derivation was based on inverse probability. In his paper "The Method of Least Squares" (Edgeworth, 1883),* Edgeworth wished to determine

> …the "probable error" incurred by taking the mean of observations as the real value. (*ibid.*, p. 367)

Edgeworth thus wished to find the probable error of the population mean from its posterior distribution. First, he assumed $x_1, x_2, \ldots, x_n$ are values from a $N\left(\xi, \frac{1}{2h^2}\right)$ distribution, where h is the precision $\left(h = 1/\left(\sigma\sqrt{2}\right)\right)$. The joint density of the data is then

$$P = \left(\frac{1}{c\sqrt{\pi}}\right)^n \exp\left\{-\frac{(x_1-\xi)^2 + (x_2-\xi)^2 + \cdots}{c^2}\right\}$$

(*ibid.*, p. 366), where $c = 1/h$ is the modulus.

By assuming a uniform prior for (ξ, h), the posterior distribution of ξ and h is

$$\frac{P}{\int_0^\infty\int_{-\infty}^\infty P\,d\xi\,dh} = \frac{\left(\frac{1}{c\sqrt{\pi}}\right)^n \exp\left\{-\frac{\sum_i (x_i-\xi)^2}{c^2}\right\}}{\int_0^\infty\int_{-\infty}^\infty P\,d\xi\,dh} = \frac{\left(\frac{1}{c\sqrt{\pi}}\right)^n \exp\left\{-\frac{\sum_i x_i^2 - 2\xi\sum_i x_i + n\xi^2}{c^2}\right\}}{\int_0^\infty\int_{-\infty}^\infty P\,d\xi\,dh}$$

$$= \frac{\frac{h^n}{\pi^{n/2}}\exp\left\{-h^2\left(\sum_i x_i^2 + n\xi^2\right)\right\}}{\int_0^\infty\int_{-\infty}^\infty P\,d\xi\,dh} \tag{4.17}$$

(*ibid.*, p. 367). In the above, Edgeworth assumed that the sample mean $\bar{x} = 0$. In the case that h is a constant, Eq. (4.17) then gives the posterior distribution of ξ as

$$\frac{P}{\int_{-\infty}^\infty P\,d\xi} = \frac{\frac{h^n}{\pi^{n/2}}\exp\left\{-h^2\left(\sum_i x_i^2 + n\xi^2\right)\right\}}{\int_{-\infty}^\infty \frac{h^n}{\pi^{n/2}}\exp\left\{-h^2\left(\sum_i x_i^2 + n\xi^2\right)\right\}d\xi} = \frac{e^{-nh^2\xi^2}}{\int_{-\infty}^\infty e^{-nh^2\xi^2}\,d\xi} = \frac{h\sqrt{n}}{\sqrt{\pi}}e^{-nh^2\xi^2}.$$

*Edgeworth's derivation is also described in Welch (1958).

Hence, the probable error of ξ is .6745 multiplied by the posterior standard deviation of ξ, that is,

$$.6745\times\frac{1}{h\sqrt{2n}}=\frac{.4769}{h\sqrt{n}}. \tag{4.18}$$

However, when h is not a constant, the posterior distribution of ξ can be obtained from Eq. (4.17) as

$$\frac{\int_0^\infty P dh}{\int_0^\infty\int_{-\infty}^\infty P d\xi dh}=\frac{\int_0^\infty \frac{h^n}{\pi^{n/2}}\exp\left\{-h^2\left(\sum_i x_i^2+n\xi^2\right)\right\}dh}{\int_0^\infty\int_{-\infty}^\infty \frac{h^n}{\pi^{n/2}}\exp\left\{-h^2\left(\sum_i x_i^2+n\xi^2\right)\right\}d\xi dh}.$$

Assuming n is even, Edgeworth wrote the above as

$$\frac{J}{\left(\sum_i x_i^2+n\xi^2\right)^{(n+1)/2}} \tag{4.19}$$

(*ibid*., p. 368), where J is a constant. When Eq. (4.19) is rewritten as

$$\frac{J\left(\sum_i x_i^2\right)^{-((n+1)/2)}}{\left\{1+\frac{n\xi^2}{\left(\sum_i x_i^2\right)}\right\}^{(n+1)/2}}$$

and compared with Eq. (4.12), it is seen that

$$\frac{n\xi}{\sqrt{\sum_i x_i^2}}$$

has a Student's t-distribution with n d.f.

Edgeworth called (4.19) a "subexponential expression" which was "incidental to this problem." This is because it arose from the choice of a particular prior for (ξ, h). Moreover, he noted that Eq. (4.18) could not be used to determine the probable error of ξ because it assumed that h was a constant. Instead, the probable error needed to be determined from Eq. (4.19).

5

THE FISHERIAN LEGACY

5.1 RONALD AYLMER FISHER (1890–1962)

Monumental and revolutionary are words that come to mind when attempting to assess Fisher's impact on mathematical statistics. Sir Ronald Aylmer Fisher* (Fig. 5.1) was born in London on February 17, 1890, and was a surviving twin. His father, George Fisher, was a fine art auctioneer. Fisher entered Harrow School in 1904 and was good at mathematics right from the start. However, his eyesight was poor so that he could not work in artificial light. This also meant that he often had to resort to mental visualization to understand mathematics. Soon he became very good at it, and this skill was to pay off in a major way throughout his statistical career. Fisher entered Gonville and Caius College, Cambridge, in 1909 and graduated in 1912. At college, he helped in the formation of a Cambridge University Eugenics Society and also met Major Leonard Darwin (1850–1943), Charles Darwin's fourth son. Leonard was the President of the Eugenics Society of London and later became an invaluable friend and mentor to Fisher.

In 1912, Fisher sent a geometric proof of the distribution of z† to Student and also published his first paper (where likelihood was used but never explicitly stated). Student was clearly impressed by Fisher's analytic proficiency and Fisher regarded Student as a major source of inspiration. A relationship of mutual admiration and respect thus grew between the two men. Meanwhile, Fisher spent a postgraduate year in the Cavendish Laboratory, Cambridge, where he studied the theory of errors under F.J.M. Stratton and statistical mechanics and quantum theory under J.H. Jeans. In 1914, Fisher was prevented from entering war service due to his poor eyesight and instead taught physics and mathematics in school. In the same year, Fisher sent his derivation of the exact distribution of

*Fisher's definitive biography has been written by his daughter Joan Fisher Box (1978). Insightful reviews of Fisher's life and work can also be found in Savage (1976), Gridgeman (1972a), Stigler (2006), Rao (1992), Efron (1998), and Healy (2003).

†See Sec. 4.3.2.

Classic Topics on the History of Modern Mathematical Statistics: From Laplace to More Recent Times, First Edition. Prakash Gorroochurn.

FIGURE 5.1 Sir Ronald Aylmer Fisher (1890–1962). Wikimedia Commons (Licensed under the Creative Commons Attribution-Share Alike 3.0 Unported license), http://commons.wikimedia.org/wiki/File:R._A._Fischer.jpg

r to Karl Pearson's (1857–1936) journal *Biometrika* for publication. Pearson accepted the paper, which appeared in May 1915. However, two years later Pearson and colleagues wrote a cooperative study (Soper et al., 1917) and attacked a portion of Fisher's paper for allegedly using inverse probability in the derivation of r. Pearson's criticism was not exactly correct and Fisher took great umbrage at it. In 1916 and 1918, two more papers of Fisher were dismissed by Pearson. In the meantime, Fisher had married Ruth Eileen Guinness in 1917. In 1919, Pearson offered him a position as Chief Statistician at the Galton Laboratory but Fisher refused:

> ...he [Fisher] recognized that nothing would be taught or published at the Galton Laboratory without Pearson's approval. Fisher required the liberty to do his own work in his own way and, by it, to win his own reputation. (Box, 1978, p. 82)

Instead, Fisher joined the Rothamsted Experimental Station where he worked on agricultural problems. There was to be a final straw: in 1920 Fisher sent yet another paper to Pearson on the probable error of the correlation coefficient. In that paper, Fisher also explained that he had not used Bayes' theorem in his earlier 1915 paper. A rejection again by Pearson resulted in Fisher's vowing that he would never submit to *Biometrika* again, a vow that he kept.

Pearson's dominance of the statistical scene, though, was to change in 1922. In that year, Fisher introduced the concept of degrees of freedom and attacked Pearson's greatest invention, the X^2 goodness-of-fit test (cf. Section 3.3.3). Fisher correctly pointed out that when parameters were estimated from the data at hand while testing for goodness of fit, the index (degrees of freedom) of the χ^2 distribution against which X^2 is compared needed to be *further* reduced by the number of parameters being estimated. Although Pearson vehemently disagreed, it soon became apparent he was fighting a losing battle.

From Rothamsted, Fisher was not only revolutionizing statistics, but he was also making equally remarkable breakthroughs in the field in genetics. Later, with J.B.S. Haldane and Sewall Wright, he united Mendelism and Darwinism in the so-called neo-Darwinian evolutionary synthesis. This gave rise to the field of theoretical population genetics. Before that, in 1922, he published a landmark paper, "On the dominance ratio" (Fisher, 1922a), dealing with a whole array of important genetics topics. In the same year, Fisher's groundbreaking statistical paper "On the mathematical foundations of theoretical statistics" (Fisher, 1922c) also came out (see Section 5.2.1). In that paper, he laid out the foundations of modern estimation theory and introduced much of the statistical vocabulary still used today. In 1925, he published the epochal book *Statistical Methods for Research Workers* (Fisher, 1925b), and in 1929 he was elected Fellow of the Royal Society. Fisher left Rothamsted in 1933 to become Galton Professor of Eugenics at University College London. In 1943, Fisher joined Cambridge as Balfour Professor of Genetics and in 1952 he was knighted.

For all his successes, Fisher engaged in countless controversies with several of his contemporaries. Starting with Karl Pearson, Fisher had famous quarrels with Jerzy Neyman, Egon Pearson, Sewall Wright, Maurice Bartlett, Student, and many others. Fisher has thus often been described a "master of polemic" (e.g., Howie, 2002, p. 80).

Many felt that he could not tolerate any criticism of his own work, although I.J. Good later noted:

> ...George Barnard told me a few years ago that Fisher was well aware of his own tendency to lose his temper, and that he regarded it as the bane of his life. I felt better disposed to Fisher after that.... (Savage, 1976, p. 492)

In 1959, Fisher immigrated to Australia and worked in the Division of Statistics at the CSIRO. Fisher died in Adelaide on July 29, 1962.

5.2 FISHER AND THE FOUNDATION OF ESTIMATION THEORY

5.2.1 Fisher's 1922 Paper: Consistency, Efficiency, and Sufficiency

5.2.1.1 Introduction As was mentioned, Fisher's paper "On the mathematical foundations of theoretical statistics" (Fisher, 1922c) was a landmark in the history of modern statistics.* The paper was exclusively concerned with estimation (hypothesis testing is not mentioned at all). One might also call it the founding paper of modern statistics because it laid out much of the statistical vocabulary and many of the concepts in estimation theory that are still used today. The concept of sufficiency was one of its most crucial components, although the notion had already been introduced two years earlier, in 1920, by Fisher himself. We now describe these topics.

Fisher started his 1922 paper with a list of 15 definitions arranged in alphabetical order. In the list, we see for the first time the terms "consistency," "efficiency," "location," "scaling," "statistic," and "sufficiency." Another fundamental term, "maximum likelihood" (ML) that was used for the first time occurred later in the paper (p. 323). "Parameter" was also used for the first time by Fisher (p. 309), but according to David (1998), that term had been used for the first time previously by Czuber (1914, p. 392). Likewise, "likelihood" also figured in the list of 15, but that term had already been used and defined in Fisher's *Metron* paper of 1921 (Fisher, 1921a, p. 24).†

Fisher then stated that one of the major reasons theoretical statistics had been neglected was because statisticians themselves had been imprecise in the formulation of statistical problems:

> ...for it is customary to apply the same name, *mean, standard deviation, correlation coefficient*, etc., both to the true value which we should like to know, but can only estimate, and to the particular value at which we happen to arrive by our methods of estimation; so also in applying the term probable error, writers sometimes would appear to suggest that the former quantity, and not merely the latter, is subject to error. (Fisher, 1922c, p. 311)

Fisher here made an important argument for a fundamental dichotomy: that of the "*hypothetical infinite population*" whose *unknown parameters* we would like to know and the *sample* from which *statistics* can be obtained as *estimators* of these parameters.

* Fisher 1922 paper is also examined in Geisser (1980, pp. 59–66; 1992, pp. 1–10).

† See Section 2.3.10, p. 264.

If $\mathbf{x} = (x_1, x_2, \ldots x_n)$ are realizations of a random variable X from a population with parameters $\theta_1, \theta_2, \ldots, \theta_p$, then any measurable function $T = T(\mathbf{X})$ is called a *statistic*. Note that the value of T depends on the data only and does not involve the unknown parameters. When a statistic is calculated with the aim of providing a numerical value for a parameter, it is called an *estimator*, and its realized value is called an *estimate*.*

Section 2 of Fisher's 1922 paper highlighted one of the fundamental aims of statistical methods:

> In order to arrive at a distinct formulation of statistical problems, it is necessary to define the task which the statistician sets himself: briefly, and in its most concrete form, the object of statistical methods is the reduction of data. A quantity of data, which usually by its mere bulk is incapable of entering the mind, is to be replaced by relatively few quantities which shall adequately represent the whole, or which, in other words, shall contain as much as possible, ideally the whole, of the relevant information contained in the original data. (*ibidem*)

How is the above task to be done? Fisher delineated three steps:

1. Problems of Specification. These arise in the choice of the mathematical form of the population.
2. Problems of Estimation. These involve the choice of methods of calculating from a sample statistical derivates, or as we shall call them statistics, which are designed to estimate the values of the parameters of the hypothetical population.
3. Problems of Distribution. These include discussions of the distribution of statistics derived from samples or in general any functions of quantities whose distribution is known. (*ibid*., p. 313)

Fisher was clearly very concerned with the second of the above, namely, the problems of estimation. He then gave three fundamental criteria that a "good" estimator should possess: *(a) consistency*, *(b) efficiency*, and *(c) sufficiency*. We will now discuss these in some detail.

5.2.1.2 *The Criterion of Consistency* Fisher defined a *consistent* estimator as one which "when applied to the whole population the derived statistic [i.e., the estimator] should be equal to the parameter." Mathematically, this means the following: suppose $F_n(x)$ is the empirical distribution function based on IID† random variables $X_1, X_2, \ldots, X_n$ each of which follows the (theoretical) distribution function $F(x \mid \theta)$; suppose also that $T_n = g\,(F_n(x))$ (where g is a function) is an estimator of θ, then T_n is consistent in Fisher's sense if $g\,(F\,(x \mid \theta)) = \theta$.

* Some scholars protested against Fisher's use of these new terms, especially his use of "statistic." For instance, writing in May 1931, the mathematician Arne Fisher (1887–1944) made the following tongue-in-cheek remark: "I am more inclined to quarrel with you over the introduction by you in statistical method of some outlanding and barbarous technical terms. They stand out like quills upon the porcupine, ready to impale the skeptical critic. Where, for instance, did you get that atrocity, *a statistic*?...". To this, Ronald Fisher replied: "I use special words as the best way of expressing special meanings. Thiele and Pearson were quite content to use the same words for what they were estimating and for their estimates of it. Hence the chaos in which they left the problem of estimation. Those of us who wish to distinguish the two ideas prefer to use different words, hence 'parameter' and 'statistic'. No one who does not feel this need is under any obligation to use them." (Bennett, 1990, pp. 311–313).

† Independent and identically distributed.

As an example, suppose $X_1, X_2, \ldots, X_n$ are IID from a $N(\mu, \theta^2)$ distribution. Since $\mathsf{E}(X-\mu)^2 = \theta^2 = g(F)$, it can be seen that $g_1(F_n) = \Sigma_i (X_i - \bar{X})^2 / n$ is Fisher consistent for θ^2 while $g_2(F_n) = \Sigma_i (X_i - \bar{X})^2 / (n-1)$ is not.

The reader will have noticed that consistency in Fisher's sense (i.e., *Fisher consistency*) is different from the way consistency is usually defined today, namely, $T = T(X_1, X_2, \ldots, X_n)$ is a consistent estimator of θ if it converges to θ in probability, that is, if $T \xrightarrow{P} \theta$. According to this definition, both $g_1(F_n)$ and $g_2(F_n)$ in the previous paragraph are consistent estimators of θ^2. Fisher regarded this definition of consistency as "much less satisfactory" than his own (Fisher, 1956b, p. 144) because:

> With respect to a function of the observations, T, defined for all possible sizes of sample this definition has a certain meaning. However, any particular method of treating a finite sample of N_1 observations may be represented as belonging to a great variety of such general functions. In particular, if T' stand for any function whatsoever of N_1 observations, and T_N for any function fulfilling the asymptotic condition of consistency, then
>
> $$\frac{1}{N}\left\{N_1 T' + \left(N - N_1\right) T_{N-N_1}\right\} \tag{5.1}$$
>
> is itself a statistic defined for all values of N, and tending asymptotically to the limit θ, yet it is recognizable when $N = N_1$ as the arbitrary function T' calculated from the finite sample. (*ibid.*, pp. 144–145)

Fisher's point in the above is that, although the estimator in (5.1) is consistent (in the sense we usually use it), it becomes equal to the arbitrary statistic T', and thus useless, when $N = N_1$. As another example, let T_n be a consistent estimator for some parameter θ. Then

$$T_n' = \begin{cases} 0 & \text{if } n \le 10^6 \\ T_n & \text{otherwise} \end{cases}$$

is also consistent but completely useless for most realistic situations, namely, when $n \le 10^6$. In spite of Fisher's example, we are often more interested in large-sample behavior and usually favor consistent (rather than Fisher consistent) estimators. Furthermore, it can be shown that for T_n to be a consistent estimator of θ, sufficient conditions are that:

- $\mathsf{E}T_n \to \theta$
- $\operatorname{var} T_n \to 0$

as $n \to \infty$.

It is interesting that Fisher preferred consistency (in his sense) to unbiasedness as a desideratum for an estimator. The reason for his indifference to unbiasedness was because unbiased estimators are not invariant to transformations:

> ...lack of bias, which since it is not invariant for functional transformation of parameters has never had the least interest for me. (Bennett, 1990, p. 196)

That is to say, Fisher's "lack of interest" in unbiasedness stemmed from the fact that if an estimator T is unbiased for a parameter θ, then for a given function g, it is not true in general that $g(T)$ will also be unbiased for $g(\theta)$.

5.2.1.3 The Criterion of Efficiency Fisher considered the following two estimators of the standard deviation σ from a normal distribution:

$$\hat{\sigma}_1 = \frac{1}{n}\sqrt{\frac{\pi}{2}}\sum_{i=1}^{n}\left|X_i - \overline{X}\right|, \tag{5.2}$$

$$\hat{\sigma}_2 = \sqrt{\frac{1}{n}\sum_{i=1}^{n}\left(X_i - \overline{X}\right)^2} \tag{5.3}$$

(Fisher, 1922c, p. 316). Both $\hat{\sigma}_1$ and $\hat{\sigma}_2$ are Fisher consistent, so we need another criterion to distinguish between the two, namely, the *criterion of efficiency* which Fisher defined as "[t]hat in large samples, when the distributions of the statistics tend to normality, that statistic is to be chosen which has the least probable error [or variance]." By assuming asymptotic *normality*, we have for large samples

$$\operatorname{var}\hat{\sigma}_1 \approx \frac{\sigma^2(\pi-2)}{2n},$$

$$\operatorname{var}\hat{\sigma}_2 \approx \frac{\sigma^2}{2n}.$$

Therefore, in large samples, $\operatorname{var}\hat{\sigma}_1 > \operatorname{var}\hat{\sigma}_2$ and $\hat{\sigma}_2$ is more efficient than $\hat{\sigma}_1$. Fisher wrote:

> many…use the first formula [in Eq. (5.2)], although the result of the second has 14 per cent. greater weight.., and the labour of increasing the number of observations by 14 per cent. can seldom be less than that of applying the more accurate formula. (*ibidem*)

5.2.1.4 The Criterion of Sufficiency Having introduced the criteria of consistency and efficiency, Fisher realized that these were large-sample properties. Fisher needed a further criterion for *finite* samples. He wrote:

> The criterion of efficiency is still to some extent incomplete, for different methods of calculation may tend to agreement for large samples, and yet differ for all finite samples. The complete criterion suggested by our work on the mean square error … is:-
>
> That the statistic chosen should summarise the whole of the relevant information supplied by the sample.
>
> This may be called the *Criterion of Sufficiency*.
>
> In mathematical language we may interpret this statement by saying that if θ be the parameter to be estimated, θ_1 a statistic which contains the whole of the information as to the value of θ, which the sample supplies, and θ_2 any other statistic, then the surface of distribution of pairs of values of θ_1 and θ_2, for a given value of θ, is such that for a given value of θ_1 the distribution of θ_2 does not involve θ_1. In other words, when θ_1 is known, knowledge of the value of θ_2 throws no further light upon the value of θ. (*ibid.*, pp. 316–317)

In the last paragraph here, Fisher gave the fundamental criterion of sufficiency. Mathematically, a statistic T is sufficient for a parameter θ if the conditional density of any other statistic T', given T, is independent of θ. In particular, the conditional density of $X_1, X_2, \ldots, X_n$, given T, is also independent of θ. Thus, T contains all the information in the sample concerning θ.

Fisher next set out to prove that "a statistic which fulfils the criterion of sufficiency will also fulfil the criterion of efficiency, when the latter is applicable" (*ibid*., p. 317). He considered two statistics θ_1 and θ_2 which, for large samples, are assumed to be BivNormal $(\theta, \theta; \sigma_1^2, \sigma_2^2; \rho)$ distributed. Their joint density is

$$\frac{1}{2\pi\sigma_1\sigma_2\sqrt{1-\rho^2}}\exp\left[-\frac{1}{1-\rho^2}\left\{\frac{(\theta_1-\theta)^2}{2\sigma_1^2}-\frac{2\rho(\theta_1-\theta)(\theta_2-\theta)}{2\sigma_1\sigma_2}+\frac{(\theta_2-\theta)^2}{2\sigma_2^2}\right\}\right].$$

The distribution of θ_1 is

$$\frac{1}{\sigma_1\sqrt{2\pi}}\exp\left\{-\frac{(\theta_1-\theta)^2}{2\sigma_1^2}\right\}$$

and the conditional density of θ_2 for a given value of θ_1 is obtained by taking the ratio of the first to the second of the above two densities:

$$\frac{1}{\sigma_2\sqrt{2\pi\left(1-\rho^2\right)}}\exp\left[-\frac{1}{2\left(1-\rho^2\right)}\left\{\frac{\rho(\theta_1-\theta)}{\sigma_1}-\frac{(\theta_2-\theta)}{\sigma_2}\right\}^2\right].$$

Now if θ_1 is sufficient for θ, then the conditional distribution of θ_2 for a given value of θ_1 will not involve θ, that is,

$$\frac{\partial}{\partial\theta}\left\{\frac{\rho(\theta_1-\theta)}{\sigma_1}-\frac{(\theta_2-\theta)}{\sigma_2}\right\}=0 \quad\Rightarrow\quad \rho=\frac{\sigma_1}{\sigma_2}$$

so that $\sigma_1 \le \sigma_2$. Fisher thus showed that θ_1 was more efficient than θ_2, and the relative efficiency of the former was measured by ρ^2. Of course, this proof is valid only within the class of estimators that are asymptotically sufficient, unbiased, and normally distributed.

5.2.2 Genesis of Sufficiency in 1920

The concept of sufficiency was a fundamental component of Fisher's 1922 paper (Fisher, 1922c) but had already been discovered in all but name by Fisher in 1920 in the paper "A mathematical examination of the methods of determining the accuracy of an observation by the mean error, and by the mean square error" (Fisher, 1920). Fisher's paper was prompted by a statement the British astrophysicist Arthur Eddington (Fig. 5.2) had made in his 1914 book *Stellar Movements and the Structure of the Universe*:

> ...just as in calculating the mean error of a series of observations it is preferable to use the simple mean residual irrespective of sign than the mean-square residual.* (Eddington, 1914, p. 147)

In his 1920 paper, Fisher examined the two quantities (estimators) Eddington had referred to, namely, (cf. Eqs. 5.2 and 5.3),

$$\hat{\sigma}_1 = \sqrt{\frac{\pi}{2}}\frac{\sum_i |x_i - \overline{x}|}{n}, \tag{5.4}$$

$$\hat{\sigma}_2 = \sqrt{\frac{\sum_i (x_i - \overline{x})^2}{n}}. \tag{5.5}$$

where the x_i's ($i=1, 2, \ldots, n$) are assumed to be random variables from a $N(m,\sigma^2)$ distribution, and $\overline{x}$ is the sample mean.

Fisher then set out to provide two arguments, one for large samples, the other for small samples, for why Eddington's claim was untrue. In his second argument, Fisher used the principle of sufficiency, although he did not name it as such, at that stage. We now examine Fisher's two arguments, paying special attention to the second one.

For his large-sample first argument, Fisher used his formidable powers of geometric reasoning and derived the joint density of $(\overline{x},\hat{\sigma}_2)$ (cf. Eq. 4.8). Fisher thus obtained the joint density of $(\overline{x},\hat{\sigma}_2)$ as proportional to

$$\exp\left\{-\frac{n(\overline{x}-m)^2}{2\sigma^2}\right\}\times\hat{\sigma}_2^{n-2}\exp\left(-\frac{n\hat{\sigma}_2^2}{2\sigma^2}\right)$$

(Fisher, 1920, p. 759), from which the independence of $\overline{x}$ and $\hat{\sigma}_2$ is apparent. Fisher then wrote the density of $\overline{x}$ as

$$\frac{\sqrt{n}}{\sigma\sqrt{2\pi}}\exp\left\{-\frac{n(\overline{x}^2-m)^2}{2\sigma^2}\right\}$$

and that of $\hat{\sigma}_2$ as

$$\frac{n^{(n-1)/2}}{2^{(n-3)/2}\Gamma\left(\frac{n-1}{2}\right)}\frac{\hat{\sigma}_2^{n-2}}{\sigma^{n-1}}\exp\left(-\frac{n\hat{\sigma}_2^2}{2\sigma^2}\right). \tag{5.6}$$

* This is contrary to the advice of most textbooks; but it can be proved to be true (*Eddington's footnote*).

FIGURE 5.2 Sir Arthur Stanley Eddington (1882–1944). Wikimedia Commons (Public Domain), http://commons.wikimedia.org/wiki/File:Arthur_Stanley_Eddington.jpg

From Eq. (5.6), Fisher obtained

$$\mathsf{E}\hat{\sigma}_2 = \sqrt{\frac{2}{n}} \cdot \frac{\Gamma\left(\frac{n}{2}\right)}{\Gamma\left(\frac{n-1}{2}\right)}\sigma, \tag{5.7}$$

$$\mathsf{E}\hat{\sigma}_2^2 = \left(1-\frac{1}{n}\right)\sigma^2. \tag{5.8}$$

Fisher noted that the distribution in (5.6) rapidly tends toward normality as n is increased. Assuming terms that are $O(n^{-2})$ or smaller can be ignored, Eq. (5.7) becomes

$$\begin{aligned}
\mathsf{E}\hat{\sigma}_2 &\approx \sqrt{\frac{2}{n}} \cdot \frac{\left(\frac{n-2}{2e}\right)^{(n-2)/2}\sqrt{2\pi(n-2)/2}}{\left(\frac{n-3}{2e}\right)^{(n-3)/2}\sqrt{2\pi(n-3)/2}}\sigma \\
&= \frac{1}{\sqrt{ne}} \cdot \frac{n-3}{\sqrt{n-2}} \cdot \sqrt{\frac{(1-2/n)^n}{(1-3/n)^n}}\sigma \\
&\approx \frac{n-3}{\sqrt{n(n-2)e}} \cdot \sqrt{\frac{e^{-2}\left(1-\frac{2}{n}\right)}{e^{-3}\left(1-\frac{9}{2n}\right)}}\sigma \\
&= \frac{n-3}{\sqrt{n(n-9/2)}}\sigma \\
&\approx \left(1-\frac{3}{n}\right)\left(1+\frac{9}{4n}\right)\sigma \\
&\approx \left(1-\frac{3}{4n}\right)\sigma.
\end{aligned}$$

Therefore, for large n, the standard error of $\hat{\sigma}_2$ is

$$\begin{aligned}
SE(\hat{\sigma}_2) &= \sqrt{\mathsf{E}\hat{\sigma}_2^2 - \mathsf{E}^2\hat{\sigma}_2} \\
&\approx \sqrt{\left(1-\frac{1}{n}\right)\sigma^2 - \left(1-\frac{3}{4n}\right)^2\sigma^2} \\
&\approx \frac{\sigma}{\sqrt{2n}}
\end{aligned} \tag{5.9}$$

(*ibid*., p. 760). By resorting to geometry again and normalizing, Fisher next obtained the density of $d_i = |x_i - \bar{x}|$ as

$$\frac{1}{\sigma}\sqrt{\frac{2n}{\pi(n-1)}}\exp\left\{-\frac{nd_i^2}{2(n-1)\sigma^2}\right\} \tag{5.10}$$

and the joint density of (d_i, d_j) as

$$\frac{2(n-1)}{\pi\sigma^2\sqrt{n(n-2)}}\exp\left\{-\frac{n-1}{2(n-2)\sigma^2}\left(d_i^2+d_j^2\right)\right\}\cosh\left\{\frac{d_i d_j}{(n-2)\sigma^2}\right\}. \tag{5.11}$$

From Eqs. (5.10) and (5.11),

$$\begin{aligned}\mathsf{E}d_i &= \sigma\sqrt{\frac{n-1}{n}}\cdot\sqrt{\frac{2}{\pi}},\\ \mathsf{E}d_i^2 &= \frac{n-1}{n}\sigma^2,\\ \mathsf{E}d_i d_j &= \frac{2\sigma^2}{n\pi}\left\{\sqrt{n(n-2)}+\sin^{-1}\left(\frac{1}{n-1}\right)\right\}.\end{aligned}$$

Using the above expectations and Eq. (5.2), Fisher obtained

$$\begin{aligned}\mathsf{E}\hat{\sigma}_1 &= \sqrt{\frac{\pi}{2}}\frac{\sum_i \mathsf{E}d_i}{n} = \sigma\sqrt{\frac{n-1}{n}},\\ \mathsf{E}\hat{\sigma}_1^2 &= \left(\frac{\pi}{2}\right)\frac{\operatorname{var}\sum_i d_i}{n^2} = \frac{(n-1)\sigma^2}{n^2}\left\{\frac{\pi}{2}+\sqrt{n(n-2)}+\sin^{-1}\left(\frac{1}{n-1}\right)\right\}.\end{aligned}$$

For large samples,

$$\begin{aligned}\mathsf{E}\hat{\sigma}_1^2 &\approx \frac{n-1}{n}\sigma^2\left\{\frac{\pi}{2n}+\left(1-\frac{2}{n}\right)^{1/2}\right\}\\ &\approx \frac{n-1}{n}\sigma^2\left(\frac{\pi}{2n}+1-\frac{1}{n}\right),\end{aligned}$$

so that the standard error of $\hat{\sigma}_1$ is

$$\begin{aligned}SE(\hat{\sigma}_1) &= \sqrt{\mathsf{E}\hat{\sigma}_1^2-\mathsf{E}^2\hat{\sigma}_1}\\ &\approx \sqrt{\frac{n-1}{n}\sigma^2\left(\frac{\pi}{2n}+1-\frac{1}{n}\right)-\frac{n-1}{n}\sigma^2}\\ &\approx \frac{\sigma}{\sqrt{n}}\sqrt{\frac{\pi-2}{2}}.\end{aligned} \tag{5.12}$$

Using Eqs. (5.9) and (5.12), Fisher was able to explain why, for large samples, $\hat{\sigma}_2$ is a superior estimator to $\hat{\sigma}_1$:

> As n is made large, therefore the standard error of σ_1 [$\hat{\sigma}_1$] tends to bear a constant ratio to that of σ_2 [$\hat{\sigma}_2$]. The former is the larger in the ratio $\sqrt{\pi-2}$; in other words, the value of the standard deviation, or probable error, obtained from the Mean Square Deviation of a sample has greater weight by 14 per cent. than that obtained from the Mean Square Deviation. To obtain

> a result of equal accuracy by the latter method, the number of observations must be increased by 14 per cent. (*ibid.*, p. 762)

Following Fisher's large-sample argument, a footnote was added to Fisher's paper by Eddington:

> Mr. Fisher kindly allows me to correct here an erroneous statement in my book, *Stellar Movements*, p. 147, footnote. I think it accords with the general experience of astronomers that, for the errors commonly occurring in practice, the mean error is a safer criterion of accuracy than the mean square error, especially if any doubtful observations have been rejected; but I was wrong in claiming a theoretical advantage for the mean error in the case of a truly Gaussian distribution...-A. S. EDDINGTON. (*ibidem*)

For his small-sample second argument, Fisher used a completely different reasoning based on the concept of sufficiency. We now describe his groundbreaking argument. Fisher's first step was to obtain the joint distribution of $(\hat{\sigma}_1, \hat{\sigma}_2)$. He pointed out that:

> When n is large the problem is simplified by the fact that both curves rapidly approach the normal form centered about the true value as mean; the only possible difference between such curves is in the standard deviation. For small values of n the case is much more complicated; failing a complete expression for the frequency surface of $\sigma_1[\hat{\sigma}_1]$ and $\sigma_2[\hat{\sigma}_2]$ in terms of n, it will be best to investigate this surface in the single case, $n=4$. This single case will be found sufficient to bring out the decisive features of the general. (*ibid.*, p. 764)

For simplicity, Fisher thus considered the case of $n=4$ observations. By using geometry, he obtained conditional, joint, and marginal densities defined across two types of spherical figures, which he called "quadrangles (Q)" (six in all) and "triangles (T)" (eight in all). The densities were:

$$f(\hat{\sigma}_1 \mid \hat{\sigma}_2) = \begin{cases} \dfrac{3}{\hat{\sigma}_2}\sqrt{\dfrac{2}{\pi}}, & (\hat{\sigma}_1,\hat{\sigma}_2) \in Q, \hat{\sigma}_2\sqrt{\pi/3} < \hat{\sigma}_1 < \hat{\sigma}_2\sqrt{\pi/2} \\ \dfrac{3}{\hat{\sigma}_2}\sqrt{\dfrac{2}{\pi}}\left(1 - \dfrac{4}{\pi}\cos^{-1}\dfrac{\hat{\sigma}_1}{\sqrt{\pi\hat{\sigma}_2^2 - 2\hat{\sigma}_1^2}}\right), & (\hat{\sigma}_1,\hat{\sigma}_2) \in Q, \hat{\sigma}_2\sqrt{\pi/4} < \hat{\sigma}_1 < \hat{\sigma}_2\sqrt{\pi/3} \\ \dfrac{8}{\hat{\sigma}_2}\sqrt{\dfrac{2}{3\pi}}, & (\hat{\sigma}_1,\hat{\sigma}_2) \in T, \hat{\sigma}_2\sqrt{\pi/3} < \hat{\sigma}_1 < \hat{\sigma}_2\sqrt{3\pi/8} \\ \dfrac{8}{\hat{\sigma}_2}\sqrt{\dfrac{2}{3\pi}}\left(1 - \dfrac{3}{\pi}\cos^{-1}\dfrac{\hat{\sigma}_1}{\sqrt{3\pi\hat{\sigma}_2^2 - 8\hat{\sigma}_1^2}}\right), & (\hat{\sigma}_1,\hat{\sigma}_2) \in T, \hat{\sigma}_2\sqrt{\pi/4} < \hat{\sigma}_1 < \hat{\sigma}_2\sqrt{\pi/3} \end{cases} \tag{5.13}$$

(*ibid.*, p. 766),

$$f(\hat{\sigma}_2 \mid \hat{\sigma}_1) = \frac{f(\hat{\sigma}_1, \hat{\sigma}_2)}{f(\hat{\sigma}_1)}, \tag{5.14}$$

where

$$f(\hat{\sigma}_1,\hat{\sigma}_2)=\begin{cases}\dfrac{48\hat{\sigma}_2}{\pi\sigma^3}e^{-2\hat{\sigma}_2^2/\sigma^2} & (\hat{\sigma}_1,\hat{\sigma}_2)\in Q, \hat{\sigma}_2\sqrt{\pi/3}<\hat{\sigma}_1<\hat{\sigma}_2\sqrt{\pi/2}\\ \dfrac{48\hat{\sigma}_2}{\pi\sigma^3}e^{-2\hat{\sigma}_2^2/\sigma^2}\left(1-\dfrac{4}{\pi}\cos^{-1}\dfrac{\hat{\sigma}_1}{\sqrt{\pi\hat{\sigma}_2^2-2\hat{\sigma}_1^2}}\right), & (\hat{\sigma}_1,\hat{\sigma}_2)\in Q, \hat{\sigma}_2\sqrt{\pi/4}<\hat{\sigma}_1<\hat{\sigma}_2\sqrt{\pi/3}\\ \dfrac{128\hat{\sigma}_2}{\pi\sqrt{3}\sigma^3}e^{-2\hat{\sigma}_2^2/\sigma^2}, & (\hat{\sigma}_1,\hat{\sigma}_2)\in T, \hat{\sigma}_2\sqrt{\pi/3}<\hat{\sigma}_1<\hat{\sigma}_2\sqrt{3\pi/8}\\ \dfrac{128\hat{\sigma}_2}{\pi\sqrt{3}\sigma^3}e^{-2\hat{\sigma}_2^2/\sigma^2}\left(1-\dfrac{3}{\pi}\cos^{-1}\dfrac{\hat{\sigma}_1}{\sqrt{3\pi\hat{\sigma}_2^2-8\hat{\sigma}_1^2}}\right), & (\hat{\sigma}_1,\hat{\sigma}_2)\in T, \hat{\sigma}_2\sqrt{\pi/4}<\hat{\sigma}_1<\hat{\sigma}_2\sqrt{\pi/3}\end{cases}$$

(*ibid.*, pp. 766–767) and

$$f(\hat{\sigma}_1)=\begin{cases}\dfrac{12}{\pi\sigma}e^{-4\hat{\sigma}_1^3/(\pi\sigma^2)}\left(1-\dfrac{4}{\pi}e^{-2\hat{\sigma}_1^2/(\pi\sigma^2)}\displaystyle\int_0^1\dfrac{e^{-2\hat{\sigma}_1^2t^2/(\pi\sigma^2)}}{1+t^2}dt\right), & (\hat{\sigma}_1,\hat{\sigma}_2)\in Q\\ \dfrac{23}{\pi\sqrt{3}\sigma}e^{-16\hat{\sigma}_1^2/(3\pi\sigma^2)}\left(1-\dfrac{3\sqrt{3}}{\pi}e^{-2\hat{\sigma}_1^2/(3\pi\sigma^2)}\displaystyle\int_0^1\dfrac{e^{-2\hat{\sigma}_1^2t^2/(\pi\sigma^2)}}{1+3t^2}dt\right), & (\hat{\sigma}_1,\hat{\sigma}_2)\in T\end{cases}$$

(*ibid.*, p. 767). Now, an examination of Eqs. (5.13) and (5.14) revealed a most remarkable peculiarity: *whereas the conditional density $f(\hat{\sigma}_2 \mid \hat{\sigma}_1)$ is a function of σ, the conditional density $f(\hat{\sigma}_1 \mid \hat{\sigma}_2)$ does not involve σ at all.* In other words:

> …, if, in seeking information as to the value of σ, we first determine $\sigma_1[\hat{\sigma}_1]$, then we can still further improve our estimate by determining $\sigma_2[\hat{\sigma}_2]$; but if we had first determined $\sigma_2[\hat{\sigma}_2]$, the frequency curve for $\sigma_1[\hat{\sigma}_1]$ being entirely independent of σ, the actual value of $\sigma_1[\hat{\sigma}_1]$ can give us no further information as to the value of σ. The whole of the information to be obtained from $\sigma_1[\hat{\sigma}_1]$ is included in that supplied by a knowledge of $\sigma_2[\hat{\sigma}_2]$.
>
> This remarkable property of $\sigma_2[\hat{\sigma}_2]$, as the methods which we have used to determine the frequency surface demonstrate, follows from the distribution of frequency density in concentric spheres over each of which $\sigma_2[\hat{\sigma}_2]$ is constant. It therefore holds if $\sigma_3[\hat{\sigma}_3]$ or any other derivate be substituted for $\sigma_1[\hat{\sigma}_1]$. If this is so, then it must be admitted that:
>
> *The whole of the information respecting σ, which a sample provides, is summed up in the value of $\sigma_2[\hat{\sigma}_2]$.** (*ibid.*, pp. 768–769)

This was the gist of Fisher's small-sample second argument in 1920, namely, that all the information on σ provided by the sample is contained in $\hat{\sigma}_2$ but not in $\hat{\sigma}_1$. In 1922, Fisher would have said that $\hat{\sigma}_2$ is a *sufficient statistic* for σ whereas $\hat{\sigma}_1$ is not, and hence $\hat{\sigma}_2$ should be preferred over $\hat{\sigma}_1$. This thus concludes our description of how Fisher first† discovered the concept of sufficiency.

* Italics are Fisher's.

† In a 1973 article (Stigler, 1973), Stigler has described how, in 1818, Laplace (1818, second supp., pp. 34–50) performed a statistical analysis similar to Fisher's and came somewhat close to discovering the concept of sufficiency. Instead of examining the sample mean deviation and sample standard deviation, Laplace compared the sample mean and sample median as estimators of the center of a normal distribution. Like Fisher, Laplace also obtained joint distributions but, unlike Fisher, did not derive any conditional distribution and was thus unable to anticipate the concept of sufficiency.

5.2.3 First Appearance of "Maximum Likelihood" in the 1922 Paper

In Section 2.3.10 (pp. 264–265), we described how Fisher first stated the principle of maximum likelihood (ML) in his very first paper (Fisher, 1912), without however using the word "likelihood." In that same section, on p. 266, we also explained how the latter word was then explicitly defined for the first time in the 1921 *Metron* paper (Fisher, 1921a). In the 1922 paper entitled "On the mathematical foundations of theoretical statistics" (Fisher, 1922c), Fisher was to give even more prominence to likelihood and ML. Fisher's motivation stemmed from the fact that the criterion of sufficiency was only a *property* of an optimal estimator and did not in itself provide a *method* for obtaining such an estimator.* That is:

> The form in which the criterion of sufficiency has been presented is not of direct assistance in the solution of problems of estimation. For it is necessary first to know the statistic concerned and its surface of distribution, with an infinite number of other statistics, before its sufficiency can be tested. For the solution of problems of estimation we require a method which for each particular problem will lead us automatically to the statistic by which the criterion of sufficiency is satisfied. Such a method is, I believe, provided by the Method of Maximum Likelihood, although I am not satisfied as to the mathematical rigour of any proof which I can put forward to that effect. (*ibid.*, p. 323)

In the above, Fisher had explicitly used the phrase "maximum likelihood" for the first time. He believed he had a proof that ML estimators (MLE) were sufficient, although he was not satisfied by its mathematical rigor. Before giving his proof, Fisher explained the method of ML as follows. Suppose a random variable X has density $f(x \mid \theta_1, \theta_2, \ldots, \theta_r)$, which depends on the parameters $\theta_1, \theta_2, \ldots, \theta_r$. Then

$$\Pr\{x < X \le x + dx\} = \Pr\{X \in dx\} = f(x \mid \theta_1, \theta_2, \ldots, \theta_r)dx.$$

Therefore, in a sample n observations, the probability that n_1 will fall in the range dx_1, n_2 will fall in the range $dx_2, \ldots, n_p$ will fall in the range dx_p is

$$P = \frac{n!}{\prod_{i=1}^{p} n_i!} \prod_{i=1}^{p} \{f(x_i \mid \theta_1, \theta_2, \ldots, \theta_r)dx_i\}^{n_i}.$$

Then:

> The method of maximum likelihood consists simply in choosing that set of values for the parameters which makes this [i.e. the above] quantity a maximum, and since in this expression the parameters are only involved in the function *f*, we have to make
>
> $$\Sigma(\log f)$$
>
> a maximum for variations of $\theta_1, \theta_2, \theta_3$ &c. In this form the method is applicable to the fitting of populations involving any number of variates, and equally to discontinuous as to continuous distributions. (*ibid.*, pp. 323–324)

* Fisher's development of maximum likelihood is also described in Stigler (2007) and Aldrich (1997).

In the foundations paper (*ibid*., p. 324), Fisher next gave a second and very powerful motivation for the use of likelihood, namely, that of *invariance under one-to-one parameter transformation*. To illustrate, Fisher considered a binomial distribution with known number of successes x out of n trials and unknown probability p of success per trial. Then, assuming a prior distribution $f(p)$, the posterior distribution of p is

$$\begin{aligned} f(p \mid x) &\propto \Pr\{X = x \mid p\} f(p) \\ &\propto p^x (1-p)^{n-x} f(p) \end{aligned}$$

At this stage, the usual practice was to appeal to Bayes' postulate (i.e., the principle of indifference): complete ignorance about a parameter is equivalent to assigning a uniform prior to it (Gorroochurn, 2012a, p. 135), that is, $f(p) = 1$ for $0 < p < 1$, so that Bayes' theorem gives

$$f(p \mid x) \propto p^x (1-p)^{n-x}. \tag{5.15}$$

Fisher strongly criticized the above method of inverse probability. To see why, let $\sin\theta = 2p - 1$ for example. The complete ignorance about p is equivalent to complete ignorance about θ, so that $f(\theta) = 1/\pi$ for $-\pi/2 < \theta < \pi/2$. Then

$$\begin{aligned} f(\theta \mid x) &\propto \Pr\{X = n \mid \theta\} f(\theta) \\ &= \Pr\{X = n \mid p\} f(\theta) \\ &\propto p^x (1-p)^{n-x} \end{aligned}$$

which implies

$$\begin{aligned} f(p \mid x) &= f(\theta \mid x) \left| \frac{dp}{d\theta} \right|^{-1} \\ &\propto p^x (1-p)^{n-x} \frac{1}{\cos\theta} \\ &\propto p^x (1-p)^{n-x} \cdot \frac{1}{p^{1/2} (1-p)^{1/2}} \\ &= p^{x-1/2} (1-p)^{n-x-1/2}. \end{aligned} \tag{5.16}$$

The above contradicted Eq. (5.15) obtained earlier. In particular, the posterior mode obtained from the two expressions (5.15) and (5.16) would be different. This showed the serious defect of inverse probability. Fisher now showed that his likelihood procedure had no such disadvantage as follows. For a binomial distribution with known number of successes x out of n trials, the likelihood of p is

$$L(p \mid x) \propto p^x (1-p)^{n-x}. \tag{5.17}$$

If we now let $\sin\theta = 2p - 1$, then the likelihood of θ is

$$L(\theta \mid x) \propto (1 + \sin\theta)^x (1 - \sin\theta)^{n-x}. \tag{5.18}$$

Differentiating the right side of Eq. (5.17) and setting to zero, we obtain the MLE

$$\hat{p} = \frac{x}{n}. \tag{5.19}$$

Likewise, from Eq. (5.18) we obtain

$$\hat{\theta} = \sin^{-1}\left(\frac{2x}{n} - 1\right). \tag{5.20}$$

Now, substituting $\theta = \sin^{-1}(2x/n - 1)$ in $\sin\theta = 2p - 1$, we get $p = x/n$. This means that whether we start from $L(p \mid x)$ or from $L(\theta \mid x)$ we end up with the same MLE $\hat{p} = x/n$, that is, the MLE is invariant to a one-to-one transformation. This example illustrates the distinction between likelihood on the one hand, being invariant to one-to-one transformations,* and inverse probability on the other hand, being affected by one-to-one transformations.

Fisher's next step in the foundations paper (Fisher, 1922c, p. 328) was now to derive the variance of the MLE θ_1 of a parameter θ. He used an asymptotic argument and assumed

$$\theta_1 \doteq N(\theta, \sigma^2)$$

for large enough samples. Then the density function of θ_1 is

$$\Phi = \frac{1}{\sigma\sqrt{2\pi}} \exp\left\{-\frac{(\theta_1 - \theta)^2}{2\sigma^2}\right\}.$$

The likelihood of any value θ is proportional to

$$\exp\left\{-\frac{(\theta_1 - \theta)^2}{2\sigma^2}\right\}$$

and attains its maximum when $\theta = \theta_1$. Moreover,

$$\begin{aligned} \frac{\partial}{\partial\theta}\log\Phi &= \frac{\theta_1 - \theta}{\sigma^2}. \\ \therefore \quad \frac{\partial^2}{\partial\theta^2}\log\Phi &= -\frac{1}{\sigma^2}. \end{aligned} \tag{5.21}$$

Now Fisher observed that Φ is the density of all samples for which the chosen statistic has the value θ_1, so that $\Phi = \sum\phi$, where ϕ is the density of a given such sample and the sum is taken over all such samples. Moreover, if f is the density of an observation from a given sample, then

$$\log\phi = C + \sum\log f, \tag{5.22}$$

*Zehna later (1966) proved that invariance of the MLE holds even when the transformation is many-to-one.

with C a constant not involving the parameters and the sum being over all observations of the given sample. By using a Taylor series expansion of f about $\theta = \theta_1$,

$$\log f(\theta) = \log f(\theta_1) + (\theta - \theta_1)\frac{\partial}{\partial\theta}\log f(\theta_1) + \frac{1}{2}(\theta - \theta_1)^2 \frac{\partial^2}{\partial\theta^2}\log f(\theta_1) + \cdots$$

Fisher wrote the above as

$$\log f = \log f_1 + a(\theta - \theta_1) + \frac{b}{2}(\theta - \theta_1)^2 + \cdots$$

(*ibidem*), where $f_1 = f(\theta_1)$, $a = \frac{\partial}{\partial\theta}\log f(\theta_1)$, and $b = \frac{\partial^2}{\partial\theta^2}\log f(\theta_1)$. Equation (5.22) then becomes

$$\begin{aligned}\log\phi &= C + \Sigma\log f_1 + (\theta - \theta_1)\Sigma a + \frac{1}{2}(\theta - \theta_1)^2 \Sigma b + \cdots \\ &= C + (\theta - \theta_1)\Sigma a + \frac{1}{2}(\theta - \theta_1)^2 \Sigma b + \cdots\end{aligned} \tag{5.23}$$

Since θ_1 is the MLE (i.e., "for optimum statistics"), $\Sigma a = 0$.* Also, from the Central Limit Theorem,

$$\frac{\Sigma b - n\mathsf{E}b}{\sqrt{n\,\mathrm{var}\,b}} \mathrel{\dot{\div}} N(0,1),$$

so that

$$\Sigma b - n\mathsf{E}b \sim O\left(n^{1/2}\right).$$

Likewise $\theta - \theta_1 \sim O(n^{-1/2})$. Therefore, "the only terms in log ϕ [from Eq. 5.23], which are not reduced without limits, as n is increased" are

$$\log\phi = C + \frac{1}{2}n(\theta - \theta_1)^2 \mathsf{E}b \tag{5.24}$$

so that

$$\phi \propto \exp\left\{\frac{1}{2}n(\theta - \theta_1)^2 \mathsf{E}b\right\}. \tag{5.25}$$

Fisher's final step was to argue that the proportionality constant in (5.25) applied to all samples which had the value θ_1, and so was also for Φ. Therefore, from Eq. (5.24),

$$\log\Phi = C' + \frac{1}{2}n(\theta - \theta_1)^2 \mathsf{E}b,$$

where $C' = C + (\theta - \theta_1)\Sigma a$, so that

$$\frac{\partial^2}{\partial\theta^2}\log\Phi = n\mathsf{E}b.$$

*Assuming that the ML estimator does not fall on the boundary of the parameter space.

Using (5.25), Fisher finally obtained

$$\begin{aligned}\operatorname{var}\theta_1 &= \sigma^2 \\ &= -\frac{1}{n\mathrm{E}b} \\ &= -\frac{1}{n\mathrm{E}\dfrac{\partial^2}{\partial\theta^2}\log f(\theta_1)}\end{aligned} \tag{5.26}$$

(*ibid*., p. 329)* and hence obtained the large-sample variance of the MLE θ_1 (see also Eq. 2.35). We shall soon revisit the above formula when discussing the prior work of Edgeworth (cf. Section 5.7.5, Eq. 5.145)

In Section 7 of the 1922 paper (Fisher, 1922c), Fisher attempted to prove that MLEs were sufficient. Unfortunately, he is incorrect here, as MLEs are not necessarily sufficient. Let us examine Fisher's argument to see where his error lies.

Fisher first denoted the MLE of a parameter θ by $\hat{\theta}$, and any other statistic by θ_1. Let the joint density of $(\hat{\theta}, \theta_1)$ be $f(\theta, \hat{\theta}, \theta_1)$. He then stated that we must have

$$\frac{\partial}{\partial\theta} f\left(\theta, \hat{\theta}, \theta_1\right)\bigg|_{\theta=\hat{\theta}} = 0. \tag{5.27}$$

Fisher noted that Eq. (5.27) was satisfied because $f(\theta, \hat{\theta}, \theta_1)$ could be written as

$$f\left(\theta, \hat{\theta}, \theta_1\right) = \phi\left(\theta, \hat{\theta}\right)\phi_1\left(\hat{\theta}, \theta_1\right) \tag{5.28}$$

(*ibid*., p. 331) (where ϕ is a function proportional to the density of $\hat{\theta}$, and ϕ_1 is a function of $\hat{\theta}$ and θ_1), so that

$$\frac{\partial f}{\partial\theta}\bigg|_{\theta=\hat{\theta}} = \frac{\partial\phi}{\partial\theta}\bigg|_{\theta=\hat{\theta}} \times \phi_1 = 0 \times \phi_1 = 0.$$

Now, in general the joint density $f(\theta, \hat{\theta}, \theta_1)$ can be written as the product of two densities as

$$f\left(\theta, \hat{\theta}, \theta_1\right) = f\left(\theta, \hat{\theta}\right) f\left(\theta, \theta_1 \mid \hat{\theta}\right). \tag{5.29}$$

Comparing Eq. (5.29) with Eq. (5.28) shows that $f(\theta, \theta_1 \mid \hat{\theta})$ does not depend on the parameter θ. Hence, Fisher concluded that the MLE $\hat{\theta}$ must also be sufficient for θ.

There are two major problems with Fisher's "proof." First, although Fisher's aim was to prove that MLEs were sufficient, his argument in fact proved the converse because it established Eq. (5.27) as a consequence of Eq. (5.28). Second, Fisher has implicitly but wrongly assumed that the sufficient statistic must be of the same dimension as the

* Fisher would later call $-\mathrm{E}\frac{\partial^2}{\partial\theta^2}\log f(\theta)$ (see Eq. 5.41) the expected information of one observation.

parameter θ. Indeed, a sufficient statistic if it exists can be of the same or of different dimension as the parameter. For example, consider (X_i, Y_i), $i=1, \ldots, n$, from the joint density

$$f(x,y\mid\theta)=\begin{cases}\exp(\theta x+y/\theta), & x>0, y>0\\ 0, & \text{elsewhere}\end{cases}$$

The likelihood is then $L\{\theta\mid(x_i,y_i)\}=\exp\left(\theta\sum_i x_i+\sum_i y_i/\theta\right)$. From this expression, it can be shown that the *bivariate* statistic $\left(\sum_i X_i, \sum_i Y_i\right)$ is sufficient for the *univariate* parameter θ. But, by setting $\partial\log L/\partial\theta=0$, it can also be shown that the MLE of θ is $\hat{\theta}=\sqrt{\sum_i Y_i/\sum_i X_i}$, which is clearly not equal to the sufficient statistic.

Therefore, the correct conclusion to be drawn from Fisher's argument is that if a statistic that has the same dimension as the parameter of interest is sufficient, then it is equal to the MLE.

It is worth noting that Fisher's factorization in Eq. (5.28) may be regarded as providing a sufficient condition for the statistic $\hat{\theta}$ to be sufficient. That the factorization is also necessary was later proved by Neyman (1935), and this became known as the *Fisher–Neyman factorization theorem* for a sufficient statistic: A statistic T is sufficient for an unknown parameter θ if and only if the likelihood function can be factorized as $L(\theta\mid\mathbf{x})=g_\theta(T)\,h(\mathbf{x})$, where g is a nonnegative function of θ and T, and h is a function of $\mathbf{x}=(x_1,\ldots,x_n)$ that does not depend on θ. Using the Fisher–Neyman factorization, we can thus make the following statement, which is close to what Fisher intended to prove in the 1922 paper: An MLE is a function of a sufficient statistic.* Of course, this does not necessarily mean that the MLE is itself sufficient (as the last example shows), although it often happens that the MLE and sufficient statistic coincide.

In some situations, there might be no sufficient statistic beyond the IID† random variables $(X_1, X_2, \ldots, X_n)$ and the order statistics $(X_{(1)}, X_{(2)}, \ldots, X_{(n)})$ (the latter two statistics being always sufficient). For example, for the Cauchy distribution

$$f(x)=\frac{1}{\pi}\cdot\frac{1}{1+(x-\theta)^2},\qquad -\infty<x<\infty; -\infty<\theta<\infty,$$

the likelihood function is

$$L(\mathbf{x})=\frac{1}{\pi^n}\prod_{i=1}^{n}\frac{1}{1+(x_i-\theta)^2}.$$

Applying the Fisher–Neyman factorization indicates that there is no lower-dimensional sufficient statistic. In particular, there is no single sufficient statistic in this case.

5.2.4 The Method of Moments and its Criticism by Fisher (Pearson, 1894; Fisher, 1912; 1922)

The method of moments was put forward by Karl Pearson for the purpose of theoretically fitting various types of curves to observational data. The method was introduced in the

* For this statement to be true, the ML estimator must be unique. See Section 5.2.7.

† Independent and identically distributed.

first of Pearson's long series of papers entitled "Mathematical contributions to the theory of evolution" (Pearson, 1894). In the first paper, we read:

> ...good results for a simple normal curve are obtained by finding the mean from the first moment, and the error of mean square from the second moment, so it seems likely that the present investigation, based on the first five or six moments of the frequency-curve, may also lead to good results. While a method of equating chosen ordinates of the given curve and those of the components leaves each equation based only on the measurements of organs of one size, the method of moments uses all the given data in the case of each equation for the unknowns, and errors in measurement will, thus, individually have less influence. (*ibid.*, p. 75)

In the above, Pearson had tried to fit two normal curves to observed data. Using the latter, he calculated the first moments and equated these to the theoretical moments. This was the method of moments and enabled Pearson to identify the two normal curves. The method of moments, though, was to play an even more important role the following year when Pearson introduced his comprehensive system of frequency curves (Pearson et al., 1895). Basing himself on the normal, binomial, and hypergeometric distributions, Pearson introduced a system of probability distributions satisfying the differential equation:

$$\frac{df}{dx} = \frac{(x-a)f}{b_0 + b_1 x + b_2 x^2}. \tag{5.30}$$

In the above, a, b_0, b_1 and b_2 are four parameters that characterize the density f. When these parameters are varied, several density functions such as the normal, gamma, beta, etc. distributions can be obtained (cf. Table 5.1).

Pearson's aim was to use the method of moments in order to identify the parameters in Eq. (5.30). In modern notation, this can be done as follows (Stuart et al., 1999, pp. 226–227). First, we note that Eq. (5.30) can be written as

$$x^r \left(b_0 + b_1 x + b_2 x^2\right) \frac{df}{dx} = x^r (x-a) f,$$

where r is a natural number. By integrating the left side of the above by parts over the support of the distribution, we have

$$\left[x^r \left(b_0 + b_1 x + b_2 x^2\right) f\right]_{-\infty}^{\infty} - \int_{-\infty}^{\infty} \left\{rb_0 x^{r-1} + (r+1)b_1 x^r + (r+2)b_2 x^{r+1}\right\} f dx = \int_{-\infty}^{\infty} \left(x^{r+1} - ax^r\right) f dx.$$

By assuming that the expression in the square brackets above vanishes at the extremities, and writing $\mu_r' = \mathsf{E}X^r$, we obtain

$$rb_0 \mu_{r-1}' + \{(r+1)b_1 - a\} \mu_r' + \{(r+2)b_2 + 1\} \mu_{r+1}' = 0, \quad k = 0,1,2,\ldots \tag{5.31}$$

TABLE 5.1 The Pearson System of Curves

No. of Type			
Main Types	**Equation**	**Origin**	**Criterion**
I	$y = y_0\left(1+\frac{x}{a_1}\right)^{\nu a_1}\left(1-\frac{x}{a_2}\right)^{\nu a_2}$	Mode (antimode)	$\kappa < 0$
IV	$y = y_0\left(1+\frac{x^2}{a^2}\right)^{-m} e^{-\nu \tan^{-1}(x/a)}$	$\frac{\nu a}{2m-2}$ after mean	$0 < \kappa < 1$
VI	$y = y_0 x^{-q_1}(x-a)^{q_2}$	a before start of curve	$\kappa > 1$
Transition Types			
Normal curve	$y = y_0 e^{-x^2/(2\sigma^2)}$	Mode (= mean)	$\kappa = 0, \beta_1 = 0, \beta_2 = 3$
II	$y = y_0\left(1-\frac{x^2}{a^2}\right)^{m}$	Mode (= mean)	$\kappa = 0, \beta_1 = 0, \beta_2 < 3$
VII	$y = y_0\left(1+\frac{x^2}{a^2}\right)^{-m}$	Mode (= mean)	$\kappa = 0, \beta_1 = 0, \beta_2 > 3$
III	$y = y_0\left(1+\frac{x}{a}\right)^{\gamma a} e^{-\gamma x}$	Mode	$2\beta_2 = 6 + 3\beta_1$
V	$y = y_0 x^{-p} e^{-\gamma/x}$	Start of curve	$\kappa = 1$
VIII	$y = y_0\left(1+\frac{x}{a}\right)^{-m}$	End of curve	$\kappa < 0, \lambda = 0,$ $5\beta_2 - 6\beta_1 - 9 < 0.$
IX	$y = y_0\left(1+\frac{x}{a}\right)^{m}$	End of curve	$\kappa < 0, \lambda = 0,$ $5\beta_2 - 6\beta_1 - 9 > 0,$ $2\beta_2 - 3\beta_1 - 6 < 0$
X	$y = y_0 e^{-x/\sigma}$	Start of curve	$\beta_1 = 4, \beta_2 = 9$
XI	$y = y_0 x^{-m}$	b before start	$\kappa > 1, \lambda = 0,$ $2\beta_2 - 3\beta_1 - 6 > 0.$
XII	$y = y_0\left\{\frac{\sigma\left(\sqrt{3+\beta_1}+\sqrt{\beta_1}\right)+x}{\sigma\left(\sqrt{3+\beta_1}-\sqrt{\beta_1}\right)-x}\right\}^{\sqrt{\frac{\beta_1}{3+\beta_1}}}$	Mean	$5\beta_2 - 6\beta_1 - 9 = 0$

From Elderton and Johnson (1969, p. 45).

$$\kappa = \frac{\beta_1(\beta_2+3)^2}{4(4\beta_2-3\beta_1)(2\beta_2-3\beta_1-6)}, \beta_1 = \frac{\mu_3^2}{\mu_2^3}, \beta_2 = \frac{\mu_4}{\mu_2^2},$$

$$\lambda = \frac{(4\beta_2-3\beta_1)(10\beta_2-12\beta_1-18)^2 - \beta_1(\beta_2+3)^2(8\beta_2-9\beta_1-12)}{(3\beta_1-2\beta_2+6)\left\{\beta_1(\beta_2+3)^2+4(4\beta_2-3\beta_1)(3\beta_1-2\beta_2+6)\right\}}$$

To determine the four parameters a, b_0, b_1, and b_2, we use the following four equations obtained from the previous difference equation:

$$\left.\begin{aligned}(b_1 - a) + (2b_2 + 1)\mu_1' &= 0\\ b_0 + (2b_1 - a)\mu_1' + (3b_2 + 1)\mu_2' &= 0\\ 2b_0\mu_1' + (3b_1 - a)\mu_2' + (4b_2 + 1)\mu_3' &= 0\\ 3b_0\mu_1' + (4b_1 - a)\mu_3' + (5b_2 + 1)\mu_4' &= 0\end{aligned}\right\} \qquad (5.32)$$

From the observed data $x_1, x_2, \ldots, x_n$, we calculate the observed moments $\hat{\mu}_j' = \left(\sum_i x_i^j\right)/n$, $j = 1,2,3,4$, and equate these to the theoretical ones. The equations in (5.32) can now be solved for a, b_0, b_1, and b_2, from which the particular distribution can be identified. This was, in essence, Pearson's approach.

Although Pearson's introduction of the method of moments was a great innovation in statistics, an important criticism of the method came in 1912 from a young Fisher (1912). Fisher criticized both the methods of least squares and moments (cf. Section 2.3.10, pp. 264–265). The former suffered from lack of invariance to rescaling of the independent variable while Fisher pointed out the latter was arbitrary. Indeed, what is the theoretical justification for choosing the *particular* four equations in (5.32) from the difference equation in (5.31)? Against these two methods, Fisher then introduced his own method of estimation, namely, the method of ML (although his did not use the term "likelihood" in 1912).

Fisher's ideas on ML were not very precise in 1912 but by 1922 had been considerably refined. In the 1922 foundations paper (Fisher, 1922c), Fisher came back to his criticism of the method of moments and his advocacy of the method of ML. Although he acknowledged that the method of moments was "without question of great practical utility," Fisher gave explicit examples where this method was unsatisfactory.

The first example considered by Fisher is that of a Cauchy distribution (a special Pearson type VII distribution), which has density

$$f(x) = \frac{1}{\pi\left\{1 + (x - m)^2\right\}}, \qquad -\infty < x < \infty,$$

where m is the parameter of the distribution. This distribution is symmetric and is centered at $x = m$. It has a similar shape to the normal distribution except that its tails are heavier. Fisher then pointed out that applying the method of moments would result in the parameter m being estimated by the sample mean $\bar{X}$. But this estimate is "entirely useless" because:

> …the distribution of the mean of such samples is in fact the same, identically, as that of a single observation. In taking the mean of 100 values of x, we are no nearer obtaining the value of m than if we had chosen any value of x out of the 100. (*ibid.*, p. 322)

Fisher then argued that, on the other hand, a better estimate of m is the sample median. This has a standard error of $\neq / (2\sqrt{n})$, which tends to zero as n becomes large.

In Section 8 of the foundations paper, Fisher used a Pearson Type III curve to calculate the efficiency of the method of moments relative to the method of ML. Let X have density

$$f(x)=\frac{1}{p!a}\left(\frac{x-m}{a}\right)^p \exp\left(-\frac{x-m}{a}\right)$$

(*ibid*., p. 332), where a, p, and m are parameters such that only m is unknown and p is a positive integer. The method of moments proceeds by setting $\mathsf{E}X=m+a(p+1)$ to $\bar{x}$, so that

$$\bar{x}=\hat{m}_\mu+a(p+1),$$

where $\hat{m}_\propto$ is the method-of-moment estimator of m. The above equation implies

$$\operatorname{var}\hat{m}_\mu=\operatorname{var}\bar{X}=\frac{\operatorname{var}X}{n}=\frac{a^2(p+1)}{n}.$$

For the method of ML, Fisher wrote the log-likelihood as

$$\begin{aligned} l &= \log\prod_{i=1}^n f(x_i) \\ &= \log\left[\left(\frac{1}{p!a}\right)^n \exp\left\{-\frac{\sum_{i=1}^n (x_i-m)}{a}\right\}\prod_{i=1}^n\left(\frac{x_i-m}{a}\right)^p\right] \\ &= -n\log a - n\log(p!)+p\sum_{i=1}^n \log\left(\frac{x_i-m}{a}\right)-\sum_{i=1}^n\frac{(x_i-m)}{a}. \end{aligned}$$

Therefore, the MLE $\hat{m}$ can be obtained by solving for m in the equation

$$\frac{\partial l}{\partial m}=p\sum_{i=1}^n\left(\frac{1}{x_i-m}\right)+\frac{n}{a}=0.$$

Using Eq. (5.26), the large-sample variance of $\hat{m}$ is

$$\begin{aligned} \operatorname{var}\hat{m} &= -\frac{1}{\mathsf{E}\dfrac{\partial^2 L}{\partial m^2}} \\ &= \frac{1}{\mathsf{E}p\sum_{i=1}^m \dfrac{1}{(x_i-m)^2}} \\ &= \frac{a^2(p-1)}{n}. \end{aligned}$$

Thus, Fisher was able to show that the relative efficiency of $\hat{m}_\mu$ with respect to $\hat{m}$ is

$$\frac{\operatorname{var} \hat{m}}{\operatorname{var} \hat{m}_\mu} = \frac{p-1}{p+1} = 1 - \frac{2}{p+1} < 1$$

(*ibid.*, p. 333). Although Fisher argued only through examples, it is true that moment estimators are not in general efficient in spite of being both consistent and asymptotically normal.

Fisher's next task in the foundations paper (*ibid.*, p. 356) was to explain why the method of moments was efficient in a small region surrounding the normal curve. Before we explain this, let us make a remark on a connection between ML and moment estimators. Consider a log-likelihood $l \equiv l(\theta_1, \theta_2, \ldots, \theta_p \mid \mathbf{x})$, which is such that

$$\frac{\partial l}{\partial \theta_j} = a_{0,j} + a_{1,j} \sum_{i=1}^{n} x_i + a_{2,j} \sum_{i=1}^{n} x_i^2 + a_{3,j} \sum_{i=1}^{n} x_i^3 + \cdots, \quad j = 1, 2, \ldots, p, \tag{5.33}$$

where $a_{0,j}, a_{1,j}, \ldots$ are constants that depend on the θ_j's. Then the MLEs of $\theta_1, \theta_2, \ldots, \theta_p$ can be obtained by solving, for $j = 1, 2, \ldots, p$,

$$a_{0,j} + a_{1,j} \sum_{i=1}^{n} x_i + a_{2,j} \sum_{i=1}^{n} x_i^2 + a_{3,j} \sum_{i=1}^{n} x_i^3 + \cdots = 0. \tag{5.34}$$

Taking expectations on both sides of the last equation,

$$a_{0,j} + a_{1,j} n\mu_1' + a_{2,j} n\mu_2' + a_{3,j} n\mu_3' + \cdots = 0. \tag{5.35}$$

Comparing Eqs. (5.34) and (5.35),

$$\mu_1' = \frac{1}{n} \sum_{i=1}^{n} x_i, \quad \mu_2' = \frac{1}{n} \sum_{i=1}^{n} x_i^2, \quad \mu_3' = \frac{1}{n} \sum_{i=1}^{n} x_i^3, \ldots$$

But the above are simply the method-of-moment equations. Therefore, when the log-likelihood equations are of the special form in (5.33), the ML and moment estimators of $\theta_1, \theta_2, \ldots, \theta_p$ coincide. Note that the density that corresponds to (5.33) is of the form

$$f\left(x \mid \theta_1, \theta_2, \ldots, \theta_p\right) = e^{b_0 + b_1 x + b_2 x^2 + b_3 x^3 + \cdots},$$

where $b_0, b_1, \ldots$ are constants that depend on the θ_j's such that $\int f(x)dx = 1$.

We now come back to Fisher's investigation of the efficiency of moment estimators in the 1922 foundations paper. His aim was to show the method of moments was efficient in a small region surrounding the normal curve. He assumed the underlying probability distribution is of the form

$$y = \exp\left\{-a^2\left(x^4 + p_1 x^3 + p_2 x^2 + p_3 x + p_4\right)\right\}. \tag{5.36}$$

Since $\int y dx = 1$, the last density has four parameters. From our previous discussion, the MLEs coincide with the moment estimators for this distribution, and the latter are therefore efficient. Next Fisher noted that Eq. (5.36) could be written as

$$y = C \exp\left(-\frac{x^2}{2\sigma^2} + k_1 \frac{x^3}{\sigma^3} + k_2 \frac{x^4}{\sigma^4}\right) \tag{5.37}$$

(where C, σ, k_1, and k_2 are constants) from which

$$\frac{d}{dx}\log y = -\frac{x}{\sigma^2}\left(1 - 3k_1 \frac{x}{\sigma} - 4k_2 \frac{x^2}{\sigma^2}\right) \approx -\frac{x}{\sigma^2\left(1 + 3k_1 \frac{x}{\sigma} + 4k_2 \frac{x^2}{\sigma^2}\right)}, \tag{5.38}$$

neglecting terms that are $O(k_1^2)$ and $O(k_2^2)$. By writing the last equation as

$$\frac{dy}{dx} \approx -\frac{xy}{\sigma^2\left(1 + 3k_1 \frac{x}{\sigma} + 4k_2 \frac{x^2}{\sigma^2}\right)}$$

and by comparing with Eq. (5.30), the above can be recognized as the (approximate) differential equation of a Pearson curve. Therefore, when k_1 and k_2 are both small, the curve in (5.37) is close to the normal curve and also to the Pearson system of curves. Since the curve in (5.37) yields efficient estimators, Fisher concluded that the method of moments is efficient in a region close to the normal curve.

5.2.5 Further Refinement of the 1922 Paper in 1925: Efficiency and Information

Fisher's 1925 paper, "Theory of statistical estimation" (Fisher, 1925c), was an important follow-up to his 1922 paper (Fisher, 1922c). The 1925 paper not only refined some aspects of the previous paper but also broke new ground by introducing the concepts of information loss and ancillarity, and tying these to efficiency. We now describe these topics.

Fisher started his paper with a "Prefatory Note" on his notion of probability as being a frequency ratio calculated from a *hypothetical infinite population* and that:

> ...the word infinite is to be taken in its proper mathematical sense as denoting the limiting conditions approached by increasing a finite number indefinitely. (Fisher, 1925c, p. 700)

Next, Fisher succinctly described the aim of the 1925 paper:

> ...When in 1921* the author put forward in the *Phil. Trans.* a paper (1)† on mathematical statistics he was principally concerned, in respect of problems of estimation, with the practical importance of making estimates of high efficiency, i.e. of using statistics which embody a large proportion of the relevant information available in the data, and which ignore, or reject along with the irrelevant information, only a small proportion of that which is

*The reference here is to the 1922 foundations paper (Fisher, 1922c).

†See numbered list of references on p. 725 (*Fisher's footnote*).

> relevant. Many of the properties of efficient statistics, such as that even moderate inefficiency of estimation will vitiate tests of goodness of fit, were at that time unknown, and the further discrimination among statistics within the efficient group, a discrimination which is essential to the advance of the theory of small samples, was left in much obscurity. Further work along the lines of the 1921 paper has, however, cleared up the main outstanding difficulties, and seems to make possible a theory of statistical estimation with some approach to logical completeness. (*ibid.*, p. 701)

Fisher was thus mainly concerned with the issue of *efficiency in small samples*. However, before investigating this problem, Fisher touched on many other important related topics. We now describe these as well as the central problem of small-sample efficiency.

The first two sections (Sections 1 and 2) of the 1925 paper were brief and repeated some of the ideas on estimation and consistency that were already discussed in the 1922 foundations paper. In accordance with his 1922 definition of "efficiency," Fisher stated in Section 3:

> ...The criterion of efficiency requires that the fixed value to which the variance of a statistic ... multiplied by n, tends, shall be as small as possible...If we know the variance of any efficient statistic and that of any other statistic under discussion, then the efficiency of the latter may be calculated from the ratio of the two values. The efficiency of a statistic represents the fraction, of the relevant information available, actually utilised, in large samples, by the statistic in question. (*ibid.*, p. 703)

In Section 4 of the 1925 paper, Fisher used a clever argument to show that Corr $(A, B) \to \sqrt{E}$ as the sample size increases, where A is an efficient statistic and B is an inefficient statistic with efficiency $E\,(0 < E \le 1)$. His argument was as follows.

Define a new statistic

$$C = \frac{\left(1-\rho\sqrt{E}\right)A + \left(E-\rho\sqrt{E}\right)B}{\left(1+E-2\rho\sqrt{E}\right)}$$

(*ibid.*, p. 705), where $\rho = \text{Corr}\,(A, B)$. Note that C will be an unbiased estimator of ρ if A and B are also unbiased. Now, for large samples, assuming that $\text{var}\,A \to \sigma^2/n$ so that $\text{var}\,B \to \sigma^2/(nE)$, we have

$$\text{var}\,C \to \frac{\left(1-\rho\sqrt{E}\right)^2\left(\frac{\sigma^2}{n}\right) + \left(E-\rho\sqrt{E}\right)^2\left(\frac{\sigma^2}{nE}\right) + 2\left(1-\rho\sqrt{E}\right)\left(E-\rho\sqrt{E}\right)\rho\sqrt{\frac{\sigma^2}{n}}\sqrt{\frac{\sigma^2}{nE}}}{\left(1+E-2\rho\sqrt{E}\right)^2}$$

$$= \frac{\sigma^2}{n}\left(\frac{1-\rho^2}{1+E-2\rho\sqrt{E}}\right).$$

Fisher wrote the above as

$$\text{var}\,C \to \frac{\sigma^2}{n}\left\{\frac{\left(1-\rho^2\right)}{\left(1-\rho^2\right)+\left(\rho-\sqrt{E}\right)^2}\right\}$$

(*ibidem*). From this equation, he reasoned that, since $\text{var}\, C \geq \text{var}\, A^* \to \sigma^2 / n$ we must have $\rho \to \sqrt{E}$ for large samples. This completes Fisher's argument.

Next, in Section 5, Fisher stated:

> ...We shall see that the method of maximum likelihood will always provide a statistic which, if normally distributed in large samples with variance falling off inversely to the sample number, will be an efficient statistic [if one exists]. (*ibid*., p. 707)

However:

> ...since the equations of maximum likelihood do not always lend themselves to direct solution, it is of importance that, starting with an inefficient estimate, we can, by a single process of approximation, obtain an efficient estimate. (*ibid*., p. 708)

Thereupon, Fisher introduced the *method of scoring* (although the term "score" was not used in the 1925 paper[†]) for the case of a Cauchy distribution:

$$f(x) = \frac{1}{\pi\left\{1+(x-m)^2\right\}}, \quad -\infty < x < \infty.$$

In the above, m is the parameter of the distribution. The log-likelihood is

$$\begin{aligned} l(\theta) &= \log \prod_{i=1}^{n} f(x_i) \\ &= \log\left\{\frac{1}{\pi^n}\prod_{i=1}^{n}\frac{1}{1+(x_i-m)^2}\right\} \\ &= -n\log\pi - \sum_{i=1}^{n}\log\left\{1+(x_i-m)^2\right\}. \end{aligned}$$

The first and second derivatives with respect to m are, respectively,

$$l'(m) = 2\sum_{i=1}^{n}\frac{(x_i-m)}{1+(x_i-m)^2},$$

$$l''(m) = 2\sum_{i=1}^{n}\frac{(x_i-m)^2-1}{\left\{1+(x_i-m)^2\right\}^2}.$$

Now, to numerically solve the *score equation* $l'(m) = 0$ with the aid of the Newton–Raphson formula, we would normally use

$$m_2 = m_1 - \frac{l'(m_1)}{l''(m_1)}, \tag{5.39}$$

*This is because A is an efficient statistic.

[†]But was first used by Fisher in 1935 (1935b).

where m_1 is an initial estimate of m, and m_2 is a second-step estimate. However, instead of Eq. (5.39), Fisher preferred to use

$$m_2 = m_1 - \frac{l'(m_1)}{\mathsf{E}l''(m_1)}, \tag{5.40}$$

with m_1 as the median and with

$$\begin{aligned}
-\mathsf{E}l''(m) &= -\int_{-\infty}^{\infty} l''(m) f(x)dx \\
&= -\int_{-\infty}^{\infty} 2\sum_{i=1}^{n} \frac{(x_i - m)^2 - 1}{\left\{1+(x_i-m)^2\right\}^3} \cdot \frac{1}{\pi\left\{1+(x_i-m)^2\right\}} dx_i \\
&= -\frac{2}{\pi}\sum_{i=1}^{n}\int_{-\infty}^{\infty} \frac{u_i^2 - 1}{\left(1+u_i^2\right)^3} du_i \quad (u_i = x_i - m) \\
&= -\frac{2}{\pi}\sum_{i=1}^{n}\left(-\frac{\pi}{4}\right) \\
&= \frac{n}{2}.
\end{aligned}$$

Thus, Eq. (5.40) becomes

$$m_2 = m_1 + \frac{4}{n}\sum_{i=1}^{n} \frac{(x_i - m_1)}{1+(x_i - m_1)^2}.$$

Fisher then observed that, while the error of estimation of m_1 is $1/\sqrt{n}$, that of m_2 is $1/n$, so that m_2 is efficient. The resulting asymptotic efficiency when the Newton–Raphson formula is used with Fisher's modification in (5.40) was later rigorously proved by Le Cam (1956).

In Section 6 of the paper, Fisher discussed the "intrinsic accuracy of error curves." He had already defined and explained this concept in the 1922 paper (Fisher, 1922c, pp. 309–310, 322, 339, 341). He now restated:

> …The variance of efficient statistics from a distribution of any form affords us a measure of an important property of the distribution itself… We may thus obtain a measure of the intrinsic accuracy of an error curve, and so compare together curves of entirely different form. If the variance of an efficient estimate derived from a large sample of n is A/n, then the intrinsic accuracy of the distribution is defined as $1/A$. (Fisher, 1925c, p. 709)

Thus, suppose $f(x \mid \theta)$ is a distribution and T is an efficient estimator of the parameter θ. For large sample sizes n, if T has variance A/n then the *intrinsic accuracy* of the distribution is given by $1/A$. Hence, in general, the intrinsic accuracy is

$$\frac{1}{n \operatorname{var} T},$$

where the sample size n is large. With $f(x \mid \theta) = y$ and using Eq. (5.26), Fisher also wrote the formula for the intrinsic accuracy as

$$-\int y \frac{\partial^2}{\partial \theta^2}(\log y)\,dx = \int \frac{1}{y}\left(\frac{\partial y}{\partial \theta}\right)^2 dx.$$

Thus, the intrinsic accuracy is also

$$-\mathsf{E}\frac{\partial^2 \log y}{\partial \theta^2} = \mathsf{E}\left(\frac{\partial \log y}{\partial \theta}\right)^2. \qquad (5.41)$$

Fisher next gave a formal definition of (expected) *information of one observation**:

> What we have spoken of as intrinsic accuracy of an error curve may equally be conceived as the [expected] amount of information of a single observation belonging to such a distribution. (*ibidem*)

Thus, Fisher defined the (expected) information to be the same as the intrinsic accuracy (see also Eq. 2.36).

In Section 7 of the 1925 paper, Fisher gave a proof of the asymptotic efficiency of the MLE (Fisher, 1925c, p. 710). In his proof, Fisher used his previous notation (See Eq. 5.22) to denote the density of an observation from a sample by f and the density of a given such sample by ϕ. The MLE of the parameter θ is now denoted by $\hat{\theta}$. Then using a Taylor series expansion about $(\theta - \hat{\theta})$ for $\partial \log \phi / \partial \theta$ in Eq. (5.22):

$$\begin{aligned}
\frac{\partial \log \phi}{\partial \theta} &= \frac{\partial}{\partial \theta}(C + \Sigma f) \\
&= \Sigma \frac{\partial f}{\partial \theta} \\
&= \Sigma \left\{ \left.\frac{\partial f}{\partial \theta}\right|_{\theta=\hat{\theta}} + (\theta - \hat{\theta}) \left.\frac{\partial^2 f}{\partial \theta^2}\right|_{\theta=\hat{\theta}} + \cdots \right\} \\
&= \Sigma (\theta - \hat{\theta}) \left.\frac{\partial^2 f}{\partial \theta^2}\right|_{\theta=\hat{\theta}} + \cdots
\end{aligned}$$

Fisher wrote the above as

$$\frac{\partial \log \phi}{\partial \theta} = -n\alpha(\theta - \hat{\theta}) \qquad (5.42)$$

(*ibid.*, p. 711), where $\frac{1}{n}\Sigma \frac{\partial^2 \log f}{\partial \theta^2} \to \mathsf{E}\frac{\partial^2 \log f}{\partial \theta^2} = -\alpha$ for large samples (α being the intrinsic accuracy or expected information) and $(\theta - \hat{\theta})$ is a small quantity of order $n^{-1/2}$. Next, let T be any other statistic with distribution Φ. Then

$$\Phi = \Sigma \phi, \qquad (5.43)$$

* The expected information is also now called the Fisher information.

the sum being taken over all samples for which the chosen statistic has the value T. If $T \sim N(\theta, \sigma^2)$ asymptotically, then using Eq. (5.21),

$$\begin{aligned} \frac{1}{\sigma^2} &= -\frac{\partial^2 \log \Phi}{\partial \theta^2} \\ &= -\frac{\partial}{\partial \theta}\left(\frac{\partial \log \Sigma \phi}{\partial \theta}\right) \\ &= \left(\frac{\Sigma \phi'}{\Sigma \phi}\right)^2 - \frac{\Sigma \phi''}{\Sigma \phi} \end{aligned} \tag{5.44}$$

(*ibidem*), where $\phi' = \partial \phi / \partial \theta$ and $\phi'' = \partial^2 \phi / \partial \theta^2$. Using Eq. (5.42),

$$\begin{aligned} \frac{\Sigma \phi'}{\Sigma \phi} &= \frac{\Sigma \phi \frac{\partial}{\partial \theta} \log \phi}{\Sigma \phi} \\ &= \frac{-n\alpha \Sigma\left(\theta - \hat{\theta}\right)\phi}{\Sigma \phi} \\ &= -n\alpha \mathrm{E}\left\{\left(\theta - \hat{\theta}\right) \mid T\right\}. \end{aligned} \tag{5.45}$$

Similarly, it can be shown that

$$\frac{\Sigma \phi''}{\Sigma \phi} = -n\alpha + n^2\alpha^2 \mathrm{E}\left\{\left(\theta - \hat{\theta}\right)^2 \mid T\right\}. \tag{5.46}$$

Using Eqs. (5.45) and (5.46) in Eq. (5.44), Fisher finally obtained

$$\frac{1}{n\sigma^2} = n\alpha^2 \mathrm{E}^2\left\{\left(\hat{\theta} - \theta\right) \mid T\right\} + \alpha - n\alpha^2 \mathrm{E}\left\{\left(\hat{\theta} - \theta\right)^2 \mid T\right\} = \alpha - n\alpha^2 \operatorname{var}\left(\hat{\theta} \mid T\right) \tag{5.47}$$

(*ibidem*). From Eq. (5.47), it can be seen that when $T = \hat{\theta}$, then var $(\hat{\theta} \mid T) = 0$ and $1/(n\sigma^2)$ achieves a maximum of α, that is, the minimum value of σ^2 is $1/(n\alpha)$. On the other hand, by writing Eq. (5.47) as

$$\operatorname{var}\left(\hat{\theta} \mid T\right) = \frac{\alpha - 1/\left(n\sigma^2\right)}{n\alpha^2},$$

it can be seen that the maximum value of $\operatorname{var}(\hat{\theta} \mid T)$ is $(\alpha - 0)/(n\alpha^2) = 1/(n\alpha)$. Thus, Fisher was able to prove that for all estimators T,

$$\operatorname{var}\left(\hat{\theta} \mid T\right) \le \operatorname{var} T = \sigma^2, \tag{5.48}$$

that is, of all estimators T of a parameter θ, the MLE is asymptotically the most efficient (has the smallest variance).

Although Fisher's proof as described above is an improvement over his 1922 proof, it is still imperfect. This is because, in his proof, Fisher treated Eq. (5.42) as being exact and T as exactly normal, although both results are only approximations.

From Section 9 onward of the 1925 paper (Fisher, 1925c), Fisher was ready to tackle the issue of efficiency in small samples. First he recalled that if T_1 is a sufficient statistic for the parameter θ, and T_2 is any other statistic, then the conditional density of T_2 given T_1 does not depend on θ, that is,

$$f(\theta,T_2 \mid T_1)=\phi_1(T_1,T_2), \tag{5.49}$$

where $\phi_1(T_1, T_2)$ is a function of T_1 and T_2 only, and $f(\theta,T_2 \mid T_1)$ is the conditional density of T_2 given T_1. Since the latter can be written as a ratio of joint and marginal densities as $f(\theta,T_2 \mid T_1)=f(\theta,T_1,T_2)/\phi(\theta,T_1)$, Eq. (5.49) becomes

$$f(\theta,T_1,T_2)=\phi(\theta,T_1)\phi_1(T_1,T_2)$$

(*ibid.*, p. 713).

Moreover, having previously defined (in Section 6 of the paper) the *intrinsic accuracy* α (or expected information) based on the distribution of an observation, Fisher next (*ibid.*, p. 714) defined the *intrinsic accuracy* of a statistic T as

$$\alpha'=\mathsf{E}\left(\frac{\partial \log f_T}{\partial\theta}\right)^2,$$

where f_T is the density of T. From Eq. (5.48), we have $\alpha'\le n\alpha$. These ideas enabled Fisher to extend the definition of efficiency of a statistic in small samples, namely, as the ratio $\alpha'/(n\alpha)$.

Fisher next (*ibid.*, p. 717) proceeded to determine the intrinsic accuracy (expected information) lost through a statistic T relative to the whole sample. Using Eq. (2.36) (with the integral changed to summation) and the notation in Eq. (5.43), the intrinsic accuracy of T and of the whole sample regarding θ are, respectively,

$$\Sigma\frac{\Phi'^2}{\Phi} \quad \text{and} \quad \Sigma\frac{\phi'^2}{\phi},$$

where Φ is the density of T, ϕ is the density of one sample which gives T, the first summation is over all values of T, the second summation is over all possible samples each giving a different value of, and the prime denotes differentiation w.r.t. θ. Using the above formulas, "if, however, a number of different samples give the same value of T, then the effect of this amalgamation will be to decrease the intrinsic accuracy [expected information] by the amount," which is equal to

$$\Sigma\left(\frac{\phi'^2}{\phi}-\frac{\Phi'^2}{\Phi}\right)=\Sigma\phi\left(\frac{\phi'}{\phi}-\frac{\Phi'}{\Phi}\right)^2$$

(*ibidem*). It can be seen that the expression above is always nonnegative, "so that the intrinsic accuracy of T can never be greater than when every possible sample yields a different value of T." The expression is zero when $\phi'/\phi=\partial L/d\theta$ (where L is the likelihood function) is constant on the set of samples for which T is constant. Now, "if these sets are the same for all values of θ," then the ML equation

$$l'(\theta)=0$$

provides a sufficient statistic.*

5.2.6 First Appearance of "Ancillary" Statistics in the 1925 Paper: Relevant Subsets, Conditional Inference, and the Likelihood Principle

5.2.6.1 First Appearance of "Ancillary" Statistics The last section of Fisher's 1925 paper (Fisher, 1925c) deals with a new concept, namely, *ancillarity*,† which has proved to be a very powerful idea in statistics, albeit somewhat controversial. This is how the concept was explained:

> When the sets of samples which for one value of θ have the same value of $\partial L/\partial\theta$, have no longer the same value for other values of θ, there exists no sufficient statistic, and some loss of information will necessarily ensue upon the substitution of a single estimate for the original data upon which it was based. (*ibid.*, p. 718)
>
> ...
>
> Since the original data cannot be replaced by a single statistic, without loss of accuracy, it is of interest to see what can be done by calculating, in addition to our estimate, an ancillary statistic which shall be available in combination with our estimate in future calculations.
>
> If our two statistics specify the values of $\partial L/\partial\theta$ and $\partial^2 L/\partial\theta^2$ for some central value of θ, such as $\hat{\theta}$, then the variance of $\partial L/\partial\theta$ over the sets of samples for which both statistics are constant, will be that of
>
> $$\frac{1}{2}\left(\theta-\hat{\theta}\right)^2\frac{\partial^3 L}{\partial\theta^3}$$
>
> which will ordinarily be of order n^{-1} at least. With the aid of such an ancillary statistic the loss of accuracy tends to zero for large samples. (*ibid.*, p. 724)

In the above, Fisher's reasoning was as follows. Suppose a nontrivial sufficient statistic for θ does not exist. Then, solving $l'(\theta)=0$ does not yield a sufficient statistic. But for any statistic T, a Taylor series expansion of $l'(\theta)$ about T gives

$$l'(\theta)=l'(T)+(\theta-T)l''(T)+\frac{1}{2}(\theta-T)^2 l'''(T)+\cdots \tag{5.50}$$

Since T is not sufficient, information will be lost, but this loss can be reduced by using more of the terms (such as $l'(T), l''(T), \ldots$) on the right side of Eq. (5.50) above. For the

* Note that, as indicated on pp. 389–390, $l'(\theta)=0$, generally, but not always, provides a sufficient statistic.

† The concepts of ancillarity and conditional inference (see later in the text) are also discussed in Schervish (1995, pp. 95–102), Ghosh (1988), Cox and Hinkley (1979, pp. 31–35), Barnett (1999, pp. 185–189), Fraser (2004), Stigler (2001), Ghosh et al. (2010), Reid (1995), Welsh (1996, pp. 157–159), Van Aarde (2009, Chapter 9), and Lehmann (2011, pp. 47–49). Ancillary statistics are also sometimes called "free" statistics, especially in the non-English literature.

case when T equals the MLE $\hat{\theta}$ (so that $l'(\hat{\theta})=0$), some lost information can be recuperated by using the value of $l''(\hat{\theta})$ in addition to $\hat{\theta}$. Indeed, using Eq. (5.50) we have

$$\text{var}\left\{l'(\theta)\,|\,l'(\hat{\theta}),l''(\hat{\theta})\right\}=\text{var}\left\{\frac{1}{2}\left(\theta-\hat{\theta}\right)^{2}l'''\left(\hat{\theta}\right)\right\}=O\left(\frac{1}{n}\right),$$

so that the loss of information by using $\hat{\theta}$ in conjunction with $l''(\hat{\theta})$ tends to zero for large samples. Fisher called $l''(\hat{\theta})$ an ancillary* statistic because, in conjunction with the MLE $\hat{\theta}$, it provided additional information about the likelihood function over and above the position of its maximum; in particular, the ancillary statistic $l''(\hat{\theta})$ provided information on the *precision* of $\hat{\theta}$.

We make two comments on Fisher's concept of ancillarity in 1925. First, Fisher provided no formal definition of an ancillary statistic and his treatment was intuitive. This explains why, in the immediate aftermath of the 1925 paper, the concept was largely ignored by others. Second, Fisher's application was restricted to the likelihood function and there was no indication as to how ancillarity could be used beyond that. Both of these limitations were dealt with in 1934 and 1935 when he next reconsidered the topic (Fisher, 1934c; 1935d). In 1935, Fisher provided for the first time something close to a definition of an ancillary statistic, and in 1934 and 1935 he extended the range of applications of the concept. Thereupon, ancillarity gained popularity and remained one of the many great lasting achievements of Fisher. We now consider these topics.

In his 1934 paper, entitled "Two new properties of maximum likelihood" (Fisher, 1934c), Fisher expanded his application of ancillary statistics. He used the latter to completely recover the information lost when the location parameter θ of the Laplace (or double exponential) distribution was estimated by the MLE T, which in this case turned out to be the sample median. The ancillary statistics used are what Fisher called *the configuration of the sample*, defined by $a_1,a_2,\ldots,a_s;a'_1,a'_2,\ldots,a'_s$† where, for $i=1,\ldots,s$,

$$\begin{cases}a_i=X_{(s+1+i)}-T & \text{for}\quad X_{(s+1+i)}>T,\\ a'_i=T-X_{(s+1+i)} & \text{for}\quad X_{(s+1+i)}<T\end{cases}\tag{5.51}$$

(*ibid*., p. 301), $X_{(i)}$ being the ith order statistic (i.e., the ith smallest value).

Now, the Laplace density is

$$\frac{1}{2}e^{-|x-\theta|},\quad -\infty<x<\infty,-\infty<\theta<\infty.\tag{5.52}$$

Assume the number of observations is $n=2s+1$, an odd number. To obtain the density of the sample median $X_{(s+1)}$, consider the three intervals $(-\infty,u],(u,u+\delta u]$ and $(u+\delta u,\infty)$. The probability that $\{u<X_{(s+1)}\le u+\delta u\}$ is then approximately the probability that

*From the Latin "ancilla," which means handmaiden.

†Hinkley (1980, p. 104) has noted that the configuration can be re-expressed in terms of $n-1$ derivatives of the log-likelihood at $\hat{\theta}$.

s observations are in each of the first and third intervals, and one observation is in the second interval, that is,

$$\frac{(2s+1)!}{s!1!s!}\left[\Pr\{X\le u\}\right]^s\left[\Pr\{u<X\le u+\delta u\}\right]^1\left[\Pr\{X>u\}\right]^s=\frac{(2s+1)!}{(s!)^2}\left(\frac{1}{2}e^{-u}\right)^s\left(\frac{1}{2}e^{-u}\delta u\right)\left(1-\frac{1}{2}e^{-u}\right)^s$$

The density of the sample median is therefore

$$f=\frac{(2s+1)!}{(s!)^2}\left(\frac{1}{2}e^{-u}\right)^s\left(1-\frac{1}{2}e^{-u}\right)^s\left(\frac{1}{2}e^{-u}\right)$$

(*ibid.*, p. 298). As s (and n) increases, using Stirling's formula and the substitution $t=u\sqrt{n}$, the above can be written as

$$f\to\frac{1}{\sqrt{2\pi}}e^{-t^2/2}.$$

Now the expected information on the mean contained in a $N(\mu,\sigma^2)$ observation, where σ^2 is known, is $1/\sigma^2$, so for the above distribution the expected information for one observation is unity. Hence, the expected information contained in the sample $x_1,x_2,\ldots,x_n$ is, for increasingly large samples,

$$n=2s+1. \tag{5.53}$$

To obtain the expected information contained in the median, that is, by using one observation from the distribution

$$f^*=\frac{(2s+1)!}{(s!)^2\,2^{2s+1}}e^{-s|u-\theta|}\left(2-e^{-|u-\theta|}\right)^2e^{-|u-\theta|}.$$ *

Fisher calculated (cf. Eq. 5.41) the expected information as

$$\mathsf{E}\left(\frac{\partial}{\partial\theta}\log f^*\right)^2=\mathsf{E}\left(s-\frac{se^{-|u-\theta|}}{2-e^{-|u-\theta|}}+1\right)^2.$$

By integration, he obtained the expected information contained in the median as

$$\begin{cases}12\left(\log 2-\dfrac{1}{2}\right) & \text{for } s=1\\[2ex] \dfrac{(s+1)(2s+1)}{(s-1)}\left\{1-\dfrac{(2s)!}{(s!)^2\,2^{2s-1}}\right\} & \text{for } s>1.\end{cases} \tag{5.54}$$

* In his paper, Fisher denoted the density f^* by df/du.

Thus, as s increases, the amount of expected information lost by using the sample median instead of the whole sample can be obtained by subtracting Eq. (5.54) from Eq. (5.53) and is

$$(2s+1)-\frac{(s+1)(2s+1)}{(s-1)}\left\{1-\frac{(2s)!}{(s!)^2 2^{2s-1}}\right\}\approx\frac{2(2s+1)}{s-1}\left\{\frac{s+1}{\sqrt{\pi\left(s+\frac{1}{4}\right)}}-1\right\}$$

$$\approx 4\left(\sqrt{\frac{s}{\pi}}-1\right)$$

(*ibid.*, p. 300). How can this lost information be recovered? Fisher proposed to *condition the distribution of the median on the ancillary statistics* $a_1,a_2,\ldots,a_s;a'_1,a'_2,\ldots,a'_s$ (cf. Eq. 5.51). First he denoted the joint density of $x_1,x_2,\ldots,x_n$ by L, where each x_i has the Laplace density in Eq. (5.52). Then, by using the transformations in Eq. (5.51) in addition to $T=T$ (T being the median), he wrote the joint density of $T,a_1,a_2,\ldots,a_s,a'_1,a'_2,\ldots,a'_s$ as

$$\text{density of}\left\{x_{(1)},x_{(2)},\ldots,x_{(n)}\right\}\times\left|\frac{\partial\left(x_{(1)},x_{(2)},\ldots,x_{(n)}\right)}{\partial\left(T,a_1,\ldots,a_s,a'_1,\ldots,a'_s\right)}\right|=n!L\times 1=n!L. \quad (5.55)$$

Therefore, the conditional density of T given $a_1,a_2,\ldots,a_s;a'_1,a'_2,\ldots,a'_s$ is

$$\frac{\text{density of}\left(T,a_1,\ldots,a_s,a'_1,\ldots,a'_s\right)}{\text{density of}\left(a_1,\ldots,a_s,a'_1,\ldots,a'_s\right)}=\frac{\text{density of}\left(T,a_1,\ldots,a_s,a'_1,\ldots,a'_s\right)}{\int\text{density of}\left(T,a_1,\ldots,a_s,a'_1,\ldots,a'_s\right)dT}.$$

Using the above formula, Fisher was able to write the conditional density of T given $a_1,a_2,\ldots,a_s;a'_1,a'_2,\ldots,a'_s$, for example, when $\theta+a'_{p-1}<T<\theta+a'_p$, as

$$f=\frac{1}{A}e^{2\left(a'_1+a'_2+\cdots+a'_{p-1}\right)}e^{-(2p-1)(T-\theta)} \quad (5.56)$$

(*ibid.*, p. 301), where

$$\int_{\theta+a'_{p-1}}^{\theta+a'_p} f dT=\frac{1}{(2p-1)}e^{2\left(a'_1+a'_2+\cdots+a'_{p-1}\right)}\left\{e^{-(2p-1)a'_{p-1}}-e^{-(2p-1)a'_p}\right\}$$

and

$$A=\sum_{p=1}^{s}\frac{1}{(2p-1)}\left[e^{2\left(a'_1+a'_2+\cdots+a'_{p-1}\right)}\left\{e^{-(2p-1)a'_{p-1}}-e^{-(2p-1)a'_p}\right\}+e^{2\left(a_1+a_2+\cdots+a_{p-1}\right)}\left\{e^{-(2p-1)a_{p-1}}-e^{-(2p-1)a_p}\right\}\right].$$

Now, to obtain the amount of information supplied by the conditional density of T given $a_1, a_2, \ldots, a_s; a'_1, a'_2, \ldots, a'_s$, Fisher needed to apply Eq. (5.41):

$$\mathsf{E}\left(\frac{\partial}{\partial \theta}\log f\right)^2, \tag{5.57}$$

where f is the above-mentioned conditional density.* For $\theta + a'_{p-1} < T < \theta + a'_p$,

$$\left(\frac{\partial}{\partial \theta}\log f\right)^2 = \left[\frac{\partial}{\partial \theta}\{-(2p-1)(T-\theta)\}\right]^2 = (2p-1)^2. \tag{5.58}$$

But the event $\theta + a'_{p-1} < T < \theta + a'_p$ corresponds to $s-p+1$ of the $2s+1$ observations lying below the true median θ. The probability of this event can therefore be obtained from a Bino($2s+1$, 1/2) distribution as

$$\frac{1}{2^{2s+1}}\frac{(2s+1)!}{(s+p)!(s-p+1)!}. \tag{5.59}$$

Therefore, the expected information in (5.57) can be obtained by taking the product of (5.58) and (5.59) and summing for all p:

$$\mathsf{E}\left(\frac{\partial}{\partial \theta}\log f\right)^2 = 2\sum_{p=1}^{s}\frac{1}{2^{2s+1}}\frac{(2s+1)!}{(s+p)!(s-p+1)!}(2p-1)^2 = 2s+1$$

(*ibid*., p. 303). Thus, the whole of the information in the sample (cf. Eq. 5.53) has been recovered! Fisher announced triumphantly:

> The process of taking account of the distribution of our estimate in samples of the particular configuration observed has therefore recovered the whole of the information available. (*ibidem*)

We next consider Fisher's 1935 paper, "The logic of inductive inference" (Fisher, 1935d). For the first time, Fisher here told us a bit more what ancillary statistics *actually were*, instead of showing us *how they worked*, as he had previously done in the 1925 and 1934 papers (Fisher, 1925c; 1934c). He stated:

> ...ancillary statistics, which themselves tell us nothing about the value of the parameter, but, instead, tell us how good an estimate we have made of it. (Fisher, 1935d, p. 48)

Thus, a statistic A is ancillary for a parameter θ if the distribution of A does not depend on θ. This is the definition that is still commonly used (e.g., Pawitan, 2001, p. 197; Casella and Berger, 2002, p. 282). However, an alternative definition that is often preferred is as follows: suppose the statistic S is minimal sufficient† for θ and can be written in the form $S = (T, A)$

* In Fisher's paper, the density f was denoted by df/dT.

† The term was first used by Lehmann and Scheffé' (1950). See pp. 469 and 629 of the current text. A statistic S is minimal sufficient for a parameter if it is a function of every other sufficient statistic. The statistic S thus represents the smallest dimensional reduction of all possible sufficient statistics. Fisher called minimal sufficient statistics "exhaustive" (Fisher, 1956b, p. 49).

TABLE 5.2 Table Used by Fisher to Illustrate the Concept of Ancillarity

	Convicted	Not Convicted	Total
Monozygotic	10	3	13
Dizygotic	2	15	17
Total	12	18	30

From Fisher (1935d, p. 48).

such that the marginal distribution of A does not depend on θ. Then A is said to be ancillary for θ and T is said to be conditionally sufficient. There are at least two reasons for preferring the alternative definition based on $S=(T,A)$. The first has to do with our natural interest in sufficient statistics. The second is because it reduces (but does not eliminate) the occurrence of nonunique ancillary statistics* for a given statistical experiment.

Fisher's application of ancillary statistics in the 1935 paper concerned the 2×2 table. He examined the data based on 30 twin pairs, where one member of each pair was known to be a criminal. The data were cross-classified based on whether the twins were monozygotic or dizygotic and on whether the other member of each pair was convicted (Table 5.2)

It is interesting how Fisher introduced the concept of ancillarity in the 2×2 table:

> To the many methods of treatment hitherto suggested for the 2×2 table the concept of ancillary information suggests this new one. Let us blot out the contents of the table, leaving only the marginal frequencies. If it be admitted that these marginal frequencies by themselves supply no information on the point at issue, namely, as to the proportionality of the frequencies in the body of the table, we may recognize the information they supply as wholly ancillary; and therefore recognize that we are concerned only with the relative probabilities of occurrence of the different ways in which the table can be filled in, subject to these marginal frequencies. (Fisher, 1935d, pp. 48–49)

In the above, Fisher stated that the margins of the 2×2 table are ancillary for the "proportionality of the frequencies in the body of the table," that is, the odds ratio. However, in his proof, he considered the common value (under the null hypothesis) of the probabilities p_1 and p_2 that the other member of a monozygotic and dizygotic twin, respectively, is convicted. Fisher's proof was as follows. First, he noted that there were 13 possible 2×2 tables with the same margins as in Table 5.2, obtained by giving the number of dizygotic convicts values from 0 to 12. He then used the binomial distribution to write the probability that $(x+1)$ are not convicted and $(12-x)$ are convicted out of the 13 monozygotic twins as

$$\binom{13}{12-x} p_1^{12-x}\left(1-p_1\right)^{1+x}.$$

Under the hypothesis that $p_2=p_1$, the probability that $(17-x)$ are not convicted and x are convicted out of the 17 dizygotic twins is

$$\binom{17}{x} p_1^{x}\left(1-p_1\right)^{17-x}.$$

*For more on nonunique ancillary statistics, see Section 5.2.8.

The probability of these two events is given by the product of their respective probabilities:

$$\binom{13}{12-x}\binom{17}{x} p_1^{12}(1-p_1)^{18}.$$

Noting that the last two terms in the above product are independent of x, Fisher next reasoned that the probability of any value of x occurring is proportional to

$$\binom{13}{12-x}\binom{17}{x}, \quad \text{for } x = 0, 1, \ldots, 12.$$

The absolute probability of any value x is then given by the hypergeometric distribution:

$$\frac{\binom{13}{12-x}\binom{17}{x}}{\sum_{x=0}^{12}\binom{13}{12-x}\binom{17}{x}} = \frac{\binom{13}{12-x}\binom{17}{x}}{\binom{30}{12}}, \quad x = 0, 1, \ldots, 12. \tag{5.60}$$

Noting that "[t]he significance of the observed departure from proportionality is therefore exactly tested by observing that a discrepancy from proportionality as great or greater than that observed, will arise, subject to the conditions specified by the ancillary information," Fisher finally calculated what later came to be known as a one-sided p-value:

$$\frac{\binom{13}{12}\binom{17}{0} + \binom{13}{11}\binom{17}{1} + \binom{13}{10}\binom{17}{2}}{\binom{30}{12}} = .000465,$$

which implies a rejection of the null hypothesis $p_1 = p_2$ (at the 5% level of significance). The above is a description of *Fisher's exact test* which, as we have just seen, was motivated by considerations of ancillarity in the 1935 paper. As can be seen from Eq. (5.60), the distribution of x conditional on both margins is independent of the common value of p_1 and p_2 under the null hypothesis.

The exact test was first put forward in the fifth edition* of Fisher's *Statistical Methods for Research Workers* (Fisher, 1934b, p. 99), although Yates has pointed out that:

> Although then unpublished, the exact form must have been known to Fisher for some years, as is indicated by a cryptic passage in an earlier paper (Fisher, 1926)†: "an exact discussion would show that [for tables with 35 entries] the average value of χ^2 should exceed unity by one part in 34." (Yates, 1984, p. 429)

* Lehmann (2011, p. 47) has wrongly stated that the exact test was first given in the *fourth* edition of Statistical Methods for Research Workers.

† (Fisher, 1926a).

Both Irwin (1935) and Yates (1934) also published the exact test almost simultaneously with Fisher, although Yates again points out that:

> ...The exact form of a binomial distribution with given **p** was of course well known, but not that of a 2×2 table with given marginal totals. This was suggested to me by Fisher.... (*ibidem*)

Let us now come back to Fisher's earlier statement that the margins of a 2×2 table should be ancillary for the "proportionality of the frequencies in the body of the table," that is, the odds ratio. Consider the cross-classification between two categorical variables as shown in Table 5.3. Denote the number of subjects in the cell lying on the ith row and jth column by X_{ij}, the sum of row i by R_i, and the sum of column j by C_j $(i, j = 1,2)$. Let $p_1 = \Pr\{N \mid M\}$ and $p_2 = \Pr\{N \mid \bar{M}\}$. Then,

$$\Pr\{X_{11} = a, X_{21} = c \mid R_1 = a+b, R_2 = c+d\} = \binom{a+b}{a} p_1^a (1-p_1)^b \binom{c+d}{c} p_2^c (1-p_2)^d.$$

To facilitate the analysis, let us reparameterize as follows. Let $\theta = \{p_1/(1-p_1)\}/\{p_2/(1-p_2)\}$ be the odds ratio and $\phi = \{p_1/(1-p_1)\}\{p_2/(1-p_2)\}$ be the odds product. Then $p_1 = \sqrt{\theta\phi}/(1+\sqrt{\theta\phi})$ and $p_2 = \sqrt{\phi}/(\sqrt{\theta}+\sqrt{\phi})$, and the above equation becomes

$$\begin{aligned}\Pr\{X_{11} = a, X_{21} = c \mid R_1 = a+b, R_2 = c+d\} &= \binom{a+b}{a}\binom{c+d}{c}\left(\frac{\sqrt{\phi\theta}}{1+\sqrt{\phi\theta}}\right)^a\left(1-\frac{\sqrt{\phi\theta}}{1+\sqrt{\phi\theta}}\right)^b \\ &\quad \times\left(\frac{\sqrt{\phi}}{\sqrt{\theta}+\sqrt{\phi}}\right)^c\left(1-\frac{\sqrt{\phi}}{\sqrt{\theta}+\sqrt{\phi}}\right)^d \\ &= \binom{a+b}{a}\binom{c+d}{c}\frac{\theta^{(a+d)/2}\phi^{(a+c)/2}}{\left(1+\sqrt{\theta\phi}\right)^{a+b}\left(\sqrt{\theta}+\sqrt{\phi}\right)^{c+d}}.\end{aligned} \quad (5.61)$$

Note that θ is the parameter of interest (in particular, when $\theta = 1$ then $p_1 = p_2$) and ϕ is a *nuisance parameter*. We also have

$$\begin{aligned}&\Pr\{C_1 = w \mid R_1 = a+b, R_2 = c+d\} \\ &= \Pr\{X_{11} + X_{21} = w \mid R_1 = a+b, R_2 = c+d\} \\ &= \sum_{i=\max(0,a-d)}^{\min(a+b,a+c)} \Pr\{X_{11} = i, X_{21} = w-i \mid R_1 = a+b, R_2 = c+d\} \\ &= \sum_{i=\max(0,a-d)}^{\min(a+b,a+c)} \Pr\{X_{11} = i \mid R_1 = a+b, R_2 = c+d\}\Pr\{X_{21} = w-i \mid R_1 = a+b, R_2 = c+d\} \\ &= \sum_{i=\max(0,a-d)}^{\min(a+b,a+c)} \binom{a+b}{i}\binom{c+d}{w-i}\left(\frac{\sqrt{\phi\theta}}{1+\sqrt{\phi\theta}}\right)^i\left(\frac{1}{1+\sqrt{\phi\theta}}\right)^{a+b-i}\left(\frac{\sqrt{\phi}}{\sqrt{\theta}+\sqrt{\phi}}\right)^{w-i}\left(\frac{\sqrt{\theta}}{\sqrt{\theta}+\sqrt{\phi}}\right)^{c+d-w+i} \\ &= \sum_{i=\max(0,a-d)}^{\min(a+b,a+c)} \binom{a+b}{i}\binom{c+d}{w-i}\frac{\theta^{i+(c+d-w)/2}\phi^{w/2}}{\left(1+\sqrt{\theta\phi}\right)^{a+b}\left(\sqrt{\theta}+\sqrt{\phi}\right)^{c+d}}.\end{aligned} \quad (5.62)$$

TABLE 5.3 Cross-classification Between two Categorical Variables

	N	Not N	Row Sum
M	a	b	$a+b$
Not M	c	d	$c+d$
Column Sum	$a+c$	$b+d$	$a+b+c+d$

From the last expression, it can be seen that the distribution of C_1 is dependent on the parameter of interest θ. Therefore, contrary to what Fisher had stated, the margins of a 2×2 are not ancillary for the odds ratio. However, the dependence of the distribution of C_1 on θ is weak and the margins convey little information about the marginal frequencies unless the marginal totals are very small (Plackett, 1977; Yates, 1984; Little, 1989). The margins of a 2×2 table are thus said to be *approximately ancillary*.

Moreover, we also have

$$\begin{aligned}
\Pr\{X_{11}=a,X_{12}=c \mid C_1=a+c,R_1=a+b,R_2=c+d\} &= \Pr\{X_{11}=a \mid C_1=a+c,R_1=a+b,R_2=c+d\} \\
&= \frac{\Pr\{X=a,Y=c \mid R_1=a+b,R_2=c+d\}}{\Pr\{C_1=a+c \mid R_1=a+b,R_2=c+d\}} \\
&= \frac{\binom{a+b}{a}\binom{c+d}{c}\theta^{(a+d)/2}\phi^{(a+d)/2}}{\sum_{i=\max(0,a-d)}^{\min(a+b,a+c)}\binom{a+b}{i}\binom{c+d}{a+c-i}\theta^{i+(c+d-a-c)/2}\phi^{(a+d)/2}} \\
&= \frac{\binom{a+b}{a}\binom{c+d}{c}\theta^{a}}{\sum_{i=\max(0,a-d)}^{\min(a+b,a+c)}\binom{a+b}{i}\binom{c+d}{a+c-i}\theta^{i}}.
\end{aligned}$$

Hence,

$$\Pr\{X_{11}=a \mid C_1=a+c,R_1=a+b,R_2=c+d\} = \frac{\binom{a+b}{a}\binom{c+d}{c}\theta^{a}}{\sum_{i=\max(0,a-d)}^{\min(a+b,a+c)}\binom{a+b}{i}\binom{c+d}{a+c-i}\theta^{i}}. \quad (5.63)$$

We note from the above that conditioning on both margins has effectively eliminated the nuisance parameter ϕ, that is, the margins are *sufficient* for ϕ. Indeed, ancillary statistics are often useful in eliminating nuisance parameters, and in such situations conditional inference has much to recommend it.

Nevertheless, Fisher's recommendation of analyzing the 2×2 table by conditioning on the margins (i.e., Fisher's exact test) has been quite controversial. There are at least three major reasons for this. First, the exact test is particularly recommended for small samples but the margins can then convey a reasonable amount of information on the odds ratio θ, so Fisher's argument based on ancillarity of the margins may seem weak. Secondly, 2×2 tables where both margins are fixed by design are rare in practice, and the use of the exact test in such situations might appear inappropriate. Finally, the exact test is particularly *conservative*, that is, it rejects a true null hypothesis much less than $\alpha\%$ of the time when conducted at a level of significance α. This happens because of the marked discreteness of the hypergeometric distribution on which the exact test is based.

5.2.6.2 Relevant Subsets. Conditional Inference Notwithstanding the difficulties listed above, there is yet another strong reason why, in many other cases, Fisher would still have us proceed conditionally, namely, that of the existence of *relevant (or recognizable) subsets** (Fisher, 1956b, p. 32). This is another one of Fisher's powerful, yet somewhat controversial, statistical legacies. The concept of relevant subsets was introduced by Fisher as follows in the context of criticizing the conventional frequency theory of probability:

> ...the information supplied by a familiar mathematical statement such as: "If a aces are thrown in n trials, the probability that the difference in absolute value between a/n and 1/6 shall exceed any positive value ε, however small, shall tend to zero as the number n is increased indefinitely", will seem not merely remote, but also incomplete and lacking in definiteness in its application to the particular throw in which he is interested. Indeed, by itself it says nothing about that throw. It is obvious, moreover, that many subsets of future throws, which may include his own, can be shown to give probabilities, in this sense, either greater or less than 1/6. Before the limiting ratio of the whole set can be accepted as applicable to a particular throw, a second condition must be satisfied, namely that before the die is cast no such subset can be recognized. This is a necessary and sufficient condition for the applicability of the limiting ratio of the entire aggregate of possible future throws as the probability of anyone particular throw. On this condition we may think of a particular throw, or of a succession of throws, as a random sample from the aggregate, which is in this sense subjectively homogeneous and without recognizable stratification. (*ibid*., pp. 32–33)

In the above, Fisher stated that a valid frequency theory of probability must not only possess the property of convergence of relative frequencies to a limiting value but must also be based on a sequence such that the same limiting value is reached in any of its subsequence. That is, the sequence must possess *no recognizable subset*. Fisher's condition of no recognizable subset is thus similar to von Mises' randomness condition in a collective† (e.g., see Keuzenkamp, 2004, p. 36).

More generally, the relevant subset is a subset of the general experimental sample space and is often provided by the ancillary statistic, conditional on which inference is made. A classic example was provided by David Cox, one of the early proponents of conditional inference. Before examining Cox's argument, we shall say a few words about him.

Sir David Roxbee Cox‡ (Fig. 5.3) was born in Birmingham, England, on July 15, 1924. He attended Handsworth Grammar School and St John's College, Cambridge. From 1944 to 1946, he worked at the Royal Aircraft Establishment, from 1946 to 1950 at the Wool Industries Research Association in Leeds. In 1949, he obtained his Ph.D. from the

*The concept of relevant subsets is also discussed in Berger and Wolpert (1988, pp. 11–13), Pace and Salvan (1997, pp. 56–58), and Welsh (1996, pp. 156–157).

†Defined as a hypothetical infinite sequence of objects such that the probability of an event is the relative frequency of that event in the collective, for example, see Gorroochurn (2012a, p. 151).

‡See also Cox's conversation with Nancy Reid (1994).

FIGURE 5.3 Sir David Roxbee Cox (b. 1924). Wikimedia Commons (Public Domain), http://commons.wikimedia.org/wiki/File:Nci-vol-8182-300_david_cox.jpg

University of Leeds, under the supervision of Henry Daniels (1912–2000) and Bernard Welch (1911–1989). Cox worked at the Statistical laboratory at Cambridge University in 1950–1956. From 1956 to 1966, he was Reader, then Professor of Statistics at Birkbeck College, London. In 1966, he was appointed Professor of Statistics at Imperial College, London, a position he held until 1988. In 1972, Cox wrote his most famous paper entitled "Regression models and life tables" (Cox, 1972), resulting in the well-known Cox Proportional Hazards Model. Cox has written many books that cover a wide spectrum of statistical fields. Of special mention is his 1974 book *Theoretical Statistics* (Cox and Hinkley, 1979), which he co-wrote with his former student David Hinkley. Although this book also contains the Neyman–Pearson theory of hypothesis testing, it is written more in the Fisherian tradition and can be considered as the definitive treatment of Fisher's ideas in mathematical statistics. Cox has received countless honors, including the Guy Medals in Silver (1961) and Gold (1973) from the Royal Statistical Society and his knighthood by Queen Elizabeth II in 1985.

We now consider Cox's classic example in favor of conditional inference. In the landmark paper "Some problems connected with statistical inference" (Cox, 1958), we can read:

> …Suppose that we are interested in the mean θ of a normal population and that, by an objective randomization device, we draw either (i) with probability ½, one observation, x, from a normal population of mean θ and variance σ_1^2, or (ii) with probability ½, one observation x, from a normal population of mean θ and variance σ_2^2, where σ_1^2, σ_2^2 are known, $\sigma_1^2 \gg \sigma_2^2$ and where we know in any particular instance which population has been sampled.
>
> The sample space formed by indefinite repetition of the experiment is clearly defined and consists of two real lines Σ_1, Σ_2, each having probability ½, and conditionally on Σ_i there is a normal distribution of mean θ and variance σ_i^2.
>
> Now suppose that we ask, accepting for the moment the conventional formulation, for a test of the null hypothesis $\theta = 0$, with size say 0.05, and with maximum power against the alternative θ', where $\theta' \approx \sigma_1 \gg \sigma_2$. (*ibid.*, p. 360)

Cox then considered two possible tests for the hypothesis $H_0: \theta = 0$ versus $H_a: \theta > 0$ (where θ is the population mean), namely, the conditional and unconditional tests. For the conditional test, calculations are made *conditionally on which population is sampled. The latter denotes the relevant subset.* If a fair coin had been tossed to determine which population to sample from, then the outcome of the coin would be the ancillary statistic.

The critical region for the conditional test is $X > c$ where

$$\begin{aligned} \Pr\{X > c \mid \Sigma_i\} &= .05 \\ 1 - \Phi\left(\frac{c}{\sigma_i}\right) &= .05 \\ \therefore \qquad c &= \sigma_i \Phi^{-1}(.95) = 1.645\sigma_i, \end{aligned} \tag{5.64}$$

where Φ is the distribution function of the N (0,1) distribution. The conditional critical region can therefore be written as

$$X > \begin{cases} 1.645\sigma_1 & \text{if the first population is sampled} \\ 1.645\sigma_2 & \text{if the second population is sampled} \end{cases} \tag{5.65}$$

For the unconditional test, the error rate of .05 is maintained by reasoning from a *frequentist perspective*: in the long run, the first and second populations will each be sampled half of the time. Therefore, the rejection region is $X > c$ whatever population is sampled, where

$$\begin{aligned} .05 &= \Pr\{X > c\} \\ &= \Pr\{X > c \mid \Sigma_1\}\Pr\{\Sigma_1\} + \Pr\{X > c \mid \Sigma_2\}\Pr\{\Sigma_2\} \\ &= \frac{1}{2}\left[\Pr\{X > c \mid \Sigma_1\} + \Pr\{X > c \mid \Sigma_2\}\right] \\ &= \frac{1}{2}\left[\Pr\left\{Z > \frac{c}{\sigma_1}\right\} + \Pr\left\{Z > \frac{c}{\sigma_2}\right\}\right] \\ &= \frac{1}{2}\left\{1 - \Phi\left(\frac{c}{\sigma_1}\right) + 1 - \Phi\left(\frac{c}{\sigma_2}\right)\right\} \end{aligned} \tag{5.66}$$

$$\therefore \quad \Phi\left(\frac{c}{\sigma_1}\right) + \Phi\left(\frac{c}{\sigma_2}\right) = 1.9.$$

For example, if $\sigma_1 = 100$ and $\sigma_2 = 1$, numerically solving the above equation gives $c \approx 128.2$. The unconditional critical region is then

$$X > 128.2 \quad \text{whichever population is sampled.}$$

Now, note that the overall $\alpha = .05$ error rate in Eq. (5.66) above is achieved by averaging two error rates, α_1 and α_2, one for each population, that is,

$$\alpha_1 = \Pr\left\{Z > \frac{c}{\sigma_1}\right\} = 1 - \Phi(1.282) = .100$$

$$\alpha_2 = \Pr\left\{Z > \frac{c}{\sigma_2}\right\} = 1 - \Phi(128.2) = .000$$

Therefore, while the long-run error rate is maintained at .05, the possible error rates are quite unreasonable for *any particular test*. For, if we were to sample for the first population, the unconditional procedure would have us operate at an error rate of .100, and if we were to sample for the second population, we would have to operate at an error rate of .000. Both of these numbers seem ridiculous and seriously discourage the unconditional approach. In the words of Cox:

> …Now if the object of the analysis is to make statements by a rule with certain specified long-run properties, the unconditional test just given is in order, although it may be doubted whether the specification of desired properties is in this case very sensible. If, however, our object is to say "what we can learn from the data that we have", the unconditional test is

> surely no good. Suppose that we know we have an observation from Σ_1. The unconditional test says that we can assign this a higher level of significance than we ordinarily do, because if we were to repeat the experiment, we might sample some quite different distribution. But this fact seems irrelevant to the interpretation of an observation which we know came from a distribution with variance σ_1^2. That is, our calculations of power, etc. should be made conditionally within the distribution known to have been sampled, i.e. if we are using tests of the conventional type, the conditional test should be chosen. (*ibid*., pp. 360–361)

It was on the basis of such examples that Fisher criticized the long-term unconditional approach of Neyman and Pearson to hypothesis testing and confidence intervals.* This is in spite of the fact that unconditional tests usually have more power than conditional ones.†

Thus, the essential characteristic of a relevant subset is its ability to substantially change the inference from a given statistical procedure. Soon after Fisher's introduction, the concept was formalized by Buehler (1959) as follows. Suppose an unconditional $100(1-\alpha)\%$ confidence interval for a parameter θ is (C_1, C_2), that is,

$$\Pr\{C_1 \le \theta \le C_2\} = 1-\alpha.$$

Then R is called a relevant subset if it is a subset of the sample space such that, for some $\varepsilon > 0$, either

$$\Pr\{C_1 \le \theta \le C_2 \mid R\} \ge 1-\alpha+\varepsilon \tag{5.67}$$

or

$$\Pr\{C_1 \le \theta \le C_2 \mid R\} \le 1-\alpha-\varepsilon. \tag{5.68}$$

In (5.67), the relevant subset is *positively biased*, whereas in (5.68) it is *negatively biased*. Thus, if a positively biased relevant subset can be identified, then higher confidence than the stated level can be achieved by proceeding *conditionally*.

A classic example of a negatively biased relevant subset was provided by Fisher (1956a) in his criticism of Welch's solution‡ (Welch, 1947) to the Behrens–Fisher problem. Fisher considered two samples of size seven each. Using a significance level of $\alpha=.1$, he showed that the probability of false rejection of Welch's test, conditional on $s_1^2/s_2^2=1$, is at least .107 for all values of σ_1/σ_2 (*ibid*., p. 58). The subset $\{(s_1^2, s_2^2): s_1^2 = s_2^2\}$ is thus a negatively biased relevant subset.

A classic example of a positively biased subset was provided by Buehler and Feddersen (1963) by building on an example provided earlier by Stein (1961). Buehler and Feddersen considered Student's t-statistic for a random sample X_1, X_2 from a normally distributed population with mean μ:

$$T = \frac{\bar{X}-\mu}{S/\sqrt{2}} = \frac{X_1+X_2-2\mu}{|X_1-X_2|} \sim t_1.$$

* See Section 5.6.4.2.

† But see the discussion in Kalbfleisch and Sprott (1974), and Severini (1995).

‡ See pp. 483–487.

Then, from t tables,

$$\Pr\{-1 \le T \le 1\} = .5.$$

However, Buehler and Feddersen showed that

$$\Pr\left\{-1 \le T \le 1 \left| \frac{|\bar{X}|}{S} \le \frac{3\sqrt{2}}{4}\right.\right\} \ge .5181.$$

Thus, the set $\{(x_1, x_2) : |\bar{x}|/s \le (3\sqrt{2})/4\}$ is a positively biased relevant subset, thus contradicting Fisher's earlier claim that no relevant subset existed for Student's t-test. A generalization of Buehler and Feddersen's result was later provided by Brown (1967). He considered a sample $X_1, X_2, \ldots, X_n$ from a $N(\mu, \sigma^2)$ distribution, where μ and σ are both unknown, from which a $100(1-\alpha)\%$ confidence interval for μ is given by

$$\left(\bar{X} - t_{n-1;\alpha/2}\frac{S}{\sqrt{n}}, \bar{X} + t_{n-1;\alpha/2}\frac{S}{\sqrt{n}}\right),$$

where $t_{n-1;\alpha/2}$ is the critical value from Student's t-distribution with $(n-1)$ degrees of freedom. For the case $n=2$ and $\alpha=.5$, unconditionally we have

$$\Pr\left\{\bar{X} - t_{1;.25}\frac{S}{\sqrt{2}} \le \mu \le \bar{X} + t_{1;.25}\frac{S}{\sqrt{2}}\right\} = \frac{1}{2}.$$

However, Brown was able to show that, conditionally,

$$\Pr\left\{\bar{X} - t_{1;.25}\frac{S}{\sqrt{2}} \le \mu \le \bar{X} + t_{1;.25}\frac{S}{\sqrt{2}} \left| \frac{|\bar{X}|}{S} \le 1 + \sqrt{2}\right.\right\} \ge \frac{2}{3}.$$

In this case, the subset $\{(x_1, x_2, \ldots, x_n) : |\bar{x}|/s \le 1 + \sqrt{2}\}$ is a positively biased relevant subset. Both of these examples will be discussed later in the context of Fisher's fiducial argument (cf. Section 5.5.2.1).

5.2.6.3 Likelihood Inference Cox's example* made a strong case for conditional inference by conditioning on the population which was actually sampled. More generally, the principle that if one of two experiments is randomly chosen and then performed, then inference about any parameter θ should be made conditional on only the experiment actually performed (and not on any experiment that could have been performed but was not) is called the *(weak) conditionality principle*. As can be inferred from Cox's example, the conditional procedure is in contradiction with usual frequentist (i.e., unconditional) methods. To drive the point further, let $\sigma_1 = 100$ and $\sigma_2 = 1$ as before. Suppose the second population was sampled and $x=2.1$ was observed. We wish to test $H_0 : \theta = 0$ versus $H_a : \theta > 0$ (where θ is the population mean). The conditional one-sided p-value is

$$\Pr\{X \ge 2.1 \mid X \sim N(0, 1^2)\} = .018.$$

* See pp. 412–416.

On the other hand, the frequentist (i.e., unconditional) one-sided p-value is

$$\frac{1}{2}\Pr\left\{X \geq 2.1 \mid X \sim N\left(0, 100^2\right)\right\} + \frac{1}{2}\Pr\left\{X \geq 2.1 \mid X \sim N\left(0, 1^2\right)\right\} = .255.$$

The conditional test leads to rejection of H_0 at the 5% level. However, we are unable to reject H_0 at the 5% level with the frequentist test.

While the conditionality principle advocates conditioning on an ancillary statistic, there is a related principle, called the *(weak) sufficiency principle*, which recommends inference should be based only on a sufficient statistic. More precisely, if T is a sufficient statistic and if $\mathbf{x}$ and $\mathbf{y}$ are two sample points such that $T(\mathbf{x}) = T(\mathbf{y})$, then the same inference should be reached concerning a parameter θ, irrespective of which one of $\mathbf{x}$ and $\mathbf{y}$ is used. Note that the conditionality principle is about *conditioning* on an ancillary statistic, while the sufficiency principle is about *using* the sufficient statistic itself for inference. Now the remarkable fact is that the conditionality and sufficiency principles together are equivalent to the so-called *likelihood principle** (LP). This was first formally shown by Allan Birnbaum (1923–1976) in a 1962 paper (Birnbaum, 1962). LP states that all the information about a parameter θ is contained in the likelihood function for θ, and that if two likelihood functions for θ (from the same or different experiments) are proportional to each other, then they must contain the same information about θ.

The origins of the LP can be traced to the works of Fisher. The following are extracts from various publications of Fisher where something close to the LP is stated:

> For the solution of problems of estimation we require a method which for each particular problem will lead us automatically to the statistic by which the criterion of sufficiency is satisfied. Such a method is, I believe, provided by the Method of Maximum Likelihood.... (Fisher, 1922c, p. 323)
>
> When sufficient statistics exist it has been shown that they will be solutions of the equations of maximum likelihood. (Fisher, 1925c, p. 714)
>
> ...the expression [likelihood]... which, when properly interpreted must contain the whole of the information respecting x which our sample of observations has to give. (Fisher, 1934c, p. 297)
>
> Objection has sometimes been made that the method of calculating Confidence Limits by setting an assigned value such as 1% on the frequency of observing 3 or less (or at the other end of observing 3 or more) is unrealistic in treating the values less than 3, which have not been observed, in exactly the same manner as the value 3, which is the one that has been observed. This feature is indeed not very defensible save as an approximation. (Fisher, 1956b, p. 56)

However, it is due to Barnard et al. (1962) and especially Birnbaum (1962) that the LP was clearly delineated. Interestingly, Fisher himself did not completely adhere to the principle. For example, the calculation of p-values under his theory of significance testing takes into account observations that were *not* actually made and therefore is *not* in accordance with the LP.

Like the conditionality principle, the LP is, not surprisingly, also at odds with frequentist reasoning. An example is as follows. Consider a binomial experiment E_1 that results in 8 successes out of 10 trials. Next consider a negative binomial experiment E_2

*The likelihood principle is also discussed in Berger and Wolpert (1988), Pawitan (2001, Chapter 7), Millar (2011, Chapter 14), Rohde (2014, Chapter 13), and Van Aarde (2009, Chapter 10).

that requires 10 trials before we get the second failure. Suppose in both experiments the success probability per trial is p. Then, the likelihood function for E_1 is

$$L_1(p) = \binom{10}{8} p^8 (1-p)^2$$

and the likelihood function for E_2 is

$$L_2(p) = \binom{9}{8} p^8 (1-p)^2.$$

Since $L_1(p) \propto L_2(p)$, application of the LP implies that both E_1 and E_2 have the same evidence about p.

However, frequentist reasoning results in a quite different conclusion. For example, if we wished to test the hypothesis $H_0 : p = .5$ versus $H_a : p > .5$, the p-value for E_1 is

$$\begin{aligned} pval_1 &= \Pr\{X \geq 8 | X \sim \text{Bino}(10,.5)\} \\ &= \sum_{x=8}^{10} \binom{10}{x} (.5)^{10} \\ &= .055, \end{aligned}$$

and the p-value for E_2 is

$$\begin{aligned} pval_2 &= \Pr\{Y \geq 10 \mid Y \sim \text{NegBino}(2,.5)\} \\ &= \sum_{y=10}^{\infty} \binom{y-1}{1} (.5)^{y-2} (1-.5)(1-.5) \\ &= \sum_{y=10}^{\infty} (y-1)(.5)^y \\ &= .020. \end{aligned}$$

Therefore, for E_1 we are unable to reject H_0 at the 5% level, whereas for E_2 we can.

5.2.7 Further Extensions: Inconsistency of MLEs (Neyman and Scott, 1948), Inadmissibility of MLEs (Stein, 1956), Nonuniqueness of MLEs (Moore, 1971)

To those who had thought that ML would be a universal estimation panacea, Neyman and Scott (1948) and Stein (1956) provided the following startling classic counterexamples:

- Neyman and Scott (1948) considered the balanced one-way ANOVA design. Consider the independent random variables $Y_{ij} \sim N(\mu_i, \sigma^2)$, where $i = 1, \ldots, n$ and $j = 1, \ldots, k$. Note that each treatment i has the same variance but a different mean μ_i. We wish to estimate σ^2. Standard calculations give the following MLEs:

$$\begin{aligned} \hat{\mu}_i &= \overline{Y}_i, \\ \hat{\sigma}^2 &= \frac{1}{nk} \sum_{i=1}^{n} \sum_{i=1}^{k} (Y_{ij} - \overline{Y}_i)^2 \end{aligned}$$

The surprising fact is that, as $n \to \infty$,

$$\hat{\sigma}^2 \xrightarrow{P} \frac{\sigma^2}{k}(k-1) \neq \sigma^2.$$

that is, the MLE $\hat{\sigma}^2$ is *not* consistent for σ^2. To explain this anomaly, first note that, as $n \to \infty$, the number n of nuisance parameters $(\mu_1, \mu_2, \ldots \mu_n)$ becomes too large compared to the number of observations (k) in each group. Therefore, k is not large enough to make the bias in $\hat{\sigma}^2$ vanish. Thus, if $\hat{\sigma}^2$ is changed to

$$\hat{\sigma}'^2 = \frac{1}{n(k-1)} \sum_{i=1}^{n} \sum_{i=1}^{k} \left(Y_{ij} - \overline{Y}_i\right)^2$$

$\hat{\sigma}'^2$ becomes consistent.

- Stein (1956) considered the random variables $X_1, \ldots, X_n$, which are independently and normally distributed with means $\xi_1, \ldots, \xi_n$ and variance unity. Then the MLE of ξ_i is X_i. Using a quadratic loss function (see Section 6.2.4), Stein showed that the MLE is admissible for $n \leq 2$ *but inadmissible* for $n \geq 3$, that is, in the latter case there is at least one other estimator with lower risk. This surprising fact is known as *Stein's paradox*. In 1961, James and Stein (1961) showed that the estimator

$$\left(1 - \frac{n-2}{\sum_{i=1}^{n} X_i^2}\right) X_i \tag{5.69}$$

has a uniformly smaller risk than the MLE X_i.* The estimator in (5.69) is known as a *shrinkage* estimator because it shrinks X_i.

In addition to the above wrinkles, MLEs can also be nonunique. Consequently, the often-made blanket statement "*An ML estimator is a function of the sufficient statistic*" is not quite accurate. This was first pointed out by Moore (1971). As an example, Moore considered IID[†] random variables $X_1, X_2, \ldots, X_n$ from Unif $(\theta - 1/2, \theta + 1/2)$ distribution. Then

$$L(\theta \mid X_1, \ldots, X_n) = \begin{cases} 1, & \text{for } \theta - 1/2 \leq x_i \leq \theta + 1/2 \\ 0, & \text{elsewhere} \end{cases}$$

L achieves a maximum of unity when

$$\theta - 1/2 \leq x_1, x_2, \ldots, x_n \leq \theta + 1/2,$$

$$\text{i.e.} \quad x_{(1)} \geq \theta - \frac{1}{2} \quad \text{and} \quad x_{(n)} \leq \theta + \frac{1}{2}$$

$$\therefore \quad x_{(n)} - \frac{1}{2} \leq \theta \leq x_{(1)} + \frac{1}{2}.$$

* Stein's paradox is further discussed in Székely (1986, pp. 77–83), Romano and Siegel (1986, pp. 208–209), and Cox and Hinkley (1979, pp. 445–451).

[†] Independent and identically distributed.

Here, $x_{(i)}$ ($i=1, 2,\ldots, n$) is the value of ith order statistic. Thus, *any* statistic $T = T(X_1, X_2, \ldots, X_n)$ such that $X_{(n)} - 1/2 \le T \le X_{(1)} + 1/2$ is an MLE for θ. As an example, Moore gave

$$\hat{\theta} = \left\{X_{(n)} - \frac{1}{2}\right\} + \left(\cos^2 X_1\right)\left\{X_{(1)} - X_{(n)} + 1\right\} \tag{5.70}$$

to be such an MLE, and there are infinitely many others. Moreover, since L can also be written as

$$L = I_{\left(x_{(n)}-1/2,\, x_{(1)}+1/2\right)}(\theta)$$

(where I is the indicator function), Fisher–Neyman factorization theorem implies that $(X_{(1)}, X_{(n)})$ is sufficient for θ. However, the MLE $\hat{\theta}$ as given in Eq. (5.70) above is not a function of $(X_{(1)}, X_{(n)})$.

The reason for this pathological result is that the MLE is a function of a sufficient statistic *only when the former is unique*. This is not the case in the current example. Moreover, within the class of the nonunique estimators, it is always possible to choose one which is a function of a sufficient statistic, for example, $\hat{\theta} = \{X_{(1)} + X_{(n)}\}/2$ in the current example.

5.2.8 Further Extensions: Nonuniqueness of Ancillaries and of Relevant Subsets (Basu, 1964)

Notwithstanding the usefulness of Fisher's concept of conditional inference, a major problem is that often neither ancillary statistics nor relevant subsets can be found on which the conditioning can be done.

In other situations, there are several ancillary statistics (each leading to a different relevant subset) and it is not always clear which one is to be preferred. The following is a classic from three examples put forward by Basu* (1964). Let the outcome when a biased die is rolled be denoted by X, where X has the distribution shown in Table 5.4. Here the parameter θ is such that $0 \le \theta \le 1$. Then, by maximizing each of the probabilities in Table 5.4, the MLE T of θ is given in Table 5.5. Note that T alone is not sufficient for θ.

TABLE 5.4 Basu's Example of Nonunique Ancillary Statistics

Score (X)	1	2	3	4	5	6
Prob.	$\frac{1-\theta}{12}$	$\frac{2-\theta}{12}$	$\frac{3-\theta}{12}$	$\frac{1+\theta}{12}$	$\frac{2+\theta}{12}$	$\frac{3+\theta}{12}$

From Basu (1964, p. 11).

TABLE 5.5 Maximum Likelihood Estimator T of θ

X	1	2	3	4	5	6
$T(X)$	0	0	0	1	1	1

From Basu (1964, p. 12).

*Basu's examples are also discussed in Cox and Hinkley (1979, pp. 33–34, 43–44, 109–110), Ghosh et al. (2010), Barnard and Sprott (1971), Cox (1971), and Fraser (1973). For more on Basu, see Section 6.1.5.

TABLE 5.6 The Six Ancillary Statistics $Y_1, \ldots, Y_6$ in Basu's Example

X	1	2	3	4	5	6
Y_1	0	1	2	0	1	2
Y_2	0	1	2	0	2	1
Y_3	0	1	2	1	0	2
Y_4	0	1	2	2	0	1
Y_5	0	1	2	1	2	0
Y_6	0	1	2	2	1	0

From Basu (1964, p. 12).

Now, each the six statistics $Y_1, \ldots, Y_6$ shown in Table 5.6 has a distribution independent of θ (moreover, when used in conjunction with T, each gives a minimal sufficient statistic). Thus, $Y_1, \ldots, Y_6$ are each ancillary.

How should inference on θ be performed? If we condition on, say Y_1, then the distribution of T is

$$(T \mid y_1 = 0) = \begin{cases} 0 & \text{prob. } \dfrac{1-\theta}{2} \\ 1 & \text{prob. } \dfrac{1+\theta}{2} \end{cases}$$

$$(T \mid y_1 = 1) = \begin{cases} 0 & \text{prob. } \dfrac{2-\theta}{4} \\ 1 & \text{prob. } \dfrac{2+\theta}{4} \end{cases}$$

$$(T \mid y_1 = 2) = \begin{cases} 0 & \text{prob. } \dfrac{3-\theta}{6} \\ 1 & \text{prob. } \dfrac{3+\theta}{6}. \end{cases}$$

On the other hand, if we condition on Y_2, then the distribution of T is

$$(T \mid y_2 = 0) = \begin{cases} 0 & \text{prob. } \dfrac{1-\theta}{2} \\ 1 & \text{prob. } \dfrac{1+\theta}{2} \end{cases}$$

$$(T \mid y_2 = 1) = \begin{cases} 0 & \text{prob. } \dfrac{2-\theta}{5} \\ 1 & \text{prob. } \dfrac{3+\theta}{5} \end{cases}$$

$$(T \mid y_2 = 2) = \begin{cases} 0 & \text{prob. } \dfrac{3-\theta}{5} \\ 1 & \text{prob. } \dfrac{2+\theta}{5}. \end{cases}$$

Similarly, conditioning on each of the other Y_i's leads to MLEs with different distributions.

Possible ways out of this conundrum were suggested by each of Barnard and Sprott (1971) and Cox (1971). Barnard and Sprott recommended the choice of the ancillary statistic that actually described the shape of the likelihood function, in accordance with Fisher's initial motivation for using ancillary statistics. Bernard and Sprott's resolution proceeds by using invariance arguments: for the above example, the transformation $gY = Y + 3(\bmod 6)$ is first defined; this induces the transformation $g^*\theta = -\theta$ on the parameter. Then the only ancillary statistic unaffected by the transformation g is Y_1. On the other hand, Cox advocated using the ancillary statistic that resulted in the *maximal* variance of the conditional information. Under this criterion, Y_1 is again the preferred ancillary statistic.

In spite of these possible resolutions (which are by no means universally accepted), Basu's example is quite disquieting and shows that care must be exercised when using ancillary statistics.

5.3 FISHER AND SIGNIFICANCE TESTING

Fisher made gigantic strides not only in estimation theory but also in significance testing. He called the latter "problems of distribution" (Fisher, 1922c) and his objective was the determination of mainly the small-sample (or exact) distribution of various statistics. Recall that Student spearheaded small-sample investigations with the groundbreaking 1908 papers (Student, 1908a; 1908b). In Fisher, Student had an ally of the highest caliber who also appreciated the importance of small-sample statistics.

Fisher's testing procedure was as follows. To test $H_0 : \theta = \theta_0$, where θ is an unknown parameter and θ_0 is a given value, Fisher would first find the distribution of an intuitively derived test statistic.* From the tails of the distribution, the probability of obtaining the value of the test statistic calculated under H_0, or a value more extreme, was obtained. This probability was later called the p-value of the test. Finally, the test was declared significant (i.e., H_0 was rejected) if the p-value was "small enough."†

5.3.1 Significance Testing for the Correlation Coefficient (Student, 1908; Soper, 1913; Fisher, 1915; 1921)

A classic example of significance testing is the test for the correlation coefficient ρ. Building on previous work by Student (1908a) and Soper (1913) (cf. p. 229), Fisher was able to derive the exact distribution of the sample correlation r in 1915 (Fisher, 1915) (cf. Section 2.3.9). He showed that when $\rho = 0$, the transformation

$$T = \frac{r}{\sqrt{1-r^2}}$$

*Formally, a test statistic T is a function of the data such that (a) the distribution of T under H_0 is known; (b) the larger the value of T, the stronger the evidence against H_0.

†Fisher recommended a cutoff of .05; see p. 430.

(*ibid.*, p. 518) has "the frequency...curves [of t] are identical with those found by 'Student' for z [with n replaced by $n-1$]." By using T as a test statistic, the hypothesis $H_0 : \rho = 0$ can therefore be tested and p-values calculated from Student's z-tables. The custom now is to define $T = r\sqrt{n-2} / \sqrt{1-r^2}$, the latter having a Student's t_{n-2} distribution under H_0.

Moreover, after an initial proposal in 1915, he was able to show in 1921 (Fisher, 1921a) (cf. Section 2.3.11) that

$$\frac{1}{2}\log\left(\frac{1+r}{1-r}\right) \dot{\sim} N\left[\frac{1}{2}\log\left(\frac{1+\rho}{1-\rho}\right), \frac{1}{n-3}\right]$$

for moderately large n. Therefore, to test $H_0 : \rho = \rho_0 \left(\neq 0\right)$, Fisher used the test statistic

$$Z = \frac{\frac{1}{2}\log\left(\frac{1+r}{1-r}\right) - \frac{1}{2}\log\left(\frac{1+\rho_0}{1-\rho_0}\right)}{1/\sqrt{n-3}} \tag{5.71}$$

which has a N (0,1) distribution under H_0. By substituting the value of r obtained from the data, the observed value z_{obs} of the test statistic can be calculated from the above so that $p-\text{value} = 2\Pr\left\{Z \geq z_{\text{obs}} \mid H_0\right\}$. In an example considered by Fisher (*ibid.*, p. 214), the hypothesis tested was $H_0 : \rho = .3000$, and the calculated correlation based on $n = 13$ pairs was r=.6259. By applying the formula in (5.71) above, we obtain $z_{\text{obs}} = 1.344$ and $p-\text{value} = 2\Pr\left\{Z \geq 1.344\right\} = .179$.* This was the first p-value† calculated by Fisher to appear in print.

5.3.2 Significance Testing for a Regression Coefficient (Fisher, 1922)

We next describe Fisher's extension of the t-test to test for the significance of regression coefficients in the 1922 paper "The goodness of fit of regression formulae, and the distribution of regression coefficients" (Fisher, 1922d). The aim of the paper was to test the goodness of fit of a linear regression model by using a form of Pearson's goodness-of-fit test that had been modified by the mathematical statistician and economist Evgeny Slutsky (1880–1948) (Slutsky, 1913) to deal with unequally weighted x-values. It is in this paper that Fisher first used the phrase "to test the significance of" (Fisher, 1922d, p. 603).

In Section 6 of the 1922 paper, Fisher considered the simple regression model:

$$y = \beta_0 + \beta_1\left(x - \bar{x}\right) + \varepsilon$$

(*ibid.*, p. 608) with least squares estimates

$$\hat{\beta}_0 = \bar{y}, \tag{5.72}$$

*This is slightly different from Fisher's p-value of .142 because he used a slightly different formula for the standard error of r.

†Although from then on Fisher would frequently calculate p-values, he did not call them as such and simply denoted them by P. According to David (1998), the term "P value" was first used by W.E. Deming in his book *Statistical Adjustment of Data* (Deming, 1943, p. 30).

$$\hat{\beta}_1 = \frac{\sum_i y_i (x_i - \bar{x})}{\sum_i (x_i - \bar{x})^2}. \tag{5.73}$$

Now the "best" estimate of var $y = \sigma^2$ is

$$s^2 = \frac{1}{n-2} \sum_i (y_i - \hat{y}_i)^2$$

where, as Fisher correctly pointed out, "the sum is divided by $(n-2)$ to allow for the two constants, used in fitting the regression line" (*ibidem*), and $\hat{y}_i = \hat{\beta}_0 + \hat{\beta}_1(x_i - \bar{x})$ is the fitted value of y_i. Then the distribution of $U = (n-2)S^2 / \sigma^2$ is

$$f_U(u) = \frac{1}{\Gamma\left(\frac{n-2}{2}\right)} \left(\frac{u}{2}\right)^{\frac{n-4}{2}} e^{-u/2}$$

(*ibidem*). Fisher also defined

$$\tau = \frac{\hat{\beta}_0 - \beta_0}{\sigma / \sqrt{n}}$$

and noted that, since from Eq. (5.72) var $\hat{\beta}_0 = \sigma^2 / n$, the random variable τ has a N (0,1) distribution. Then, since S and $\hat{\beta}_0$ (and hence U and τ) are independent, "following 'Student,' we find the distribution of a quantity completely calculable from the sample, namely,"

$$Z = \frac{\tau}{\sqrt{U}} = \frac{\left(\hat{\beta}_0 - \beta_0\right)\sqrt{n}}{\sqrt{\sum_i (y_i - \hat{y}_i)^2}} \tag{5.74}$$

(*ibidem*). Then, making the transformation

$$\begin{aligned} Z &= \tau / \sqrt{U}, \\ U &= U, \end{aligned} \tag{5.75}$$

we have

$$f(z,u) = f(\tau,u) \left| J\left(\frac{\tau,u}{z,u}\right) \right| = \frac{1}{2^{(n/2)-1}\Gamma\left(\frac{n-2}{2}\right)} e^{-u/2} u^{(n/2)-2} \cdot \frac{1}{\sqrt{2\pi}} e^{-uz^2/2} \cdot \sqrt{u}.$$

Integrating u out, Fisher finally obtained

$$f_z(z) = \int_0^\infty f(z,u)\,du = \frac{1}{\sqrt{\pi}} \frac{\Gamma\left(\frac{n-1}{2}\right)}{\Gamma\left(\frac{n-2}{2}\right)} \frac{1}{\left(1+z^2\right)^{(n-1)/2}}$$

(*ibid*., p. 609), which is the same Pearson Type VII distribution obtained by Student in Eq. (4.6), with n replaced by $(n-1)$. Had Fisher used $Z = \tau / \sqrt{U/(n-2)}$ in (5.74), the distribution of Z would have been t_{n-2}. Fisher also noted that since

$$\operatorname{var} \hat{\beta}_1 = \frac{\sigma^2}{\sum_i (x_i - \bar{x})^2},$$

we can also let

$$Z = \frac{\left(\hat{\beta}_1 - \beta_1\right)\sqrt{\sum_i (x_i - \bar{x})^2}}{\sqrt{\sum_i (y_i - \hat{y}_i)^2}} \tag{5.76}$$

and "we arrive at the same distribution as before" (*ibidem*). Generalizing the above, Fisher considered the multiple linear regression model

$$y = \beta_1 x_1 + \beta_2 x_2 + \cdots + \beta_p x_p + \varepsilon$$

and defined the determinant

$$\Delta = \begin{vmatrix} \sum x_1^2 & \sum x_1 x_2 & \ldots & \sum x_1 x_p \\ \sum x_1 x_2 & \sum x_2^2 & \ldots & \sum x_2 x_p \\ \vdots & \vdots & & \vdots \\ \sum x_1 x_p & \sum x_2 x_p & \ldots & \sum x_p^2 \end{vmatrix}.$$

The variance of $\hat{\beta}_i$, the least squares estimator of β_i, is given by

$$\operatorname{var} \hat{\beta}_i = \frac{\sigma^2 \Delta_{ii}}{\Delta},$$

where Δ_{ii} is the (i,i)th minor. In analogy with Eqs. (5.74) and (5.76), the Z statistic defined by

$$Z = \frac{\left(\hat{\beta}_i - \beta_i\right)\sqrt{\Delta}}{\sqrt{\sum_i (y_i - \hat{y}_i)^2}\sqrt{\Delta_{ii}}}$$

will have the Type VII distribution:

$$f_z(z) = \frac{1}{\sqrt{\pi}} \frac{\Gamma\left(\frac{n-p}{2}\right)}{\Gamma\left(\frac{n-p-1}{2}\right)} \frac{1}{\left(1+z^2\right)^{(n-p)/2}}$$

(*ibid*., p. 610).

5.3.3 Significance Testing Using the Two-Sample *t*-test Assuming a Common Population Variance (Fisher, 1922)

Another remarkable feature of Fisher's 1922 paper (Fisher, 1922d) was the first explicit statement of the two-sample *t*-test under the assumption of normality and a common (unknown) population variance. Suppose $\bar{x}$ and $\bar{y}$ are the means of two samples of size n_1 and n_2, respectively, and "we wish to test if the means are in sufficient agreement to warrant the belief that the samples are drawn from the same population." Fisher then defined

$$\begin{aligned} Z &= \frac{(\bar{x}-\bar{y})}{\sqrt{\sum_i (x_i-\bar{x})^2 + \sum_i (y_i-\bar{y})^2}} \cdot \sqrt{\frac{n_1 n_2}{n_1+n_2}} \\ &= \frac{(\bar{x}-\bar{y})}{s_p\sqrt{n_1+n_2-2}\sqrt{\frac{1}{n_1}+\frac{1}{n_2}}} \end{aligned}$$

(*ibid*., p. 610), where $s_p^2 = \left\{\sum_i (x_i-\bar{x})^2 + \sum_i (y_i-\bar{y})^2\right\} / (n_1+n_2-2)$ is the pooled estimated variance. He then stated that the distribution of Z is

$$f_z(z) = \frac{1}{\sqrt{\pi}} \cdot \frac{\Gamma\left(\frac{n_1+n_2-1}{2}\right)}{\Gamma\left(\frac{n_1+n_2-2}{2}\right)} \frac{1}{\left(1+z^2\right)^{(n_1+n_2-1)/2}}.$$

under $H_0 : \mu_1 - \mu_2 = 0$. Again, the above is a Pearson Type VII distribution. The common practice nowadays is to define $T = (\bar{x}-\bar{y})/\{s_p\sqrt{1/n_1+1/n_2}\}$, which has a Student's $t_{n_1+n_2-2}$ distribution under H_0.

Although the two-sample *t*-test was first explicitly stated in the 1922 paper, Seal has noted that:

> [Fisher] had, in fact, already proved this in (1921)* for $n_1 = n_2 = n$ by considering the so-called intraclass correlation coefficient calculated from the $\binom{n}{2}$ pairs of observations that can be formed by taking one member from each sample. (Seal, 1967)

As straightforward as the *t*-test for two population means under a common unknown population variance was, the case when the unknown population variances were different

* On the "probable error" of a coefficient of correlation deduced from a small sample. Metron 1, N. 4, 3–32. (*Author's note*).

turned out to be a much more complex and controversial affair. This is known as the Behrens–Fisher problem and will be discussed in detail in Sections 5.5.2.3–5.5.2.6.

5.3.4 Significance Testing for Two Population Variances (Fisher, 1924)

Having shown how to test the significance of population means, correlations, and regression coefficients, it was natural for Fisher to next turn to population variances. In 1924, Fisher presented a paper* (Fisher, 1924a) at the International Congress of Mathematics in Toronto. In Section 3 of the paper, Fisher introduced the following problem:

> We have two estimates s_1^2 and s_2^2 derived from the two small samples, and we wish to know, for example, if the variances are significantly different. (*ibid.*, p. 807)

He further explained why using $s_1^2 - s_2^2$ for a test of $H_0 : \sigma_1^2 = \sigma_2^2$ was not appropriate when the samples were small:

> If we introduce hypothetical true values σ_1^2 and σ_2^2 we could theoretically calculate in terms of σ_1 and σ_2, how often $s_1^2 - s_2^2$ (or $s_1 - s_2$) would exceed its observed value. The probability would of course involve the hypothetical σ_1 and σ_2, and our formulae could not be applied unless we were willing to substitute the observed values s_1^2 and s_2^2 for σ_1^2 and σ_2^2; but such a substitution, though quite legitimate with large samples, for which the errors are small, becomes extremely misleading for small samples; the probability derived from such a substitution would be far from exact. The only exact treatment is to eliminate the unknown quantities σ_1 and σ_2 from the distribution by replacing the distribution of s by that of log s, and so deriving the distribution of $\log s_1/s_2$. Whereas the sampling errors in s_1 are proportional to σ_1, the sampling errors of $\log s_1$ depend only upon the size of the sample from which s_1 was calculated. (*ibid.*, pp. 807–808)

Thereupon, Fisher defined

$$\nu_1 s_1^2 = \sigma_1^2 X_1^2, \quad \nu_2 s_2^2 = \sigma_2^2 X_2^2,$$
$$e^{2z} = \frac{s_1^2}{s_2^2} = \frac{\sigma_1^2}{\sigma_2^2} \cdot \frac{\nu_2 X_1^2}{\nu_1 X_2^2}$$

(*ibid.*, p. 808). Although Fisher did not demonstrate this, it can be shown that (e.g., see Kendall, 1945, p. 249) the distribution of $z = \frac{1}{2}\log(s_1^2 / s_2^2)$ is

$$f(z) = \frac{\sigma_1^{\nu_1}\sigma_2^{\nu_1}\nu_1^{\nu_1/2}\nu_2^{\nu_2/2}}{B\left(\frac{\nu_1}{2}, \frac{\nu_2}{2}\right)} \cdot \frac{e^{\nu_1 z}}{\left(\frac{\nu_1}{\sigma_1^2}e^{2z} + \frac{\nu_2}{\sigma_2^2}\right)^{(\nu_1+\nu_2)/2}}.$$

where $\nu_1 = n_1 - 1$ and $\nu_2 = n_2 - 1$ are the degrees of freedom associated with s_1^2 and s_2^2, respectively, and B (.,.) is the beta function. Under $H_0 : \sigma_1^2 = \sigma_2^2$, we obtain the so-called Fisher's z-distribution

*This paper was published much later in 1928.

$$f(z)=\frac{\nu_1^{\nu_1/2}\nu_2^{\nu_2/2}}{B\left(\frac{\nu_1}{2},\frac{\nu_2}{2}\right)}\cdot\frac{e^{\nu_1 z}}{\left(\nu_1 e^{2z}+\nu_2\right)^{(\nu_1+\nu_2)/2}},$$

which is independent of σ_1^2 and σ_2^2. The same z-statistic given here was also included in Fisher's widely influential book *Statistical Methods for Research Workers* (Fisher, 1925b, p. 192).

As the reader will no doubt realize, the current practice is to use the test statistic

$$F=\frac{s_1^2}{s_2^2}$$

for a test of $H_0:\sigma_1^2=\sigma_2^2$, instead of Fisher's original $z=\frac{1}{2}\log(s_1^2/s_2^2)$. P.C. Mahalanobis (1883–1972) was the first to make use of tabulated values of the *variance ratio* in a 1932 paper published in *The Indian Journal of Agricultural Science* (Mahalanobis, 1932). He was soon followed by George W. Snedecor (1881–1974) in a 1934 monograph (Snedecor, 1934, p. 15). Snedecor does not seem to have been aware of Mahalanobis' paper. Whereas Mahalanobis used the symbol x for the variance ratio, Snedecor chose the symbol F in honor of Fisher.* It is the latter symbol that has stuck. The F-distribution with ν_1 and ν_2 degrees of freedom is given by

$$f(u)=\frac{\nu_1^{\nu_1/2}\nu_2^{\nu_2/2}}{B\left(\frac{\nu_1}{2},\frac{\nu_2}{2}\right)}\cdot\frac{u^{(\nu_1/2)-1}}{\left(\nu_1 u+\nu_2\right)^{(\nu_1+\nu_2)/2}}.$$

The F-statistic was to play an even more prominent role in Fisher's analysis of variance (ANOVA), as we shall describe in Section 5.4.

5.3.5 *Statistical Methods for Research Workers (Fisher, 1925)*

Fisher's first book, *Statistical Methods for Research Workers (SMRW)*† (Fisher, 1925b), was written to give the reader the results of his statistical research over the last few years, mainly in small-sample statistics. Fisher started writing SMRW in the summer of 1923 and almost finished by the middle of 1924. Using the minimal amount of mathematics, the book was originally to be called *Statistics for Biological Research Workers*. In all, it went through 14 editions,‡ the last one being in 1970 (Fisher died in 1962). The contents of the first edition are as follows:

I. INTRODUCTORY
II. DIAGRAMS

*Fisher didn't exactly agree with the use of the symbol F. In a letter dated February 16, 1938, Fisher informed Snedecor: "… I am afraid I have a difficulty which you may not know of in assigning the particular symbol F to the variance ratio, and that is, that a table of this value derived from my table of z was published in India, I think in Sankhya, before yours. The author, if I remember right, used a different symbol—probably x." (Bennett, 1990, p. 323).

† SMRW is also described in Pearce (1992) and Edwards (2005).

‡ 2nd 1928, 3rd 1930, 4th 1932, 5th 1934, 6th 1936, 7th 1938, 8th 1941, 9th 1944, 10th 1946, 11th 1950, 12th 1954, 13th 1958, 14th 1970.

III. DISTRIBUTIONS
IV. TESTS OF GOODNESS OF FIT, INDEPENDENCE AND HOMOGENEITY; WITH TABLE OF X^2
V. TESTS OF SIGNIFICANCE OF MEANS, DIFFERENCES OF MEANS, AND REGRESSION COEFFICIENTS
VI. THE CORRELATION COEFFICIENT
VII. INTRACLASS CORRELATIONS AND THE ANALYSIS OF VARIANCE
VIII. FURTHER APPLICATIONS OF THE ANALYSIS OF VARIANCE

The introductory chapter is made up of six sections where many of ideas of Fisher's 1922 paper "On the foundations of theoretical statistics" (Fisher, 1922c),* are repeated. We are told in Section **1: The Scope of Statistics** that:

> ...Statistics may be regarded as (i;) the study of **populations**, (ii.) as the study of **variation**, (iii.) as the study of methods of the **reduction of data.** (Fisher, 1925b, p. 1)

In Section **2: General Method, Calculation of Statistics**, Fisher introduced his fundamental distinction between parameters and statistics. He then reiterated the three types of problems faced by the statistician: (1) Problems of **Specification**, (2) Problems of **Estimation**, and (3) Problems of **Distribution**. Having introduced these problems, Fisher next repeated his usual condemnation of Bayesian methods:

> ...I have sustained elsewhere, that the theory of inverse probability is founded upon an error, and must be wholly rejected. (*ibid.*, p. 10)

Probability, we are told by Fisher, is not appropriate for inference on populations. A different quantity is needed, namely, *likelihood*. These ideas go back to Fisher's 1921 *Metron* paper (Fisher, 1921a). In Section **3**: **The Quantification of Satisfactory Statistics**, Fisher revisited his concepts of consistent, efficient, and sufficient statistics (see Section 5.2.1) and stated that efficient statistics could be found by the method of ML. Section **4**: **Scope of the Book** stated the rationale behind Fisher's book:

> The prime object of this book is to put into the hands of research workers, and especially of biologists, the means of applying statistical tests accurately to numerical data accumulated in their own laboratories or available in the literature. (Fisher, 1925b, p. 16)

Section **5: Mathematical Tables** explains the tables to be used with the book and Section **6** gives an application of ML to a problem in genetics.

Of special note among the other chapters of SMRW is that, in Chapter IV, Fisher for the first time recommended a p-value of less than .05† to declare statistical significance:

> If P is between .1 and .9 there is certainly no reason to suspect the hypothesis tested. If it is below .02 it is strongly indicated that the hypothesis fails to account for the whole of the

* See Section 5.2.1.

† The origins of the .05 level of significance are discussed in Cowles and Davis (1982).

> facts. We shall not often be astray if we draw a conventional line at .05, and consider that higher values of χ^2 indicate a real discrepancy. (*ibid.*, p. 79)

Equally noteworthy is the fact that ANOVA was introduced in a chapter (Chapter VII) largely devoted to the intraclass correlation. The reason for this is explained in Section 5.4.1. Chapter VIII further extended ANOVA to treat regression problems (see end of Section 5.4.1).

SMRW met with only a lukewarm reception from the statistical experts. Two major problems with the book, as some reviewers saw it, were its strong didactic style coupled with a complete lack of mathematical details. The absence of the latter was especially felt to be a major drawback since several of the methods Fisher had introduced in SMRW were new to both biologists and statisticians. A notable exception to some of the negative reviews SMRW received was the one given by Harold Hotelling who concluded that:

> ...[t]he author's work is of revolutionary importance and should be far better known in this country [the United States]. (Hotelling, 1927, p. 2)

Hotelling was right, because SMRW proved to be a truly revolutionary book and was widely read. It changed the practice of statistics among a wide spectrum of scientific workers including biologists, social scientists, psychologists, and so on. It can safely be said that Fisher's method of significance testing took off in the wake of SMRW.

5.4 ANOVA AND THE DESIGN OF EXPERIMENTS

Fisher revolutionized not only the world of statistical theory but also that of experimental design. Before he entered the scene, experiments lacked a sound theoretical basis and were conducted in an *ad hoc* manner. There was no systematic way of estimating experimental errors, as can be seen in works on experimentation in the early twentieth century (e.g., Wood and Stratton, 1910; Mercer and Hall, 1911).

Sound experimental techniques such as randomization, replication, and blocking were used well before Fisher (Hall, 2007) but in a confused way. With Fisher, all these came in full light. Starting in the mid-20s, he developed various experimental techniques all linked by one underlying thread, namely, analysis of variance (ANOVA).*

5.4.1 Birth and Development of ANOVA (Fisher and Mackenzie, 1923; Fisher, 1925)

ANOVA is one of the most enduring and fundamental of Fisher's countless contributions. The term "variance" itself was first used by Fisher in his 1918 genetics paper "The correlation between relatives on the supposition of Mendelian inheritance" (Fisher, 1918) and was motivated by the additive property it enjoyed:

> ...When there are two independent causes of variability capable of producing in an otherwise uniform population distributions with standard deviations† σ_1 and σ_2, it is

*A survey of Fisher's 1922–1926 contributions to the design of experiments is given by Box (1980), Preece (1990), and Street (1990).

†The term "standard deviation" was first used by Pearson (1894, p. 80).

> found that the distribution, when both causes act together, had a standard deviation $\sqrt{\sigma_1^2 + \sigma_2^2}$. It is therefore desirable in analyzing the causes of variability to deal with the square of the standard deviation as the measure of variability. We shall term this quantity the Variance of the normal population to which it refers, and we now ascribe to the constituent causes fractions and percentages of the total variance which they together produce. (*ibid.*, p. 399)

In the above we find not only the first use of "variance," but also, in the second part of the last sentence, the germ of the notion of ANOVA.

Fisher first used the term "analysis of variance" in another genetics paper published the following year, entitled "The causes of human variability" (Fisher, 1919). This occurred in the context of finding the proportion of the total variance in a phenotype that is due to heritable causes:

> There is, then, in this analysis of variance no indication of any other than innate and heritable factors at work, and a strong probability that whatever non-heritable factors are at work, including errors of measurement, do not contribute more than 5 per cent. and perhaps much less, to the total variance. (*ibid.*, p. 7)

In 1919, having refused a position at the Galton Laboratory by Karl Pearson, Fisher joined the Rothamsted Experimental Station. In the aftermath of World War I, there was increased emphasis on research into food production. Fisher seems to have enjoyed the opportunity to work at Rothamsted and, starting from 1921, he published a series of papers under the general title "Studies in crop variation." In the first of these papers (Fisher, 1921b), Fisher introduced the concept of orthogonal polynomials. This technique involved dividing group sum of squares into independent components. Student had previously used polynomials as a tool to remove time trends in data (Fisher, 1916), but with the orthogonality feature, Fisher was able to assess the importance of each polynomial *independently* of the other.

Having fitted orthogonal polynomials in "Studies in crop variation I," Fisher was now ready for the next step. In the 1923 paper "Studies in crop variation II," in collaboration with W.A. Mackenzie, Fisher introduced the first ANOVA table (Fisher and Mackenzie, 1923) (Fig. 5.4). Data were obtained by T. Eden of Rothamsted Station from a $2 \times 12 \times 3$ factorial experiment involving the response to manure (absent vs. present) of 12 different varieties of potatoes under three conditions (no potash, potassium sulfate, and potassium chloride). The experiment made use of a split–split plot design: the field was first divided into two blocks, one with dung, the other without; each block was then split into 36 plots involving 12 varieties of potatoes in three replicates; each plot was further split into three subplots, one receiving potassium sulfate, one potassium chloride, and the third receiving no potassium. The correct ANOVA table is therefore of the form shown in Table 5.7.

Two important facts emerge from the last ANOVA table. First, since a single replicate was used for dung within each block, there is no estimate for the within-block error and the effect of dung cannot be tested. Second, there are two separate estimates of error, one between plot, and the other within plot.

32

STUDIES IN CROP VARIATION.

II. THE MANURIAL RESPONSE OF DIFFERENT POTATO VARIETIES.

By R. A. FISHER, M.A. and W. A. MACKENZIE, B.Sc.
Rothamsted Experimental Station, Harpenden.

(With Two Charts.)

1. Introductory.

It is not infrequently assumed that varieties of cultivated plants differ not only in their suitability to different climatic and soil conditions, but in their response to different manures. Since the experimental error of field experiments is often underestimated, this supposition affords a means of explaining discrepancies between the results of manurial experiments conducted with different varieties; in the absence of experimental evidence adequate to prove or disprove the supposed differences between varieties in their response to manures such explanations cannot be definitely set aside, although we very often suspect that the discrepancies are in reality due to the normal errors of field experiments.

On the other hand, if important differences exist in the manurial response of varieties a great complication is introduced into both variety and manurial tests; and the practical application of the results of past tests becomes attended with considerable hazard. Only if such differences are non-existent, or quite unimportant, can variety tests conducted with a single manurial treatment give conclusive evidence as to the relative value of different varieties, or manurial tests conducted with a single variety give conclusive evidence as to the relative value of different manures.

In a recent experiment at Rothamsted twelve potato varieties were tested with six manurial treatments, and since the tests were carried out in triplicate the normal error may be evaluated with some accuracy. There is thus afforded a basis for comparing the discrepancies between the different varieties with those to be expected if all varieties had responded alike to the differences in manurial treatment. Both the general response to manurial treatment, and the general differences in the yield of varieties were well marked, so that the data are well suited to the present enquiry.

Journal of Agricultural Science, 13: 311-320, (1923).

FIGURE 5.4 First page of Fisher and Mackenzie's "Studies in crop variation. II. The manurial response of different potato varieties." First page in Fisher and Mackenzie (1923). Courtesy of Cambridge University Press

TABLE 5.7 Correct ANOVA Table for Fisher and Mackenzie's Split–Split Plot Design

Source of Variation	Degrees of Freedom
Block	1
Variety	11
Block × variety	11
Between-plot error	47
Potassium	2
Potassium × variety	22
Potassium × block	2
Potassium × variety × block	22
Within-plot error	94
Total	**212**

TABLE 5.8 Incorrect ANOVA Table From Fisher and Mackenzie's "Studies in Crop Variation. II. The Manurial Response of Different Potato Varieties"

Variation due to	Degrees of Freedom	Sum of Squares	Mean Square	Standard Deviation
Manuring	5	6,158	1231.6	35.09
Variety	11	2,843	258.5	16.07
Deviation from summation formula	55	981	17.84	4.22
Variation between parallel plots	141	1,758	12.47	3.53
Total	**212**	**11,740**	—	—

From Fisher and Mackenzie (1923, p. 316).

However, Fisher and Mackenzie were not cognizant of these subtleties when they wrote their first ANOVA paper in 1923. The ANOVA table (see Table 5.8) that they gave in their paper was incorrect for at least two reasons:

1. They wrongly assumed there were six manuring treatments (three potash dressings with and without dung).
2. They failed to distinguish the two separate sources of error and instead combined them into one (SS = 1758 on 47 + 94 = 141 d.f.).

In any case, Fisher and Mackenzie's aim was to test the manuring x variety interaction, which they denoted in their ANOVA table by "Deviations from summation formula." The distribution of s_1^2 / s_2^2 under $H_0 : \sigma_1^2 = \sigma_2^2$ had not been discovered yet (cf. Section 5.3.4), and they instead used

$$\log s_i^2 \sim N\left(\log \sigma_i^2, \frac{1}{2\nu_i} \right),$$

where ν_i is the number of degrees of freedom associated with the sample variance $s_i^2 (i=1,2)$. To test $H_0:\sigma_1^2=\sigma_2^2$, Fisher and Mackenzie therefore calculated $\log(s_1^2/s_2^2)=\log 4.22-\log 3.53=.1785$ and $1/282+1/110=.01264$. They concluded:

> ...The difference in the logarithm thus exceeds its standard error, but not sufficiently to be significant. (*ibid.*, p. 316)

More precisely, the normal deviate is .1785/.1124 = 1.591 with one-sided *p*-value = .056 so that the result fails to be significant at the 5% level.*

Readers today may find it surprising that Fisher's treatment of ANOVA in his book *Statistical Methods for Research Workers* (Fisher, 1925b) occurred in a chapter largely devoted to the intraclass correlation. But as Box recalls:

> ...problems that today would be dealt with by the analysis of variance, were usually thought of as problems in correlation. If there were n families and k siblings in each, one could calculate the intraclass correlation among siblings by averaging correlations. But since with k siblings, there would be $k(k-1)/2$ pairs to average over, the direct method was tedious. J. A. Harris (1913)† had proposed an abbreviated method of calculation for the intraclass correlation, which simply involved the identity between the sum of squares of deviations from the average over all families and the total of the corresponding sums of squares between and within families. Fisher used this method to estimate not correlations but the contributions of component variances to the total variance. (Box, 1978, pp. 100–101)

In Fisher's own words:

> A very great simplification is introduced into questions involving intraclass correlation when we recognise that in such cases the correlation merely measures the relative importance of two groups of factors causing variation. (Fisher, 1925b, p. 188)

Indeed, we explained in Section 2.3.8 how Fisher showed that the determination of the intraclass correlation can be viewed as essentially a problem of ANOVA (involving components of variance). In particular, consider n classes each with k values, and let the quantity being analyzed be denoted by x_{ij} ($i=1, 2, \ldots, n$; $j=1, 2, \ldots, k$). Fisher then provided the fundamental ANOVA identity:

$$\sum_{i=1}^{n}\sum_{j=1}^{k}\left(x_{ij}-\bar{x}\right)^2 = k\sum_{i=1}^{n}\left(\bar{x}_i-\bar{x}\right)^2+\sum_{i=1}^{n}\sum_{j=1}^{k}\left(x_{ij}-\bar{x}_i\right)^2 \tag{5.77}$$

(*ibid.*, p. 188), where the quantities from left to right in the above are the total, between-class, and within-class sum of squares, respectively; $\bar{x}_i$ is the mean of class i; and $\bar{x}$ is the grand mean.

*At about the same time that "Studies of crop variation II" was published, Fisher communicated his ideas on ANOVA to Student (1923). Fisher used ANOVA to provide a pooled estimate of error in a design involving blocking, and his full derivation was provided as a footnote in Student's paper (pp. 283–284).

† Harris J.A., 1913 On the calculation of intraclass and interclass coefficients of correlation from class moments when the number of possible combinations is large. Biometrika 9: 446–472 (*Author's Note*).

The last identity may be proved by writing

$$\begin{aligned}
\sum_{i=1}^{n}\sum_{j=1}^{k}\left(x_{ij}-\bar{x}\right)^2 &= \sum_{i=1}^{n}\sum_{j=1}^{k}\left(x_{ij}-\bar{x}_i+\bar{x}_i-\bar{x}\right)^2 \\
&= \sum_{i=1}^{n}\sum_{j=1}^{k}\left(x_{ij}-\bar{x}_i\right)^2+\sum_{i=1}^{n}\sum_{j=1}^{k}\left(\bar{x}_i-\bar{x}\right)^2+2\sum_{i=1}^{n}\sum_{j=1}^{k}\left(x_{ij}-\bar{x}_i\right)\left(\bar{x}_i-\bar{x}\right) \\
&= \sum_{i=1}^{n}\sum_{j=1}^{k}\left(x_{ij}-\bar{x}_i\right)^2+\sum_{i=1}^{n}\sum_{j=1}^{k}\left(\bar{x}_i-\bar{x}\right)^2+2\sum_{i=1}^{n}\left(\bar{x}_i-\bar{x}\right)\sum_{j=1}^{n}\left(x_{ij}-\bar{x}_i\right) \\
&= \sum_{i=1}^{n}\sum_{j=1}^{k}\left(x_{ij}-\bar{x}_i\right)^2+\sum_{i=1}^{n}\sum_{j=1}^{k}\left(\bar{x}_i-\bar{x}\right)^2+2(0)\sum_{j=1}^{k}\left(x_{ij}-\bar{x}_i\right) \\
&= \sum_{i=1}^{n}\sum_{j=1}^{k}\left(x_{ij}-\bar{x}_i\right)^2+k\sum_{i=1}^{n}\left(\bar{x}_i-\bar{x}\right)^2.
\end{aligned}$$

Now, the between-class and within-class variances are given, respectively, by

$$s_b^2=\frac{\sum_{i=1}^{n}\left(\bar{x}_i-\bar{x}\right)^2}{n-1},$$

$$s_w^2=\frac{\sum_{i=1}^{n}\left(x_{ij}-\bar{x}_i\right)^2}{n(k-1)}.$$

The (sample) intraclass correlation is then

$$r_{\mathrm{IC}}=\frac{s_b^2-s_w^2}{s_b^2+(k-1)s_w^2}$$

(cf. Eq. 2.74). Whereas in the 1923 paper (Fisher and Mackenzie, 1923) Fisher had used a normal approximation for the distribution of $\log(s^2)$, by 1925 Fisher had already obtained and tabulated the specific distribution of the z-statistic defined by

$$z=\frac{1}{2}\log\left(\frac{s_1^2}{s_2^2}\right)=\log\left(\frac{s_1}{s_2}\right) \tag{5.78}$$

(cf. Section 5.3.4). Fisher's z-statistic for the class effect in an ANOVA table is then

$$z=\frac{1}{2}\log\left(\frac{s_b^2}{s_w^2}\right)=\frac{1}{2}\log\left\{\frac{1+(k-1)r_{\mathrm{IC}}}{1-r_{\mathrm{IC}}}\right\}. \tag{5.79}$$

The z-statistic has degrees of freedom corresponding to the two variances s_b^2 and s_w^2, namely, $n-1$ and $nk-n$. If we define $F=s_b^2/s_w^2(=e^{2z})$, then

$$F=\frac{1+(k-1)r_{\mathrm{IC}}}{1-r_{\mathrm{IC}}}$$

with degrees of freedom $n-1$ and $nk-n$. From this last formula, it can be seen that when $r_{IC}=0$ then $F=1$, so that an intraclass correlation significantly different from zero is equivalent to an F-statistic significantly different from unity.

The specific distribution of z in Eq. (5.78) had been obtained by Fisher in 1924 (1924a).* As we mentioned before, P.C. Mahalanobis was the first to make use of tabulated values of the *variance ratio* $(=s_1^2/s_2^2)$ in 1932 (Mahalanobis, 1932). He was soon followed by George W. Snedecor (1934, p. 15). Mahalanobis used the symbol x for the variance ratio, but Snedecor opted for F (in honor of Fisher), a symbol that has stuck.

While the ANOVA in "Studies in crop variation II" (Fisher and Mackenzie, 1923) was, as we pointed out, incorrect, Fisher gave a correct analysis soon after in his book *Statistical Methods for Research Workers* (Fisher, 1925b, pp. 203–209). But before we examine this, let us consider the very first ANOVA example Fisher gave in his book:

> In an experiment on the accuracy of counting soil bacteria, a soil sample was divided into four parallel samples, and from each of these after dilution seven plates were inoculated. The number of colonies on each plate is shown below [see Table 5.9]. Do the results from the four samples agree within the limits of random sampling? In other words, is the whole set of 28 values homogeneous, or is there any perceptible intraclass correlation? (*ibid.*, p. 194)

Thereupon, by using Eq. (5.77), Fisher gave the ANOVA shown in Table 5.10. Therefore:

> The table shows that $P=.05$ for $z=1.0781$, so that the observed difference, .3224, is really very moderate, and quite insignificant. The whole set of 28 values appears to be homogeneous with variance about 57.07. (*ibid.*, pp. 195–196)

Let us now examine how Fisher gave the correct analysis in his book for the potato example previously considered in "Studies of crop variation II" (Fisher and Mackenzie, 1923). Having realized that the effect of dung could not be tested, since there was a single

TABLE 5.9 Data for First ANOVA Example in Fisher's *Statistical Methods for Research Workers*

	Sample			
Plate	I	II	III	IV
1	72	74	78	69
2	69	72	74	67
3	63	70	70	66
4	59	69	58	64
5	59	66	58	62
6	53	58	56	58
7	51	52	56	54
Mean	60.86	65.86	64.28	62.86

From Fisher (1925b, p. 195).

* See also Section 5.3.4.

TABLE 5.10 Table for First ANOVA Example in Fisher's *Statistical Methods for Research Workers*

Source	d.f.	Sum of Squares	Mean Square	S.D.	log(S.D.)
Within Classes	24	1446	60.25	7.762	2.0493
Between Classes	3	94.86	31.62	5.623	1.7269
Total	**27**	**1540.9**	57.07	7.55	−.3224
					(Difference) = z

From Fisher (1925b, p. 195).

TABLE 5.11 First ANOVA Table From *Statistical Methods for Research Workers* for the Potato Example

Source	d.f.	Sum of Squares	Mean Square	log(S.D.)
Between varieties	11	43.638	3.967	.689
Between potash dressings for the same variety	24	17.440	.727	−.159
Within potash dressings	72	10.620		
Total	**107**	**71.699**		

From Fisher (1925b, p. 207).

TABLE 5.12 Second ANOVA Table From *Statistical Methods for Research Workers* for the Potato Example

Source	d.f.	Sum of Squares	Mean Square
Between potash dressings	2	.350	.175
Differential response of varieties	22	2.191	.100
Differential response in potash dressings with the same variety	48	8.080	.168
Total	**72**	**10.620**	

From Fisher (1925b, p. 208).

replicate, Fisher used only half of the field this time for the analysis. He performed several tests using Tables 5.11 and 5.12.

Fisher first tested for the effect of variety. From Table 5.11, he obtained $z = .689 + .159 = .85$, which is more than twice the 5% value he had tabulated: "the effect of variety is therefore very significant." The next test was that of potash dressings from the same variety. Fisher compared the value .727 with 24 d.f. (Table 5.11) against the value .168 with 48 d.f. (Table 5.12). He obtained $z = .7315$, while the 5% table value is about .28: "the evidence for unequal fertility of the different patches [potash dressings] is therefore unmistakable." Likewise, Fisher tested for the differential response among varieties and for the effect of potash dressings and found both to be nonsignificant.

The last two chapters of Fisher's *Statistical Methods for Research Workers* (Fisher, 1925b) were a real display of the ubiquity and power of ANOVA. Here Fisher showed how various previously used as well as new techniques could all be viewed through the single lens of ANOVA.

After having considered the intraclass correlation, ANOVA per se, and the split–split plot experiment of "Studies in crop variation II," the next step for Fisher was to show how the problem of fitting regression lines could also be viewed as one of ANOVA.

As a first example, Fisher considered the case of simple linear regression. An independent variable x is measured at a different values such that, at $x = x_p (p = 1,\ldots,a)$, n_p measurements of y are made. Then Fisher gave the following decomposition of sum of squares:

$$\Sigma(y-\bar{y})^2 = \Sigma n_p(\bar{y}_p - \bar{y})^2 + \Sigma\Sigma(y-\bar{y}_p)^2 \tag{5.80}$$

(*ibid.*, p. 213), where $\bar{y}_p$ is the mean of the values of y at $x = x_p$, and $\bar{y}$ is the grand mean of all values of y. To illustrate, Fisher considered the influence of temperature (x) on the number of eye facts (y) in *Drosophila melanogaster*. The data obtained by A.H. Hersh had nine different temperatures (arrays) with varying number of y-values for each temperature (823 values in all). Fisher obtained the ANOVA shown in Table 5.13.

He then noted:

> The variance within the arrays is thus only about 4.7; the variance between the arrays will be made up of a part which can be represented by a linear regression, and of a part which represents the deviations of the observed means of arrays from a straight line. (*ibid.*, p. 216)

To test for departure from linear regression, Fisher calculated the regression sum of squares:

$$SS_{reg} = \frac{\{\Sigma(x-\bar{x})(y-\bar{y})\}^2}{\Sigma(x-\bar{x})^2} = 11{,}974.$$

He then decomposed the between-array sum of squares (=12,370) in Table 5.13 into the regression sum of squares (=11,974) and the sum of squares (=12,370−11,974=396) due to deviation from a regression line (see Table 5.14).

From Tables 5.13 and 5.14, Fisher obtained

$$z = \frac{1}{2}\log\left(\frac{s_1^2}{s_2^2}\right) = \frac{1}{2}\log\left(\frac{56.6}{4.708}\right) = 1.2434.$$

TABLE 5.13 First ANOVA Table for Regression Example in Fisher's *Statistical Methods for Research Workers*

Variance	d.f.	Sum of Squares	Mean Square
Between arrays	8	12,370	…
Within arrays	814	3,832	4.708
Total	**822**	**10,620**	

From Fisher (1925b, p. 216).

TABLE 5.14 Second ANOVA Table for Regression Example in Fisher's *Statistical Methods for Research Workers*

Variance Between Arrays due to	d.f.	Sum of Squares	Mean Square
Linear regression	1	11,974	...
Deviations from regression	7	396	56.6
Total	**8**	**12,370**	

From Fisher (1925b, p. 217).

TABLE 5.15 ANOVA Table for Multiple Linear Regression

Variance due to	d.f.	Sum of Squares
Regression function	p	SS_{reg}
Deviations from regression	$n-p-1$	SS_{dev}
Total	$\boldsymbol{n-1}$	$\mathbf{SS_{tot}}$

Since the 5% value is .35, he concluded:

> ...There can therefore be no question of the statistical significance of the deviations from the straight line, although the latter accounts for the greater part of the variation. (*ibid.*, p. 218)

For multiple regression, Fisher considered a fitted regression line of the form

$$Y = b_1 x_1 + b_2 x_2 + \cdots + b_p x_p,$$

where $b_1, b_2, \ldots, b_p$ are the estimated regression coefficients, Y is the fitted value of the deviation y of the outcome from its average value, and x_i ($i=1, \ldots, p$) is the deviation of the ith predictor from its average value. To test for the regression effect, Fisher gave the ANOVA shown in Table 5.15 (*ibid.*, p. 222).

In the above, n is the total number of observations and

$$\begin{aligned} SS_{tot} &= \Sigma y^2, \\ SS_{dev} &= \Sigma (y-Y)^2, \\ SS_{reg} &= b_1 \Sigma x_1 y + b_2 \Sigma x_2 y + \cdots + b_p \Sigma x_p y = SS_{tot} - SS_{dev}. \end{aligned}$$

The test for regression is then carried out by calculating

$$z = \frac{1}{2} \log \left\{ \frac{SS_{reg} / p}{SS_{dev} / (n-p-1)} \right\},$$

and comparing with z-tables as before.

5.4.2 Randomization, Replication, and Blocking (Fisher, 1925; 1926), Latin Square (Fisher, 1925), Analysis of Covariance (Fisher, 1932)

5.4.2.1 Randomization Fisher remained a fierce advocate of randomization, replication, and blocking as the three cornerstones of a well-planned experiment. Already in the "Studies in crop variation II" paper (Fisher and Mackenzie, 1923), the importance of these three components was recognized. Replication and blocking were explicitly used. As Cochran has correctly pointed out (Cochran, 1980, p. 18), no randomization was used in the split–split plot experiment. But Fisher seems to have been aware of its importance for he wrote in the 1923 paper that:

> ...Thus the sum of all the squares of deviations from the general mean may be divided up into two parts: one measures the variation within the triplicates, while the other measures the variation between triplicates differently treated. Further, if all the plots are undifferentiated, as if the numbers had been mixed up and written down in *random order**, the average value of each of the two parts is proportional to the number of degrees of freedom in the variation of which it is compared. (Fisher and Mackenzie, 1923, p. 315)

As his ideas on the foremost importance of randomization were getting cemented,† Fisher was more forceful in *Statistical Methods for Research Workers*:

> ...For our test of significance to be valid the difference in fertility between plots chosen as parallels must be truly representative of the differences between plots with different treatment; and we cannot assume that this is the case if our plots have been chosen in any way according to a prearranged system; for the systematic arrangement of our plots may have, and tests with the results of uniformity trials show that it often does have, features in common with the systematic variation of fertility, and thus the test of significance is wholly vitiated. (Fisher, 1925b, p. 224)

The term "randomization" itself was first used by Fisher in the 1926 paper "The arrangement of field experiments" (Fisher, 1926b, p. 510). In the same paper, Fisher properly delineated why randomization was important:

> One way of making sure that a valid estimate of error will be obtained is to arrange the plots deliberately at random, so that no distinction can creep in between pairs of plots treated alike and pairs treated differently; in such a case an estimate of error, derived in the usual way from the variations of sets of plots treated alike, may be applied to test the significance of the observed difference between the averages of plots treated differently.
>
> The estimate of error is valid, because, if we imagine a large number of different results obtained by different random arrangements, the ratio of the real to the estimated error, calculated afresh for each of these arrangements, will be actually distributed in the theoretical distribution by which the significance of the result is tested. (*ibid.*, pp. 506–507)

In the above, Fisher's argument was that randomization simulated the effect of independence in the observed values of the outcome and yielded a valid estimate of error. Thus, the variance ratio calculated in ANOVA had the correct theoretical distribution, making tests of significance valid.

*Italics are ours.

† Fisher's advocacy of randomization is thoroughly discussed in Hall (2007).

Fisher's advocacy of randomization thus relied on his understanding of the underlying distribution theory. But his adamancy about this principle was such that, in one of the less fortunate episodes in the history of statistics, it resulted in a dispute with his longtime friend and colleague, Student. We shall examine the Fisher–Student dispute on randomization in Section 5.4.3.

5.4.2.2 Replication Fisher's second *sine qua non* for a sound experiment was *replication*. As we explained in the previous section, "Studies in crop variation II" (Fisher and Mackenzie, 1923) involved replication but not completely so. The field was first divided into two blocks, one with dung, the other without; each block was then split into 36 plots involving 12 varieties of potatoes in three replicates; each plot was further split into three subplots, one receiving potassium sulfate, one potassium chloride, and the third receiving no potassium. But since the dung factor appeared in a single replicate, no estimate for the standard error of its average effect could be calculated. Fisher made no comment on this shortcoming in the 1923 paper but soon realized the importance of proper replication. In the book *Statistical Methods for Research Workers*, we can read:

> ...The first requirement which governs all well-planned experiments is that the experiment should yield not only a comparison of different manures, treatments, varieties, etc., but also a means of testing the significance of such differences as are observed. Consequently all treatments must at least be duplicated, and preferably further replicated, in order that a comparison of replicates may be used as a standard with which to compare the observed differences. (Fisher, 1925b, p. 224)

In the landmark paper "Arrangement of field experiments," Fisher was even more specific:

> ...It would be exceedingly inconvenient if every field trial had to be preceded by a succession of even ten uniformity trials; consequently, since the only purpose of these trials is to provide an estimate of the standard error, means have been devised for obtaining such an estimate from the actual yields of the trial year.
>
> The method adopted is that of replication.... (Fisher, 1926b, pp. 505–506)

But Fisher was quick to point that, although replication enabled the standard error to be estimated:

> One way of making sure that a valid estimate of error will be obtained is to arrange the plots deliberately at random.... (*ibid*., p. 506)

5.4.2.3 Blocking Fisher's third dictum for a good experiment was *blocking*. As was mentioned at the beginning of this section, blocking was used in the 1923 paper (Fisher and Mackenzie, 1923) but without randomization. The use of blocking in a randomized experiment first took place in Fisher's book *Statistical Methods for Research Workers* (Fisher, 1925b). Let us examine how Fisher introduced this concept. In his book, Fisher considered a completely randomized experiment in which five different treatments were applied to 20 strips of land. In one such layout (see Table 5.16), the data were analyzed and (using the formulas in Eq. 5.77) displayed in a one-way ANOVA table (see Table 5.17).

TABLE 5.16 Data for Completely Randomized Design

B	C	A	C	E	E	E	A	D	A
3504	3430	3376	3334	3253	3314	3287	3361	3404	3366
B	C	B	D	D	B	A	D	C	E
3416	3291	3244	3210	3168	3195	3330	3118	3029	3085

From Fisher (1925b, p. 225).
The letters A, B, …, E stand for the five different treatments.

TABLE 5.17 One-way ANOVA for Completely Randomized Design

Variance due to	d.f.	Sum of Squares	Mean Square	Standard Deviation
Treatment	4	58,725	14,681	121.1
Experimental error	15	231,041	15,403	124.1
	19	**289,766**	**15,251**	**123.5**

From Fisher (1925b, p. 226).

Fisher next made the following observation:

> While adhering to the essential condition that the errors by which the observed values are affected shall be a random sample of the errors which contribute to our estimate of experimental error, it is still possible to eliminate much of the effect of soil heterogeneity, and so increase the accuracy of our observations by laying restrictions on the order in which the strips are arranged. As an illustration of a method which is widely applicable, we may divide the 20 strips into 5 blocks, and impose the condition that each treatment shall occur once in each block; we shall then be able to separate the variance into three parts representing (i.) local differences between blocks, (ii.) differences due to treatment, (iii.) experimental errors; and if the five treatments are arranged at random within each block, our estimate of experimental error will be an unbiassed [*sic*] estimate of the actual errors in the differences due to treatment. (*ibid.*, pp. 226–227)

Fisher's point in the above was that the experimental error could be greatly reduced,* but still be unbiased, by dividing the 20 strips of land into blocks (here five blocks were used). Each block contained each treatment only once and randomization was done within each block. This is the "randomized block" design, a term that was first used by Fisher one year later in "The arrangement of field experiments." (Fisher, 1926b, p. 509). By using one such layout (see Table 5.18), the original data were reanalyzed and displayed as a two-way ANOVA (see Table 5.19)

The sums of squares in Table 5.19 can be obtained from Eq. (5.77) by using the formula for SS_t to calculate the total sum of squares, and the formula for SS_b to calculate both (i) "local differences" and (ii) "treatment." The only difference is that, for (i), the data are arranged by block, as in Table 5.18, while for (ii) the data are arranged by treatment. The sum of squares due to experimental error is then the difference between the first and the sum of the last two sums of squares.

*A reduction in experimental error means that the design has more *power* to test for the treatment effect.

TABLE 5.18 Data for Randomized Block Experiment

	A	E	C	D	B
Block 1	3504	3430	3376	3334	3253
	C	B	E	D	A
Block 2	3314	3287	3361	3404	3366
	A	D	E	B	C
Block 3	3416	3291	3244	3210	3168
	C	E	B	A	D
Block 4	3195	3330	3118	3029	3085

From Fisher (1925b, p. 227).

TABLE 5.19 Two-way ANOVA for Randomized Block Design

Variance due to	d.f.	Sum of Squares	Mean Square	Standard Deviation
Local differences	3	154,483	51,494	...
Treatment	4	40,859	10,215	...
Experimental error	12	94,424	7,869	88.7
	19	**289,766**	**15,251**	

From Fisher (1925b, p. 227).
Fisher used "Local differences" to denote differences between blocks.

From Tables 5.17 and 5.19, it is seen that the experimental error has been reduced from 15,403 to 7,869. Fisher commented that:

> The local differences between the blocks are very significant, so that the accuracy of our comparisons is much improved, in fact the remaining variance is reduced almost to 55 per cent of its previous value. (Fisher, 1925b, p. 227)

5.4.2.4 Latin Square The randomized block design can thus be considered as an improvement over the completely randomized design since the former results in a reduced experimental error. However, Fisher noted that in some cases an even further reduction in experimental error was possible:

> The method of laying restrictions on the distribution of the plots and eliminating the corresponding degrees of freedom from the variance is, however, capable of some extension in suitably planned experiments. In a block of 25 plots arranged in 5 rows and 5 columns, to be used for testing 5 treatments, we can arrange that each treatment occurs once in each row, and also once in each column, while allowing free scope to chance in the distribution subject to these restrictions. (*ibid.*, p. 229)

Thus, if each treatment appears just once not only in each row but also in each column, a further reduction in experimental error can be achieved. Note that some randomization is retained as, although each treatment appears only once in each row and column, the *order* in which this occurs is still random. Fisher called this design the "Latin Square" design, a term first used by Euler in 1782 (Euler, 1782, p. 90).* As an illustration, Fisher considered

* See the article by Kendall (1948) for an early history of the Latin Square.

TABLE 5.20 Data for Latin Square Design

						Total of Row
	D	E	C	B	A	
	376	371	355	356	335	**1793**
	B	D	E	A	C	
	316	338	336	332	332	**1678**
	C	A	B	D	E	
	326	326	335	330	330	**1660**
	E	B	A	C	D	
	317	342	330	336	336	**1653**
	A	C	D	E	B	
	321	332	317	306	306	**1594**
Total of Column	**1656**	**1710**	**1673**	**1700**	**1639**	**8378**

From Fisher (1925b, p. 230).

TABLE 5.21 ANOVA Table for Latin Square

Differences between	d.f.	Sum of Squares	Mean Square	Standard Deviation
Rows	4	4240.24	...	...
Columns	4	701.84	...	...
Treatments	4	330.24	130.3 (Treatments and Remainder combined)	11.41
Remainder	12	1754.32		
	24	**7026.64**	**292.8**	**17.11**

From Fisher (1925b, p. 230).

the root weights for mangolds in 25 plots, with five treatments appearing once in each row and column (see Table 5.20). The ANOVA for the Latin Square is shown in Table 5.21.

The sums of squares in Table 5.21 can be obtained from Eq. (5.77) by using the formula for SS_t to calculate the total sum of squares, and the formula for SS_b to calculate each of (i) "rows," (ii) "columns," and (iii) "treatment." The only difference is that, for (i), the row sums in Table 5.20 are used, for (ii) the column sums in Table 5.20 are used, while for (iii) the data are arranged by treatment. The sum of squares due to experimental error is then the difference between the first and the sum of the last three sums of squares.

From Table 5.21, Fisher concluded:

> By eliminating the soil differences between different rows and columns, the mean square has been reduced to less than half, and the value of the experiment as a means of detecting differences due to treatment is therefore more than doubled. (Fisher, 1925b, p. 230)

5.4.2.5 Analysis of Covariance In the fourth edition of *Statistical Methods for Research Workers* (Fisher, 1932, p. 249), Fisher introduced yet another technique for further error reduction, namely, *analysis of covariance (ANCOVA)*. This is how the method was introduced:

> ...[I]n nutritional experiments the growth rates of males and females may be distinctly different, while nevertheless both sexes may be equally capable of showing the advantage of

TABLE 5.22 Preliminary Yields of Tea Plots

88	102	91	88	**369**
94	110	109	118	**431**
109	105	115	94	**423**
88	102	91	96	**377**
379	**419**	**406**	**396**	**1600**

From Fisher (1932, p. 252).

TABLE 5.23 Experimental Yields of Tea Plots

90	93	85	81	**349**
93	106	114	121	**434**
114	106	111	93	**424**
92	107	92	102	**393**
389	**412**	**402**	**397**	**1600**

From Fisher (1932, p. 252).

> one diet over another. The effect of sex, on the growth rates compared, will, therefore, be eliminated by assigning the same proportion of males to each experimental treatment, and, what is more often neglected, eliminating the average difference between the sexes from the estimate of error. (*ibid*., p. 250)

Thus, to adjust for the effect of gender in an experiment with several treatments, one has to make sure that the same proportion of males is assigned to each treatment. This results in reduced experimental error. Similarly, the effects of other factors can be adjusted. As an example, Fisher considered the data provided by T. Eden concerning pluckings from 16 plots of tea bushes in Sri Lanka. Both the preliminary yields (Table 5.22) and experimental yields (Table 5.23) are shown.

Fisher assumed a Latin Square design in four treatments, but since no actual treatment differences were applied, the ANOVA table for the experimental yields is given in Table 5.24. The residual variance is 97.22. Fisher then noted that this variance could be further reduced by choosing sets of plots in which the preliminary yields were nearly equalized. However, a more practical method would be to incorporate a linear regression into the ANOVA (hence the term analysis of *covariance*). This has two advantages over the equalization strategy, namely:

> ...the advantage of eliminating differences between rows and columns (or blocks) would often have to be sacrificed to equalisation, and that such equalisation as would be possible would always be inexact. (*ibid*., p. 254)

The idea behind ANCOVA is that the difference between any two experimental plots needs to be adjusted by subtracting the expected difference obtained from preliminary plots treated alike. To obtain the expected difference, a regression of the experimental yields (Y) is performed on the preliminary yields (X). The regression coefficient of such a

TABLE 5.24 Analysis of Experimental Yields

	d.f.	Sum of Squares	Mean Square
Rows	3	1095.5	…
Columns	3	69.5	…
Error	9	875.0	97.22
	15	**2040.0**	**136.00**

From Fisher (1932, p. 253).

TABLE 5.25 Analysis of Preliminary Yields

	d.f.	Sum of Squares
Rows	3	745.0
Columns	3	213.5
Error	9	567.5
	15	**1526.0**

From Fisher (1932, p. 255).

TABLE 5.26 Sum of Squares and Products (*x* and *y* Are Deviations From Their Respective Row Means for "Rows" and Deviations From Their Respective Column Means for "Columns")

	d.f.	x	xy	y
Rows	3	745.0	837.00	1095.5
Columns	3	213.5	120.75	69.5
Error	9	567.5	654.25	875.0
	15	**1526.0**	**1612.00**	**2040.0**

From Fisher (1932, p. 256).

regression is the ratio of the covariance between X and Y to the variance of X. The latter variance can be obtained from the ANOVA in Table 5.25.

The final step in the analysis is to construct an ANCOVA table by using the products of yields of the preliminary and experimental results instead of sum of squares. The ANCOVA table can be readily obtained by first writing the sum of squares shown in Table 5.26.

From Table 5.26, the regression coefficient of Y on X is estimated as $\hat{\beta}_1 = 654.25 / 567.7 = 1.1529$. Since the adjusted sum of squares is the sum of

$$\left(y - \hat{\beta}_1 x\right)^2 = \left(\hat{\beta}_1^2\right)x^2 + \left(-2\hat{\beta}_1\right)xy + y^2,$$

the ANCOVA is finally obtained by multiplying each set of row entries in Table 5.26 by $\hat{\beta}_1^2$, $-2\hat{\beta}_1$, and 1, and then adding the products, as shown in Table 5.27.

TABLE 5.27 ANCOVA Table

	d.f.	Sum of Squares	Mean Square
Rows	3	155.8	51.93
Columns	3	74.8	24.93
Error	8	120.7	15.09
	14	**351.3**	**15.09**

From Fisher (1932, p. 257).

Comparing Table 5.27 against Table 5.24, three points can be made:

1. The mean square error has decreased from 97.22 to 15.09.
2. The degrees of freedom for error has decreased from 9 to 8 due to the one degree of freedom lost by estimating the regression coefficient.
3. The large sum of squares for rows has decreased substantially from 1095.5 to 155.8, implying that a large fraction of the row variability in the experimental yields was due to preliminary differences.

5.4.3 Controversy with Student on Randomization (1936–1937)

Fisher and Student shared a long relationship of mutual respect and admiration. Fisher (who was almost 15 years younger than Student) would never lose an opportunity in his papers to highlight the groundbreaking nature of Student's 1908 work in small samples. Similarly, Student often expressed admiration for the mathematical sophistication of Fisher's treatment of statistical problems, especially of small-sample statistics. Of the two men, Fisher undoubtedly had the more fiery temper and had already been embroiled in several controversies. But Student had been very wise not to take sides in these controversies and often shared amicable relationships with both disputing parties.

It therefore came as a big surprise when Fisher and Student got involved in a major disagreement concerning the merits of randomization in 1936.* The controversy started just one year before Student's death, but the seeds of discord between the two men had already been sown as far back as in 1923. We now describe these topics.

To understand the story better, we should go back to the 1890s when Edwin S Beaven proposed two systematic designs for comparative experiments, namely, the half-drill strip design (later called by Student as the ABBA design) and the chessboard design (a variant of the Knight's move design). These designs are illustrated in Figure 5.5.

The designs were adopted by Student who wrote an appendix to Mercer and Hall's classic 1911 paper (Mercer and Hall, 1911) and recommended systematic designs. In the 1920s, Student became particularly interested in the half-drill strip design and used it in his 1923 paper (Student, 1923). In that paper (*ibid.*, p. 285), he called the half-drill strip design "the most accurate method yet devised for field trials by which two varieties are compared." Undoubtedly, the design had certain definite advantages. A major one was its

* The Fisher–Student controversy is also discussed in Picard (1980, pp. 51–58), Pearson et al. (1990, pp. 54–60), and Senn (2004).

(a)

A A A...A

B B B...B

B B B...B

A A A...A

A A A...A

B B B...B

B B B...B

A A A...A

...

(b)

A	B	C	D	E
D	E	A	B	C
B	C	D	E	A
E	A	B	C	D
C	D	E	A	B

(c)

A	F	C	H	E	B	G	D
B	G	D	A	F	C	H	E
C	H	E	B	G	D	A	F
D	B	F	C	H	E	B	G
E	A	G	D	A	F	C	H

FIGURE 5.5 Half-drill strip, Knight's move, and chessboard designs. (a) Half-drill strip (ABBA) design for two treatments (the strips are along the rows and the ABBA "sandwiches" are along the columns), (b) Knight's move with five treatments and five replicates. The name Knight's move comes from the fact that each treatment is repeated by moving one down and two across. This particular design is also known as the Knut Vik Square in honor of the Norwegian Knut Vik who presented it in 1924, and (c) Beaven's chessboard for eight treatments.

ease of implementation. One half of the seed drill was filled with A, the other half with B. As the drill moved across the field (from left to right in Figure 5.5a, then making a head turn at the right end, then from right to left, and so on), the rows of crops had the perfectly balanced ABBA pattern. This also meant that if there was a fertility gradient along the columns the symmetry of the ABBA design ensured that there was no resulting bias. A design such as ABABABA... would not be able to counter such a gradient. Even randomization did not guarantee the perfect balance possessed by the ABBA design.

At the same time as Student was advocating the usefulness of his systematic design, Fisher wrote his breakthrough 1923 ANOVA paper in collaboration with Mackenzie (Fisher and Mackenzie, 1923). Although randomization was formally introduced later in 1926 (Fisher, 1926b) (see Section 5.4.2.1), Fisher had already started thinking about randomization in 1923. In that same year, in his correspondence with Fisher, Student started advocating systematic designs to him. For instance, in an April 16, 1923, letter to Fisher:

> ...my impression is that on occasion one has saved half the space of random placing by Chessboarding and that one has always gained something.... (Gosset, 1970, April 16, 1923 letter)

It can also be seen that, right from the start, Student did not hide his disapproval of Fisher's randomization ideas:

> ...I don't expect to convince you but I don't agree with your controlled randomness. You would want a large lunatic asylum for the operators who are apt to make mistakes enough even at present.
>
> I quite agree that such an experiment as the 6×6 of the Irish plots is not at all good when systematically arranged but when you replicate the sets of six often enough the thing becomes random again. If you say anything about Student in your preface you should I think make a note of his disagreement with the practical part of the thing: of course he agrees in theory. (Gosset, 1970, October 20, 1924 letter)

At this stage, Student and Fisher had what can safely be called a mere disagreement. However, things were to take a more dramatic turn in 1936 when Student presented a paper, "Co-operation in large scale experiments" (Gosset, 1936), to the Royal Statistical Society. Fisher was one of the attendants.

At the very start of his paper, Student made these provocative statements:

> Finally, about fifteen years ago, Professor Fisher introduced the principle of randomizing the position of the plots in the various systems of randomized blocks and Latin squares with which many of you are familiar. This enabled us to obtain a certainly valid estimate of the variability of our results, though usually at the expense of increasing that variability when compared with balanced arrangements. (*ibid.*, p. 115)

A few pages later, another punch was delivered:

> Hence, since the tendency of deliberate randomizing is to increase the error, a balanced arrangement like the half-drill strip [ABBA design] is best if otherwise convenient. (*ibid.*, p. 118)

TABLE 5.28 Student's ANOVA Table for the ABBA Design

	Degrees of Freedom	Sum of Squares
Treatment	1	$\frac{\left(\Sigma\overline{AB}-\Sigma\overline{BA}\right)^2}{n}$
Error	$n-2$	$\frac{\Sigma(A-B)^2}{n}-\frac{\left(\Sigma\overline{AB}-\Sigma\overline{BA}\right)^2}{n}$
Total	$n-1$	$\frac{\Sigma(A-B)^2}{n}$

From Gosset (1936, p. 121).

Thus, Student's major contention was that randomization resulted in a higher experimental error than a balanced arrangement such as the ABBA design. In an appendix to the paper, Student then gave an ANOVA table (Table 5.28) for two treatments A and B in an ABBA design. In the table, $\Sigma\overline{AB}$ is the sum of $A-B$ for all AB comparisons, $\Sigma\overline{BA}$ is the sum of $A-B$ for all BA comparisons, and n is the number of pairs.

All of this must have distressed Fisher, especially when the latter had previously (see Section 5.4.2.1) successfully argued that no valid estimate of error was possible unless randomization was used. In his discussion of the paper, he said:

> What is my own disappointment may, I hope, be to the benefit of others. If Mr. Gosset has chosen to give his time to more familiar and elementary aspects of the subject, we can at least recognize the justification that these elementary aspects are still occasionally misunderstood, even to a ridiculous extent. (*ibid.*, p. 122)

Fisher then explained why "modern" (meaning "his") methods were important:

> The advantage of modern methods lies in their capacity to recognize the genuineness of discrepant results obtained at different places, or at different times, and so to perceive the limitations of generalizations which might otherwise be accepted uncritically. (*ibid.*, p. 123)

Next he criticized the half-drill strip (ABBA design) and Student's ANOVA table (Table 5.28):

> ...If the error [in the split-drill design] is estimated in the way he suggests, which is not a method which has been always used, he thinks the error will be slightly over-estimated. This may be so on the average, but this is not incompatible with the error being sometimes largely under-estimated, and a systematic bias in the estimation of error in either direction is a serious drawback...The serious fact is that the actual errors of the split-drill method are always unknown, and though the result of a trial may be ornamented by the addition of the standard error, estimated by some plausible process, such estimates can never be scientifically on the same level as are standard errors of known validity. (*ibid.*, p. 124)

In the above, Fisher explained that the errors of the ABBA design could either be systematically overestimated or underestimated, and that in either case they were *not* statistically valid estimates of the true but unknown errors. Thus, tests of significance could not be applied to Student's ANOVA table.

In his reply to Fisher's criticism, Student was cordial but still unyielding on the issue of systematic designs:

> ...That is an old matter of controversy between Professor Fisher and myself. He says to me, "Your half-drill strips have no validity and conclusions cannot be drawn from them"; I say to him, "Your errors are so large that no conclusions are drawn." Neither of these criticisms is true, and the one is about as good as the other. (*ibid.*, p. 136)

That might indeed have been an old matter, but it did not stop there. In that same year, Barbacki and Fisher published a paper with the highly suggestive title "A test of the supposed precision of systematic arrangements" (Barbacki and Fisher, 1936). Right at the start, the authors stated:

> ...Recently...in connexion with Beavan's split drill method of testing cereal varieties "Student" has claimed explicitly that higher precision is attainable with systematic than with randomized arrangements.
>
> The only method of testing such an assertion is by the direct application of the two alternative methods to yields harvested in half-drill strips from a trial using only a single variety. (*ibid.*, p. 189)

With the ABBA design shown in Table 5.29, Barbacki and Fisher calculated the differences (A − B) in the yields between the half-drill strips as shown in Table 5.30. The authors next randomized the half-drill strips by randomly choosing between ABBA and BAAB. The differences in yield for one such randomization is shown in Table 5.31.

TABLE 5.29 ABBA Design with 16 Strips and 12 Sections per Strip used by Barbacki and Fisher (1936, p. 190)

	(i)	(ii)	(iii)	(iv)	(v)	(vi)	(vii)	(viii)	(ix)	(x)	(xi)	(xii)
A	4410	4035	3865	3640	3650	3985	3490	3330	3358	3712	3487	3781
B	3950	3865	3295	2960	2925	3685	3400	3040	2889	3195	3496	3576
B	4185	4075	3325	2860	2965	3770	3240	2735	2764	3460	3273	3442
A	3785	3515	3255	2815	2630	3295	2875	2630	2775	3040	2940	3152
A	3870	3780	3660	2980	2650	3250	2925	2915	2933	3277	3042	3363
B	3910	3690	3705	3050	2910	3630	2985	3130	2986	3040	2778	3123
B	3890	3695	3720	2990	2970	3315	2910	2985	2851	2635	2906	3081
A	4190	3970	4335	3350	3325	3870	3120	3015	3097	2909	2936	3628
A	4170	4070	4455	3610	3365	3460	2970	2855	2877	2834	3020	3632
B	4015	4480	4730	3805	3375	3545	3080	2810	2794	2974	2770	3805
B	4150	4755	5065	4125	3550	3740	3425	2690	2789	2810	2895	3695
A	4190	4740	5265	4415	3675	3965	3685	3030	2782	2904	3080	2798
A	4095	5075	5495	4270	3760	4010	3695	3255	2759	3118	3287	3547
B	3805	4360	4415	3870	3585	3785	4025	3300	3199	3407	3473	3572
B	4005	4225	3840	3800	3780	3780	4025	3710	3564	3616	3539	3853
A	3700	4325	3550	3455	3540	3660	3980	3705	3577	3759	3558	3673

The numbers are the yields in grams of grain for 1500 15 ft. rows of wheat.

TABLE 5.30 Differences (A − B) for the Design in Table 5.29

	(i)	(ii)	(iii)	(iv)	(v)	(vi)	(vii)	(viii)	(ix)	(x)	(xi)	(xii)
	460	170	570	680	725	300	90	290	469	517	−9	205
	−400	−560	−70	−45	−335	−475	−365	−105	11	−420	−333	−290
	−40	90	−45	−70	−260	−380	−60	−215	−53	237	264	240
	300	275	615	360	355	555	210	30	246	274	30	547
	155	−410	−275	−195	−10	−85	−110	45	83	−140	250	−173
	40	−15	200	290	125	225	260	340	−7	94	185	103
	290	715	1080	400	175	225	−330	−45	−440	−289	−186	−25
	−305	100	−290	−345	−240	−120	−45	−5	13	143	19	−180
Total	500	365	1785	1075	535	245	−350	335	322	416	220	427

From Barbacki and Fisher (1936, p. 190).

TABLE 5.31 Differences (A − B − B + A) for One Particular Randomization of Half-Drill Strips

	(i)	(ii)	(iii)	(iv)	(v)	(vi)	(vii)	(viii)	(ix)	(x)	(xi)	(xii)
	60	−390	500	635	390	−175	−275	185	480	97	−342	−85
	260	365	570	290	95	175	150	−185	193	511	294	787
	195	−425	−75	95	115	140	150	385	76	−46	435	−70
	−15	815	790	55	−65	105	−375	−50	−427	−146	−167	−205
Total	500	365	1785	1075	535	245	−350	335	322	416	220	427

From Barbacki and Fisher (1936, p. 190).

TABLE 5.32 ANOVA Table for ABBA Design

	Degrees of Freedom	Sum of Squares	Mean Square
Varieties	1	719,076	719,076
Estimated error	47	4,819,203	102,536
Total	48		

From Barbacki and Fisher (1936, p. 191).

Now the systematic and randomized designs could be compared. The results are as follows. For the ABBA design, the difference in yields between A and B is 339,535 − 333,660 = 5,875. The mean yield is 336,598 so that the actual error of the ABBA design is (5,875/336,598) × 100 = 1.745%. For the randomized design, the sum of squares of the 48 differences is 5,538,279 so that the standard error is $\sqrt{5,538,279} = 2353.35$. The percentage estimated error is therefore (2,353.35/336,598) × 100 = .699%. Finally, using the ANOVA table for the systematic design (see Table 5.32), they calculated the standard error of the difference between the total yields as $\sqrt{48 \times 102,536} = 2,218.50$ so that the percentage estimated error of the ABBA design is (2,218.50/336,598) × 100 = .659%. From these numbers, Barbacki and Fisher made two conclusions:

1. The real error of the systematic design is *higher* than the estimated error of the randomized design (1.745% vs. .699%).

2. The estimated error of the systematic design is *lower* than the estimated error of the randomized design (.659% vs. .699%).

Hence, "the test of significance is vitiated for both reasons."

Barbacki and Fisher next also randomized AB pairs instead of ABBA sandwiches, but the conclusion was the same. Barbacki and Fisher then observed that "there is no more reason for estimating the errors from pairs than from sandwiches," so that "the error estimated from a systematic design is ambiguous, and the experimenter has an arbitrary choice between several widely different estimates."

Now it was Student's turn to fire back, which he did in a *Nature* article published the same year (Student, 1936):

> ...There is a good deal in the [Barbacki & Fisher's] paper with which I am not in agreement and with which I hope to deal elsewhere, but a letter from a friend of mine in Australia, who had heard at second-hand that Fisher's "results showed not only that the halfdrill strip failed to give a valid estimate of error but was less accurate", shows that it would be better not to let such rumours get a start, for they are quite unfounded. (*ibid.*, p. 971)

Student's main criticism of the Barbacki and Fisher paper was that the authors' calculations assumed that the 192 sections in Table 5.29 were independent. In fact, Student maintained, the sections of a strip were highly correlated so that the whole strip should be used as the unit in the calculation. Based on the latter assumption, he calculated the estimated error of the ABBA design as 2.37%. This could be compared to the actual difference of 1.75%, so "the difference between two things which should be the same within the error of random sampling is in fact no more than .75 times the standard error."

Of course, Student's *Nature* paper elicited another *Nature* paper by Fisher published in the same year (Fisher, 1936). Regarding Student's criticism of his (and Barbacki's) estimation of the error of the ABBA design, Fisher retorted:

> It is not noticed, apparently, by "Student" that Dr. Barbacki and I criticize systematic arrangements in general on the ground that "the experimenter has an arbitrary choice between several widely different estimates". The two we show to be misleading correspond with the two arrangements which we also test. (*ibid.*, p. 1101)

Then he observed that Student's error estimate of 2.37%, if it was correct, would be further evidence of the inappropriateness of the ABBA design, since the randomized design has an error of only .669%:

> ...If "Student's" estimate is right, the randomized experiment is worth as much as the average of eleven such systematic experiments. (*ibidem*)

Therefore:

> ...what becomes of the claim that randomization tends to increase the error, or that experimenters can usefully try to diminish it by adopting regular balanced arrangements? (*ibidem*)

The final rejoinder in the debate was a paper by Student who died (on October 16, 1937) before the manuscript was completed. It was therefore published posthumously (Student,

1938) with some additional notes from Neyman and Egon Pearson, with whom (!) Student had apparently discussed the matter a fortnight before his death. In his paper, Student maintained that balanced designs have higher precision (smaller standard errors) than randomized ones. Then he considered the possibility of a bad randomization:

> And this brings me to a question which has often interested me. Suppose there are two treatments to be randomized—I take two for simplicity only—and suppose that by the luck of the draw they come to be arranged in a very unbalanced manner, say AAAABBBB: is it seriously contended that the risk should be accepted of spoiling the experiment owing to the bias which will affect the mean if there is the usual fertility slope? (*ibid.*, p. 366)

Student's third point was that Fisher's previous conclusion that "the randomized experiment is worth as much as the average of 11 such systematic experiments" was not the fault of the ABBA design but rather the large size of the strips that were used for the calculations:

> Now one of the things that was noticed when uniformity trials first began I was that the same piece of land laid out in large plots gave a very much larger error than if subdivided into small plots, and since half-drill strips were in this trial twelve times as large as "sheaf weights".... (*ibid.*, p. 369)

Therefore:

> Prof. Fisher's conclusion naturally follows since he is not comparing like with like. (*ibidem*)

Student next considered Barbacki and Fisher's claim that "the error estimated from a systematic design is ambiguous, and the experimenter has an arbitrary choice between several widely different estimates." To this, Student answered that his preference was to choose the method that was most profitable when designing the experiment. Finally, Student proposed a double ABBA design (i.e., ABBA both horizontally and vertically) for the data Barbacki and Fisher had previously analyzed. His estimate of the standard error was then similar to that of the randomized sandwiches.

Thus ended the most unlikely dispute in the history of statistics, with Student and Fisher each holding on to their respective positions. In a later letter to Jeffreys, Fisher commented on the fight as follows:

> So far as I can judge, "Student" and I would have differed quite inappreciably on randomisation if we had seen enough of each other to know exactly what the other meant, and if he had not felt in duty bound, not only to extol the merits, but also to deny the defects of Beaven's half drill strip system. (Bennett, 1990, p. 271)

Fisher also revealed in his letter that he had been taken aback by Student's vehement criticism of randomization:

> ...until the last two years I had really thought that "Student" accepted all that I had put forward on behalf of randomization. (*ibidem*)

In case the reader is still wondering which of the half-drill strip and randomized design is preferable, consider that the former is hardly ever used nowadays while the latter is usually the gold standard.

5.4.4 *Design of Experiments (Fisher, 1935)*

Fisher's second book, *The Design of Experiments* (DOE) (Fisher, 1935a), was the natural successor to his first book *Statistical Methods for Research Workers* (Fisher, 1925b). The latter dealt mainly with significance testing, but the last two chapters introduced the planning of experiments through various ANOVA designs. The subject of planning was to receive a full treatment in DOE. In all, the book went through nine editions,* the last one occurring in 1971. The contents of the first edition of DOE are as follows:

I. INTRODUCTION
II. THE PRINCIPLES OF EXPERIMENTATION, ILLUSTRATED BY A PSYCHO-PHYSICAL EXPERIMENT
III. A HISTORICAL EXPERIMENT ON GROWTH RATE
IV. AN AGRICULTURAL EXPERIMENT IN RANDOMISED BLOCKS
V. THE LATIN SQUARE
VI. THE FACTORIAL DESIGN IN EXPERIMENTATION
VII. CONFOUNDING
VIII. SPECIAL CASES OF PARTIAL CONFOUNDING
IX. THE INCREASE OF PRECISION BY CONCOMITANT MEASUREMENTS. STATISTICAL CONTROL
X. THE GENERALISATION OF NULL HYPOTHESES. FIDUCIAL PROBABILITY
XI. THE MEASUREMENT OF AMOUNT OF INFORMATION IN GENERAL

Chapter I of DOE contains some introductory remarks on experiments in general, followed by a discussion of inductive inference. We can also find Fisher's familiar rejection of inverse probability. Chapter II starts with the description of what is now famously known as the "Lady Tasting Tea" experiment†:

> A LADY declares that by tasting a cup of tea made with milk she can discriminate whether the milk or the tea infusion was first added to the cup. We will consider the problem of designing an experiment by means of which this assertion can be tested... Our experiment consists in mixing eight cups of tea, four in one way and four in the other, and presenting them to the subject for judgment in a random order. The subject has been told in advance of what the test will consist, namely that she will be asked to taste eight cups, that these shall be four of each kind, and that they shall be presented to her in a random order, that is in an order not determined arbitrarily by human choice...Her task is to divide the 8 cups into two sets of 4, agreeing, if possible, with the treatments received. (*ibid.*, pp. 13–14)

* Second 1937, third 1942, fourth 1947, fifth 1949, sixth 1951, seventh 1960, eighth 1966, ninth 1971.

† An excellent discussion of the "Lady Tasting Tea" is provided by Gridgeman (1959). See also Salsburg (2001, Chapter 1) and Gorroochurn (2012a, Prob. 26).

Fisher then noted that there are $\binom{8}{4} = 70$ ways of choosing a group of 4 objects out of 8 and if the lady had no discrimination skill she would have only one way of correctly dividing the cups, that is, "with a frequency that would approach 1 in 70 more nearly the more of the test is repeated." A statistical test of the lady's discrimination ability will thus have a p-value of $1/70 \approx .0143$.

In the next pages of Chapter II, Fisher used the phrase "null hypothesis" for the first time and explained it as follows:

> Our examination of the possible results of the experiment has therefore led us to a statistical test of significance, by which these results are divided into two classes with opposed interpretations...The two classes of results which are distinguished by our test of significance are, on the one hand, those which show a significant discrepancy from a certain hypothesis; namely, in this case, the hypothesis that the judgments given are in no way influenced by the order in which the ingredients have been added; and on the other hand, results which show no significant discrepancy from this hypothesis. This hypothesis, which may or may not be impugned, by the result of an experiment, is again characteristic of all experimentation. Much confusion would often be avoided if it were explicitly formulated when the experiment is designed. In relation to any experiment we may speak of this hypothesis as the "null hypothesis," and it should be noted that the, null hypothesis is never proved or established, but is possibly disproved, in the course of experimentation. Every experiment may be said to exist only in order to give the facts a chance of disproving the null hypothesis. (*ibid.*, pp. 18–19)

Fisher's insistence in the above that "a null hypothesis is never proved" was to be one of his major arguments against the Neyman–Pearson's theory of hypothesis testing (see Section 5.6.4.3). The last sentence "Every experiment may be said..." is one of the often-quoted famous phrases uttered by Fisher.

Chapter II also contains an important section (Section 11) dealing with the sensitivity of an experiment:

> By increasing the size of the experiment, we can render it more sensitive, meaning by this that it will allow of the detection of a lower degree of sensory discrimination, or, in other words, of a quantitatively smaller departure from the null hypothesis. Since in every case the experiment is capable of disproving, but never of proving this hypothesis, we may say that the value of the experiment is increased whenever it permits the null hypothesis to be more readily disproved. (*ibid.*, p. 25)

The above is nothing but a statement of Neyman–Pearson's concept of power. The only difference is that, in Fisher's framework, there is no explicit alternative hypothesis and therefore power cannot be quantified.

Chapter III contains an example illustrating the use of the paired t-test by using the data reported in Darwin's 1876 book (Darwin, 1876). The last section in that chapter gives Fisher's first permutation test.

Fisher next considered randomized blocks (Chapter IV) and Latin Squares (Chapter V), both of which had previously been introduced in *Statistical Methods for Research Workers* (Fisher, 1925b). Chapter VI is an extension of these designs to factorial experiments.

Chapters VII and VIII give the first complete treatments of full and partial confounding. The last two chapters deal with fiducial probability and information, respectively.

In 1935, Fisher's position was quite different from that in 1925. The power and ubiquity of his methods had now been recognized. It was natural then that DOE got a much better reception than *Statistical Methods* had in 1925.

5.5 FISHER AND PROBABILITY

It is easier to describe what Fisher thought probability ought *not* to be than to delineate what he thought it is. This is because Fisher never completely and unequivocally exposed his own conception of probability, although he often wrote about it. For sure, Fisher was adamantly against inverse probability. To counter this concept, he formally introduced likelihood in the early twenties. At around the same time, he defined probability as a frequency ratio in a *hypothetical infinite population*.* This was close, but no identical, to the frequency definition of probability. Later, against the persistence of inverse probability, he created fiducial probability as an alternative. In the early thirties, Fisher got into a dispute with Jeffreys about the nature of probability and was forced to re-examine his own understanding of this concept. In his later years, against the backdrop of the increasing popularity of the Neyman–Pearson's overt frequentist approach, Fisher introduced the concept of relevant subsets to define probability and edged toward an epistemic ("degree of belief") concept of the latter. We now examine these topics.

5.5.1 Formation of Probability Ideas: Likelihood, Hypothetical Infinite Populations, Rejection of Inverse Probability

We first encounter probability in Fisher's writings in his very first paper, "On an absolute criterion for fitting frequency curves" (Fisher, 1912) (cf. Section 2.3.10). In this paper, Fisher criticized the methods of least squares and moments, and announced his own "criterion" for estimation. The latter was an informal version of the method of maximum likelihood (ML), but no mention of "likelihood" was made at that stage. Later in the paper, Fisher described how his method could be used to estimate the mean (m) and the precision (h) in a normal distribution. He then stated:

> Corresponding to any pair of values, m and h, we can find the value of P, and the inverse probability system may be represented by the surface traced out by a point at a height P above the point on a plane, of which m and h are the coordinates. (*ibid.*, p. 158)

In the above, Fisher used the phrase "inverse probability system," but as it turned out, he did not actually use inverse probability in the statement of his criterion.† Little did

*Jeffreys (1961, p. 370) has pointed out that the notion of a hypothetical infinite population occurred earlier in statistical physics in the writings of Willard Gibbs.

†The following reveals that Fisher had Bayesian training right from school: "In the latter half of the nineteenth century the theory of inverse probability was rejected more decisively by Venn and by Chrystal, but so retentive is the tradition of mathematical teaching that I may myself say that I learned it at school as an integral part of the subject, and for some years now saw no question to question its validity" (Fisher, 1936, p. 248).

Fisher know that his reference to inverse probability would be the reason for his first quarrel with Karl Pearson.

Moving on, in the 1915 paper, "Frequency distribution of the values of the correlation coefficient in samples from an indefinitely large population" (Fisher, 1915), Fisher used ML (again, "likelihood" was not mentioned) to derive an estimate of the population correlation coefficient from a bivariate normal distribution (cf. Eq. 2.79). In that paper, he stated:

> I have given elsewhere* a criterion, independent of scaling, suitable for obtaining the relation between an observed correlation of a sample and the most probable value of the correlation of the whole population. (*ibid.*, p. 520)

Pearson and his associates attacked Fisher's work in a paper (Soper et al., 1917), which was intended as an appendix to Student's and Fisher's own, but without notifying the latter. They accused Fisher of having made "indiscriminate use of Bayes' Theorem" in his derivation of the "most likely value of ρ," and in particular for having used a uniform prior for ρ.

Having been subjected to Pearson's criticism, which Fisher must have resented, it is not surprising that the latter forcefully rejected inverse probability from then on.† His first such statement occurred in a 1921 paper published in the journal *Metron* (Fisher, 1921a), where he also explicitly delineated his concept of likelihood:

> My treatment of this problem differs radically from that of BAYES. BAYES (1763) attempted to find, by observing a sample, the actual *probability* that the population value lay in any given range. In the present instance the complete solution of this problem would be to find the probability integral of the distribution of r. Such a problem is indeterminate without knowing the statistical mechanism under which different values of r come into existence; it cannot be solved from the data supplied by a sample, or any number of samples, of the population. What we can find from a sample is the likelihood of any particular value of r, if we define the likelihood as a quantity proportional to the probability that, from a population having that particular value of r, a sample having the observed value r, should be obtained.
>
> So defined, probability and likelihood are quantities of an entirely different nature. (*ibid.*, p. 24)

Fisher thus pointed out that although probability and likelihood were proportional to each other, they were nevertheless fundamentally different. Bayes' theorem attempted to determine the probabilities of parameters (or hypotheses). Fisher believed that, in the absence of legitimate prior probabilities, this task was impossible. According to him, probabilities therefore should be used for events or observations not for hypotheses. On the other hand, likelihood circumvented the need for prior probabilities and was thus perfectly suited to deal with hypotheses.

Fisher's opposition to Bayes' theorem stemmed from his rejection of the principle of indifference (cf. Section 1.3), which was frequently used in the application of the theorem:

> ...BAYES adds a scholium the purport of which would seem to be that in the absence of all knowledge save that supplied by the sample, it is reasonable to assume this particular a priori

* R.A. Fisher, "On an absolute criterion for fitting frequency curves," *Messenger of Mathematics*, February, 1912. (*Fisher's footnote*).

† Fisher's rejection of inverse probability is also discussed in Zabell (1989a).

distribution of p. The result, the datum, and the postulate implied by the scholium, have all been somewhat loosely spoken of as BAYES' Theorem.

The postulate would, if true, be of great importance in bringing an immense variety of questions within the domain of probability. It is, however, evidently extremely arbitrary. Apart from evolving a vitally important piece of knowledge, that of the exact form of the distribution of values of p, out of an assumption of complete ignorance, it is not even a unique solution. (Fisher, 1922c, pp. 324–325)

Fisher's point in the above was that complete ignorance about a parameter did not legitimize the assignment of a uniform distribution to it. Moreover, even the latter task could not be done in a unique manner.

The last point was illustrated by Fisher by considering an example that is described in Section 1.3.4.

Having defined the appropriate quantity (i.e., likelihood) to use when dealing with hypotheses, Fisher's next task was to define the corresponding measure, namely, probability, when dealing with events or observations. His definition depended on the notion of a *hypothetical infinite population*, which he introduced as follows:

...briefly, and in its most concrete form, the object of statistical methods is the reduction of data...This object is accomplished by constructing a hypothetical infinite population, of which the actual data are regarded as constituting a random sample. The law of distribution of this hypothetical population is specified by relatively few parameters, which are sufficient to describe it exhaustively in respect to all quantities under discussion. Any information given by the sample, which is of use in estimating the values of these parameters, is relevant information. (*ibid.*, p. 311)

Fisher's hypothetical infinite population can thus be regarded as an "infinite urn" model. He later clarified that:

...the word infinite is to be taken in its proper mathematical sense as denoting the limiting conditions approached by increasing a finite number indefinitely. (Fisher, 1925c, p. 700)

In the 1922 foundations paper, Fisher next explained his concept of probability:

When we speak of the *probability* of certain object fulfilling a certain condition, we imagine all such objects to be divided into two classes, according as they do or do not fulfil the condition. This is the only characteristic in them of which we take cognizance. For this reason probability is the most elementary of statistical concepts. (Fisher, 1922c, p. 312)

Fisher was now ready to enunciate his definition:

[Probability] is a parameter which specifies a simple dichotomy in an infinite hypothetical population, and it represents neither more or less than the frequency ratio which we imagine such a population to exhibit. (*ibidem*)

According to Fisher, probability was thus a frequency ratio derived from a hypothetical infinite population. As an example:

> ...when we say that the probability of throwing a five with a die is one-sixth, we must not be taken to mean that of any six throws with that die one and one only will necessarily be a five; or that of any six million throws, exactly one million will be fives; but that of a hypothetical population of an infinite number of throws, with the die in its original condition, exactly one-sixth will be fives. Our statement will not then contain any false assumption about the actual die, as that it will not wear out with continued use, or any notion of approximation, as in estimating the probability from a finite sample, although this notion may be logically developed once the meaning of probability is apprehended. (*ibidem*)

In the above, Fisher argued that by making use of an *hypothetical* population the need for making false assumptions, such as that the die should remain in its original condition, is obviated. Moreover, since the population is assumed to be *infinite*, there is no need to approximate probability.

It can be seen that Fisher's use of the concept of a *hypothetical infinite population* to define probability was meant to draw a contrast with the usual frequentist approach of defining probability from an *experimental procedure* repeated ad infinitum. Thus, although Fisher was a frequentist, his frequentism was slightly different from those of either Venn or von Mises. He did not espouse the limiting frequency definition of probability, as he made clear himself in a later paper:

> ...My own definition is not based on the limit of frequencies, if by this [is meant] experimental frequencies, for I believe we have no knowledge of the existence of such limits. (Fisher, 1935d, p. 81)

Indeed, as Fisher pointed out in the above, one of the problems with the limiting frequency concept of probability is that, although the existence of a limit is common, it is not assured. In the paper "Theory of statistical estimation" (Fisher, 1925c), Fisher gave other advantages of his hypothetical infinite population concept:

> The idea of an infinite hypothetical population is, I believe, implicit in all statements involving mathematical probability. If, in a Mendelian experiment, we say that the probability is one half that a mouse born of a certain mating shall be white, we must conceive of our mouse as one of an infinite population of mice which might have been produced by that mating. The population must be infinite for in sampling from a finite population the fact of one mouse being white would affect the probability of others being white, and this is not the hypothesis which we wish to consider; moreover, the probability may not always be a rational number. Being infinite the population is clearly hypothetical, for not only must the actual number produced by any parents be finite, but we might wish to consider the possibility that the probability should depend on the age of the parents, or their nutritional conditions. We can, however, imagine an unlimited number of mice produced upon the conditions of our experiment, that is, by similar parents, of the same age, in the same environment. The proportion of white mice in this imaginary population appears to be the actual meaning to be assigned to our statement of probability. Briefly, the hypothetical population is the conceptual resultant of the conditions which we are studying.... (*ibid.*, p. 700)

Here, Fisher pointed out that sampling from a hypothetical infinite population guaranteed that the successive draws were independent of each other. This is not the case with the frequency concept which uses sampling from a finite population. Moreover, in an infinite hypothetical population, every proportion is possible, as opposed to sampling from a sample of finite size n where the only frequencies possible are multiples of $1/n$.

Notwithstanding some of the advantages of the hypothetical infinite model, it has none of the popularity of the limiting frequency concept. In his book *Scientific Explanation*, Braithwaite has argued that defining probability using a hypothetical infinite population contains an intrinsic contradiction:

> Since the notion of the proportion of black balls among the infinity of balls in the bag cannot be taken literally, to make the probability p of the theory correspond with this proportion in the model is not very illuminating. What happens in the minds of those using this bag model is, I believe, that when they start by making the probability correspond to the proportion of balls in the bag, they are thinking of the bag as finite so that the proportion has a literal meaning. The size of the bag then has to become infinite in order than that the proportion should not be altered by the removal of one ball, and in order that as many balls as are desired can be drawn from the bag without the bag becoming exhausted. This confusion of thought as to the size of the bag is imposed by the attempt to use drawings *without replacement* of balls from a bag as a model for the theory of probability; to serve this purpose the bag must be both finite and infinite. The model suggested by the hypothetical-infinite population view is therefore inappropriate, through being self-contradictory. (Braithwaite, 1955, p. 129)

5.5.2 Fiducial Probability and the Behrens-Fisher Problem

5.5.2.1 The Fiducial Argument (1930) Of the many statistical innovations Fisher made, fiducial probability is one of the very few that has found little favor among statisticians. Yet Fisher regarded it as a major achievement since, he believed, fiducial probability would supplant inverse probability as a tool for inference directly on parameters (and hypotheses). However, none of this happened, and fiducial probability has remained a very controversial subject. We now examine these topics.

Fisher himself credited the original fiducial idea to E.J. Maskell* (Fisher, 1935a, p. 196). Box has further pointed out that the concept was first made public by J.O. Irwin in 1929 while giving a talk at the British Association meeting (Box, 1978, p. 254). Irwin was then an assistant at Rothamsted to Fisher and the latter had suggested that the idea of fiducial

*In 1929, while testing for the difference between two means using the t-test, Maskell had written: "In general, however, the exact probability for each individual difference is immaterial: we wish usually to know what differences we may accept and what should be rejected as not established. For this purpose it is convenient to calculate, from the standard error and the table of t, values which may be called '*significant differences*', which would be exceeded by chance only once in twenty trials ($P=0.05$) or, if we wish for greater certainty, only once in one hundred trials ($P=0.01$)" (Maskell, 1929, p. 7). See also Bennett (1990, p. 212).

probability might be included in Irwin's presentation. One year later, Fisher formally introduced the fiducial argument* in the paper "Inverse probability" (Fisher, 1930).

Fisher started his paper with yet another condemnation of inverse probability. He pointed out that Bayes himself had been circumspect in the application of this method but that Laplace on the other hand was much to be blamed for its uncritical implementation:

> I know only one case in mathematics of a doctrine which has been accepted and developed by the most eminent men of their time, and is now perhaps accepted by men now living, which at the same time has appeared to a succession of sound writers to be fundamentally false and devoid of foundation. Yet that is quite exactly the position in respect of inverse probability. Bayes, who seems to have first attempted to apply the notion of probability, not only to effects in relation to their causes but also to causes in relation to their effects, invented a theory, and evidently doubted its soundness, for he did not publish it during his life. It was posthumously published by Price, who seems to have felt no doubt of its soundness. It and its applications must have made great headway during the next 20 years, for Laplace takes for granted in a highly generalised form what Bayes tentatively wished to postulate in a special case. (Fisher, 1930, p. 528)

Concluding his argument on inverse probability, Fisher said:

> ...Inverse probability has, I believe, survived so long in spite of its unsatisfactory basis, because its critics have until recent times put forward nothing to replace it as a rational theory of learning by experience. (*ibid.*, p. 531)

Fisher's aim was therefore to provide the "rational theory" that had been lacking. Before doing that, Fisher reviewed his method of maximum likelihood (ML), which he was careful to point out was *not* a probability (and it certainly had "no logical connection with inverse probability at all"), but rather a method of statistical estimation. Finally came the fiducial part. First, Fisher declared triumphantly:

> There are, however, certain cases in which statements in terms of probability can be made with respect to the parameters of the population.... (*ibid.*, p. 532)

Next, he considered the maximum likelihood estimator (MLE) T of a parameter θ such that

$$P = \Pr\{T \le t \mid \theta\} = F(t, \theta),$$

where F is a function.† Fisher then made the perfectly valid statement:

> If now we give to P any particular value such as .95, we have a relationship between the statistic T and the parameter θ, such that T is the 95 per cent. value corresponding to a given

*Fisher's fiducial argument is also described in Pedersen (1978), Zabell (1992), Aldrich (2000), Geisser (2006, pp. 149–161), Kalbfleisch (1985, pp. 314–321), Lehmann (2011, pp. 78–80), Buehler (1980), Dempster (1964), Edwards (1976), Neyman (1941), Barnard (1995), Rao (1973, pp. 339–340), Van Aarde (2009, Chapter 14), and Seidenfeld (1979, Chapter 4; 1992).

†Fisher later insisted that T should instead be sufficient.

> θ, and this relationship implies the perfectly objective fact that in 5 per cent. of samples T will exceed the 95 per cent. value corresponding to the actual value of θ in the population from which it is drawn.... (*ibid.*, p. 533)

That is, suppose the Pth percentile of T for a given θ is t_P. Then

$$\Pr\{T \le t_P \mid \theta\} = F(t_P, \theta) = P. \tag{5.81}$$

Thus, for a given P, there exists a functional relationship between t and θ, and we can write $t_P = t_P(\theta)$, where $t_p(\theta)$ is usually an increasing function of θ. Now came the fiducial argument:

> ...To any value of T there will moreover be usually a particular value of θ to which it bears this relationship; we may call this the "fiducial 5 per cent. value of θ" corresponding to a given T. If, as usually if not always happens, T increases with θ for all possible values, we may express the relationship by saying that the true value of θ will be less than the fiducial 5 per cent. value corresponding to the observed value of T in exactly 5 trials in 100. By constructing a table of corresponding values, we may know as soon as T is calculated what is the fiducial 5 per cent. value of θ, and that the true value of θ will be less than this value in just 5 per cent. of trials. This then is a definite probability statement about the unknown parameter θ, which is true irrespective of any assumption as to its a priori distribution. (*ibidem*)

In the above, Fisher stated that for a given $T=t$, there should similarly exist a function $\theta_p(t)$ such that $F\{t, \theta_P(t)\} = P$. The fiducial argument then claims that, since $t_p(\theta)$ is usually an increasing function of θ, we must also have

$$\Pr\{\theta \ge \theta_P(t)\} = P. \tag{5.82}$$

If we try to reconstruct Fisher's reasoning, we first see that Eq. (5.81) leads to the perfectly valid relation:

$$P = \Pr\{T \le t_P(\theta)\} = \Pr\{\theta \ge t_P^{-1}(T)\} = \Pr\{\theta \ge \theta_P(T)\},$$

where $t_P^{-1}() \equiv \theta_P()$. Now, as is apparent from Eq. (5.82), Fisher's fiducial argument consists of insisting that the above is true even for a given value t of T. The fiducial argument thus makes the following probabilistic equivalence*:

$$\Pr\{T \le t_P(\theta)\} \equiv \Pr\{\theta \ge \theta_P(t)\}. \tag{5.83}$$

*Fisher's reasoning can be exemplified as follows: suppose $X_1, X_2, \ldots, X_n$ is a random sample from a normal population with unknown mean μ and known variance σ^2. Then the sample mean $\bar{X} \sim N(\mu, \sigma^2/n)$ and $Z = \sqrt{n}(\bar{X} - \mu)/\sigma$ where $Z \sim N(0,1)$. Therefore, using the fiducial argument, we have $\Pr\{Z \le z_P\} = \Pr\{\mu \ge \bar{x} - z_P\sigma/\sqrt{n}\} = \Pr\{\mu \le \bar{x} + z_P\sigma/\sqrt{n}\}$. Hence, the fiducial density of μ evaluated at $\bar{x} + z_P\sigma/\sqrt{n}$ is the $N(0,1)$ density, that is, the fiducial density of μ evaluated at u is $\frac{\sqrt{n}}{\sigma\sqrt{2\pi}}\exp\left\{-\frac{1}{2}\left(\frac{u-\bar{x}}{\sigma/\sqrt{n}}\right)^2\right\}$, which is the $N(\bar{x}, \sigma^2/n)$ distribution.

TABLE 5.33 Part of Table Used by Fisher to Illustrate Fiducial Intervals for ρ

Fiducial 5% ρ	95% r
.761594	.980916
.800499	.991770
.833655	.993335
.861723	.994593
.885352	.995608
.905148	.996427

From Fisher (1930, p. 533).

Fisher was therefore able to achieve his aim of assigning probabilities to parameters. But he warned that (5.82) is not a probability in the usual sense of the word, since it does not refer to a hypothetical event (i.e., there are no random variables involved). It is a *fiducial** probability regarding θ and has the following frequency interpretation: suppose, for a given P, the percentiles corresponding to $\theta_1, \theta_2, \ldots, \theta_n$ are denoted by $t_1, t_2, \ldots, t_n$, then, for a given θ^* and in the limit as $n \to \infty$, the inequality $\theta^* \le \theta_i$ will be true for exactly the same proportion P of times as the inequality $t_i \le t_P\,(\theta^*)$ will be true.

Continuing with the 1930 paper, Fisher gave a table (see Table 5.33) of 95% values of the sample correlation r for given values of the population correlation ρ and of fiducial 5% values for ρ for a given r. These values were obtained from Fisher's formula for the exact distribution of r obtained in 1915 (Fisher, 1915)† and were compiled by Miss F.E. Allen. For a given $\rho = \rho_0$, the 95% value r_0 of r is obtained by solving

$$\Pr\{r \le r_0 \mid \rho = \rho_0\} = .95.$$

Using the fiducial argument, ρ_0 is also the fiducial 5% value in the sense that

$$\Pr\{\rho \le \rho_0 \mid r = r_0\} = .05.$$

For example:

> ...if a value $r = .99$ were obtained from the sample, we should have a fiducial 5 per cent. ρ equal to about .765. The value of ρ can then only be less than .765 in the event that r has exceeded its 95 per cent. point, an event which is known to occur just once in 20 trials. In this sense ρ has a probability of just 1 in 20 of being less than .765. (Fisher, 1930, p. 534)

After the previous example, Fisher gave the following expression for the fiducial density of θ:

$$f(\theta) = -\frac{\partial}{\partial\theta} F(t, \theta). \tag{5.84}$$

* "Fiducial" means "based on trust."

† See also Eq. (2.76).

This equation can be obtained from Eq. (5.82) by writing

$$\Pr\{\theta \geq \theta_P\} = P = F(t, \theta_P).$$

Therefore,

$$f(\theta_P) = \frac{\partial}{\partial \theta_P} \Pr\{\theta < \theta_P\} = -\frac{\partial}{\partial \theta_P} F(t, \theta_P),$$

which gives Eq. (5.84).

We now make some further comments on the equivalence in (5.83). Since, on the right side of the latter relation, we are calculating the probability that θ exceeds or equals θ_p, it would seem that the parameter θ is now behaving as a random variable. Many have thus argued that Fisher's fiducial argument is simply a disguised form of Bayesian inference.* Fisher would have vehemently protested against such a charge, but the fact remains that he wished to make probability statements about θ without giving the latter the status of a random variable. This apparent contradictory state of affairs prompted Jimmie Savage (1917–1971) to famously say that fiducial probability was really:

> ...a bold attempt to make the Bayesian omelet without breaking the Bayesian eggs. (Savage, 1961, p. 578)

Savage was not the only one to cast doubts on the validity of the assertion in (5.83). In 1940, for example, Fisher had an interesting exchange with the French mathematician Maurice Fréchet (1878–1973).† We have chosen to present this exchange because of the clarity with which it presents the issue at heart.

Referring to Fisher's 1930 paper (Fisher, 1930), Fréchet explained in a letter dated January 21, 1940:

> ...The difficulty arises in a very short sentence which looks as obvious as the other ones but of which the meaning has a capital importance...
>
> Page 533: "we may express the relationship by saying that the true value of θ will be less than the fiducial 5 per cent value corresponding to the observed value of T *in exactly 5 trials in* 100". Now the five per cent and the 5 trials in 100 refer to two probabilities of the same event, it is true, but the populations where these probabilities are computed are extremely different. In the first one θ is fixed whereas T is a random variable and the table has been computed on this assumption. In the second one T is fixed, θ is a random variable and the first table which is still used (or the corresponding formula) has not been computed under this second assumption. (Bennett, 1990, p. 120)

In the above, Fréchet's point was that, on the left side of (5.83), T is a random variable and θ is a constant. On the other hand, the converse is true on the right side of the same

*A discussion of fiducial probability from a Bayesian perspective can be found in Barnard (1987) and Seidenfeld (1992).

†For more on Maurice Fréchet, see p. 607.

relation. Therefore the probabilities on the two sides of (5.83) refer to different random variables and hence different populations, so it does not make sense to use the same table of critical values for both.

In his reply, Fisher disagreed with Fréchet that, on the right side of (5.83), T is a constant and θ is a random variable:

> You say, in your third paragraph: "In the second case T is fixed, θ is a random variable, etc." If this were so, then the probability statement under consideration would be precisely a statement of inverse probability and, as such, as we both agree, probably not true, and certainly not known to be true.
>
> ...in the fiducial statement we are not considering T to be fixed. If, on the contrary, we remember that T will vary from sample to sample, and with it the corresponding fiducial 5% value of θ, then it will be true that the true value of θ will be less than the fiducial 5% value in exactly five trials in 100. (*ibid.*, p. 121)

However, Fréchet correctly observed that indeed on the right side of (5.83), one needs to work with a given value of T.* For the probability statement to make sense, θ would then have to be treated as a random variable. In the words of Fréchet:

> Now, it may be understood that, as you say, T will vary from sample to sample and with it the corresponding fiducial 5% value of θ. Still I expect that everybody will understand that the probability 5 trials in 100 is computed in the population where T gets a fixed value...
>
> But, even if we leave that, it remains that identification is admitted between probabilities of the same event in 2 populations which are different and essentially different *because in one, θ is fixed* (and this is implicitly supposed in the computation of the corresponding first probability) whereas *in the second, θ is supposed to be a random variable.*† (*ibid.*, p. 122)

Seeing that it would be impossible to convince Fréchet, Fisher made the following desperate statement:

> I doubt if we shall be able to get to a clearer understanding of the problem of fiducial probability, unless you are willing to accept it as a fact which I demonstrated to you by quotation in my last letter, that for the population of cases relative to which a fiducial probability is defined, the value of any relevant statistic T is not regarded as fixed. (*ibid.*, p. 124)

In the above, Fisher effectively said if Fréchet could not understand it he would have to just accept it as a fact! This is far from a convincing argument.

About 10 more letters were exchanged between Fréchet and Fisher on the matter without an eventual resolution of the issue.

5.5.2.2 Neyman's Confidence Intervals (1934) Thus, as innovative as Fisher's 1930 paper was, the fiducial argument in it was only that, an argument. Fisher attempted no systematic development of a theory of fiducial probability, and he also offered no reason why the MLE was chosen for the statistic T. Consequently, considerable doubt and confusion continued to exist on the true nature of fiducial probability. However, more details soon trickled and statisticians were offered further insights into the fiducial argument.

* The example in the footnote of p. 464 also illustrates this fact.

† All italics are Fréchet's.

Perhaps few today realize that the theory of confidence intervals, which remains one of the cornerstones of modern statistical theory, was first put forward by Jerzy Neyman in 1934 as possibly an extension of Fisher's fiducial argument, which is now all but defunct.* Neyman's paper was entitled "On the two different aspects of the representative method" (Neyman, 1934) and was the English version of a pamphlet already published in Polish. We next consider some portions of this paper in relation to the fiducial argument and examine how they forced Fisher to further clarify his argument.

The start of Neyman's paper expressed the state of affairs regarding the current understanding of Fisher's fiducial argument:

> ...an approach to problems of this type has been suggested by Professor R.A. Fisher which removes the difficulties involved in the lack of knowledge of the a priori probability law.† Unfortunately the papers referred to have been misunderstood and the validity of statements they contain formally questioned. This I think is due largely to the very condensed form of explaining ideas used by R.A. Fisher, and perhaps also to a somewhat difficult method of attacking the problem. (*ibid.*, p. 562)

Later in the paper, Neyman delineated the problem of interval estimation:

> The new form of the problem of estimation of the collective character [parameter] θ may be stated as follows: given any positive number $\varepsilon < t$, to associate with any possible value of x an interval
>
> $$\theta_1(x) < \theta_2(x) \tag{1}$$
>
> such that if we accept the rule of stating that the unknown value of the collective character θ is contained within the limits
>
> $$\theta_1(x') \leq \theta \leq \theta_2(x') \tag{2}$$
>
> every time the actual sampling provides us with the value $x = x'$, the probability of our being wrong is less than or at most equal to $1 - \varepsilon$, and this whatever the probability law a priori, $\varphi(\theta)$.
>
> The value of ε, chosen in a quite arbitrary manner, I propose to call the "confidence coefficient." If we choose, for instance, $\varepsilon = .99$ and find for every possible x the intervals $[\theta_1(x), \theta_2(x)]$ having the properties defined, we could roughly describe the position by saying that we have 99 per cent. confidence in the fact that θ is contained between $\theta_1(x)$ and $\theta_2(x)$. (*ibid.*, pp. 589–590)

Although Neyman did not make use of the fiducial argument in his paper, the following shows he identified his method of confidence intervals with Fisher's fiducial limits:

> The numbers $\theta_1(x)$ and $\theta_2(x)$ are what R.A. Fisher calls the fiducial limits of θ. Since the word "fiducial" has been associated with the concept of "fiducial probability" which has caused the misunderstandings I have already referred to, and which in reality cannot be distinguished from the ordinary concept of probability, I prefer to avoid the term and call the intervals $[\theta_1(x), \theta_2(x)]$ the confidence intervals, corresponding to the confidence coefficient ε. (*ibid.*, p. 590)

* But revived to some extent in modified form as "structural inference" by Fraser (1968). See also Barnett (1999, pp. 313–319).

† R.A. Fisher: Proc. Camb. Phil. Soc., Vol. XXVI, Part 4, Vol. XXVIII, Part 3, and Proc. Roy. Soc., A. Vol. CXXXIX *(Neyman's footnote)*.

Fisher too thought that this was the case* for in the discussion to Neyman's paper his comments were written as:

> The particular aspect of this work, of which Dr. Neyman's paper was a notable illustration, was the deduction of what Dr. Fisher had called fiducial probability. (*ibid.*, p. 617)

However, Fisher was quick to point out certain caveats in Neyman's work, and these further elucidated the fiducial argument. First, contrary to his 1930 assumption that the statistic T should be an MLE, Fisher now stated that T ought to be sufficient:

> ...he [Fisher] would apply the fiducial argument, or rather would claim unique validity for its results, only in those cases for which the problem of estimation proper had been completely solved, i.e. either when there existed a statistic of the kind called sufficient, which in itself contained the whole of the information supplied by the data, or when, though there was no sufficient statistic, yet the whole of the information could be utilized in the form of ancillary information. Both these cases were fortunately of common occurrence, but the limitation seemed to be a necessary one, if they were to avoid drawing from the same body of data statements of fiducial probability which were in apparent contradiction. (*ibid.*, pp. 617–618)

The importance of sufficiency in fiducial inference stems from the fact that a sufficient statistic T contains all the information about the parameter θ. Since there exists several sufficient statistics for θ, it is more appropriate to use a *minimal sufficient statistic*. Thus, one of the sufficient conditions for fiducial inference in the one-parameter case is the existence of a minimal sufficient statistic.

In his discussion of Neyman's 1934 paper, the second point raised by Fisher concerned the appropriateness of using the fiducial argument when the underlying distribution was discrete:

> Dr. Fisher had limited his application to continuous distributions, hoping, with more confidence in this case, that the limitation might later be removed. Dr. Neyman removed this limitation, but at the expense of replacing inferences that stated the exact value of the fiducial probability by inequalities, which asserted that it was not less than some assigned value. This also was somewhat a wide departure, for it raised the question whether exact statements of probability were really impossible, and if they were, whether the inequality arrived at was really the closest inequality to be derived by a valid argument from the data. (*ibid.*, p. 618)

In the above, Fisher correctly pointed out that it was not possible to achieve an exact fiducial probability when the underlying distribution was discrete. This is generally the case although it is possible to circumvent the problem in some cases by transforming a discrete random variable to a continuous one.

Fisher's third point concerned Neyman's extension of the inference to several parameters:

> Thirdly, Dr. Neyman proposed to extend the fiducial argument from cases where there was only a single unknown parameter, to cases in which there were several. Here, again, there might be serious difficulties in respect to the mutual consistency of the different inferences

* It was only later realized that, although confidence and fiducial limits frequently coincided, they were in fact conceptually and mathematically different.

> to be drawn; for, with a single parameter, it could be shown that all the inferences might be summarized in a single probability distribution for that parameter, and that, for this reason, all were mutually consistent; but it had not yet been shown that when the parameters were more than one any such equivalent frequency distribution could be established. (*ibidem*)

In the above, Fisher pointed to the risk of mutually contradictory inferences when the fiducial argument was applied to multiple parameters in a given probability system. Indeed, the extension of the fiducial argument to multiple parameters has proved very challenging.

As a parting note on Neyman's paper, we wish to draw the reader's attention to the cordiality of the exchange between Fisher and Neyman in 1934. For example, at one instant Fisher said:

> Dr. Neyman claimed to have generalized the argument of fiducial probability, and he had every reason to be proud of the line of argument he had developed for its perfect clarity. (*ibidem*)

Neyman was equally kind when, for instance, he said:

> ...If I call a distribution "Student's" distribution, it means clearly that I attribute its discovery to "Student," and not to anybody else. This does not prevent me from recognizing and appreciating the work of Professor Fisher concerning the same distribution. (*ibid.*, p. 625)

However, in 1935, the relationship between the two men took a dramatic turn resulting in a major ongoing feud, as we shall soon describe (see Section 5.6).

5.5.2.3 The Behrens-Fisher Problem (1935) Fisher's 1930 paper (Fisher, 1930) had thus created some confusion about the exact nature of fiducial probability. Neyman's 1934 paper (Neyman, 1934) had suggested that fiducial limits perhaps coincided with confidence limits and therefore enjoyed the desirable frequentist properties of the latter. Fisher's 1935 paper, "The fiducial argument in statistical inference" (Fisher, 1935c), was meant to extend the fiducial argument to two population means but really ended up exposing holes in it.

The 1935 paper started with the derivation of the fiducial limits for a normal population with mean μ when the population variance is unknown. If the sample mean and sample variance are denoted by $\bar{X}$ and S^2, respectively, and n is the sample size, then

$$\Pr\left\{\frac{\bar{X}-\mu}{S/\sqrt{n}} < t_{.95} \middle| \mu\right\} = .95,$$

where $t_{.95}$ is the 95th percentile of the Student's t-distribution with $n-1$ degrees of freedom. From the above, we obtain

$$\Pr\left\{\mu < \bar{X} - t_{.95}\frac{S}{\sqrt{n}}\right\} = .05.$$

Applying the fiducial argument, for observed $\bar{x}$ and s^2,

$$\Pr\left\{\mu < \bar{x} - t_{.95}\frac{s}{\sqrt{n}}\right\} = .05.$$

Therefore, the fiducial 5% limit for μ is given by $\mu < \bar{x} - t_{.95}s / \sqrt{n}$* (*ibid*., p. 392).

It is noteworthy that the random variable $T = \sqrt{n}(\bar{X} - \mu)/S$ used in deriving the fiducial distribution of μ has a distribution that does not depend on any population parameters. T is therefore said to be *pivotal* (or a *pivot*), a term first used by Fisher in 1941 (Fisher, 1941, p. 147). It might seem that the existence of a pivot would be sufficient to derive a valid fiducial distribution, but Fisher was quick to point out this was not the case:

> It is necessary to emphasize also that statements similar to those of fiducial probability can only represent the true state of knowledge derivable from the sample, if the statistics used contain the whole of the relevant information which the sample provides. If, for example, an estimate s', derived from the mean error, had been used in place of one derived from the mean square error, and a quantity t' had been defined by the equation
>
> $$t' = \frac{(\bar{x} - \mu)\sqrt{n}}{s'},$$
>
> the distribution of t', like that of t [T], would depend only on the size of the sample [i.e. would be pivotal]; and probability statements accurate for t' could be expressed in terms of μ. The probability distribution for μ obtained in this way would, of course, differ from that obtained from t [T], and the probability statements derived from the two distributions would be discrepant. There is, however, in the light of the theory of estimation, no difficulty in choosing between such inconsistent results, for it has been proved that, whereas s' uses only a portion of the information utilised by s, on the contrary, s utilises the whole of the information used by s', or indeed by any alternative estimate. To use s', therefore, in place of s would be logically equivalent to rejecting arbitrarily a portion of the observational data, and basing probability statements upon the remainder as though it had been the whole. (Fisher, 1935c, pp. 392–393)

Fisher's point in the above was that a pivotal statistic that was not sufficient would not lead to a valid fiducial inference. However, subsequent investigations revealed that even if a pivotal statistic was sufficient, unique fiducial statements might still not be possible (Stone, 2006, p. 2304).

After having derived the fiducial limits for the mean in the one-sample problem, Fisher next considered the fiducial limits for the difference of the means of two independent normal populations with unknown and unequal variances. From the first population, suppose a sample of n observations $x_1, x_2, \ldots, x_n$ is obtained with mean $\bar{x}$ and with estimated variance of the mean s^2, where

$$s^2 = \frac{\sum_i (x_i - \bar{x})^2}{n(n-1)}.$$

*Fisher later remarked that Student's 1908 paper, where the t-distribution was introduced, contained an early example of fiducial reasoning: "'Student' was immediately concerned only with developing an exact test of significance. That he was, in the 1908 paper, thinking of probability statements respecting μ cannot, however, be doubted. In the case $n=2$, for example, he points out that the quartiles of the distribution of μ lie at the two observed points. This conclusion is notable as illustrating a type of inference entirely independent of the form of the distribution sampled; for if μ stands for the median of such a distribution, it follows, since the three probabilities that two independent observations (a) should both exceed the median, (b) should lie on each side of it, and (c) should both fall short of it, must occur in the frequency ratio 1 : 2 : 1, that we may infer that the two observations are the quartiles of the fiducial distribution of the median" (Fisher, 1939a, p. 4).

Since the fiducial distribution of the population mean μ is of the Student type, Fisher wrote

$$\mu = \bar{x} + st,$$

where t has Student's distribution with $n-1$ degrees of freedom. Similarly, for the second population,

$$\mu' = \bar{x}' + s't',$$

where t' has Student's distribution with $n'-1$ degrees of freedom. Writing $\mu' - \mu = \delta$ and $\bar{x}' - \bar{x} = d$, Fisher thus obtained

$$\varepsilon = \delta - d = s't' - st.$$

Now,

> …since s' and s are known, the quantity represented on the right [above] has a known distribution, though not one which has been fully tabulated. (*ibid.*, p. 396)

Furthermore, once the tabulated values were obtained,

> …we may use $d/\sqrt{s^2 + s'^2}$ to test the hypothesis that δ has the chosen value zero. This is, in fact, the exact test of significance for the difference, d, between the observed means, equivalent to that given in 1929 by W.-V. Behrens. (*ibid.*, p. 397)

Indeed, the test statistic

$$W = \frac{\delta - d}{\sqrt{s^2 + s'^2}} \tag{5.85}$$

had previously been proposed by the German mathematician W.V. Behrens in 1929 (Behrens, 1929). Behrens had worked on the problem (assuming the same sample size in each group), albeit from a non-fiducial perspective,* and his solution had been approved by Fisher right from the start (Bennett, 1990, p. 54). The problem of comparing the means of two independent normal populations with different unknown variances for general sample sizes thus came to be known as the Behrens–Fisher problem.†

Note that, from Eq. (5.85), we can write the Behrens–Fisher test statistic as

$$W = \frac{s't' - st}{\sqrt{s^2 + s'^2}} = t'\cos\theta - t\sin\theta,$$

where $\tan\theta = s/s'$.

*Breny has been critical of Behrens' own approach and has called it "pseudo-conditional" (Breny, 1955, pp. 111–112).

†In his papers and correspondences, Fisher himself most of the time referred to the problem as the "Behrens problem." Indeed, Yates states: "…Fisher in a letter to C.I. Bliss shortly before his death expressed the view that it should be referred to as the 'Behrens' test." (Yates, 1964, p. 348). The Behrens–Fisher problem is also discussed in Wallace (1980, pp. 119–138), Weerahandi (2003, pp. 174–181), Lehmann (2011, pp. 49–51), Scheffé (1970), Barnard (1984; 1995), and Welsh (1996, pp. 150–152).

At Fisher's request, 5% and 1% values of the distribution of W were tabulated by Sukhatme (1938) for values of the angle θ differing by 15° and the 25 combinations of degrees of freedom for t and t' chosen from $6, 8, 12, 24, \infty$.

5.5.2.4 Controversy with Bartlett (1936–1939) With his 1935 paper, Fisher thus seemed to have solved the problem of two means with different unknown variances. However, only one year later, a young Maurice Bartlett wrote a paper (Bartlett, 1936), which would stand as the first major blow to Fisher's fiducial argument. Let us say a few words about Bartlett before considering his paper.

Maurice Stevenson Bartlett* (Fig. 5.6) was one of the greatest statisticians of the twentieth century. He was born on March 18, 1910 in Chiswick, London. His father was a clerk of modest means. Bartlett did his secondary education at Latymer Upper School after winning a scholarship. It was at school, as he later recalled (Bartlett, 1982, p. 42), that his interest in probability began when he read a chapter in Hall and Knight's book *Algebra*. In 1929, Bartlett obtained a scholarship at Queen's College, Cambridge, to study mathematics. In 1932, he achieved the rank of wrangler (with distinction) and published his first paper with his teacher John Wishart (1898–1956) on the distribution of second-order moments in a normal system (Wishart and Bartlett, 1932). After graduating, Bartlett did an extra year as the first postgraduate student of Wishart. They co-wrote a second paper, which was awarded the Raleigh Prize for 1933 (Wishart and Bartlett, 1933). During his postgraduate, Bartlett also read Ramsey's book *Foundations of Mathematics and Other Logical Essays* (Ramsey, 1931), which included a famous chapter on probability (see Section 6.3.1). In 1933, Bartlett was offered a position as Assistant Lecturer by Egon Pearson at University College London. This was an exciting time for Bartlett to learn more statistics since Fisher was also at University College London as the new Galton Professor and Neyman had recently joined Pearson. However, in 1934, Bartlett accepted the position of statistician at the Imperial Chemical Industries (ICI). This was a very productive research period for him as he published papers in a wide variety of fields. Of special interest to Bartlett was the field of stochastic processes, about which he learned a lot by interacting with Jo Moyal. Bartlett became Chair of Mathematical Statistics at the University of Manchester in 1947. In 1955, he published his most important book *An Introduction to Stochastic Processes* (Bartlett, 1955). Bartlett received many accolades during his lifetime, including the Guy Silver Medal in 1952 and the Guy Gold Medal in 1969 by the Royal Statistical Society. Bartlett passed away at Exmouth in 2002, at the age of 91.

As was mentioned before, Bartlett's 1936 paper (Bartlett, 1936) was the first major attack on Fisher's fiducial argument. In this paper, Bartlett considered the Behrens–Fisher test for the case $n = n' = 2$. For the latter case, he also proposed a test statistic that consisted of choosing one of the following two statistics at random:

$$\begin{aligned} T &= \frac{x_1 + x_2 - x_1' - x_2'}{|x_1 + x_2| + |x_1' - x_2'|}, \\ T' &= \frac{x_1 + x_2 - x_1' - x_2'}{\big||x_1 + x_2| - |x_1' - x_2'|\big|}. \end{aligned} \tag{5.86}$$

* Bartlett's life is also described in Bartlett (1982) and Whittle (2004).

FIGURE 5.6 Maurice Stevenson Bartlett (1910–2002). From Olkin (1989, p. 152). Courtesy of Institute of Mathematical Statistics

Here, x_1, x_2 and x_1', x_2' are observations in the first and second samples, respectively, and T corresponds to the Behrens–Fisher test for $n=2$.

Bartlett showed that, for $n = n' = 2$, the Behrens–Fisher test gave a fiducial probability (that $\mu > \mu'$), which was less for the case $s/s' = 1$* than when $s/s' = 0$ or ∞. This was problematic because the test should have had more power when $s/s' = 1$ since $\sigma^2 + \sigma'^2$ was then more efficiently estimated. Bartlett also showed that the width of the fiducial interval for $\mu' - \mu$, based on the Behrens–Fisher test and assuming $n = n'$, depended on the factor

$$\lambda = \sqrt{\frac{\phi + s^2/s'^2}{1 + s^2/s'^2} \cdot \frac{1+\phi}{2\phi}}$$

(*ibid.*, p. 565), where $\phi = \sigma^2 / \sigma'^2$. Thus, the $100(1-\alpha)\%$ fiducial interval for $\mu' - \mu$ does not enclose the true mean difference in a proportion $(1-\alpha)$ of cases in the long run, since the interval depends on the value of the nuisance parameter ϕ. The Behrens–Fisher test and the corresponding critical region are thus said to be *non-similar*† (to the whole sample space) with respect to the nuisance parameter. This fact casted a long shadow of doubt on the validity of the fiducial method. On the other hand, Bartlett's test statistic (which consisted of choosing one of the statistics in (5.86) at random) turned out to be similar.

Bartlett's 1936 paper was a source of major ongoing discord between himself and Fisher. The latter resigned from the *Cambridge Philosophical Society* when it refused to publish Fisher's reply to Bartlett's paper. In a letter dated November 24, 1938, to H. Jeffreys (with whom Fisher had previously been embroiled in a major disagreement over the nature of probability) Fisher described the incident as follows:

> The situation was that M.S. Bartlett thought he had detected an error in a paper of mine dealing with a test of significance originally put forward by W.U. Behrens. Bartlett's paper purporting to expose this error was published by the Society, without notification by him or them to me as the author criticised; they thereby took the ordinary risk of receiving a reply. The Referee objected to my reply on the ground that I had chosen a degenerate case, although this case was not chosen by me, but by Bartlett, and reiterated a number of facts which I had demonstrated and emphasised in my note, as though I had overlooked them and as though they were opposed to my point of view. Nevertheless, Mr. Wilson and his Committee decided to refuse publication of my answer in which I gave my grounds for dissenting from Bartlett's conclusions. This course was, I think, particularly objectionable as I had been myself responsible for first putting forward the fiducial argument, and had, therefore, to continue to shoulder the responsibility of preventing its being misunderstood, as I think Bartlett had done.
>
> The excuse was put forward that I had misrepresented Bartlett, but when Wilson was challenged to cite the particular passage which was thought to be a misrepresentation, with a view to my modifying it if it really seemed to be such, he failed to produce any case. (Bennett, 1990, pp. 349–350)

Eventually, Fisher was able to publish his reply (Fisher, 1937a) in the journal *Annals of Eugenics*. In his note, he criticized Bartlett's randomized test in (5.86):

> Bartlett's test of significance appears to require, either that the knowledge of whether the larger value, T', or the smaller value, T, had been chosen by lot has been obliterated from the experimenter's mind, or that he is supposed to accept the judgment of significance with

* Here, s and s' are the standard deviations of the first and second samples, respectively, with the sample size as divisor in each case.

† The term "similar" was first introduced in 1933 by Neyman and Pearson (1933). See also Section 5.6.1.2.

> the same confidence, when he knows that chance has chosen T', and that T is much smaller, as he might feel if T also had been large ...The alternative test of significance proposed involves, when the variance ratio of the two populations sampled is unknown, the choice by lot between the value T, used in Behrens' test, and a second value T', which reverses the order of significance of different possible sets of observations. High values of T' are not, therefore, by themselves evidence of inequality of the means. (*ibid.*, p. 375)

Indeed, since T' is always larger than T, it is possible for the same data set to give rise to totally different levels of significance, depending on which test statistic was chosen.

In his rejoinder (Bartlett, 1939), Bartlett explained that he proposed randomization between T and T' not as a *viable* test of significance but only to highlight the deficiency of the Behrens–Fisher solution (i.e., using T alone):

> ...It is sufficient to note that the distribution [of Bartlett's randomized test] certainly provides us with an exact inference of fiducial type, as Fisher himself confirmed ([9], p. 375); and this inference clashes with the apparent inference to be drawn from the Behrens-Fisher solution. In general it is of course true that different distributions might validly lead to different inferences of fiducial type, but here the distributions are sufficiently similar mathematically for it to be possible to assert that they cannot both be correct. (*ibid.*, pp. 135–136)

Thereupon, Bartlett reiterated his previous major rebuke that the Behrens–Fisher test was not similar (and therefore did not maintain the nominal significance level). Concerning the latter criticism, Fisher was quick to brush it away on several occasions:

> The fact that choosing at random T and T' will give us a quantity which, on the null hypothesis, is distributed in "Student's" distribution is, thus, insufficient to justify its use as a test of significance. (Fisher, 1939b, p. 385)
>
> ...
>
> It was obvious from the first, and particularly emphasised by the present writer, that Behrens' test rejects a smaller proportion of such repeated samples than the proportion specified by the level of significance, for the sufficient reason that the variance ratio of the populations sampled was unknown. (Fisher, 1950, p. 35.173a)
>
> ...
>
> This circumstance, indeed, caused me no surprise...but was eagerly seized upon by M.S. Bartlett, as though it were a defect in the test of significance of the composite hypothesis, that in special cases the criterion of rejection is less frequently attained by chance than in others. On reflexion I do not think one should expect anything else.... (Fisher, 1956b, p. 96)

5.5.2.5 Welch's Approximations (1938, 1947) Notwithstanding Fisher's dismissal of the issue, many statisticians agreed with Bartlett and started looking for confidence level-based similar regions. But before studying these attempts, let us consider Welch's important paper of 1938 (Welch, 1938).

Bernard Lewis Welch* was born in Sunderland, England, on January 9, 1911. He was the youngest of four brothers and did his early education at Bede School. From there, he obtained a scholarship to Brasenose College, Oxford. In 1933, Welch graduated with

*Welch's life is also described in Mardia (1990).

first class in mathematics. He then became interested in mathematical statistics and joined University College London, where Fisher and Egon Pearson were Chairs of Eugenics and Statistics, respectively. Although Welch was interested in theoretical statistics, he was also determined to apply statistics to industrial and agricultural problems. During 1939–1946, Welch served as Scientific Officer on the Ordnance Board of the Ministry of Supply. He then returned to academic life by becoming Reader in Statistics at the University of Leeds. Welch was appointed Chair of Statistics in 1968 and retired in 1976. He was an unassuming person and loathed controversy. He was apparently distressed during the disagreement, which we shall soon describe, he had with R.A. Fisher. Welch died on December 29, 1989.

Welch 1938 paper* (Welch, 1938) examined how the relationship between the underlying population variances (σ_1^2 and σ_2^2) affected the performance of two commonly used test statistics for testing the difference between the means of two independent normal populations:

$$u = \frac{(\bar{x}_1 - \bar{x}_2)}{\sqrt{\dfrac{\Sigma_1 + \Sigma_2}{(n_1 + n_2 - 2)}\left(\dfrac{1}{n_1} + \dfrac{1}{n_2}\right)}}, \tag{5.87}$$

$$v = \frac{(\bar{x}_1 - \bar{x}_2)}{\sqrt{\dfrac{\Sigma_1}{n_1(n_1 - 1)} + \dfrac{\Sigma_2}{n_2(n_2 - 1)}}}, \tag{5.88}$$

where, for $i = 1,2$, $\bar{x}_i$ is the mean of sample i, $\Sigma_i = \sum_j (x_{ij} - \bar{x}_i)^2$, x_{ij} is the jth observation in sample i, and n_i is the size of sample i. Note that, in the above, u is the test statistic commonly used when $\sigma_1^2 = \sigma_2^2$ and, under the latter condition, has a Student's t-distribution with $n_1 + n_2 - 2$ degrees of freedom. On the other hand, v is the Behrens–Fisher test statistic in Eq. (5.85) (calculated under $\delta = 0$) and "may be referred to the normal probability table if the samples are large enough, but for small samples it does not yield an exact test and it is not clear how it may best be made to furnish approximations." Next, for arbitrary σ_1^2 and σ_2^2, let

$$X' = \frac{(\bar{x}_1 - \bar{x}_2)}{\sqrt{\dfrac{\sigma_1^2}{n_1} + \dfrac{\sigma_2^2}{n_2}}} \sim N(0,1),$$

$$X_1^2 = \frac{\Sigma_1}{\sigma_1^2} \sim \chi_{n_1 - 1}^2,$$

$$X_2^2 = \frac{\Sigma_2}{\sigma_2^2} \sim \chi_{n_2 - 1}^2.$$

* See also Stuart et al. (1999, pp. 145–147).

Using the last equations, Eq. (5.87) becomes

$$u=\frac{X'\sqrt{\dfrac{\sigma_1^2}{n_1}+\dfrac{\sigma_2^2}{n_2}}}{\sqrt{\dfrac{X_1^2\sigma_1^2+X_2^2\sigma_2^2}{(n_1+n_2-2)}\left(\dfrac{1}{n_1}+\dfrac{1}{n_2}\right)}}.$$

This is then written as

$$u=\frac{X'}{\sqrt{aX_1^2+bX_2^2}}=\frac{X'}{\sqrt{w}}, \tag{5.89}$$

where

$$a=\frac{\sigma_1^2\left(\dfrac{1}{n_1}+\dfrac{1}{n_2}\right)}{(n_1+n_2-2)\left(\dfrac{\sigma_1^2}{n_1}+\dfrac{\sigma_2^2}{n_2}\right)},\quad b=\frac{\sigma_2^2\left(\dfrac{1}{n_1}+\dfrac{1}{n_2}\right)}{(n_1+n_2-2)\left(\dfrac{\sigma_1^2}{n_1}+\dfrac{\sigma_2^2}{n_2}\right)} \tag{5.90}$$

(*ibid*., p. 352) and, for constants f and g, w/g is to be approximated by a χ_f^2-distribution. The latter approximation implies

$$\mathsf{E}w\approx gf,\quad \operatorname{var} w\approx 2g^2f.$$

From $w=aX_1^2+bX_2^2$ in Eq. (5.89),

$$\mathsf{E}w=a(n_1-1)+b(n_2-1),\quad \operatorname{var} w=2\{a^2(n_1-1)+b^2(n_2-1)\}.$$

Solving for f and g in the last two sets of equations,

$$g\approx\frac{a^2(n_1-1)+b^2(n_2-1)}{a(n_1-1)+b(n_2-1)},\quad f\approx\frac{\{a(n_1-1)+b(n_2-1)\}^2}{a^2(n_1-1)+b^2(n_2-1)}. \tag{5.91}$$

Finally note that, since $w/g \mathrel{\dot{\sim}} \chi_f^2$, the test statistic u in Eq. (5.89) can be written as

$$u=\frac{1}{\sqrt{gf}}\cdot\frac{X'}{\sqrt{w/(gf)}}=\frac{1}{\sqrt{gf}}T\quad\text{where } T\mathrel{\dot{\sim}} t_f.$$

Thus, for arbitrary σ_1^2 and σ_2^2, the test statistic u is approximately* distributed as a factor $1/\sqrt{gf}$ of a Student's t-distribution with f degrees of freedom, where by substituting for f and g in Eq. (5.91), and for a and b in Eq. (5.90),

* The distribution is exact when $\sigma_1^2=\sigma_2^2$.

$$f \approx \frac{\left\{(n_1-1)\sigma_1^2+(n_2-1)\sigma_2^2\right\}^2}{(n_1-1)\sigma_1^4+(n_2-1)\sigma_2^4},$$

$$\frac{1}{\sqrt{gf}} \approx \sqrt{\frac{(n_1+n_2-2)\left(\dfrac{\sigma_1^2}{n_1}+\dfrac{\sigma_2^2}{n_2}\right)}{\left(\dfrac{1}{n_1}+\dfrac{1}{n_2}\right)\left\{(n_1-1)\sigma_1^2+(n_2-1)\sigma_2^2\right\}}} \tag{5.92}$$

(*ibid*., p. 353).

Similarly, for the statistic v, Welch used the form in (5.89)

$$v = \frac{X'}{\sqrt{a^* X_1^2 + b^* X_2^2}} = \frac{X'}{\sqrt{w^*}}$$

where

$$a^* = \frac{\dfrac{\sigma_1^2}{n_1(n_1-1)}}{\left(\dfrac{\sigma_1^2}{n_1}+\dfrac{\sigma_2^2}{n_2}\right)}, \quad b^* = \frac{\dfrac{\sigma_2^2}{n_2(n_2-1)}}{\left(\dfrac{\sigma_1^2}{n_1}+\dfrac{\sigma_2^2}{n_2}\right)} \tag{5.93}$$

and, for constants f^* and g^*, w^*/g^* is to be approximated by a $\chi^2_{f^*}$ distribution. By equating means and variances,

$$g^* \approx \frac{a^{*2}(n_1-1)+b^{*2}(n_2-1)}{a^*(n_1-1)+b^*(n_2-1)}, \quad f^* \approx \frac{\left\{a^*(n_1-1)+b^*(n_2-1)\right\}^2}{a^{*2}(n_1-1)+b^{*2}(n_2-1)}.$$

Moreover,

$$v = \frac{1}{\sqrt{g^* f^*}} \cdot \frac{X'}{\sqrt{w^*/(g^* f^*)}} = \frac{1}{\sqrt{g^* f^*}} T^* \quad \text{where } T^* \doteq t_{f^*}.$$

Thus, for arbitrary σ_1^2 and σ_2^2, the test statistic v is approximately distributed as a factor $1/\sqrt{g^* f^*}$ of a Student's t-distribution with f^* degrees of freedom, where

$$f^* \approx \frac{\left(\dfrac{\sigma_1^2}{n_1}+\dfrac{\sigma_2^2}{n_2}\right)^2}{\dfrac{\sigma_1^4}{n_1^2(n_1-1)}+\dfrac{\sigma_2^4}{n_2^2(n_2-1)}},$$

$$\frac{1}{\sqrt{g^* f^*}} \approx 1 \tag{5.94}$$

(*ibidem*).*

* It can be shown that $\min(n_1-1, n_2-1) \le f^* \le n_1+n_2-2$ (e.g., Brownlee, 1965, p. 301).

From Eqs. (5.92) and (5.94), it is seen that, for the statistic u, the nuisance parameter σ_1^2 / σ_2^2 occurs both in the effective degrees of freedom f and in the approximate test statistic itself; on the other hand, for the statistic v, the nuisance parameter occurs only in the effective degrees of freedom. Therefore,

> If it is known that $\sigma_1 = \sigma_2$ then there can be no doubt that u is a better, more sensitive, criterion than v. If, however, there exists the possibility that σ_1 and σ_2 differ, then u may give very misleading results and it will be safer to use v. (*ibid.*, p. 355)

Welch also found that, when the sample sizes were the same, u was much less adversely affected* by $\sigma_1^2 \neq \sigma_2^2$ than when the sample sizes were different. Therefore, when the sample sizes were the same, u could still be used even though $\sigma_1^2 \neq \sigma_2^2$.

It is interesting that simulations have confirmed Welch's findings (e.g., Stapleton, 2008, pp. 251–252) and these results hold true even today.

It would seem intuitive that, when the approximately t-distributed statistic v was used, its degrees of freedom could be calculated from Eq. (5.94) by estimating the population variances using the sample variances. Welch himself was more cautious:

> When there is no very precise *a priori* information about $\theta \left[= \sigma_1^2 / \sigma_2^2 \right]$ available it might seem permissible to use the ratio Σ_1/Σ_2 from the samples to estimate θ and hence f. Complications arise, however, owing to the fact that the distribution of v is not independent of that of Σ_1/Σ_2. (*ibid.*, p. 358)

However, F.E. Satterthwaite (1946) was later able to show that it was reasonably accurate to use Eq. (5.94) with the population variances replaced by their sample counterparts:

$$f^{**} \approx \frac{\left(s_1^2 / n_1 + s_2^2 / n_2\right)^2}{\dfrac{\left(s_1^2 / n_1\right)^2}{n_1 - 1} + \dfrac{\left(s_2^2 / n_2\right)^2}{n_2 - 1}}, \tag{5.95}$$

where $s_i^2 = \Sigma_i / (n_i - 1)$ for $i = 1, 2$. In recognition of H. Fairfield Smith who first proposed the above approximation (Smith, 1936), the testing of means from two independent normal populations using the approximately t-distributed test statistic v with f^{**} degrees of freedom is called the *Smith–Satterthwaite* test.†

Following his 1938 paper (Welch, 1938), Welch further refined his method in the 1947 paper "The generalization of 'Student's' problem when several different population variances are involved" (Welch, 1947).‡ Welch couched his analysis in terms of k normal populations but it was the $k = 2$ case that was of particular interest to him, so it is the latter case that we shall consider here.

*The main adverse effect here is the failure of the test to maintain the nominal error rate. Welch called such tests "invalid" or "biased." The "bias" of a test is used in a very different sense today (see end of Section 5.6.1.2).

†Although the names Welch, Welch–Satterthwaite, and Smith–Welch–Satterthwaite are also used.

‡See also Stuart et al. (1999, pp. 147–148).

Essentially, Welch's aim was to find suitable critical values for the statistic v in Eq. (5.88). He started by seeking a statistic $h\left(s_1^2, s_2^2\right)$ such that

$$\Pr\left\{y-\eta < h\left(s_1^2, s_2^2, P\right)\right\} = P \tag{5.96}$$

(*ibid.*, p. 28), where $y = \bar{x}_1 - \bar{x}_2, \eta = \mu_1 - \mu_2$, and P is a given probability. Furthermore, let

$$j\left(s_1^2, s_2^2, P\right) = \Pr\left\{y-\eta < h\left(s_1^2, s_2^2, P\right) \middle| s_1^2, s_2^2\right\}. \tag{5.97}$$

Then,

$$j\left(s_1^2, s_2^2, P\right) = \Pr\left\{\frac{y-\eta}{\sqrt{\sigma_1^2/n_1+\sigma_2^2/n_2}} < \frac{h\left(s_1^2, s_2^2, P\right)}{\sqrt{\sigma_1^2/n_1+\sigma_2^2/n_2}} \middle| s_1^2, s_2^2\right\} = I\left\{\frac{h\left(s_1^2, s_2^2, P\right)}{\sqrt{\sigma_1^2/n_1+\sigma_2^2/n_2}}\right\},$$

where $I(\bullet)$ is the normal probability integral. Comparing the two probabilities in Eqs. (5.96) and (5.97),

$$\iint I\left\{\frac{h\left(s_1^2, s_2^2, P\right)}{\sqrt{\sigma_1^2/n_1+\sigma_2^2/n_2}}\right\} p\left(s_1^2\right) p\left(s_2^2\right) ds_1^2 ds_2^2 = P, \tag{5.98}$$

where $p(\cdot)$ is the probability density of each of s_1^2 and s_2^2, the latter two random variables being assumed to be independent of each other. The next step is to expand I in a Taylor series about $\left(\sigma_1^2, \sigma_2^2\right)$:

$$I\left\{\frac{h\left(s_1^2, s_2^2, P\right)}{\sqrt{\sigma_1^2/n_1+\sigma_2^2/n_2}}\right\} = \exp\left\{\sum_{i=1}^{2}\left(s_i^2-\sigma_i^2\right)\partial_i\right\} I\left\{\frac{h\left(w_1, w_2, P\right)}{\sqrt{\sigma_1^2/n_1+\sigma_2^2/n_2}}\right\}, \tag{5.99}$$

where

$$\partial_i^r I\left\{\frac{h\left(w_1, w_2, P\right)}{\sqrt{\sigma_1^2/n_1+\sigma_2^2/n_2}}\right\} \equiv \left[\frac{\partial^r}{\partial w_i^r} I\left\{\frac{h\left(w_1, w_2, P\right)}{\sqrt{\sigma_1^2/n_1+\sigma_2^2/n_2}}\right\}\right]_{w_1=\sigma_1^2, w_2=\sigma_2^2}.$$

Substituting (5.99) into Eq. (5.98), Welch obtained

$$\prod_{i=1}^{2}\left[\int \exp\left\{\left(s_i^2-\sigma_i^2\right)\partial_i\right\} p\left(s_i^2\right) ds_i^2\right] I\left\{\frac{h\left(w_1, w_2, P\right)}{\sqrt{\sigma_1^2/n_1+\sigma_2^2/n_2}}\right\} = P. \tag{5.100}$$

Now, since

$$p\left(s_i^2\right) ds_i^2 = \frac{1}{\Gamma\left(\frac{n_i-1}{2}\right)}\left\{\frac{\left(n_i-1\right)s_i^2}{2\sigma_i^2}\right\}^{\frac{n_i-1}{2}} \exp\left\{-\frac{1}{2}\frac{\left(n_i-1\right)s_i^2}{\sigma_i^2}\right\} d\left\{\frac{\left(n_i-1\right)s_i^2}{2\sigma_i^2}\right\},$$

Equation (5.100) becomes

$$\begin{aligned}
P &= \prod_{i=1}^{2}\left(1-\frac{2\sigma_i^2\partial_i}{n_i-1}\right)^{-(n_i-1)/2}\exp\left(-\sigma_i^2\partial_i\right)I\left\{\frac{h(w_1,w_2,P)}{\sqrt{\sigma_1^2/n_1+\sigma_2^2/n_2}}\right\}\\
&= \exp\left\{-\sum_i\sigma_i^2\partial_i-\frac{1}{2}\sum_i(n_i-1)\log\left(1-\frac{2\sigma_i^2\partial_i}{n_i-1}\right)\right\}I\left\{\frac{h(w_1,w_2,P)}{\sqrt{\sigma_1^2/n_1+\sigma_2^2/n_2}}\right\}\\
&= \exp\left\{\sum_i\frac{\sigma_i^4\partial_i^2}{n_i-1}+\frac{4}{3}\sum_i\frac{\sigma_i^6\partial_i^3}{(n_i-1)^2}+2\sum_i\frac{\sigma_i^8\partial_i^4}{(n_i-1)^3}+\cdots\right\}I\left\{\frac{h(w_1,w_2,P)}{\sqrt{\sigma_1^2/n_1+\sigma_2^2/n_2}}\right\}\\
&= \left[1+\sum_i\frac{\sigma_i^4\partial_i^2}{n_i-1}+\left\{\frac{4}{3}\sum_i\frac{\sigma_i^6\partial_i^3}{(n_i-1)^2}+\frac{1}{2}\left(\sum_i\frac{\sigma_i^4\partial_i^2}{n_i-1}\right)^2\right\}+\cdots\right]I\left\{\frac{h(w_1,w_2,P)}{\sqrt{\sigma_1^2/n_1+\sigma_2^2/n_2}}\right\}
\end{aligned}$$

(*ibid*., p. 29). The above equation can be solved for any given P to obtain $h\left(\sigma_1^2,\sigma_2^2,P\right)$ and hence $h\left(s_1^2,s_2^2,P\right)$ of Eq. (5.96). From $h\left(s_1^2,s_2^2,P\right)$, the desired critical value c_p for the statistic v in Eq. (5.88) can be calculated from

$$c_p=\frac{h\left(s_1^2,s_2^2,P\right)}{\sqrt{s_1^2/n_1+s_2^2/n_2}}. \tag{5.101}$$

Welch also gave a series solution for $h\left(s_1^2,s_2^2,P\right)$:

$$\begin{aligned}
h\left(s_1^2,s_2^2,P\right)=\xi\sqrt{\frac{s_1^2}{n_1}+\frac{s_2^2}{n_2}}\Bigg[1&+\frac{(1+\xi^2)}{4}\frac{\left(\frac{s_1^4/n_1^2}{n_1-1}+\frac{s_2^4/n_2^2}{n_2-1}\right)}{\left(s_1^2/n_1+s_2^2/n_2\right)^2}-\frac{(1+\xi^2)}{2}\frac{\left\{\frac{s_1^4/n_1^2}{(n_1-1)^2}+\frac{s_2^4/n_2^2}{(n_2-1)^2}\right\}}{\left(s_1^2/n_1+s_2^2/n_2\right)^2}\\
&+\frac{(3+5\xi^2+\xi^4)}{3}\frac{\left\{\frac{s_1^6/n_1^3}{(n_1-1)^2}+\frac{s_2^6/n_2^3}{(n_2-1)^2}\right\}}{\left(s_1^2/n_1+s_2^2/n_2\right)^3}-\frac{(15+32\xi^2+9\xi^4)}{32}\frac{\left(\frac{s_1^4/n_1^2}{n_1-1}+\frac{s_2^4/n_2^2}{n_2-1}\right)^2}{\left(s_1^2/n_1+s_2^2/n_2\right)^4}+\cdots\Bigg]
\end{aligned} \tag{5.102}$$

(*ibid*., p. 31), where $\xi=I^{-1}(P)$. Note that the above is a series to order $1/(n_i-1)^2$. Welch further noted that, for the one-sample case, Eq. (5.102) gives the series expansion of Student's *t:*

$$t_P=\frac{h(s,p)}{s/\sqrt{n}}=\xi\left\{1+\frac{(1+\xi^2)}{4(n-1)}+\frac{(3+16\xi^2+5\xi^4)}{96(n-1)^2}+\cdots\right\},$$

where t_p is the pth percentile of Student's distribution with $n-1$ degrees of freedom. The latter result had been obtained earlier by Fisher (1941, p. 151).

Building on Welch's work, A. Aspin (1948) further expanded $h\left(s_1^2, s_2^2, P\right)$ to terms of the order $1/(n_i-1)^4$ in Eq. (5.102) and suggested that this expanded expression be used to calculate the critical values in Eq. (5.101). A further paper by Aspin and Welch (1949) gave small tables of the fourth-order critical values for various degrees of freedom, test sizes, and sample variance ratios. The t-test for means from two normal populations based on Aspin's critical values is known as the *Welch–Aspin* test. We note that, although the test does not depend on the nuisance parameter σ_1^2/σ_2^2, it is based on *approximate* values of $h\left(s_1^2, s_2^2, P\right)$. The Welch–Aspin test and the corresponding confidence intervals are therefore *approximately (or asymptotically) similar.*

Lee and Gurland (1975) have shown that, for small-sample sizes (<10), the nominal significance level is better controlled with the fourth-order critical value than with either the second-order one in (5.102) or the critical value based on the approximate degrees of freedom in (5.95). However, all three procedures controlled the Type I error rate reasonably well.

Lee and Gurland have also reviewed other confidence level-based solutions of the Behrens–Fisher problem.

5.5.2.6 Criticism of Welch's Solution (1956) Although the Welch–Aspin test was not an exact solution to the Behrens–Fisher problem, it was approximately similar and did a good job of controlling the nominal significance rate. Quite naturally, the test gained favor among many statisticians and its critical values were later tabulated in Pearson and Hartley's *Biometrika Tables for Statisticians* (Pearson and Hartley, 1954). However, in the introduction section of their book, the authors were particularly critical of Fisher's own fiducial solution to the Behrens–Fisher problem:

> …One of the consequences of the use of the Fisher-Behrens test is that, if the null hypothesis is true, the probability (in the usual direct sense) of rejecting it will not equal the figure specified as the level of significance. Fisher, indeed, does not consider this to be a drawback. It should be noted, however, that the table given in the present volume is not related to the Fisher-Behrens test in any way. It is the result of an attempt to produce a test which does satisfy the condition that the probability of rejection of the hypothesis tested (or of the truth of a confidence statement) will be equal to a specified figure. (*ibid.*, p. 26)

Now such criticism did not usually go without a swift retaliation from Fisher, and this case proved no exception. In a paper published just one year later in *The Journal of the Royal Statistical Society* (Fisher, 1956a), Fisher mounted a major attack on the Welch–Aspin test, as we now describe.

Fisher considered two samples of sizes n_1+1 and n_2+1 obtained from two independent normal populations. His aim was to derive, "of those samples giving an assigned value for the ratio s_2/s_1," the exact sampling distribution of

$$d = \frac{\bar{x}_1 - \bar{x}_2}{\sqrt{s_1^2 + s_2^2}}, \tag{5.103}$$

where $\bar{x}_i$ ($i=1, 2$) is the mean of the ith sample and

$$s_1^2 = \frac{1}{n_1(n_1+1)}\Sigma(x_1-\bar{x}_1)^2, \quad s_2^2 = \frac{1}{n_2(n_2+1)}\Sigma(x_2-\bar{x}_2)^2$$

are the estimated variances of the sample means. Then the joint density of (s_1^2, s_2^2) is $f(s_1^2, s_2^2)$ where

$$f(s_1^2, s_2^2)ds_1^2 ds_2^2 = \frac{1}{\Gamma\left(\frac{n_1}{2}\right)}\left(\frac{n_1 s_1^2}{2\sigma_1^2}\right)^{\frac{n_1-2}{2}} \exp\left(-\frac{n_1 s_1^2}{2\sigma_1^2}\right)\frac{d(n_1 s_1^2)}{2\sigma_1^2}$$
$$\times\frac{1}{\Gamma\left(\frac{n_2}{2}\right)}\left(\frac{n_2 s_2^2}{2\sigma_2^2}\right)^{\frac{n_2-2}{2}} \exp\left(-\frac{n_2 s_2^2}{2\sigma_2^2}\right)\frac{d(n_2 s_2^2)}{2\sigma_2^2}$$

(*ibid*., p. 56). In the above, σ_1^2, σ_2^2 stand for the variances of the sample means. Fisher now made the substitutions

$$s_1^2 = u,$$
$$s_2^2 = u\cot^2\theta$$

and found the conditional density of u given θ (i.e., given s_2/s_1) to be

$$\frac{1}{\Gamma(n_1+n_2)}\left\{\frac{u}{2}\left(\frac{n_1}{\sigma_1^2}+\frac{n_2\cot^2\theta}{\sigma_2^2}\right)\right\}^{\frac{n_1+n_2-2}{2}} \exp\left\{-\frac{u}{2}\left(\frac{n_1}{\sigma_1^2}+\frac{n_2\cot^2\theta}{\sigma_2^2}\right)\right\}\times\frac{1}{2}\left(\frac{n_1}{\sigma_1^2}+\frac{n_2\cot^2\theta}{\sigma_2^2}\right).$$

Therefore,

$$u\left(\frac{n_1}{\sigma_1^2}+\frac{n_2\cot^2\theta}{\sigma_2^2}\right)|\theta \sim \chi^2_{n_1+n_2},$$

which implies, for given θ,

$$s_1^2+s_2^2 = (1+\cot^2\theta)u = \frac{\operatorname{cosec}^2\theta}{\left(\frac{n_1}{\sigma_1^2}+\frac{n_2\cot^2\theta}{\sigma_2^2}\right)}X^2, \quad \text{where} \quad X^2 \sim \chi^2_{n_1+n_2} \tag{5.104}$$

(*ibid*., p. 57). The next step for Fisher was to let

$$e^z = \frac{\sigma_2/\sigma_1}{s_2/s_1}.$$

Therefore, conditional on s_2/s_1, the numerator in (5.103), that is, $\bar{x}_1-\bar{x}_2$, is normally distributed with mean 0 and variance

$$\begin{aligned}\operatorname{var}(\bar{x}_1-\bar{x}_2)&=\sigma_1^2+\sigma_2^2\\&=\sigma_1^2+\sigma_1^2\left(\frac{s_2^2}{s_1^2}\right)e^{2z}\\&=\sigma_1^2\left(1+e^{2z}\cot^2\theta\right).\end{aligned}$$

That is,

$$W=\frac{\bar{x}_1-\bar{x}_2}{\sigma_1\sqrt{1+e^{2z}\cot^2\theta}}\sim N(0,1).$$

On the other hand, from (5.104), the square of the denominator in (5.103) can be written as

$$\begin{aligned}s_1^2+s_2^2&=\frac{\operatorname{cosec}^2\theta}{\left(\frac{n_1}{\sigma_1^2}+\frac{n_2\cot^2\theta}{\sigma_2^2}\right)}X^2\\&=\sigma_1^2X^2\frac{\operatorname{cosec}^2\theta}{n_1+n_2\left(\frac{\sigma_1^2}{\sigma_2^2}\right)\cot^2\theta}\\&=\sigma_1^2X^2\frac{\operatorname{cosec}^2\theta}{n_1+n_2\left(\frac{s_1^2}{s_2^2}e^{-2z}\right)\cot^2\theta}\\&=\sigma_1^2X^2\frac{\operatorname{cosec}^2\theta}{n_1+n_2e^{-2z}}.\end{aligned}$$

Hence, with $T\sim t_{n_1+n_2}$,

$$\begin{aligned}\frac{d^2}{T^2}&=\frac{(\bar{x}_1-\bar{x}_2)^2/(s_1^2+s_2^2)}{W^2\Big/\frac{X^2}{(n_1+n_2)}}\\&=\frac{(\bar{x}_1-\bar{x}_2)^2/(s_1^2+s_2^2)}{\frac{(\bar{x}_1-\bar{x}_2)^2}{\sigma_1^2(1+e^{2z}\cot^2\theta)}\Big/\frac{(s_1^2+s_2^2)(n_1+n_2e^{-2z})/(\sigma_1^2\operatorname{cosec}^2\theta)}{n_1+n_2}}\\&=\frac{(1+e^{2z}\cot^2\theta)(n_1+n_2e^{-2z})}{(n_1+n_2)\operatorname{cosec}^2\theta}\\&=\frac{(\cosh z+\cos 2\theta\sinh z)(n_1e^{z}+n_2e^{-z})}{n_1+n_2}.\end{aligned}$$

Fisher considered the "doubly simple case" of $n_1 = n_2$ and $s_1 = s_2$, so that

$$\frac{d^2}{T^2} = (\cosh z + \cos 90° \sinh z)\frac{(e^z + e^{-z})}{2} \quad \Rightarrow \quad \frac{d}{T} = \cosh z \geq 1.$$

Therefore,

> ...whatever the value of z may be for the populations from which the samples are taken, the value of d which is exceeded by chance in 5 per cent or 1 per cent of trials cannot be less than the value of $t_{n_1+n_2}$ at these points. (*ibid*., p. 58)

This means that, for any value a,

$$\Pr\left\{|d| > a \left| \frac{s_2}{s_1} = 1 \right.\right\} \geq \Pr\{|T| > a\}.$$

For illustration, Fisher took $n_1 + 1 = n_2 + 1 = 7$, $\alpha = .1$. The Welch–Aspin critical value is then $a = 1.74$ so that

$$\Pr\left\{|d| > 1.74 \left| \frac{s_2}{s_1} = 1 \right.\right\} \geq 2\Pr\{T > 1.74\} = .107.$$

Now this is a remarkable result for it implies that the rejection rate of the conditional Welch–Aspin test is greater than the nominal significance level. It would seem as if the criticism leveled by Pearson and Hartley against the Behrens–Fisher fiducial test to the effect that it failed to control the nominal significance level was much weaker now, given that the same could be said of the Welch–Aspin test. This last sentence would have been true if not for the fact that, as the reader will have observed, Fisher did not show a defect in the Welch–Aspin per se, rather in a *conditional* version of it.

Indeed, what Fisher had effectively done was to show the existence of a negatively biased relevant subset,* namely $s_1 = s_2$, for the Welch–Aspin test. For Fisher, since s_1/s_2 was a relevant subset, any valid inference had to condition on it, which is precisely what the Behrens–Fisher test did.

However, the conditional test was not the test that was put forward by Welch, who questioned the validity of conditioning on s_2/s_1 in his reply to Fisher's criticism:

> ...He [Fisher] is just expressing the *opinion* that a statistician concerned with this problem, should restrict himself solely to probabilities evaluated for a class of samples which has $\left[c = \frac{s_1^2 / n_1}{s_1^2 / n_1 + s_2^2 / n_2}\right]$ fixed. With this opinion I simply disagree. (Welch, 1956, p. 298)

* See Eq. (5.68) and the discussion thereby.

To make matters worse for Fisher, Buehler and Feddersen (1963) later showed that a relevant subset* also existed for Student's t-test (the latter coincided with Fisher's own fiducial t-test). This was in spite of Fisher's earlier claim that:

> ...it might be thought that some feature calculable from the observations could define a relevant subset [for the fiducial t-test]. However, it has been shown that for Normal samples the two statistics $\bar{x}$ and s^2 are *exhaustive*, meaning that the sampling distribution of any such function, conditional upon given values of $\bar{x}$ and s shall be completely independent of the true mean and variance of the population sampled. There is therefore no relevant and recognizable subset, and all the conditions for a genuine statement of probability are satisfied. (Fisher, 1959, p. 26)

Buehler and Feddersen concluded:

> ...Thus Fisher's fiducial intervals based on Student's t could be criticized in very nearly the same way as Fisher himself has criticized the Welch solution. (Buehler and Feddersen, 1963, p. 1098)

By now the reader will probably wonder whether the Behrens–Fisher problem does have an exact nonrandomized similar confidence-based solution. In 1940, Wilks had stated that such a solution would be difficult to find (Wilks, 1940). A more definitive answer was later provided by the Soviet mathematician Yuri. V. Linnik (1915–1972). The latter showed that no similar test that used sufficient statistics in a smooth way existed (Linnik, 1968). Nonetheless, most of the confidence-based tests we have mentioned in this section are approximately similar.

5.5.3 Clash with Jeffreys on the Nature of Probability (1932–1934)

Sir Harold Jeffreys† (Fig. 5.7) was a world authority in geophysics and also made important contributions to probability. He was born on April 22, 1891 in Fatfield, Tyne and Wear, England. His was the only child of his parents, who were both schoolteachers (when Jeffreys was born, his father was headmaster). After finishing his primary education in Fatfield, Jeffreys obtained a Durham County minor scholarship to Rutherford College in Newcastle upon Tyne and was taught mathematics, English, and chemistry. Jeffreys' family moved to Birtley in 1905. Two years later, Jeffreys entered Armstrong College (now University of Newcastle) and was taught mathematics, physics, chemistry and, for one year, geology. He graduated in 1910 with first-class marks and a distinction in mathematics. Thereupon, he entered St John's College, Cambridge, to read for the Mathematical Tripos. Although his first two years there were difficult, he gained the college Adams Memorial Prize in 1912 for an essay on precession and mutation. In his third year, he was a wrangler. Jeffreys' college scholarship was extended for one more year and he began research. He obtained a D.Sc. in 1917 and then joined the Meteorological Office as a Senior Professional Assistant. In London, Jeffreys met Sir Arthur Holmes who influenced his geological thinking. Two years later, while still in London, Jeffreys wrote

*Which in this case was positively biased.

†Jefrreys' life is also described in Lindley (1989), Smith (1989), Cook (1990), and Swirles (1992).

FIGURE 5.7 Sir Harold Jeffreys (1891–1989). Wikipedia, http://en.wikipedia.org/wiki/Harold_Jeffreys

important papers in probability, scientific method, and semiology with his friend Dorothy Wrinch, who was then a lecturer at University College London. Jeffreys returned to Cambridge in 1922 as College lecturer in mathematics. In 1926, he was appointed to a university lectureship, and later he became the Plumian Professor of Astronomy. Jeffreys was a polymath who wrote more than 170 papers and several major books. Among the latter, two classics that ought to be mentioned are his 1924 book *The Earth, It's Origin, History and Physical Constitution* (Jeffreys, 1924) and his 1939 book *Theory of Probability* (Jeffreys, 1939). The former went through six editions and the latter was a major boost to the Bayesian revival in the forties and fifties. Although Jeffreys was a poor lecturer, his writing was excellent. His scientific outlook was very much influenced by Karl Pearson's *Grammar of Science* (Pearson, 1892), which had been introduced to him by his botanist friend, E.P. Farrow, in 1914. But Jeffreys was opposed to Karl Popper's falsification theory.* Jeffreys was knighted in 1953. He retired in 1958 and passed away much later on March 18, 1989, Cambridge.

Our main interest here is in Jeffreys' views on probability and how these clashed with those of Fisher. Unlike Fisher, who was a frequentist, Jeffreys belonged to the objective Bayesian School and viewed probability as an objective measure of belief. Jeffreys thus differed from a subjective Bayesian like de Finetti (cf. Section 6.3.2) who regarded probability as completely personal (subjective). For Jeffreys, probability ought to be the same for all rational persons. This is the logical interpretation of probability and was popularized before by Keynes (1921).

Jeffrey's first work on probability goes back to the period 1919–1923 when he published three papers (Wrinch and Jeffreys, 1919; 1921; 1923) in collaboration with Dorothy Wrinch. In their papers, Wrinch and Jeffreys described probability as an extension of logic and rejected the frequentist definition. They also argued that, for a given phenomenon, the number of possible quantitative laws was theoretically infinite. The laws could be ranked in decreasing order of simplicity, with the simplest law having the highest probability.

The Fisher–Jeffreys dispute[†] took place during the years 1932–1934 in the wake of important events in the lives of both Fisher and Jeffreys. Fisher had recently (Fisher, 1930) put forward what was arguably his most ambitious idea to date, namely, fiducial probability, and was trying to popularize the fiducial method as a valid means of assigning probabilities to hypotheses and population parameters. On the other hand, Jeffreys had just published his important book, *Scientific Inference* (Jeffreys, 1931), which not only contained various topics in physics but also in logical probability and the theory of errors.

In *Scientific Inference* (*ibid*., p. 66), Jeffreys studied the Bayesian problem of finding an appropriate joint prior for the mean and standard deviation of a normal distribution. Assuming a normal law of error, he considered a true value x which was observed as $x+\xi$, where ξ is the error of observation. Let the precision of ξ be h, where $h = 1/\left(\sigma\sqrt{2}\right)$

*In particular, Popper upheld that simple postulates were easier to falsify and therefore less probable. Jeffreys disagreed and thought simpler theories were more probable. See, for example, Keuzenkamp and McAleer (1995).

[†]The Fisher–Jeffreys dispute is also described in Lane (1980), Bartlett (1933), Howie (2002, Chapter 5), and Aldrich (2005).

and σ is its standard deviation. In order to obtain the posterior distribution of x and h, Jeffreys needed an appropriate prior $P(x, h)$. Concerning the prior for x:

> ...In most cases the prior probability of x is nearly uniformly distributed, at any rate over a range several times that covered by the observations. We are initially prepared for values of x over a wide range, and the purpose of making observations at all is to permit a considerable reduction of this range. (*ibidem*)

Thus, Jeffreys chose the uniform distribution for the prior of x, that is, $p_x \propto 1$. Concerning the prior for h, however:

> ...The position is different with regard to h. Initially we may have no special views about the probability of one value of h rather than another, but we do at least know that negative values are excluded, since they would imply negative probabilities. Again, x is not usually in fact a number; it is usually a length or an interval of time, and h is a reciprocal of whatever kind of magnitude x is, while the standard error [deviation] σ is the same kind of quantity as x. There seems to be no special reason for measuring the precision in terms of h rather than σ, and their product is constant, so that
>
> $$d \log h + d \log \sigma = 0. \tag{5.105}$$
>
> If then $P(x, h)dh$ is proportional to dh/h or $d\sigma/\sigma$, an ambiguity is removed. It means that the probability of a value of σ or h within a definite range is proportional to the increase of its logarithm. (*ibid.*, pp. 66–67)

In the above, Jeffreys explained it was unlikely that h would have a uniform prior. Moreover, by virtue of Eq. (5.105), the prior would have the same form whether we used h or σ if we allowed a prior for h of the type

$$p_h \propto 1/h. \tag{5.106}$$

This is because then the prior for σ would be

$$p_\sigma = p_h \left|\frac{dh}{d\sigma}\right| \propto \frac{1}{h} \cdot \frac{1}{\sigma^2} \propto \frac{1}{\sigma}.$$

This type of prior, which remained invariant under transformations of the parameter, would later be called a *Jeffreys' prior*. Thus, by using the two priors $p_x \propto 1$ and $p_h \propto 1/h$, and by implicitly assuming their independence, Jeffreys wrote the joint prior $P(x, h)$ as

$$P(x, h) \propto \frac{1}{h}.$$

One year later, in 1932, Jeffreys wrote a paper, "On the theory of errors and least-squares" (Jeffreys, 1932), where he attempted to provide an alternative justification for the prior in (5.106). Jeffreys considered the problem of predicting a third observation given the first two, and it was his solution to this problem that ignited a fight between himself and Fisher. But before we describe the fight, let us first examine Jeffreys' alternative justification for the prior in (5.106).

Jeffreys started his argument as follows:

> ...The normal law of error is supposed to hold, but the true value x and the precision constant h are unknown. Two measures are made: what is the probability that the third observation will lie between them? The answer is easily seen to be one-third. For the law says nothing about the order of occurrence of errors of different amounts, and therefore the middle one in magnitude is equally likely to be the first, second, or third made (provided, of course, that we know nothing about the probable range of error already). Now let us see what distribution of prior probability will give this result. (*ibid.*, pp. 48–49)

Jeffreys' strategy was therefore to investigate what prior would lead to an answer of 1/3 for the probability that the third observation would lie between the first two given observations. This is not very different from the principle used by Gauss hundreds of years ago when he applied the postulate of the arithmetic mean to derive the normal distribution (see Section 1.5.3.2). Jeffreys derived the prior as follows.

First Jeffreys assumed that the prior for x to be uniform and the prior for h to be f_h, so that joint prior $P(x, h) = f_h$. Next, he wrote the joint probability density of two independent observations $\pm a$, given x and h, as

$$\frac{h}{\sqrt{\pi}}e^{-h^2(x-a)^2}\cdot\frac{h}{\sqrt{\pi}}e^{-h^2(x+a)^2}=\frac{h^2}{\pi}e^{-2h^2\left(x^2+a^2\right)}$$

(*ibid.*, p. 49). If the joint posterior of (x, h) is denoted by $I(x, h)$, then

$$I(x,h)\propto f_h h^2 e^{-2h^2\left(x^2+a^2\right)}.$$

Now the density of a third observation x_3 is

$$\begin{aligned}\int_0^\infty\int_{-\infty}^\infty \frac{h}{\sqrt{\pi}}e^{-h^2(x_3-x)^2}I(x,h)\,dxdh &\propto \int_0^\infty\int_{-\infty}^\infty f_h h^3\exp\left[-h^2\left\{3\left(x-\frac{x_3}{3}\right)^2+2a^2+\frac{2x_3^2}{3}\right\}\right]dxdh\\ &\propto \int_0^\infty f_h h^2\exp\left\{-2h^2\left(a^2+\frac{x_3^2}{3}\right)\right\}dh.\end{aligned} \tag{5.107}$$

Hence, the probability that x_3 lies between two numbers b and c is the integral with respect to x_3 of the above, between b and c, and is proportional to

$$\int_0^\infty f_h h e^{-2h^2a^2}\left\{\operatorname{erf}\left(hc\sqrt{\frac{2}{3}}\right)-\operatorname{erf}\left(hb\sqrt{\frac{2}{3}}\right)\right\}dh, \tag{5.108}$$

where

$$\operatorname{erf}(z)=\frac{2}{\sqrt{\pi}}\int_0^z e^{-u^2}du.$$

Now, Jeffreys had postulated that the conditional probability was 1/3 that $-a < x_3 < a$ for given a. This means that the conditional probability was 1/6 that $0 < x_3 < a$ and was $1/6 + 1/3 = 1/2$ that $0 < x_3 < \infty$, that is, the first probability was one-third the second. Therefore, by substituting in Eq. (5.108):

$$\int_0^\infty f_h h e^{-2h^2a^2} \operatorname{erf}\left(ha\sqrt{\frac{2}{3}} \right) dh = \frac{1}{3}\int_0^\infty f_h h e^{-2h^2a^2}\, dh. \tag{5.109}$$

Concerning Eq. (5.109), Jeffreys explained:

> This is to hold for all values of a. If we take ha as a new variable, and $f(h)$ [f_h] is a power of h, a cancels from the equation; the problem then reduces to determining what power of h we must choose. If, on the other hand, $f(h)$ [f_h] is not a power of h, a does not cancel throughout, and (5) [5.109] cannot hold for all values of a. Hence $f(h)$ [f_h] is a power of h, and we proceed to investigate whether, if we take it proportional to h^{-1}, (5) [5.109] will hold. (*ibid*., pp. 49–50)

Thereupon, Jeffreys proceeded to show that Eq. (5.109) was verified for

$$f_h \propto h^{-1}. \tag{5.110}$$

Furthermore, he argued that $hf_h \propto h^\gamma$ was satisfied only for $\gamma = 0$. Hence, by postulating 1/3 for the probability that a third observation lies between the first two given observations, Jeffreys was able to obtain the same prior as in Eq. (5.106), namely, $f_h \propto 1/h$.

As has been effectively argued by Lane (1980, p. 152), Jeffreys' Bayesian analysis achieved two major feats:

1. First, Jeffreys had shown how a *unique* prior could be developed without recourse to arbitrary reasoning.
2. Second, Jeffreys' prior was *not constant* and thus different from that prescribed by Bayes' postulate.

Jeffreys' 1932 paper naturally did not go unnoticed by Fisher who, as we explained in Section 5.5.1, held very strong anti-Bayesian views. Fisher also probably saw an opportunity to demonstrate the superiority of his own fiducial method in solving the problem Jeffreys had tackled. In a paper published in the same year and purporting to criticize Jeffreys' method, Fisher started by writing:

> Jeffreys considers the question: What distribution *a priori* should be assumed for the value of h, regarding it as a variate varying from population to population of the ensemble of populations which might have been sampled? (Fisher, 1933, p. 343)

The reader can at once see from the above that Fisher either misunderstood the problem or purposely chose to transcribe it in his own frequentist terms. This is because, in Jeffreys' problem, the prior distribution of h was not the result of variation across an ensemble of populations, rather it represented a state of prior belief about h.

Given the way in which Jeffreys' problem was represented by Fisher, there was no hope for the latter to convince the former of anything and vice versa. As we shall now see, this is exactly what eventually happened.

After having given *his* version of Jeffreys' problem, Fisher severely criticized the latter's derivation of the prior in (5.110):

> That there should be a method of evolving such a piece of information by mathematical reasoning only, without recourse to observational material, would be in all respects remarkable, especially since the same principle of reasoning should, presumably, be applicable to obtain the distribution a prior of other statistical parameters. The proof can, however, scarcely in any case establish all that is claimed. (*ibidem*)

Fisher then gave an example from "an artificially constructed series of populations" for which the prior of h was $f_h = \alpha e^{-\alpha h}$, where α is a constant. In this case, "Jeffreys's reasoning would certainly lead to a false conclusion." Fisher's aim here was to provide a counterexample to Jeffreys' claim to have derived $f_h \propto 1/h$ as the *unique* prior for h.

Fisher was now ready to explain what he thought was wrong with Jeffreys' reasoning:

> ...we may note at once that, for any particular population, the probability will generally be larger when the first two observations are far apart than when they are near together. This is important since, as will be seen, the fallacy of Jeffreys's argument consists just in assuming that the probability shall be 1/3, independently of the distance apart of the first two observation. (*ibid*., p. 344)

However, as both Bartlett (1933, p. 526) and Lane (1980, p. 153) have observed, the probability that Fisher considered in the last quotation and which he thought Jeffreys had used is

$$\Pr\{-a \le x_3 \le a \mid a, h, x\},$$

where $\pm a$ and x_3 are the first three observations, h is the precision, and x is the true value of the observations. But in fact Jeffreys had considered the probability

$$\Pr\{-a \le x_3 \le a \mid a\}. \tag{5.111}$$

Fisher also argued that the true probability that should have been calculated was

$$\Pr\{-a \le x_3 \le a \mid h, x\}. \tag{5.112}$$

Fisher showed that the last probability was 1/3, irrespective of the prior for h and x. His demonstration was as follows.

Fisher considered a population with mean μ and precision h. Then, the joint density of the first two observations x_1 and x_2 is

$$\frac{h^2}{\pi}\exp\left[-h^2\left\{(x_1-\mu)^2+(x_2-\mu)^2\right\}\right]$$

(Fisher, 1933, p. 345). If the larger of the observations is written as $u+v$ and the smaller is written as $u-v$, the previous joint density becomes

$$\frac{2h^2}{\pi}\exp\left[-2h^2\left\{(u-\mu)^2+v^2\right\}\right].$$

Assuming the prior of h is f_h, the joint density of u, v, and h is

$$\frac{2h^2 f_h}{\pi}\exp\left[-2h^2\left\{(u-\mu)^2+v^2\right\}\right]$$

and that u, v, h, and the third observation x_3 is

$$\frac{2h^3 f_h}{\pi^{3/2}}\exp\left[-2h^2\left\{(u-\mu)^2+v^2\right\}-h^2(x_3-\mu)^2\right].$$

Fisher next wrote $x_3=u+c$ so that c determined whether the third observation lied between the first two. Integrating with respect to u from $-\infty$ to ∞, he obtained

$$\frac{2h^2 f_h}{\pi\sqrt{3}}\exp\left(-2h^2v^2-\frac{2}{3}h^2c^2\right) \tag{5.113}$$

which corresponds to Jeffreys' (5.107). Fisher noted that "μ, the mean of the population, has disappeared, showing that its value, and therefore its distribution, a priori is irrelevant." Now, $-v<c<v$ is the event that the third observation lies between the first two. The probability of the latter event, for a given v, can be obtained by integrating Eq. (5.113) over c from $-v$ to v and then integrating over h from 0 to ∞. Fisher wrote the first integral as

$$\frac{2hf_h}{\sqrt{2\pi}}\alpha\left(hv\sqrt{\frac{2}{3}}\right)e^{-2h^2v^2},$$

where

$$\alpha(x)=\frac{1}{\sqrt{\pi}}\int_{-x}^{x}e^{-u^2}du.$$

On the other hand the integral of Eq. (5.113) over c from $-\infty$ to ∞ is

$$\frac{2hf_h}{\sqrt{2\pi}}e^{-2h^2v^2}.$$

Fisher was thus able to obtain Jeffreys' Eq. (5.109) in the form

$$\int_0^\infty \frac{2hf_h}{\sqrt{2\pi}}\alpha\left(hv\sqrt{\frac{2}{3}}\right)e^{-2h^2v^2}dh=\frac{1}{3}\int_0^\infty\frac{2hf_h}{\sqrt{2\pi}}e^{-2h^2v^2}dh. \tag{5.114}$$

Now Fisher stated that it was not for a particular v but "*on average for all values of* v,* that the probability is exactly one-third." He therefore proceeded to integrate the *integrands* on both sides of Eq. (5.114) over v from 0 to ∞ and obtained

$$\frac{f_h}{3} = \frac{1}{3} f_h.$$

Therefore, "the probability is just one-third is assured, irrespective of h, and therefore for every frequency element of that variate independently" (*ibid.*, p. 346). Fisher also pointed out that Jeffreys obtained $f_h \propto 1/h$ only because in the latter case the integrals in Eq. (5.114), which are over h, coincided with those over v, which is what Fisher used. This completes Fisher's demonstration.

However, as we mentioned earlier, Fisher overlooked the fact that his computed probabilities were of the type (5.112), whereas the probabilities Jeffreys was interested in were of the type (5.111). By the same token, Fisher's analysis, unlike Jeffreys', was strictly non-Bayesian since he did not actually integrate with respect to any of the parameters (i.e., the mean and the precision).

In the final part of his paper, Fisher considered the fiducial distribution of σ. He noted that the latter distribution coincided with the posterior distribution of σ using Jeffreys' prior:

> ...For this particular distribution *a priori* makes the statements of inverse and of fiducial probability numerically the same, and so allows their logical distinctness to be slurred over. (*ibid.*, p. 348)

Concerning fiducial probability, Fisher was quick to point out that

> ...Probability statements of this type are logically entirely distinct from inverse probability statements, and remain true whatever the distribution a priori of σ may actually be... The probabilities differ in referring to different populations; that of the fiducial probability is the population of all possible random samples, that of the inverse probability is a group of samples selected to resemble that actually observed. (*ibid.*, pp. 347–348)

Having seen his previous paper misrepresented by Fisher, it came as no surprise when Jeffreys wrote a strong rebuttal of the points raised by the latter (Jeffreys, 1933). Jeffreys saw that his disagreement with Fisher was fundamentally about their different concepts of probability. His paper was therefore more or less a defense of why probability should be only logically interpreted. He started with this clarification:

> ...By "probability" I mean probability and not frequency, as Fisher seems to think, seeing that he introduces the latter word in restating my argument. (*ibid.*, p. 523)

He maintained that probability had a fundamental role in scientific inference:

> I believe that the correct solution is in terms of probability; that our laws themselves, and further observational results obtainable from them, are never proved, what is proved being that they have definite degrees of reliability on the knowledge available. (*ibid.*, p. 524)

* Italics are Fisher's.

But he insisted that all inductive inference should be based on some *a priori* assumption:

> *There is therefore no possibility of drawing any inductive inference without an a priori assumption, and if any method appears to avoid such an assumption it must be either erroneous in principle or involve some a priori assumption that the author has not stated and possibly has not noticed.** (*ibid.*, p. 525)

Jeffreys maintained that the fundamental rule for eliciting an *a priori* distribution was the principle of indifference ("non-sufficient reason"):

> ...according to which propositions mutually exclusive on the same data must receive equal probabilities if there is nothing to enable us to choose between them. (*ibid.*, p. 528)

He further stated that an *a priori* distribution must satisfy certain conditions and can be tested in two ways:

> ...We may consider it as such, examining what distribution describes our *a priori* knowledge; or we may investigate what distribution is consistent with facts otherwise known about the posterior probability on certain types of data. (*ibid.*, p. 530)

Indeed, as we have seen, these were the two ways Jeffreys employed to derive the prior distribution in Eq. (5.106). It is important to note that the prior here is not completely subjective, rather it is the only one that can be admitted by a rational person:

> ...*The fact is that my distribution is the only distribution of prior probability that is consistent with complete previous ignorance of the value of h.*† (*ibid.*, p. 531)

Such an attitude is in sharp contrast to de Finetti's subjectivism (see Section 6.3.2), whereby two people with the same information could invoke different priors. For Jeffreys, probability should be logically and objectively derived. He explained its logical interpretation as follows:

> ...We introduce the idea of a relation between one proposition p and another proposition (or aggregate of propositions) q, expressing the degree of knowledge concerning p provided by q. This may amount to implication (q implies p) or inconsistency (q implies not-p), as extreme cases. This relation is called "the probability of p given q." The fundamental assumption of the theory is that probabilities are comparable in terms of the relations "more probable than" and "less probable than," which are transitive. With this assumption we can proceed, by a series of conventions, involving no further hypothesis, to the measurement of probabilities by numbers from 0 to 1 and to the fundamental laws connecting probabilities. From these the formal development follows deductively. The principle of Inverse Probability, which Fisher and others' habitually mention as an unwarranted hypothesis, becomes on this view a theorem. (*ibid.*, pp. 527–528)

* Italics are Jeffreys'.

† Italics are Jeffreys'.

It was through this concept of probability that Jeffreys proposed to resolve the various paradoxes, discussed in Section 1.3.4, associated with the principle of indifference. Indeed, Fisher had been particularly critical of this principle. For example, Jeffreys considered a problem discussed in Keynes' book *A Treatise on Probability*:

> ...If, to take an example, we have no information whatever as to the area or population of the countries of the world, a man is as likely to be an inhabitant of Great Britain as of France, there being no reason to prefer one alternative to the other. He is also as likely to be an inhabitant of Ireland as of France. And on the same principle he is as likely to be an inhabitant of the British Isles as of France. (Keynes, 1921, p. 44)

Keynes had pointed out that this led to a contradiction. For the three statements in the last quotation can be written as follows:

$$\begin{aligned} \Pr\{GB\} &= \Pr\{F\}, \\ \Pr\{I\} &= \Pr\{F\}, \\ \Pr\{BI\} &= \Pr\{F\}, \end{aligned}$$

where GB is the event that a person lives in Great Britain and so on. But

$$\begin{aligned} \Pr\{BI\} &= \Pr\{GB\} + \Pr\{I\} \\ &= \Pr\{F\} + \Pr\{F\} \\ &= 2\Pr\{F\}, \end{aligned}$$

which contradicts the third statement that $\Pr\{BI\} = \Pr\{F\}$. However, Jeffreys argued that there was no contradiction and that the fallacy in Keynes' reasoning consisted of:

> ...supposing that the probability of a proportion is function of that proposition and nothing else, instead of an expression of our state of knowledge relative to particular data. (Jeffreys, 1933, p. 529)

Thus, if a third person had a state of knowledge d that Great Britain, Ireland, and France are three different countries, then

$$\Pr\{GB \mid d\} = \Pr\{I \mid d\} = \Pr\{F \mid d\},$$

so that

$$\Pr\{BI \mid d\} = \Pr\{GB \mid d\} + \Pr\{I \mid d\} = 2\Pr\{F \mid d\}. \tag{5.115}$$

On the other hand, if the third person had a state of knowledge d' that the British Isles is one country, of which Great Britain and Ireland are divisions, then

$$\Pr\{GB \mid d'\} = \Pr\{I \mid d'\} = \frac{1}{2}\Pr\{BI \mid d'\}$$

and

$$\Pr\{BI \mid d'\} = \Pr\{F \mid d'\}. \tag{5.116}$$

Therefore, according to Jeffreys, there was no contradiction between Eqs. (5.115) and (5.116) since they referred to conditioning on different states of knowledge d and d'.

After having defended his logical interpretation and use of probability, Jeffreys was now ready to attack the points made by Fisher in his 1933 paper (Fisher, 1933, p. 343). Consistent with his logical use of probability as illustrated in Eqs. (5.115) and (5.116), his first counterpoint was:

> Fisher considers that it is an objection to my theory that with different assumptions I should have got a different answer, and thereby misses the entire point of the theory. Had I begun with different assumptions I should have been discussing a different problem. (Jeffreys, 1933, p. 530)

Concerning Fisher's counterexample that $f_h = \alpha e^{-\alpha h}$ could be a plausible prior of h derived from "an artificially constructed series of populations," thus contradicting Jeffreys' $f_h \propto 1/h$, the latter pointed that $f_h = \alpha e^{-\alpha h}$ would imply previous knowledge that the probability was $1/e$ that h exceeds the value $1/a$. This, Jeffreys said, would be inconsistent with the search for an appropriate form for f_h: "if it is known, it is known, and there is no more to be said."

Fisher had also cynically qualified Jeffreys search for the form of f_h by mathematical reasoning only as "remarkable." To this, Jeffreys retorted that not only Fisher was unaware that he had himself used several *a priori* hypotheses in constructing a theory of statistical inference but also that he had got one particular *a priori* hypothesis wrong! Here Jeffreys alluded to Fisher's definition of probability, as used in the latter's foundations paper of 1922:

> ...when we say that the probability of throwing a five with a die is one-sixth, we must not be taken to mean that of any six throws with that die one and one only will necessarily be a five; or that of any six million throws, exactly one million will be fives; but that of a hypothetical population of an infinite number of throws, with the die in its original condition, exactly one-sixth will be fives. (Fisher, 1922c, p. 312)

Concerning the latter definition, this is what Jeffreys had to say:

> ...That a mathematician of Dr. Fisher's ability should commit himself to the statement that the ratio of two infinite numbers has an exact value can only be regarded as astonishing. (Jeffreys, 1933, p. 533)

Fisher had also claimed that fiducial probability referred to the "populations of all possible samples" while inverse probability referred to "a group of samples selected to resemble that actually observed." Jeffreys considered this to be in fact a drawback of the fiducial argument:

> ...it may be asked why this [i.e. taking all possible samples] should be thought to have much relevance to any particular sample. (*ibid.*, p. 534)

We make two final points on Jeffreys' rejoinder to Fisher. First, throughout his paper Jeffreys referred to *his* concept of probability as *the* theory of probability. We thus see that, if Fisher was adamant about interpreting probability in frequentist terms, Jeffreys was equally unyielding in casting it in logical terms. Second, although Jeffreys was notoriously poor at articulation, his writing was crisp and clear. This was in contrast to Fisher's often condensed and abstruse style of writing.

However, Jeffreys' forceful paper did not prevent Fisher from submitting a second rebuttal on December 14, 1933, to the journal *Proceedings of the Royal Society of London*. Seeing that there was no chance of a resolution of the issue, the Mathematical Committee of the Royal Society wrote to Fisher on February 1934. The Committee asked Fisher to submit his rejoinder together with a final rejoinder of the same length from Jeffreys, but only after each had seen the other's manuscript. Thereupon, Fisher wrote to Jeffreys and both men exchanged cordial letters until Jeffreys submitted both papers to the Royal Society on April 27, 1934. They were printed together in the *Proceedings*.

Fisher's paper, entitled "Probability, likelihood, and quantity of information in the logic of uncertain inference" (Fisher, 1934a), contained several criticisms of Jeffreys' previous paper. First, Fisher's severely criticized Jeffreys' definition of probability.

> Jeffreys' definition of probability is subjective and psychological…it resembles the more expressive phrase used by Keynes, "the degree of rational belief"…Obviously no mathematical theory can really be based on such verbal statements. Any such theory which purports to be based upon them must in reality be derived from the supplementary assumptions and definitions subsequently introduced, a "series of conventions, involving no further hypotheses," in Jeffreys' explanatory phrase. (*ibid*., p. 3)

Second, Fisher pointed that Jeffreys' logical probability suffered from

> …the difficulty of establishing the laws of mathematical probability, without basing the notion of probability, on the concept of frequency, for which these laws are really true, and from which they were originally derived. (*ibidem*)

Fisher's third point was that Jeffreys'

> …rejection also of frequency as an observational measure of probability…makes it impossible for any of his deductions to be verified experimentally. (*ibid*., p. 4)

Fisher's fourth criticism of Jeffreys' concept of probability concerned the latter's adoption of the principle of indifference as a fundamental rule. Fisher found fault with Jeffreys' resolution of the paradox raised by Keynes concerning the inhabitants of the British Isles, etc. Jeffrey's resolution rested on the verbal use of the word "country" and ignored Keynes' admonishment that:

> …It is not plausible to maintain, when we are considering the relative populations of different areas, that the number of names of sub-divisions which are within our knowledge, is, in the absence of any evidence as to their size, a piece of relevant evidence. (Keynes, 1921, p. 44)

The point in the above is that a principle for which just knowing the number of names in a subdivision affects our probability cannot be valid. Therefore, Fisher maintained that the principle of indifference "leads to inconsistencies which seem to be ineradicable." In summary, it was Jeffreys' definition of probability, not his, that was defective.

After having rebuked Jeffreys' definition of probability, Fisher complained that

> ...It is difficult to understand the difficulty expressed by Jeffreys as to the definition of probability, when incommensurable, as the limit of the ratio of two numbers, when these both become infinite or increase without limit. (Fisher, 1934a, p. 7)

Thereupon, Fisher further elaborated on his concept of an infinite hypothetical population. All sampling properties of the latter could be obtained as limits of those of its finite counterparts. However, Fisher pointed that frequency ratios *did* exist for his infinite hypothetical population model because all repetitions were done under identical conditions and were independent of each other. This was in contrast to a limiting ratio for *an experimental procedure* whereby the existence of a limit was not guaranteed.

In the final section of Fisher's paper, he provided a statistical alternative to Jeffreys' use of the degree of knowledge provide by q concerning p in his definition of logical probability. Fisher pointed to his own work where he had used the concept of information calculated from the likelihood function. He disagreed with Jeffreys that probability was the fundamental quantity to be used in scientific inference:

> ...it is now too late, in view of what has already been done in the mathematics of inductive reasoning, to accept the assumption that a single quantity, whether "probability" or some other word be used to name it, can provide a measure of "degree of knowledge" in all cases in which uncertain inference is possible. (*ibid.*, p. 8)

The second paper in the joint submission was Jeffreys' paper, entitled "Probability and the scientific method" (Jeffreys, 1934). This was essentially an argument for the necessity of an *a priori* element in scientific inference. To Fisher's first point that Jeffreys' definition of probability was defective, the latter explained that he had:

> ...originally [been] somewhat attracted by the wish to define probability in terms of frequency, but found that the existing theory of Venn failed in its objects. It avoided no *a priori* hypothesis, several having been used but not stated, and its results, when interpreted in terms of the definition, were not in a practically applicable form. (*ibid.*, p. 9)

Concerning Fisher's second point that it was difficult to establish the laws of probability using Jeffreys' logical interpretation, the latter replied that

> ...It is not claimed that either the laws of probability or the assessment of prior probabilities can be proved, either by logic or by experiment. If they could be proved, the belief that inductive inference is meaningless would be disproved, and this seems impossible. All we can say is that we do believe that inference is possible, and therefore take this as an axiom. It is a question whether a self-consistent theory can be constructed with a given set of postulates, but enough seems to have been done now to entitle us to answer this question in the affirmative. (*ibid.*, pp. 10–11)

Fisher's third point had been that Jeffreys' probability was such that it was impossible for any of its deductions to be experimentally verified. To this, Jeffreys explained:

...Prior probabilities could logically be assigned in any way; they must in practice be assigned to correspond as closely as possible to our actual state of knowledge, and in such a way that the sort of general laws that we, in fact, consider capable of being established can acquire high probabilities as a result of sufficient experimental verification. (*ibid*., p. 11)

Regarding Fisher's fourth criticism that Jeffreys had ignored Keynes' admonishment that the principle of indifference could not be defended if it could be affected merely by some verbal usage, Jeffreys pleaded:

With regard to the quotation from Keynes, I did not overlook Keynes's answer. I do not agree with it. If the word "country" has any intelligible meaning, I think it is a very relevant piece of information whether Great Britain and Ireland are two different countries or two parts of the same country. (*ibid*., p. 14)

He further continued to maintain that Fisher's use of an infinite hypothetical population to define probability implied that "no probability would have a definite value."

The Fisher–Jeffreys debate thus ended, with each man vindicating his own concept of probability but without having been able to convince the other. However, the exchange was not totally futile for two reasons. First, it forced each man to further examine and clarify his own concept of probability. Second, it compelled each one to consider an alternative and perhaps viable viewpoint on such a fundamental concept as probability. It is noteworthy that, despite the sometimes heated exchange between the two men in their papers, they maintained a cordial relationship in their private correspondence and exchanged friendly letters years after their debate on probability. It seems that Jeffreys even regretted the incident. In his major book, *Theory of Probability*,* this is what he had to say:

...The general agreement between Professor R.A. Fisher and myself has been indicated already in many places. The apparent differences have been much exaggerated owing to a rather unfortunate discussion some years ago, which was full of misunderstandings on both sides. Fisher thought that a prior probability based on ignorance was meant to be a statement of a known frequency, whereas it was meant merely to be a formal way of stating that ignorance, and I had been insisting for several years that no probability is simply a frequency. I thought that he was attacking the "Student" rule, of which my result for the general least squares problem was an extension; at the time, to my regret, I had not read "Student's" papers and it was not till considerably later that I saw the intimate relation between his methods and mine. This discussion no longer, in my opinion, needs any attention.... (Jeffreys, 1939, p. 323)

As for Fisher, even during his debate with Jeffreys, he had privately admitted (in a letter dated February 26, 1934) to the latter that:

...I am glad you think the differences in our points of view do not go deep as one might judge. (Bennett, 1990, p. 151)

*This is the same book referring to which Fisher reportedly said about Jeffreys: "He [Jeffreys] makes a logical mistake [of adopting Bayes' postulate] on the first page which invalidates all the 395 formulae in his book" (Box, 1978, p. 441).

In fact, Fisher's view on probability underwent some change in later years. One of his major griefs was the repeated sampling (frequentist) interpretation of probability that had been put forward by Neyman and Pearson in their treatment of hypothesis testing and confidence intervals (see Section 5.6.4.2). Fisher condemned this approach in no uncertain terms:

> The operative properties of an acceptance procedure [i.e. Neyman-Pearson hypothesis testing]...are ascertained practically or conceptually by applying it to a series of successive similar samples from the same source of supply, and determining the frequencies of the various possible results...The root of the difficulty of carrying over the idea from the field of acceptance procedures to that of tests of significance is that, where acceptance procedures are appropriate, the source of supply has an objective reality ... whereas if we possess a unique sample in student's sense on which significance tests are to be performed, there is always...a multiplicity of populations to each of which we can legitimately regard our sample as belonging; so that the phrase "repeated sampling from the same population" does not enable us to determine which population is to be used to define the probability level, *for no one of them has objective reality, all being products of the statistician's imagination.** (Fisher, 1955, pp. 70–71)

What transpires from the above, even more than his criticism of the Neyman–Pearson approach, is Fisher's closeness to Jeffreys' view on probability (a view he had debated in the 1930s!).

5.6 FISHER VERSUS NEYMAN–PEARSON: CLASH OF THE TITANS

Few disputes in the history of statistics have been as acrimonious and protracted as the one opposing Ronald Fisher (1890–1962) on one side, and Jerzy Neyman (1894–1981) and Egon Pearson (1885–1980) on the other side. The dispute between Fisher and Neyman was even more vitriolic and continued well after the latter had moved to the United States in the late 30s. The ongoing fight caught the attention of the entire statistical community, not only because it involved three of the greatest statisticians of the time but also because it touched the philosophy of statistical inference at its very core: how should statistical hypotheses be tested? Fisher advocated an approach based on *inductive reasoning (or inference)* while Neyman favored *inductive behavior*. We now examine these topics.

5.6.1 The Neyman-Pearson Collaboration

5.6.1.1 The Creation of a New Paradigm for Hypothesis Testing in 1926 Neyman[†] was born Jerzy Splawa-Neyman (Fig. 5.8) from Polish parents on April 16, 1894, in Bendery, Moldavia, then part of Russia. Neyman studied mathematics at the University of

* Italics are ours.

[†] Neyman's biography has been written by Constance Reid (1998). His life is also described in Fienberg and Tanur (2001) and Lehmann (1994).

FIGURE 5.8 Jerzy Neyman (1894–1981). Wikimedia Commons (Licensed under the Creative Commons Attribution-Share Alike 2.0 Germany license), http://commons.wikimedia.org/wiki/File:Jerzy_Neyman2.jpg

Kharkov from 1912 to 1917. His instructors included the famous probabilist Sergei Bernstein (1880–1968) and his earliest papers were in the field of measure theory. He received his Master's degree in 1920 and then became a lecturer. In the summer of 1921, Neyman went to Bydgoszcz in northern Poland, as part of an exchange program between Russia and Poland, and continued his mathematical studies culminating in a doctoral dissertation in 1923 on statistical problems in agricultural experimentation. Owing to his interest in statistics, Neyman obtained a postdoctoral position under Karl Pearson at University College London, in 1925. However, the next several years were difficult for both him and his wife owing to constant financial hardships.

A key member of Karl Pearson's laboratory was his son Egon Sharpe Pearson (Fig. 5.9) who had moved in his father's Department of Applied Statistics in 1921. Egon obtained his degree in mathematics from Cambridge in 1919 and was still a teaching assistant in 1925 when Neyman came to University College. Whereas Neyman was at the time primarily interested in Karl Pearson's work, especially the latter's *Grammar of Science* (Pearson, 1892), Egon was attracted by the modern revolution in statistics that was being single-handedly ushered by Fisher, especially the latter's recently published book *Statistical Methods for Research Workers* (Fisher, 1925b). Until now, Egon had been trained by his father in large-sample statistics (which Egon called "Mark I" statistics), much of which has been discovered by Karl Pearson himself. However, Fisher's publications and book turned Egon's attention to small-sample statistics (which he called "Mark II" statistics):

> ... in 1925–6, I was in a state of puzzlement, and realized that, if I was to continue an academic career as a mathematical statistician, I must construct for myself what might be termed a statistical philosophy, which would have to combine what I accepted from K. P.'s large sample tradition with the newer ideas of Fisher. (Pearson et al., 1990, p. 77)

An apparently innocent statement made by his father caused Egon to probe further. In 1924, E.C. Rhodes read a paper, "On the problem whether two given samples can be supposed to have been drawn from the same population" (Rhodes, 1924) in which he (wrongly) argued that three tests with degrees of freedom 1, 2, and 3 all yielded the same expression for Pearson's X^2 test statistic. In his reply to this article, Karl Pearson said:

> ...I think it extremely likely that different tests may give different results. The statistician has to apply of such tests a number reasonable for the purpose he has in hand; and will, I should say, always be guided in rejected or accepting common origin by the most stringent of these tests. (Pearson, 1924, p. 249)

By "most stringest" in the above, Karl Pearson was alluding to the test with the smallest p-value. However, this recommendation was not quite acceptable to Egon who wished to find:

> ...a more logical principle to follow in choice among alternative statistical tests. (Pearson et al., 1990, p. 77)

Thereupon, on April 7, 1926, Egon sought the advice of Student and wrote him a letter to enquire about the logical basis of the Student's ratio. In his reply, Student made

FIGURE 5.9 Egon Sharpe Pearson (1885–1980). Front photo in Bartlett (1981). Courtesy of Oxford University Press

some statements that can be viewed as the building blocks of what later came to be known as the Neyman–Pearson (NP) theory:

> In your large samples with a known normal distribution you are able to find the chance that the mean of a random sample will lie at any given distance from the mean of the population. (Personally I'm inclined to think your cases are best considered as mine to the limit n large.) That doesn't in itself necessarily prove that the sample is not drawn randomly from the population even if the chance is very small, say 0.00001: What it does is to show that if there is any alternative hypothesis which will explain the occurrence of the sample with a more reasonable probability, say 0.05 (such as that it belongs to a different population or that the sample wasn't random or whatever will do the trick) you will be very much more inclined to consider that the original hypothesis is not true. (Pearson, 1966, p. 7)

In the above, for the first time Student explicitly considered the notion of an *alternative hypothesis* in addition to the original (null) hypothesis and also addressed the issue of the *probabilities of the observed sample under the two hypotheses (or likelihoods of the two hypotheses)*. As we shall see, the notions of an alternative hypothesis and of relative likelihoods are the key ingredients of the NP theory of hypothesis testing. Moreover, none of them form part of Fisher's theory of significance testing.

Egon was very inspired by Student's reply, and from that point his ideas on the likelihood ratio (LR) started to take shape. However, he needed somebody with higher mathematical abilities than himself to further explore these ideas. Soon, Egon decided that Neyman would be the right person* probably because:

> ...Neyman was so "fresh" to statistics. The other English statisticians would all have had preferences between Mark I and Mark II statistics. (Reid, 1998, p. 62)

In late November 1926, Egon sent a long series of notes on his LR ideas to Neyman while the latter was in Paris attending the lectures of Émile Borel, Henry Lebesgue, Jacques Hadamard, and Paul Lévy. Neyman was quite receptive to Egon's ideas but his response of December 9, 1926, indicates that he completely missed the point of the letter:

> I think that to have the possibility of testing it is necessary to adopte [*sic*] such a principle as Students, but it seems to me also that this principle is equivalent to the principle leading to the inverse probabilities and that it must be generalised. I think we must have curage [*sic*] enough to adopte this principle, because without it, every test seems to be impossible. (*ibid.*, p. 70)

In the above, Neyman referred to the principle of inverse probability and stated that, unless the principle was adopted, every test would be impossible. However, as he soon realized, with LR none of that was needed.

In the spring of 1927, Egon visited Neyman in France to start preparing a paper. The latter was published in *Biometrika* in 1928 and was entitled "On the use and interpretation of certain test criteria for purposes of statistical inference: Part I" (Neyman and Pearson, 1928a). Neyman was the first author of the paper but the fundamental ideas in it were mainly Egon's.

*The story of the Neyman–Pearson collaboration is also described in Reid (1998, pp. 65–119) and Pearson (1966).

The 1928 paper (Fig. 5.10) started by what may be described as a sketch of Fisher's procedure for significance testing:

> One of the most common as well as most important problems which arise in the interpretation of statistical results, is that of deciding whether or not a particular sample may be judged as likely to have been randomly drawn from a certain population, whose form may be either completely or only partially specified. We may term Hypothesis A the hypothesis that the population from which the sample Σ has been randomly drawn is that specified, namely Π. In general the method of procedure is to apply certain tests or criteria, the results of which will enable the investigator to decide with a greater or less degree of confidence whether to accept or reject Hypothesis A, or, as is often the case, will show him that further data are required before a decision can be reached. (*ibid.*, p. 175)

However, in the next page, there is a departure from Fisher by explicitly introducing the notion of an alternative hypothesis:

> …however strong may be the *à priori* evidence in favour of Π, there would be no problem at all to answer if we were not prepared to consider the possibility of alternative hypotheses as to the population sampled. (*ibid.*, p. 176)

Neyman and Pearson next made the concept of an alternative hypothesis even more explicit and also introduced the notions of two types of errors:

> Σ must either have been drawn randomly from Π or from Π', where the latter is some other population which may have any one of an infinite variety of forms differing only slightly or very greatly from Π. The nature of the problem is such that it is impossible to find criteria which will distinguish exactly between these alternatives, and whatever method we adopt two sources of error must arise:
>
> (1) Sometimes, when Hypothesis A is rejected, Σ will in fact have been drawn from Π.
> (2) More often, in accepting Hypothesis A, Σ will really have been drawn from Π'. (*ibid.*, p. 177)

These two types of errors could not be completely eliminated; instead they had to be controlled:

> In the long run of statistical experience the frequency of the first source of error (or in a single instance its probability) can be controlled by choosing as a discriminating contour, one outside which the frequency of occurrence of samples from Π is very small-say, 5 in 100 or 5 in 1000…
>
> The second source of error is more difficult to control, but if wrong judgments cannot be avoided, their seriousness will at any rate be diminished if on the whole Hypothesis A is wrongly accepted only in cases where the true sampled population, Π', differs but slightly from Π. It is not of course possible to determine Π', but making use of some clearly defined conception of probability we may determine a "probable" or "likely" form of it, and hence fix the contours so that in moving "inwards" across them the difference between Π and the population from which it is "most likely" that Σ has been sampled should become less and less. This choice also implies that on moving "outwards" across the contours, other hypotheses as to the population sampled become more and more likely than Hypothesis A. (*ibidem*)

ON THE USE AND INTERPRETATION OF CERTAIN TEST CRITERIA FOR PURPOSES OF STATISTICAL INFERENCE. PART I.

By J. NEYMAN, Ph.D. and E. S. PEARSON, D.Sc.

CONTENTS.

I. Introductory.

One of the most common as well as most important problems which arise in the interpretation of statistical results, is that of deciding whether or not a particular sample may be judged as likely to have been randomly drawn from a certain population, whose form may be either completely or only partially specified. We may term Hypothesis A the hypothesis that the population from which the sample Σ has been randomly drawn is that specified, namely Π. In general the method of procedure is to apply certain tests or criteria, the results of which will enable the investigator to decide with a greater or less degree of confidence whether to accept or reject Hypothesis A, or, as is often the case, will show him that further data are required before a decision can be reached. At first sight the problem may be thought to be a simple one, but upon fuller examination one

FIGURE 5.10 First page of Neyman and Pearson's 1928 paper (1928a, p. 175). Courtesy of Oxford University Press

We note that the two types of errors (later dubbed "Type I" and "Type II" errors) formed an integral part of the NP testing procedure. These two errors were interpreted as frequencies in "the long run of statistical experience." In contrast, neither type explicitly existed in Fisher's testing procedure.* This is because Fisher did not view hypothesis testing as a *decision* procedure to accept one of Π and Π′ such that each decision carried its own probability of error. Rather, for Fisher, hypothesis testing was an *inferential* procedure whereby the sample provided a probability value (p-value), which measured its degree of support of the hypothesis being tested.

Neyman and Pearson now introduced the LR, which they called the "criterion of likelihood":

> Suppose there to be two hypotheses regarding the population from which a given sample has been drawn; there will be one density field for Hypothesis A corresponding to Π, and another for Hypothesis A′ corresponding to Π′…we measure the chance of drawing the n variates $[x_1, x_2, \ldots, x_n]$ in the range
>
> $$x_1 \pm \frac{1}{2}h, \ldots, x_n \pm \frac{1}{2}h,$$
>
> by $$f(x_1)f(x_2)\ldots f(x_n)h^n \qquad \text{(vx)}$$
>
> …We shall therefore take the ratio of two expressions such as (xv), with a common h, that is to say the ratio of the point densities at P in the two fields, as a measure for the comparison of Hypotheses A and A′. (*ibid*., pp. 184–185)

As an application of LR, Neyman and Pearson considered the example of testing the hypothesis A that the mean and standard deviation of a normal population are known to be a and σ, respectively, against the alternative that both of these values (a' and σ') are unknown. Suppose the random sample $X_1, X_2, \ldots, X_n$ has mean $\bar{X}$ and variance $s^2 = \frac{1}{n}\sum_i (X_i - \bar{X})^2$. Then the likelihood of a' and σ' is

$$D = \frac{1}{\left(\sqrt{2\pi}\sigma'\right)^n} \exp\left\{\frac{1}{2\sigma^2}\sum_i \left(X_i - \bar{X}\right)^2\right\} = \frac{1}{\left(\sqrt{2\pi}\sigma'\right)^n} e^{-\frac{n}{2\sigma'^2}\left\{\left(\bar{X}-a'\right)^2 + s^2\right\}}. \qquad (5.117)$$

To find the Π′ for which D is maximum, we solve

$$\frac{\partial D}{\partial a'} = 0 \;\Rightarrow\; a' = \bar{X},$$

$$\frac{\partial D}{\partial \sigma'} = 0 \;\Rightarrow\; \frac{n}{\left(\sigma'\right)^{n+1}} + \frac{n}{\sigma'^n}\frac{\left\{\left(\bar{X}-a'\right)^2 + s^2\right\}}{\sigma'^2} = 0 \;\Rightarrow\; \sigma'^2 = \left(\bar{X}-a'\right)^2 + s^2.$$

*Although the notion of a Type I error is implied.

Substituting $a' = \bar{X}$ from the first equation into the second, we obtain $\sigma' = s$. Hence, the maximum likelihood (ML) is

$$D_M = \frac{1}{\left(\sqrt{2\pi}\sigma'\right)^n} e^{\frac{-n}{2\sigma'^2}\left\{\left(\bar{X}-a'\right)^2 + s^2\right\}} \Bigg|_{a'=\bar{X},\sigma'=s} = \frac{1}{\left(\sqrt{2\pi}s\right)^n} e^{\frac{-n}{2}}. \tag{5.118}$$

Thus, Neyman and Pearson were able to write the LR as

$$\lambda = \frac{\text{Likelihood of } \Pi}{\text{Likelihood of } \Pi'(\max)} = \frac{\dfrac{1}{\left(\sqrt{2\pi}\sigma\right)^n} e^{\frac{-n}{2\sigma^2}\left(m^2+s^2\right)}}{\dfrac{1}{\left(\sqrt{2\pi}s\right)^n} e^{\frac{-n}{2}}} = \left(\frac{s}{\sigma}\right)^n e^{-\frac{n}{2}\left(\frac{m^2+s^2}{\sigma^2}-1\right)} \tag{5.119}$$

(*ibid*., p. 187), where $m = \bar{X} - a$. The above ratio remains constant on a set of contours with equation

$$\frac{\left(m^2+s^2\right)}{\sigma^2} - \log\frac{s^2}{\sigma^2} = 1 - \frac{2}{n}\log\lambda = k\log 10,$$

where k is a constant that determines the particular contour. We note that λ decreases as we move outward from one contour to another. To use the λ-test, the probability level P_λ corresponding to a given λ is calculated by double integrating the joint density of m and s^2 over the region ω, which is located outside the contour defined by the given λ:

$$P_\lambda = \int_\omega \frac{n^{n/2}}{\sqrt{\pi}\,2^{(n-2)/2}\,\Gamma\left(\dfrac{n-1}{2}\right)} S^{n-2} e^{-\frac{n}{2}\left(M^2+S^2\right)} dMdS$$

(*ibid*., pp. 188 and 233), where $M = m/\sigma$ and $S = s/\sigma$. Then:

> …If then we only reject Hypothesis A when P_λ is, let us say, $\leq .01$, we can control the error of form (1) [i.e. the Type I error], while the use of the contours of likelihood will minimise as far as possible the effect of errors of form (2) [i.e. the Type II error]. (*ibid*., p. 188)

In the appendix to their paper, the authors provided tables of P_λ for $n = 3$ to 50.

In the next section of their paper, Neyman and Pearson discussed the testing of the hypothesis B that Σ had been drawn from a population with known mean a and unspecified standard deviation against the alternative that both the mean and standard deviation are unknown. This corresponds to Student's test. However, the authors noted that:

> …B is really a multiple hypothesis concerning the sub-universe of normal populations, $M(\Pi)$, with means at a and with varying standard deviations. It only becomes precise upon definition of the manner in which σ is distributed within this sub-universe, that is to say, upon defining the a priori probability distribution of σ. (*ibid*., p. 189)

Neyman and Pearson thus contemplated using inverse probability to solve Student's problem. This was undoubtedly Neyman's idea since he had initially viewed the problem from the point of inverse probability and was still attached to that principle. On the other hand, Pearson had since the start been of the firm opinion that inverse probability

> ...had been forever discredited by Fisher in his 1922 paper in the *Philosophical Transactions* of the Royal Society. (Reid, 1998, p. 79)

However, Neyman and Pearson then postponed the inverse probability approach to a later section of their paper, since they realized that "the criterion of likelihood will probably be of service" in solving Student's problem. Thereupon, they decided to use the λ-test in Eq. (5.119) *with the numerator* (in addition to the denominator) *also maximized*. To do this, D in Eq. (5.117) is differentiated with respect to σ' only and set to zero, resulting in $\sigma'^2 = m^2 + s^2$ where $m = \bar{X} - a$. By further replacing a' by a, one obtains the ML of Π as

$$D_0 = \frac{1}{(2\pi)^{n/2}} \frac{1}{(m^2 + s^2)^{n/2}} e^{-n/2}.$$

Using the ML under the alternative as given in (5.118), LR becomes

$$\lambda' = \frac{D_0}{D_M} = \left(\frac{s^2}{m^2 + s^2} \right)^{n/2} = (1 + z^2)^{-n/2} \tag{5.120}$$

(Neyman and Pearson, 1928a, p. 190), where $z = m / s$. The λ'-test (with both numerator and denominator maximized) was later called the LR test. We note that the larger the value of z, the smaller the value of λ', and the lower the evidence in favor of hypothesis B. The value of z can thus be used as a criterion for deciding between hypothesis B and the alternative. Note that Student's test under Fisher's framework is based on the same z (and is equal to $z\sqrt{n-1}$):

> Student's test was originally devised to allow for the use of s instead of σ in testing the significance of a deviation in the sample mean. We have above another interpretation of it as a method of testing the sample as a whole, based upon the criterion of likelihood, which will be valid as long as that criterion can be employed. (*ibidem*)

In Section 7 of their paper (*ibid*., p. 192), Neyman and Pearson considered a solution to Student's problem based on inverse probability. As we pointed out before, this approach was undoubtedly contributed by Neyman rather than Egon. Moreover, the approach is no longer used within the NP framework.

The next application of LR of interest is the testing of the hypothesis that two given samples come from the two populations with the same (unknown) mean a and same (unknown) standard deviation σ against the alternative that the (unknown) means are different (a_1 and a_2) but the (unknown) standard deviations are the same (σ'). Then the likelihood of the first hypothesis is maximized when

$$a = \frac{n_1\bar{x}_1 + n_2\bar{x}_2}{n_1 + n_2}, \quad \sigma^2 = \frac{1}{n_1 + n_2} \left\{ n_1 s_1^2 + n_2 s_2^2 + \frac{n_1 n_2}{n_1 + n_2} (\bar{x}_1 - \bar{x}_2)^2 \right\},$$

where $n_i, \bar{x}_i$ and s_i^2 are the sample size, sample mean, and sample variance of sample i ($i=1, 2$). Then the ML of the first hypothesis is

$$L(\max) \propto \frac{1}{\sigma^{n_1+n_2}} e^{-\frac{n_1+n_2}{2}}$$

Similarly, like likelihood of the second (i.e., alternative) hypothesis is maximized when

$$a_1 = \bar{x}_1, \quad a_2 = \bar{x}_2, \quad \sigma'^2 = \frac{n_1 s_1^2 + n_2 s_2^2}{n_1 + n_2}$$

and the ML of the second hypothesis is

$$L'(\max) \propto \frac{1}{(\sigma')^{n_1+n_2}} e^{-\frac{n_1+n_2}{2}}.$$

Therefore, the λ'-test is

$$\frac{L(\max)}{L'(\max)} = \left(\frac{\sigma'}{\sigma}\right)^{n_1+n_2} = \left(1+z^2\right)^{-\frac{n_1+n_2}{2}},$$

where

$$z = \frac{\bar{x}_1 - \bar{x}_2}{\sqrt{n_1 s_1^2 + n_2 s_2^2}} \sqrt{\frac{n_1 n_2}{n_1 + n_2}}$$

(*ibid.*, p. 207). As z increases, the lower the evidence in support of the hypothesis that the two populations have the same means. Therefore, z can be used as a criterion for deciding between the two hypotheses and is also the basis of the test statistic $(= z\sqrt{n_1 + n_2 - 2})$ under Fisher's framework.

Part II of the Neyman–Pearson paper (Neyman and Pearson, 1928b) came out in December of the same year. Here, the authors for the first time drew a distinction between simple and composite hypotheses:

> a simple hypothesis is that Σ has been sampled from a population for which $a = a_1, \sigma = \sigma_1$, while a composite hypothesis is that in the population $a = a_1$, but σ may have any value. (*ibid.*, p. 264)

Note that Neyman and Pearson had previously dealt with a composite hypothesis (Hypothesis B) in Part I when they had solved Student's problem. At that time, they had called Hypothesis B a "multiple hypothesis." In Part II, the notion was now formalized. Moreover, the authors now made it explicit that, while a simple hypothesis required a λ–test of the type in (5.119), a composite hypothesis required a λ'–test of the type in (5.120). They wrote the latter test in new notation as

$$\lambda_1 = \frac{C(\omega_{\max})}{C(\Omega_{\max})} \tag{5.121}$$

(*ibid.*, p. 265), where $C(\omega_{max})$ is the ML of the composite hypothesis being tested and $C(\Omega_{max})$ is the ML of the alternative. The authors recalled that, for Student's problem, $\lambda_1 = (1+z^2)^{-n/2}$ (see Eq. 5.120).

Neyman and Pearson next set out to solve Karl Pearson's X^2 goodness-of-fit problem* as follows. Consider k classes that are observed with frequencies $n_1, n_2, \ldots, n_k$. Then, assuming a multinomial distribution for the observed frequencies, the likelihood function is

$$C = \frac{N!}{n_1! \cdots n_k!}(p_1)^{n_1} \cdots (p_k)^{n_k}$$

(*ibidem*), where $p_1, p_2, \ldots, p_k$ are the probabilities of the classes and $n_1 + n_2 + \cdots + n_k = N$. By writing $m_s = p_s N$ and $\delta n_s = n_s - m_s$, the above becomes

$$C = \frac{N!}{N^N}\frac{(\tilde{m}_1)^{n_1}}{n_1!}\cdots\frac{(\tilde{m}_k)^{n_k}}{n_k!}. \tag{5.122}$$

Assuming the frequencies $n_1, n_2, \ldots, n_k$ are large enough, Stirling's approximation can be applied to the factorials resulting in

$$C = \frac{1}{\left(\sqrt{2\pi}\right)^{k-1}}\sqrt{\frac{N}{\tilde{m}_1 \cdots \tilde{m}_k}}\exp\left\{-\frac{1}{2}\sum_{s=1}^{k}\left(\frac{\delta^2 n_s}{\tilde{m}_s} - \frac{1}{3}\frac{\delta^3 n_s}{\tilde{m}_s^2} + \frac{\delta n_s}{\tilde{m}_s} - \frac{1}{2}\frac{\delta^2 n_s}{\tilde{m}_s^2} + \cdots\right)\right\}.$$

Now, as n_s increases, $\tilde{m}_s$ also increases and all terms in the exponential above, except for the first, tend to zero. The mass function above can then be approximated by the following density function

$$D = D_0 e^{-\frac{1}{2}X^2},$$

where

$$X^2 = \sum_{s=1}^{k}\left(\frac{\delta^2 n_s}{\tilde{m}_s}\right) = \sum_{s=1}^{k}\frac{(n_s - \tilde{m}_s)^2}{\tilde{m}_s}$$

(*ibid.*, p. 266). To test the simple hypothesis that the probability of class s is known to be p_s for $s = 1, \ldots, k$ against the alternative that class probability is unknown, Neyman and Pearson used the λ-test in (5.119) which they re-wrote in new notation as

$$\lambda = \frac{C}{C(\Omega_{max})}. \tag{5.123}$$

* See Section 3.2.2.

Here, C is given by Eq. (5.122) and $C(\Omega_{max})$ is obtained by maximizing C subject to the condition $p_1 + p_2 + \cdots + p_k = 1$. When this is done, we obtain $p_s = n_s / N$, so that

$$C(\Omega_{max}) = \frac{N!}{n_1! \ldots n_k!}\left(\frac{n_1}{N}\right)^{n_1} \cdots \left(\frac{n_k}{N}\right)^{n_k}.$$

Therefore,

$$\lambda = \frac{C}{C(\Omega_{max})} = \left(\frac{\tilde{m}_1}{n_1}\right)^{n_1} \cdots \left(\frac{\tilde{m}_k}{n_k}\right)^{n_k}$$

and

$$\begin{aligned}
\log \lambda &= -n_1 \log\left(\frac{n_1}{\tilde{m}_1}\right) - \cdots - n_k \log\left(\frac{n_k}{\tilde{m}_k}\right) \\
&= -n_1 \log\left(1 + \frac{\delta n_1}{\tilde{m}_1}\right) - \cdots - n_k \log\left(1 + \frac{\delta n_k}{\tilde{m}_k}\right) \\
&= -\sum_{s=1}^{k} (\tilde{m}_s + \delta n_s)\left\{\frac{\delta n_s}{\tilde{m}_s} - \frac{1}{2}\left(\frac{\delta n_s}{\tilde{m}_s}\right)^2 + \frac{1}{3}\left(\frac{\delta n_s}{\tilde{m}_s}\right)^3 - \cdots\right\} \\
&= -\sum_{s=1}^{k} \left(\delta n_s + \frac{1}{2}\frac{\delta^2 n_s}{\tilde{m}_s} - \frac{1}{6}\frac{\delta^3 n_s}{\tilde{m}_s^2} - \cdots\right).
\end{aligned}$$

Since the first sum in the above is zero and the third sum is small, Neyman and Pearson finally obtained

$$\lambda = e^{-\frac{1}{2}X^2}$$

(*ibid*., p. 267). As X^2 increases, λ decreases so that there is less support for the hypothesis being tested. Hence, X^2 can be used as a criterion for deciding between the two hypotheses and is in fact the test statistic used by Karl Pearson.

5.6.1.2 The 'Big Paper' of 1933 With the two 1928 papers, Neyman and Pearson had made a major breakthrough in the theory of hypothesis testing. With the LR test, they were able to give a formal justification for the tests used by Fisher and others. However, Neyman was still attached to inverse probability methods. In the summer of 1929, Neyman planned a Part III, which would be coauthored by Egon and which he would present at the ISI meeting to be held in Poland. However, because the paper contained inverse probability methods used in conjunction with the LR test, Egon refused to coauthor the paper.

The fundamental ideas in the first two papers thus originated mainly from Egon. However, the situation soon changed in 1930. Egon seemed to be sold on the idea that LR was justified since it resulted in the usual tests used by Fisher and others, and therefore did not require further justification. On the other hand, Neyman wanted a more fundamental justification for it. Thus, in a letter dated February 1, 1930, addressed to Egon, Neyman stated:

> ...It *seems* that we can have an experimental proof that the principle of likelihood "est fait pour quelque chose" [is of some value]...If we show that the frequency of accepting a false hypothesis is minimum when we use λ tests, I think it will be quite a thing! (Reid, 1998, p. 92)

Then, one late evening in February 1930, Neyman had an epiphany whereby most of the ideas of what later came to be known as the NP fundamental lemma took shape. In his February 20, 1930, letter to Egon, he wrote the essentials of that lemma:

> At present I am working on a variation calculus problem connected with the likelihood method. The results already obtained are a vigorous [rigorous?] argument in favour of the likelihood method. I considerably forget the variation calculus and until the present time I have only results for samples of two. But in all cases considered I have found the following:
>
> We test a simple hypothesis H_0 concerning the value of *some* character $a = a_0$, and wish to find a contour $\phi(x_1, x_2, \ldots, x_n) = c$ such that:
>
> (i) The probability $P(\phi, a_0)$ of a sample, Σ, lying inside the contour (which probability is determined by H_0) is $P(\phi, a_0) = \alpha$ where α is a certain fixed value, say 0.01 (this is for controlling the errors of rejecting a true hypothesis).
> (ii) The probability $P(\phi, a_1)$ determined by some other hypothesis that $a = a_1 \neq a_0$ of the sample lying inside the same contour be a maximum.
>
> Using such contours and rejecting H_0 when Σ is inside $\phi = \text{const}$, we are sure that a true hypothesis is rejected with a frequency less than α, and that if H_0 is false and the true hypothesis is, say H', then most often the observed sample will be inside $\phi = \text{const}$ and hence H_0 will be rejected. I feel you start to be angry as you think I am attacking the likelihood method! Be quiet! In all cases I have considered the $\phi = \text{const}$ contours are the λ-contours! (Pearson, 1966, pp. 17–18)

In the above, Neyman considered the test of a simple hypothesis against a simple alternative. Then, for a given probability of wrongly accepting the first hypothesis, what type of test would have the maximum probability of correctly rejecting that hypothesis? For all cases considered by Neyman, he had found that such a "best" test coincided with their LR test. This was a very remarkable statement that foreshadowed what would become one of the most influential theories of hypothesis testing.

No less remarkable was the fact that Neyman's February 20 letter also contained a reference to confidence intervals, which was to later prove vital in Neyman's work on estimation.

On March 24, 1930, Neyman wrote a letter to Egon containing the proof of the NP lemma. In that letter, he also responded to a counterexample Egon had sent him, purporting to show that the LR test was *not* the best test. To this, Neyman said that, *if* a best exists then it is the LR test. However, there was many situations where a best test does not exist. In 1931, he was able to resolve the issue more fully. In the case a uniformly best test did not exist, Neyman imposed the condition of *unbiasedness** whence a best test could be found.

* See later in the text.

Neyman and Pearson were now ready to publish their next paper, which Neyman called the "Big Paper." In late December 1931, Neyman wrote to Fisher informing him of their project. The latter showed interest in reading the manuscript, even saying on February 12, 1932 that:

> ...It is quite probable that if the work is submitted to the Royal Society, I might be asked to act as referee, and in that case I certainly shall not refuse.* (Bennett, 1990, p. 189)

The Big Paper, entitled "On the problem of the most efficient tests of statistical hypotheses"† (Neyman and Pearson, 1933), was presented to the Royal Society by none other than Karl Pearson on November 11, 1932 and was published the next year.

The start of the 1933 paper contained a very important paragraph, which represented the core philosophy behind NP hypothesis testing:

> ...Without hoping to know whether each separate hypothesis is true or false, we may search for rules to govern our behaviour with regard to them, in following which we insure that, in the long run of experience, we shall not be too often wrong. Here, for example, would be such a "rule of behaviour": to decide whether a hypothesis, H, of a given type be rejected or not, calculate a specified character, x, of the observed facts; if $x > x_0$ reject H, if $x \leq x_0$ accept H. Such a rule tells us nothing as to whether in a particular case H is true when $x \leq x_0$ or false when $x > x_0$. But it may often be proved that if we behave according to such a rule, then in the long run we shall reject H when it is true not more, say, than once in a hundred times, and in addition we may have evidence that we shall reject H sufficiently often when it is false. (Neyman and Pearson, 1933, p. 291)

In the above, the reader should pay particular attention to "rules to govern our behavior" and to "in the long run of experience." Both of these are hallmarks of NP hypothesis testing and stand in sharp contrast to Fisher's philosophy of significance testing.

Neyman and Pearson next provided the motivation behind their current paper:

> In earlier papers we have suggested that the criterion appropriate for testing a given hypothesis could be obtained by applying the principle of likelihood. This principle was put forward on intuitive grounds after the consideration of a variety of simple cases. It was subsequently found to link together a great number of statistical tests already in use, besides suggesting certain new methods of attack. It was clear, however, in using it that we were still handling a tool not fully understood, and it is the purpose of the present investigation to widen, and we believe simplify, certain of the conceptions previously introduced. (*ibid.*, p. 293)

After reviewing the λ–test for simple hypotheses and the λ_1 – test for composite hypotheses, the authors considered (*ibid.*, p. 297) the issue of choosing an appropriate rejection (or critical) region when testing a particular hypothesis H_0. They made the following remarks. First is it easy to keep the error of the first kind (i.e., rejecting H_0 when

*When Constance Reid was preparing the biography of Neyman, she wrote to the librarian of the Royal Society and was informed that Fisher did not actually referee the NP paper: the only referee was A.C. Aitken (Reid, 1998, pp. 103–104).

†The 1933 paper is also described in Lehmann (1992a).

it is true) as low as one pleases. Once this is done, the second type of error (i.e., accepting H_0 when it is false) needs to be dealt with. Here one needs to choose the critical region for which there is maximum probability of rejecting H_0 when it is false. They called such a region as the "best critical region." Now the best critical region with regard to a particular alternative H_1 will not necessarily be also the best critical region with regard to another alternative H_2. In some cases, there will a common family of best critical regions for H_0 with regard to the set of all admissible alternatives. In other cases, such a common family does not exist, and an additional principle needs to be introduced to obtain a "good critical region." Such a good critical region is the envelope of best critical regions obtained from the individual alternative hypotheses in the set. Finally, there are also best critical regions in the case of composite hypotheses, but here again, an additional principle needs to be introduced for this to be possible.

All the above ideas were next developed. Neyman and Pearson first (*ibid.*, p. 298) considered the problem of finding the best critical region for testing the simple hypothesis H_0 against the simple alternative H_1. Let ω_0 be this region and suppose $P_i(\omega_0)$ be the probability of rejecting H_0 given that H_i ($i=0$, 1) is true. Then the aim is to maximize $P_1(\omega_0)$ subject to the condition that

$$P_0(\omega_0)=\varepsilon, \tag{5.124}$$

where ε is an assigned value for the error of the first kind. The maximization is performed by using the calculus of variations and unconditionally minimizing

$$P_0(\omega_0)-kP_1(\omega_0)=\int_{\omega_0}\cdots\int\{p_0(x_1,x_2,\ldots,x_n)-kp_1(x_1,x_2,\ldots,x_n)\}\,dx_1dx_2\ldots dx_n,$$

where p_0 is the likelihood of H_0, p_1 is the likelihood of H_1, $x_1,x_2,\ldots,x_n$ is the observed sample, and k is a constant determined by the condition in (5.124). Now, for $P_0(\omega_0)-kP_1(\omega_0)$ to have a minimum, it is necessary that

$$p_0(x_1,x_2,\ldots,x_n)=kp_1(x_1,x_2,\ldots,x_n)$$

on the boundary of ω_0. For the region inside ω_0, the necessary condition is

$$p_0(x_1,x_2,\ldots,x_n)\le kp_1(x_1,x_2,\ldots,x_n) \tag{5.125}$$

(*ibid.*, p. 300) and for the region outside ω_0 the necessary condition is

$$p_0(x_1,x_2,\ldots,x_n)>kp_1(x_1,x_2,\ldots,x_n).$$

Equation (5.125) corresponds to the criterion provided by the LR for the test of a simple hypothesis against a simple alternative.

Neyman and Pearson next showed that the condition in (5.125) is also sufficient for ω_0 to be the best critical region, as follows. Let ω_1 be any other region such that $P_0(\omega_1)=\varepsilon$ and let ω_{01} the common part of the regions ω_0 and ω_1 (Fig. 5.11).

From Figure 5.11 and using $P_0(\omega_1)=\varepsilon$,

$$P_0(\omega_0-\omega_{01})=\varepsilon-P_0(\omega_{01})=P_0(\omega_1-\omega_{01})$$

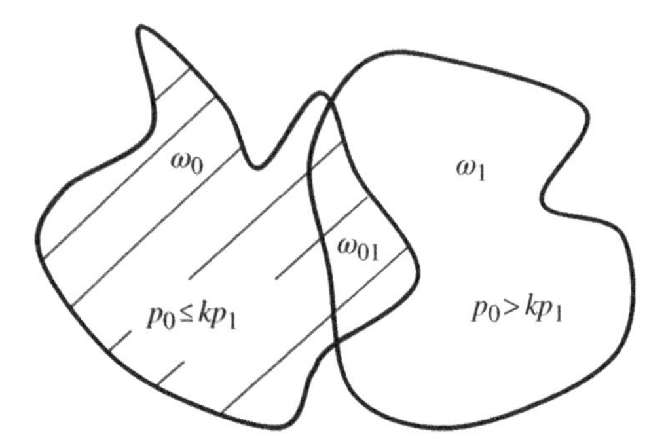

FIGURE 5.11 Neyman and Pearson's demonstration of the sufficiency of the condition $p_0 \le kp_1$

Equation (5.125) therefore implies that

$$kP_1(\omega_0 - \omega_{01}) \ge P_0(\omega_0 - \omega_{01}) = P_0(\omega_1 - \omega_{01}) > kP_1(\omega_1 - \omega_{01}),$$

the last inequality being due to the region $\omega_1 - \omega_{01}$ lying outside ω_0. The above thus gives $kP_1(\omega_0 - \omega_{01}) > kP_1(\omega_1 - \omega_{01})$. Adding $kP_1(\omega_{01})$ on both sides of the latter, we eventually obtain

$$P_1(\omega_0) > P_1(\omega_1).$$

Hence, ω_0 is the best region and sufficiency is proved.

With Eq. (5.125), Neyman and Pearson had thus proved what later came to be known as the NP fundamental lemma: *for a simple hypothesis H_0 against a simple alternative H_1, then for a given probability of wrongly rejecting H_0 (i.e. Type I error), the best critical region (i.e. the region leading to the most powerful test) is that provided by the likelihood ratio.*

Moreover, when the alternative H_1 was composite and consisted of a family of simple alternatives, the authors observed that the envelopes of the family $p_0 = kp_1$ that bound the best critical regions coincided with the families of constant likelihood λ defined by (5.123):

$$p_0 = \lambda p(\Omega_{\max}).$$

After some illustrative examples, Neyman and Pearson now (*ibid.*, p. 312) considered the case of a composite hypothesis H_0'. How should the best critical region be determined in that case? Suppose the likelihood of this hypothesis, $p(x_1, x_2, \ldots, x_d)$ depends on the $c+d$ parameters. If H_0' has c degrees of freedom then the values of d of these parameters are specified and c unspecified. Let the parameters be written as

$$\alpha^{(1)}, \alpha^{(2)}, \ldots, \alpha^{(c)}; \alpha_0^{(c+1)}, \ldots, \alpha_0^{(c+d)}.$$

Then a necessary condition for a critical region ω suitable for testing H_0' is that

$$P_0(\omega) = \int_{\omega}\cdots\int p_0(x_1, x_2, \ldots, x_n)\, dx_1 dx_2 \cdots dx_n = \text{constant} = \varepsilon. \qquad (5.126)$$

In particular, $P_0(\omega)$ should be independent of $\alpha^{(1)}, \alpha^{(2)}, \ldots, \alpha^{(c)}$. Thus, this is the condition introduced by the authors to deal with composite hypotheses. They further wrote:

> If this condition is satisfied we shall speak of ω as a region of "size" ε, similar to W [the sample space] with regard to the c parameters $\alpha^{(1)}, \alpha^{(2)}, \ldots \alpha^{(c)}$. (*ibid.*, p. 313)

Two terms should hold our attention in the above. First, Neyman and Pearson defined a critical region ω satisfying (5.126) as one of *size* ε. Second, since in that equation, $P_0(\omega)$ is independent of $\alpha^{(1)}, \alpha^{(2)}, \ldots, \alpha^{(c)}$, the region ω is said to be *similar* to the sample space with respect to the unspecified parameters. The search of a best critical region when H_0' is composite should therefore be made within the subset of *similar regions of size* ε.

After having considered similar regions, Neyman and Pearson finally (*ibid.*, p. 317) came to the thorny issue of how to proceed when no single best critical region existed for all classes of possible alternative, that is, when there was no *uniformly best critical region*. Such situations arise when the alternative is of the form $H_1: \theta \neq \theta_0$, where θ_0 is a known value of the parameter θ. In these situations, the authors proceeded by considering the two cases $\theta < \theta_0$ and $\theta > \theta_0$ separately and finding a best critical region for each case. However, with such a procedure, the consequence is that the best critical region for $\theta < \theta_0$ is not best for $\theta > \theta_0$, so that there is still no uniformly best critical region. Thus, the authors concluded their paper by saying that:

> The question of the choice of a "good critical region" for testing a hypothesis, when there is no common best critical region with regard to every alternative admissible hypothesis, remains open. (*ibid.*, p. 337)

However, as early as 1931, Neyman had an intuitive idea of how to proceed when a single best test did not exist. The idea came while he was in Poland:

> I think I have got the point! Suppose we have to test a simple hypothesis H_0 with regard to a class of alternatives C with no common B.C.R [best critical region]. It would be no good to use a critical region ω, having the following property: the class of alternatives contains a hypothesis, say H_1, such that ε_1 [the probability of rejection under H_1] is $<\varepsilon$. In fact, doing so we shall accept H_0 with larger frequency when it is false (and the true hypothesis is H_1) than when it is true....
>
> Therefore, the good critical region ω_0 should be chosen in such a way that [the probability of rejection under H_1 is $\geq$ than that under H_0]. (Lehmann, 2011, p. 38)

This was the germ of the idea of *unbiasedness* for statistical tests, that is, tests for which the power was always greater than the Type I error. Such tests made sense because they rejected H_0 more often when the latter was false than when it was true. Thus, just like for composite hypotheses, the idea was to search for best tests within the class of similar tests, so with two-sided alternatives the trick was to find best tests within the class of unbiased tests.

The concept of unbiased tests was further developed by Neyman and Pearson in their 1936 paper "Contributions to the theory of testing statistical hypotheses. I" (Neyman and Pearson, 1936).

5.6.2 Warm Relationships in 1926–1934

Perhaps few today realize that the relationship between Fisher on one side, and Neyman and Pearson on the other, got off to a reasonably good start. When the latter two started their collaboration, Fisher was the leading statistician of the time. Neyman had considerable admiration for Fisher, and Egon, although he did somewhat resent the verbal attacks of Fisher against his father, could not help but respect the pioneering works of Fisher. It may also be surprising that the initial fight between the two sides was not about hypothesis testing, a topic which would antagonize them throughout their lives. Rather it involved the issue of the Latin Square. We now describe these topics.

There is no evidence that, when he came to England in 1925, Neyman was aware of the works of Fisher. However, by the spring of 1926 Neyman had heard sufficiently of the latter when he made a request through Student to meet Fisher. This is how Student described Neyman to Fisher:

> He is fonder of algebra than correlation tables and is the only person except yourself I have heard talk about maximum likelyhood [*sic*] as if he enjoyed it. (Reid, 1998, p. 59)

Neyman and Fisher met for the first time on July 20, 1926. That Neyman initially had high regard for Fisher's work is reflected in a letter he wrote to the latter in 1929. At the start of the letter, Neyman wrote:

> My dear Dr. Fisher, you are probably informed that a good deal of time Dr. Pearson and myself we are studying your work. I write to say how it is exciting for me. Very often I am not able to follow your arguments and even the theorems seem at the beginning to be false. Then Dr. Pearson or myself we are looking for a proof and finally the theorem appears to be perfectly correct. (*ibid*., p. 84)

The letter was written to inform Fisher about the NP work in relation to the testing of hypotheses and also to ask his opinion on some inverse probability ideas Neyman wished to incorporate in a potential Part III paper. Not only was the letter not sent but Egon also refused to coauthor the paper with Neyman because:

> ...[I]f they published the proposed paper, with its admission of inverse probability, they would find themselves involved in a disagreement with Fisher, who had come out decisively against it. (*ibidem*)

Given Karl's Pearson highly antagonistic relationship with Fisher, Egon was eager to avoid any further controversy with the latter.

On his part, Fisher also tried his best to be accommodating to the NP team. As we observed in the last section, he had blessed their "Big Paper" of 1933 (Neyman and Pearson, 1933) before it was published and had even stated he would gladly referee it if asked. Fisher also made a small correction to the "Big Paper" and admitted that he found the question of significance testing of immense philosophical importance. The relationship between him and Neyman got so warm that Fisher invited the latter to the Rothamsted Station on several occasions.

In 1934, Neyman presented his paper "On the two different aspects of the representative method" (Neyman, 1934). This contained a discussion of confidence intervals, which both Neyman and Fisher at the time thought was an extension of the fiducial argument. In the discussion of the paper, both men exchanged very complimentary remarks, which were described in Section 5.5.2.2.*

In 1934 still, Fisher published a paper entitled "Two new properties of mathematical likelihood" (Fisher, 1934c), which is of major historical importance in relation to the later Fisher–NP dispute. This is because the paper is the only one whereby Fisher explicitly accepted the existence of an alternative hypothesis as a legitimate concept in the theory of hypothesis testing:

> ...Neyman and Pearson introduce the notion that any chosen test of a hypothesis H_0 is more powerful than any other equivalent test, with regard to an alternative hypothesis H_1, when it rejects H_0 in a set of samples having an assigned aggregate frequency ε when H_0 is true, and the greatest possible aggregate frequency when H_1 is true.
>
> If any group of samples can be found within the region of rejection whose probability of occurrence on the hypothesis H_1 is less than that of any other group of samples outside the region, but is not less on the hypothesis H_0, then the test can evidently be made more powerful by substituting the one group for the other. Consequently, for the most powerful test possible the ratio of the probabilities of occurrence on the hypothesis H_0 to that on the hypothesis H_1 is less in all samples in the region of rejection than in any sample outside it. For samples involving continuous variation the region of rejection will be bounded by contours for which this ratio is constant. The regions of rejection will then be required in which the likelihood of H_0 bears to the likelihood, of H_1, a ratio less than some fixed value defining the contour. The test of significance is termed uniformly most powerful with regard to a class of alternative hypotheses if this property holds with respect to all of them. (*ibid.*, pp. 294–295)

The above should be contrasted with Fisher's later statement about the NP approach to hypothesis testing.

The last positive interaction between NP and Fisher was during the famous incident involving the latter and Arthur Bowley (1869–1957) on December 18, 1934, when Fisher read his paper "The logic of inductive inference" (Fisher, 1935d). During the discussion, Bowley was very critical, at times even disparaging, of Fisher's paper:

> I found the treatment to be very obscure. I took it as a week-end problem, and first tried it as an acrostic, but I found that I could not satisfy all the "lights." I tried it then as a cross-word puzzle, but I have not the facility of Sir Josiah Stamp for solving such conundrums. Next I took it as an anagram, remembering that Hooke stated his law of elasticity in that form, but when I found that there were only two vowels to eleven consonants, some of which were Greek capitals, I came to the conclusion that it might be Polish or Russian, and therefore best left to Dr. Neyman or Dr. Isserlis. Finally, I thought it must be a cypher, and after a great deal of investigation, decided that Professor Fisher had hidden the key in former papers, as is his custom, and I gave it up. (*ibid.*, p. 56)

Bowley thereafter pointed that an important formula that Fisher used in his paper (and which Fisher had in fact obtained as early as 1922, see Eq. 5.26) had previously been

* See p. 470.

obtained by Edgeworth in one of his "Genuine Inverse Method" papers (see Eq. 5.145). Bowley's implication clearly was that either Fisher was not sufficiently aware of the literature or that he had intentionally failed to give credit to Edgeworth. After additional incisive comments by Isserlis and Wolf (who had been brought in by Bowley to critique the logical parts of Fisher's paper), one person who came to Fisher's defense was none other than Egon. The latter's comments were written as follows:

> ...At the beginning of the paper Professor Fisher had said that he regarded the essential effect of the general body of researches in mathematical statistics during the last fifteen years to be fundamentally a reconstruction of logical rather than mathematical ideas. Dr. Pearson agreed with that statement, but he rather gathered that Professor Bowley and Dr. Isserlis did not. It seemed to him that Professor Fisher had contributed to this development of logical ideas something which was definitely lacking before. When these ideas were fully understood, whether there was final agreement or not with his particular terminology and the details of his theory of inductive inference, it would be realized that statistical science owed a very great deal to the stimulus Professor Fisher had provided in many directions. (*ibid.*, p. 65)

Egon was supported by (of course) Neyman:

> Professor Fisher's papers are interesting not only because of the many important problems stated and solved, but, perhaps still more, because they contain so many hints and questions which the author did not have time or perhaps did not care to solve himself. Therefore, going through this paper one has not only to follow the writer's ideas, but is, as it were, compelled to think of many other problems, not directly discussed there. (*ibid.*, p. 73)

Finally, Fisher concluded very nicely:

> ...However true it may be that Professor Bowley is left very much where he was, the quotations show at least that Dr. Neyman and myself have not been left in his company. (*ibid.*, p. 77)

5.6.3 1935: The Latin Square and the Start of an Ongoing Dispute

The promising relationship between Fisher and NP took a turn for the worse when, on March 28, 1935, Neyman presented the paper "Statistical problems in agricultural experimentation" (Neyman and Iwaszkiewicz, 1935) to the Section on Industrial and Agricultural Research of the Royal Society.* In his paper, Neyman compared the randomized block and Latin Square that had been developed previously by Fisher (see Sections 5.4.2.3–5.4.2.4). Neyman first stated:

> In the early stages of pushing forward with their application it has not been wise to raise any questions as to the absolute accuracy of the two arrangements whose statistical soundness obviously surpassed that of all previous work. However, by now the new methods seem to be sufficiently established, and I have thought it useful to discuss at some length the nature of the experimental errors involved. (*ibid.*, p. 109)

*The Fisher versus Neyman–Pearson dispute is also described in Howie (2002, pp. 172–191), Lenhard (2006), Lehmann (1993; 2011, Chapter 4), Spanos (1999, pp. 720–727), Gigerenzer et al. (1989, pp. 90–106), and Hubbard and Bayarri (2003). A recent discussion with particular emphasis on Neyman's 1935 paper is provided by Sabbaghi and Rubin (2014).

As we shall see, Fisher was to find much fault with the last statement.

Neyman then proceeded by building an explicit linear model for the randomized block and Latin Square, both of which had been treated before rather heuristically by Fisher. Neyman's analysis was as follows.

Considering the randomized block, let $X_{ij}(k)$ be the mean yield on "object" (or treatment) k on block i and plot j. Let $x_{ij}(k)$ be the corresponding observed yield so that

$$x_{ij}(k) = X_{ij}(k) + \varepsilon_{ij}(k), \tag{5.127}$$

(*ibid.*, p. 110) where $\varepsilon_{ij}(k)$ is the "technical error" in the particular yield $x_{ij}(k)$. Now $X_{ij}(k)$ can be further decomposed as follows:

$$X_{ij}(k) = X_{..}(k) + \{X_{i.}(k) - X_{..}(k)\} + \{X_{ij}(k) - X_{i.}(k)\} = X_{..}(k) + B_i(k) + \eta_{ij}(k),$$

where $X_{..}(k)$ is the mean yield of treatment k in all plots, $X_{i.}(k)$ is the mean yield in all plots of block i, $B_i(k) = X_{i.}(k) - X_{..}(k)$, and $\eta_{ij}(k) = X_{ij}(k) - X_{i.}(k)$. Equation (5.127) then becomes

$$x_{ij}(k) = X_{..}(k) + B_i(k) + \eta_{ij}(k) + \varepsilon_{ij}(k). \tag{5.128}$$

The mean difference in two treatments k and k' is

$$\Delta = X_{..}(k) - X_{..}(k'),$$

and the observed difference is

$$d = x_{..}(k) - x_{..}(k').$$

Neyman wished to investigate the conditions under which $\mathsf{E}d = \Delta$. If there are n' repetitions of each treatment in the field, then the average yield of treatment k is

$$x_{..}(k) = \frac{1}{n'} \sum_k x_{ij}(k).$$

Therefore, from Eq. (5.128),

$$\mathsf{E}x_{..}(k) = X_{..}(k) + \frac{1}{n'} \sum_k \eta_{ij}(k),$$

since $\sum_{i=1}^{n'} B_i(k) = 0$ and $\mathsf{E}\varepsilon_{ij}(k) = 0$. There is a similar expression to the above for $\mathsf{E}x_{..}(k')$, so that

$$\mathsf{E}d = \Delta + \frac{1}{n'} \left\{ \sum_k \eta_{ij}(k) - \sum_{k'} \eta_{ij}(k') \right\} \tag{5.129}$$

(*ibid.*, p. 111). Now, $\eta_{ij}(k) = X_{ij}(k) - X_{i.}(k)$ is the deviation of the yield from block i and plot j from the mean yield of block i and is not zero. Therefore, the expression inside the brackets of Eq. (5.129) will not usually be zero and $\mathsf{E}d \neq \Delta$. However, Neyman noted that:

> …[t]he difficulty has been overcome by the device proposed by R.A. Fisher, which consists in making η's random variables with the mean equal to zero. (*ibid.*, p. 112)

The device Neyman referred to here is that of randomizing plots within blocks. Still, the fact that the η's have zero mean does not imply that, for a particular experiment, $\sum_k \eta_i(k)$ and $\sum_{k'} \eta_i(k')$ are both exactly zero, but only that they are approximately zero for reasonably large k and k'.

Turning next to the Latin Square, Neyman used $X_{ij}(k)$ to denote the mean yield on treatment k on row i and column j. Like in the previous case, $X_{ij}(k)$ can be decomposed:

$$\begin{aligned} X_{ij}(k) &= X_{..}(k) + \{X_{i.}(k) - X_{..}(k)\} + \{X_{.j}(k) - X_{..}(k)\} + \eta_{ij}(k) \\ &= X_{..}(k) + R_i(k) + C_j(k) + \eta_{ij}(k) \end{aligned}$$

(*ibidem*), where

$$\eta_{ij}(k) = X_{ij}(k) + X_{i.}(k) + X_{.j}(k) + X_{..}(k),$$

$R_i(k) = X_{i.}(k) - X_{..}(k)$, and $C_j(k) = X_{.j}(k) - X_{..}(k)$. In analogy with Eq. (5.128), we have

$$x_{ij}(k) = X_{..}(k) + R_i(k) + C_j(k) + \eta_{ij}(k) + \varepsilon_{ij}(k). \tag{5.130}$$

As before, let Δ and d be the mean and observed difference in yield between two treatments k and k'. Since there are as many rows and columns and each treatment occurs only once in each row and column in a Latin Square, we have $\sum_k R_i(k) = \sum_k C_j(k) = 0$. Therefore,

$$d = \Delta + \frac{1}{n'}\sum_k \eta_{ij}(k) - \frac{1}{n'}\sum_{k'} \eta_{ij}(k') + \frac{1}{n'}\sum_k \varepsilon_{ij}(k) - \frac{1}{n'}\sum_{k'} \varepsilon_{ij}(k')$$

and

$$\mathsf{E}d = \Delta + \frac{1}{n'}\sum_k \eta_{ij}(k) - \frac{1}{n'}\sum_{k'} \eta_{ij}(k')$$

(*ibid*., p. 113). Again, it can be seen that $\mathsf{E}d \neq \Delta$, but this difficulty is overcome by Fisher through randomizing the soil errors $\eta_{ij}(k)$. That is, plots are assigned to different treatments at random such that each treatment occurs only once in each row and column. This implies that the η's can be regarded as random variables with mean zero. But, as Neyman observed, there are two random variables in Eq. (5.130), namely, $\eta_{ij}(k)$ and $\varepsilon_{ij}(k)$. These two error components are quite different:

> It may be easily assumed that $\varepsilon_{ij}(k)$ is independent of, say, $\varepsilon_{rs}(k')$, corresponding to some other plot, and is varying normally about zero. Now if we take into consideration any two η's we shall find that they are dependent and that the dependency is rather complex. Besides, there is only a finite number of possible values for each η. It follows that the conditions under which the application of the z distribution is legitimate are not strictly satisfied. (*ibid*., p. 114)

Neyman therefore concluded:

> In the case of the Randomized Blocks the position is somewhat more favourable to the z test, while in the case of the Latin Square this test seems to be biased, showing the tendency to discover differentiation when it does not exist. (*ibidem*)

Neyman then gave some numerical examples which showed that the randomized block was more efficient than the Latin Square.

Fisher was furious. He opened the discussion of Neyman's paper by the following two sentences:

> PROFESSOR R.A. FISHER, in opening the discussion, said he had hoped that Dr. Neyman's paper would be on a subject with which the author was fully acquainted, and on which he could speak with authority, as in the case of his address to the Society delivered last summer. Since seeing the paper, he had come to the conclusion that Dr. Neyman had been somewhat unwise in his choice of topics. (*ibid.*, p. 154)

Fisher complained that Neyman's analysis was erroneous:

> Apart from its theoretical defects, Dr. Neyman appeared also to have discovered that it [the Latin Square] was, contrary to general belief, a less precise method of experimentation than was supplied by Randomized Blocks, even in those cases in which it had hitherto been regarded as the more precise design. (*ibid.*, p. 155)

Then, referring to the "extraordinary" preliminary statement made by Neyman,* he reproached the latter "for concealing...his discoveries...from public knowledge until such time as the method should be widely adopted in practice!"

In concluding his comments, Fisher referred to a statement made by Neyman with regard to the Latin Square. Neyman had said:

> This bias vanishes...when the objects compared are reacting to differences in soil fertility in exactly the same manner...This is not always true. (*ibid.*, p. 153)

Fisher retorted (p. 157) that, in fact, this was always true *under the null hypothesis*. Thus, he claimed, Neyman was confusing estimation with significance testing: by finding bias in estimates he had thought this would undermine significance tests. Fisher concluded by bringing Egon into the fray:

> ...Were it not for the persistent efforts which Dr. Neyman and Dr. Pearson had made to treat what they speak of as problems of estimation, by means merely of tests of significance, he had no doubt that Dr. Neyman would not have been in any danger of falling into the series of misunderstandings which his paper revealed. (*ibidem*)

Egon's reaction was understandably that of irritation:

> DR. PEARSON said that while he knew there was a widespread belief in Professor Fisher's infallibility, he must, in the first place, beg leave to question the wisdom of accusing a fellow-worker of incompetence without, at the same time, showing that he had succeeded in mastering his argument. He felt sure that Dr. Neyman would be able himself to answer the criticisms that had been raised, but there was a rather more general point touching on the whole field of statistical enquiry on which he would like to comment. Professor Fisher had

*See p. 522.

on more than one occasion thrown out the suggestion that while some other statisticians were academic he and his collaborators were practical men; Dr. Pearson believed that attempts to dub as academic enquiries into the underlying principles upon which practical tests were based, showed a serious loss of perspective. (*ibid.*, p. 170)

In his reply to Fisher, Neyman stated:

...Professor Fisher ascribed to me statements, and intentions which I never dreamed of...The problem I considered is ...as follows: "Our purpose in the field experiment consists in comparing numbers such as $\ddot{X}(k)$, or the average true yields which our objects are able to give when applied to the whole field." It is seen that this problem is essentially different from what Professor Fisher suggested. (*ibid.*, pp. 172–173)

Neyman thus explained that he had been interested in average yields across the whole field rather than the yield for a particular plot. But Fisher would have none of this:

It [i.e. testing for a particular plot] may be foolish, but this is what the z test was designed for, and the only purpose for which it has been used. (*ibid.*, p. 173)

In retrospect, it is seen that Neyman and Fisher were arguing from different vantage points, although it seems that the latter missed the point the former was trying to make. However, a direct consequence of the meeting of March 28, 1935, was that the relationship between Fisher and NP was completely and irreversibility destroyed. This is how Neyman recalled Fisher's reaction to him one week after the meeting:

...he [Fisher] said that, as I know, he had published a book—and that's *Statistical Methods for Research Workers*—and he is upstairs from me so he knows something about my lectures—that from time to time I mention his ideas, this and that—and that this would be quite appropriate if I were not here in the college but, say, in California...—but if I am going to be at University College, then this is not acceptable to him. And then I said, "Do you mean that if I am here, I should just lecture using your book?" And then he gave an affirmative answer. Yes, that's what he expected. And I said, "Sorry, no. I cannot promise that." And then he said, "Well, if so, then from now on I shall oppose you in all my capacities." And then he enumerated—member of the Royal Society and so forth. There were quite a few. Then he left. Banged the door. (Reid, 1998, p. 126)

In all fairness, the later "opposition" was not always from Fisher but often from Neyman himself. But the situation at University College became very tense, especially since the Department of Eugenics, chaired by Fisher, was located just above the Department of Applied Statistics, headed by Egon and of which Neyman was then a member.

Several incidents subsequently occurred between Fisher and NP that soured their relationship even more. For example, shortly after Karl Pearson's death in 1936, Fisher wrote a scathing paper entitled "Professor Karl Pearson and the Method of Moments" where he criticized Pearson for being a "clumsy mathematician" with an "arrogant temper." Fisher also explained that he had decided to settle the issue of Pearsonian methods only after Pearson's death because "during his last years Pearson's intimates have earnestly represented that irritation and controversy might be dangerous to his health." However, owing to his respect for Pearson, Neyman took great umbrage at Fisher's paper. He quickly wrote a reply (Neyman, 1938a) accusing the "regular campaign carried by Prof. R.A. Fisher to discredit the work of the late Karl Pearson."

Even when Neyman moved to Berkeley, University of California, in August 1938, there was no let-up in the fight. In his review of Neyman's book, *Lectures and Conferences on Mathematical Statistics* (Neyman, 1938c), Fisher brushed the book aside for not having "enough original material to justify publication as a book, and too much that is trivial" (Fisher, 1938). Neyman later "returned the favor" when Fisher's *Contributions to Mathematical Statistics* (Fisher, 1950) was published:

> A very able "manipulative" mathematician, Fisher enjoys a real mastery in evaluating complicated multiple integrals.... (Neyman, 1951, p. 406)

However, Neyman was wrong in the next statement he made:

> ...three major concepts were introduced by Fisher and consistently propagandized by him in a number of publications. These are mathematical likelihood as a measure of the confidence in a hypothesis, sufficient statistics, and fiducial probability. Unfortunately, in conceptual mathematical statistics Fisher was much less successful than in manipulatory, and of the three above concepts only one, that of a sufficient statistic, continues to be of substantial interest. (*ibid.*, p. 407)

Stating that only sufficient statistics continue to be of substantial interest completely misrepresents the huge importance of likelihood in statistical theory.

Neyman's third point was much more legitimate when he repeated Bowley's previous criticism that Fisher had failed to give due credit to Edgeworth's pioneering work on maximum likelihood:

> ...Since the results of Edgeworth (1908) described above form a very substantial part of Fisher's paper of 1922, one would think that the prefatory note used with this paper (or, perhaps, with the paper of 1925) republished in the present volume would contain some reference to Edgeworth. It does not...No mention is made by Fisher that, some thirteen years previously, the same idea "emerged" in the mind of Edgeworth (loc. cit., 678) the only difference being that Edgeworth's measure is the reciprocal of that of Fisher and does not bear the picturesque label of "amount of information." (*ibidem*)

When Fisher heard of Neyman's scathing review, he said to H. Gray:

> Neyman is, judging by my own experience, a malicious mischief-maker. Probably by now this is sufficiently realised in California. I would not suggest to anyone to engage in scientific controversy with him, for I think that scientific discussion is only profitable when good faith can be assumed in the common aim of getting at the truth. (Bennett, 1990, p. 138)

It was now Fisher's turn to be wrong in his above negative assessment of Neyman's position at Berkeley, California. This is because the Statistics Department was completely revolutionized and became a first-class place for statisticians from all around the world under the leadership of Neyman. But Fisher was somewhat more credible, though not quite correct, in his response concerning Edgeworth's priority on ML:

> Edgeworth's paper of 1908 has, of course, been long familiar to me, and to other English statisticians. No one could now read it without realising that the author was profoundly confused. I should say, for my own part, that he certainly had an inkling of what I later demonstrated. The view that, in any proper sense, he anticipated me is made difficult by a number of verifiable facts...He based his argument on Bayesian inverse probability; my results are free from this assumption and represent an entirely different approach. (*ibid.*, pp. 138–139)

It is true that Edgeworth's writings were often abstruse, although he did not use inverse probability in all of his work on ML (see Section 5.7.5).

5.6.4 Fisher's Criticisms (1955, 1956, 1960)

5.6.4.1 Introduction Faced with the rise of the NP theory of hypothesis testing, especially in the late 50s, Fisher made valiant efforts to discredit it and promote his own theory of significance testing. Three main sources where Fisher's views are expounded on the subject are the two papers "Statistical methods and scientific induction" (Fisher, 1955) and "Scientific thought and the refinement of human reasoning" (Fisher, 1960) and the book *Statistical Methods and Scientific Inference* (Fisher, 1956b).

5.6.4.2 Repeated Sampling One of Fisher's major griefs against the NP theory was the latter's interpretation of hypothesis testing and confidence intervals as procedures to be considered over the long run in repeated sampling (i.e., as procedures which are frequentist or unconditional). In their very first joint paper, Neyman and Pearson made it clear that the two types of errors were to be interpreted as long-run frequencies, for example:

> In the long run of statistical experience the frequency of the first source of error (or in a single instance its probability) can be controlled by choosing as a discriminating contour, one outside which the frequency of occurrence of samples from Π is very small-say, 5 in 100 or 5 in 1000.... (Neyman and Pearson, 1928a, p. 177)

A common example Fisher liked to use to criticize the repeated sampling viewpoint is based on an incident that opposed him to George Barnard (Fig. 5.12) in 1945. The latter had published an article entitled "A new test for 2×2 tables" (Barnard, 1945) in *Nature* purporting to show a more powerful test than Fisher's exact test for 2×2 tables (see Table 5.34).

In Table 5.34, the rows sums m and n have been fixed in advance and the hypothesis to be tested is that of an association between the attributes A and B, and P and not-P. Fisher's exact test is based on the hypergeometric probability

$$\frac{m!n!r!s!}{N!a!b!c!d!}. \tag{5.131}$$

Now let p_1 be the probability that A has P and p_2 be the probability that B has P. Then the hypothesis tested is $H(p) \equiv p_1 = p_2 = p$. Barnard next proceeded by representing

FIGURE 5.12 George Alfred Barnard (1915–2002). From DeGroot (1988, p. 197). Courtesy of Institute of Mathematical Statistics

TABLE 5.34 2×2 Table for Barnard's Example

	P	Not-P	Total
A	a	c	m
B	b	d	n
Total	r	s	N

Table 5.34 geometrically by the point (a, b) in the xy-plane with only integer coordinates allowed. Since m and n are fixed, the sample space is the set of integer points on the boundaries or inside the rectangle bounded by the x- and y-axes and the lines $x=m$, $y=n$.

Barnard now assigned to each point (a, b) a weight $W(a,b,p)$, calculated under $H(p)$ as the product of two binomial probabilities:

$$W(a,b,p)=\binom{m}{a}p^{a}(1-p)^{c}\times\binom{n}{b}p^{b}(1-p)^{d}=\frac{m!n!}{a!b!c!d!}p^{a+b}(1-p)^{c+d}.$$

A valid test with significance level α is then a selection of a region R inside the rectangle such that

$$\max_{0\le p\le 1}\sum_{R}W(a,b,p)\le\alpha.$$

Since the above is not sufficient to determine R uniquely, Barnard imposed the additional conditions that R should consist of as many points as possible and should lie away from the diagonal (of the rectangle) that passed through the origin (i.e., the complement of R should be convex, symmetrical, and minimal).

Taking $m=n=3$ and $\alpha=1/32$, Barnard found that R consists of the two points (3, 0) and (0, 3). On the other hand, the level of significance with Fisher's exact in (5.131) is 1/10. Therefore, if the point (3, 0) was observed, Barnard's test* would give a one-sided p-value of 1/64, while Fisher's exact test would give a larger p-value of 1/20. "Thus the new test is more powerful than Fisher's."

In that same year, Fisher wrote a reply to *Nature* on Barnard's paper and argued that Barnard's unconditional test was inappropriate (Fisher, 1945). He also criticized the notion of repeated sampling on which the NP concept of power (= 1 − Type II error) was based. Fisher's reasoning was as follows. He considered the situation when the point (3, 0) was observed. There are in all 20 tables for which all margins are fixed to three. Only one is more or as extreme as the one observed, namely, (3, 0) itself, hence the p-value is $p=1/20$ for the conditional test (i.e., Fisher's exact test). Now the unconditional test does not fix all margins and allows for 44 extra tables in addition to the previous 20. However, many of these extra tables are uninformative. Fisher gave examples of two such tables (see Table 5.35).

*Barnard's unconditional test was later called the CSM test, the name coming from the Company Sergeant Major in Barnard's Guard unit at the time (Yates, 1984, p. 450).

TABLE 5.35 Two Uninformative Tables that are Used in Barnard's Unconditional Test

	Die	Survive		Die	Survive
Experimental	3	0	Experimental	0	3
Control	3	0	Control	0	3

In these two cases, all subjects either die or survive and therefore contribute no evidence with respect to a potential association between experimental condition and survival. Yet these two cases are included in the unconditional test. Fisher thus concluded:

> In my view the notion of defining the level of significance by "repeated sampling of the same population" is misleading in the theory of small samples just because it allows of the uncritical inclusion in the denominator of material irrelevant to a critical judgment of what has been observed. In 2 of the 64 cases enumerated above, all animals die or all survive. The fact that such an unhelpful outcome me as these might occur, or must occur with a certain probability, is surely no reason for enhancing our judgment of significance in cases where it has not occurred. (*ibidem*)

So convincing was Fisher's argument and other examples he sent to Barnard that the latter decided to retract his unconditional test a few years later (Barnard, 1949) in favor of Fisher's exact test, admitting that "Professor Fisher was right after all."

Fisher often used the Barnard incident to attack the NP insistence on repeated sampling. For example, in his book *Statistical Methods and Scientific Inference*, he said:

> Professor Barnard has since then frankly avowed that further reflection has led him to the same conclusion as Yates and Fisher…It is, therefore, obvious that he had at first been misled by the form of argument developed by Neyman and Pearson…It is to be feared, therefore, that the principles of Neyman and Pearson's "Theory of Testing Hypotheses" are liable to mislead those who follow them into much wasted effort and disappointment, and that its authors are not inclined to warn students of these dangers. (Fisher, 1956b, p. 88)

Another example of the advantage of Fisher's conditional approach compared to the NP unconditional approach is provided in Section 5.2.6.2.

If statistical tests were not to be viewed "in the long run of statistical experience," then how should they be viewed? Fisher's answer was that:

> …no one uses the rule of determining the level of significance by successive sampling from the population of all random samples of N pairs of values, but, ever since the right approach was indicated (Fisher 1922), *the selection of all random samples having a constant value A, equal to that actually observed in the sample under test, is what has in fact been used.** (Fisher, 1955, pp. 71–72)

Thus, according to Fisher, each observed sample should be regarded as unique and compared only to other samples similar in all relevant respects to the one observed. A prototypal example is Fisher's exact test, where inference is made by comparing to other tables with the same marginal frequencies, that is, by conditioning on both margins of the given table.

* Italics are ours.

However, even Fisher's conditional approach to testing can run into difficulties. As we mentioned in Section 5.2.8, often ancillary statistics or relevant subsets cannot be found on which conditional inference can be done. On other occasions, there are several ancillary statistics to choose from, and it is not always clear on how to proceed.

5.6.4.3 Type II Errors Fisher's second major criticism of NP hypothesis testing, which he called "acceptance procedures," pertains to the issue of "errors of the second kind" (Type II errors). Fisher stated:

> The phrase "Errors of the second kind", although apparently only a harmless piece of technical jargon, is useful as indicating the type of mental confusion in which it was coined. (*ibid.*, p. 73)

He argued that Type II errors could not in fact be calculated:

> ...A well-designed acceptance procedure is one which attempts to minimize the losses entailed by such events. To do this one must take account of the costliness of each type of error, if errors they should be called, and in similar terms of the costliness of the testing process; it must take account also of the frequencies of each type of event. For this reason probability *a priori*, or rather knowledge based on past experience, of the frequencies with which lots of different quality are offered, is of great importance; whereas, in scientific research, or in the process of "learning by experience", such knowledge *a priori* is almost always absent or negligible. (*ibidem*)

Fisher also pointed out that, since the NP procedure advocated the acceptance of one hypothesis, the investigator might mistakenly believe that an irreversible decision had been reached:

> The fashion of speaking of a null hypothesis as "accepted when false", whenever a test of significance gives us no strong reason for rejecting it, and when in fact it is in some way imperfect, shows real ignorance of the research workers' attitude, by suggesting that in such a case he has come to an irreversible decision. (*ibidem*)

Not only is acceptance irreversible in the NP framework, but the whole process of scientific inference becomes a mechanical procedure:

> In an acceptance procedure...acceptance is irreversible, whether the evidence for it was strong or weak. It is the result of applying mechanically rules laid down in advance; no *thought* is given to the particular case, and the tester's state of mind, or his capacity of *learning*, is inoperative. (*ibid.*, pp. 73–74)

On the other hand, in Fisher's significance testing, a hypothesis is never accepted and all conclusions are provisional:

> By contrast, the conclusions drawn by a scientific worker from a test of significance are provisional, and involve an intelligent attempt to understand the experimental situation. (*ibid.*, p. 74)

By denying that a particular hypothesis could ever be accepted, Fisher thus in a sense reiterated one of the major tenets of the philosopher Karl Popper (1902–1994): *falsification*, that is, a hypothesis (or theory) can be corroborated but *never confirmed* since it is still open to countless chances of falsification.*

Neyman of course took exception to Fisher's dismissal of the issue of Type II error (and power). In his reply to Fisher, he enumerated three advantages of the concepts of Type II error and power:

> First, these concepts serve as a basis for the *deduction* of tests that are the most powerful either absolutely or compared with a specified class of tests. Second they serve as a means of comparison and evaluation of two or more suggested alternative tests. In particular, the power of nonparametric tests is frequently evaluated for this specific purpose. Third, the numerical values of probabilities of errors of the second kind are most useful for deciding whether or not the failure of a test to reject a given hypothesis could be interpreted as any sort of "confirmation" of this hypothesis. (Neyman, 1956, p. 290)

Neyman thus contended that "the concepts have penetrated modern statistical thinking" deeply. As to the legitimacy of the alternative hypothesis, Neyman said that:

> ...ordinarily, when a scientist designs an experiment he, consciously or subconsciously, has in mind at least a general outline of the set of admissible hypotheses, against which a particular hypothesis is to be tested. (*ibidem*)

He pointed that Fisher himself had been aware of the issue of power when he "contemplates factorial field trials with six replicates rather than with three" in the second edition of his book "Design of Experiments" (Fisher, 1937b).

Neyman's point above that Fisher appreciated the need for sufficient power in a statistical test is evident from Fisher's inclusion of a whole section entitled "The Sensitiveness of an Experiment. Effect of Enlargement and Repetition" in the first edition of the above-mentioned book. In that section, we can also read:

> ...By increasing the size of the experiment, we can render it more sensitive, meaning by this that it will allow of the detection of a lower degree of sensory discrimination, or, in other words, of a quantitatively smaller departure from the null hypothesis. Since in every case the experiment is capable of disproving, but never of proving this hypothesis, we may say that the value of the experiment is increased whenever it permits the null hypothesis to be more readily disproved. (Fisher, 1935a, p. 25)

Although Fisher did not use the term "power," he was in fact referring to that concept when talking about the sensitiveness of an experiment. Thus, it seems that Fisher was not against the need for "well-powered" experiments, rather he objected to the formalization of the notions of power and of alternative hypotheses.

*"In so far as a scientific statement speaks about reality, it must be falsifiable: and in so far as it is not falsifiable, it does not speak about reality" (Popper, 1959, p. 314). Einstein had also made a similar statement earlier in 1921.

Neyman did not explicitly address the criticism regarding "accepting" particular hypotheses (under the NP framework) in his rebuttal to Fisher, but he had earlier offered the following caveat:

> The terms "accepting" and "rejecting" a statistical hypothesis are very convenient and are well established. It is important, however, to keep their exact meaning in mind and to discard various additional implications which may be suggested by intuition. Thus, to accept a hypothesis H means only to decide to take action A rather than action B. This does not mean that we necessarily believe that the hypothesis H is true. (Neyman, 1950, p. 259)

We are thus admonished by Neyman that to accept a hypothesis H is not to actually believe that H is true but only to decide to act as if H was true.

5.6.4.4 Inductive Behavior Probability and statistics both involve the process of scientific induction whereby we attempt to make inference from particular cases to more general situations. But what is the exact nature of the inductive process involved? As early as 1933, Neyman had espoused the notion that statistical hypotheses endow us with *rules of behavior*:

> Without hoping to know whether each separate hypothesis is true or false, we may search for rules to govern our behaviour with regard to them, in following which we insure that, in the long run of experience, we shall not be too often wrong. (Neyman and Pearson, 1933, p. 291)

On the other hand, Fisher believed that statistical hypotheses involved the process of *inductive reasoning*, which should be contrasted with deductive reasoning as follows:

> In deductive reasoning all knowledge obtainable is already latent in the postulates. Rigour is needed to prevent the successive inferences growing less and less accurate as we proceed. The conclusions are never more accurate than the data. In inductive reasoning we are performing part of the process by which new knowledge is created. The conclusions normally grow more and more accurate as more data are included. It should never be true, though it is still often *said*, that the conclusions are no more accurate than the data on which they are based. Statistical data are always erroneous, in greater or less degree. The study of inductive reasoning is the study of the embryology of knowledge, of the processes by means of which truth is extracted from its native ore in which it is fused with much error. (Fisher, 1935d, p. 54)

Thus, according to Neyman, statistical hypotheses result in specific decisions that govern how we should behave in light of the decisions reached. On the other hand, Fisher's doctrine was that statistical hypotheses are a process of continually obtaining knowledge without ever definitely accepting a hypothesis.

After the Fisher–NP breakup in 1935, Neyman was to give his philosophy of statistical inference a more definite form and called it *inductive behavior*. The latter concept was formally introduced and contrasted with Fisher's inductive reasoning in Neyman's article in French entitled "L'estimation statistique traitée comme un problème classique de probabilité":

> Some authors attached the term "inductive reasoning" to statistical methods. If, after having made observations and calculated the limits $[8.04 \le \lambda' \le 11.96]$ the physician decides to assert that $8.04 \le \lambda' \le 11.96$, it seems to us that the process which has led to this assertion

> cannot be called inductive reasoning...But to decide to "assert" does not mean to "know" nor does it mean to "believe". It is a willful act which is preceded by some experience or some deductive reasoning, just like to take a life insurance even if one hopes to live long...If one wants a special term to describe these methods and in particular to describe the decision to assert that the inequalities (90) [i.e. $8.04 \leq \lambda' \leq 11.96$] exist, one may perhaps propose "inductive behavior". (Neyman, 1938b, 1967 reprint, p. 352)

In his book, *First Course in Probability and Statistics*, Neyman later gave a much more explicit exposition of his concept of inductive behavior on the very first page:

> [Inductive behavior] may be used to denote the adjustment of our behavior to limited amounts of observation. The adjustment is partly conscious and partly subconscious. The conscious part is based on certain rules (if I see this happening, then I do that) which we call rules of inductive behavior. In establishing these rules, the theory of probability and statistics both play an important role, and there is a considerable amount of reasoning involved. As usual, however, the reasoning is all deductive.
>
> ...
>
> Human progress is based on "permanencies" or, rather, on our ability to detect permanencies both in the objects surrounding us and in changes in these objects which we describe as phenomena...With many phenomena certain permanencies appear quite stable. This created the habit of regulating our actions in regard to some observed events by referring to the permanencies which at the particular moment seem to be established. This is what we call inductive behavior. (Neyman, 1950, p. 1)

Thus, when one accepts a particular hypothesis under the NP framework, one has expressed commitment to a particular permanency. Once this commitment is made, one creates rules of behavior based on deductive reasoning. This is the essence of inductive behavior. For example, in Neyman's terminology, the stability of long-term frequencies is a permanency. Based on the latter, the rules of probability can be developed by using deductive reasoning.

Later in his book, Neyman stated:

> *When the frequency function $p_X(e)$ is uncertain and when there are only two actions contemplated, the desirability of which depends on the nature of $p_X(e)$, then every rule of inductive behavior which determines the choice between the two possible actions, in accordance with the observed values of X, is called a test of a statistical hypothesis.** (*ibid.*, p. 259)

Thus, according to Neyman, a test of a statistical hypothesis is a rule of inductive behavior.

Fisher disputed the validity of Neyman's philosophy of inductive behavior, as described above. He charged that Neyman did not actually believe in the inductive process for his reasoning was only deductive:

> When, therefore, Neyman denies the existence of inductive reasoning he is merely expressing a verbal preference. For him "reasoning" means what "deductive reasoning" means to others. He does not tell us what in his vocabulary stands for inductive reasoning, for he does not clearly understand what that is. (Fisher, 1955, p. 74)

* Italics are Neyman's.

Furthermore, he accused Neyman of putting forward inductive behavior in "an effort to assimilate a test of significance to an acceptance procedure."

5.6.4.5 Conclusion In spite of Fisher's criticisms and Egon Pearson's own reluctance vis-à-vis inductive behavior,* the NP framework for hypothesis testing gained increasing popularity. Starting from 1939, Abraham Wald extended the NP theory to one of statistical decision functions (see Section 6.2). The major tenets of the NP theory continue to hold today. Concerning Fisher, it can safely be said that his theory of estimation is more popular than his theory of significance testing.

5.7 MAXIMUM LIKELIHOOD BEFORE FISHER

Before Fisher, several other mathematicians touched on the concept of likelihood in some way or another.† These include Johann Heinrich Lambert (1728–1777), Joseph-Louis Lagrange (1736–1813), Daniel Bernoulli (1700–1782), Robert Adrain (1775–1843), and especially Francis Ysidro Edgeworth (1845–1926). We now briefly describe the relevant contributions of each of these mathematicians. In each of the works examined, the reader will see that direct, rather than inverse, probability is used to model a set of observations and is then maximized, hence the relationship to maximum likelihood (ML).

5.7.1 Lambert and the Multinomial Distribution (1760)

Johann Heinrich Lambert‡ (Fig. 5.13) was a versatile Swiss scientist who made contributions to mathematics, probability, photometry, and philosophy. Lambert was born in Mulhouse, Alsace (now part of France) on August 26, 1728. His father was a tailor and Lambert had to leave school at the age of 12 to help him. Much of his early education was therefore self-taught. At the age of 20, he became tutor to the sons of a noble Swiss family. During the 10 years he spent with the latter, he was able to study, publish his own work, and travel in Europe to meet some of the great scientists of the time. In 1758, Lambert unsuccessfully sought a permanent position as Chair at the University of Göttingen. He then went to Zurich where he made astronomical observations with Gessner, was elected a member of the Physical Society of Zurich, and published *Die freye Perspektive*.¶ In 1764, he arrived in Berlin to be made member of the Prussian Academy of Sciences. But it is reported that his strange appearance and behavior delayed his appointment. Concerning Lambert's intellect, Gray has offered the following insights:

> ...He was no genius, but a man of great intelligence and imagination with occasional sparks of brilliance; he aimed at clarifying the fundamentals of his subjects, but was never

*Egon Pearson published his own rebuttal of Fisher (1955) in a paper entitled "Statistical concepts in their relation to reality" (Pearson, 1955). Pearson's paper, although it has been overlooked, is a most interesting read, not least because it shows some important nuances between his and Neyman's philosophies of hypothesis testing. Regarding Neyman's inductive behavior, this is what Pearson had to say in his paper: "[T]his is Professor Neyman's field rather than mine" (*ibid*., p. 207). See also Mayo (1996, Chapter 11).

†A survey of the use of likelihood before Fisher is also given in Edwards (1974) and Stigler (2007).

‡Lambert's life is also described in Gray (1978), Scriba (1973), Shafer (2006), and Daw (1980).

¶Free Perspective.

FIGURE 5.13 Johann Heinrich Lambert (1728–1777). Wikimedia Commons (Public Domain), http://commons.wikimedia.org/wiki/File:JHLambert.jpg

profoundly original either in his mathematics or in his science. The fascination of his work for us lies in the composite nature of his endeavour: he not only propounded a philosophy which showed significant modifications of current ideas, but also suggested a methodology of science in harmony with his philosophy. Further he was a practicing scientist who achieved at least a limited degree of success in his investigations. (Gray, 1978, p. 14)

Given the breadth of his knowledge, Lambert wrote many papers and books in German, Latin, and French. In mathematics, he was the first to demonstrate the irrationality of π and e. In physics, he is remembered for the law of cometary orbits and for the cosine law of photometry. Lambert's main philosophical work, *Neues Organon** (Lambert, 1764), dealt with the rules for distinguishing subjective and objective appearances. He also corresponded with the famous philosopher Immanuel Kant (1724–1804), author of the *Critique of Pure Reason* (Kant, 1781). Lambert died on September 25, 1777, in Berlin at the age of 49.

Lambert was 30 years old when he began writing his classic book *Photometria* (Lambert, 1760) (Fig. 5.14). In 2001, the book was translated in English and commented by David L. DiLaura (Lambert and DiLaura, 2001). In Section 295 of his book, Lambert made these interesting remarks:

Since in individual experiments in Photometry, as in countless others, aberrations are not equally frequent, another method is provided for determining the mean quantity from a finite number of them, so that the probability is greatest that it, of all, will differ least from the true quantity. (*ibid*., 102)

Lambert's above statement can be regarded as one of the earliest statements of the principle of ML.†

Thereupon, Lambert considered (Fig. 5.15) an unknown quantity AC, which is measured by experimentation. The four possible errors ("aberrations") that can be made are denoted by CP, CQ, CR, and CS. The latter correspond to the observations AP, AQ, AR, AS and have respective probabilities PN, QM, RL, and SK. Now suppose that, in a particular experiment, observations AP, AQ, AR, and AS have been made with frequencies n, m, l, and k, respectively. Lambert's aim was to maximize the probability of occurrence of the four observations. He wrote the probability of the observed sample as

$$N = \frac{\left[1\cdot 2\cdot 3\cdot 4\cdots(n+m+l+k)\right]PN^n QM^m RL^l SK^k}{(1\cdot 2\cdot 3\cdots n)(1\cdot 2\cdot 3\cdots m)(1\cdot 2\cdot 3\cdots l)(1\cdot 2\cdot 3\cdots k)} \tag{5.132}$$

(*ibid*., p. 103). Viewed as a function of PN, QM, RL, and SK, N is nowadays called the likelihood and is based on a multinomial distribution for the observed frequencies.

Moreover,

Since in general that case is most probable which occurs most frequently of all, it should be that

$$PN^n \cdot QM^m \cdot RL^l \cdot SK^k = \text{the greatest number.}$$

(*ibid*., p. 104)

* New Organon.

† Lambert's work is also described in Sheynin (1966; 1971, pp. 250–252) and Hald (1998, pp. 79–83).

I. H. LAMBERT
ACADEMIAE SCIENTIARVM ELECTORALIS BOICAE; ET SOCIETATIS PHYSICO-MEDICAE BASILIENSIS MEMBRI, REGIAE SOCIETATI SCIENTIARUM GOETINGENSI COMMERCIO LITERARIO ADIVNCTI

PHOTOMETRIA
SIVE
DE
MENSVRA ET GRADIBVS
LVMINIS,
COLORVM ET VMBRAE

AUGUSTAE VINDELICORUM,
Sumptibus VIDVAE EBERHARDI KLETT
Typis CHRISTOPHORI PETRI DETLEFFSEN.
MDCCLX.

FIGURE 5.14 Title page of Lambert's *Photometria* (Lambert, 1760)

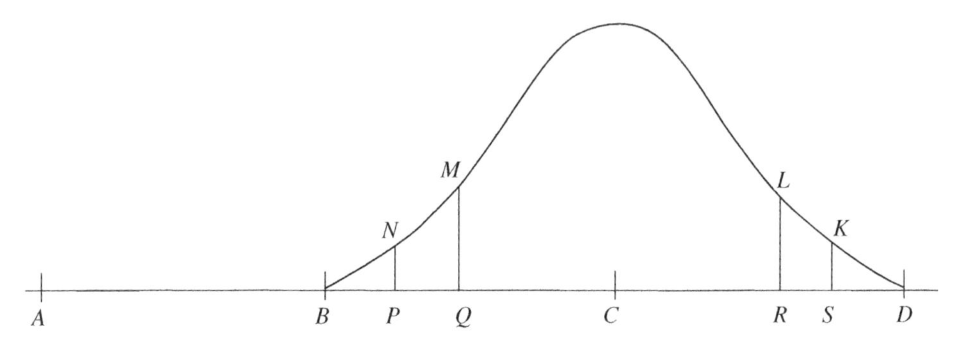

FIGURE 5.15 Lambert's maximum likelihood procedure

This is an explicit statement of the ML principle. Lambert wished to maximize N, hence $n \log N$, and therefore wrote

$$n \log(PN) + m\log(QM) + l \log(RL) + k \log(SK) = \text{maximum number.}$$

By taking differentials in the above and setting to zero, he obtained

$$n\frac{d(PN)}{PN} + m\frac{d(QM)}{QM} + l\frac{d(RL)}{RL} + k\frac{d(SK)}{SK} = 0.$$

The above is then written as

$$\frac{n}{\nu} + \frac{m}{\mu} + \frac{l}{\lambda} + \frac{k}{\chi} = 0, \qquad (5.133)$$

where $\nu = PN / d(PN)$, $\mu = QM / d(QM)$, and so on. Lambert next considered the case when only two observations AQ and AR were made, such that $n = k = 0$ and $m = l = 1$. Equation (5.133) becomes

$$\frac{1}{\mu} + \frac{1}{\lambda} = 0 \qquad (5.134)$$

(*ibid*., p. 105), so that $\lambda = \mu$ in magnitude. The latter condition holds only when C is equidistant from Q and R, so that

$$AC = \frac{AQ + AR}{2}$$

(*ibidem*), that is, AC is the arithmetic mean of AQ and AR.

Today, Lambert's problem in its original version is readily solved as follows. Denoting the log-likelihood, as in Lambert's solution, by

$$\begin{aligned} \log L &= n \log PN + m \log QM + l \log RL + k \log SK \\ &= n \log PN + m \log QM + l \log RL + k \log(1 - PN - QM - RL), \end{aligned}$$

we set

$$\frac{\partial \log L}{\partial PN} = \frac{\partial \log L}{\partial QM} = \frac{\partial \log L}{\partial RL} = 0.$$

The above are the ML equations, which give

$$\frac{n}{PN} - \frac{k}{1 - PN - QM - RL} = 0,$$
$$\frac{m}{QM} - \frac{k}{1 - PN - QM - RL} = 0,$$
$$\frac{l}{RL} - \frac{k}{1 - PN - QM - RL} = 0.$$

The solution of the above yields estimates of the probabilities *PN*, *QM*, *RL*, *SK*:

$$\widehat{PN} = \frac{n}{n+m+l+k},$$
$$\widehat{QM} = \frac{m}{n+m+l+k},$$
$$\widehat{RL} = \frac{l}{n+m+l+k},$$
$$\widehat{SK} = \frac{k}{n+m+l+k}.$$

Thus, the ML estimates of the probabilities are simply the respective observed frequencies.

5.7.2 Lagrange on the Average of Several Measurements (1776)

The second mathematician we shall consider in the context of ML is Joseph-Louis Lagrange. The latter's life has been described in Section 1.4.4. Our interest here is in "Mémoire sur l'utilité de la méthode de prendre le milieu entre les résultats de plusieurs observations*" (Lagrange, 1776). This memoir treats 11 problems having to do with the analysis of the errors arising from a set of observations. In each case, the distribution of the errors is given, and solutions are presented to problems such as the probability of the sum of errors being zero, the probability of the sum of errors lying between specified limits, and so on. *Problem VI* is of special interest because its solution involves an elementary application of the principle of ML.†

* Memoir on the usefulness of the method of taking the mean of the results of many observations.

† Lagrange's work is also described in Hald (1998, pp. 41–43).

Problem VI

19. I assume that we have verified some instrument, and having repeated the same verification several times, we have found different errors, each one of which is repeated a certain number of times; it is required to find what error must be taken for the correction of the instrument. (Lagrange, 1776, Oeuvres 2, p. 200)

In this problem, errors $p, q, r, \ldots$ have been observed $\alpha, \beta, \gamma, \ldots$ times in n repeated measurements ("*n vérifications*") so that $\alpha + \beta + \gamma + \cdots = n$. It is given that the "number of cases" that give the respective errors are $a, b, c, \ldots$. This means that the probabilities of the errors $p, q, r, \ldots$ are, respectively,

$$\frac{a}{a+b+c+\cdots}, \frac{b}{a+b+c+\cdots}, \frac{c}{a+b+c+\cdots}, \cdots \tag{5.135}$$

Lagrange's aim was to determine the single correction, based on the observed errors, that should be applied.

Lagrange denoted the probability of the errors $p, q, r, \ldots$ by the multinomial distribution

$$\frac{N a^{\alpha} b^{\beta} c^{\gamma} \ldots}{\left(a+b+c+\cdots\right)^{n}} \tag{5.136}$$

where

$$N = \frac{1 \cdot 2 \cdot 3 \cdots n}{1 \cdot 2 \cdot 3 \cdots \alpha \times 1 \cdot 2 \cdot 3 \cdots \beta \times 1 \cdot 2 \cdot 3 \cdots \gamma \times \cdots}$$

He now wished to determine the values of $a, b, c, \ldots$ that maximized the probability in (5.136). In modern terminology, he wanted to determine the ML estimates of the probabilities in (5.135). To do this, Lagrange referred to *Problem IV* (*ibid*., p. 195) where he used a clever algebraic reasoning (instead of the calculus), as follows. Considering the case of three errors p, q, and r, he let

$$M = \frac{1 \cdot 2 \cdot 3 \cdots n \times a^{\alpha} b^{\beta} c^{\gamma}}{1 \cdot 2 \cdot 3 \cdots \alpha \times 1 \cdot 2 \cdot 3 \cdots \beta \times 1 \cdot 2 \cdot 3 \cdots \gamma}$$

(*ibid*., p. 196). Then, assuming M was already a maximum, varying the exponents α, β, γ would make M decrease. Therefore increasing α to $\alpha + 1$ while at the same time decreasing* β to $\beta - 1$ changes the above expression to

$$\frac{\beta}{\alpha+1} \cdot \frac{aM}{b}$$

*If one exponent is increased by unity, another one needs also to be decreased by unity since $\alpha + \beta + \gamma = n$, a constant.

and we must have

$$\frac{\beta}{\alpha+1}\cdot\frac{aM}{b}<M.$$

Hence,

$$\frac{\beta}{\alpha+1}\cdot\frac{a}{b}<1. \qquad (5.137)$$

Lagrange now reasoned that if he had instead *decreased* α to $\alpha-1$ while at the same time increased β to $\beta+1$, a similar reasoning to the above would have led to

$$\frac{\alpha}{\beta+1}\cdot\frac{b}{a}<1. \qquad (5.138)$$

Equations (5.137) and (5.138) are both satisfied when

$$\frac{\alpha}{\beta+1}<\frac{a}{b}<\frac{\alpha+1}{\beta} \quad \Rightarrow \quad \frac{\alpha}{\beta}=\frac{a}{b}.$$

Similarly, $\alpha/\gamma=a/c$. Thus, if M is a maximum, then

$$\alpha=\rho a, \quad \beta=\rho b, \quad \gamma=\rho c,$$

where ρ is a constant. But since $\alpha+\beta+\gamma=n$, we have $\rho=n/(a+b+c)$ and Lagrange finally obtained

$$\alpha=\frac{na}{a+b+c}, \quad \beta=\frac{nb}{a+b+c}, \quad \gamma=\frac{nc}{a+b+c} \qquad (5.139)$$

(*ibid.*, p. 197).

Coming back to *Problem VI*, Lagrange used the above result to state that the distribution in (5.136) is maximized when

$$\alpha=\frac{na}{a+b+c+\cdots},$$
$$\beta=\frac{nb}{a+b+c+\cdots},$$
$$\gamma=\frac{nc}{a+b+c+\cdots},$$
$$\ldots$$

(*ibid.*, p. 201). By writing the "total number of cases" as $s=a+b+c+\cdots$, he expressed the above as

$$a=\frac{s\alpha}{n}, b=\frac{s\beta}{n}, c=\frac{s\gamma}{n}, \ldots$$

Although he undoubtedly recognized it, Lagrange did not state that the last result implied that M was maximized when probabilities of the various errors were estimated by their respective observed relative frequencies.

The above analysis also enabled Lagrange to answer his initial question, namely, how the average of the measurements taken by the instrument should be corrected. The required correction should be

$$\frac{ap+bq+cr+\cdots}{a+b+c+\cdots}=\frac{\alpha p+\beta q+\gamma r+\cdots}{n}$$

(*ibidem*), that is, the mean error across all n observations.

5.7.3 Daniel Bernoulli on the Choice of the Average Among Several Observations (1778)

In 1778, Daniel Bernoulli* (Fig. 5.16) published a paper that contains an early use of the principle of ML. Daniel Bernoulli was born on February 8, 1700, in Groningen, Netherlands. His father was the famous mathematician John Bernoulli, who was then Professor at Groningen. John was a bitter rival of his elder brother James, one of the founding fathers of the calculus of probabilities. In 1705, John moved to Basel, Switzerland, and took over the Chair of Mathematics, the latter being vacant owing to James' death. Daniel thus received his education in Basel. He studied philosophy and logic, passed his baccalaureate in 1715, and obtained his Master's degree in 1716. He was also taught mathematics by his father and his older brother Nicholas II. Daniel's father tried to force him to be a commercial apprentice, but when he refused, he was allowed to study medicine. Daniel completed his doctorate in 1721 on the mechanics of breathing. In 1724, he published his book *Exercitationes quaedam mathematicae*† while he was in Venice. The publication made him famous and he was invited to the St Petersburg Academy. Both Daniel and Nicholas II went to the Russian capital, but suddenly Nicholas II died there. Although Daniel was much affected by the loss of his dear brother, his years in St Petersburg were very productive. He wrote the outlines of his future book *Hydrodynamica* and also an original treatise on probability. Daniel returned to Basel in 1733, where he successively held chairs in anatomy and botany, in physiology, and in physics. Daniel's *Hydrodynamica* came out in 1738 and contained the famous Bernoulli's principle‡ that has immortalized his name in physics. However, in that same year, Daniel also published a memoir entitled "Specimen Theoriae Novae de Mensura Sortis", where he introduced the concept of utility to solve the St Petersburg paradox, (e.g., see Gorroochurn, 2012a, pp. 108–118). This paper turned out to be his most important contribution to the theory of probability and was in part responsible for the development of more sophisticated theories of utility and risk in the early twentieth century. Daniel had antagonistic relationships personally with his father and professionally with Jean le Rond d'Alembert (1717–1783) (see Section 5.8.4). It seems that John was irked by his own son's success as a mathematician. When he heard that Daniel was writing the *Hydrodynamica*, John wrote his own

* Daniel Bernoulli's life is also described in Gani (2001) and Straub (1970).

† Mathematical Exercises.

‡ Bernoulli's principle: As the velocity of a fluid increases, the pressure it exerts decreases (*ceteris paribus*).

FIGURE 5.16 Daniel Bernoulli (1700–1782). Wikimedia Commons (Public Domain), http://commons.wikimedia.org/wiki/Daniel_Bernoulli#/media/File:Daniel_Bernoulli_001.jpg

Hydraulica in which he copied large chunks from his son's work. Worse, he falsified the date of his book so that it would appear to predate Daniel's. On another occasion, when both won the 1734 Prize from the Paris Academy, John was furious that he had to share the prize with Daniel.* Concerning d'Alembert, Daniel had his most bitter dispute with him regarding the merits of inoculation against smallpox. The two also disagreed on the St Petersburg paradox and on the inclination of the planes of planetary orbits. Notwithstanding these controversies, Daniel was one of the most brilliant members of the prestigious Bernoulli family. He passed away on March 17, 1782, leaving a rich legacy of scientific work.

Daniel's use of the principle of ML was published in the paper entitled "Diiudicatio maxime probabilis plurium obseruationum discrepantium atque verisimillima inductio inde formanda"† (Bernoulli, 1778) (Fig. 5.17).‡ The memoir was translated in English by C.G. Allen and published by M.G. Kendall in 1961. Right at the start of the paper, Daniel made his motivation clear:

> I choose to propound ... doubts that I have sometimes entertained about the universally accepted rule for handling several slightly discrepant observations of the same event. By this rule the observations are added together and the sum divided by the number of observations; the quotient is then accepted as the true value of the required quantity, until better and more certain information is obtained. In this way, if the several observations can be considered as having, as it were, the same weight, the center of gravity is accepted as the true position of the objects under investigation. This rule agrees with that used in the theory of probability when all errors of observation are considered equally likely. (Bernoulli, 1778, English translation p. 3)

Daniel thus (wrongly) believed that the sample mean was a good estimator for the center of a set of observations only in the case of a uniform distribution, that is, only when "the observations have the same weight." Next, Daniel made a statement that is a rudimentary version of the principle of ML:

> ...of all the innumerable ways of dealing with errors of observation one should choose the one that has the highest degree of probability for the complex of observations as a whole. (*ibid.*, p. 4)

Daniel next sought a distribution for the set of observations $x_1, x_2, \ldots, x_n$ such that:

> Observations will ... be more numerous and indeed more probable near to the center of forces; at the same time they will be less numerous in proportion to their distance from that center. (*ibidem*)

Daniel chose a semicircular distribution with radius r as an example of such a distribution. He denoted the observations from smallest to largest as $A, A+a, A+b, A+c, \ldots$ and the distance between the center of the semicircle and the smallest observation by x (Fig. 5.18).

* Daniel won the prize a total of 10 times, a feat rivaled only by Euler who won it 12 times.

† The most probable choice between several discrepant observations and the formation therefrom of the most likely induction.

‡ Daniel Bernoulli's derivation is also described in Hald (1998, pp. 83–87).

DIIVDICATIO
MAXIME PROBABILIS
PLVRIVM OBSERVATIONVM DISCREPANTIVM
ATQVE
VERISIMILLIMA INDVCTIO INDE FORMANDA.

Auctore

DANIELE BERNOVLLI.

§. 1.

Aſtronomis potiſſimum, genti ſagaciſſime ſcrupuloſae, diiudicandas proponam haeſitationes, quas mihi aliquoties feci, de regula, ad quam confugiunt omnes, quoties plures de eadem re factas prae ſe habent obſeruationes aliquantulum inter ſe diſcrepantes; ſcilicet obſeruationes tunc omnes in vnam colligunt ſummam, quam poſtmodum diuidunt per obſeruationum numerum; quod a diuiſione oritur, pro vera accipiunt quantitate quaeſita, donec aliunde meliora et certiora fuerint edocti. Hoc modo ſi ſingulae obſeruationes eiusdem veluti ponderis cenſeantur, incidunt in centrum grauitatis,

A 2 quod

FIGURE 5.17 Original Latin version of Daniel Bernoulli's article "Diiudicatio maxime probabilis plurium obseruationum discrepantium atque verisimillima inductio inde formanda" (Bernoulli, 1778)

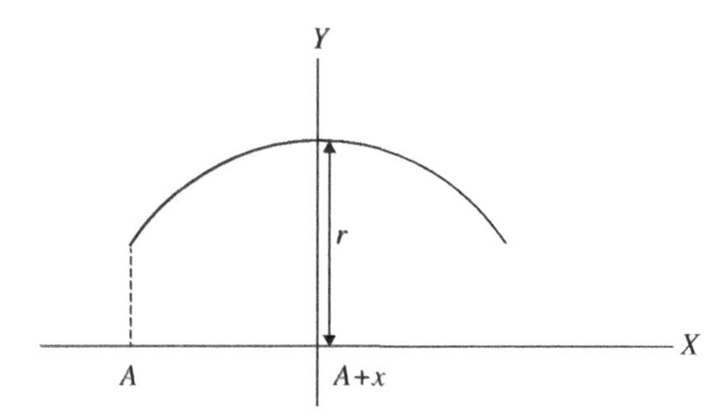

FIGURE 5.18 Semi-circular distribution considered by Daniel Bernoulli in his article "Diiudicatio maxime probabilis plurium obseruationum discrepantium atque verisimillima inductio inde formanda" (Bernoulli, 1778)

The sum $A+x$ is "the quantity which is most probably to be assumed on the basis of all the observations," that is, $A+x$ is the mode of the distribution.

We note that for n observations $A, A+a, A+b, A+c, \cdots$, the sample mean is

$$\frac{A+(A+a)+(A+b)+(A+c)+\cdots}{n} = A+\frac{a+b+c+\cdots}{n}.$$

Although the above was the "common sense" estimator of $A+x$, Daniel wished to prove that it did not correspond to the ML solution.

Daniel's probability curve in the XY-plane is given by

$$Y=\sqrt{r^2-(X-A-x)^2}, \quad A \le X \le A+2x. \tag{5.140}$$

Note that Y is not normalized (does not integrate to unity) and thus is not a proper probability density. However, this fact does not affect the search for ML solutions. The probability curve evaluated at $X{=}A$ is $\sqrt{r^2-x^2}$, at $X{=}A+a$ is $\sqrt{r^2-(a-x)^2}$, at $X{=}A+b$ is $\sqrt{r^2-(b-x)^2}$, and so on. Hence, "by the rules of the theory of probability," the probability density of the given sample is proportional to

$$\sqrt{r^2-x^2}\times\sqrt{r^2-(x-a)^2}\times\sqrt{r^2-(x-b)^2}\times\sqrt{r^2-(x-c)^2}\times\cdots \tag{5.141}$$

(*ibid*., p. 7). Viewed as a function of the parameter x, the above can be recognized as the likelihood function.

Daniel's aim was now to find the value of x that maximized (5.141). Since this was computationally inconvenient, he maximized the square of the expression instead (the maximum of both expressions occur at the same value(s) of x).

Daniel first considered the case of a single observation A. Then the square of (5.141) becomes r^2-x^2, which is maximized when

$$\frac{d}{dx}\left(r^2-x^2\right)=0 \quad \Rightarrow \quad x=0.$$

Therefore the ML solution is $A+x=A$, and "our hypothesis agrees with the common (arithmetic mean) one."

For the case of two observations, A and $A+a$, the square of (5.141) is $(r^2-x^2)\{r^2-(x-a)^2\}$. The latter is maximized when

$$\frac{d}{dx}\left(r^2-x^2\right)\left\{r^2-\left(x-a\right)^2\right\}=0 \quad \Rightarrow \quad x=\frac{a}{2}$$

is the "only useful root" (*ibid.*, p. 8). Therefore, the ML solution is $A+x=A+a/2$. This coincides with the arithmetic mean of A and $A+a$, and "this also is the teaching of the common hypothesis."

However, when there are three observations, A, $A+a$, and $A+b$, the square of (5.141) is $(r^2-x^2)\{r^2-(x-a)^2\}\{r^2-(x-b)^2\}$. This is maximized when

$$\begin{aligned}&\left(-2a^2br^2-2ab^2r^2+2ar^4+2br^4\right)+\left(-2a^2b^2+4a^2r^2+8abr^2+4b^2r^2-6r^4\right)x\\&+\left(6a^2b+6ab^2-12ar^2-12br^2\right)x^2+\left(-4a^2-16ab-4b^2+12r^2\right)x^3\\&+10\left(a+b\right)x^4-6x^5=0\end{aligned} \tag{5.142}$$

(*ibidem*). Now the "teaching of the common hypothesis" would lead us to estimate x from $A+x=(A+A+a+A+b)/3$ as $x=(a+b)/3$. Daniel then proceeded to show that, although there were situations when the solution of the quintic in (5.142) corresponded to the common (arithmetic mean) solution, this was not the case in general.

For example, Daniel observed that, when $r\to\infty$, Eq. (5.142) reduced to $2ar^4+2br^4-6r^4x=0$ and the common solution $x=(a+b)/3$ is obtained (*ibid.*, p. 9). Other cases considered by Daniel where the two solutions matched were $b=2a$ and $b=-a$.

For other cases, Daniel gave numerical examples that showed discrepancies between the two solutions. As a first example, he considered the three observations A, $A+.2$, and $A+1$, that is, $a=.2$ and $b=1$. The common solution is $x=.4$. With $a=.2$, $b=1$, and $r=1$ in Eq. (5.142), the MLE is $x=.4427$. This value "exceeds the commonly accepted value by more than a tenth" and "is due to the fact that the middle observation is much nearer to the first than the third." Using some more numerical examples, Daniel concluded that:

> ...the excess will be changed to a defect of the middle observation is nearer to the third than to the first, and the nearer the middle observation to the mean between the two extreme observations, the smaller will be this defect. (*ibidem*)

Daniel thus showed that the sample mean does not coincide with the ML solution. But his implication that the sample mean is suboptimal unless the distribution is uniform is wrong. The sample mean does not correspond to the ML solution because, in this case, the latter is biased and not because the former is deficient.

5.7.4 Adrain's Two Derivations of the Normal Law (1808)

Robert Adrain's maximization of direct probability to derive the normal distribution is described in Section 1.5.2.

5.7.5 Edgeworth and the Genuine Inverse Method (1908, 1909)

We finally turn to Francis Ysidro Edgeworth. Of all the mathematicians we have considered in this section, none can perhaps make a stronger claim to have anticipated Fisher's method of ML than Edgeworth. The latter's relevant work occurred in a set of four papers, "On the probable errors of frequency-constants" (Edgeworth, 1908; 1909), and has been studied in detail by Pratt (1976). Edgeworth considered a sample of independent and identically distributed observations and his method was based on maximizing a posterior distribution based on a uniform prior, which he called the "genuine inverse method." We note that the posterior mode thus obtained, being based on a *uniform* prior, coincides with the usual MLE.

First Edgeworth derived the variance of the posterior distribution as follows* (Edgeworth, 1908, pp. 500–504). He assumed an observation x_i ($i=1, 2, \ldots, n$) had a density proportional to $\exp\{\psi(x_i - x)\}$, where ψ is an arbitrary function and x is a parameter.† Then, assuming a uniform prior for x, its posterior density given a set of observations $x_1, x_2, \ldots, x_n$ can be written as

$$y = H\exp\{\psi(x_1 - x) + \psi(x_2 - x) + \cdots\} \tag{5.143}$$

(*ibid*., p. 500), where H is a normalizing constant. The estimated value x' (the *quaesitum*) of the parameter x that maximizes y must satisfy

$$\frac{d}{dx}\Sigma\psi(x_t - x)\bigg|_{x=x'} = 0.$$

Now, let $x = x' + \mathbf{x}$ and expand $\Sigma\psi(x_t - x)$ in powers of $\mathbf{x}$ as follows:

$$\begin{aligned}\Sigma\psi(x_t - x) &= \Sigma\psi(x_t - x' - \mathbf{x}) \\ &= \Sigma\psi(x_t - x') - \mathbf{x}\frac{d}{dx}\Sigma\psi(x_t - x)\bigg|_{x=x'} + \frac{\mathbf{x}^2}{2}\frac{d^2}{dx^2}\Sigma\psi(x_t - x)\bigg|_{x=x'} + \cdots \\ &= \Sigma\psi(x_t - x') + \frac{\mathbf{x}^2}{2}\Sigma\psi''(x_t - x'),\end{aligned}$$

* The same method had been used before by Laplace.

† The term "parameter" was introduced much later by Fisher (1922c); at the time of Edgeworth's writing, it was called "frequency constant."

ignoring terms in $\mathbf{x}^3$ and higher powers. Equation (5.143) then gives the posterior distribution as

$$\begin{aligned} y &= H\exp\left\{\Sigma\psi(x_t - x') + \frac{\mathbf{x}^2}{2}\Sigma\psi''(x_t - x')\right\} \\ &= H\exp\{\Sigma\psi(x_t - x')\}\times\exp\left\{\frac{\mathbf{x}^2}{2}\Sigma\psi''(x_t - x')\right\} \\ &= H_0\exp\left\{\frac{(x - x')^2}{2}\Sigma\psi''(x_t - x')\right\}. \end{aligned}$$

Edgeworth recognized the above as implying that the posterior distribution of x is normal with mean x' and inverse modulus squared equal to

$$-\frac{1}{2}\Sigma\psi''(x) \tag{5.144}$$

(*ibid*., p. 504). Since the modulus is $\sqrt{2}$ times the standard deviation, Eq. (5.144) means that the posterior variance of x is

$$\operatorname{var} x = -\frac{1}{\Sigma\psi''(x)}.$$

Edgeworth's result in (5.144) implies another important formula. From Eq. (5.143), $\Sigma\psi = \Sigma\log f - n\log H$, where f is the density of each x_i, so that $\Sigma\psi'' = d^2\,\Sigma\log f/dx^2$. Therefore,

$$\operatorname{var} x = -\frac{1}{\sum\frac{d^2\log f}{dx^2}} = -\frac{1}{n\mathsf{E}\frac{d^2\log f}{dx^2}} \tag{5.145}$$

for large n. This is the same as Fisher's formula in (5.26) except that Edgeworth's method is based on inverse probability while Fisher's is based on direct sampling theory.*

Edgeworth was now ready to prove a statement he had made earlier:

> [the posterior mode] not only…is the *most probable*† value, but also that it is the *most advantageous* value (in the classical sense of the term above explained with reference to our *first* scruple), provided that the number of observations is so large that the law of distribution for the *quaesitum* does not differ significantly from a normal curve of error. (*ibid*., p. 391)

* It would make no sense to evaluate the variance of a parameter under sampling theory (as the answer would always be zero).

† All italics in this quotation are Edgeworth's.

Edgeworth interpreted "most advantageous" in the sense of *minimum* mean square error:

> ...the Mean Square of deviation from the true point...is less for the formula given by the inversion proper than it is for the Arithmetic Mean, and by parity of reasoning, for any other rival method, say $\chi(x_1, x_2, \ldots, x_n)$. (*ibid.*, p. 507)

He thus wished to prove that

$$\mathsf{E}(x'-x)^2 \leq \mathsf{E}(\hat{\chi}-x)^2, \tag{5.146}$$

where x' is the estimator (the *quaesitum*) of the parameter x that maximizes the posterior density in (5.143), and $\hat{\chi}$ is any other estimator of x such as $\hat{\chi}=\bar{x}$, the sample mean. The proof of the above was supplied by Professor Augustus E.H. Love (1863–1940) at Edgeworth's request and is in the appendix to the third paper (Edgeworth, 1908, pp. 662–665). The proof is as follows.

First, Eq. (5.145) is used to write the left side of Eq. (5.146) as

$$\mathsf{E}(x'-x)^2 = -\frac{1}{n\mathsf{E}\dfrac{d^2\log f}{dx^2}} = \frac{1}{n\displaystyle\int_{-\infty}^{\infty}\frac{(df/dx)^2}{f}dx}.$$

The above is true since

$$\begin{aligned}\mathsf{E}\frac{d^2\log f}{dx^2} &= \mathsf{E}\frac{\partial}{d\theta}\left(\frac{1}{f}\frac{df}{dx}\right)\\ &= \mathsf{E}\left\{-\frac{1}{f^2}\left(\frac{df}{dx}\right)^2 + \frac{1}{f}\left(\frac{d^2 f}{dx^2}\right)\right\}\\ &= -\mathsf{E}\frac{1}{f^2}\left(\frac{df}{dx}\right)^2\\ &= -\int_a^b \frac{1}{f}\left(\frac{df}{dx}\right)^2 dx.\end{aligned}$$

(*ibid.*, p. 663). In the above, it has been assumed that df/dx is independent of x so that $d^2 f/dx^2 = 0$. It is also assumed that $\mathsf{E}d^2\log f/d\theta^2 < \infty$ and that f and df/dx are both continuous functions that vanish at $x=a$ and $x=b$.

Second, using $\hat{\chi}=\bar{x}$, the right side of Eq. (5.146) is

$$\mathsf{E}(\bar{x}-x)^2 = \operatorname{var}\bar{x} = \frac{1}{n}\left\{\int_{-\infty}^{\infty} x^2 f dx - \left(\int_{-\infty}^{\infty} x f dx\right)^2\right\}.$$

By assuming $\int_{-\infty}^{\infty} xfdx = 0$, the above becomes

$$\mathsf{E}(\bar{x}-x)^2 = \frac{1}{n}\int_{-\infty}^{\infty} x^2 f dx.$$

What remains to be proved is therefore obtained from Eq. (5.146) as

$$\frac{1}{\int_a^b \frac{(df/dx)^2}{f}dx} \le \int_a^b x^2 f dx \tag{5.147}$$

(*ibid.*, p. 664). Love's proof was based on the Cauchy–Schwarz inequality as follows. Suppose u and v are two real functions that are integrable between $x = a$ and $x = b$. Then, the Cauchy–Schwarz inequality states that

$$\left(\int_a^b uv dx\right)^2 \le \left(\int_a^b u^2 dx\right)\left(\int_a^b v^2 dx\right). \tag{5.148}$$

In the above, let

$$u = \frac{1}{\sqrt{f}}\frac{df}{dx},$$
$$v = x\sqrt{f}.$$

Equation (5.148) then becomes

$$\begin{aligned}\left\{\int_a^b \frac{1}{f}\left(\frac{df}{dx}\right)^2 dx\right\}\left\{\int_a^b x^2 f dx\right\} &\ge \left\{\int_a^b x\frac{df}{dx}dx\right\}^2 \\ &= \left([xf]_a^b - \int_a^b f dx\right)^2 \\ &= (0-1)^2 \\ &= 1\end{aligned}$$

(*ibidem*). Equation (5.147) (and hence Eq. 5.146) is thus proved.

In the 1909 Addendum, Edgeworth's work edged even closer to ML. As we shall see, here Edgeworth *explicitly* departed from inverse probability and faced the estimation problem from a purely sampling theory approach. At the start of the Addendum, Edgeworth explained:

> The general theorem…may itself likewise be proved by a direct method free from the speculative character which attaches to inverse probability. For this purpose we no longer take our stand on a particular set of observations in order thence to remount *a posterior* to the originating cause; rather we watch in prior experience the distribution of observations and determine that function of which the several values, each formed from a large set of observations, hover with minimum dispersion about the true value of some constant represented by a symmetrical function of the observations. (Edgeworth, 1909, p. 82)

Thereupon, Edgeworth considered an estimator χ that had a normal distribution with mean x' when the sample size n was large. Moreover, χ belonged "to the comprehensive class (of functions constituting averages), which is defined by the equation (solved for χ)"

$$\sum_t \theta\left(\frac{x_t - \chi}{c}\right) = 0 \tag{5.149}$$

(*ibid*., p. 82), where θ is a function and c is a measure of scale. The modern term for χ is "*M*-estimator." Edgeworth wished to obtain the asymptotic variance of χ, which he did as follows. First he defined

$$\xi = \frac{\chi - x'}{c}$$

and

$$\mathbf{x}_t = \frac{x_t - x'}{c}.$$

Equation (5.149) then becomes

$$\sum_t \theta(\mathbf{x}_t - \xi) = 0. \tag{5.150}$$

Then, using a Taylor series expansion for $\theta(\mathbf{x}_t - \xi)$ about $\mathbf{x}_t$, and noting that ξ is typically small, the above becomes

$$\sum_t \{\theta(\mathbf{x}_t) - \xi\theta'(\mathbf{x}_t) + \cdots\} = \sum_t \theta(\mathbf{x}_t) - \xi \sum_t \theta'(\mathbf{x}_t) = 0,$$

where terms quadratic in ξ and higher powers have been ignored. From the above,

$$\xi = \frac{\sum_t \theta(\mathbf{x}_t)}{\sum_t \theta'(\mathbf{x}_t)}.$$

Therefore,

$$\operatorname{var}\xi = \frac{\sum_t \operatorname{var}\theta(\mathbf{x}_t)}{\left\{\sum_t \theta'(\mathbf{x}_t)\right\}^2} = \frac{n\{\mathsf{E}\theta^2(\mathbf{x}_t) - \mathsf{E}^2\theta(\mathbf{x}_t)\}}{\left\{\sum_t \theta'(\mathbf{x}_t)\right\}^2}.$$

By the law of large numbers, we have $\left\{\sum_t \theta'(\mathbf{x}_t)\right\}/n \approx \mathsf{E}\theta'(\mathbf{x}_t)$ for large n. The variance above thus becomes

$$\operatorname{var}\xi = \frac{\mathsf{E}\theta^2 - \mathsf{E}^2\theta}{\mathsf{E}^2\theta'} = \frac{\int \theta^2 y d\mathbf{x} - \left(\int \theta y d\mathbf{x}\right)^2}{\left(\int \theta' y d\mathbf{x}\right)^2},$$

where y is the density of $\mathbf{x}_t$. By assuming $\mathsf{E}\theta = 0$, Edgeworth finally obtained

$$\operatorname{var}\xi = \frac{\int \theta^2 y d\mathbf{x}}{\left(\int \theta' y d\mathbf{x}\right)^2} \tag{5.151}$$

(*ibid*., p. 83). Edgeworth next used the calculus of variations to show that the variance in Eq. (5.151) was minimized when

$$\theta \propto \frac{1}{y}\frac{dy}{dx}$$

(*ibid*., p. 84).

We note that, since $\theta \propto y'/y$, Eq. (5.150) is simply the ML equation, namely,

$$\sum_{t} \frac{y'(\mathbf{x}_t - \xi)}{y(\mathbf{x}_t - \xi)} = 0.$$

Therefore, Edgeworth had effectively proved that, within the class of M-estimators, the MLE had the smallest variance asymptotically.

From the above, the reader will hopefully realize how close Edgeworth was to ML. However, Fisher seems to have been unaware of these early contributions when he later formally developed likelihood theory. Hald has thus commented that:

> …it is strange that Fisher does not refer to Edgeworth; he seems to have overlooked Edgeworth's papers not only in 1912 but also in 1922. (Hald, 1998, p. 717)

It must the admitted, though, that Edgeworth's abstruse style of writing did not help in popularizing his work.

5.8 SIGNIFICANCE TESTING BEFORE FISHER

In this section, we outline some of the attempts that were made at significance testing prior to Fisher's groundbreaking contributions.

5.8.1 Arbuthnot on Divine Providence: The First Published Test of a Statistical Hypothesis (1710)

John Arbuthnot* (Fig. 5.19) was born in Kincardineshire, Scotland, on April 29, 1667. His father was an Episcopal clergyman. Arbuthnot entered Marischal College, Aberdeen, at the age of 14 and obtained an M.A. in medicine in 1685. After his father's death in 1691, Arbuthnot left for England and initially taught mathematics to make a living. During that time, he also famously translated and annotated Christiaan Huygens' (1629–1695) book *De Ratiociniis in ludo aleae*† (Huygens, 1657). The English translation was published in 1692 as *On the Laws of Chance* (Arbuthnot, 1692). In 1696, Arbuthnot obtained his Doctor's degree in Medicine in St. Andrews and thereafter decided to settle in London. Soon he became well known as a competent physician and a man of learning.

*Arbuthnot's life is also described in Pearson (1978, pp. 127–129), Shoesmith (2006, pp. 203–207), and Freudenthal (1970).

†On the Calculations in Games of Chance.

FIGURE 5.19 John Arbuthnot (1667–1735). Wikimedia Commons (Public Domain), http://commons.wikimedia.org/wiki/John_Arbuthnot#/media/File:Arbuthnot_John_Kneller.jpg

After a few scientific publications, Arbuthnot was elected a Fellow of the Royal Society (1704) as well as the Royal College of Physicians (1710). In 1705, Arbuthnot became Physician Extraordinary to Queen Anne. He was also famous for his comical satirical writings. Arbuthnot passed away at London on February 27, 1735.

In 1710, Arbuthnot communicated a memoir entitled "An Argument for Divine Providence taken from the constant regularity of the births of both sexes" (Arbuthnot, 1710) (Fig. 5.20) to the Royal Society. This paper is generally regarded as containing the first published test of a statistical hypothesis.*

Arbuthnot started his paper by claiming that the exact balance between the numbers of men and women was "not the effect of chance but Divine Providence, working for a good end."

To prove his point, he modeled births by considering a set of n fair dice, each of which had two sides labeled M (for male) and F (for female). Then the "number of chances" for all combinations of males and females are given by the coefficients of the following binomial expansion:

$$(M+F)^n = M^n + \frac{n}{1}\times M^{n-1}F + \frac{n}{1}\times\frac{n-1}{2}\times M^{n-2}F^2 + \frac{n}{1}\times\frac{n-1}{2}\times\frac{n-2}{3}\times M^{n-3}F^3 + \cdots \quad (5.152)$$

(*ibid.*, p. 187). Thus, "the number of chances" for n males is unity; for $n-1$ males and 1 female is $n(n-1)/2$ and so on. In modern statistical language, for the set of n fair dice, the number of ways of obtaining k M's and $n-k$ F's is given by the coefficient of $M^k F^{n-k}$ in the expansion of $(M+F)^n$, this coefficient being

$$\binom{n}{k} = \frac{n(n-1)\cdots(n-k+1)}{k!}.$$

Moreover, the probability of obtaining k M's (and $n-k$ F's) is equal to the above coefficient divided 2^n, that is,

$$\frac{1}{2^n}\binom{n}{k}.$$

Continuing with his analysis, Arbuthnot next assumed n was an even number. Then, if somebody wished to obtain an equal number of M's and F's:

> ...his Lot is to the Sum of all the Chances, as the coefficient of the middle Term is to the power of 2 raised to an exponent equal to the Number of Dice.... (*ibidem*)

To find the middle term in (5.152):

> ...continue the Series $\frac{n}{1}\times\frac{n-1}{2}\times\frac{n-2}{2}\times\cdots$ till the number of terms are equal to $\frac{1}{2}n$. (*ibidem*)

*Arbuthnot's problem is also discussed in Pearson (1978, pp. 131–133), Shoesmith (1985; 1987), Hald (1990, Chapter 17), Bellhouse (2002), Hacking (1965, pp. 75–81), Chatterjee (2003, pp. 174–175), and Samueli (2012, Chapter 11).

(186)

II. *An Argument for Divine Providence, taken from the conſtant Regularity obſerv'd in the Births of both Sexes. By Dr.* John Arbuthnott, *Phyſitian in Ordinary to Her Majeſty, and Fellow of the College of Phyſitians and the Royal Society.*

AMong innumerable Footſteps of Divine Providence to be found in the Works of Nature, there is a very remarkable one to be obſerved in the exact Ballance that is maintained, between the Numbers of Men and Women; for by this means it is provided, that the Species may never fail, nor periſh, ſince every Male may have its Female, and of a proportionable Age. This Equality of Males and Females is not the Effect of Chance but Divine Providence, working for a good End, which I thus demonſtrate:

Let there be a Die of Two ſides, M and F, (which denote Croſs and Pile), now to find all the Chances of any determinate Number of ſuch Dice, let the Binome M+F be raiſed to the Power, whoſe Exponent is the Number of Dice given; the Coefficients of the Terms will ſhew all the Chances ſought. For Example, in Two Dice of Two ſides M+F the Chances are $M^2+2MF+F^2$, that is, One Chance for M double, One for F double, and Two for M ſingle and F ſingle; in Four ſuch Dice there are Chances $M^4+4M^3F+6M^2F^2+4MF^3+F^4$, that is, One Chance for M quadruple, One for F quadruple, Four for triple M and ſingle F, Four for ſingle M and triple F, and Six for M double and F double; and univerſally, if the Number of Dice be *n*, all their Chances will be expreſſed in this Series

M^n+

FIGURE 5.20 First page of Arbuthnot's 1710 memoir (Arbuthnot, 1710)

Thus, the probability of $n/2\,M$'s is

$$\frac{1}{2^n}\binom{n}{n/2}.$$

Next Arbuthnot argued that for large n there was a very small probability of obtaining an equal number of M's and F's:

> It is visible from what has been said, that with a very great Number of Dice... (supposing M to denote Male and F Female) ... there would be but a small part of all the possible Chances, for its happening at any assignable time, that an equal Number of Males and Females should be born. (*ibidem*)

Arbuthnot's observation above is correct as it can be shown that the probability above tends to zero as n becomes infinitely large.

Up to now, Arbuthnot showed a thorough familiarity with the binomial distribution. However, Arbuthnot next contended that, assuming again that the dice were fair, it was also very unlikely that there should be *approximately* the same number of M's and F's. He stated that the probability was also small of not observing a considerable excess of M's over F's, or vice versa, that is, it was quite likely that there *should* be such an excess:

> But it is very improbable (if mere Chance govern'd) that they would never reach as far as the Extremities. (*ibid.*, p. 188)

Arbuthnot's above reasoning is incorrect, as a simple calculation can show. For example, letting X_n denote the number of M's ("successes") in a Bino(n, 1/2) distribution,

$$\Pr\{X_{82}=41\}=\frac{1}{2^{82}}\binom{82}{41}=8.78\times10^{-2},$$

$$\Pr\{X_{82}\geq 60\}=\frac{1}{2^{82}}\sum_{i=60}^{82}\binom{82}{i}=1.62\times10^{-5}.$$

There is thus an even *smaller* probability of observing at least 60 M's compared to 41 M's in a toss of 82 fair dice. Nevertheless, Arbuthnot's erroneous reasoning was used to make his first argument for some divine force at work rather than mere chance, as follows:

> But this Event is wisely prevented by the wise Oeconomy [*sic*] of Nature; and to judge of the wisdom of the Contrivance, we must observe that the external Accidents to which are Males subject (who must seek their Food with danger) do make a great havock of them, and that this loss exceeds far that of the other Sex, occasioned by Diseases incident to it, as Experience convinces us. To repair that Loss, provident Nature, by the Disposal of its wife Creator, brings forth more Males than Females; and that in almost a constant proportion. (*ibidem*)

Arbuthnot's above argument was based on the false premise that there should be a *considerable* excess of one sex under chance alone. Since in actual life a considerable excess is prevented, he argued there must exist a divine force other than chance. Moreover, such a force wisely made sure there was a consistent *slight* excess of males over females across the years, in almost constant proportion, since males are more prone to danger and death. As a case in point, Arbuthnot showed a table (see Table 5.36) of the christenings in London for 82 consecutive years from 1629 to 1710. For each of these years, there was an excess of males over females, that is, each year was a "male year." However, in each year the excess was only slight.

Arbuthnot next proceeded to give a second statistical argument to prove his claim in favor of "Divine Providence." Assuming that the probability of a "male year" (a year with more male births than female births) is 1/2, he used logarithms to calculate the probability of observing 82 male years out of 82 years as

$$\frac{1}{4{,}836{,}000{,}000{,}000{,}000{,}000{,}000{,}000}$$

(*ibidem*).* Note that the above is simply

$$\Pr\{X_{82} = 82\} = \frac{1}{2^{82}}\binom{82}{82} \approx 2.07 \times 10^{-25}.$$

In face of such a small probability, Arbuthnot concluded:

> ...if A wager with B, not only that the Number of Males shall exceed that of Females, every Year, but that this Excess shall happen in a constant Proportion, and the Difference lye within fix'd limits; and this not only for 82 Years, but for Ages of Ages, and not only at *London*, but all over the World; (which 'tis highly probable is Fact, and designed that every Male may have a Female of the same Country and suitable Age) then A's Chance will be near an infinitely small Quantity, at least less than any assignable Fraction. From whence it follows, that it is Art, not Chance, that governs. (*ibid*., pp. 188–189)

Arbuthnot's point in the above was that since, under the assumption of chance alone, the probability of 82 consecutive male years was so small, the hypothesis of chance must be rejected in favor of divine providence or "art." Although Arbuthnot's second statistical argument is valid, many would take issue with his subsequent conclusion regarding divine providence. Arbuthnot's test certainly suggests that it is very unlikely that the probability of a male year is half. This could be explained today by evolutionary forces such as natural selection. However, Arbuthnot lived at a time when religion and fear of God were deeply entrenched in society, and there was more than a century to wait before Darwin's *The Origin of Species* would come and radically transform the social fabric.

It should also be noted that Arbuthnot's test belongs to the distribution-free class of tests since it makes no assumption about the underlying distribution of the number of births of either gender. The distribution-free test that he actually used is today known as the *sign test*.

*Arbuthnot's fraction should contain an extra zero in the denominator. The correct fraction is given here.

TABLE 5.36 Christenings in London for 82 Consecutive Years

	Christened.			Christened.	
Anno.	Males	Females	Anno.	Males	Females
1629	5218	4683	1648	3363	3181
30	4858	4457	49	3079	2746
31	4432	4102	50	2890	2722
32	4994	4590	51	3231	2840
33	5158	4839	52	3220	2908
34	5035	4820	53	3196	2959
35	5106	4928	54	3441	3179
36	4917	4605	55	3655	3349
37	4703	4457	56	3668	3382
38	5359	4952	57	3396	3289
39	5366	4784	58	3157	3013
40	5518	5332	59	3209	2781
41	5470	5200	60	3724	3247
42	5460	4910	61	4748	4107
43	4793	4617	62	5216	4803
44	4107	3997	63	5411	4881
45	4047	3919	64	6041	5681
46	3768	3395	65	5114	4858
47	3796	3536	66	4678	4319

	Christened.			Christened.	
Anno.	Males	Females	Anno.	Males	Females
1667	5616	5322	1689	7604	7167
68	6073	5560	90	7909	7302
69	6506	5829	91	7662	7392
70	6278	5719	92	7602	7316
71	6449	6061	93	7676	7483
72	6443	6120	94	6985	6647
73	6073	5822	95	7263	6713
74	6113	5738	96	7632	7229
75	6058	5717	97	8062	7767
76	6552	5847	98	8426	7626
77	6423	6203	99	7911	7452
78	6568	6033	1700	7578	7061
79	6247	6041	1701	8102	7514
80	6548	6299	1702	8031	7656
81	6822	6533	1703	7765	7683
82	6909	6744	1704	6113	5738
83	7577	7158	1705	8366	7779
84	7575	7127	1706	7952	7417
85	7484	7246	1707	8379	7687
86	7575	7119	1708	8239	7623
87	7737	7214	1709	7840	7380
88	7487	7101	1710	7640	7288

From Arbuthnot (1710, pp. 189–190).

Arbuthnot's paper was quite a sensation and elicited reactions from a wide range of personalities. For instance, Hacking has noted:

> Arbuthnot's argument for divine providence was instantly adopted by theologians and preached, midst more familiar nonsense, from the pulpits of Oxford and Munich for the rest of the century. (Hacking, 1965, p. 77)

Of course, several mathematicians of the time also got involved, among whom we shall consider Willem Jacob 's Gravesande (1688–1742) and Nicholas Bernoulli (1687–1759).

5.8.2 's Gravesande on the Arbuthnot Problem (1712)

Willem 's Gravesande (Fig. 5.21) was a Dutch mathematician and physicist, and one of the most brilliant scientists of his time. According to Karl Pearson (Pearson, 1978, p. 301), he must have been around 29 when he was asked by his colleague Bernard Nieuwentyt (1654–1718) to weigh on the Arbuthnot problem when Nieuwentyt was writing his book *Het regt gebruik der Wereldbeschouwingen* (Nieuwentyt, 1715).* 's Gravesande's calculations were included in the book and also appeared in the *Oeuvres Philosophiques et Mathématiques de Mr G.F. 's Gravesande* (Allamand, 1774, pp. 221–248) (Fig. 5.22).

's Gravesande's solution was as follows. First, reproducing the table in Arbuthnot's paper, he calculated the mean number of births for the 82 years as 11,429. Next he found that the least difference between males and females occurred in 1703, when 7765 males and 7683 females were christened. When this difference was adjusted for a population of 11,429, 's Gravesande obtained 5,745† males and 5684 females. Similarly, the greatest difference between males and females occurred in 1661, and when adjusted gave 6128 males and 5301 females. Since all 82 years, when adjusted for the population size of 11,429, gave between 5,745 and 6,128 males, 's Gravesande now set out to calculate:

> ...the odds in favor of A, who has wagered against B, that of 11429 children who will be born in a year, the number of males will lie between the two limits 5745 & 6128. (*ibid.*, p. 230)

Assuming an equal probability for males and females, 's Gravesande's calculated the required odds by using the recursion

$$\binom{n}{x+1} = \binom{n}{x}\frac{n-x}{x+1}$$

and tabulating values of

$$10^5 \frac{\binom{n}{x}}{\binom{n}{5{,}715}}$$

* The English translation of the book read: "The Religious Philosopher: Or, the Right Use of Contemplating the Works of the Creator."

† $7{,}765 \times 11{,}429/(7{,}765+7{,}683) = 5{,}745$. Similarly for all other adjustments.

FIGURE 5.21 Willem Jacob 's Gravesande (1688–1742). Wikimedia Commons (Public Domain), http://commons.wikimedia.org/wiki/File:Willemsgravesande.jpg

DÉMONSTRATION Mathématique du ſoin que Dieu prend de diriger ce qui ſe paſſe dans ce monde, tirée du nombre des Garçons & des Filles qui naiſſent journellement (*).

La contemplation de cet Univers, & de ce que nous y voyons arriver tous les jours, nous fournit une infinité de preuves de l'exiſtence d'un Etre, qui non ſeulement a créé les cieux & la terre, & en a réglé le cours en les aſſujettiſſant à des loix fixes & immuables, mais qui encore dirige continuellement ce qui s'y paſſe. Les créatures qui nous paroiſſent être de la plus petite conſéquence, les évènemens qui ſemblent à peine mériter notre attention, pourroient fournir des raiſons capables de fermer la bouche aux Athées les plus ſubtils, & de démontrer l'exiſtence d'un Dieu, ſi on les propoſoit de manière à en faire ſentir toute la force. 1.

Le nombre des Enfans qui naiſſent en eſt un exemple; peu de gens font reflexion à ce qu'il nous offre de remarquable; & qui conſiſte en ce qu'il nait à peu près autant de garçons que de filles, mais de façon cependant que le nombre de ceux-là ſurpaſſe toujours un peu le nombre de celles-ci. Ce ſeul fait, examiné avec attention, prouve démonſtrativement que la naiſſance des Enfans eſt dirigée par un Etre intelligent, de qui elle dépend.

Je vais travailler à mettre cette preuve dans tout ſon jour. Pour cela je me bornerai aux Enfans nés dans la Ville de Londres, & cela ſeulement pendant l'eſpace de quatre-vingt-deux ans, ſçavoir depuis le commencement de 1629. juſqu'à la fin de 1710. J'en donnerai ici la liſte tirée des régîtres des Enfans qu'on y bâtiſe: régîtres qu'on conſerve dans les Egliſes de cette Ville; & je laiſſerai juger au Lecteur du nouveau dégré de force qu'acquerroit cette preuve, ſi l'on appliquoit les calculs que je vais faire à tout un pays, & à une plus longue ſuite d'années. 2.

C A-

(*) On trouve dans le N°. 328. des *Tranſactions Philoſophiques* de la Société Royale de Londres, une Lettre du Dr. *Arbuthnot*, dans laquelle il démontre que la régularité qu'on obſerve dans la naiſſance des Enfans des deux ſexes, ne ſauroit être l'effet du hazard, & qu'elle eſt une preuve de la Providence divine. Cette Lettre fit beaucoup de bruit: bien des gens ne convenoient pas de la force de cette preuve. Cela engagea Mr. *'s Graveſande* à l'examiner avec attention, & convaincu de ſa ſolidité, il en donna cette démonſtration, qu'il ſe contenta de communiquer à ſes amis, ſans la faire imprimer. Mr. *B. Nieuwentyt* fut un de ceux à qui il en fit part, & l'on en trouve le réſultat dans l'Ouvrage de cet Auteur, intitulé *l'Exiſtence de Dieu démontrée par les merveilles de la création. pag.* 176.

Ee 3

FIGURE 5.22 's Gravesande's article on Arbuthnot's Problem, taken from the *Oeuvres Philosophiques et Mathématiques de Mr G.F. 's Gravesande* (Allamand, 1774, p. 221)

from $x=5{,}715$ to 5,973 for $n=11{,}429$. He was thus able to obtain the odds in favor of A as

$$\frac{3{,}849{,}150}{9{,}347{,}650} \tag{5.153}$$

(*ibid*., p. 235). This implies that the probability that the number of males will be between 5,745 and 6,128 is $3{,}849{,}150/13{,}196{,}800 \approx .2917$.

's Gravesande now wished to calculate the probability that the number of males lies between 5745 and 7128 in each of 82 years (assuming the same probability for each year). Writing the odds in (5.153) as $a:b$ where $a=1$ and $b=2\dfrac{32{,}987}{76{,}983}$, the required probability is

$$\frac{a^{82}}{(a+b)^{82}} = \frac{1}{(a+b)^{82}}.$$

's Gravesande calculated

$$(a+b)^{82} = 75{,}598{,}215{,}229{,}552{,}469{,}135{,}802{,}469{,}135{,}802{,}469{,}135{,}802{,}469$$

(*ibidem*) (giving all 44 digits!) and concluded that there was only one chance in this number* for the number of males to lie between 5745 and 7128 in each of 82 years. This is an even smaller probability than that previously obtained by Arbuthnot. 's Gravesande clearly shared Arbuthnot belief in "art, not chance," for he said:

> What to make of what we have just said? It is that the one who has created the Skies and the Earth, governs everything that goes on not only by the general laws that he has established since the start of their existence, but also by special laws whose effect is felt daily. Only an intelligent Being can give birth to boys and girls in precisely the amount needed for each and other, so that everything remains in order, despite the phenomenal probability that opposes itself to it, if one pays attention to only what follows for general and physical laws. (*ibid*., p. 236)

5.8.3 Nicholas Bernoulli on the Arbuthnot Problem: Disagreement with 's Gravesande and Improvement of James Bernoulli's Theorem (1712)

Nicholas Bernoulli (1687–1759), the gifted nephew of James and John Bernoulli, met 's Gravesande in 1712 in The Hague while Nicholas was touring the Netherlands, England, and France. Nicholas discussed the Arbuthnot problem with 's Gravesande and wrote the latter three letters (dated September 30, November 9, and December 30, 1712) on the subject. These can be found in the *Oeuvres Philosophiques et Mathématiques de Mr G.F. 's Gravesande* (Allamand, 1774, pp. 236–248) (Fig. 5.23). Nicholas also commented on the Arbuthnot problem in two (dated October 11, 1712 and January 23, 1713) of his famous letters to Pierre Rémond de Montmort (1678–1719) while the latter was preparing the second edition of his *Essay d'Analyse sur les Jeux de Hazard* (Montmort, 1713, pp. 371–375, 388–394). All these letters were written just before the posthumous release

* The probability turns out to be approximately 1.32×10^{-44}.

DE LA PROVIDENCE. 237

gularité qu'on obſerve dans le nombre des garçons & des filles qui naiſſent chaque année. J'ai remarqué qu'on ne convenoit pas généralement de la validité de cette preuve. Mr. *Nicolas Bernoulli* qui, dans un voyage qu'il fit en Hollande, s'en entretint avec Mr. *'s Graveſande*, fut du nombre de ceux qui ne convenoient point de la force de cet argument. Voici ce qu'il lui en écrivit de Londres dans une Lettre dattée le 30 7bre 1712.

„ J'ai diſputé fort au long avec Mrs. *Craig* & *Burnet*, ſur cet argu-
„ ment pour la Providence divine tiré de la régularité qu'on obſerve dans
„ les nombres des mâles & des femelles, qui naiſſent chaque année à Lon-
„ dres, & dont vous m'avez parlé, lorsque j'ai eu l'honneur de vous ſaluer
„ à la Haye. Je leur ai enfin démontré que dans un grand nombre de
„ jettons à deux faces (croix & pile) jettés en l'air, il y a une fort gran-
„ de probabilité que presque la moitié ſeront croix, & la moitié pile, &
„ que par conſéquent il ne doit pas paſſer pour un miracle, qu'on voit
„ qu'il naît chaque année presque un égal nombre de mâles & de femel-
„ les, & qu'au contraire ce devroit être un miracle ſi ceci n'arrivoit pas.
„ La faute de Mr. *Arbuthnot* conſiſte en ce qu'il a pris cette égalité trop
„ préciſe, & qu'il n'a point obſervé que dans ſon catalogue de 82 ans les
„ limites ſont ſi grandes, qu'il y a une très grande probabilité que le nom-
„ bre des mâles & des femelles tombera entre ces limites. J'ai trouvé en
„ faiſant le calcul qu'il eſt plus de 7000 fois plus probable que ce nom-
„ bre tombera entre ces limites qu'au dehors. Mr. *Burnet* m'a dit que
„ Mr. *Nieuwentyt* doit imprimer ce prétendu argument dans un Livre
„ qu'il va donner au public, c'eſt pourquoi je crois que Mr. *Nieuwentyt*
„ ſera bien aiſe d'être averti de l'invalidité de cet argument."

Mr. *'s Graveſande* prit la défenſe du Dr. *Arbuthnot*, dans une réponſe qu'il fit à Mr. *Bernoulli*, dont je n'ai point trouvé la copie parmi ſes manuſcrits, mais dont on pourra conjecturer le contenu, par la replique de Mr. *Bernoulli*, dans une Lettre écrite de la Haye en datte du 9 Novembre 1712, & que voici:

„ J'aurois fort ſouhaité que nos ſentimens ſur le prétendu argument de
„ la Providence divine euſſent été les mêmes. Vous dites dans la Lettre
„ que vous m'avez fait l'honneur de m'envoyer à Londres, que c'eſt la
„ même choſe, que ſi l'on vouloit trouver le ſort de celui qui parieroit
„ qu'un jetton tombera 82 fois de ſuite ſur la même face. Je vous prie,
„ Monſieur, de reflechir encore un peu ſur cela; je ſuis perſuadé que vous
„ trouverez aiſément que vous vous êtes trompé. Vous n'avez qu'à con-
„ ſidérer ces deux points. 1°. Que ce n'eſt pas un jetton, qu'on jette 82
„ fois en l'air, mais que ce ſont pluſieurs, par ex. 1000. 2°. Qu'entre
„ ces 1000 jettons, jettés 82 fois en l'air, il y a toujours *presque* la moi-

Gg 3 tié,

FIGURE 5.23 Nicholas Bernoulli's comments on the Arbuthnot's Problem, taken from the *Oeuvres Philosophiques et Mathématiques de Mr G.F. 's Gravesande* (Allamand, 1774, p. 237)

of James Bernoulli's epochal *Ars Conjectandi* (Bernoulli, 1713) whose preface was written by none other than Nicholas himself.

Nicholas vehemently rejected Arbuthnot's analysis and ensuing inference about divine intervention. His views were clearly described in his very first letter of September 30, 1712:

> I had a lengthy argument with Mr Craig and Mr Burnet, on the argument in favor of Divine Providence following from the regularity which one observes in the number of males & females, who are born every year in London…I have finally showed to them that in a large number of counters with two sides (heads & tails), there is a very high probability that almost half will be heads, & half tails, & consequently there is no miracle if one sees that every year there is an almost equal number of male & female births, & that on the contrary there would be a miracle if this did not happen. Mr Arbuthnot's error consists in that he has taken this equality too precisely, & that he has failed to observe that in his bills of 82 years the limits are so wide, that there is a very high probability that the number of males & females will fall in these limits…. (Allamand, 1774, p. 237)

Nicholas backed his claim with calculations. First he assumed that the yearly number of births was constant at 14,000. Using this adjusted population size, he calculated the mean number of male births as 7,237 and the probability of a male birth as 7,237/14,000 ($\approx$.5169). Now, Nicholas wished to determine the odds that the number of male births in any year fell within a given number from 7237. Nicholas chose this number to be 200. To obtain the required odds, Nicholas used a method that represents a sharpening of James Bernoulli's law of large numbers. Because of its importance in the history of probability and statistics, Nicholas' analysis will be described in detail here.*

In the first stage of his analysis, Nicholas considered the binomial expansion of $(M+F)^n$ in increasing power of F, where M, F, and n are all positive integers. Then the $(F+1)$th term is

$$\frac{n}{1}\times\frac{n-1}{2}\times\frac{n-2}{3}\times\cdots\times\frac{n-F+1}{F}\times M^{n-F}F^F \tag{5.154}$$

and the $(F-L+1)$th term is

$$\frac{n}{1}\times\frac{n-1}{2}\times\frac{n-2}{3}\times\cdots\times\frac{n-F+L+1}{(F-L)}\times M^{n-F+L}F^{F-L} \tag{5.155}$$

(*ibid*., p. 239). The ratio of the $(F+1)$th term to the $(F-L+1)$th term is

$$\frac{n-F+L}{F-L+1}\times\frac{n-F+L-1}{F-L+2}\times\frac{n-F+L-2}{F-L+3}\times\cdots\times\frac{n-F+1}{F}\times\left(\frac{F}{M}\right)^L.$$

Substituting $n-F=M$, the above becomes

$$\frac{M+L}{F-L+1}\times\frac{M+L-1}{F-L+2}\times\frac{M+L-2}{F-L+3}\times\cdots\times\frac{M+1}{F}\times\left(\frac{F}{M}\right)^L. \tag{5.156}$$

* Nicholas Bernoulli's derivation is also described in Yushkevich (1987) and Hald (1984).

$$\ldots,\underbrace{T_L''',\ldots,T_2''',T_1'''}_{\text{Class 3}},\underbrace{T_L'',\ldots,T_2'',T_1''}_{\text{Class 2}},\underbrace{T_L',\ldots,T_2',T_1'}_{\text{Class 1}},T_{F+1},\ldots$$

$$\begin{cases} \dfrac{T_L'}{T_L''} > \cdots > \dfrac{T_2'}{T_2''} > \dfrac{T_1'}{T_1''} \\ \dfrac{T_L''}{T_L'''} > \cdots > \dfrac{T_2''}{T_2'''} > \dfrac{T_1''}{T_1'''}. \end{cases}$$

FIGURE 5.24 Nicholas Bernoulli's classification of the terms before the $(F+1)$th term in the binomial expansion of $(M+F)^n$

Hence also, the ratio of the Fth term to the $(F-L)$th term is

$$\frac{M+L+1}{F-L}\times\frac{M+L}{F-L+1}\times\frac{M+L-1}{F-L+2}\times\cdots\times\frac{M+2}{F-1}\times\left(\frac{F}{M}\right)^L. \tag{5.157}$$

Now, the ratio in (5.157) is larger than that in (5.156) because each term in the latter is larger than the corresponding one in the former. Using this observation, Nicholas proceeded to define a number of classes, each one consisting of L terms, that preceded the $(F+1)$th term. By the previous reasoning, reading from the left of the $(F+1)$th term, the ratio of each term in a given class to the corresponding term in the next class will be larger than the ratio of the previous terms in the two classes. Nicholas' reasoning is shown in Figure 5.24.

Therefore, the sum of all the terms in a given class will be larger than that of the next class, that is, sum of terms in class 1 > sum of terms in class 2 > sum of terms in class 3 > ⋯ Now suppose the sum of the terms in class 1 is s, and let the ratio of the $(F+1)$th term to the $(F-L+1)$th term in (5.156) be m (>1). Then the sum of the terms in class 2 will be less than S/m, that of the terms in class 3 will be less than S/m^2, and so on. Hence, the sum of all terms in classes 2, 3, … will be less than

$$\frac{S}{m}+\frac{S}{m^2}+\cdots=\frac{S}{m}\cdot\frac{1}{1-1/m}=\frac{S}{m-1}.$$

Therefore, the sum of the terms in class 1 ($(F-L+1)$th to Fth terms) will be larger than the sum of the terms in classes 2, 3, … (1st to $(F-L)$th terms) by a factor greater than $m-1$ (*ibid*., p. 240).

In the second stage of his analysis, Nicholas came back to the ratio in (5.156):

$$\underbrace{\frac{M+L}{F-L+1}\times\frac{M+L-1}{F-L+2}\times\frac{M+L-2}{F-L+3}\times\cdots\times\frac{M+1}{F}}_{P}\times\left(\frac{F}{M}\right)^L.$$

He assumed that the successive terms in the product P here are in geometric progression ("*progressione geometrica*"). He noted that this assumption was very close to reality, especially when M and F were both large. Under this assumption, the logarithms of the terms in P are in arithmetic progression ("*progressione aritmetica*"), and the sum of such an arithmetic progression is the product of half the number of terms in the progression, and the sum of the first and last terms in it, that is, the value of log (P) is

$$\frac{L}{2}\left(\log\frac{M+L}{F-L+1}+\log\frac{M+1}{F}\right).$$

Hence, the logarithm of the ratio in (5.156) is

$$\begin{aligned}\frac{L}{2}\left(\log\frac{M+L}{F-L+1}+\log\frac{M+1}{F}\right)+L\log\left(\frac{F}{M}\right)&=\frac{L}{2}\left(\log\frac{M+L}{F-L+1}+\log\frac{M+1}{F}+\log\frac{F}{M}+\log\frac{F}{M}\right)\\&=\frac{L}{2}\left(\log\frac{M+L}{F-L+1}+\log\frac{M+1}{M}+\log\frac{F}{M}\right)\end{aligned}$$

and the ratio itself is

$$m=\left(\frac{M+L}{F-L+1}\times\frac{M+1}{M}\times\frac{F}{M}\right)^{L/2} \tag{5.158}$$

(*ibid*., p. 240). Thus, from the result in the first stage of his analysis, Nicholas reasoned that the sum of the $(F-L+1)$th to Fth terms in the binomial expansion of $(M+F)^n$ will be larger than the sum of the first to $(F-L)$th terms by a factor greater than

$$\left(\frac{M+L}{F-L+1}\times\frac{M+1}{M}\times\frac{F}{M}\right)^{L/2}-1.$$

By interchanging M and F, this also means that the sum of the $(F+2)$th to $(F+L+1)$ th terms will be greater than the sum of the $(F+L+2)$th to last terms by a factor larger than

$$\left(\frac{F+L}{M-L+1}\times\frac{F+1}{F}\times\frac{M}{F}\right)^{L/2}-1.$$

Hence, the ratio of the sum of the $(F-L+1)$th to $(F+L+1)$th terms, excluding the middle $(F+1)$th term, to the sum of all remaining terms at both ends will be greater than

$$\min\left\{\left(\frac{M+L}{F-L+1}\times\frac{M+1}{M}\times\frac{F}{M}\right)^{L/2},\left(\frac{F+L}{M-L+1}\times\frac{F+1}{F}\times\frac{M}{F}\right)^{L/2}\right\}-1 \tag{5.159}$$

(*ibid.*, p. 241). This is Nicholas' final result. His analysis is similar to his uncle's (James Bernoulli) except for the way in which they dealt with the ratio in (5.156). Nicholas was able to approximate the ratio by (5.158) whereas James had only sought a lower bound for it.

Nicholas applied his result to Arbuthnot's problem as follows. Recall that Nicholas had taken the yearly number of births to be 14,000 and the mean number of male births as 7,237. He wished to calculate the odds in favor of the number of male births to be between 7,037 and 7,437. Therefore, he took $M=7{,}237$, $F=140{,}000-7{,}237=6{,}763$, and $L=200$. From (5.159), Nicholas obtained

$$\min\left(306\frac{9}{10},304\right)-1=303.$$

Thus, the required odds are greater than 303 to 1, and the corresponding probability is greater than 303/304 (*ibid.*, p. 242).

Nicholas also calculated the probability that the number of male births to be between 7037 and 7437 for 100 consecutive years as greater than $(303/304)^{100}\approx 1000/1389$.* With this number, he was able to refute Arbuthnot's first statistical argument that it was also very unlikely that there should be *approximately* the same number of M's and F's. However, Nicholas' calculation does not counter Arbuthnot's second statistical argument that there is an extremely small probability that there is an excess of males for 82 consecutive years, assuming an equal probability for each sex to exceed the other in a given year.

In a cordial letter, 's Gravesande expressed his disagreement with Nicholas. He said that the very fact that Nicholas chose a probability of male birth of $p=7{,}237/1{,}4000\neq 1/2$ was evidence enough for divine providence:

> If chance brings people, there is at each birth the same amount of probability that a boy will be born, as there is that a girl will be born: chance is not as farsighted to put more probability on one side than the other.... (*ibid.*, p. 244)

Thereupon, 's Gravesande repeated his previous mathematical calculations in favor of Arbuthnot's argument for divine providence. In the December 30, 1772 letter to 's Gravesande, Nicholas clarified his point:

> Mr. Arbuthnot's argument consists of two things: 1°. in that, assuming an equality in the births of girls & boys, there is little probability that the number of boys & girls can be found within limits close to equality: 2°. that there is little probability that the number of boys will exceed a large number of successive times the number of girls. It is the first argument that I refute, not the second. (*ibid.*, p. 247)

In the January 23, 1713 letter to Montmort (1713, pp. 388–394), Nicholas repeated his previous calculations with some minor changes. The total number of yearly births was

* $(303/304)^{100}\approx .7192$, $1000/1389\approx .7199$.

still 14,000 but the probability of a male birth was now changed to 18/35. The expected number of male births is then 7200. Repeating his previous calculations, Nicholas obtained the probability that the number of male births fell within 163 of the mean number as greater than 2179/2229 ($\approx$.9776).

In summary, through his calculations Nicholas was able to refute Arbuthnot's first statistical argument that it was unlikely that there should be approximately the same number of male and female years. However, he acknowledged he had no objections to Arbuthnot's second argument that "there is little probability that the number of boys will exceed a large number of successive times the number of girls." While Nicholas rejected divine intervention, 's Gravesande upheld it and further contended that if one were to assume a probability of a male birth different from half (as Nicholas did), this assumption itself, if it were true, was enough evidence of divine providence. The latter was the same argument that Abraham de Moivre (1667–1754) added in the third edition of the *Doctrine of Chances* (de Moivre, 1756, pp. 252–253).

5.8.4 Daniel Bernoulli on the Inclination of the Planes of the Planetary Orbits (1735). Criticism by d'Alembert (1767)

Daniel Bernoulli's (1700–1782) life has been described in Section 5.7.3 In 1732, the Paris Academy of Sciences set a prize for an answer to the following problem: "What is the physical cause of the inclination of the planes of the planetary orbits with the plane of the solar equator, and why are the inclinations of these orbits different from each other?" No one was able to answer the question to the judges' satisfaction and the problem was posed again in 1734 with double the prize. This time, Daniel and his father John both came up with a satisfactory answer and the prize was shared between the two.* The memoirs of both mathematicians appeared in the *Recueil des Pièces qui ont Remporté le Prix de l'Académie Royale des Sciences* (Bernoulli, 1735) (Fig. 5.25). It is the memoir of Daniel that is of interest here for it is the first application of probability theory to a problem of astronomy. Two versions of his memoir are given in the *Recueil*: one in the original Latin, the second translated in French by Daniel himself, with "minor additions or clarifications." Daniel wished to address two issues:

1. The mutual inclinations of the planetary orbits
2. The inclinations of the planetary orbits to the plane of the solar equator

Regarding the first issue, Daniel announced his strategy explicitly as follows:

> Of all the planetary Orbits I will look for the two that cut each other at the greatest angle; after which I will calculate the probability that all other Orbits are contained by chance within the limits of these two Orbits. We shall see that this probability is so small, that it should be considered a moral impossibility. (Bernoulli, 1735, p. 96)

* Much to John's irritation that his son had been equal to him. See also Section 5.7.3.

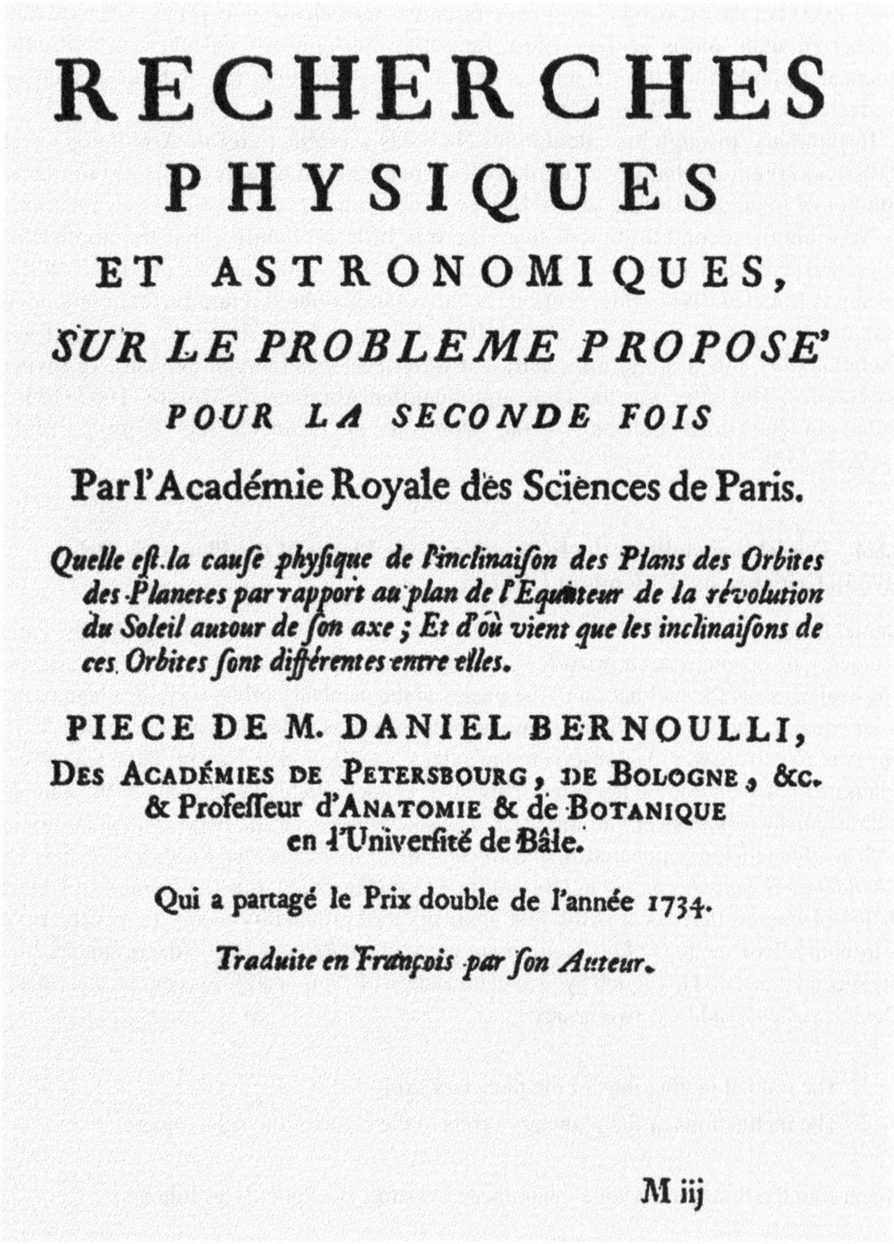

RECHERCHES
PHYSIQUES
ET ASTRONOMIQUES,

SUR LE PROBLEME PROPOSE'
POUR LA SECONDE FOIS

Par l'Académie Royale des Sciences de Paris.

Quelle eſt la cauſe phyſique de l'inclinaiſon des Plans des Orbites des Planetes par rapport au plan de l'Equateur de la révolution du Soleil autour de ſon axe; Et d'où vient que les inclinaiſons de ces Orbites ſont différentes entre elles.

PIECE DE M. DANIEL BERNOULLI,
DES ACADÉMIES DE PETERSBOURG, DE BOLOGNE, &c.
& Profeſſeur d'ANATOMIE & de BOTANIQUE
en l'Univerſité de Bâle.

Qui a partagé le Prix double de l'année 1734.

Traduite en François par ſon Auteur.

M iij

FIGURE 5.25 French translated version of Daniel Bernoulli's prize-winning article (Bernoulli, 1735)

Using Kepler's Rudolphine Tables, Daniel found the maximum angle of 6°54′ was that between Mercury and Earth. Now the area of the zonal belt on a sphere of radius R subtended by a small angle θ is approximately

$$(R\sin\theta)(2\pi R) = 2\pi R^2 \sin\theta.$$

Moreover the area of a sphere of radius R is $4\pi R^2$. Therefore, assuming pure chance, the probability of an orbit lying in the zone subtended by an the angle 6°54′ is approximately

$$\frac{2\pi R^2 \sin\left(60\tfrac{54}{60} \times \pi / 180\right)}{4\pi R^2} \approx \frac{1}{17}$$

(*ibid.*, p. 97). Hence, for the six planets Daniel considered, the probability that the inclinations of all five of the planes to one plane will be less than 6°54′ is approximately

$$\frac{1}{17^5} = \frac{1}{1{,}419{,}856} \tag{5.160}$$

(*ibidem*). In the French version of his memoir, Daniel also considered the case of all planes intersecting in a common line. Then the angles can be taken to be uniformly distributed between 0° and 90°. Hence, the probability that the inclinations of all five of the planes to one plane will be less than 6°54′ is, in this case, approximately

$$\left(\frac{6\frac{54}{60}}{90}\right)^5 \approx \frac{1}{13^5} = \frac{1}{371{,}293}. \tag{5.161}$$

Finally, for the second issue (namely, the inclinations of the planetary orbits to the plane of the solar equator), Daniel took the solar equator as the plane of reference. He found that the largest inclination of the plane of any orbit to the reference plane was 7°30′ (and interestingly is that between Earth and the Sun). Hence, the probability that all of the *six* inclinations will be less than 7°30′ is

$$\left(\frac{7\frac{30}{60}}{90}\right)^6 \approx \frac{1}{12^6} = \frac{1}{2{,}985{,}984}. \tag{5.162}$$

Concerning the three probabilities he had calculated in (5.160)–(5.162), Bernoulli commented:

> All these methods, although quite different, do not give numbers which are extremely unequal. Nevertheless, I will prefer the number given in the first case. (*ibid.*, p. 99)

In any case, the three calculated probabilities are so small that Daniel believed he had shown "how ridiculous it is to attribute to pure chance the tight positions of the Orbits."

We shall next discuss two noteworthy critiques of Daniel Bernoulli's analysis, namely, those of the French mathematician Jean le Rond d'Alembert (Fig. 5.26) and the mathematical historian Isaac Todhunter (Fig. 5.27).

D'Alembert and Daniel frequently clashed on scientific issues, and these were not restricted to the calculus of probabilities. Thus, when d'Alembert beat Daniel for the prize on the theory of winds offered by the Berlin Academy of Sciences in 1746, Daniel expressed his frustration in a letter to Euler 4 years later:

> After one has read his [d'Alembert's] paper, one knows no more about the winds than before. (Fuss, 1843, p. 650)

Further disagreements between the two occurred on the St Petersburg paradox and the problem of inoculation against smallpox, and have been well documented in the literature (e.g., see Gouraud, 1848, pp. 57–59; Bradley, 1971; Pearson, 1978, pp. 537–551; Daston, 1988, pp. 76–91; Paty, 1988).

Concerning Daniel's work on planetary orbits, d'Alembert first expressed his opposition in his *Mélanges de Littérature, d'Histoire et de Philosophie*:

> ...But I say, mathematically speaking, it is equally possible, either that the five planets deviate as little as they do on the ecliptic plane [i.e. the plane of the earth's orbit], or that they take an altogether different arrangement, which makes them deviate much more, & dispersed like comets under all possible angles with the ecliptic; nevertheless, nobody dares to ask why comets do not have limits in the inclination, & one asks why planets do? What could be the reason? If not because one considers as very likely, & almost as self-evident, that a combination where regularity and a certain design appears, is not the effect of chance, although mathematically speaking, it is as likely as any other combination with no order or regularity, and to which for this reason one would not think look for a cause.
>
> If one throws a die with seventeen faces five consecutive times, and that in each of the five times a six occurs, M. Bernoulli can show that, there are precisely the same odds as in the case of the planets, that a six will not occur thus. However, I ask him if he will look for a reason to this event, or he will not. If he does not at all, and that he considers it the result if chance, why does he look for a reason in the arrangement of planets, which is precisely the same case?...
>
> I will first conclude that if regular effects due to chance are not completely impossible, physically speaking, they are at least much more likely the effect of an intelligent and regular cause, than non-symmetric and irregular effects; I will secondly conclude, that if there is at the very outside, & even physically speaking, no combination that is possible, the physical likelihood that all these combinations (so long as one supposes the effect of pure chance) will not be the same, although their mathematical likelihood could be exactly the same. (d'Alembert, 1767, pp. 295–301)

In the above, D'Alembert made a forceful argument against Daniel's analysis. But one sees that he did not dispute Daniel's calculations nor did he deny the existence of a common cause or design. His grievance concerned Daniel's use of mathematical probability to study such matters. D'Alembert asked, since, mathematically speaking, *every particular configuration* of the planets was unique and had the same small probability,

FIGURE 5.26 Jean le Rond d'Alembert (1717–1783). Wikimedia Commons (Public Domain), http://commons.wikimedia.org/wiki/File:Jean_Le_Rond_d%27Alembert,_by_French_school.jpg

FIGURE 5.27 Isaac Todhunter (1820–1884). Wikimedia Commons (Public Domain), http://commons.wikimedia.org/wiki/Category:Isaac_Todhunter#/media/File:Todhunter_Isaac.jpg

why would Daniel then search for a cause when there was regularity and would most likely react differently if there were no regularity? D'Alembert used the analogy of a 17-sided die to make his point. If this die was thrown five times and each time there was a six, one would look for a cause. However, if there was any other particular sequence of numbers, one would not, although in both cases the mathematical probability was the same. Therefore, d'Alembert believed that the calculus of probabilities did not enable one to distinguish cases of regularity, which would point to some kind of design, from cases of irregularity, which would point to pure chance.

D'Alembert thus invoked his concepts of metaphysical (or mathematical) and physical probabilities (e.g., see Gorroochurn, 2012a, pp. 112–113) to discredit Daniel's analysis. These concepts go back to his 1761 *Opuscule Mathématiques*, where we can read:

> One must distinguish between what is metaphysically possible, & what is physically possible. In the first class are all things whose existence is not absurd; in the second are all things whose existence not only has nothing that is absurd, but also has nothing that is very extraordinary either, and that is not in the everyday course of events. It is metaphysically possible, that we obtain a double-six with two dice, one hundred times in a row; but this is physically impossible, because this has never happened, & will never happen. (d'Alembert, 1761, p. 10)

D'Alembert's argument against Daniel boiled down to his saying that a regular arrangement of the planets was physically impossible whereas an irregular arrangement was physically possible. However, since both arrangements had the same metaphysical probabilities, the latter could not be used to distinguish between the two.

But how are we to calculate physical probabilities? D'Alembert never gave any rigorous guidelines and his views about physical and mathematical possibilities are now obsolete. Nevertheless we still need to grapple with his argument that both a regular and any irregular arrangement of planets have the same small mathematical probabilities; therefore, why should only the first case not be regarded as fortuitous? An answer to this question has been provided by Gower (1987) by appealing to van Fraassen's concept of *favoring* as used by the latter in his book *The Scientific Image* (van Fraassen, 1980, pp. 141–157). According to van Fraassen, there are two different ways in which a hypothesis can favor an observation. First, it may be because the hypothesis can favor no other *alternative observation* as well or better, and second it may be because no other *alternative hypothesis* can favor the observation as well or better. D'Alembert's arguments pertained to the first case. He contended that the hypothesis of design cannot favor the regular disposition of planets because it also favors other irregular disposition of planets. However, Daniel's thinking was more along the lines of the second case. His argument was that the hypothesis of design *does* favor the regular disposition of planets because there is no other alternative hypothesis that can favor the disposition as well or better. Gower has thus summarized the issue as follows:

> It would appear, then, that d'Alembert's arguments, reasonable though they may be, do not show that Bernouilli's [*sic*] calculations were based on faulty principles. Bernoulli was, in effect, concerned with the extent to which particular hypotheses about the planets "probability", "favour", or are "confirmed" by actual observations... D'Alembert, by contrast, was intent on comparisons between the extent to which one particular hypothesis—the hypothesis of random

> distribution—probabilified, or favoured, different possible observations. He claimed, quite correctly, that an observation cannot be used to support a hypothesis if every other possible observation would be probabilified by, or would confirm, that hypothesis to the same extent...
>
> What d'Alembert failed to notice, however, is that his view is not, as he thought, inconsistent with that of Bernoulli...Whatever "peculiar significance" attaches to the actual disposition derives not from the fact of its improbability in the absence of a known law which would favour it, but rather from the fact of its high probability in the presence of a justifiable conjecture which favours it. (Gower, 1987, pp. 453–454)

A second, and less fundamental, criticism of Daniel's analysis concerns the following comment made by Isaac Todhunter (1820–1884) in his seminal book, *A History of the Mathematical Theory of Probability from the Time of Pascal to that of Laplace*:

> It is difficult to see why in the first of the three preceding calculations Daniel Bernoulli took 1/17 instead of 2/17; that is why he compared his zone with the surface of a sphere instead of with the surface of a hemisphere. (Todhunter, 1865, p. 223)

Gower (1987) has again disagreed with this criticism and has correctly pointed out that Daniel was concerned with the maximal mutual inclination of *any* two orbits, not with the maximal inclination of any orbit to that of the ecliptic (i.e., the orbit of the earth). If the latter had been the case then Todhunter would have been correct in his criticism. Hence, it seems that Daniel was right in comparing the zonal belt with the surface of a sphere rather than that of a hemisphere.

5.8.5 Michell on the Random Distribution of Stars (1767): Clash Between Herschel and Forbes (1849)

5.8.5.1 Michell on the Random Distribution of Stars (1767) John Michell (1724–1793) was an English natural philosopher and geologist whose interests also included astronomy. In the latter field, he is known for being the first to realistically estimate the distance of the stars and for discovering the existence of physical double stars. It is the last topic that shall concern us, for it deals with significance testing.

In the 1767 paper entitled "An inquiry into the probable parallax and magnitude of the fixed stars from the quantity of light which they afford us, and the particular circumstances of their situation" (Michell, 1767) (Fig. 5.28),* Michell wrote:

> It has always been usual with Astronomers to dispose the fixed stars into constellations: this has been done for the sake of remembering and distinguishing them, and therefore it has in general been done merely arbitrarily, and with this view only; nature herself however seems to have distinguished them into groups. What I mean is, that, from the apparent situation of the stars in the heavens, there is the highest probability, that, either by the original act of the Creator, or in consequence of some general law (such perhaps as gravity) they are collected together in great numbers in some parts of space, whilst in others there are either few or none.
>
> The argument, I intend to make use of, in order to prove this, is of that kind, which infers either design, or some general law, from a general analogy, and the greatness of the odds against things having been in the present situation, if it was not owing to some such cause. (*ibid.*, p. 243)

* Michell's derivation is also described in Hald (1998, pp. 70–74).

[234]

Received April 24, 1767.

XXVII. *An Inquiry into the probable Parallax, and Magnitude of the fixed Stars, from the Quantity of Light which they afford us, and the particular Circumſtances of their Situation, by the Rev.* John Michell, *B. D. F. R. S.*

Read May 7, and 14, 1767.

THOUGH no man can at preſent doubt, that the want of a ſenſible parallax in the fixed ſtars, is owing to their immenſe diſtance, yet it may not perhaps be diſagreeable to ſee, that this diſtance is farther confirmed by other circumſtances; for let us ſuppoſe them to be, at a medium, equal in magnitude and natural brightneſs to the ſun, to which they ſeem in all reſpects to be analagous. And, having laid this down as a foundation to build upon, let us inquire what would be the parallax of the ſun, if he were to be removed ſo far from us, as to make the quantity of the light, which we ſhould then receive from him, no more than equal to that of the fixed ſtars. In order to do this with accuracy, it would be proper to compare the quantity of light; which we at preſent receive from him, with that of the fixed ſtars, by ſome ſuch methods, as are made uſe of by Monſieur Bouguer

FIGURE 5.28 First page of Michell's article "An inquiry into the probable parallax, and magnitude of the fixed stars, from the quantity of light which they afford us, and the particular circumstances of their situation" (Michell, 1767)

Here, Michell referred to the widespread belief of the times that bright stars* were *randomly scattered* in the celestial sphere, thus forming different constellations. Michell intended to disprove the randomness assumption by using probabilistic arguments. He first used the example of double stars to make his point. Double stars had been known of well before Michell's times. However, they had been thought of as *optical pairs*, that is, they appeared close due to chance alignment, although they were far from each other. Michell wished to argue that the frequency of double stars with such small angular separation was much higher than would be expected by chance alone.

Accordingly, Michell considered two stars and calculated the chance of one star being within an angle of, say, 1° of the other by using geometric probability. Assuming a uniform distribution of each star in space, he divided the area of a small circle of angular radius 1°† by the area of the celestial sphere (assumed to have radius R), thus obtaining a probability of

$$\frac{\pi\left(\pi R/180\right)^2}{4\pi R^2}=7.615\times10^{-5}\approx\frac{1}{13{,}131} \tag{5.163}$$

(*ibid*., p. 244). Therefore, the probability of one star *not* being within 1° of the other is 13,130/13,131. Next, Michell argued that if there was a total of n stars then the probability that *none* of them would be within 1° of a given star is, by the product rule for independent events, equal to

$$\left(\frac{13{,}130}{13{,}131}\right)^n \tag{5.164}$$

(*ibidem*). Moreover, Michell also reasoned that the probability that none of the stars will be within 1° of *any other star* is

$$\left\{\left(\frac{13{,}130}{13{,}131}\right)^n\right\}^n=\left(\frac{13{,}130}{13{,}131}\right)^{n^2}. \tag{5.165}$$

Clearly, Michell had again assumed independence in this last calculation, but this time his assumption is incorrect. For example, the conditional probability that none of n stars will be within 1° of star A given that none of n stars will be within 1° of star B is larger than the unconditional probability that none of n stars will be within 1° of star A. This is because the conditioning event implies that star B is not within 1° of star A. Moreover, as Gower (1982) has pointed out, even if nonindependence had escaped Michell, he could have realized that the probability that none of the stars will be within 1° of any other star should be zero for $n \geq 13{,}131$ (and thus (5.165) is wrong). This is because, from (5.164),

* The brightness of a star was used by Michell and his contemporaries as an indication of its distance. Brightness was classified into six "magnitudes," the first magnitude being the brightest and the last magnitude being the faintest to the naked eye. For more details, see McCormmach (2012, pp. 128–129).

† By an angular radius of 1°, we mean a radius equal to the length of the arc subtended by 1° on a great circle (of radius R) of the celestial sphere. The length of such an arc is therefore equal to $\pi R/180$.

we can view the celestial sphere as having 13,131 square units of area and the small circle of angular radius 1° as having one square unit of area. When $n \geq 13{,}131$, all the area on the celestial sphere is taken up and the (geometric) probability that none of the stars will be within 1° of any other star is zero.

As a first illustration of his calculations, Michell considered the double star β-Capricorni. He noted that there was less than 3⅓′ between the pair and that there are about 230 stars of equal magnitude. By using a reasoning similar to the above, Michell obtained the probability that none of the 230 stars would be within 3⅓′ of each other as

$$\left\{1-\frac{\pi\left(\frac{3\frac{1}{3}}{60}\times\pi R/180\right)^2}{4\pi R^2}\right\}^{230^2}\approx\frac{80}{81}.$$

(*ibid*., p. 246). As a second illustration, Michell considered the Pleiades galactic cluster in Taurus. The six brightest stars in the Pleiades are Taygeta, Electra, Merope, Alcyone, Atlas, and Maya. Taking the latter as the reference, the angles of the others were, respectively, 11′, 19½′, 24½′, 27′, and 49′. Also, Michell stated there were about 1500 stars at least as bright as the faintest among the six. The joint probability that one star will be within 11′ of Maya, and one will be within 19½′ of Maya, and so on, could then be calculated as

$$\xi=\left[1-\left\{1-\frac{\pi\left(\frac{11}{60}\times\pi R/180\right)^2}{4\pi R^2}\right\}^{1500}\right]\left[1-\left\{1-\frac{\pi\left(\frac{19\frac{1}{2}}{60}\times\pi R/180\right)^2}{4\pi R^2}\right\}^{1500}\right]\times$$
$$\left[1-\left\{1-\frac{\pi\left(\frac{24\frac{1}{2}}{60}\times\pi R/180\right)^2}{4\pi R^2}\right\}^{1500}\right]\left[1-\left\{1-\frac{\pi\left(\frac{27}{60}\times\pi R/180\right)^2}{4\pi R^2}\right\}^{1500}\right]\times$$
$$\left[1-\left\{1-\frac{\pi\left(\frac{49}{60}\times\pi R/180\right)^2}{4\pi R^2}\right\}^{1500}\right].$$

The probability that no six stars out of the 1500 will be as close to each other as the ones observed was then

$$(1-\xi)^{1500}\approx\frac{496{,}000}{496{,}001}. \tag{5.166}$$

Michell thus found that:

> …the odds to be near 500000 to 1, that no six stars, out of that number [1500] scattered at random, in the whole heavens, would be within so small a distance from each other, as the Pleiades are. (*ibidem*)

He stated that similar results could be found for other stars as well as those stars "which appear double, treble, &c. when seen through telescopes," and concluded:

> We may from hence, therefore, with the highest probability conclude (the odds against the contrary opinion being many million millions to one) that the stars are really collected together in clusters in some places, where they form a kind of systems, whilst in others there are either few or none of them, to whatever cause this may be owing, whether to their mutual gravitation, or to some other law or appointment of the Creator. And the natural conclusion from hence is, that it is highly probable in particular, and next to a certainty in general, that such double stars, &c. as appear to consist of two or more stars placed very near together, do really consist of stars placed near together, and under the influence of some general law, whenever the probability is very great, that there would not have been any such stars so near together, if all those, that are not less bright than themselves, had been scattered at random through the whole heavens. (*ibid.*, pp. 249–250)

A careful reading of the above shows that Michell was making reference to the *probability of the hypothesis* that seemingly close stars are indeed physically close. Now, such a reference is in line with Bayesian or inverse probability. Consequently, much controversy subsequently occurred on whether Michell was really using a Bayesian argument. Dale (1999) has summarized the issue as follows:

> Had Michell contented himself with stopping before the last quotation, his work would in all probability have been seen as an early significance test, and we should have been spared much of the ensuing controversy. But the passage quoted above suggests strongly that Michell thought the strength of his argument to be measurable, and his work came to be seen as an application of inverse probability. (*ibid.*, p. 90)

However, nowhere in his calculations did Michell in fact use or refer to Bayes' theorem, and he seems to have been using probability informally.

5.8.5.2 Clash Between Herschel and Forbes (1849) A slightly different kind of debate was also to take place in the aftermath of Michell's paper. The fight here was not on inverse probability per se, rather it essentially involved the nature of randomness and the appropriateness of a uniform distribution. The two main protagonists were John Herschel (Fig. 5.29) and James Forbes (Fig. 5.30). Prior to the controversy, John Herschel's father, William Herschel (1738–1822), had empirically confirmed Michell's hypothesis on double stars in 1803:

> I shall therefore now proceed to give an account of a series of observations on double stars, comprehending a period of about 25 years, which, if I am not mistaken, will go to prove, that many of them are not merely double in appearance, but must be allowed to be real binary combinations of two stars, intimately held together by the bond of mutual attraction. (Herschel, 1803, p. 340)

FIGURE 5.29 Sir John Frederick William Herschel (1792–1871). Wikimedia Commons (Public Domain), http://commons.wikimedia.org/wiki/File:Herschel_sitzend.jpg

FIGURE 5.30 James David Forbes (1809–1868). Wikimedia Commons (Public Domain), http://commons.wikimedia.org/wiki/Category:James_David_Forbes#/media/File:James_David_Forbes.png

Then, in his 1849 book *Outlines of Astronomy*, John Herschel more or less approved of Michell's analysis:

> Mitchell [*sic*], in 1767, applying the rules for the calculation of probabilities to the case of the six brightest stars in the group called the Pleiades, found the odds to be 500000 to 1 against their proximity being the mere result of a random scattering of 1500 stars (which he supposed to be the total number of stars of that magnitude in the celestial sphere) over the heavens ... The conclusion of a physical connection of some kind or other is therefore unavoidable. (Herschel, 1849, pp. 564–565)

Regarding Michell's use of 1500 stars, Herschel added in a footnote:

> This number is considerably too small, and in consequence, Mitchell's odds in this case materially overrated. But enough will remain, if this be rectified, fully to bear out his argument. (*ibid.*, p. 564)

Shortly after the publication of Herschel's book, James Forbes wrote the following criticism:

> Nearly a century ago, Mitchell [*sic*] computed the chances to the 500,000 to 1 against the stars composing the group of the Pleiades being *fortuitously* concentrated within the small apparent space with they occupy; and he thence infers the probability of a physical connexion between them. Struve has pushed this consideration much further...
>
> Now I confess my inability to attach any idea to what would be the distribution of stars or of anything else, if "fortuitously scattered," much more must I regard with doubt and hesitation an attempt to assign a numerical value to the antecedent probability of any given arrangement or grouping whatever. An equable spacing of the stars over the sky would seem to me to be far more inconsistent with a total absence of Law or Principle, than the existence of spaces of comparative condensation, including binary or more numerous groups, as well as of regions of great paucity of stars. (Forbes, 1849, pp. 132–133)

Forbes' main argument in the above was that if the stars were truly "fortuitously scattered" then it did not make sense to assign a uniform distribution, and indeed any other distribution, to them. This was because such distributions were based on certain laws, and the latter fact was contrary to the very notion of randomness. Thus, any attempt to "assign a numerical value to the antecedent probability of any given arrangement or grouping whatever" must be regarded with "doubt and hesitation." As far as Forbes was concerned, Michell's equating of randomness with a uniform distribution invalidated his calculations.

Herschel's response did not have to wait for too long:

> ...a singular misconception of the true incidence of the argument from probability which has prevailed in a quarter where we should least have expected to meet it. The scattering of the stars over the heavens, does it offer any indication of law? In particular, in the apparent proximity of the stars called "double," do we recognise the influence of any *tendency to proximity*, pointing to a cause exceptional to the abstract law of probability resulting from equality of chances *as respects the area occupied by each star*?...
>
> Such we conceive to be the nature of the argument for a physical connexion between the individuals of a double star prior to the direct observation of their orbital motion round each

> other. To us it appears conclusive; and if objected to on the ground that every attempt to assign a numerical value to the antecedent probability of any given arrangement or grouping of fortuitously scattered bodies must be doubtful,* we reply, that if this be admitted as an argument, there remains no possibility of applying the theory of probabilities to any registered facts whatever. We set out with a certain hypothesis as to the chances: granting which, we calculate the probability, not of one certain definite arrangement, which is of no importance whatsoever, but of certain *ratios* being found to subsist between the cases in certain predicaments, on an average of great numbers. Interrogating Nature, we find these ratios contradicted by appeal to her facts; and we pronounce accordingly on the hypothesis. It may, perhaps, be urged that the scattering of the stars is *un fait accompli*, and that their actual distribution being just as possible as any other, can have no *à priori* probability. In reply to this, we point to our target, and ask whether the same reasoning does not apply equally to that case? When we reason on the result of a trial which, in the nature of things, cannot be repeated, we must agree to place ourselves, in idea, at an epoch antecedent to it. On the inspection of a given state of numbers, we are called on to hold up our hands on the affirmative or negative side of the question, Bias or No bias? In this case who can hesitate? (Herschel, 1850, pp. 36–37)

In the above, Herschel disagreed with Forbes and argued that we *could* assign an antecedent probability to a random arrangement. Herschel contended that if antecedent probabilities could *not* be assigned, then there would be no hope of "applying the theory of probabilities to any registered facts whatever."

Soon after, Forbes came back and further elaborated his previous argument:

> Let us look at the case straightforwardly. Suppose two luminaries in the heavens, as two dots on a sheet of paper of known dimensions and figure. Is it or is it not common sense to say, that the position of one luminary or one dot being given, the law of "random scattering" can assign any other position in the field as more or less probable or improbable for the second. If one star be in one pole of the heavens, is the most probable position of the second in the opposite pole, or is it 90° from it, or is it 30°, or is it in any assignable position whatever? I think not. Every part of the field, even that close to the star A, is all equally probable allotment for the star B, if we are guided by no predetermining hypothesis or preconception whatever. Whence, then, this astounding probability of 18180 to 1 that B shall not be within a degree of A? If it express anything, what does it express?
>
> What the probability 13130/13131 expresses may be thus illustrated.† Suppose it to be known that two comets exist at once in the sky, of which one is telescopic, and its position has not been indicated. The probability in question represents the *expectation* which will exist in the mind of a person possessed of this partial information, that if he takes a telescope with a field of 2° in diameter, and fixes the cross wires upon the larger comet, the telescopic one shall *not* be somewhere in the field, which includes 1/13131 of the whole spherical area. (Forbes, 1850, pp. 405–406)

In the first paragraph of the above, we find Forbes more or less reiterating d'Alembert's argument against Daniel Bernoulli (see Section 5.8.4). Forbes argued that, under true randomness (i.e., when "we are guided by no predetermining hypothesis or preconception whatever"), every configuration of the stars was as likely as the other, so why should we try

* London, Ed. and Dub. Philosoph. Magazine, &c., Aug, 184 (*Herschel's footnote*).

† This illustration is different from that given in the original paper (*Forbes' footnote*).

to infer any particular cause from a given configuration? Furthermore, since Michell's value of 13,130/13,131 for the probability that one star was not within 1° of the other was based on the assumption of a uniform law, Forbes believed it was not valid. The question then was, according to Forbes, what did this probability represent? In the second paragraph of the last quotation, he stated that the number 13,130/13,131 represented an *expectation* that "will exist in a person's mind." Forbes here showed that, contrary to Michell's, his view of probability was *subjective*, that is, was related to our degree of belief (Gower, 1982).

A modern take on the Michell–Herschel–Forbes debate would be that Michell's analysis was very much in the spirit of Fisher's significance testing. Thus, Michell's probabilities were not those of the *hypothesis* of randomness. Instead they were those of the *observed configuration* of stars (or a more extreme configuration) calculated under the (null) hypothesis of randomness. For example, Michell's value of 490,000/496,001 in (5.166) is the probability that of no six stars out of 1500 would be as close or closer to the ones observed, under the assumption of randomness. Therefore, on the basis of the observations made, it was plausible that the assumption of randomness did not hold. This certainly did *not* imply that there was a probability of 496,000/496,001 that the hypothesis of randomness did not hold. By using language that appeared to place probabilities on hypotheses, Michell might have contributed much to the ensuing controversy. On the other hand, Forbes' criticism of Michell and Herschel seems misplaced when he criticized the former for assuming a uniform distribution in his calculations. However, Forbes' arguments were insightful. As Gower has said:

> Forbes' forceful criticisms, though they failed to touch the heart of Michell's reasoning, encouraged nascent doubts about the capacity of such methods to yield determinate conclusions. (Gower, 1982, p. 160)

We shall next describe Fisher's take on Michell's problem. In discussing the simple test of significance in his book *Statistical Methods and Scientific Inference* (Fisher, 1956b, p. 38), Fisher complained that he found "the details of Michell's calculation obscure." He then gave his own analysis of the problem by using the Poisson approximation as follows. First, Fisher took the fraction of the celestial share that lied within a circle of a minutes as approximately (cf. Eq. 5.163)

$$p=\frac{\pi\left\{a\pi R/(60\times 180)\right\}^2}{4\pi R^2}=\left(\frac{a}{6875.5}\right)^2.$$

Taking $a=49'$ (the greatest angle between Maya and the five other stars), he obtained

$$p=\left(\frac{1}{140.316}\right)^2=\frac{1}{19,689}.$$

Next, out of the 1499 stars other than Maia, the expected number within a circle of radius 49′ was

$$m=\frac{1,499}{19,689}=\frac{1}{13.1345}=.07613.$$

Now, the actual number of stars which lie within a circle of radius 49′ has the following Poisson distribution:

$$e^{-.07613}\frac{(.07613)^x}{x!}, \quad x=0,1,2,\cdots$$

Therefore, the probability that 5 stars fall within a circle of radius 49′ was

$$e^{-.07613}\frac{(.07613)^5}{5!}=1.975\times10^{-8}$$

or about 1 in 50,000,000. Finally, "since 1500 stars had each this probability of being the center of such a close cluster of 6, although these probabilities were not strictly independent, the probability that among them any one fulfils the condition" was approximately

$$\frac{1500}{50,000,000}\approx\frac{1}{33,000}.$$

Fisher concluded:

> Michell arrived at a chance of only 1 in 500,000, but the higher probability obtained by the calculations indicated above is amply low enough to exclude at a high level of significance any theory involving a random distribution. (*ibid.*, p. 39)

5.8.6 Laplace on the Mean Inclination of the Orbit of Comets (1776)

Laplace's first test of significance took place in his 1776 *Mémoire sur l'Inclinaison moyenne des orbites* (Laplace, 1776a). Laplace investigated the mean inclination of the orbits of comets with respect to the ecliptic. Details can be found in Section 1.2.4

5.8.7 Edgeworth's "Methods of Statistics" (1884)

It has often been said that Edgeworth anticipated many, if not most, of the ideas of modern mathematical statistics, and the case of significance testing is no exception. Edgeworth's relevant work took place in the paper "Methods of statistics" (Edgeworth, 1885) where we can read:

> The science of Means comprises two main problems: 1. To find how far the difference between any proposed Means is accidental or indicative of a law? 2. To find what is the best kind of Mean… An example of the first problem is afforded by some recent experiments in so called "psychiatric research." One person chooses a suit of cards. Another person makes a guess as to what the choice has been. Many hundred such choices and guesses having been recorded, it has been found that the proportion of successful guesses considerably exceeds the figure which would have been the most probable supposing chance to be the only agency at work, namely ¼. *E.g.*, in 1,833 trails the number of successful guesses exceeds 458, the

> quarter of the total number, by 52. The first problem investigates how far the difference between the average above stated and the results usually obtained in similar experience where pure chance reigns is a significant difference; indicative of the working of a law other than chance, or merely accidental. (*ibid.*, p. 182)

In the above, Edgeworth effectively formulated the one-sample problem of significance for a proportion and became the first to use the word "significant" in its statistical sense.

Later in the paper (*ibid.*, pp. 187 and 195), Edgeworth considered the following two-sample problem for means: the mean height of 2315 criminals differs from the mean height of 8585 members of the general adult male population by about 2 in. The modulus* (the standard deviation multiplied by $\sqrt{2}$) for population i ($i=1, 2$) is estimated by the formula

$$c_i = \sqrt{\frac{2\sum_{i=1}^{n_i}\left(x_{ij} - \overline{x}_i\right)^2}{n_i}},$$

where n_i is the size of the ith sample, x_{ij} is the jth observation for the ith sample, and $\overline{x}_i$ is the mean of the ith sample. For his example, Edgeworth used $c_1^2 = c_2^2 = 13$ so that the modulus of the difference between the two means is

$$\sqrt{\frac{c_1^2}{n_1} + \frac{c_2^2}{n_2}} = \sqrt{\frac{13}{8585} + \frac{13}{2315}} \approx .08.$$

Edgeworth then took three times the above modulus (i.e., .25), which was much less than the observed difference of 2, so he concluded the latter "is not accidental, but indicates a law" (*ibid.*, p. 196).

Edgeworth's criterion for significance is thus

$$\frac{\overline{x}_1 - \overline{x}_2}{\sqrt{\frac{\sigma_1^2}{n_1} + \frac{\sigma_2^2}{n_2}}} > 3\sqrt{2}$$

where $\sigma_i = c_i / \sqrt{2}$. The two-sided p-value of the test is therefore

$$2\Pr\left\{Z > 3\sqrt{2} \mid Z \sim N(0, 1)\right\} = .00002.$$

Edgeworth's paper also contained various other examples of significance testing.

* Edgeworth also called the square of the modulus the "fluctuation."

5.8.8 Karl Pearson's Chi-squared Goodness-of-Fit Test (1900)

Karl Pearson's test of significance, the chi-squared goodness-of-fit test, took place in the groundbreaking paper "On the criterion that a given system of deviations from the probable in the case of a correlated system of variables is such that it can be reasonably supposed to have arisen from random sampling" (Pearson, 1900b). Details of the test have been given in Section 3.2.2.

5.8.9 Student's Small-Sample Statistics (1908)

Student's 1908 paper (Student, 1908b) was a watershed in the history statistics because it pioneered the use of exact (or small-sample) statistics for the purpose of inference. Details of Student's work can be found in Section 4.2.

PART THREE: FROM DARMOIS TO ROBBINS

6

BEYOND FISHER AND NEYMAN–PEARSON

6.1 EXTENSIONS TO THE THEORY OF ESTIMATION

In his statistical work, Fisher was much more guided by his intuition than with concerns about mathematical rigor. Often his papers would contain proofs that were found wanting by his contemporaries. On one occasion at least, Fisher himself acknowledged that he was not satisfied with a proof because it lacked mathematical rigor. This occurred in the foundation paper of 1922 (Fisher, 1922c) where Fisher attempted to prove that maximum likelihood (ML) estimates are sufficient (see Section 5.2.3). This is what Fisher had to say concerning his proof:

> ...For my own part I should gladly have withheld publication until a rigorously complete proof could have been formulated; but the number and variety of the new results which the method discloses press for publication, and at the same time I am not insensible of the advantage which accrues to Applied Mathematics from the co-operation of the Pure Mathematician, and this co-operation is not infrequently called forth by the very imperfections of writers on Applied Mathematics. (*ibid.*, p. 15)

The above shows that Fisher acknowledged the importance of mathematical rigor, but he was not prepared to withhold publication just because his proofs lacked rigor. In fact, Fisher was opposed to the "mathematization of statistics," that is, to the introduction of so much formalism that the subject became overly abstract. In a 1957 letter to David Finney, he stated:

> ...There is surely a lifetime's work in front of me there, but it is less agreeable than doing a bit towards clarifying the situation in one of the natural sciences, where the application of statistical methods is not bogged down with all the pedantic verbiage that seems to be necessary in mathematical departments, where it seems to amount to no more than finding excuses for replacing known competent methods by less adequate ones. (Bennett, 1990, p. 327)

Classic Topics on the History of Modern Mathematical Statistics: From Laplace to More Recent Times, First Edition. Prakash Gorroochurn.

Many would argue that excess abstraction and rigor were precisely what happened to probability to some extent, when Lebesgue's measure theory was introduced into probability at the start of twentieth century. Perhaps Fisher had seen how abstract probability had become, and had been dismayed by it.

However, in spite of Fisher's reservations, the fact remains that no coherent theory of statistics (and of any other mathematical science for that matter) can be built without due attention to rigor. This was recognized by several mathematicians who followed Fisher, and they set out to fill the gaps in Fisher's work. In what follows, we shall examine the work of some of these mathematicians and also of others who extended his work.

6.1.1 Distributions Admitting a Sufficient Statistic

6.1.1.1 Fisher (1934) In 1934, Fisher pointed to the existence of an important class of distributions that admitted sufficient statistics (Fisher, 1934c). This class was later called the *exponential family of distributions* and constitutes an important aspect of current statistical theory. Fisher's argument was as follows.

Fisher first implicitly assumed that the support of the density of each observation is independent of the unknown parameter θ. He then denoted the solution of the maximum likelihood (ML) equation by

$$\phi(T) = A$$

(*ibid*., p. 293), where ϕ is a function of the sufficient statistic T, and A is a symmetric function of the observations but does not involve θ. Now, for the ML equation to be of the above form, the derivative of the log-likelihood ($\partial \log L / \partial \theta$) must have been of the form

$$C\psi'(\theta)\{A - \phi(\theta)\} = C\{A\psi'(\theta) - \phi(\theta)\psi'(\theta)\},$$

where $\psi(\theta)$ is a differentiable function of θ. Moreover, using the factorization condition for a sufficient statistic (see end of Section 5.2.3), Fisher assumed that C must either be a constant or a function of A. From the last equation, by integration, $\log L$ must be of the form

$$CA\psi(\theta) - C\int \phi(\theta)\, d\psi(\theta) + B,$$

where B is a function of the observations only. Now, "$\log L$ is the sum of expressions involving each observation singly," so Fisher concluded that $C = n$. By further writing

$$CA = \Sigma X, \quad B = \Sigma X_1,$$

where X, X_1 are functions of the observations, Fisher obtained the final form of L as

$$e^{-n\int \phi(\theta) d\psi(\theta)} e^{\psi(\theta)\Sigma X} e^{\Sigma X_1}$$

(*ibid.*, p. 294). We note that this implies that the density of each observation x is of the form

$$\exp\{a(\theta)b(x)+c(\theta)+d(x)\} \tag{6.1}$$

where $a(\theta), c(\theta)$ are functions of θ only, $b(x), d(x)$ are functions of x only, and the support of the density is independent of θ. The above is the exponential family of distributions, of which the normal, binomial, Poisson, and many more distributions are members. Therefore, distributions admitting a sufficient statistic belong to the exponential family.

6.1.1.2 Darmois (1935) Fisher did not provide the explicit form in (6.1) in his paper. However, one year later, Darmois filled the gap with the paper "Sur les lois de probabilité à estimation exhaustive" (Darmois, 1935).

Georges Darmois* (Fig. 6.1) was born on June 24, 1888, in Epley, France. He did his first studies at the Collège de Toul and Lycée de Nancy. He was then admitted to the École Normale Supérieure, where his brother had been admitted two years before. After graduating in mathematics, he studied geometry. During World War I, his duties geared him toward problems of ballistics and then of the transmission of sound. His doctorate was thus in these fields. Darmois' interest in statistics started in 1923 when he had to teach the subject at Nancy. In 1925, at the request of Émile Borel, Darmois also taught mathematical statistics at the Institute of Statistics of the University of Paris. Darmois' interest in mathematical statistics culminated in the publication of his treatise *Statistique Mathématique* (Darmois, 1928) in 1928. Apart from being one of the first modern French books in the field, *Statistique Mathématique* also contains an early use of the technique of characteristic functions (*ibid.*, p. 44).

Darmois' research in mathematical statistics was influenced by Fisher, and in particular Darmois studied sufficient statistics in great detail. In 1946, Darmois was made Director of the Institute of Statistics at the University of Paris. In 1953, he became Director of the International Statistical Institute, a position he held until his death in 1960.

Our main interest here is in Darmois' 1935 paper (Darmois, 1935). Darmois started by referring to the fecundity of Fisher's ideas in the theory of estimation. He then reviewed the basic properties of sufficient estimators ("*estimation exhaustive*"), which were the subject of interest of his current paper. He explained the problem he wished to solve as follows:

> ...If we represent the set of independent random variables $x_1, x_2, \ldots, x_n$ by a point M in n-dimensional space, then the probability law of M must have a particular structure. If a is indeed a sufficient estimator, which is a function of $x_1, x_2, \ldots, x_n$, then for a given value of a the point M must lie on a hyperspace Σ; the necessary and sufficient condition for a to be a sufficient estimator is that the probability law of M lying on Σ no longer depends on α [the parameter]. (*ibid.*, p. 1265)

The question then is, what is the general form of the probability law of M for which such a sufficient estimator a exists? Without any demonstration, Darmois gave the form of the probability density $f(x;\alpha)$ as

$$f = e^{u(x)v(\alpha)+r(x)+s(\alpha)}$$

* Darmois' life is also described in Bru (2001a, pp. 382–385), Roy (1961), and Dugué (1961).

FIGURE 6.1 Georges Darmois (1888–1960). Back photo in Dugué (1960). Courtesy of John Wiley & Sons, Inc.

(*ibidem*), where $s(\alpha)$, $v(\alpha)$ are functions of α only, $u(x)$, $r(x)$ are functions of x only, and the support of $f(x;\alpha)$ is independent of α. Since the density integrates to unity, he also noted that $s(\alpha)$ can be obtained once the other three functions are specified.

When there are k variables $x, y, z, \ldots$ and h parameters $\alpha, \beta, \ldots$, Darmois gave the analogous form of $\log f$ as

$$\log f = U_1(x,y,z,\ldots)A_1(\alpha,\beta,\ldots)+\cdots+U_h(x,y,z,\ldots)A_h(\alpha,\beta,\ldots)+V(x,y,z,\ldots)+B(\alpha,\beta,\ldots),$$

where $U_1,\ldots,U_h, V$ are functions of $x, y, z, \ldots$, the quantities $A_1,\ldots,A_h, B$ are functions of $\alpha, \beta, \ldots$, and the support of f is independent of $\alpha, \beta, \ldots$

6.1.1.3 Koopman (1936) A rigorous proof of the exponential form was first provided by Koopman in 1936 in the paper "On distributions admitting a sufficient statistic" (Koopman, 1936).

Bernard Osgood Koopman* was born in Paris on January 19, 1900. He did his early studies in Lycée Montaigne. The year 1912 was marked by the divorce of his parents, Koopman's serious illness, and his father's declining health. After his father passed away in 1914, Koopman migrated to the United States with his mother and sister in 1914. In 1918, Koopman entered Harvard and studied electrodynamics with Emory Chaffee, thermodynamics with Percy Bridgman (whose daughter he married later in 1948, after his wife passed away in 1946), and mathematics with George Birkhoff (1884–1944). Shortly after earning his Ph.D., he joined the Mathematics Department at Columbia University in 1927. During World War II, Koopman joined the Anti-Submarine Warfare Operations Research Group in Washington, D.C., to work for the US Navy. There, he and several colleagues developed techniques to hunt U-boats. The theoretical work was responsible for the later development of search theory, now a field in operations research. Apart from operations research, Koopman was also interested in the pure side of mathematics and probability. He became Head of the Mathematics Department at Columbia University in 1951. Koopman retired in 1968 and passed away many years later, in 1981.

In his 1936 paper (Koopman, 1936), Koopman considered a random sample of independent observations $x_1,\ldots,x_n$ such that each x_i has density function $f(\theta_1,\ldots,\theta_\nu,x)$, where $\theta_1,\ldots,\theta_\nu$ are unknown parameters of the distribution and the support of f does not depend on the parameters. Let $\phi_j(x_1,\ldots,x_n)$ be an estimator of θ_j such that ϕ_j is continuous in $\Re_n$. Then Koopman proved that the necessary condition for $(\phi_1,\ldots,\phi_\nu)$ to form a set of sufficient statistics for the distribution is that

$$f\left(\theta_1,\ldots,\theta_\nu,x\right)=\exp\left(\sum_{k=1}^{\mu}\Theta_k X_k+\Theta+X\right) \tag{6.2}$$

(*ibid.*, p. 402), where Θ_k, Θ are analytic functions of $(\theta_1,\ldots,\theta_\nu)$, X_k, X are analytic functions of x, and $\mu \le \nu$.

In his proof, Koopman considered the cases $\nu=2$ and $n=3$ and used the following definition of sufficiency. The distribution $f(\theta_1,\ldots,\theta_\nu,x)$ admits the sufficient statistics $\phi_j(x_1,\ldots,x_n)$ $(j=1,\ldots,\nu)$ if the equations

$$\phi_j\left(x_1,\ldots,x_n\right)=\phi_j\left(x_1',\ldots,x_n'\right) \tag{6.3}$$

* Koopman's life is also described in Morse (1982).

imply

$$\frac{\prod_{i=1}^{n} f(\theta_1,\ldots,\theta_v,x_i)}{\prod_{i=1}^{n} f(\theta_1,\ldots,\theta_v,x_i')} = \frac{\prod_{i=1}^{n} f(\theta_1',\ldots,\theta_v',x_i)}{\prod_{i=1}^{n} f(\theta_1',\ldots,\theta_v',x_i')} \tag{6.4}$$

(*ibid.*, p. 400). We note that Koopman's definition expresses the intuitive notion that if the values of the sufficient statistics are the same for two data sets $(x_1,\ldots,x_n)$ and $(x_1',\ldots,x_n')$, then the relative joint density of the two data sets is independent of the parameters $(\theta_1,\ldots,\theta_v)$, that is, each joint density carries the same amount of information regarding the parameters. This statement is also known as the *sufficiency principle*.

Koopman's proof of (6.2) was as follows. By successively setting $(\theta_1',\theta_2')=(a_{11},a_{12})$, (a_{21},a_{22}), (a_{31},a_{32}) in Eq. (6.4) and taking logarithms,

$$\sum_{i=1}^{n} \log \frac{f(\theta_1,\theta_2,x_i)}{f(a_{j1},a_{j2},x_i)} = \sum_{i=1}^{n} \log \frac{f(\theta_1,\theta_2,x_i')}{f(a_{j1},a_{j2},x_i')} \quad \text{for } j=1,2,3 \tag{6.5}$$

(*ibid.*, p. 403). Since ϕ_1,ϕ_2 form a system of sufficient statistics, the above equations are a consequence of Eq. (6.3), namely,

$$\begin{aligned} \phi_1(x_1,x_2,x_3) &= \phi_1(x_1',x_2',x_3'), \\ \phi_2(x_1,x_2,x_3) &= \phi_2(x_1',x_2',x_3'). \end{aligned} \tag{6.6}$$

Geometrically this means that the locus of (x_1,x_2,x_3) in (6.6) through each given point (x_1',x_2',x_3') must be a subset of the locus of (x_1,x_2,x_3) in (6.5) through the same point. Koopman then argued that the Jacobian of the left members of the latter equation with respect to x_1,x_2,x_3 must therefore be identically zero, for otherwise the preservation of dimensionality under homeomorphism is violated.

Now consider the Jacobian matrix

$$\mathbf{M}_3 = \left\| M_{ij} \right\|_{(i,j=1,2,3)} = \left\| \frac{\partial}{\partial x_i} \log \frac{f(\theta_1,\theta_2,x_i)}{f(a_{j1},a_{j2},x_i)} \right\|_{(i,j=1,2,3)},$$

together with the upper left-hand minors

$$\mathbf{M}_2 = \left\| M_{ij} \right\|_{(i,j=1,2)}, \quad \mathbf{M}_1 = \left\| M_{11} \right\|$$

(*ibid.*, p. 404). Furthermore, let ρ be the identical rank of $\mathbf{M}_3$, that is, the order of the nonidentically vanishing determinant of highest order in $\mathbf{M}_3$. Then the possible values of ρ are 0, 1, and 2. Koopman wrote $\rho = \det \mathbf{M}_\rho$ and considered each possible value separately:

- $\rho = 0$. In this case $\det \mathbf{M}_1 = 0$, so that

$$\frac{\partial}{\partial x_1} \log \frac{f(\theta_1,\theta_2,x_1)}{f(a_{11},a_{12},x_1)} = 0. \tag{6.7}$$

Integrating this with respect to x_1 from b to x,

$$\log\frac{f(\theta_1,\theta_2,x)}{f(a_{11},a_{12},x)}=\log\frac{f(\theta_1,\theta_2,b)}{f(a_{11},a_{12},b)}.$$

Solving for $f(\theta_1,\theta_2,x)$, the desired form in (6.2) is obtained, with $\mu=0$ and

$$\Theta(\theta_1,\theta_2)=\log\frac{f(\theta_1,\theta_2,b)}{f(a_{11},a_{12,},b)},$$
$$X(x)=\log f(a_{11},a_{12},x).$$

- $\rho=1$. In this case $\det\mathbf{M}_1\neq 0$, but $\det\mathbf{M}_2=0$. The latter determinant is still zero if the first column of $\mathbf{M}_2$ is subtracted from the second. Once this is done, $\partial/\partial x_1$ is taken outside the determinant and the resulting equation is integrated with respect to x_1 from b to x. Then

$$\begin{vmatrix} \log\dfrac{f(\theta_1,\theta_2,x)}{f(a_{11},a_{12},x)} & \log\dfrac{f(a_{11},a_{12},x)}{f(a_{21},a_{22},x)} \\ \dfrac{\partial}{\partial x_2}\log\dfrac{f(\theta_1,\theta_2,x_2)}{f(a_{11},a_{12},x_2)} & \dfrac{\partial}{\partial x_2}\log\dfrac{f(a_{11},a_{12},x_2)}{f(a_{21},a_{22},x_2)} \end{vmatrix}$$
$$=\begin{vmatrix} \log\dfrac{f(\theta_1,\theta_2,b)}{f(a_{11},a_{12},b)} & \log\dfrac{f(a_{11},a_{12},b)}{f(a_{21},a_{22},b)} \\ \dfrac{\partial}{\partial x_2}\log\dfrac{f(\theta_1,\theta_2,x_2)}{f(a_{11},a_{12},x_2)} & \dfrac{\partial}{\partial x_2}\log\dfrac{f(a_{11},a_{12},x_2)}{f(a_{21},a_{22},x_2)} \end{vmatrix}$$

  As for the previous case, the above can be solved for $f(\theta_1,\theta_2,x)$ and the form in (6.2) established, with $\mu=1$. However, for this to happen it is necessary that an x_2 can be found such that

$$\frac{\partial}{\partial x_2}\log\frac{f(a_{11},a_{12},x_2)}{f(a_{21},a_{22},x_2)}\neq 0.$$

  Koopman argued that such an x_2 must exist, for otherwise upon multiplying the above by -1 and replacing a_{21},a_{22},x_2 by θ_1,θ_2,x_1, we would be led to Eq. (6.7), that is, $\det\mathbf{M}_1=0$. But this possibility has already been excluded for the case $\rho=1$.
- $\rho=2$. In this case $\det\mathbf{M}_2\neq 0$, but $\det\mathbf{M}_3=0$. The argument used here is the same as for the $\rho=1$ case.

Having shown the exponential form for all three cases $\rho=0,1,\text{and }2$, Koopman thus proved Eq. (6.2).

6.1.1.4 Pitman (1936) A few months after Koopman's paper, Pitman published a paper entitled "Sufficient statistics and intrinsic accuracy" (Pitman, 1936) where he gave an independent proof of the exponential family.

Edwin James George Pitman* (Fig. 6.2) was born in Melbourne, Australia, on October 29, 1897. He was brilliant in mathematics right from a young age. In his final year of school, he obtained the Wyselaskie and Dixson Scholarships in Mathematics, and also a scholarship in Ormond College, University of Melbourne. He graduated from the latter with first-class honors. He was acting Professor of Mathematics at the University of New Zealand in 1922–1923 and then tutor in mathematics and physics at the Melbourne in 1924–1925. Pitman was appointed Professor and Chair of Mathematics at the University of Tasmania in 1926, a position he held till his retirement in 1962. Pitman himself recalled that his first encounter with statistics was at the age of 23 while he was a student at the University of Melbourne:

> ...I decided then and there that statistics was the sort of thing that I was not interested in, and would never have to bother about. (Pitman, 1982, p. 112)

But two years after his appointment as Chair, an experimenter (R.A. Scott) from the State Department of Agriculture asked Pitman a statistical question. Pitman was given Fisher's recently published *Statistical Methods for Research Workers* (Fisher, 1925b) to consult. This episode effectively launched Pitman's career into statistics as he learned more about Fisher's methods. In 1936–1939, Pitman published a series of papers in leading statistical journals, resulting in the development of the Fisher–Pitman permutation test.†

In 1948, Pitman introduced the concept of asymptotic relative efficiency while lecturing at the University of North Carolina and at Columbia University. However, he had already taught this concept to his students in Tasmania in 1946. Pitman was elected Fellow of the Australian Academy of Science in 1954. He passed away in Hobart, Tasmania, on July 21, 1993.

Pitman's demonstration of the exponential family in 1936 (Pitman, 1936) was along the lines of Fisher's earlier argument, as we now describe.

Pitman considered a random sample of independent observations $x_1, x_2, \ldots, x_n$. Each $x_r (r = 1, 2, \ldots, n)$ has probability density $f(x_r, \theta)$, where θ is an unknown parameter of interest and the support of f does not depend on θ. Let the likelihood of θ be ϕ and the log-likelihood be L, that is,

$$\phi = \prod f(x_r, \theta),$$
$$L = \log \phi = \sum \log f(x_r, \theta).$$

Assuming T_1 to be a sufficient statistic, Pitman then wrote

$$\frac{\phi'}{\phi} = \frac{\partial L}{\partial \theta} = \sum \frac{f'(x_r, \theta)}{f(x_r, \theta)} = \Phi(T_1, \theta) \tag{6.8}$$

(*ibid*., p. 569), where Φ is a function of T_1 and θ only, and $f'(x_r, \theta) = \partial f(x_r, \theta)/\partial \theta$. We see that the last part of the equality above is a consequence of the Fisher–Neyman factorization of ϕ (see end of Section 5.2.3). Next, since T_1 is a function of the data only, let

$$T_1 = \psi\left(\sum g(x_r)\right) = \psi(T),$$

* Pitman's life is also described in Pitman (1982, pp. 111–125), Williams (2001), and Sprent (1994).

† See also Berry et al. (2014, pp. 78–82).

FIGURE 6.2 Edwin James George Pitman (1897–1993). From Williams (1994). Courtesy of the Australian Academy of Science and CSIRO Publishing

where ψ, g are arbitrary functions and $T = \Sigma g(x_r)$ is a statistic. From Eq. (6.8), $\partial L / \partial\theta$ must be a function of T and θ only, say,

$$\frac{\partial L}{\partial\theta} = H(T,\theta). \tag{6.9}$$

Then, using the chain rule,

$$\begin{aligned}\frac{\partial^2 L}{\partial\theta\partial x_r} &= \frac{\partial}{\partial x_r}H(T,\theta)\\ &= \frac{\partial}{\partial T}H(T,\theta)\frac{\partial T}{\partial x_r}\\ &= H_T'(T,\theta)\frac{\partial T}{\partial x_r},\end{aligned}$$

where $H_T' = \partial H / \partial T$. Now, the left side of the above is a function of x_r and θ only, and $\partial T / \partial x_r$ is a function of x_r only. Therefore, $H_T'(T,\theta)$ must be a function of x_r and θ only. But since T is a symmetric function of the x_r's, $H_T'(T,\theta)$ must be a function of θ only. With this in mind, integrating $H_T'(T,\theta)$ with respect to θ, we have

$$H(T,\theta) = p(\theta)T + q(\theta),$$

where p and q are arbitrary functions of θ. Combining this result with Eqs. (6.8) and (6.9),

$$\begin{aligned}\sum\frac{f'(x_r,\theta)}{f(x_r,\theta)} &= H(T,\theta) = p(\theta)T + q(\theta).\\ \Rightarrow\quad \sum\frac{f'(x_r,\theta)}{f(x_r,\theta)} &= p(\theta)\sum g(x_r) + q(\theta),\\ \frac{f'(x_r,\theta)}{f(x_r,\theta)} &= p(\theta)g(x_r) + \frac{q(\theta)}{n},\\ \log f(x_r,\theta) &= p_1(\theta)g(x_r) + \frac{q_1(\theta)}{n} + k(x_r),\end{aligned}$$

where $p_1(\theta) = \int p(\theta)d\theta$, $q_1(\theta) = \int q(\theta)d\theta$, and $k(x_r)$ is a function of x_r. Thus, Pitman obtained the following exponential form for the density $f(x_r, \theta)$:

$$f(x_r,\theta) = u(\theta)h(x_r)\exp\{p_1(\theta)g(x_r)\}$$

(*ibid.*, p. 570), where $u(\theta) = \exp\{q_1(\theta)/n\}$ and $h(x_r) = \exp\{k(x_r)\}$. The exponential family of distributions was previously known as the *Darmois–Koopman–Pitman* family (or a permutation thereof).

6.1.2 The Cramér–Rao Inequality

6.1.2.1 Introduction The Cramér–Rao (CR) inequality is an important extension of Fisher's theory of statistical estimation. It provides a lower bound for the variance of an unbiased estimator of a function of a parameter. According to the inequality, under the regularity conditions that the likelihood function (i) has a support that does not depend on the unknown parameter of interest and (ii) can be differentiated with respect to the latter, we have

$$\operatorname{var} t \geq \frac{\{\tau'(\theta)\}^2}{\mathsf{E}\left(\dfrac{\partial \log L}{\partial \theta}\right)^2} = \frac{\{\tau'(\theta)\}^2}{-\mathsf{E}\dfrac{\partial^2 \log L}{\partial \theta^2}}, \tag{6.10}$$

where t is an unbiased estimator of $\tau(\theta)$ (a function of the parameter θ) and L is the likelihood function. Since the denominator is the expected information in the sample, the expression above is also called the *information inequality*. The right-hand side of the inequality is called the Cramér–Rao lower bound (CRLB). Note that some estimators *do not* attain the CRLB, depending on the underlying distribution of the observations. For some distributions, there is a least attainable variance that is higher than the CRLB.

Note also that Eq. (6.10) is based on direct (rather than inverse) probability and is different from a similar formula obtained by Edgeworth (see Eq. 5.145). Moreover, Eq. (6.10) applies to all sample sizes and is thus also different from a similar formula obtained by Fisher (see Eq. 5.26).

6.1.2.2 Aitken & Silverstone (1942) Although Eq. (6.10) is usually credited to the independent works of Cramér and Rao, it had been previously discovered in various forms by other mathematicians. In fact, the CR inequality was first implicitly given by Aitken in 1942 while working with his student Silverstone and first explicitly given by Fréchet in 1943. We shall soon examine each of these works, starting with the 1942 paper. But first let us say a few words regarding Aitken.

Alexander Craig Aitken* was born in Dunedin, New Zealand, on April 1, 1885. He was the eldest of seven children and his father was a farmer. In 1908, Aitken attended Otago Boys' High School and in 1913 he won a Junior University Scholarship to Otago University. Aitken wished to study languages and mathematics with a view to becoming a teacher. However, in 1915 he enlisted as a private in the New Zealand Infantry and served in Gallipoli, Egypt, and France. In 1916, Aitken was wounded in the battle of the Somme and returned to New Zealand in 1917. He rejoined Otago University, but there was no mathematics professor there. The upshot was that Aitken graduated in 1920 with First-Class Honors in Languages and Literature, but only Second-Class Honors in Mathematics. After his graduation, he married a fellow student, Mary Winifred Betts, and taught languages in Otago Boys' High School until 1923. In that year, Aitken was given a scholarship to study under E.T. Whittaker in Edinburgh, Scotland. Aitken's doctoral dissertation, which was on the graduation of observational data, was a superlative work, and consequently Aitken was awarded the higher degree of D.Sc. instead of a Ph.D. In 1932 he co-wrote the book *An Introduction to the Theory of Canonical Matrices* with Turnbull (Aitken and Turnbull, 1932), and in 1946 he succeeded Whittaker as the Chair of Mathematics. The 1932 book, together with Aitken's earlier 1931 paper (Aitken, 1931), also marked the introduction of matrix algebra into statistics (David, 2006). Apart from being a renowned teacher, Aitken had an extraordinary memory and a real passion for music. He passed away on November 3, 1967, in Edinburgh.

*Aitken's life is also described in Kendall (1968), Whittaker and Bartlett (1968), and Watson (2006, pp. 101–103).

In their paper, "On the estimation of statistical parameters" (Aitken and Silverstone, 1942), Aitken and Silverstone explained that their approach was motivated by the theory of least squares. In that theory, the normal equations can be obtained either (A) by assuming a normal distribution for the errors and maximizing their joint probability density, or (B) by seeking the unbiased linear combinations that minimizes the error variance. Similarly, in Fisher's theory of maximum likelihood, the usual route is to start from the likelihood and then maximize it. However, the authors wished to investigate approach (B): assume the existence of an unbiased estimator and seek the solution that minimizes its variance. Their analysis was as follows.

Let $\phi(x,\theta)$ be the density of x and let θ be an unknown parameter that is to be estimated from a random sample of independent observations $x_1, x_2, \ldots, x_n$. It is assumed that the support of ϕ is independent of θ and that ϕ is uniformly differentiable with respect to θ. The aim is to estimate θ through a statistic $t = t(x_1, x_2, \ldots, x_n)$ subject to the two conditions:

1. $\int \cdots \iint t\Phi dx_1 dx_2 \ldots dx_n = \theta$, and
2. $\int \cdots \iint (t-\theta)^2 \Phi dx_1 dx_2 \ldots dx_n = \text{minimum}$,

where Φ is the joint density of $x_1, x_2, \ldots, x_n$. The solution to the problem is obtained from the calculus of variations and was given by Aitken and Silverstone as

$$(t-\theta)\Phi - \lambda(\theta)\Phi_\theta = 0 \tag{6.11}$$

(*ibid*., p. 188), where $\Phi_\theta = \partial\Phi/\partial\theta$ and $\lambda(\theta)$ (the Lagrange function) is independent of $x_1, x_2, \ldots, x_n$. We note that the above solution can be obtained by writing the first condition as

$$\int \ldots \iint t\Phi_\theta dx_1 dx_2 \ldots dx_n = 1,$$

so that it is required to unconditionally minimize

$$\int \cdots \iint \left\{ (t-\theta)^2 \Phi - 2\lambda(\theta)\Phi_\theta \right\} dx_1 dx_2 \ldots dx_n.$$

This has solution

$$\frac{\partial}{\partial t}\left\{ (t-\theta)^2 \Phi - 2\lambda(\theta) t \Phi_\theta \right\} = 0,$$

whence (6.11) is obtained.

Since

$$(t-\theta) = \lambda(\theta)\frac{\partial\Phi}{\Phi\partial\theta} = \lambda(\theta)\frac{\partial \log \Phi}{\partial\theta},$$

Aitken and Silverstone concluded that an estimator t of θ satisfying the two aforementioned conditions exists as long as $\partial \log\Phi/\partial\theta$ can be written in form

$$\frac{\partial \log\Phi}{\partial\theta} = \frac{t-\theta}{\lambda(\theta)} \tag{6.12}$$

(*ibidem*). The authors noted that although it is sometimes not possible to obtain the above form for an estimator of θ, it is possible to do so for a function $\tau(\theta)$ of θ. So they modified their previous conclusion by stating that an estimator t of $\tau(\theta)$ satisfying the two aforementioned conditions exists as long as $\partial \log\Phi/\partial\theta$ can be written in form

$$\frac{\partial \log \Phi}{\partial \theta}=\frac{t-\tau(\theta)}{\lambda(\theta)}.$$

Coming back to Eq. (6.12), how should the minimum variance v of t be found? We have

$$\begin{aligned} v &= \int\cdots\iint (t-\theta)^2\, \Phi dx_1 dx_2 \ldots dx_n \\ &= \int\cdots\iint \lambda^2 \left(\frac{\partial \log \Phi}{\partial \theta}\right)^2 \Phi dx_1 dx_2 \ldots dx_n \\ &= \lambda^2 \int\cdots\iint \left(\frac{\Phi_\theta}{\Phi}\right)^2 \Phi dx_1 dx_2 \ldots dx_n \end{aligned}$$

(*ibid.*, p. 190–191). Now,

$$\frac{\partial^2 \log \Phi}{\partial \theta^2}=\frac{\partial}{\partial \theta}\left(\frac{\Phi_\theta}{\Phi}\right)=\Phi^{-1}\frac{\partial^2 \Phi}{\partial \theta^2}-\left(\frac{\Phi_\theta}{\Phi}\right)^2 \Rightarrow \left(\frac{\Phi_\theta}{\Phi}\right)^2=\Phi^{-1}\frac{\partial^2 \Phi}{\partial \theta^2}-\frac{\partial^2 \log \Phi}{\partial \theta^2}.$$

Therefore,

$$v=\lambda^2 \int\cdots\iint \left(\Phi^{-1}\frac{\partial^2 \Phi}{\partial \theta^2}-\frac{\partial^2 \log \Phi}{\partial \theta^2}\right)\Phi dx_1 dx_2 \ldots dx_n.$$

Now since $\int\cdots\iint \Phi dx_1 dx_2 \ldots dx_n = 1$, we have $\int\cdots\iint (\partial^2\Phi/\partial\theta^2) dx_1 dx_2 \ldots dx_n = 0$, and the above expression becomes

$$v=-\lambda^2 \int\cdots\iint \frac{\partial^2 \log \Phi}{\partial \theta^2}\Phi dx_1 dx_2 \ldots dx_n. \tag{6.13}$$

Now, from Eq. (6.12)

$$\frac{\partial^2 \log \Phi}{\partial \theta^2}=(t-\theta)\frac{\partial}{\partial \theta}\left(\frac{1}{\lambda}\right)-\frac{1}{\lambda},$$

so that

$$\begin{aligned} v &= -\lambda^2 \int\cdots\iint \left\{(t-\theta)\frac{\partial}{\partial \theta}\left(\frac{1}{\lambda}\right)-\frac{1}{\lambda}\right\}\Phi dx_1 dx_2 \ldots dx_n \\ &= -\lambda^2 \int\cdots\iint (t-\theta)\frac{\partial}{\partial \theta}\left(\frac{1}{\lambda}\right)\Phi dx_1 dx_2 \ldots dx_n + \lambda^2 \int\cdots\iint \frac{1}{\lambda}\Phi dx_1 dx_2 \ldots dx_n \\ &= -\lambda^2 \frac{\partial}{\partial \theta}\left(\frac{1}{\lambda}\right)(0)+\frac{\lambda^2}{\lambda}(1) \\ &= \lambda. \end{aligned}$$

Thus, the minimum variance v is the function λ in (6.12). But it is somewhat cumbersome to find v in this way since we must first arrange $\partial \log\Phi / \partial\theta$ in the form of Eq. (6.12). We can do better and obtain an explicit formula from Eq. (6.13):

$$\begin{aligned}
v &= -v^2 \int \cdots \iint \frac{\partial^2 \log \Phi}{\partial \theta^2} \Phi dx_1 dx_2 \ldots dx_n \\
\Rightarrow \quad v &= \frac{1}{-\int \cdots \iint \frac{\partial^2 \log \Phi}{\partial \theta^2} \Phi dx_1 dx_2 \ldots dx_n} \\
&= \frac{1}{-\mathsf{E} \frac{\partial^2 \log \Phi}{\partial \theta^2}} \\
&= \frac{1}{-n\mathsf{E} \frac{\partial^2 \log \phi}{\partial \theta^2}}
\end{aligned} \tag{6.14}$$

(*ibid*., p. 191). The authors then noted:

> …that v^{-1} is the mean value of $-\partial^2 L / \partial\theta^2$, where $L = \log\Phi$. This last is a well-known result of R.A. Fisher, which here we see holding accurately in finite samples, provided that $\lambda(\theta)$ exists, and that $\partial L / \partial\theta = (t-\theta)/\lambda(\theta)$. (*ibidem*)

However, no mention of Fisher's information (which is equal to the denominator of v) is made.

Aiken's and Silverstone's formula for v thus coincides with Fisher's formula in (5.26) for the large-sample variance of the ML estimator. Therefore, asymptotically, ML estimators are unbiased and have the minimum variance out of all unbiased estimators.

The formula for v also implies that the variance of *any* unbiased estimator t^* of θ satisfies

$$\operatorname{var} t^* \geq \frac{1}{-n\mathsf{E} \frac{\partial^2 \log \phi}{\partial \theta^2}}.$$

This is the same formula as (6.10) for the case $\tau(\theta) = \theta$.

In Section 6 (*ibid*., p. 192) of their paper, Aitken and Silverstone also investigated the type of distributions that admitted the minimum variance in (6.14). Remembering that t is independent of θ, Eq. (6.12) is integrated with respect to θ to obtain

$$\log \Phi = \mu(\theta) t + v(\theta) + C.$$

In the above,

$$\mu(\theta) = \int \frac{d\theta}{\lambda(\theta)}, \quad v(\theta) = -\int \frac{\theta}{\lambda(\theta)} d\theta,$$

and C is a function of the data only. Hence, $\phi(x;\theta)$ must be of the form

$$\phi(x;\theta)=\exp\{\mu(\theta)t(x)+F(\theta)+f(x)\} \tag{6.15}$$

(*ibidem*), where f and F are arbitrary functions of x and θ, respectively. We note that the above is the exponential family of distributions (see Eq. 6.1), a family that almost exclusively admits sufficient statistics. Hence, the exponential family is the one that allows unbiased estimators to have the minimum variance in (6.14).

As we mentioned before, Aitken and Silverstone had mentioned in their paper that sometimes it was not possible to obtain the form in (6.12) for an estimator t of θ. In that case, it was more reasonable to estimate a function $\tau(\theta)$ of θ, so that Eq. (6.12) becomes

$$\frac{\partial \log \Phi}{\partial \theta}=\frac{t-\tau(\theta)}{\lambda(\theta)}.$$

The authors now (*ibidem*) addressed the issue of the function $\tau(\theta)$ to be estimated. From Eq. (6.15),

$$\begin{aligned}
&\phi(x;\theta)=\exp\{\mu(\theta)t(x)+F(\theta)+f(x)\}.\\
&\Rightarrow \quad \Phi=\phi(x_1;\theta)\phi(x_2;\theta)\ldots\phi(x_n;\theta)=\exp\left(\mu\sum t+nF+\sum f\right),\\
&\frac{\partial}{\partial\tau}\log\Phi=\frac{\partial\mu}{\partial\tau}\sum t+n\frac{\partial F}{\partial\tau}=n\frac{\partial\mu}{\partial\tau}\left(\frac{\sum t}{n}+\frac{\partial F}{\partial\mu}\right).
\end{aligned}$$

If we define

$$\tau\equiv-\frac{\partial F}{\partial\mu}=-\frac{\dfrac{\partial F}{\partial\theta}}{\dfrac{\partial\mu}{\partial\theta}},$$

then the last but one equation can be written as

$$\frac{\partial}{\partial\tau}\log\Phi=\frac{\sum t/n-\tau}{\lambda(\tau)}$$

(*ibidem*), where $\lambda(\tau)=(n\partial\mu/\partial\tau)^{-1}$. This is in the form (6.12). Thus, the function of θ that should be estimated is $\tau=-\partial F/\partial\mu$, the estimator of τ is $\sum t/n$, and the variance of $\sum t/n$ is $(n\partial\mu/\partial\tau)^{-1}$.

6.1.2.3 Fréchet (1943) Apparently unaware of Aitken and Silverstone's paper, Fréchet published an article in 1943 entitled "Sur l'extension de certaines évaluations statistiques au cas de petits échantillons" (Fréchet, 1943), where the CR inequality was first explicitly given. Before examining Fréchet's work, we shall say a few words about his life.

Maurice Fréchet* (Fig. 6.3) was born in Maligny, France, on September 2, 1878, in a Protestant family. Soon after, the latter moved to Paris and Fréchet studied at the

* Fréchet's life is also described in Kendall (1977) and Dugué (1974).

FIGURE 6.3 Maurice Fréchet (1878–1973). Wikimedia Commons (Public Domain), http://commons.wikimedia.org/wiki/File:Frechet.jpeg

Lycée Buffon, where he was taught by Jacques Hadamard (1865–1963). In 1900, Fréchet was admitted to the École Normale Supérieure. His superlative 1906 thesis dealt with abstract metric spaces. In 1907–1908, he worked as schoolteacher at Besançon and Nantes. Between 1910 and 1919, he became Professor at the University of Poitiers. During World War I, Fréchet worked as an Anglo-French interpreter. He then served as Professor in the Faculty of Sciences in Strasbourg, where he taught mathematics, statistics, actuarial science, and nomography. Fréchet was made Professor at the Sorbonne's Rockefeller Foundation in late 1928. In the same year, he published his famous book *Les Espaces Abstraits* (Fréchet, 1928), which developed the ideas of his thesis. Fréchet also wrote several books in probability, of which *Recherches Théoriques Modernes sur le Calcul des Probabilités* (Fréchet, 1938) should be mentioned. He later succeeded Émile Borel as the Chair of the Calculus of Probabilities and Mathematical Physics in 1941. In 1956, Fréchet was elected to the *Académie des Sciences*. He died in 1973 in Paris, at the age of 94.

In his 1943 paper (Fréchet, 1943), Fréchet considered a probability density $f(x,\theta)$, where θ is a parameter to be estimated by an unbiased estimator T based on a sample of size n. He then explicitly showed that, under regularity conditions, the standard deviation of T satisfies

$$\sigma_T \geq \frac{1}{\sigma_A \sqrt{n}} \tag{6.16}$$

(*ibid.*, p. 185, Eq. 10), where σ_A is the standard deviation of the random variable

$$A = \frac{1}{f(x,\theta)} \frac{\partial f(x,\theta)}{\partial \theta} = \frac{\partial \log f(x,\theta)}{\partial \theta}.$$

To prove the result in (6.16), Fréchet proceeded as follows. The joint density of a random sample of independent observations $x_1, x_2, \ldots, x_n$ is $\prod f(x_i,\theta)$, so that

$$\int_{-\infty}^{\infty} \cdots \int_{-\infty}^{\infty} \left\{ \prod f(x_i,\theta) dx_i \right\} = 1.$$

Differentiating the above with respect to θ,

$$\int_{-\infty}^{\infty} \cdots \int_{-\infty}^{\infty} \left\{ \sum A(x_i) \right\} \left\{ \prod f(x_i,\theta) dx_i \right\} = 0.$$

This is the same as $\mathcal{M}_\theta \sum A(x_i) = 0$, where $\mathcal{M}_\theta$ denotes expectation. Therefore, with $U = \sum A(x_i)$,

$$\mathcal{M}_\theta U = 0$$

(*ibid.*, p. 187). Now since T is unbiased,

$$\int_{-\infty}^{\infty} \cdots \int_{-\infty}^{\infty} (T - \theta) \left\{ \prod f(x_i,\theta) dx_i \right\} = 0.$$

Differentiating the last expression with respect to θ,

$$\int_{-\infty}^{\infty}\cdots\int_{-\infty}^{\infty}\left[(T-\theta)\frac{\partial}{\partial\theta}\left\{\prod f(x_i,\theta)\,dx_i\right\}+\left\{\prod f(x_i,\theta)dx_i\right\}\frac{\partial}{\partial\theta}(T-\theta)\right]=0$$

$$\int_{-\infty}^{\infty}\cdots\int_{-\infty}^{\infty}\left[(T-\theta)\left\{\sum A(x_i)\right\}\left\{\prod f(x_i,\theta)dx_i\right\}\right]+(1)\,(0-1)=0$$

$$\int_{-\infty}^{\infty}\cdots\int_{-\infty}^{\infty}\left[(T-\theta)\left\{\sum A(x_i)\right\}\left\{\prod f(x_i,\theta)dx_i\right\}\right]-1=0$$

Thus, writing $\mathcal{M}_\Theta$ as $\mathcal{M}$

$$\mathcal{M}(T-\theta)U=1,$$

which is the same as

$$\mathcal{M}(T-\mathcal{M}T)\,(U-\mathcal{M}U)=1$$

(*ibid*., pp. 187–188). Now, from the Cauchy–Schwarz inequality*

$$\mathcal{M}^2\left(T-\mathcal{M}T\right)\left(U-\mathcal{M}U\right)\le\mathcal{M}\left(T-\mathcal{M}T\right)^2\mathcal{M}\left(U-\mathcal{M}U\right)^2$$

$$\therefore\quad 1\le\left(\sigma_T\sigma_U\right)^2,$$

where σ_U is the standard deviation of U. Since $U=\sum A(x_i)$, we have $\sigma_U^2=\sum\sigma_A^2=n\sigma_A^2$, so that the above becomes

$$\sigma_T\ge\frac{1}{\sigma_A\sqrt{n}}$$

(*ibid*., p. 188). Eq. (6.16) is thus proved.

We note that

$$\sigma_A^2=\mathcal{M}\left\{\frac{\partial\log f(x,\theta)}{\partial\theta}\right\}^2-\mathcal{M}^2\frac{\partial\log f(x,\theta)}{\partial\theta}=\mathcal{M}\left\{\frac{\partial\log f(x,\theta)}{\partial\theta}\right\}^2$$

since, by Eq. (2.37), the expectation of $A=\partial\log f(x,\theta)/\partial\theta$ is zero. Therefore, Eq. (6.16) becomes

$$\sigma_T^2\ge\frac{1}{n\mathcal{M}\left\{\dfrac{\partial\log f(x,\theta)}{\partial\theta}\right\}^2}.$$

This is the same formula as (6.10) for the case $\tau(\theta)=\theta$.

Fréchet did not mention Aitken and Silverstone in his paper, but he referred to Doob (1936), whom he credited for a related form of the inequality.

*This inequality here also means that the square of the correlation between U and T must be less than or equal to unity.

6.1.2.4 Rao (1945) The third proof we shall consider is one of the two more well-known proofs of the inequality. It was provided by Rao in the landmark paper "Information and the accuracy attainable in the estimation of statistical parameters" (Rao, 1945). Rao's proof is essentially the same provided earlier by Fréchet, but his proof seems to be independent of the latter. Before examining Rao's paper, we shall say a few words about his life.

Calyampudi Radhakrishna Rao* (Fig. 6.4) was born on September 10, 1920, in Hadagali (now in Karnataka State), India. He studied in schools in Andhra Pradesh and in 1940 received his master's degree in mathematics at Andhra University, Visakhapatnam. Soon he started training at the Indian Statistical Institute (ISI), but then joined Calcutta University for a master's degree in statistics. At Calcutta University, he was taught by K.R. Nair, S.N. Roy, and R.C. Bose. In 1943, Rao received his Master's degree and a gold medal. Five years later, he married Bhargavi, who herself had two Master's degrees, one in history and the other in psychology. In August 1946, Rao sailed to England to work on a project at the Museum of Anthropology and Archeology at Cambridge University. Two years later, he received his Ph.D. from King's College under the supervision of R.A. Fisher. The title of Rao's dissertation was "Statistical Problems of Biological Classification." By that time, Rao had already made some of the greatest statistical discoveries associated with his name, including the CR inequality, Rao–Blackwell theorem, and the Rao score test. For the next 30 years, Rao worked at the ISI. In 1979, he moved to the United States and joined the University of Pittsburgh. In 1988, he moved to Pennsylvania State University where he stayed until his retirement at the age of 81. Rao is still active in the field, having published about 400 papers and several books, among which *Linear Statistical Inference and its Applications* (Rao, 1973) deserves special mention.

Our interest here is in Rao's 1945 paper (Rao, 1945). Rao started by mentioning the Gauss–Markov problem of least squares, namely, that of finding the linear function of observations whose expectation is a given linear function of parameters and whose variance is a minimum. Rao credited Markov for posing this problem, but made no mention of Gauss' earlier solution (see Section 1.5.6). Rao then referred to Fisher's concepts of consistency, efficiency, and sufficiency. He noted that a major merit of the method of maximum likelihood (ML) was that, of all unbiased estimators following a normal distribution, the ML estimator† had the least variance. Furthermore, even when the estimators were not normally distributed, the variance was still minimum for large-sample sizes. Finally, Rao cited Aitken and Silverstone's paper‡ (Aitken and Silverstone, 1942) where the problem of finding unbiased estimators with minimum variance for finite samples was first tackled.

In his paper, Rao wished to obtain explicit inequalities relating Fisher's information and the variances (and covariances) of estimators. He considered the joint probability density $\phi(x_1,x_2,\ldots,x_n;\theta)$ of a sample of n observations, where the parameter θ was to be estimated by some function $t=f(x_1,x_2,\ldots,x_n)$. Ideally, the best estimator would be one for which

$$\Pr\{\theta-\lambda_1<t<\theta+\lambda_2\}\geq\Pr\{\theta-\lambda_1<t'<\theta+\lambda_2\},$$

* Rao's life is also described in Bera (2003) and Agarwal and Sen (2014, p. 428).

† Note that this does not mean that ML estimators are unbiased. In fact, they are often biased.

‡ While citing this paper, Rao mentioned Aitken only.

FIGURE 6.4 Calyampudi Radhakrishna Rao (1920–). Wikimedia Commons (Licensed under the Creative Commons Attribution-Share Alike 4.0 International, 3.0 Unported, 2.5 Generic, 2.0 Generic and 1.0 Generic license), http://commons.wikimedia.org/wiki/File:Calyampudi_Radhakrishna_Rao_at_ISI_Chennai.JPG

where t' is any other estimator and λ_1, λ_2 are positive real numbers. However, a more reasonable and less stringent condition is

$$\mathsf{E}(t-\theta)^2 \le \mathsf{E}(t'-\theta)^2$$

(Rao, 1945, 1992 reprint, p. 236). Additionally, a joint condition can be added to the above that $\mathsf{E}t = \theta$, that is, that t is an unbiased estimator of θ. Then

$$\int \cdots \int t\phi dx_1 dx_2 \ldots dx_n = \theta.$$

Differentiating with respect to θ under the integral sign,

$$\int \cdots \int t \frac{d\phi}{d\theta} dx_1 dx_2 \ldots dx_n = 1,$$

where regularity conditions have been assumed. The last equation implies that the covariance of t and $d\phi/(\phi d\theta)$ is unity (by Eq. 2.37, the expectation of $d\phi/(\phi d\theta)$ is zero). Next, Rao applied the Cauchy–Schwarz inequality ("the square of the covariance of two variates is not greater than the product of the variances of the variates") to write

$$V(t)V\left(\frac{1}{\phi}\frac{d\phi}{d\theta}\right) \ge \left\{C\left(t, \frac{1}{\phi}\frac{d\phi}{d\theta}\right)\right\}^2,$$

where V and C denote variance and covariance, respectively. Therefore,

$$V(t) \ge \frac{1}{V\left(\frac{1}{\phi}\frac{d\phi}{d\theta}\right)}.$$

But

$$V\left(\frac{1}{\phi}\frac{d\phi}{d\theta}\right) = -\mathsf{E}\frac{d^2 \log \phi}{d\theta^2} = I$$

is Fisher's intrinsic accuracy (expected information). Hence, Rao finally obtained

$$V(t) \ge \frac{1}{I} \tag{6.17}$$

(*ibid*., p. 237). Therefore,

> ...*the variance of any unbiassed [sic] estimate of θ is greater than the inverse of I which is defined independently of any method of estimation.** The assumption of the normality of the distribution function of the estimate is not necessary. (*ibidem*)

* Italics are Rao's.

In other words, no asymptotic arguments are used in the derivation of (6.17). The latter is the same formula as (6.10) for the case $\tau(\theta)=\theta$. Rao also noted that if the interest was in estimating a function $f(\theta)$ (instead of θ itself), then Eq. (6.17) should be changed to

$$V(t)\geq\frac{\{f'(\theta)\}^2}{I}.$$

This is (6.10) with $\tau\equiv f$.

Further in his paper (*ibid*., p. 238), Rao reviewed the results already obtained by Aitken and Silverstone* regarding the exponential form and the function $\tau(\theta)$ to be estimated when the form in (6.12) was not possible for θ.

Rao next considered the extension of (6.17) to the case of several parameters $\theta_1,\theta_2,\ldots,\theta_q$.† The corresponding inequality then is

$$\left|V_{rs}-I^{rs}\right|\geq 0,\quad r,s=1,2,\ldots,i;\ i=1,2,\ldots,q$$

(*ibid*., p. 240), where $V_{rs}=\mathsf{E}(t_r-\theta_r)\,(t_s-\theta_s)$ and I^{rs} is the (r, s)th entry in the inverse of the information matrix $\mathbf{I}=\{I_{rs}\}$, where

$$I_{rs}=-\mathsf{E}\frac{\partial^2\log\phi}{\partial\theta_r\partial\theta_s}.$$

6.1.2.5 Cramér (1946) The last proof the CR inequality we shall consider is the one provided by Cramér in his influential book *Mathematical Methods of Statistics* (Cramér, 1946).

Harald Cramér‡ (Fig. 6.5) was born on September 25, 1893, in Stockholm, Sweden. He did his higher studies at the University of Stockholm in 1912 and showed early interest in chemistry and mathematics. During 1913–1914, as a research assistant in biochemistry, Cramér published his first paper with Hans von Euler (1983–1964), later a Nobel laureate. But he soon turned his attention solely to mathematics and became a student of the influential Swedish mathematician Mittag Leffler. Cramér received his Ph.D. in 1917 for his dissertation on Dirichlet series. He became an Assistant Professor of Mathematics in 1917, and over the next 7 years he published about 20 papers in analytic number theory. Through this early research, Cramér got good training in Fourier integrals that he later used in his fundamental contributions to probability. Cramér also took up a position as an actuary with the Svenska Life Insurance company, a job that increased his interest in probability and statistics. In 1929, he was appointed Chair of a newly created Department of Actuarial Mathematics and Mathematical Statistics. His 1937 book *Random Variables and Probability Distributions* (Cramér, 1937) was one of the first English books to teach probability from Kolmogorov's axiomatic viewpoint. However, his masterpiece was the 1946 book *Mathematical Methods of Statistics* (Cramér, 1946)§ which used the measure-theoretic probabilistic framework of Kolmogorov to give a rigorous treatment of the statistical ideas of Fisher, Neyman, and Pearson.

* See pp. 603–607

† Previously, Darmois had also considered this case using the notion of *concentration ellipsoids* (Darmois, 1945).

‡ Cramér's life is also described in Blom (1987) and Leadbetter (2001, pp. 439–443).

§ The book was first published in 1945 in Sweden by Almqvist and Wiksells.

FIGURE 6.5 Harald Cramér (1893–1985). Wikimedia Commons (Licensed under the Creative Commons Attribution-Share Alike 2.0 Germany license), http://commons.wikimedia.org/wiki/File:Harald_Cram%C3%A9r.jpeg

Cramér was made president of Stockholm University in 1950 and died on October 5, 1985, having spent almost all of his professional life in Stockholm.

Our interest here is in Cramér's proof of the CR inequality. In his 1946 book (Cramér, 1946, p. 478), Cramér considered a random sample of n independent observations $x_1, x_2, \ldots, x_n$, where the density of $x_i (i = 1, 2, \ldots, n)$ is $f(x_i; \alpha)$, α being a parameter of interest. Let $\alpha^* = \alpha^*(x_1, x_2, \ldots, x_n)$ be a continuous estimator of α such that $\partial\alpha^*/\partial x_i$ is continuous, except possibly for a finite number of points. Moreover, suppose

$$\mathsf{E}\alpha^* = \alpha + b(\alpha), \tag{6.18}$$

so that $b(\alpha)$ is the bias of α^*. Note that the equation $\alpha^* = c$ defines a set of hypersurfaces in $\mathfrak{R}_n$ for different values of c. Furthermore, a point in $\mathfrak{R}_n$ is uniquely determined through both α^*, which determines the particular hyperspace, and the $(n-1)$ "local coordinates" $\xi_1, \xi_2, \ldots \xi_{n-1}$, which fix the point on the chosen hyperspace. Now, the joint density of $\alpha^*, \xi_1, \xi_2, \ldots \xi_{n-1}$ is given by

$$f(x_1;\alpha) f(x_2;\alpha) \cdots f(x_n;\alpha) |J|,$$

where J is the Jacobian of the transformation. The above joint density can also be written as

$$g(\alpha^*;\alpha) h(\xi_1, \xi_2, \ldots, \xi_{n-1} \mid \alpha^*;\alpha),$$

where $g(\alpha^*;\alpha)$ is the density of α^* and $h(\xi_1, \xi_2, \ldots, \xi_{n-1} \mid \alpha^*;\alpha)$ is the conditional density of $\xi_1, \xi_2, \ldots, \xi_{n-1}$ given α^*. Equating the last two expressions, Cramér obtained

$$f(x_1;\alpha) f(x_2;\alpha) \cdots f(x_n;\alpha) |J| = g(\alpha^*;\alpha) h(\xi_1, \xi_2, \ldots, \xi_{n-1} \mid \alpha^*;\alpha) \tag{6.19}$$

(*ibid*., p. 479). This is the same as

$$\begin{aligned} & f(x_1;\alpha) f(x_2;\alpha) \cdots f(x_n;\alpha) dx_1 dx_2 \ldots dx_n \\ &= g(\alpha^*;\alpha) h(\xi_1, \xi_2, \ldots, \xi_{n-1} \mid \alpha^*;\alpha) d\alpha^* \, d\xi_1 d\xi_2 \ldots d\xi_{n-1}. \end{aligned} \tag{6.20}$$

Next, since $f(x;\alpha) \equiv f(x_i;\alpha)$ and $h(\xi_1, \xi_2, \ldots, \xi_{n-1} \mid \alpha^*;\alpha)$ are probability densities,

$$\int_{-\infty}^{\infty} f(x;\alpha) dx = \int_{-\infty}^{\infty} \cdots \int_{-\infty}^{\infty} h(\xi_1, \xi_2, \ldots, \xi_{n-1} \mid \alpha^*;\alpha) d\xi_1 d\xi_2 \ldots d\xi_{n-1} = 1.$$

Assuming the integrands can be differentiated with respect to α, the above implies

$$\begin{aligned} & \int_{-\infty}^{\infty} \frac{\partial f(x;\alpha)}{\partial\alpha} dx = \int_{-\infty}^{\infty} \cdots \int_{-\infty}^{\infty} \frac{\partial h(\xi_1, \xi_2, \ldots, \xi_{n-1} \mid \alpha^*;\alpha)}{\partial\alpha} d\xi_1 d\xi_2 \ldots d\xi_{n-1} = 0. \\ \therefore \; & \int_{-\infty}^{\infty} \frac{\partial \log f}{\partial\alpha} f(x;\alpha) dx = \int_{\infty}^{\infty} \cdots \int_{-\infty}^{\infty} \frac{\partial \log h}{\partial\alpha} h(\xi_1, \xi_2, \ldots, \xi_{n-1} \mid \alpha^*;\alpha) d\xi_1 d\xi_2 \ldots d\xi_{n-1} = 0. \end{aligned} \tag{6.21}$$

Coming back to Eq. (6.19), taking logarithms, and differentiating with respect to α,

$$\sum_{i=1}^{n}\frac{\partial\log f(x_i;\alpha)}{\partial\alpha}=\frac{\partial\log g}{\partial\alpha}+\frac{\partial\log h}{\partial\alpha},$$

since J is independent of α. The next step is to square the above, multiply by Eq. (6.20), and integrate over the whole space:

$$\begin{aligned}&\int_{-\infty}^{\infty}\cdots\int_{-\infty}^{\infty}\left\{\sum_{i=1}^{n}\frac{\partial\log f(x_i;\alpha)}{\partial\alpha}\right\}^2 f(x_1;\alpha)f(x_2;\alpha)\cdots f(x_n;\alpha)dx_1dx_2\ldots dx_n\\&=\int_{-\infty}^{\infty}\cdots\int_{-\infty}^{\infty}\left(\frac{\partial\log g}{\partial\alpha}+\frac{\partial\log h}{\partial\alpha}\right)^2 g(\alpha^*;\alpha)h(\xi_1,\xi_2,\ldots,\xi_{n-1}\mid\alpha^*;\alpha)d\alpha^*\,d\xi_1d\xi_2\ldots d\xi_{n-1}.\end{aligned}$$

From Eq. (6.20), the integral of the product of any two different derivatives is zero, so the above simplifies to

$$\begin{aligned}n\int_{-\infty}^{\infty}\left(\frac{\partial\log f}{\partial\alpha}\right)^2 fdx&=\int_{-\infty}^{\infty}\left(\frac{\partial\log g}{\partial\alpha}\right)^2 g(\alpha^*;\alpha)d\alpha^*+\int_{-\infty}^{\infty}g\,d\alpha^*\int_{-\infty}^{\infty}\cdots\int_{-\infty}^{\infty}\left(\frac{\partial\log h}{\partial\alpha}\right)^2 h\,d\xi_1d\xi_2\ldots d\xi_{n-1}\\&\geq\int_{-\infty}^{\infty}\left(\frac{\partial\log g}{\partial\alpha}\right)^2 g(\alpha^*;\alpha)d\alpha^*\end{aligned}\tag{6.22}$$

(*ibid*., p. 481). Cramér credited the inequality above to Dugué (1937). In the final step, Cramér appealed to a lemma he had previously (Cramér, 1946, pp. 475–476) established: let the expectation in (6.18) be written as

$$\mathsf{E}\alpha^*=\int_{-\infty}^{\infty}\alpha^*g(\alpha^*;\alpha)d\alpha^*=\psi(\alpha).$$

Then,

$$\begin{aligned}\frac{d\psi}{d\alpha}&=\int_{-\infty}^{\infty}\alpha^*\frac{\partial g}{\partial\alpha}d\alpha^*\\&=\int_{-\infty}^{\infty}(\alpha^*-\alpha)\frac{\partial g}{\partial\alpha}d\alpha^*\\&=\int_{-\infty}^{\infty}\underbrace{(\alpha^*-\alpha)\sqrt{g}}\,\underbrace{\frac{\partial\log g}{\partial\alpha}\sqrt{g}}\,d\alpha^*.\end{aligned}$$

Applying the Cauchy–Schwarz inequality to the above, we have

$$\left(\frac{d\psi}{d\alpha}\right)^2\leq\int_{-\infty}^{\infty}(\alpha^*-\alpha)^2 gd\alpha^*\int_{-\infty}^{\infty}\left(\frac{\partial\log g}{\partial\alpha}\right)^2 gd\alpha^*.\tag{6.23}$$

Combining Eqs. (6.22) and (6.23),

$$n\int_{-\infty}^{\infty}\left(\frac{\partial \log f}{\partial \alpha}\right)^2 fdx \geq \int_{-\infty}^{\infty}\left(\frac{\partial \log g}{\partial \alpha}\right)^2 g(\alpha^*;\alpha)d\alpha^* \geq \frac{\left(\dfrac{d\psi}{d\alpha}\right)^2}{\int_{-\infty}^{\infty}(\alpha^*-\alpha)^2 gd\alpha^*}.$$

$$\therefore \quad \int_{-\infty}^{\infty}(\alpha^*-\alpha)^2 gd\alpha^* \geq \frac{\left(\dfrac{d\psi}{d\alpha}\right)^2}{n\int_{-\infty}^{\infty}\left(\dfrac{\partial \log f}{\partial \alpha}\right)^2 fdx}.$$

Since the left-hand side of the above is the variance of α^* and $\psi = \alpha + b(\alpha)$ from (6.18), Cramér finally obtained the following inequality for $D^2(\alpha^*) \equiv \text{var}\,\alpha^*$:

$$D^2(\alpha^*) \geq \frac{\left(1+\dfrac{db}{d\alpha}\right)^2}{n\int_{-\infty}^{\infty}\left(\dfrac{\partial \log f}{\partial \alpha}\right)^2 fdx}.$$

For the particular case that α^* is unbiased, we have $b(\alpha) = 0$, and the above becomes

$$D^2(\alpha^*) \geq \frac{1}{n\int_{-\infty}^{\infty}\left(\dfrac{\partial \log f}{\partial \alpha}\right)^2 fdx}$$

(*ibid*., p. 480). This is the same formula as (6.10) for the case $\tau(\theta) = \theta$.

Apart from the first proof by Aitken and Silverstone, all other proofs made use of the Cauchy–Schwarz inequality, with the proofs of Rao and Fréchet being the simplest. Aitken and Silverstone did not give the CR inequality explicitly; this was first done by Fréchet. Cramér was the last to have independently discovered the inequality, but the latter is usually known by his name first. The reason for this is not hard to understand. Both Fréchet's and Rao's papers were published during the war period in foreign journals not readily available to the Anglo-Saxon statistical community. On the other hand, Cramér's book, apart from being a landmark, was published by the prestigious Princeton University Press. But if credit is to be given to the first discoverer, the inequality should be called Fréchet's inequality.*

6.1.3 The Rao–Blackwell Theorem

6.1.3.1 Rao (1945) In the same paper (Rao, 1945) that the CR inequality was proved, Rao demonstrated another powerful theorem that addressed the following important question: given an initial unbiased estimator t for a parameter θ, is there a way of "improving" upon this estimator, that is, can we obtain another unbiased estimator that has a smaller variance than the first? To this question, Rao answered in the affirmative in

*As do Van der Waerden (1968, p. 160) and Sverdrup (1967, p. 73), for instance.

what later came to be known as the Rao–Blackwell theorem. The theorem states that: suppose $x_1, \ldots, x_n$ are observations from a random sample, each with density $f(x;\theta)$*; if t is an unbiased estimator for the parameter θ and T is a sufficient statistic, then the new estimator $\mathsf{E}(t \mid T)$ is also unbiased and its variance is at most that of t.

Rao's proof was as follows. Consider the joint density $\phi(x_1, \ldots, x_n;\theta)$ of the observations $x_1, \ldots, x_n$ such that the statistic T is sufficient θ. Using the Fisher–Neyman factorization theorem (see end of Section 5.2.3), Rao wrote

$$\phi(x_1, \ldots, x_n;\theta) = \Phi(T,\theta)\psi(x_1, \ldots, x_n)$$

(Rao, 1945, 1992 reprint, p. 238), where ψ is a function of the data only and $\Phi(T,\theta)$ is the probability density of T. Now, if t is an unbiased estimator of θ, then

$$\theta = \int \cdots \int t\phi \, dx_1 dx_2 \ldots dx_n. \tag{6.24}$$

Rao then stated that there should be a function $f(T)$ that is also unbiased for θ, that is,

$$\theta = \int f(T)\Phi(T,\theta)\,dT. \tag{6.25}$$

To see the reasoning behind this last statement, consider Eq. (6.24). If we change variables from $x_1, x_2, \ldots, x_n$ to $T, x_2, \ldots, x_n$ in the multiple integral, then using $\phi = \Phi\psi$ we have

$$\theta = \int \cdots \int t\Phi(T,\theta)\psi(x_1, \ldots, x_n)|J|\,dT dx_2 \ldots dx_n,$$

where J is the Jacobian of the transformation. Integrating out $x_2, \ldots, x_n$, we can write

$$\theta = \int f(T)\Phi(T,\theta)\,dT,$$

which is Eq. (6.25).†

To finish his proof of the Rao–Blackwell theorem, Rao next compared the variance of t with that of $f(T)$:

$$\begin{aligned}\int \cdots \int (t-\theta)^2 \phi \, dx_1 dx_2 \ldots dx_n &= \int \cdots \int \{t - f(T) + f(T) - \theta\}^2 \phi \, dx_1 dx_2 \ldots dx_n \\ &= \int \cdots \int \{t - f(T)\}^2 \phi \, dx_1 dx_2 \ldots dx_n + \int \{f(T) - \theta\}^2 \Phi(T,\theta)\,dT \\ &\geq \int \{f(T) - \theta\}^2 \Phi(T,\theta)\,dT,\end{aligned}$$

so that, with V as the variance,

$$V\{f(T)\} \leq V(t)$$

(*ibidem*). Therefore,

> ...*if a sufficient statistic and an unbiassed [sic] estimate exist for θ, then the best unbiassed estimate of θ is an explicit function of the sufficient statistic.*‡ (*ibidem*)

* Note that there is no restriction here that the support of $f(x;\theta)$ should be independent of θ.

† Furthermore, by using the last two equations together with $\phi = \Phi\psi$, it is seen that $f(T) = \mathbf{E}(t|T)$, although the latter expression was not *explicitly* given by Rao.

‡ Italics are Rao's.

That is, given an unbiased estimator, another unbiased estimator with smaller variance can be obtained by taking a function of a sufficient statistic.

6.1.3.2 Blackwell (1947) Two years after Rao's paper, Blackwell independently gave a more detailed treatment of the topic of improving unbiased estimators through sufficient statistics. Before describing his derivation, let us say a few words about Blackwell.

David Harold Blackwell* (Fig. 6.6) was one of the most distinguished figures of mathematical statistics and the first African American to be inducted to the National Academy of Sciences. Blackwell was born in Centralia, Illinois, on April 24, 1919. In those days, the United States was marred by segregation and racial discrimination. Blackwell's father worked for the Illinois Central Railroad as a hostler. From 1925 to 1935, Blackwell attended Centralia public schools. But his parents made sure to enroll him in integrated (rather than racially segregated) schools, so that, as Blackwell later recalled:

> ...I was not even aware of these problems—I had no sense of being discriminated against. My parents protected us from it and I didn't encounter enough of it in the schools to notice it. (Davis et al., 2011, p. 15)

Blackwell did his higher education in 1935 at the University of Illinois where he took all the undergraduate mathematics courses. He received his Master's degree in 1939. Blackwell was 22 years old when he received his Ph.D. under Joseph L. Doob (1910–2004). The title of his thesis was "Properties of Markov Chains." Doob seems to have had a tremendous influence on him. In 1941, Blackwell was awarded a Rosenwald Fellowship at the Institute for Advanced Studies in Princeton. In that year, he met John von Neumann, Samuel Wilks, and Jimmie Savage, all of whom influenced his interests and future research. While in Princeton, Blackwell wrote more than a hundred letters of applications to black colleges. Eventually he received three job offers and accepted the position of instructor at Southern University in Baton Rouge. From 1944 to 1955, Blackwell was at the Mathematics Department at Howard University, Washington, D.C. In the period 1946–1950, he worked at the RAND Corporation, where he became a full-fledged Bayesian. In 1954, Blackwell co-wrote the influential book *Games and Statistical Decisions* (Blackwell and Girshick, 1954) with Meyer A Girshick. The next year, he was appointed Full Professor at the University of California, Berkeley, and then became Chair of the Department in 1957. Blackwell retired in 1988 and passed away in Berkeley on July 8, 2010, aged 91.

In his paper "Conditional expectation and unbiased sequential estimation" (Blackwell, 1947), Blackwell started by proving the following two basic relationships:

$$\mathsf{E}\{f(x)\mathsf{E}(y\,|\,x)\}=\mathsf{E}\{f(x)y\} \tag{6.26}$$

$$\text{var}\,\mathsf{E}(y\,|\,x)\leq \text{var}\,y \tag{6.27}$$

(*ibid*., p. 105). In the above, $f(x)$ is a simple function of x and the relationships are valid as long as $\mathsf{E}\{f(x)y\}$ and $\text{var}\,y$ are both finite. Blackwell first proved that the left side of

* Blackwell's life is also described in Brillinger (2011) and Agwu et al. (2003).

FIGURE 6.6 David Harold Blackwell (1919–2010). Wikimedia Commons (Licensed under the Creative Commons Attribution-Share Alike 2.0 Germany license), http://commons.wikimedia.org/wiki/File:David_Blackwell.jpg

Eq. (6.26) is finite and is equal to the right side of that equation as follows. For the function $f(x)$, note that there is a sequence of simple functions $f_n(x)$ such that $f_n(x) \to f(x)$ and $|f_n(x)| \leq |f(x)|$. Let $\mathsf{E}\{y \mid x\} \equiv \mathsf{E}_x y$, $f_n(x) \equiv f_n$, and $f(x) \equiv f$. Then, using the basic inequality $\mathsf{E}\,|y| \geq |\mathsf{E}y|$, we have

$$\left|f_n \mathsf{E}_x y\right| \leq \left|\mathsf{E}_x (f_n y)\right| \leq \mathsf{E}_x \left|f_n y\right| \leq \mathsf{E}_x \left|fy\right|. \tag{6.28}$$

Blackwell then wrote

$$\mathsf{E}(f_n \mathsf{E}_x y) = \mathsf{E}(f_n y). \tag{6.29}$$

To see why the above is true, note that

$$\begin{aligned}
\mathsf{E}(f_n \mathsf{E}_x y) &= \int (f_n \mathsf{E}_x y) f_x dx \\
&= \int f_n \left(\int y f_{y|x} dy \right) f_x dx \\
&= \int f_n \left(\int y \frac{f_{xy}}{f_x} dy \right) f_x dx \\
&= \int f_n \left(\int y f_{xy} dy \right) dx \\
&= \iint (f_n y) f_{xy} dy dx \\
&= \mathsf{E}(f_n y),
\end{aligned}$$

where $f_{xy}, f_x, f_{y|x}$ denote the joint densities of (x, y), the marginal density of x, and the conditional density of y given x, respectively.

Continuing with Blackwell's proof, since $f_n \mathsf{E}_x y$ and $f_n y$ are bounded in absolute value by the summable functions $\mathsf{E}_x\,|fy|$ and $|fy|$, respectively, application of Lebesgue's dominated convergence theorem* implies that:

- $\mathsf{E}(f\mathsf{E}_x y)$ is finite (by Eq. 6.28).
- $\mathsf{E}(f_n \mathsf{E}_x y) \to \mathsf{E}(f\mathsf{E}_x y) = \mathsf{E}(fy)$ (by Eq. 6.29).

Hence, Eq. (6.26) is established.

Using his proof of (6.26), Blackwell next made the following statement, which is essentially the Rao–Blackwell theorem:

> ...if u is a sufficient statistic for a parameter θ and f is any unbiased estimate for θ, then $\mathrm{E}(f \mid u)$ (which, since u is a sufficient statistic, is a function of u independent of θ) is an unbiased estimate for θ... The interesting fact is that the estimate $\mathrm{E}(f \mid u)$ is always a better estimate for [θ] than f in the sense of having a smaller variance, unless f is already a function of u only, in which case the two estimates f and $\mathrm{E}(f \mid u)$ clearly coincide. (*ibid.*, p. 106)

The first part of the statement above, that $\mathsf{E}(f \mid u)$ is unbiased for θ, can be shown to be true by taking $f(x) = 1, y = f$ and $x = u$ in (6.26). Thus, if u is a sufficient statistic and f is

*This states that if $X_1, X_2, \ldots$ are random variables such that $|X_n| \leq Y$ (where $\mathsf{E}Y$ is finite) and $X_n \to X$ as $n \to \infty$, then $\mathsf{E}X$ is finite and $\mathsf{E}X_n \to \mathsf{E}X$ as $n \to \infty$.

an unbiased estimator, then $\mathsf{E}(f \mid u)$ is both unbiased and an improvement over f in the sense of the former having a lower variance than the latter.

To prove the part on $\mathsf{E}(f \mid u)$ having a variance less or equal to that of f, Blackwell first proved Eq. (6.27) as follows. If we take $f = 1$ in (6.26), we see that $\mathsf{E}_x y$ and y both have the same expected value. Let this common expectation be m. Assuming var $\mathsf{E}_x y$ is finite, then

$$\begin{aligned} \operatorname{var} y &= \mathsf{E}(y-m)^2 \\ &= \mathsf{E}\{(y-\mathsf{E}_x y)+(\mathsf{E}_x y-m)\}^2 \\ &= \mathsf{E}(y-\mathsf{E}_x y)^2 + \operatorname{var}\mathsf{E}_x y, \end{aligned} \tag{6.30}$$

since the expectation of the product term is zero, that is, $\mathsf{E}(y-\mathsf{E}_x y)(\mathsf{E}_x y - m) = \mathsf{E}\{y(\mathsf{E}_x y - m)\} - \mathsf{E}\{\mathsf{E}_x y(\mathsf{E}_x y - m)\} = 0$ (by taking $f = \mathsf{E}_x y - m$ in Eq. 6.26). Next, from Eq. (6.30), we have

$$\operatorname{var} y \geq \operatorname{var}\mathsf{E}_x y.$$

The last part is to show that var $\mathsf{E}_x y$ is indeed finite. Blackwell showed this by using the Cauchy–Schwarz inequality

$$\mathsf{E}_x^2(fg) \leq \mathsf{E}_x^2(f)\mathsf{E}_x^2(g)$$

and substituting $f = y,\ g = 1$ in the latter. Equation (6.27) is thus completely proved.

Now, this proof can be used to demonstrate Blackwell's earlier statement that $\mathsf{E}(f \mid u)$ has a variance less or equal to that of f. For this we only have to take $x = u,\ y = f$ in Eq. (6.27). Since it has been shown that $\mathsf{E}(f \mid u)$ is also unbiased for θ, Rao–Blackwell theorem is thus proved.*

It is noteworthy that the theorem was at first (and still sometimes) called Blackwell's theorem, without any mention of Rao.† Similarly, the technique of improving estimation by using the expectation of an unbiased estimator conditional on a sufficient statistic was called (only) Blackwellization. Rao later recounted the frustrating experience of asserting his priority over the result:

> ...When I told Berkson that I have priority over Blackwell for the result..., he said that Raoization does not sound so nice, so he called it Rao–Blackwellization. A few years later, L. K. Schmetterer wrote a book on mathematical statistics where he attributed my result to Blackwell only. D.V. Lindley, in a review of the book, referred to the result as Blackwell theorem. When I wrote to Lindley that I proved this theorem earlier he replied on the following lines:
>
> Of course, the result is there in your 1945 paper, but you were not aware of the importance of the result as you have not mentioned it in the introduction to the paper.

* The famous probabilist Kolmogorov also gave a proof of the theorem in 1950 (Kolmogorov, 1950) by referring to Blackwell's paper and extending the definition of sufficiency. The Rao–Blackwell is thus sometimes, especially in the Russian literature, called the Rao–Blackwell–Kolmogorov theorem.

† For example, see the books by Fisz (1963, p. 477) and Schmetterer (1974, p. 277).

I replied to Lindley saying that it was my first research paper and I was not aware that introduction is generally written for the benefit of those who do not want to read the paper. (Rao, 2001, p. 399)

6.1.4 The Lehmann–Scheffé Theorem

6.1.4.1 Introduction We have seen how, given an initial unbiased estimator, the Rao–Blackwell theorem enables one to obtain another unbiased estimator with a smaller variance than the first, by conditioning on a sufficient statistic. As useful as the Rao–Blackwell theorem is, there is one lingering question it does not answer. Can we still do better? That is, is there another function of the sufficient statistic that is both unbiased and has still a smaller variance? To this question, Lehmann and Scheffé answered in the negative under the important condition of completeness, which the authors introduced in their paper "Completeness, similar regions, and unbiased estimation—Part I" (Lehmann and Scheffé, 1950). Before considering this paper, we shall say a few words about Lehmann and Scheffé.

Erich Leo Lehmann* (Fig. 6.7) was one of the most renowned statisticians of the current and last centuries. He was born in Strasbourg, France, on November 20, 1917. Lehmann's family soon moved to Frankfurt where his father was a prominent lawyer. When Hitler came in power in 1933, they left for Switzerland, where Lehmann did his high school and then studied mathematics for two years at the University of Zürich. In 1938, upon his father's suggestion, Lehmann went to Trinity College, Cambridge, to study mathematics. In 1940, again influenced by his father, Lehmann went to New York. He was then advised by Richard Courant to study at the University of California, Berkeley. Lehmann received his Master's degree in mathematics in 1942. He then switched to statistics and worked as a Teaching Assistant in Neyman's Statistical Laboratory in 1942–1944 and 1945–1946. In 1946, Lehmann obtained his Ph.D. in Mathematics under the supervision of George Pólya (1887–1985). The title of his dissertation was "Optimum Tests of a Certain Class of Hypotheses Specifying the Value of a Correlation Coefficient," and the Berkeley thesis examiner was Jerzy Neyman. In the same year, Lehmann published his first paper (Lehmann, 1946) in the *Comptes Rendus de l'Académie des Sciences*, his paper being communicated to the Academy by none other than Émile Borel. Still in 1946, Lehmann became instructor in mathematics at Berkeley, and then in 1947 he was promoted to Assistant Professor of Mathematics. In 1948–1949, Neyman asked Lehmann to teach a graduate course in mathematical statistics. One of the student, Colin Blyth, took notes of the lectures, which were mainly on testing hypotheses. Blyth took another set of notes on statistical estimation from Lehmann's summer classes. These notes were later turned into Lehmann's masterpiece, *Testing Statistical Hypotheses* (Lehmann, 1959). This book is viewed by many as the definitive treatment of the Neyman–Pearson (NP) theory of hypothesis testing. During the period 1950–1951, Lehmann was a Visiting Associate Professor at Columbia University and then Lecturer at Princeton University. During this time, he met Ted Anderson (1918–), Howard Levene (1914–2003), Henry Scheffé (1907–1977), Abraham Wald (1902–1950), and Jacob Wolfowitz (1910–1981). In 1954, Lehmann was made Professor of Statistics at Berkeley when the Statistics Department was created.

*Lehmann's life is also described in Brillinger (2010), van Zwet (2011), and Rojo (2011).

FIGURE 6.7 Erich Leo Lehmann (1917–2009). http://magazine.amstat.org/blog/2010/02/01/obitsfeb2010/. Courtesy of Peter Bickel and Juliet Popper Shaffer

He received many professional awards, including three Guggenheim Fellowships (1955, 1966, and 1980) and Wald Memorial Lecturer (1964), and is an elected member of the American Academy of Arts and Science (1975) and a member of the US National Academy of Sciences (1978). Lehmann passed away on September 12, 2009, at the age of 91.

Henry Scheffé* (Fig. 6.8) was born in New York City on April 11, 1907, and grew up there. He graduated from high school in 1924 and then studied electrical engineering at the Cooper Union Free Night School for the Advancement of Science and Art and at the Polytechnic Institute of Brooklyn. In the years 1924–1928, he also worked as a Technical Assistant at the Bell Telephone Laboratories. In 1928, he joined the University of Wisconsin. There he switched to mathematics and did his B.A., and then his Ph.D. in 1935, under the supervision of Rudolf E. Langer[†] (1894–1968). The title of Scheffé's dissertation was "Asymptotic Solutions of Certain Linear Differential Equations in Which the Coefficient of the Parameter May Have a Zero," which was published the next year in the *Transactions of the American Mathematical Society* (Scheffé, 1936). After teaching mathematics for a number of years, Scheffé switched to statistics and wrote his first statistics paper in 1942 (Scheffé, 1942). Soon he moved to Princeton, New Jersey, and got more deeply involved in statistics. In 1944, Scheffé was elected Fellow of the Institute of Mathematical Statistics. Four years later, he joined Columbia University as Associate Professor of Mathematical Statistics, becoming Chair of the Statistics Department in 1951–1953. Apart from several outstanding papers in mathematical statistics, Scheffé's most important work was undoubtedly his classic 1959 book *Analysis of Variance* (Scheffé, 1959). This book was written while Scheffé was at the University of California, Berkeley, and continues to serve as a reference to statisticians. Scheffé died on July 5, 1977, at the age of 70, as a result of injuries he sustained that day while riding his bicycle to Berkeley.

6.1.4.2 The Lehmann-Scheffé Theorem. Completeness (1950) We now return to Lehmann and Scheffé's 1950 paper (Lehmann and Scheffé, 1950). The authors introduced completeness as a unification of the concept of similar regions, first put forward by Neyman and Pearson (1933), with that of (unbiased) estimation, introduced by Fisher (1922c)[‡]:

> The aim of this paper is the study of two classical problems of mathematical statistics, the problems of similar regions and of unbiased estimation. The reason for studying these two problems is that both are concerned with a family of measures and that essentially the same condition [of completeness] on this family insures a very simple solution to both. (Lehmann and Scheffé, 1950, p. 305)

The definition of completeness soon followed:

> …we introduce the notion of completeness of a family of measures $M^x = \{M_\theta^x \mid \theta \in \omega\}$. The family M^x is said to be complete if
>
> $$\int f(x) dM_\theta^x(x) = 0 \quad \text{for all} \quad \theta \in \omega$$
>
> implies $f(x) = 0$ except on a set N with $M_\theta^x(N) = 0$. (*ibid.*, p. 306)

* Scheffé's life is also described in Brillinger (1978) and Lehmann (2006, pp. 7472–7474).

† The latter became the President of the Mathematical Association of America (MAA) in 1949–1950.

‡ Note, however, that Fisher did not explicitly use "bias" in the 1922 paper, although he did refer to the concept of consistency of estimators, along with many other properties of the latter.

FIGURE 6.8 Henry Scheffé (1907–1977). Wikimedia Commons (Free Art License), http://commons.wikimedia.org/wiki/File:Henry_Scheffe.jpeg

In other words, consider the family of distributions $\{F_X(x;\theta)\}$ for a statistic X. Then the family is said to be complete if, for any function $f(X)$, we have

$$\mathsf{E}_\theta\{f(X)\}=0 \;\Rightarrow\; f(X)=0 \text{ with probability 1} \tag{6.31}$$

for all θ.* Thus, the *only* unbiased estimator of zero is zero itself (and no other statistic). Often, instead of saying that the statistic X belongs to a family of distributions that is complete, we shall simply say that the statistic X is complete.†

It is seen that completeness has to do with the *uniqueness* of an unbiased estimator of θ. For, suppose there are two such unbiased estimators, $h_1(X)$ and $h_2(X)$. Then

$$\begin{aligned}\mathsf{E}h_1 &= \theta,\\ \mathsf{E}h_2 &= \theta.\\ \Rightarrow \mathsf{E}(h_1-h_2) &= 0.\end{aligned}$$

But since the family is complete, $h_1-h_2=0$, that is, $h_1=h_2$ almost surely.

Coming back to what was mentioned at the start of this section, what does completeness have to do with the Rao–Blackwell theorem? Lehmann and Scheffé's answer was as follows. Let $\mathsf{P}^x=\{P_\theta^x \mid \theta\in\omega\}$ denote a family of distributions and let V be the class of all real-valued statistics. Suppose there exists a V in V such that $\mathsf{E}_\theta V=g(\theta)$ (i.e., suppose $g(\theta)$ is *estimable*‡). Then

> THEOREM 5.1 If there exists a sufficient statistic T for P^x such that P^t is complete then every estimable function has a minimum variance estimate, and a statistic V in V is a minimum variance estimate of its expected value if and only if it is a function of T (a.e. P^x). (*ibid.*, p. 321)

The above is the Lehmann–Scheffé theorem. It states that, supposing an unbiased estimator for $g(\theta)$ exists and is a function of a complete sufficient statistic, then the estimator will uniquely have minimum variance.

The Rao–Blackwell theorem states that if t is an unbiased estimator for $g(\theta)$ and T is a sufficient statistic, then the new estimator $\mathsf{E}(t \mid T)$ is unbiased and its variance is at most that of t. The Lehmann–Scheffé theorem complements the Rao–Blackwell theorem by stating that if, in addition, the statistic T is complete, then $\mathsf{E}(t \mid T)$ is the unique estimator with minimum variance.§

*As an example, let $X_1,\ldots,X_n \sim N(0,\sigma^2)$, then $T_1=X_1$ is not complete for σ^2 but $T_2=X_1^2$ is. Thus, the family $\{f_X : X\sim N(0,\sigma^2)\}$ is not complete, but the family $\{f_U : \sqrt{U}\sim N(0,\sigma^2)\}$ is. Further, let $Y_1,\ldots,Y_n \sim N(\mu,1)$, where $\mu\neq 0$. Then $T_3=Y_1$ is complete (in contrast to T_1).

†Lindgren (1993, p. 267) has pointed out that the use of "complete" in statistical estimation is similar to its use in linear algebra: a set of vectors $\{\mathbf{u}_n\}$ is complete if the only vector with zero components in all directions $\mathbf{u}_n$ is the zero vector, that is, $\mathbf{u}_n\cdot\mathbf{V}=0$ for all n implies $\mathbf{V}=0$.

‡The concept is due to Bose (1944).

§Note that achieving minimum variance should not be confused with attaining the Cramér–Rao lower bound (CRLB): the minimum variance will either be equal to or greater than the CRLB. Consider the following example (Bickel and Doksum, 1977, p. 147, Problem 4.3.3). Suppose $X\sim\text{Poisson}(\lambda)$ and let the parameter of interest be $e^{-\lambda}$. The statistic T, where $T=1$ if $X=0$ and $T=0$ otherwise, is unbiased and complete sufficient for $e^{-\lambda}$. Therefore, T is the unique minimum variance unbiased estimator, but $\operatorname{var} T = e^{-\lambda}(1-e^{-\lambda}) > \lambda e^{-2\lambda} = \text{CRLB}$.

To prove their theorem, Lehmann–Scheffé proceeded as follows. Suppose $g(\theta)$ is estimable, that is, there exists a statistic V such that $\mathsf{E}_\theta V = g(\theta)$. Let $\psi = \psi(T)$ be also unbiased for $g(\theta)$, where T is a sufficient statistic. Now suppose V' is any other unbiased estimator of $g(\theta)$ such that $\psi'(t) = \mathsf{E}(V' \mid t)$. By the Rao–Blackwell theorem, $\psi'(T)$ is unbiased and $\text{var}_\theta \psi'(T) \le \text{var}_\theta V'$. Since $\mathsf{E}_\theta(\psi - \psi') = 0$, it follows by completeness that $\psi = \psi'$. Therefore, $\text{var}_\theta \psi = \text{var}_\theta \psi' \le \text{var}_\theta V'$, and ψ is the unique minimum variance estimator of $g(\theta)$. Hence, the Lehmann–Scheffé theorem is proved.

6.1.4.3 Minimal Sufficiency and Bounded Complete Sufficiency (1950) Lehmann and Scheffé's paper (1950) also introduced the important concept of *minimal sufficiency*:

> A sufficient statistic T is said to be minimal if T is a function of all other sufficient statistics. (*ibid.*, p. 307)

To understand the above definition, note that if T' is a sufficient statistic, then every one-to-one function of T' is also sufficient. Thus, a parameter θ could potentially have an infinite number of sufficient statistics. The question then is, is there a class of sufficient statistics that summarize the data with maximum reduction? The answer is yes, by choosing a sufficient statistic that summarizes every other sufficient statistic through a function of the latter. Such a chosen statistic is called a *minimal sufficient* statistic.* Note that a minimal sufficient statistic is not unique, as every one-to-one function of a given minimal sufficient statistic is also minimal sufficient.†

We can also think of a minimal sufficient statistic as follows. The original data $\{X_1, \ldots, X_n\}$ is itself sufficient for the parameter θ and induces a *partition* of the sample space Ω. By taking a function h of the data such that the resulting function is also sufficient, a coarser partition of Ω is obtained. By taking a further sufficient function of h, an even coarser partition of Ω is obtained. By continuing to take functions (which may be different each time), we obtain further coarser partitions of Ω but also run the risk of ending with a resulting function that is no longer sufficient. A function (or statistic) that induces the *coarsest partition* of Ω while still being sufficient is called a minimal sufficient statistic.

One reason Lehmann and Scheffé introduced minimal sufficient statistics is that they wanted to draw a connection between the latter and their previously introduced complete sufficient statistics. In particular, are *complete sufficient* and *minimal sufficient* statistics equivalent to each other? Lehmann and Scheffé showed that under the weaker condition of *bounded completeness* (rather than completeness), a sufficient statistic was also minimal sufficient.

A boundedly complete statistic is one for which (6.31) holds only when the function f is bounded, that is, there exists an $M < \infty$ such that $|f| < M$. It can be shown that all complete statistics are also boundedly complete, but not vice versa.

Note that the converse of what Lehmann and Scheffé proved is not true in general: *a sufficient statistic that is minimal sufficient is not necessarily complete*. In their

*For example, if $X_1, \ldots, X_n \sim N(\mu, \sigma^2)$, then the order statistics $X_{(1)}, \ldots, X_{(n)}$ are sufficient for (μ, σ^2) but *not* minimal sufficient. On the other hand, $(\Sigma X_i, \Sigma X_i^2)$ is minimal sufficient.

†For example, if $X_1, \ldots, X_n \sim N(\mu, \sigma^2)$, then $(\Sigma X_i, \Sigma X_i^2)$ and $\{\Sigma X_i, \Sigma(X_i - \bar{X})^2\}$ are both minimal sufficient for (μ, σ^2).

paper (*ibid*., p. 312), the authors gave an example to illustrate this fact. Let X be a discrete random variable such that

$$\Pr\{X=-1\}=\theta$$
$$\Pr\{X=x\}=(1-\theta)^2\theta^x, \quad x=0,1,2,\ldots$$

where $0<\theta<1$. Now, for a sample size of one (i.e., a single realization of X), X must invariably be minimal sufficient for θ. But at the same time, it can be shown that $\mathsf{E}X=0$. Thus, X is not complete.

The final question we shall address is the following: how can minimal sufficient statistics be obtained? Recall that the Fisher–Neyman factorization* provided a means to identify *sufficient* statistics. Is there a similar way for *minimal sufficient statistics?* Lehmann and Scheffé's answer (*ibid*., p. 328) was as follows: let $\mathbf{X}=(X_1,\ldots,X_n)$ be a sample with joint probability density† $f(\mathbf{x}\mid\theta)$. Then suppose there is a statistic $T(\mathbf{X})$ such that, for every two sample points $\mathbf{x}$ and $\mathbf{y}$, the ratio $f(\mathbf{x}\mid\theta)/f(\mathbf{y}\mid\theta)$ does not depend on θ if and only if $T(\mathbf{x})=T(\mathbf{y})$. Then $T(\mathbf{X})$ is a minimal sufficient statistic. As an example, Lehmann and Scheffé considered the case $X_1,\ldots,X_n \sim$ Bernoulli (p), where $0<p<1$. Then for two sample points $\mathbf{x}$ and $\mathbf{y}$,

$$\frac{f(\mathbf{x}\mid\theta)}{f(\mathbf{y}\mid\theta)}=\frac{\prod_{i=1}^{n}p^{x_i}(1-p)^{1-x_i}}{\prod_{i=1}^{n}p^{y_i}(1-p)^{1-y_i}}=\left(\frac{p}{1-p}\right)^{\left(\sum_{i=1}^{n}x_i-\sum_{i=1}^{n}y_i\right)}.$$

It is seen that the above ratio is independent of p if and only if $\sum_{i=1}^{n}x_i=\sum_{i=1}^{n}y_i$. Hence, $T(\mathbf{X})=\sum_{i=1}^{n}X_i$ is minimal sufficient for p.

6.1.5 The Ancillarity–Completeness–Sufficiency Connection: Basu's Theorem (1955)

In 1955, Basu discovered a deceptively simple relationship (or rather, lack of) between ancillary and complete sufficient statistics (Basu, 1955). Before explaining this we shall say a few words about Basu.

Debabrata Basu‡ was born on July 5, 1924, in Dhaka, which is now in Bangladesh. He took a course in statistics at Dhaka University as part of the Undergraduate Honors Program in mathematics. At around 1945, Basu received his Master's degree in Mathematics from Dhaka University and taught briefly there during 1947–1948. Because of the turbulent political times, he then moved to Calcutta where he worked as an actuary in an insurance company for some time. In 1950, he joined the Indian Statistical Institute (ISI) as a Ph.D.

* See p. 390.

† Or joint probability mass function, if the random variables are discrete.

‡ Basu's life is also described in Ghosh (2002), Ghosh and Pathak (1992, pp. i–ii), Ghosh (1988, pp. xvii–xviii), and DasGupta (2011, pp. vii–ix).

student under the supervision of C.R. Rao. Basu's Ph.D. thesis was entitled "Contributions to the Theory of Statistical Inference" and dealt mainly with decision theory. He submitted his thesis to Calcutta University in 1953 and went to Berkeley as a Fulbright Scholar. At Berkeley, he met Neyman and attended many of the latter's lectures. In 1954, Basu returned to India "duly influenced" by the NP theory. During the winter of 1954–1955, he met Fisher while the latter was visiting India. It was from Fisher that Basu learned several paradoxes in conditional inference, which not only made him question the validity of both the Fisherian and Neyman–Pearsonian schools but also to discover some countexamples of his own.* Basu thus wrote important papers on conditional inference and ancillarity. He also went through a process of conversion to the Bayesian paradigm and was fully transformed in 1968. Basu was Professor Emeritus at the ISI and at Florida State University and taught at those institutions for many years. He visited universities all around the world. Basu retired in 1986 and passed away in Calcutta on March 24, 2001, at the age of 76.

Let us now come back to Basu's 1955 paper (Basu, 1955). Although Basu's key theorem in that paper dealt with ancillary statistics, the word "ancillary" did not appear in the paper. Instead he identified such a statistic T_1 as one that is independent of the parameter θ, in the sense that the marginal distribution $P_\theta^{T_1}(B)$ of T_1 is the same for all $\theta \in \Omega$ and for every B, where B is an element of the Borel field $\boldsymbol{B}$ (i.e., B is an interval on the real line). Basu also noted that for two statistics T_1 and T with measurable spaces $(\boldsymbol{I}_1, \boldsymbol{B}_1)$ and $(\boldsymbol{I}, \boldsymbol{B})$ (where $\boldsymbol{I}, \boldsymbol{I}_1$ represent the real line), we have

$$P_\theta^{T_1}(B_1) = P_\theta\left(T_1^{-1}B_1\right) = \int_{\boldsymbol{I}} f_\theta\left(T_1^{-1}B_1 \mid t\right) dP_\theta^T \tag{6.32}$$

(*ibid*., p. 378). The above is simply the law of total probability with $f_\theta\left(T_1^{-1}B_1 \mid t\right)$ denoting the conditional probability of the event $T_1^{-1}B_1$ given $T = t$.

After these preliminaries, Basu was ready to announce his celebrated theorem:

> If T be a boundedly complete sufficient statistic then any statistic T_1 which is independent of θ is stochastically independent of T. (*ibidem*)

Basu's theorem thus states that a boundedly complete sufficient statistic for a parameter θ is independent of any ancillary statistic. Since a complete statistic is also boundedly complete, the theorem also implies that a complete sufficient statistic for a parameter θ is independent of any ancillary statistic.

To prove his theorem, Basu noted that, since T is sufficient, the integrand on the right side of (6.32) is independent of θ. The integrand is also essentially bounded. Moreover, on the left side of (6.32), the distribution $P_\theta^{T_1}(B)$ does not depend on θ (by the definition of T_1). Therefore, let the left side of (6.32) be a constant k, and let the right side be written as $\mathsf{E} f_\theta(T_1^{-1}B_1 \mid t)$, where the expectation is with respect to the distribution of T. Therefore, we have $\mathsf{E}\{f_\theta(T_1^{-1}B_1 \mid t) - k\} = 0$. The bounded completeness of T implies that $f_\theta(T_1^{-1}B_1 \mid t) = k = P_\theta(T_1^{-1}B_1)$ and hence T_1 is independent of T.

* See Section 5.2.8.

Basu's theorem can be used with great efficiency to prove that, for $X_1,\ldots,X_n \sim N(\mu,\sigma^2)$, the two statistics

$$T = \bar{X},$$

$$T_1 = S^2 = \frac{\sum_{i=1}^{n}\left(X_i - \bar{X}\right)^2}{n-1}$$

are independent of each other. We only have to note that, for a fixed $\sigma^2 = \sigma_0^2$, $\bar{X}$ is complete sufficient for μ and S^2 is ancillary for μ. Hence, by Basu's theorem, $\bar{X}$ and S^2 are independent. But this conclusion is true for *any* fixed $\sigma^2 = \sigma_0^2$; hence it is true for all μ and σ^2. Note that this proof is much simpler than the one through Helmert's transformations (see Section 3.4.3).*

6.1.6 Further Extensions: Sharpening of the CR Inequality (Bhattacharyya, 1946), Variance Inequality without Regularity Assumptions (Chapman and Robbins, 1951)

Following the derivations of Cramér, Rao, and others, several improvements of the basic Cramér-Rao (CR) inequality were made. We mention two classic ones:

1. We stated before that there can exist minimum variance unbiased estimators that have variances larger than the Cramér-Rao lower bound (CRLB). In such cases, the CR inequality is not sharp. However, a method of addressing this drawback was provided by Bhattacharyya in 1946 (Bhattacharyya, 1946). Bhattacharyya's solution assumed that the kth (k being a positive integer) derivative with respect to θ of the likelihood function L exists and that the kth derivative with respect to θ of each of $\int\cdots\int L dx_1 \ldots dx_n$ and $\int\cdots\int tL dx_1 \ldots dx_n$ (where t is an unbiased estimator of θ) can be taken under the integral sign. Furthermore, let

   $$v_{ij} = \operatorname{cov}\left(S_i, S_j\right) \quad i,j = 1,2,\ldots,k,$$

   where

   $$S_i = \frac{1}{L}\frac{\partial^i L}{\partial \theta^i}.$$

*But one needs to proceed carefully when applying Basu's theorem, because if we start by assuming that both μ and σ^2 are unknown, then neither is $\bar{X}$ sufficient, nor is S^2 ancillary, for (μ, σ^2). See also Mukhopadhyay (2000, pp. 325–326).

Bhattacharyya's lower bound (BLB) for the variance of the unbiased estimator t is then given by

$$\operatorname{var} t \geq \frac{\begin{vmatrix} v_{22} & v_{23} & \cdots & v_{2k} \\ v_{32} & v_{33} & \cdots & v_{3k} \\ & & \vdots & \\ v_{k2} & v_{k3} & \cdots & v_{kk} \end{vmatrix}}{\begin{vmatrix} v_{11} & v_{12} & \cdots & v_{1k} \\ v_{21} & v_{22} & \cdots & v_{2k} \\ & & \vdots & \\ v_{k1} & v_{k2} & \cdots & v_{kk} \end{vmatrix}}.$$

Bhattacharyya's method works by taking $k=2, 3, \ldots$.* Then a series of lower bounds is obtained with each one being sharper than the previous one and converging to the minimum variance.

2. An even sharper variance inequality that imposed *no regularity condition* was obtained by Chapman and Robbins (1951). Let t be an unbiased estimator of $\tau(\theta)$, where τ is a function of the parameter $\theta \in \Theta$. Further, let $\theta_0 \in \Theta$ be any fixed value of θ such that for sufficiently small $h \neq 0$, $\theta_0 + h \in \Theta$ and $L(\theta_0) = 0 \Rightarrow L(\theta_0 + h) = 0$, where L is the likelihood function. Then

$$\operatorname{var}_{\theta_0} t \geq \sup_h \frac{\{\tau(\theta_0+h)-\tau(\theta_0)\}^2}{\mathsf{E}_{\theta_0}\left\{\dfrac{L(\theta_0+h)}{L(\theta_0)}-1\right\}^2} = \sup_h \frac{\{\tau(\theta_0+h)-\tau(\theta_0)\}^2}{\mathsf{E}_{\theta_0}\left\{\dfrac{L(\theta_0+h)}{L(\theta_0)}\right\}^2 - 1}. \tag{6.33}$$

We give an example to show how the Chapman–Robbins method works. Consider the uniform distribution

$$f_\theta(x) = \begin{cases} \dfrac{1}{\theta}, & 0 < x < \theta \\ 0, & \text{elsewhere} \end{cases}.$$

We are required to find the Chapman–Robbins lower bound for an unbiased estimator of θ. First note that the support of X depends on θ and the regularity conditions do not hold here. We have $L(\theta) = 1/\theta^n$. Now $L(\theta_0) = 0 \Rightarrow L(\theta_0 + h) = 0$ is equivalent to saying that $x > \theta_0 \Rightarrow x > \theta_0 + h$. This implies that $h < 0$. Also since $\theta_0 + h \in \Theta$,

* Note that, for $k=1$, we define BLB=CRLB.

we have $\theta_0 + h > 0$. Hence, $-\theta_0 < h < 0$. Now that we have obtained the bounds for h, we take $\tau(\theta_0) = \theta_0$ and apply Eq. (6.33):

$$\begin{aligned}
\text{var}_{\theta_0} t &\geq \sup_{-\theta_0<h<0} \frac{h^2}{\mathsf{E}_{\theta_0}\left\{\dfrac{L(\theta_0+h)}{L(\theta_0)}\right\}^2 - 1} \\
&= \sup_{-\theta_0<h<0} \frac{h^2}{\mathsf{E}_{\theta_0}\left\{\dfrac{\theta_0^n}{(\theta_0+h)^n}\right\}^2 - 1} \\
&= \sup_{-\theta_0<h<0} \frac{h^2}{\left(\dfrac{\theta_0^{2n}}{(\theta_0+h)^{2n}} \displaystyle\int_0^{\theta_0+h} \cdots \int_0^{\theta_0+h} \frac{1}{\theta_0^n} dx_1 \ldots dx_n\right) - 1} \\
&= \sup_{-\theta_0<h<0} \frac{h^2 (\theta_0+h)^n}{\theta_0^n - (\theta_0+h)^n}.
\end{aligned}$$

The Chapman–Robbins lower bound for an unbiased estimator of θ when, for example, $n = 1$ is $\sup_{-\theta<h<0} \{-h(\theta+h)\} = \theta^2/4$.

6.2 ESTIMATION AND HYPOTHESIS TESTING UNDER A SINGLE FRAMEWORK: WALD'S STATISTICAL DECISION THEORY (1950)

6.2.1 Wald's Life

Abraham Wald* (Fig. 6.9) was another towering figure in statistics, besides Fisher, Neyman, and Pearson. His major work, the theory of statistical decisions, is essentially an extension of the Neyman-Pearson (NP) theory, and the combined approach is usually called the Neyman–Pearson–Wald theory. Wald was born in Cluj, which at that time belonged to Hungary (after World War I, it became part of Romania), on October 31, 1902. He was part of five children in a family of Orthodox Jews. Wald attended the University of Vienna, where he became acquainted with the mathematicians Hans Hahn (1879–1934) and Carl Menger (1902–1985). He settled in Vienna in 1930 and worked under Menger in geometry. In 1932, he worked in economics at the Austrian Institute for Business Cycle Research. In 1938, Wald accepted an invitation from the Cowles Commission to do research in econometrics in the United States. This turned out to be a good decision for him on two grounds. First, the situation in Austria had taken a turn for the worse in 1938, with its annexation (*Anschluss*) to Nazi Germany. Second, Wald got the opportunity to work at Columbia University under Harold Hotelling (1895–1973), one of the leading statisticians in the United States at the time.

* Wald's life is also described in Wolfowitz (1952) and Weiss (2006, pp. 9018–9020).

FIGURE 6.9 Abraham Wald (1902–1950). Wikimedia Commons (Public Domain), http://en.wikipedia.org/wiki/File:Abraham_Wald_in_his_youth.jpg

Wald was keen to stay in the United States and wrote to Jerzy Neyman in August 1938 about a potential position at Berkeley. In her biography of Neyman, Reid explains:

> ...but Neyman, having come himself so recently to Berkeley, had been unable to do anything for him then. In March 1939, however, when Wald wrote again—"I would be happy to be at the same university as you are and to have the opportunity to discuss different statistical problems with you"—Neyman pressed Evans [the then Chair of Mathematics] to add the younger man to the mathematics faculty, but without success. Ultimately Wald wrote that, although he would still have preferred to be at the same institution as Neyman, he was accepting the position which Hotelling had been able to arrange for him at Columbia. (Reid, 1998, p. 165)

Wald thus stayed at Columbia and continued his research in statistics under Hotelling. Soon, a stream of important research papers were authored by him. Wald started gaining recognition for being a leading statistician of his time, but that was short lived: on December 13, 1950, the airplane that carried both him and his wife crashed in India while Wald was on an invited tour of Indian universities. Thus "died this great statistician, whose work had changed the whole course and emphasis of statistics" (Wolfowitz, 1952, p. 1).

6.2.2 Statistical Decision Theory: Nonrandomized and Randomized Decision Functions, Risk Functions, Admissibility, Bayes, and Minimax Decision Functions

Wald's most important work was in statistical decision theory and stemmed from the realization that both the problems of statistical estimation (mainly from Fisher's work) and of hypothesis testing (mainly from Neyman and Pearson's work) could be viewed as a general problem of making decisions under uncertainty. The decision in question could thus be to choose a particular estimator or reject a particular hypothesis. An important aspect of Wald's work in statistical decision theory was the game-theoretic framework he adopted from von Neumann and Morgenstern (1944).

One of Wald's first statistical papers (Wald, 1939) was precisely on the topic of statistical decision theory. That paper was written before von Neumann and Morgenstern's book and does not contain any game-theoretic ideas. However, the 1939 paper has rightly been called by Wolfowitz (1952) as possibly Wald's most important paper, since it foreshadowed many of the decision-theoretic concepts that were to form the backbone of Wald's theory.

It is important to bear in mind that although Wald's *statistical* decision theory shared many similarities with the decision-theoretic ideas of both Ramsey earlier (see Section 6.3.1) and Savage later (see Section 6.3.3) it also had a major difference. The latter two approached the subject of decisions from a purely epistemic viewpoint. On the other hand, Wald's approach was distinctly frequentist and similar to those of Neyman and Pearson. But that does not mean Wald never used Bayesian methods in his statistical decision theory. As we shall soon see, he did in fact use Bayes' theory, but that was only to compare the latter's performance against frequentist methods.

The definite repository of Wald's decision-theoretic ideas is undoubtedly his major book *Statistical Decision Functions* (Wald, 1950). We shall now examine this book, especially its first chapter where most of the essential ideas can be found.

In the preface of Wald's book, we find an interesting paragraph that gives the motivation behind his theory:

> Until about ten years ago, the available statistical theories, except for a few scattered results, were restricted in two important respects: (1) experimentation was assumed to be carried out in a single stage; (2) the decision problems were restricted to problems of testing a hypothesis, and that of point and interval estimation. The general theory, as given in this book, is freed from both of these restrictions. It allows for multi-stage experimentation and includes the general multi-decision problem. (*ibid.*, p. v)

In (1) above, Wald thus wished to use decision theory to deal with sequential analysis, which is another field in which he made major contributions. In (2), he referred to the ability of his theory to treat problems of both hypothesis testing and estimation as general decision problems.

At the start of the first chapter, Wald noted that statistical decision problems arose when we are faced with alternative choices, one of which must be made, depending on the distribution of the random variables $\{X_i\}$ generating the data. Since the decisions of interest were made at the end of the experiment, he used d^t to denote a particular terminal decision from the decision space D^t. Moreover, suppose d^e is a particular decision from the space D^e of all decisions at the first stage of an experiment performed in k stages, and let $D = D^t \cup D^e$. Then *a nonrandomized decision function* is a function $d(x; s_1, \ldots, s_k)$ such that

1. it is a single-valued function defined for all positive integral values k, for any sample point x, and for any finite disjoint sets $s_1, \ldots, s_k$ of positive integers;
2. the value of $d(x; s_1, \ldots, s_k)$ is independent of the coordinates x_i of x for which the integer i is not contained in any of the sets $s_1, \ldots, s_k$;
3. it is a constant when $k=0$ [we shall denote this constant by $d(0)$];
4. for $k \geq 1$, the value of the function $d(x; s_1, \ldots, s_k)$ may be any element of $D_{i_1,\ldots,i_r}$, where the set $\{i_1, \ldots, i_r\}$ is the set-theoretical sum of $s_1, \ldots, s_k$;
5. for $k=0$, the value of $d(x; s_1, \ldots, s_k)$, i.e., the value $d(0)$, may be any element of D. (*ibid.*, p. 6)

Thus, essentially, a (nonrandomized) decision function is a mapping from $X_1, \ldots, X_n$ to D. To understand how the above definition works, consider a particular experiment. If $d(0) \in D^t$, the experiment is not carried out and the terminal decision $d(0)$ is chosen. On the other hand, if $d(0) = s_1 = (i_1, \ldots, i_r) \in D^e$, then observations are made on $X_{i_1}, \ldots, X_{i_r}$ and the value $d(x; s_1)$ is computed. If $d(x; s_1) \in D^t$, then the experiment is stopped and the terminal decision $d(x; s_1)$ is chosen. On the other hand, if $d(x; s_1) = s_2 = (j_1, \ldots, j_r) \in D^e$, then observations are made on $X_{j_1}, \ldots, X_{j_r}$ and the value $d(x; s_1, s_2)$ is computed, and so on.

For theoretical reasons, Wald also needed the concept of a *randomized* decision function. The latter was defined as a function $\delta(x; s_1, \ldots, s_k)$ whose values belonged to the space Δ of all probability measures δ such that

1. it is a single-valued function defined for any positive integer k, for any finite disjoint sets $s_1, \ldots, s_k$ of positive integers, and for any sample point x;
2. it is a constant $\delta(0)$ when $k=0$;

3. $\delta(x; s_1, \ldots, s_k)$ may be any element of $\Delta_{i_1,\ldots,i_r}$ if $r \geq 1$, and $\delta(0)$ is an element of Δ where $\{i_1, \ldots, i_r\}$ is the set-theoretical sum of $s_1, \ldots, s_k$;
4. the value of $\delta(x; s_1, \ldots, s_k)$ is independent of the coordinates x_i of x for which the integer i is not contained in any of the sets $s_1, \ldots, s_k$. (*ibid.*, p. 7)

It is therefore seen that, whereas a nonrandomized decision function is a rule for a *specific* decision, a randomized decision function is a probability distribution for *several* decisions. Thus, the first decision function can be regarded as a special case of the second, such that the probability distribution assigns a value of unity for a particular decision and values of zero for other decisions.

Next, Wald explained that an experimenter's choice of a particular decision function may depend on two factors:

1. the relative degree of preference given to the various elements d' of D' when the true distribution F of X is known, and
2. the cost of experimentation. (*ibid.*, p. 8)

Wald denoted the degree of preference by the nonnegative weight function $W(F, d')$, which expresses the loss incurred by making the decision d' when the true distribution of X is F. As an example, consider a lot of N units of a manufactured product. Let d_1' and d_2' be the decisions to respectively accept and reject the lot. Suppose the true proportion of defective units is known to be p, the only parameter that determines F. Then, for example, a simple weight function $W^*(p, d')$ can be created by choosing two values p_1 and p_2 such that

$$W^*\left(p, d_1'\right) = \begin{cases} 0 & \text{if} \quad p \leq p_2 \\ 1 & \text{if} \quad p > p_2 \end{cases}$$

and

$$W^*\left(p, d_2'\right) = \begin{cases} 1 & \text{if} \quad p \leq p_1 \\ 0 & \text{if} \quad p > p_1 \end{cases}.$$

On the other hand, the cost function for an experiment carried out in k stages was denoted by $c(x; s_1, \ldots, s_k)$. An example of a simple cost function is when c is proportional to the number of observations.

Given $\{X_i\}$, $W(F, d')$, and $c(x; s_1, \ldots, s_k)$, the fundamental problem was then to choose a particular decision function $\delta(x; s_1, \ldots, s_k)$. Concerning the latter function, Wald made the following noteworthy remark:

> The adoption of a particular decision function by the experimenter may be termed "inductive behavior," since it determines uniquely the procedure for carrying out the experiment and for making a terminal decision. Thus the above decision problem may be called the problem of inductive behavior. (*ibid.*, p. 10)

Thus, Wald's adoption of inductive *behavior* (rather than inductive *reasoning*; see Section 5.6.4.4) as a working statistical philosophy shows that he aligned himself much more with Neyman than with Fisher.*

Now, how can several statistical decision functions be compared (or ordered) so that an optimal choice can then be made? Wald next set out to answer this question. First let

$$\delta(x;s) \equiv \delta(x;s_1,\ldots,s_k),$$
$$c(x;s) \equiv c(x;s_1,\ldots,s_k).$$

For any $D^* \subseteq D$, let $\delta(D^* \mid x;s)$ denote the probability that the decision d made will be contained in D^* when the decision function δ is used, x is the observed sample, and r stages of the experiment have been carried out. Now, suppose the decision function $\delta(y;s)$ is adopted and $x = \{x_i\}$ is the observed sample. Then the probability that the experiment will be carried out in k stages, the first stage in accordance with d_1^e, the second in accordance with $d_2^e,\ldots$, the kth stage in accordance with d_k^e, and that the terminal decision will be an element of $\bar{D}^t \subseteq D^t$ is given by

$$p\left(d_1^e,d_2^e,\ldots,d_k^e,\bar{D}^t \mid x,\delta\right) = \delta\left(d_1^e \mid 0\right)\delta\left(d_2^e \mid x;d_1^e\right)$$
$$\times\,\delta\left(d_3^e \mid x;d_1^e,d_2^e\right)\cdots\delta\left(d_k^e \mid x;d_1^e,\ldots,d_{k-1}^e\right)\delta\left(\bar{D}^t \mid x,d_1^e,\ldots,d_k^e\right).$$

The probability of $d_1^e d_2^e,\ldots,d_k^e,\bar{D}^t$ conditional only on δ (i.e., for *any* sample point x) is

$$q\left(d_1^e,d_2^e,\ldots,d_k^e,\bar{D}^t \mid F,\delta\right) = \int_M p\left(d_1^e,d_2^e,\ldots,d_k^e,\bar{D}^t \mid x,\delta\right)dF(x),$$

where M is the totality of all sequences x. Therefore, the conditional probability that the terminal decision will be an element of $\bar{D}^t$ is

$$P\left(\bar{D}^t \mid F,\delta\right) = \sum_{k=0}^{\infty}\sum_{d_1^e,d_2^e,\ldots,d_k^e} q\left(d_1^e,d_2^e,\ldots,d_k^e,\bar{D}^t \mid F,\delta\right).$$

With the above development, Wald finally defined the *risk function* as the expectation of the loss $W(F,d^t)$ with respect to the probability measure $P(\bar{D}^t \mid F,\delta)$, that is,

$$r_1(F,\delta) = \int_{D^t} W\left(F,d^t\right)dP\left(\bar{D}^t \mid F,\delta\right) \tag{6.34}$$

(*ibid.*, p. 12). Similarly, the expectation of the cost function is

$$r_2(F,\delta) = \sum_{k=1}^{\infty}\sum_{d_1^e,d_2^e,\ldots,d_k^e}\int_M c\left(x;d_1^e,d_2^e,\ldots,d_k^e\right)p\left(d_1^e,d_2^e,\ldots,d_k^e,D^t \mid x,\delta\right)dF(x).$$

*Thus, Wald also was not spared from Fisher's criticisms, for example, "It is, of course, also to be suspected that those authors, such as Neyman and Wald, who have treated these tests with little regard to their purpose in the natural sciences, may not have been more successful in the application of their ideas to the needs of acceptance procedures" (Fisher, 1956b, p. 77).

The total risk is then

$$r(F,\delta) = r_1(F,\delta) + r_2(F,\delta).$$

The function $r(F,\delta)$ was simply called the *risk function*.

Having obtained the risk function, Wald was now able to compare different decision functions:

> ...It seems reasonable to judge the merit of any given decision function δ_0 for purposes of inductive behavior entirely on the basis of the risk function $r(F,\delta_0)$ associated with it. This already permits a partial ordering of the decision functions as to their suitability for purposes of inductive behavior. (*ibidem*)

Thus, the decision function δ_1 will be preferred to the decision function δ_2 if $r(F,\delta_1) \le r(F,\delta_2)$ for all $F \in \Omega$ and $r(F,\delta_1) < r(F,\delta_2)$ for at least one $F \in \Omega$.

The next important concept introduced by Wald was that of an *admissible* decision function:

> A decision function δ will be said to be admissible if there exists no other decision function δ^* which is uniformly better than δ. (*ibid.*, p. 15)

Thus, the decision function δ is admissible if *there exists no other decision function* δ^* such that $r(F,\delta^*) \le r(F,\delta)$ for all $F \in \Omega$ and $r(F,\delta^*) < r(F,\delta)$ for at least one $F \in \Omega$.

Furthermore, a class C of decisions functions is said to be complete if for any $\delta \notin C$ there exists at least one element $\delta^* \in C$ such that δ^* is uniformly better than δ. A complete class C is said to be *minimal complete* if no proper subset of C is complete. The relationship between minimal complete classes and admissible decisions is that if a minimal complete class exists, then it must be equal to the class of all admissible decision functions.

The above describes *properties* of decision functions. Wald now addressed the issue of how to *actually obtain* decision functions. These would naturally have to satisfy certain optimality conditions. Wald considered two particular types of decision functions:

1. Bayes decision functions
2. Minimax decision functions

We note that the second (i.e., the minimax) criterion had previously been extensively used by von Neumann and Morgenstern in their book (von Neumann and Morgenstern, 1944).

Let us now examine Wald's use of the Bayes criterion. Wald proceeded by using a prior distribution ξ on Ω, where as before $F \in \Omega$. For example, if the distribution function F is completely determined by the parameter θ, then ξ is the prior distribution of θ. The value of the decision function $r(F,\delta)$ averaged across the prior ξ is given by

$$r^*(\xi,\delta) = \int_{\Omega} r(F,\delta)\,d\xi.$$

Wald then defined a *Bayes decision function (or a Bayes solution)* as a decision function δ_0 that minimizes the Bayes risk $r^*(\xi,\delta)$, that is,

$$r^*(\xi,\delta_0) \le r^*(\xi,\delta)$$

(Wald, 1950, p. 16). Note that $r^*(\xi,\delta)$ is called *a* Bayes solution, not *the* Bayes solution, since $r^*(\xi,\delta)$ depends on the particular prior ξ.

Wald was well aware of the difficulty of choosing a suitable prior, but included Bayes solutions because, perhaps surprisingly, the latter belong to a complete class:

> ...The main reason for discussing Bayes solutions here is that they enter into some of the basic results in Chapter 3. It will be shown there that under certain rather weak conditions the class of all Bayes solutions corresponding to all possible a priori distributions is a complete class. (*ibidem*)

Bayes solutions (when unique) are thus admissible, that is, there is no other decision function with at most its risk for all $F \in \Omega$ and no other decision function with less than its risk for at least one $F \in \Omega$.

Having considered Bayes decision functions, Wald next examined *minimax* solutions. A decision function δ_0 is minimax if it *min*imizes the *max*imum of $r(F,\delta)$ with respect to F, that is,

$$\sup_F r(F,\delta_0) \le \sup_F r(F,\delta)$$

(*ibid*., p. 18). A minimax solution thus "chooses the best from the worst" and has been criticized for being overly pessimistic. However, apart from other theoretical advantages, one of its key attributes is that it circumvents the use of any prior:

> In the general theory of decision functions...much attention is given to the theory of minimax solutions for two reasons: (1) a minimax solution seems, in general, to be a reasonable solution of the decision problem when an a priori distribution in Ω does not exist or is unknown to the experimenter; (2) the theory of minimax solutions plays an important role in deriving the basic results concerning complete classes of decision functions. (*ibidem*)

Wald later showed that, under some weak conditions, a minimax solution is a Bayes solution relative to the least favorable prior, that is, relative to a prior ξ_0, such that

$$\inf_\delta r(\xi_0,\delta) \le \inf_\delta r(\xi,\delta).$$

6.2.3 Hypothesis Testing as a Statistical Decision Problem

Having so far developed the basic theory of statistical decision functions, Wald now wished to connect it to the Neyman-Pearson (NP) theory of hypothesis testing. In particular, he proceeded to show that the latter can be viewed as a special case of a general class of decision-making problems.

In NP theory, δ is a nonrandomized decision function and D^t consists of the two elements d_1^t and d_2^t, that is, respectively, to accept and reject the null hypothesis H. Now, the set of all sample points $x=(x_1,\ldots,x_N)$ for which we decide to reject H is called the critical (rejection) region. *The choice of a critical region in NP theory corresponds to the choice of a decision function in Wald's theory*. Thus, if the critical region is $R\subseteq M$, then the decision function $d(x_1,\ldots,x_N)$ can be written as

$$d(x_1,\ldots,x_N)=\begin{cases} d_1^t & \text{if} \quad x\notin R \\ d_2^t & \text{if} \quad x\in R \end{cases}$$

where $x_1,\ldots,x_N$ denote the data.

Moreover, in NP theory, the probabilities of rejecting H when it is false and true are called the power and size of R, respectively. These two probabilities correspond to particular risks in Wald's theory, as we now show. Let $H:F\in\omega$. Then we define

$$W(F,d_1^t)=\begin{cases} 0 & \text{if} \quad F\in\omega \\ 1 & \text{if} \quad F\notin\omega \end{cases}$$

$$W(F,d_2^t)=\begin{cases} 1 & \text{if} \quad F\in\omega \\ 0 & \text{if} \quad F\notin\omega \end{cases}$$

(*ibid*., p. 20). $W(F,d_1^t)$ and $W(F,d_2^t)$ above are the two possible values of the weight function $W(F,d^t)$. Then the risk in Eq. (6.34) gives the size of R when $F\in\omega$ and the power of R when $F\notin\omega$. Note that the expected cost of experimentation has been ignored here since the experiment is performed in a single stage.

6.2.4 Estimation as a Statistical Decision Problem

Wald next turned his attention to the problem of point estimation, as developed by Fisher. Here also, Wald showed that point estimation can be viewed as a problem of decision making.

In point estimation, δ is a nonrandomized decision function and can be represented by $d(x_1,x_2,\ldots,x_N)$, which is an element d^t of D^t. Let Ω be a finite-parameter family of distribution functions F. For simplicity, suppose F is determined by a single parameter θ. Wald also denoted the loss function by $W(\theta,\theta^*)$, where θ^* is a chosen estimator of θ. An important example is the quadratic loss function

$$W(\theta,\theta^*)=(\theta^*-\theta)^2$$

(*ibid*., p. 22). Then, from Eq. (6.34), the risk is $\mathsf{E}_\theta W=\mathsf{E}_\theta(\theta^*-\theta)^2$. Since the risk equals the variance when θ^* is unbiased, Wald noted:

> A great deal of the literature on point estimation is devoted to the study of unbiased estimators with minimum variance, which are called efficient estimators. This theory can be regarded as a special case of the general decision theory when $W(\theta,\theta^*)$ is given by $(\theta^*-\theta)^2$. (*ibidem*)

As an example of another loss function, Wald mentioned a particular estimator introduced previously by Pitman (1939), now called a *Pitman estimator*. Here an estimator θ^* is sought such that

$$\Pr\{|\theta^*(x_1,\ldots,x_N)-\theta|\le c\}, \quad \text{where } c>0,$$

is maximized. This corresponds to the loss function

$$W(\theta,\theta^*) = \begin{cases} 0 & \text{when} \quad |\theta-\theta^*| \le c \\ 1 & \text{when} \quad |\theta-\theta^*| > c \end{cases}$$

Wald also showed that the theory of confidence intervals, as developed by Neyman, was a special case of a decision-making problem. For simplicity, let Ω be a one-parameter family of distribution functions. Suppose the parameter is θ and let C be a given class of intervals. For $I \in C$, let d_I^t be the terminal decision that I is chosen to be an interval estimate of θ. The weight function can thus be represented by $W(\theta, I)$. For example,

$$W(\theta, I) = \begin{cases} 1 & \text{when} \quad \theta \notin I \\ 0 & \text{when} \quad \theta \in I \end{cases}$$

(Wald, 1950, pp. 23–24). The risk $\mathsf{E}_\theta W$ is then equal to the probability that the chosen I will *not* cover the true θ. By setting the last probability to $1-\gamma\,(0 \le \gamma \le 1)$, one can obtain a confidence interval I. The latter will be a confidence interval with confidence coefficient γ, that is, a confidence interval with probability γ of including the true θ.

6.2.5 Statistical Decision as a Two-Person Zero-Sum Game

Wald showed that his statistical decision theory was at core a two-person zero-sum game. Thus, statistical decisions can be approached by using the theory of games developed earlier by von Neumann and Morgenstern (1944). Game theory is the mathematical study of situations when there are conflicts of interest. A game is any situation where two or more players who do not have identical interests can affect the outcome. The players attempt to maximum their individual utility, and their moves may or may not be random. A two-person zero-sum game means that whatever one player loses, the other wins (i.e., the system is closed).

Since any game with a finite number of moves can be converted into an equivalent "normalized game" (i.e., in a payoff matrix form), Wald examined the latter form. Consider a game with two players such that when player 1 chooses point $a \in A$ and player 2 chooses point $b \in B$, then player 2 pays an amount $K(a,b)$ to player 1. Any $a \in A$ is a *pure strategy* for player 1 and any $b \in B$ is a *pure strategy* for player 2. On the other hand, if a is chosen according to a chance mechanism such that $\Pr\{a \in \alpha\} = \xi(\alpha)$ for an interval α, then the probability distribution ξ is a mixed strategy for player 1. Similarly, η is a mixed strategy for player 2 if $\Pr\{b \in \beta\} = \eta(\beta)$ for an interval β. The expected value of the outcome $K(a,b)$ is then

$$K^*(\xi,\eta) = \int_B \int_A K(a,b)\,d\xi\,d\eta$$

(Wald, 1950, p. 25). One sees that a pure strategy is a special case of a mixed strategy: let the probability measures ξ and η assign probability of one to a and b respectively in a mixed strategy, and we obtain a pure strategy.

Writing $K^*(\xi,\eta)$ simply as $K(\xi,\eta)$, a strategy ξ_0 is *minimax* for player 1 if

$$\inf_\eta K(\xi_0,\eta) \ge \inf_\eta K(\xi,\eta) \quad \text{for all } \xi.$$

Similarly, a strategy η_0 is minimax for player 2 if

$$\sup_{\xi} K(\xi,\eta_0) \le \sup_{\xi} K(\xi,\eta) \quad \text{for all } \eta.$$

On the other hand, a strategy η_0 of player 2 is *minimal* relative to a strategy ξ of player 1 if

$$K(\xi,\eta_0) = \min_{\eta} K(\xi,\eta)$$

Further, a strategy η_1 of player 2 is said to be *uniformly better* than the strategy η_2 if $K(\xi,\eta_1) \le K(\xi,\eta_2)$ for all ξ and if $K(\xi,\eta_1) < K(\xi,\eta_2)$ for at least one ξ. A strategy for player i ($i = 1, 2$) is said to be *admissible* if there is no uniformly better strategy for player i. A class C of strategies for player i ($i = 1, 2$) will be *complete* if for any strategy not in C there exists a strategy in C that is uniformly better.

Having given the above basic terminology for a two-person zero-sum game, Wald next stated that the decision problem can be viewed as such a game, with the players being the Nature and the experimenter:

> In a decision problem the experimenter wishes to minimize the risk $r(F,\delta)$. The risk, however, depends on two variables F and δ, and the experimenter can choose only the decision function δ but not the true distribution F. The true distribution F, we may say, is chosen by Nature, and Nature's choice is unknown to the experimenter. Thus the situation that arises here is very similar to that of a two-person game. (*ibid.*, p. 26)

By showing that each of the concepts in his decision theory had a counterpart in game theory, Wald was able to make the link between the two theories (see Table 6.1).

TABLE 6.1 Correspondence Between Two-Person Game and Decision Problem

Two-Person Game	Decision Problem
Player 1	Nature
Player 2	Experimenter
Pure strategy a of player 1	Choice of true distribution F by Nature
Space A of pure strategies of player 1	Space Ω
Pure strategy b of player 2	Choice of decision function δ by experimenter
Outcome $K(a, b)$	Risk $r(F,\delta)$
Mixed strategy ξ of player 1	A priori distribution ξ in Ω
Mixed strategy η of player 2	Probability measure ξ in D
Outcome $K(\xi,\eta)$	$r(\xi,\eta) = \int_D \int_\Omega r(F,\delta)\,d\xi\,d\eta$
Minimax strategy of player 2	Minimax solution of decision problem
Minimax player of player 1	Least favorable a priori distribution in Ω
Minimal strategy of player 2	Bayes solution
Admissible strategy of player 2	Admissible decision function

From Wald (1950, p. 27).

6.3 THE BAYESIAN REVIVAL

After Laplace's death, there was an increasing realization within the mathematical community of the deficiencies of inverse probability,* which relied on an epistemic interpretation of probability. A major reason for this was the reliance on the principle of indifference to use a uniform prior in conjunction with Bayes' theorem. In Section 1.3, we explained how the principle of indifference can often lead to absurdities. In spite of the problems associated with inverse probability, it continued to be used by all of Edgeworth, Karl Pearson, Student, Bowley, and others. Even Fisher referred to inverse probability in his very first paper (Fisher, 1912). An exhaustive examination of the use of inverse probability right from Thomas Bayes up to Karl Pearson is given in Dale's book (1999). An engaging, nontechnical history of Bayes' theorem can be found in the book by McGrayne (2011).

What was the reason for the continued use of inverse probability? The reason is not difficult to figure: right up into the 1920s, there was no alternative probabilistic framework upon which statistical inference could be performed. Writing in 1925, the mathematician Julian Coolidge thus offered the following candid assessment:

> It is perfectly evident that Bayes' principle is open to very grave question, and should only be used with the greatest caution. The difficulty lies with the *a priori* probabilities. We generally have no real line on them, so take them all equal...
>
> Why not, then, reject the formula outright? Because, defective as it is, Bayes' formula is the only thing we have to answer certain important questions which do arise in the calculus of probability...Therefore we use Bayes' formula with a sigh, as the only thing available under the circumstances. (Coolidge, 1925, pp. 99–100)

However, in the mid-1920s, Fisher put forward frequentist theories of estimation and significance testing. There was a consequent decline in inverse probability. Nevertheless, despite of its decline, it never died. Thus, mathematicians such as Jeffreys† continued to advocate inverse probability as *the* method of inference.

In 1928–1930, the Italian mathematician Bruno de Finetti gave further vindication to the Bayesian method through some brilliant work, culminating in one of the truly beautiful theorems of statistics, the so-called de Finetti representation theorem. But before De Finetti, there was Frank Ramsey, who laid the foundations for the first serious subjective theory of probability in 1926. Ramsey's theory has been rightly qualified by Bernardo and Smith, as by many other scholars, as a "revolutionary landmark in the history of ideas" (Bernardo and Smith, 2000, p. 84). Building on the works of Ramsey and de Finetti, Jimmie Savage later brought Bayesianism to the forefront and made the latter a serious contender for a coherent theory of statistics. A further breakthrough in the application of Bayesian methods was later made by Herbert Robbins in 1956, though it is debatable whether the latter's work was strictly within the Bayesian framework.

*An excellent article detailing the decline of inverse probability after Laplace is provided in Kamlah (1987).

†See Section 5.5.3.

6.3.1 Ramsey (1926): Degree of Belief, Ethically Neutral Propositions, Ramsey's Representation Theorem, Calibrating the Utility Scale, Measuring Degree of Belief, and The Dutch Book

Frank Plumpton Ramsey* (Fig. 6.10) was born in Cambridge, England, on February 22, 1903. He was the elder son of a Cambridge mathematician, and much of his mathematical and philosophical background was shaped by the Cambridge intellectual community in the 1920s. Ramsey read Whitehead and Russell's *Principia Mathematica* at an early age and received a degree in mathematics from Trinity College in 1923. In 1926, he was appointed University Lecturer in Mathematics. Ramsey developed jaundice after an abdominal operation and died in Cambridge in 1930, at the age of 26.

One of Ramsey's earliest publications was a critique of John Maynard Keynes' landmark book *A Treatise on Probability* (Keynes, 1921). Keynes was a mentor to Ramsey and had put forward a very influential logical theory of probability in his book. We shall soon describe some of Ramsey's criticisms of Keynes' theory, but for now we note that, in spite of being influenced by Keynes (as well as Russell), Ramsey was closest to Wittgenstein, both personally and philosophically. Ramsey's other writings include "The Foundations of Mathematics," "Universals," "Mathematical Logic," "Facts and Propositions," and especially "Truth and Probability" (which was written in 1926). These works as well as other writings were published in 1931 (Ramsey, 1931) and show the originality of Ramsey's thoughts.

Undoubtedly, Ramsey's most original work can be found in the essay "Truth and Probability"†, where he laid the foundations of a subjective theory of probability as well as of decision theory. Unfortunately, this landmark essay did not gain much attention until the 1950s, when it was rediscovered by several scholars, among whom Jimmie Savage was one of the most prominent. We shall now examine Ramsey's "Truth and Probability."

Ramsey started his essay by commenting on the frequency interpretation of probability, that was spearheaded by the likes of John Venn and Richard von Mises. Ramsey stated that, although he himself did not believe that the frequency interpretation was correct, he was willing to temporarily admit that

> ...Suppose we start with the mathematical calculus, and ask, not as before what interpretation of it is most convenient to the pure mathematicism, but what interpretation gives results of greatest value to science in general, then it may be that the answer is again an interpretation in terms of frequency.... (*ibid.*, p. 159)

Ramsey's conciliatory attitude toward the frequency theory ought to be contrasted against that of his immediate successor, Bruno de Finetti, who was categorically opposed to any interpretation of probability other than the subjective one (see Section 6.3.2). But having admitted the usefulness of the frequency theory,

> ...it still remains the case that we have the authority both of ordinary language and of many great thinkers for discussing under the heading of probability what appears to be quite a different subject, *the logic of partial belief*‡.... (*ibidem*)

* Of the several books written on Ramsey, we mention those by Dokic and Engel (2003), Sahlin (1990), Frápolli (2004), and Lillehammer and Mellor (2005).

† This essay also features as a chapter in Kyburg and Smokler's *Studies in Subjective Probability* (Ramsey, 1964, pp. 61–92)

‡ Italics are ours.

FIGURE 6.10 Frank Plumpton Ramsey (1903–1930). Wikipedia, http://en.wikipedia.org/wiki/Frank_P._Ramsey

Ramsey's objective was thus to put forward a theory of subjective probability, which he called "a logic of partial belief."* However, before that, he mounted a major criticism against Keynes' logical theory. According to the latter, probability is a logical relation between two propositions, one being the premise and the other the conclusion. If c is conclusion and p is the premise, then the probability $\Pr\{c \mid p\}$ measures the extent to which p implies (or entails) c. Ramsey's first major criticism of Keynes' theory was that, for many pairs of propositions, $\Pr\{c \mid p\}$ was too elusive to be determined. For example,

> ...if...we take the simplest possible pairs of propositions such as "This is red" and "That is blue" or "This is red" and "That is red", whose logical relations should surely be easiest to see, no one, I think, pretends to be sure what is the probability relation which connects them.... (*ibid.*, p. 162)

Ramsey's second major point was that Keynes' prescription of measuring probability as a logical relation between two propositions was against the natural manner in which degrees of belief were construed by individuals:

> ...no one estimating a degree of probability simply contemplates the two propositions supposed to be related by it; he always considers *inter alia* his own actual or hypothetical degree of belief. (*ibid.*, p. 163)

Having criticized Keynes' theory of probability, it remained for Ramsey to offer an alternative. How should degrees of belief be measured? A first possibility would be to use the intensity of the feeling of conviction, but Ramsey discarded this as being impractical. He now offered a second, more reasonable possibility:

> ...degree of a belief is a causal property of it, which we can express vaguely as the extent to which we are prepared to act on it. (*ibid.*, p. 169)

Thus, *one's degree of belief can be measured by the extent to which one is prepared to act on it*. Ramsey mentioned that this idea had been used by Russell in his book *The Analysis of Mind* (Russell, 1921). In fact the idea goes back to Hume (e.g., Dokic and Engel, 2003, p. 7). But what kind of actions did Ramsey have in mind? He then went on to give a definition of subjective probability that was later independently repeated by de Finetti and that is still used today:

> The old-established way of measuring a person's belief is to propose a bet, and see what are the lowest odds which he will accept. (Ramsey, 1931, p. 172)

Thus, if *I* am willing to bet on an event when its probability of happening is as low as P, but no lower than this value, then *my* degree of belief in the event is P. This is the same as saying that I am willing to bet up to $\$PS(S > 0, 0 \le P \le 1)$ on the event for a return of $\$S$.

But Ramsey noted that a bet was not sufficient to decide on one's course of action

*Ramsey's logic or theory of partial belief, which we shall next consider, is also described in Eells (1982, pp. 65–71), Jeffrey (1983, Chapter 3), Sahlin (1990, Chapter 1), Dokic and Engel (2003), and Parmigiani and Inoue (2009, pp. 76–81).

TABLE 6.2 Utilities ($U_1 \ldots U_4$) of Outcomes for Two Possible Actions (α, β) Undertaken under Two Possible Conditions (p, not ($\sim$) p)[a]

	Condition p (Prob. P)	Condition $\sim p$ (Prob. $1-P$)	Expected Utility
Action α	U_1	U_2	$PU_1+(1-P)U_2$
Action β	U_3	U_4	$PU_3+(1-P)U_4$

[a] The action that is finally taken is the one with higher expected utility. A more general situation can be considered whereby the probability of the condition is dependent on the action. Then we would have a different probability P_i ($i=1, \ldots, 4$) corresponding to each utility U_i.

> ...partly because of the diminishing marginal utility of money, partly because the person may have a special eagerness or reluctance to bet, because he either enjoys or dislikes excitement or for any other reason.... (*ibid.*, p. 172)

Thus, Ramsey wished to take into account, just like Daniel Bernoulli had done hundreds of years ago, the *marginal utility* of an amount when deciding on one's course of action:

> In order therefore to construct a theory of quantities of belief which shall be both general and more exact, I propose to take as a basis a general psychological theory...that we act in the way we think most likely to realize the objects of our desires, so that a person's actions are completely determined by his desires and opinions. (*ibid.*, p. 173)

Therefore, one should act by taking into account not only one's *degree of belief* ("opinion") but also one's *utility or desirability* ("desires") for the outcomes, and such action should maximize our expected utility. Thus, in the above, Ramsey stated a crude form of what later came to be known as the *principle of maximum expected utility*. The latter is illustrated in the utility matrix* shown in Table 6.2.

We give an example. Let us take actions α and β to respectively be "revised whole syllabus for an exam" and "revised only first half of syllabus," and let p be the condition "exam questions were taken from only the second half of the syllabus." Then U_3 is the utility of the outcome "failed the exam," and the other utilities similarly depend on the corresponding possible outcomes.

To proceed further, Ramsey needed to find some proposition (or condition) p that did not change the utilities resulting from of a given action. Such a proposition is said to be *ethically neutral*, and in Table 6.2 it would imply that $U_1 = U_2$ and $U_3 = U_4$. We shall shortly show how the concept can be used to *calibrate* the utility scale. This is how Ramsey introduced the idea of an ethically neutral proposition:

> There is first a difficulty which must be dealt with; the propositions like p in the above case which are used as conditions in the options offered may be such that their truth or falsify is an object of desire to the subject. This will be found to complicate the problem, and we have to assume that there are propositions for which this is not the case, which we shall call ethically neutral. More precisely an atomic proposition p is called ethically neutral if two possible worlds differing only in regard to the truth of p are always of equal value. (*ibid.*, p. 177)

*For the time being we shall assume that utilities for various outcomes exist. The existence of utilities will be made explicit later through Ramsey's axioms.

To give a concrete example, suppose I live in New York. Consider the action

$$\alpha : \text{I go out today}$$

Now consider the two propositions

$$p_1 : \text{it is raining in New York}$$
$$p_2 : \text{it is raining in Melbourne}$$

Then p_1 is *not* ethically neutral because the utility (or desirability) of my going out would definitely depend on whether it rained or not. On the other hand, p_2 is ethically neutral since whether or not it is raining in Melbourne has no effect on the utility of my going out in New York.

Of all ethically neutral propositions, those with probability 1/2 are of special importance:

> ...The subject is said to have belief of degree 1/2 in such a proposition p if he has no preference between the options (1) α if p is true, β if p is false, and (2) α if p is false, β if p is true, but has a preference between α and β simply. (*ibidem*)

Let us now show why indifference to the two actions

$$\alpha \text{ if } p \text{ is true}, \beta \text{ if } p \text{ is false}$$
$$\alpha \text{ if } p \text{ is false}, \beta \text{ if } p \text{ is true}$$

(where p is ethically neutral and α, β are not equally preferred) implies that the probability of p is 1/2. Consider Table 6.3, where the utilities resulting from the actions α and β are, respectively, x and $y(x \neq y)$, given p.

Since one is indifferent to the two actions,

$$Px+(1-P)y = Py+(1-P)x$$
$$x(1-2P) = y(1-2P).$$

Since $x \neq y$, the above is true only if $1-2P=0$. Hence, $P=1/2$, which is what we wanted to show. For the rain example above, an ethically neutral condition with probability 1/2 would be "the toss of a fair coin gives a head."

Having defined an ethically neutral proposition with probability 1/2, Ramsey now showed how it could be used to define the difference in value (or preference) between two outcomes. He first stated:

TABLE 6.3 Utilities for A Particular Ethically Neutral Proposition p

	Proposition p (Prob. P)	Proposition $\sim p$ (Prob. $1-P$)	Expected Utility
Action α if p, else β	x	y	$Px+(1-P)y$
Action β if p, else α	y	x	$Py+(1-P)x$

TABLE 6.4 Utility Implication of Ramsey's Definition of Value Difference

	Condition p (Prob. 1/2)	Condition ~p (Prob. 1/2)	Expected Utility
Action α if p, else δ	$U(\alpha)$	$U(\delta)$	$\{U(\alpha)+U(\delta)\}/2$
Action β if p, else γ	$U(\beta)$	$U(\gamma)$	$\{U(\beta)+U(\gamma)\}/2$

> Belief of degree 1/2 as thus defined can be used to measure values numerically in the following way. We have to explain what is mea nt by the difference in value between α and β being equal to that between γ and δ; and we define this to mean that, if p is an ethically neutral proposition believed to degree 1/2, the subject has no preference between the options (1) α if p is true, δ if p is false, and (2) β if p is true, γ if p is false. (*ibid*., p. 178)

In the above, Ramsey used the same Greek symbols for both actions and the outcomes, that is, he extended the possible actions to include the outcomes themselves. Let $d(\alpha,\beta)$ be the value distance between the outcomes α and β. Then, in the above, Ramsey *defined* $d(\alpha,\beta)=d(\gamma,\delta)$ if one is indifferent to the two actions:

$$\alpha \text{ if } p, \text{ else } \delta$$
$$\beta \text{ if } p, \text{ else } \gamma,$$

where p is ethically neutral. To see why this definition makes sense, consider Table 6.4 where we have denoted the actions by the outcomes themselves.

Since one is indifferent to the two actions, we have

$$d(\alpha,\beta)=d(\gamma,\delta)\Rightarrow\frac{U(\alpha)+U(\delta)}{2}=\frac{U(\beta)+U(\gamma)}{2}\Rightarrow U(\alpha)-U(\beta)=U(\gamma)-U(\delta).$$

Therefore, equality in value distances translates into an equality in utility differences. This is a reasonable conclusion and gives legitimacy to the use of utilities. However, the definition does not guarantee the existence of utilities. Moreover, we do not yet know how to assign the latter to various outcomes. Both tasks would be made possible through the following axioms that Ramsey next stated:

> Axioms.
>
> 1. There is an ethically neutral proposition p believed to degree 1/2.
> 2. If p, q are such propositions and the option
> α if p, δ if not-p is equivalent to β if p, γ if not-p
>   then α if q, δ if not-q is equivalent to β if q, γ if not-q.
>   *Def.* In the above case we say $\alpha\beta=\gamma\delta$.
>   *Theorems.* If $\alpha\beta=\gamma\delta$,
>   then $\beta\alpha=\delta\gamma$, $\alpha\gamma=\beta\delta$, $\gamma\alpha=\delta\beta$.
>
> (2a) If $\alpha\beta=\gamma\delta$, then $\alpha>\beta$ is equivalent to $\gamma>\delta$
> and $\alpha=\beta$ is equivalent to $\gamma=\delta$.
>
> 3. If option A is equivalent to option B and B to C then A to C.
>   *Theorem.* If $\alpha\beta=\gamma\delta$ and $\beta\eta=\zeta\gamma$,
>   then $\alpha\eta=\zeta\delta$

4. If $\alpha\beta = \gamma\delta$, $\gamma\delta = \eta\zeta$, then $\alpha\beta = \eta\zeta$.
5. $(\alpha, \beta, \gamma)\ E!\ (\hat{}x)\ (\alpha x = \beta\gamma)$
6. $(\alpha, \beta)\ E!\ (\hat{}x)\ (\alpha x = x\beta)$.
7. Axiom of continuity: Any progression has a limit (ordinal).
8. Axiom of Archimedes.

(*ibid.*, pp. 178–179)

Here, Axiom 1 *postulates the existence* of an ethically neutral proposition with probability 1/2. Axiom 1 is thus said to be an *ontological* axiom. In Axiom 2, the value distance $d(\alpha,\beta)$ is represented by $\alpha\beta$. The axiom states that if $d(\alpha,\beta) = d(\gamma,\delta)$ then $d(\beta,\alpha) = d(\delta,\gamma)$, $d(\alpha,\gamma) = d(\beta,\delta)$, and $d(\gamma,\alpha) = d(\delta,\beta)$. Axiom 2a states that, given $d(\alpha,\beta) = d(\gamma,\delta)$, if α is preferred to β (i.e., $\alpha > \beta$), then γ is preferred to δ. Axioms 3 and 4 are transitivity axioms. Axiom 5 states that if α, β, and γ are three outcomes, then there exists a unique outcome x such that $d(\alpha,x) = d(\gamma,\delta)$. Axiom 6 states that if α and β are two outcomes, then there exists a unique outcome x such that $d(\alpha,x) = d(x,\beta)$. Axiom 7 implies that every bounded nondecreasing (or nonincreasing) set of numbers tends to a limit. Axiom 8 implies that no matter how small $d(\alpha,\beta)$ is and no matter how large $d(\gamma,\delta)$ is, there exists an integer N such that $N \times d(\alpha,\beta) \geq d(\gamma,\delta)$.

Having given the eight axioms above, Ramsey stated that he had "defined a way of measuring value" (*ibid.*, p. 179). The existence of what we have hitherto called the utility U of an outcome thus follows from the axioms. That is, if $d(\alpha,\beta) = d(\gamma,\delta)$, then there exists a utility function such that $U(\alpha) - U(\beta) = U(\gamma) - U(\delta)$. This is known as Ramsey's *representation theorem*: a subject's preferences can be represented by a utility function (up to a linear transformation).

With Ramsey's axioms at our disposal, let us now show through a concrete example how utilities can be measured. This proceeds by calibrating the utility scale. Suppose a subject wishes to watch a movie on a Sunday afternoon and she has choices between the three types: Action, Spy, and Crime. Her preferences are Crime>Action>Spy. We can arbitrarily create a utility scale by assigning $U(\text{Crime}) = 1$ and $U(\text{Spy}) = 0$. On this scale, how can we measure U(Action)? The "trick" is to first use an ethically neutral proposition with probability 1/2 (Axiom 1), such as p: fair coin gives head. Next, consider the action

$$\alpha : \text{Crime if head, else Spy}$$

Now, in Axiom 6, let us take $x = \beta$. Therefore, for any α, there should be another action x such that $d(\alpha,x) = 0$, that is, such that the subject is indifferent between α and x. Suppose this x for the subject is "Fantasy" (watching a fantasy movie). The situation is illustrated in Table 6.5.

Since the subject is indifferent between α and x, we see that $U(\text{Fantasy}) = 1/2$. Next, the question is: does the subject prefer Action to Fantasy? Suppose the answer is no. Then we have $0 \leq U(\text{Action}) \leq 1/2$. This is our first estimate for U(Action). We now make this estimate more precise. Consider the action

$$\beta : \text{Fantasy if head, else Spy}$$

TABLE 6.5 Calibrating a Subject's Utility Scale (First Estimate)

	Head (Prob. 1/2)	Tail (Prob. 1/2)	Expected Utility
α: Crime if head, else Spy	1	0	1/2
x: Fantasy	U(Fantasy)	U(Fantasy)	U(Fantasy)

TABLE 6.6 Calibrating a Subject's Utility Scale (Second Estimate)

	Head (Prob. 1/2)	Tail (Prob. 1/2)	Expected Utility
β: Fantasy if head, else Spy	1/2	0	1/4
y: Drama	U(Drama)	U(Drama)	U(Drama)

TABLE 6.7 Ramsey's Definition of the Subjective Probability P of a Proposition p (Which is not Necessarily Ethically Neutral)

	p (Prob. P)	$\sim p$ (Prob. $1-P$)	Expected Utility
β if p, else γ	β	γ	$P\beta+(1-P)\gamma$
α	α	α	α

Again, appealing to Axiom 6, there should be another action y such that $d(\beta,y)=0$. Suppose such a y for the subject is "Drama." The situation is illustrated in Table 6.6.

From the table, we see that $U(\text{Drama})=1/4$. Again, the question to the subject is: does she prefer Action to Drama? Suppose the answer is yes. Then we have $1/4 \le U(\text{Action}) \le 1/2$. This is our second estimate for U(Action) and is more precise than the first. By going on in this way, we can thus narrow down the value of U(Action) as precisely as we wish. If at any stage, $U(\text{Movie Type A}) \le U(\text{Action}) \le U(\text{Movie Type B})$, then the interval can be further narrowed down by considering the two actions

$$\gamma : \text{Movie Type A if head, else Movie Type B}$$
$$z : \text{Movie Type C}$$

where the subject is indifferent between γ and z. The actions $x, y, \ldots$ that mark the ½ point, ¼ point… of the utility scale serve to calibrate the latter.

Ramsey's next major task was to show how subjective probabilities (degrees of belief) could be assigned to propositions. His definition was as follows:

> If the option of α for certain is indifferent with that of β if p is true and γ if p is false, we can define the subject's degree of belief in p as the ratio of the difference between α and γ to that between β and γ. (*ibid*., p. 179)

To understand Ramsey's definition, consider Table 6.7. Since the subject is indifferent between the two actions,

$$P\beta+(1-P)\gamma=\alpha \Rightarrow P=\frac{\alpha-\gamma}{\beta-\gamma},$$

TABLE 6.8 Ramsey's Definition of Conditional Probability

	p & q	p & ~q	~p & q	~p & ~q
α if q, β if ~q	α	β	α	β
γ if p & q, δ if ~p & q, β if ~q	γ	β	δ	β

which is Ramsey's definition. Continuing with the movie example, let p be the proposition "subject's friend will watch movie with her," and suppose the subject is indifferent between the two actions:

$$\zeta : \text{Crime if } p, \text{ else Spy}$$
$$\delta : \text{Fantasy}$$

Then, $\alpha = U(\text{Fantasy}) = 1/2$, $\gamma = U(\text{Spy}) = 0$, and $\beta = U(\text{Crime}) = 1$. Hence, the subjective probability of p is

$$P = \frac{\alpha - \gamma}{\beta - \gamma} = \frac{1}{2}.$$

Thus, based on one's utilities for various actions, Ramsey achieved the major task of assigning subjective probabilities to various propositions (or events).

Ramsey's next task was to use the idea of bets to define conditional probability, which he called "the degree of belief in p given q." He noted that the latter

> does not mean the degree of belief in "If p then q", or that in "p entails q" or that which the subject would have in p if he knew q, or that which he ought to have. It roughly expresses the odds at which he would now bet on p, the bet only to be valid if q is true. (*ibid.*, p. 180)

Ramsey's definition of the degree of belief in p given q was

> Suppose the subject indifferent between the options (1) α if q true, β if q false, (2) γ if p true and q true, δ if p false and q true, β if q false. Then the degree of his belief in p given q is the ratio of the difference between α and δ to that between γ and δ. (*ibidem*)

To understand this definition, consider Table 6.8. Since the subject is indifferent between the two actions, their expected utilities are the same, that is,

$$\begin{aligned}
\alpha \Pr\{p\&q\} + \beta \Pr\{p\&\sim q\} + \alpha \Pr\{\sim p\&q\} + \beta \Pr\{\sim p\&\sim q\} &= \gamma \Pr\{p\&q\} + \beta \Pr\{p\&\sim q\} \\
&\quad + \delta \Pr\{\sim p\&q\} + \beta \Pr\{\sim p\&\sim q\} \\
(\alpha - \gamma)\Pr\{p\&q\} + (\alpha - \delta)\Pr\{\sim p\&q\} &= 0 \\
(\alpha - \gamma)\Pr\{p\&q\} + (\alpha - \delta)(\Pr\{q\} - \Pr\{p\&q\}) &= 0 \\
(\delta - \gamma)\Pr\{p\&q\} &= (\delta - \alpha)\Pr\{q\} \\
\therefore \quad \Pr\{p \mid q\} = \frac{\Pr\{p\&q\}}{\Pr\{q\}} &= \frac{\delta - \alpha}{\delta - \gamma}.
\end{aligned}$$

Having defined conditional probability, Ramsey stated that all previous definitions and axioms could prove the fundamental laws of probable belief:

(1) Degree of belief in p + degree of belief in $\sim p = 1$.
(2) Degree of belief in p given q + degree of belief in $\sim p$ given $q = 1$.
(3) Degree of belief in (p and q) = degree of belief in p × degree of belief in q given p.
(4) Degree of belief in (p and q) + degree of belief in (p and $\sim q$) = degree of belief in p.

(*ibid.*, p. 181)

We see that these laws correspond exactly to those of the usual probability calculus. Moreover, any system of belief that violated these laws would be inconsistent (incoherent) in the sense that it would violate at least one of Ramsey's axioms. In Ramsey's words,

> Having degrees of belief obeying the laws of probability implies a further measure of consistency, namely such a consistency between the odds acceptable on different propositions as shall prevent a book being made against you. (*ibid.*, p. 183)

In the above, Ramsey introduced the notion of a "book," now called a "Dutch Book," that is, a systems of bets that guarantees loss. Any system of belief that is not coherent results in a Dutch Book.

Let us show how, for example, the probability law in (1) above follows from the Dutch Book argument. Suppose player Q believes that the proposition p has a probability P_1 and the proposition $\sim p$ has a probability P_2. What should the relationship between P_1 and P_2 be so that Q does not incur a Dutch Book? Based on her beliefs, Q bets:

- $\$P_1S$ for a payment of $\$S$ when p occurs.
- $\$P_2S$ for a payment of $\$S$ when $\sim p$ occurs.

The total amount Q bets is $\$(P_1 + P_2)S$. If p occurs Q gets $\$S$ back. Similarly, if $\sim p$ occurs Q gets $\$S$ back. Thus, whatever happens, Q gets $\$S$. To avoid a Dutch Book, we must have $S \geq (P_1 + P_2)S$, that is,

$$P_1 + P_2 \leq 1. \tag{6.35}$$

Now, based on her beliefs, Q can also pay:

- $\$S$ to somebody who bets $\$P_1S$ on p
- $\$S$ to somebody else who bets $\$P_2S$ on $\sim p$

The total amount received by Q is $\$(P_1 + P_2)S$. If p occurs Q pays $\$S$. Similarly, if $\sim p$ occurs, Q pays $\$S$ as well. Thus, whatever happens, Q pays $\$S$. To avoid a Dutch Book, we must have $(P_1 + P_2)S \geq S$, that is,

$$P_1 + P_2 \geq 1. \tag{6.36}$$

By using Eqs. (6.35) and (6.36), we see that the only way for P to avoid a Dutch Book is to have $P_1 + P_2 = 1$, that is, $\Pr\{p\} + \Pr\{\sim p\} = 1$, in accordance with the probability law in (1).

6.3.2 De Finetti (1937): The Subjective Theory, Exchangeability, De Finetti's Representation Theorem, Solution to the Problem of Induction, and Prevision

Bruno de Finetti* (Fig. 6.11) was born on June 13, 1906, in Innsbruck, Austria, where his father was working as an engineer. Both his parents were Italian. De Finetti did his early school in his mother's hometown Trento, a formerly Austrian town annexed to Italy after World War I. He graduated in mathematics from Milan University in 1927. While in Milan, his interest in probability was aroused when he read a short article on Mendel's law written by the biologist Carlo Foà (de Finetti, 1982, p. 4). De Finetti wrote a paper on the topic and sent it to Foà. The latter passed it to his colleagues and also to the famous Corrado Gini (1884–1965), president of the newly created National Census Bureau in Rome. Gini was very impressed and published the article in his journal *Metron*. He also offered de Finetti a position at the Bureau after he graduated. De Finetti thus went to Rome after graduating and stayed there until 1931. In 1930 alone, he published 17 papers (Lindley, 1986). Afterward, he worked for an insurance company in Trieste until 1946. He held chairs both at Trieste University and the University of Rome.

De Finetti was well known for his extreme subjectivist views on probability and for his utter denial of the existence of objective probabilities. His famous aphorism "PROBABILITY DOES NOT EXIST" featured in the preface of his two-volume opus *Theory of Probability: A Critical Introductory Treatment* (de Finetti, 1974), meaning that probability is not an objective quantity "out there" that needs to be discovered. Rather it is a subjective relationship between an observer and her surrounding (i.e., it changes across observers). De Finetti's work in probability and statistics was abundant, and his 1937 paper (which we shall examine next) was a landmark for subjectivists. De Finetti died in Rome in 1985.

De Finetti was unaware of Ramsey's (1926) prior work on subjective probability† but seems to have written a complete exposition of his views on subjective probability in 1928. These were published in 1930 (de Finetti, 1930a; 1930b; 1930c) and therefore anteceded the posthumous publication of Ramsey's work in 1931.

De Finetti's 1937 paper‡ was originally entitled "La prévision: Ses lois logiques, ses sources subjectives" (de Finetti, 1937). This was later translated in English by Henry Kyburg as "Foresight: Its logical laws, its subjective sources" (de Finetti, 1964). De Finetti admitted at the start that his viewpoint "may be considered the extreme of subjectivist solutions" and stated his aims as follows:

> ...The aim of the first lecture will be to show how the logical laws of the theory of probability can be rigorously established within the subjectivistic point of view. (*ibid.*, p. 99)

He then considered a well-defined event and gave four axioms upon which his subjective theory of probability could be built:

> [AXIOM 1:] one uncertain event can only appear to us (a) equally probable, (b) more probable, or (c) less probable than another...[AXIOM 2:] an uncertain event always seems to us more

*De Finetti's life is also described in Lindley (1986) and Cifarelli and Regazzini (1996).

†In de Finetti's own words, "I did not know of Ramsey's work before 1937..." (de Finetti, 1964, p. 102, footnote 4(a)).

‡De Finetti's paper is also discussed in Barlow (1992, pp. 127–133).

FIGURE 6.11 Bruno de Finetti (1906–1985). Wikipedia L'Enciclopedia libera (Public Domain), http://it.wikipedia.org/wiki/File:Bruno_de_Finetti.jpg

> probable than an impossible event and less probable than a necessary event...[AXIOM 3:] when we judge an event E' more probable than an event E, which is itself judged more probable than an event E'', the event E' can only appear more probable than E'' (transitive property)...[AXIOM 4:] inequalities are preserved in logical sums: if E is incompatible with E_1 and with E_2, then $E_1 \vee E$ will be more or less probable than $E_2 \vee E$, or they will be equally probable, according to whether E_1 is more or less probable than E_2, or they are equally probable. (*ibid.*, p. 100)

A fifth axiom regarding conditional probability would shortly follow, but for the time being de Finetti noted that, by starting from the above purely qualitative axioms, one could arrive at a quantitative measure of probability and then at the theorem of total probability* from which the whole calculus of probabilities could be built.

However, it is also possible to

> ...give a direct, quantitative, numerical definition of the degree of probability attributed by a given individual to a given event, in such a fashion that the whole theory of probability can be deduced immediately from the very natural condition having an obvious meaning. (*ibid.*, p. 101)

The natural condition in question is that under which a subject would be disposed to bet on an event. The idea of assessing one's probability based on one's disposition to bet had been put forward by Ramsey before (Ramsey, 1931)[†], but in 1937 de Finetti was not aware of Ramsey's work. Probability can thus be defined as follows:

> Let us suppose that an individual is obliged to evaluate the rate p at which he would be ready to exchange the possession of an arbitrary sum S (positive or negative) dependent on the occurrence of a given event E, for the possession of the sum pS; we will say by definition that this number p is the measure of the degree of probability attributed by the individual considered to the event E, or, more simply, that p is the probability of E. (de Finetti, 1964, p. 102)

For example, if I am prepared to accept a sure sum of \$50 in lieu of an uncertain sum of \$200, conditional on the occurrence of an event E, then I would say that *my* probability of the event E was 50/200 = .25.

We see that de Finetti's definition is the same as saying that if an individual is willing to bet an amount pS on an event for a return of an amount S if the event occurs, then her probability for the event is p.[‡] We also see that, with the above subjective definition of probability, de Finetti was able to assign probabilities to one-off events, a task not possible with the frequency theory (e.g., see Gorroochurn, 2012a, pp. 150–151).

An important caveat regarding the above definition of probability in terms of betting disposition was that, as de Finetti pointed out, an individual should *not* be able to bet in such a way as to be assured of gaining. This is the condition of *coherence* without which the calculus of probabilities cannot be deduced and is similar to Ramsey's Dutch Book argument.[§]

*De Finetti here did not mean the familiar law of total probability in terms of conditional events, but rather additivity for mutually exclusive and exhaustive events. See further in the text.

[†]See p. 648.

[‡]Thus, in the above example, if I am willing to bet \$50 on an event for a return of \$200, then my probability for the event is .25.

[§]See p. 655.

De Finetti next (1964, p. 103) proceeded to show how his definition of probability led to the "theorem of total probability" and then to finite additivity. Because finite additivity is so fundamental in de Finetti's theory and is one of the major differences with Kolmogorov's axioms, we shall describe his proof. Let $E_1, E_2, \ldots, E_n$ be incompatible (i.e., mutually exclusive) events that form a complete class (i.e., that are exhaustive). Then if one bets $p_1S_1, p_2S_2, \ldots, p_nS_n$ on the respective events $E_1, E_2, \ldots, E_n$ for returns of $S_1, S_2, \ldots, S_n$, then the gain on the hth bet ($h=1, 2, \ldots, n$) is

$$G_h = S_h - \sum_{i=1}^{n} p_i S_i.$$

Regarding $S_1, S_2, \ldots, S_n$ as unknowns, the above gives the following linear system of equations:

$$\begin{aligned} G_1 &= (1-p_1)S_1 - p_2S_2 - \cdots - p_nS_n, \\ G_2 &= -p_1S_1 + (1-p_2)S_2 - \cdots - p_nS_n, \\ &\ldots \\ G_n &= -p_1S_1 - p_2S_2 - \cdots + (1-p_n)S_n. \end{aligned} \tag{6.37}$$

The determinant of the above system is

$$\begin{vmatrix} 1-p_1 & -p_2 & \cdots & -p_n \\ -p_1 & 1-p_2 & & -p_n \\ \vdots & & & \\ -p_1 & -p_2 & & 1-p_n \end{vmatrix} = 1-(p_1+p_2+\cdots+p_n).$$

Now, if the above determinant is not zero, it is possible to obtain arbitrary values of $G_1, G_2, \ldots, G_n$ for given values of $S_1, S_2, \ldots, S_n$. In particular it is possible to make all G's positive, which violates coherence. Therefore, for the coherence condition to be maintained, one needs

$$p_1 + p_2 + \cdots + p_n = 1$$

(*ibid*., p. 104). Thus, the sum of probabilities of n mutually exclusive and exhaustive events must be unity. This is de Finetti's theorem of total probability. As a corollary and generalization to it,

> the probability of the logical sum of n incompatible events is the sum of their probabilities. (*ibidem*)

The above is now called the principle of finite additivity. We see that the latter is a consequence of de Finetti's axioms and the principle of coherence. Finite additivity can be contrasted with countable (or σ–) additivity, that is, additivity across an *infinite* sequence of mutually exclusive events. Countable additivity, as it appears in Kolmogorov's axiomatic system, is not the corollary to some theorem. Rather it has the status of an axiom, being equivalent to the axiom of continuity (Gorroochurn, 2012a, p. 204).

De Finetti next made two statements that would form the hallmarks of his subjective theory of probability. First

> ...a complete class of incompatible events E_1, E_2, ..., E_n being given, all the assignments of probability that attribute to $p_1, p_2, \ldots, p_n$ any values whatever, which are non-negative and have a sum equal to unity, are admissible assignments: each of these evaluations corresponds to a coherent opinion, to an opinion legitimate in itself, and every individual is free to adopt that one of these opinions which he prefers, or, to put it more plainly, that which he *feels*. (de Finetti, 1964, p. 104)

Thus, according to de Finetti's subjective theory, every individual is *completely free* to assign her own probabilities, depending on her opinion, as long as the axioms and coherence are adhered to. This is the reason why de Finetti's theory of probability is often said (as he himself admitted) to be one of extreme subjectivism.

De Finetti's second statement relates to the issue of countable additivity:

> When the events considered are infinite in number, our definition introduces no new difficulty: **P** is a probability function for the infinite class of events E when it is probability function for all finite subclasses of E. This conclusion implies that the theorem of total probability cannot be extended to the case of an infinite or even denumerable number of events.... (*ibid.*, p. 108)

Thus, in de Finetti's theory, additivity is only finite.

De Finetti's next task was to obtain the multiplication theorem for probabilities. But first he had to define an axiom for conditional probabilities. This would be the fifth axiom of his theory. Consider the "tri-event" E' conditioned on E'', denoted as $E' \mid E''$. This tri-event can have three values:

1. True, if E' and E'' are true
2. False, if E'' is true and E' is false
3. Zero, if E'' is false

Then

> [Axiom 5:] If E' and E'' are contained in E, $E'|E$ is more or less probable than (or is equal in probability to) $E''|E$, according to whether E' is more or less probable than (or equal in probability to) E''. (*ibid.*, p. 109)

On the other hand, we note that under the Kolmogorov axiomatic system the conditional probability $\mathbf{P}(E' \mid E'')$ is perfectly legitimate but the conditional event $E' \mid E''$ has no meaning (Pfeiffer, 1990, p. 51; Hailperin, 1996, p. 253). In any case, let $E' \subset E''$. Proceeding as in de Finetti's theorem of total probability, suppose one bets amounts $p'S', p''S'', pS$ on the respective events $E', E'', E' \mid E''$ for returns of S', S'', S, then the gains are:

- If E' occurs, $G_1 = (S' - p'S') + (S'' - p''S'') + (S - pS) = (1 - p')S' + (1 - p'')S'' + (1 - p)S$.
- If E'' occurs but E' does not occur, $G_2 = -p'S' + (1 - p'')S'' - pS$.
- If E'' does not occur, $G_3 = -p'S' - p''S''$.

Treating S', S'', S as unknowns, we obtain a linear system similar to the one before (see Eq. 6.37) with determinant

$$\begin{vmatrix} 1-p' & 1-p'' & 1-p \\ -p' & 1-p'' & -p \\ -p' & -p'' & 0 \end{vmatrix} = p' - pp''.$$

Using the same arguments as before, coherence implies that $p' = p''$. This will still be true if we do away with the assumption $E' \subset E''$ and consider $E' \cdot E''$ instead of E'. De Finetti thus obtained the multiplication theorem for probabilities:

$$\mathbf{P}(E' \cdot E'') = \mathbf{P}(E')\mathbf{P}(E'' \mid E') \tag{6.38}$$

(de Finetti, 1964, p. 109). By writing $\mathbf{P}(E' \cdot E'')$ as $\mathbf{P}(E'' \cdot E')$ and applying the above formula leads to Bayes' theorem

$$\mathbf{P}(E'' \mid E') = \frac{\mathbf{P}(E'')\mathbf{P}(E' \mid E'')}{\mathbf{P}(E')}.$$

In Chapter 3 of the 1937 paper, de Finetti introduced the concept of *exchangeable* events in order to replace the stronger condition of independence. Exchangeable events had at first been called *equivalent* events ("*événements équivalents*")* by de Finetti and had before him been studied by William Johnson under the name "permutation postulate" (Johnson, 1924, p. 183).

De Finetti's initial motivation was to solve the problem of induction (see Section 1.3.3), that is, can we use the frequency of past occurrences of an event to predict a future occurrence?

> Why are we obliged in the majority of problems to evaluate a probability according to the observation of a frequency? This is a question of the relations between the observation of past frequencies and the prediction of future frequencies which we have left hanging, but which presents itself anew under a somewhat modified form when we ask ourselves if a prediction of frequency can be in a certain sense confirmed or refuted by experience. The question we pose ourselves now includes in reality the problem of reasoning by induction. (de Finetti, 1964, p. 118)

As an example, de Finetti considered an irregular coin that was tossed n times such that the first r tosses were heads and the last $s\ (=n-r)$ tosses were tails. Denote the latter event by A. If a head on the ith toss is represented by E_i, then

$$A = E_{i_1} E_{i_2} \ldots E_{i_r} \bar{E}_{j_1} \bar{E}_{j_2} \ldots \bar{E}_{j_s}. \tag{6.39}$$

Moreover, the probability that the $(n+1)$th toss will be a head is given by Eq. (6.38) as

$$\mathbf{P}(E_{n+1} \mid A) = \frac{\mathbf{P}(A \cdot E_{n+1})}{\mathbf{P}(A)}. \tag{6.40}$$

*De Finetti later stated that the change to "exchangeable" was suggested by Maurice Fréchet in 1939.

An objectivist would easily simplify this formula by assuming *independence* of the tosses: $\mathbf{P}(E_{n+1} \mid A)$ would be simply $\mathbf{P}(E_{n+1})$, the probability of heads in a given toss. However, for a subjectivist, the successive accumulation of data provides increasing information about the future, so that the assumption of independence of the tosses is not tenable. Therefore, (6.40) cannot be simplified for the subjectivist, and de Finetti therefore sought "the simplest conditions which define the events," namely, *exchangeability*. The concept was introduced as follows. Let $\omega_r^{(n)}$ be the probability of obtaining a total of r heads and $n-r$ tails. Then $\omega_r^{(n)}$ is the sum of the probabilities of the $\binom{n}{r}$ distinct ways in which the result can be obtained. Regarding each of these distinct ways,

> ...[i]n general, different probabilities will be assigned, depending on the order, whether it is supposed that one toss has an influence on the one which follows it immediately, or whether the exterior circumstances are supposed to vary, etc.; nevertheless it is particularly interesting to study the case where the probability does not depend on the order of the trials. In this case every result having the same frequency r/n on n trials has the same probability, which is $\omega_r^{(n)}/\binom{n}{r}$; if this condition is satisfied, we will say that the events of the class being considered, e.g., the different tosses in the example of tossing coins, are *exchangeable* (in relation to our judgment of probability). (*ibid*., p. 121)

Thus, a sequence of events is deemed to be exchangeable if its probability does not depend on the order in which the events occur.

Therefore, each of the $\binom{n}{r}$ distinct ways in which r heads and $n-r$ tails can be obtained will have the same probability, namely,

$$\frac{\omega_r^{(n)}}{\binom{n}{r}}.$$

The right side of Eq. (6.40) becomes

$$\mathbf{P}(A) = \frac{\omega_r^{(n)}}{\binom{n}{r}}, \quad \mathbf{P}(A \cdot E) = \frac{\omega_{r+1}^{(n+1)}}{\binom{n+1}{r+1}},$$

so that

$$\mathbf{P}(E_{n+1} \mid A) = \frac{r+1}{n+1}\left\{\frac{\omega_{r+1}^{(n+1)}}{\omega_r^{(n)}}\right\} \tag{6.41}$$

(*ibid*., p. 122). De Finetti came back to the above formula later in his continued solution to the problem of induction. But before that, he went on to prove his celebrated representation theorem* as follows. First, let X_i be the indicator variable for the event E_i (= ith toss is a head), that is,

*The de Finetti representation theorem had previously been communicated to the International Congress of Mathematicians held in Bologna in 1928 (de Finetti, 1930b). However, in the 1937 paper, de Finetti also gave the proof of the theorem.

$$X_i = \begin{cases} 1 & \text{if } E_i \text{ occurs} \\ 0 & \text{otherwise} \end{cases}$$

Then the indicator for $\bar{E}_i$ is $1-X_i$ and the indicator the event A in (6.39) is

$$X_{i_1}X_{i_2}\ldots X_{i_r}\left(1-X_{j_1}\right)\left(1-X_{j_2}\right)\cdots\left(1-X_{j_s}\right) = X_{i_1}X_{i_2}\ldots X_{i_r} - \sum_{h=1}^{s} X_{i_1}X_{i_2}\ldots X_{i_r}X_{j_h}$$
$$+\sum_{k,h=1}^{s} X_{i_1}X_{i_2}\ldots X_{i_r}X_{j_h}X_{j_k} - \cdots + (-1)^s X_1X_2\ldots X_n.$$

Taking the expectation of the above and remembering that the expectation of an indicator variable is the probability of the associated event, de Finetti obtained for the probability of the event A,

$$\frac{\omega_r^{(n)}}{\binom{n}{r}} = \omega_r - \binom{s}{1}\omega_{r+1} + \binom{s}{2}\omega_{r+2} - \cdots (-1)^s\,\omega_{r+s} = (-1)^s\,\Delta^s\omega_r \tag{6.42}$$

(*ibid.*, p. 124), where $\omega_k \equiv \omega_k^{(k)}$ and $\Delta a_i \equiv a_{i+1} - a_i$ (so that $\Delta^2 a_i = a_{i+2} - 2a_{i+1} + a_i$, and so on). Now, define

$$Y_h = \frac{X_1 + X_2 + \cdots + X_h}{h},$$
$$\Phi_n(\xi) = \mathbf{P}(Y_n \le \xi),$$
$$\Phi(\xi) = \lim_{n\to\infty}\Phi_n(\xi).$$

In the above, the limiting distribution function $\Phi(\xi)$ exists as a consequence of the law of large numbers. Note that we are now effectively dealing with the *infinite* exchangeable sequence $X_1, X_2, \ldots$ Now, $\Phi(\xi)$ is completely determined once all its moments m_n $(n=1, 2, \ldots)$ are known, where

$$m_n = \int_0^1 \xi^n\, d\Phi(\xi).$$

But since the expectation of an indicator variable is the probability of the associated event,

$$m_n = \omega_n.$$

Therefore, Eq. (6.42) leads to

$$\omega_r^{(n)} = (-1)^s\binom{n}{r}\Delta^s\omega_r = (-1)^s\binom{n}{r}\Delta^s m_r.$$

Since m_r is the rth moment of the distribution $\Phi(\xi)$, de Finetti was able to write the last expression as

$$\omega_r^{(n)} = \binom{n}{r} \int_0^1 \xi^r (1-\xi)^s \, d\Phi(\xi) \tag{6.43}$$

(*ibid.*, p. 128), which is the *de Finetti representation theorem.** It shows that a sequence of independent Bernoulli trials each with objective probability ξ can be reduced to (or represented by) a sequence of exchangeable trials when appropriately weighted by the prior of ξ:

> ...the nebulous and unsatisfactory definition of "independent events with fixed but unknown probability" should be replaced by that of "exchangeable events." (*ibid.*, p. 142)

The representation theorem thus does away with the objective probability ξ in favor of the subjective probability $\omega_r^{(n)}$, which is the only justifiable probability for de Finetti. The theorem also shows that the prior distribution of ξ, namely $\Phi(\xi)$, is a legitimate quantity.

Equipped with the representation theorem, de Finetti was now ready to take up the problem of induction from where he left off in (6.41). He stated the solution to the problem of induction in qualitative terms as follows:

> ...a rich enough experience leads us always to consider as probable future frequencies or distributions close to those which have been observed. (*ibidem*)

In other words, having observed with frequency r/n a past event, it is likely that the future probability of that event will be close to the past frequency. To prove this statement, de Finetti first denoted the limiting distribution of r/n by $\bar{\Phi}(\xi)$. Using the representation theorem in (6.43),

$$d\bar{\Phi}(\xi) = \alpha \xi^r (1-\xi)^s \, d\Phi(\xi),$$

where α is a constant such that $\alpha \int_0^1 \xi^r (1-\xi)^s d\Phi = 1$. The probability of an additional event is then

$$p_r^{(n)} = \int_0^1 \xi d\bar{\Phi} = \int_0^1 \xi \cdot \alpha \xi^r (1-\xi)^s \, d\Phi$$

(*ibid.*, p. 143). One thus sees that $p_r^{(n)}$ is the mean of ξ weighted by $\alpha\xi^r(1-\xi)^s d\Phi$. Since the maximum of $\xi^r(1-\xi)^s$ is $r/(r+s) = r/n$, "the ξ around the maximum...are evidently strengthened more and more."

*Note that, as it stands, the theorem does not hold if the underlying sequence is only *finitely* exchangeable. In the latter case, finite forms of the representation theorem can be obtained (e.g., see Diaconis (1977)). Exchangeability and de Finetti's representation theorem are also discussed in Torretti (1990, pp. 212–218), Zabell (2005, pp. 3–13), Press (2009, pp. 238–242), Schervish (1995, pp. 24–52), Dale (1985), Jackman (2009, pp. 39–46), Singpurwalla (2006, Chapter 3), and Cooke (1991, pp. 110–115).

Using Eq. (6.41), de Finetti provided a more rigorous demonstration of the last statement as follows. Since $\mathbf{P}(E_{n+1} \mid A) = p_r^{(n)}$, we have

$$p_r^{(n)} = \frac{r+1}{n+1}\left\{\frac{\omega_{r+1}^{(n+1)}}{\omega_r^{(n)}}\right\}.$$

Now,

$$\omega_r^{(n)} = \frac{s+1}{n+1}\omega_r^{(n+1)} + \frac{r+1}{n+1}\omega_{r+1}^{(n+1)}.$$

The last but one equation then becomes

$$p_r^{(n)} = \frac{(r+1)\omega_{r+1}^{(n+1)}}{(s+1)\omega_r^{(n+1)} + (r+1)\omega_{r+1}^{(n+1)}} = \frac{r+1}{n+2+(s+1)\left\{\frac{\omega_r^{(n+1)}}{\omega_{r+1}^{(n+1)}} - 1\right\}} \tag{6.44}$$

(*ibid.*, p. 144). Note that by applying the principle of indifference, one has $\omega_r^{(n)} = 1/(n+1)$, so that

$$p_r^{(n)} = \frac{r+1}{n+2},$$

which is Laplace's rule of succession (see Eq. 1.6). In general (i.e., without applying the principle of indifference), de Finetti concluded from Eq. (6.44) that

> In any case, $p_r^{(n)}$ is close to $\frac{r+1}{n+2}$, and hence close to the frequency r/n, if the ratio differs little from unity; it thus suffices to admit this condition in order to justify easily the influence of observation on prediction in the case of exchangeable events. (*ibid.*, p. 145)

De Finetti was thus able to find a solution to the problem of induction. His solution had the advantage that it did *not* rely on the principle of indifference and thus did *not* make use of a uniform prior. Rather it was based on the more reasonable notion of exchangeability.

In this last section, we examine the word "prévision," which was used in the original (French) 1937 paper (de Finetti, 1937). In his English translation of the original article, Kyburg translated "La prévision" in the title as "Foresight," but everywhere else in the paper "prévision" was translated as "prediction." Although de Finetti himself stuck to the latter translation in his 1972 book *Probability, Induction and Statistics* (de Finetti, 1972), in 1974 he drew a contrast between "prediction" and "prevision" in his book *Theory of Probability: A Critical Introductory Treatment* (de Finetti, 1974). Let us now examine the concept of prevision, as de Finetti later explained it, and its relation to probability. This examination is important inasmuch as it gives a deeper insight into de Finetti's views on probability.

In 1974, de Finetti said, '*Prevision, not prediction*' (*ibid.*, Vol. 1, p. 70). He explained:

> To make a *prediction* would mean (using the term in the sense we propose) to venture to try to "guess", among the possible alternatives, the one that will occur. This is an attempt often made, not only by would-be magicians and prophets, but also by experts and such like who are inclined to precast the future in the forge of their fantasies. To make a "prediction", therefore, would not entail leaving the domain of the logic of certainty, but simply including the statements and data which we assume ourselves capable of guessing, along with the ascertained truths and the collected data. (*ibidem*)

On the other hand,

> *Prevision*, in the sense in which we have said we want to use this word, does not involve guessing anything. It does not assert—as prediction does—something that might turn out to be true or false, by transforming (over-optimistically) the uncertainty into a claimed, but worthless, certainty. It acknowledges (as should be obvious) that what is uncertain is uncertain: in so far as statements are concerned, all that can be said beyond what is said by the logic of certainty is illegitimate…[Prevision] consists in considering, after careful reflection, all the possible alternatives, in order to distribute among them, in the way which will appear most appropriate, one's own expectations, one's own sensations of probability. (*ibid.*, Vol. 1, pp. 71–72)

Thus, according to de Finetti, prediction is the act of guessing which one of several alternatives *will occur*. As such, it carries overtones of certainty and is not suited for de Finetti's purposes. On the other hand, prevision is a subjective measure of expectation in the face of the uncertainty, with full acknowledgment of the latter in one's assessment.

Quantitatively, consider a random gain X. *My* specification of a sure gain that is equivalent to X is defined to be *my* price of X, denoted by $\mathbf{P}(X)$.* We see that $\mathbf{P}(X)$ is numerically equal to the mathematical expectation of X. The definition can be generalized to *any* random quantity Y by writing $Y = aX$, where a is a nonzero constant. We now speak of the *prevision* of Y, defined as

$$\mathbf{P}(Y) = a\mathbf{P}(X)$$

(*ibid.*, Vol. 1, p. 75). In the case of an event, note that any event E is associated with a random (indicator) variable I_E where

$$I_E = \begin{cases} 1 & \text{if } E \text{ occurs} \\ 0 & \text{otherwise} \end{cases}.$$

The *probability* of E is then defined as the prevision of I_E, that is, $\mathbf{P}(I_E)$. However, de Finetti denoted both the event and its associated indicator variable by the same letter E, so we shall denote the probability of the event E by $\mathbf{P}(E)$. Therefore,

*For example, let $X = 1$, 2, or 10 units. Suppose I am willing to accept a minimum sure gain of 2 units in lieu of the uncertain gain of 1, 2, or 10 units. Then my price of X is 2 units, that is, $\mathbf{P}(X) = 2$.

> The probability $\mathbf{P}(E)$ that You attribute to an event E is therefore the certain gain p which You judge equivalent to a unit gain conditional on the occurrence of E: in order to express it in a dimensionally correct way, it is preferable to take pS equivalent to S conditional on E, where S is any amount whatsoever, one Lira or one million, \$20 or £75. Since the possible values for a possible event E satisfy $\inf E=0$ and $\sup E=1$, for such an event we have $0 \leq \mathbf{P}(E) \leq 1$, while necessarily $\mathbf{P}(E)=0$ for the impossible event, and $\mathbf{P}(E)=1$ for the certain event. (*ibidem*)

One thus sees that probability is defined in terms of prevision (which is numerically equal to mathematical expectation). This is the reverse of current practice, whereby mathematical expectation is defined in terms of probability. Prevision is thus the primitive (or fundamental quantity) in de Finetti's theoretical framework.

6.3.3 Savage (1954): The Seven Postulates, Qualitative Probability, Quantitative Personal Probability, Savage's Representation Theorem, and Expected Utility

Leonard Jimmie Savage* (Fig. 6.12) was one of the most original and influential thinkers in statistics. Jimmie Savage, as he was most commonly known, was born in Detroit, Michigan, on November 20, 1917. His father was in real estate business, while his mother was a nurse. Throughout his life, Savage suffered from poor eyesight. He received his bachelor's degree in science in 1938 and his Ph.D. degree from the University of Michigan in 1941. His Ph.D. was in the application of vectorial methods in metric geometry. In 1944 he spent a brief period of time with a group that was to have a major influence on him. This was the Statistical Research Group at Columbia University and included the likes of Churchill Eisenhart, Milton Friedman, Abraham Girshick, Harold Hotelling, Frederick Mosteller, Abraham Wald, W. Allen Wallis, and Jacob Wolfowitz. All these figures were to become leading statisticians, and Savage's interactions with them developed his own interest in statistics. In 1947, Savage went to the University of Chicago, and in 1949 he co-founded the Statistics Department at Chicago with W. Allen Wallis. Savage became Chair in 1957 but left in 1960 for the University of Michigan. In 1964, Savage became Eugene Higgins Professor at Yale University. He died in 1971 at the age of 53.

Of all the interactions Savage had with mathematicians, his association with the probabilist Bruno de Finetti was the most fruitful. Savage was much influenced by the subjective viewpoint of de Finetti, and the two met and collaborated on many projects.

Although Savage made many contributions to statistics, his major one is contained in his opus *The Foundations of Statistics* (Savage, 1954). Drawing on the decision-theoretic work of Ramsey (1964), the subjectivist viewpoint of de Finetti (1937), and the utility theory of von Neumann and Morgenstern (1944), Savage built a sophisticated theory of statistics† founded on a subjective (or "personal," as Savage called it) theory of probability. This was done mainly in the first seven chapters of his book.

* Savage's life is also described in Lindley (1980; 2006).

† Savages's theory is also described in Kreps (1988, pp. 127–137), Fishburn (1970, pp. 191–210), Shafer (1986, pp. 193–234), Eells (1982, pp. 71–82), Luce and Raiffa (1957, pp. 302–304), Joyce (1999, Chapter 3), and Parmigiani and Inoue (2009, pp. 81–91).

FIGURE 6.12 Leonard Jimmie Savage (1917–1971). Book cover of "*The Writings of Leonard Jimmie Savage – A Memorial Selection*" *by L.J. Savage (1981)*. Courtesy of Institute of Mathematical Statistics

TABLE 6.9 Consequences in Egg Example Given by Savage (1954, p. 14)

	State	
Act	Good	Rotten
Break into bowl	Six-egg omelet	No omelet and five good eggs destroyed
Break into saucer	Six-egg omelet and a saucer to wash	Five-egg omelet and a saucer to wash
Throw away	Five-egg omelet and a good egg destroyed	Five-egg omelet

We now describe the terminology used by Savage before moving on to his theory:

- The "world": the object about which the person is concerned
- A "state" (of the world): a description of the world, leaving no relevant aspect undescribed ("state" corresponds to "proposition" in Section 6.3.1)
- The "true" state (of the world): the state that does in fact obtain, that is, the true description of the world
- An "event": a set of states
- An "act": a decision
- A "consequence": anything that can happen to a person as a result of her act ("consequence" corresponds to "outcome" in Section 6.3.1)

To illustrate these definitions, Savage considered the following example:

> ...Your wife has just broken five good eggs into it bowl when you come in and volunteer to finish making the omelet. A sixth egg, which for some reason must either be used for the omelet or wasted altogether, lies unbroken beside the bowl. You must decide, what to do with this unbroken egg. Perhaps it is not too great an oversimplification to say that you must decide among three acts only, namely, to break it into the bowl containing the other five, to break it into a saucer for inspection, or to throw it away without inspection. Depending on the state of the egg, each of these three acts will have some consequence of concern to you.... (Savage, 1954, pp. 13–14)

He then gave some possible consequences, as shown in Table 6.9.

In general, once an act **f** is made then, depending on the state s, this results in a consequence $f(s)$, for example, $f(\text{good})=$six-egg omelet.

Next, Savage gave an axiomatic system upon which his theory of probability would be built. First he needed the concept of a *simple ordering*. In general, the relation $\leq\cdot$ is a simple ordering among the set of elements $x, y, z, \ldots$ if and only if

- for any x, y, either $x \leq\cdot y$ or $y \leq\cdot x$; and
- for any x, y, z, if $x \leq\cdot y$ and $y \leq\cdot z$, then $x \leq\cdot z$.

Using the symbol $\leq$ to denote "is not preferred to," Savage enunciated his first postulate (*ibid.*, p. 18) as follows:

POSTULATE 1: The relation $\leq$ is a simple ordering among acts.

For his second postulate, Savage needed the *sure-thing* principle:

> ...If the person would not prefer **f** to **g**, either knowing that the event B obtained, or knowing that the event $\sim B$ obtained, then he does not prefer **f** to **g**. Moreover (provided he does not regard B as virtually impossible) if he would definitely prefer **g** to **f**, knowing that B obtained, and, if he would not prefer **f** to **g**, knowing that B did not obtain, then he definitely prefers **g** to **f**. (*ibid.*, p. 21)

We can use the sure-thing principle to explain *conditional preference*. Consider the statement "**f** is not preferred to **g** (i.e., $\mathbf{f} \leq \mathbf{g}$) given B." This means the following. Suppose actions **f** and **g** are modified to **f**′ and **g**′, respectively, such that they agree outside B (i.e., $\mathbf{f}'(s) = \mathbf{g}'(s)$ for all $s \in \sim B$). However, given B, **f**′ is not preferred to **g**′.

The only problem in the above explanation is that it may not hold for all permitted modifications. In order for this not to happen, Savage gave the following postulate (*ibid.*, p. 23):

POSTULATE 2: If **f**, **g**, and **f**′, **g**′ are such that:

1. In $\sim B$, **f** agrees with **g**, and **f**′ agrees with **g**′.
2. In B, **f** agrees with **f**′, and **g** agrees with **g**′.
3. If $\mathbf{f} \leq \mathbf{g}$,
   then $\mathbf{f}' \leq \mathbf{g}'$.

The above postulate is called the *independence postulate*: if two acts **f** and **f**′ agree in $\sim B$, then the preference between them should depend on how they differ in B, *not* on how they agree in $\sim B$ (Fig. 6.13a and b). Of all Savage's postulates, the independence postulate has probably been the most controversial. Allais and Ellsberg have each given examples where the postulate is violated, but Morgenstern has insisted that the postulate is reasonable (Shafer, 1986).

Continuing with Savage's theory, the next concept is that of a *null* (or virtually impossible) event: B is null if and only if, for every **f** and **g**, we have $\mathbf{f} \doteq \mathbf{g}$ (where $\doteq$ means "indifferent to"). Using this concept, we have the third postulate (Savage, 1954, p. 26):

POSTULATE 3: If $\mathbf{f} \doteq \mathbf{g}$, $\mathbf{f}' \doteq \mathbf{g}'$ and B is not null, then $\mathbf{f} \leq \mathbf{f}'$ given B, if and only if $\mathbf{g} \leq \mathbf{g}'$.

The essence of the third postulate is that conditional preferences on acts and marginal preferences on acts are identical.

Savage now introduced the concept of a *prize*. To offer a prize given the event A means to make available to the person an act $\mathbf{f}_A$ such that

$$f_A(s) = \begin{cases} f & \text{for } s \in A \\ f' & \text{for } s \in \sim A \end{cases}$$

(a)

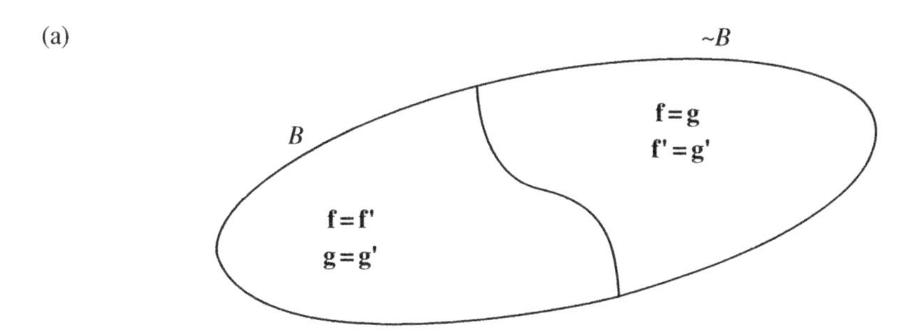

(b)

	B		~B
	Event α	Event β	Event $\sim\alpha \cap \sim\beta$
f	a	b	x
g	c	d	x
f'	a	b	y
g'	c	d	y

FIGURE 6.13 (a) Savage's second postulate. (b) Illustration of the second postulate: the acts **f** and **g** are modified to **f′** and **g′**, respectively; since both pairs agree on ~B, preference between **f** and **g** (and **f′** and **g′**) should be based on B, not on the utility values x and y

where f is preferred over f' (i.e., $f' < f$). However, it is important that "on which of two events the person will choose to stake a given prize does not depend on the prize itself." This is guaranteed by the fourth postulate (*ibid.*, p. 31):

Postulate 4: If f, f', g, g'; A, B; $\mathbf{f}_A$, $\mathbf{f}_B$, $\mathbf{g}_A$, $\mathbf{g}_B$ are such that:

1. $f' < f, g' < g$;

2a. $f_A(s)=\begin{cases} f & \text{for } s \in A \\ f' & \text{for } s \in \sim A \end{cases};\quad g_A(s)=\begin{cases} g & \text{for } s \in A \\ g' & \text{for } s \in \sim A \end{cases};$

2b. $f_B(s)=\begin{cases} f & \text{for } s \in B \\ f' & \text{for } s \in \sim B \end{cases};\quad g_B(s)=\begin{cases} g & \text{for } s \in B \\ g' & \text{for } s \in \sim B \end{cases};$

3. $\mathbf{f}_A \le \mathbf{f}_B$;

then $\mathbf{g}_A \le \mathbf{g}_B$.

In the above, the act $\mathbf{f}_A$ results in a "good" outcome f if A occurs and a "bad" outcome f' if ~A occurs; the act $\mathbf{f}_B$ results in the same outcomes depending on whether B occurs or not; and it is also given that the person does not prefer $\mathbf{f}_A$ to $\mathbf{f}_B$. In Savage's terminology, this means that *the person believes that event A is not more probable than event B*. Postulate 4 therefore essentially means that the person's belief is maintained whether the good outcome is *f* or *g*, that is, it does not depend on the *magnitude* of the outcome.

Savage's next axiom had to do with the postulate there was *always* a worthwhile prize:

POSTULATE 5: There is at least one pair of consequences f, f' such that $f' < f$.

With the first five postulates, Savage was now ready to pass from *preferences of events* to *personal probabilities*. He now defined:

A relation $\leq \cdot$ between events is a **qualitative probability** if and only if, for all events *B*, *C*, *D*,

1. $\leq \cdot$ is a simple ordering,
2. $B \leq \cdot C$, if and only if $B \cup D \leq \cdot C \cup D$, provided $B \cap D = C \cap D = 0$,
3. $0 \leq \cdot B, 0 < \cdot S$ [*S* is the universal event].

(*ibid.*, p. 32)

The five postulates imply that *the relation* $\leq$ *applied to events is a qualitative probability*. In fact, $\leq$ can be taken to mean "not more probable than."

Having defined qualitative probability, it now remained for Savage to link the latter to *quantitative personal probability* (or probability measure). First

A **probability measure** on a set *S* is a function *P*(*B*) attaching to each $B \subset C$ a real number such that:

1. $P(B) \geq 0$ is a simple ordering,
2. If $B \cap C = 0$, $P(B \cup C) = P(B) + P(C)$,
3. $P(S) = 1$ [*S* is the universal event].

(*ibid.*, p. 33)

The above are the usual (i.e., Kolmogorov's) axioms of probability, except for 2. Here, like de Finetti, Savage assumed finite (but not countable) additivity.

But to be able to make a connection between qualitative probability $\leq \cdot$ and quantitative personal probability *P*, Savage needed an additional axiom that the universal event *S* could be partitioned into an arbitrarily large number of equivalent sets. He noted that this was similar to an assumption previously made by de Finetti (1937). Savage's sixth axiom therefore was (Savage, 1954, p. 39):

POSTULATE 6: If $\mathbf{g} < \mathbf{h}$ and *f* is any consequence, then there exists a partition of *S* such that if **g** or **h** is so modified on anyone element of the partition as to take the value *f* at every *s* there, other values being undisturbed, then the modified **g** remains less than **h**, or **g** remains less than the modified **h**, as the case may require.

It is now possible to make the connection between qualitative probability and quantitative probability: suppose *S* has an associated probability measure *P* and a qualitative probability $\leq \cdot$; then for any events *B* and *C*,

- If $B \leq \cdot C \Leftrightarrow P(B) \leq P(C)$, then P **(strictly) agree**s with $\leq \cdot$.
- If $B \leq \cdot C \Rightarrow P(B) \leq P(C)$, then P **almost agrees** with $\leq \cdot$.

With these six postulates, Savage was thus able to move from preferences (applied to events) to qualitative probabilities and then to probability measures (or quantitative personal probabilities). Thus, once postulate 6 is accepted, "there will scarcely again be any need to refer directly to qualitative probability."

Savage later made two statements that showed the proximity of his own viewpoint on probabilities with that of de Finetti. Concerning quantitative personal probabilities, he first said:

> the concept of personal probability … is, except possibly for slight modifications, *the only probability concept essential to science** and other activities that call upon probability.

(*ibid*., p. 56)

The above statement shows that Savage's absolute insistence of the personalistic interpretation of probability is not different from de Finetti's radicalism concerning the same interpretation.

Savage's second statement concerned the interpretation of quantitative personal probabilities:

> the personalistic view…insists that probability is concerned with consistent action in the face of uncertainty. (*ibidem*)

In the above, Savage introduced the concept of acts. Consistency in acts is similar to de Finetti's (and Ramsey's) concepts of coherence.

To fully develop his theory, Savage next concentrated on acts. Savage wished to achieve a "far-reaching arithmetization of comparison among acts," and for this he appealed to the concept of utility (U). Essentially, the idea was to attach to each consequence f a number $U(f)$ such that

$$\mathbf{f} \leq \mathbf{g} \quad \Leftrightarrow \quad \mathrm{E}U(\mathbf{f}) \leq \mathrm{E}U(\mathbf{g}). \tag{6.45}$$

provided $U(\mathbf{f})$ and $U(\mathbf{g})$ are bounded (i.e., acts are limited with probability one to a finite number of consequences). The above can be taken as the definition of utility, which Savage showed to exist by using the same proof as in the appendix of von Neumann and Morgenstern's book (von Neumann and Morgenstern, 1944).

Let us explore Eq. (6.45) further. Let an act that has a finite number of consequences (with probability one) be called a *gamble*. Consider now two gambles $\boldsymbol{f}$ and $\boldsymbol{g}$ for which there exist respective partitions B_i and C_i of the event space such that

* Italics are ours.

$$
\begin{aligned}
P(B_i) &= \rho_i \quad \text{where} \quad \sum \rho_i = 1, \\
P(C_i) &= \sigma_i \quad \text{where} \quad \sum \sigma_i = 1, \\
f(s) &= f_i \quad \text{for} \quad s \in B_i, \\
g(s) &= g_j \quad \text{for} \quad s \in C_j.
\end{aligned}
$$

Then the gambles $\boldsymbol{f}$ and $\boldsymbol{g}$ can be, respectively, written as $\boldsymbol{f} = \sum \rho_i f_i$ and $\boldsymbol{g} = \sum \sigma_i g_i$. Equation (6.45) can therefore be written as

$$
\boldsymbol{f} \leq \boldsymbol{g} \quad \Leftrightarrow \quad \sum \rho_i U(f_i) \leq \sum \sigma_i U(g_i) \tag{6.46}
$$

(Savage, 1954, p. 73). The above is a fundamental result in Savage's theory and is called the *Savage representation theorem*. It shows how, by starting with a simple set of postulates, one act is preferred to the other if and only if the expected utility of the first is greater than that of the second.*

There was one last postulate Savage needed to complete his theory. His aim was to be able to extend his theory to acts with an *infinite* number of consequences so that Eq. (6.46) would still be valid. For this, he needed:

POSTULATE 7: If $\mathbf{f} \leq (\geq)\ g(s)$ given B for every $s \in B$, then $\mathbf{f} \leq (\geq)\ \mathbf{g}$ given B

(*ibid*., p. 77). The above postulate is the *dominance* postulate: if a person prefers every outcome of one act to the corresponding outcome of another act, then the person prefers the first act to the second.

This completes Savage's theory. We note that whereas Ramsey proceeded from utilities to probabilities, Savage's development was in the opposite direction. Although improvements on Savage's work were later made, notably by Richard Jeffrey (1983), his contribution remains nothing short of seminal. In an obituary, Dennis Lindley called Savage quite deservedly as "the Euclid of statistics…[for his] formal description of inference and decision-making based on the standard mathematical method of axioms and theorems deduced from them" (Lindley, 1980, p. 8).

6.3.4 A Breakthrough in "Bayesian" Methods: Robbins' Empirical Bayes (1956)

Whereas the contributions of Ramsey, de Finetti, and Savage were mainly concerned with the foundational aspects of modern Bayesian theory, an application-oriented breakthrough was made by Herbert Robbins in 1956, in the paper "An empirical Bayes approach to statistics" (Robbins, 1956). However, a deeper analysis of Robbins' approach shows that it also touches the core of the Bayesian philosophy.

Herbert Ellis Robbins† (Fig. 6.14) was born on January 12, 1915, in New Castle, Pennsylvania. He joined Harvard College at the age of 16 and became interested in mathematics under the influence of Marston Morse. Robbins received the A.B. summa cum laude

* According to von Neumann–Morgenstern theory, the utility function is *unique* up to a linear transformation.

† Robbins' life is also described in Lai and Siegmund (1986). See also Robbins' interview by Page (1984).

FIGURE 6.14 Herbert Ellis Robbins (1915–2001). Wikipedia, http://en.wikipedia.org/wiki/Herbert_Robbins

in 1935 and his Ph.D. 3 years later, both from Harvard. His doctoral thesis was in the field of combinatorial topology and was written under the supervision of Hassler Whitney. After graduating, Robbins worked for a year at the Institute for Advanced Study at Princeton as assistant to Marston Morse. The next 3 years were then spent as instructor in mathematics at New York University. It was around that time, in 1939, that Robbins was approached by Richard Courant to collaborate on the book *What Is Mathematics?* (Courant and Robbins, 1941), which became a classic*. In 1941, Robbins enlisted in the Navy and, in his own words:

> ...It was in the Navy, in a rather strange way, that my future career in statistics originated. I was reading in a room, close to two naval officers who were discussing the problem of bombing accuracy. In no way could I keep from overhearing their conversation: "We're dropping lots of bombs on an airstrip in order to knock it out, but the bomb impacts overlap in a random manner, and it doesn't do any good to obliterate the same area seventeen times. Once is enough." They were trying to decide how many bombs were necessary to knock out maybe 90% of an area, taking into account the randomness of impact patterns. The two officers suspected that some research groups working on the problem were probably dropping poker chips on the floor in order to trace them out and measure the total area they covered. Anyway, I finally stopped trying to read and asked myself what really does happen when you do that? Having scribbled something on a piece of paper, I walked over to the officers and offered them a suggestion for attacking the problem. Since I wasn't engaged in war research, they were not empowered to discuss it with me. So I wrote up a short note and sent it off to one of the two officers.... (Page, 1984, pp. 8–10)

In any case, Robbins' solution to the problem which the officers were discussing was later published (Robbins, 1944) and was his first paper in statistics (more precisely, in geometric probability).

In 1946, Robbins was offered the position of Associate Professor by Harold Hotelling at the University of North Carolina at Chapel Hill. He accepted the offer and, in the next 6 years, he was able to make several important statistical contributions. From 1953 onward, Robbins joined Columbia University as Professor and Chair of the Department of Mathematical Statistics. His major achievements were the creation of the empirical Bayes methodology in 1955, the theory of power-one sets, and the development of sequential methods of estimation. Robbins was a member of the National Academy of Sciences and the American Academy of Arts and Sciences. He passed away in New Jersey on February 12, 2001, at the age of 86.

In his 1956 paper† (Robbins, 1956), Robbins considered a discrete random variable with probability mass function

$$p(x \mid \lambda) = \Pr\{X = x \mid \Lambda = \lambda\}.$$

In the above, the parameter Λ is also a random variable and has distribution function

$$G(\lambda) = \Pr\{\Lambda \le \lambda\}.$$

*But that collaboration turned out to be controversial, since Robbins' name did not initially appear in the title page of the book, although Robbins managed to change that before the book came out. For more details, see Chapter 22 of Reid's book *Courant* (Reid, 1996)

†Robbins' paper is also described in Good (1992). See also Kiefer and Moore (1976).

The unconditional (or marginal) distribution of X is then

$$p_G(x) = \Pr\{X = x\} = \int p(x \mid \lambda)\, dG(\lambda). \tag{6.47}$$

Now the Bayes estimator of Λ under quadratic loss is simply the mean of the posterior distribution and is given by

$$\phi_G(x) = \frac{\int \lambda p(x \mid \lambda)\, dG(\lambda)}{\int p(x \mid \lambda)\, dG(\lambda)} \tag{6.48}$$

(*ibid.*, p. 158). As we are reminded by Robbins, the evaluation of $\phi_G(x)$ in the above takes us back to the usual thorny issue of how to obtain a suitable prior G for Λ.

Robbins next considered the hypothetical situation where "the problem of estimating Λ from an observed value X is going to occur with a fixed and known $p(x|\lambda)$ and a fixed but unknown $G(\lambda)$." Suppose the sequence so generated is denoted by

$$(\Lambda_1, X_1), (\Lambda_2, X_2), \ldots, (\Lambda_n, X_n).$$

If we wanted to estimate the unknown Λ_n from an observed X_n and *if the previous values* $\Lambda_1, \ldots, \Lambda_{n-1}$ *were known*, then we could obtain an empirical distribution function for Λ by using

$$G_{n-1}(\lambda) = \frac{\text{no. of terms } \Lambda_1, \ldots, \Lambda_{n-1} \text{ which are} \leq \lambda}{n-1}.$$

The latter could then be plugged in Eq. (6.48) to obtain an estimate of Λ_n. Robbins admitted that it was unusual for $\Lambda_1, \ldots, \Lambda_{n-1}$ to be known, but that

> ...*However, in many cases the previous values* $X_1, \ldots, X_{n-1}$ *will be available to the experimenter at the moment when* Λ_n *is to be estimated,** and the question then arises whether it is possible to infer from the set of values $X_1, \ldots, X_{n-1}$ the approximate form of the unknown G, or at least, in the present case of quadratic estimation, to approximate the value of the functional of G [in Eq. (6.48)]. (*ibid.*, pp. 158–159)

As a first step toward the estimation of Λ_n, Robbins then proposed to estimate the unconditional distribution of X in (6.47) by the empirical distribution function given by

$$p_n(x) = \frac{\text{no. of terms } X_1, \ldots, X_n \text{ which are equal to } x}{n}$$

(*ibid.*, p. 159). Moreover, he noted that $p_n(x) \xrightarrow{P} p_G(x)$ irrespective of the prior G. Then, depending on the form of $p(x \mid \lambda)$ and on the class to which G belonged, it was possible in several cases to obtain an approximation to G or at least to Λ_n in (6.48).

As a concrete example, Robbins considered the Poisson distribution:

$$p(x \mid \lambda) = \frac{e^{-\lambda}\lambda^x}{x!} \quad x = 0, 1, \ldots; \lambda > 0.$$

* Italics are Robbins'.

The unconditional distribution of X in (6.47) is then

$$p_G(x) = \frac{\int_0^\infty e^{-\lambda}\lambda^x dG(\lambda)}{x!} \tag{6.49}$$

and the posterior mean in (6.48) is

$$\phi_G(x) = \frac{\int_0^\infty e^{-\lambda}\lambda^{x+1} dG(\lambda)}{\int_0^\infty e^{-\lambda}\lambda^x dG(\lambda)}.$$

The last two equations give the "fundamental relation"

$$\phi_G(x) = (x+1)\frac{p_G(x+1)}{p_G(x)}.$$

In view of the fact that $p_G(x)$ can be estimated by $p_n(x)$, Robbins defined

$$\phi_n(x) = (x+1)\frac{p_n(x+1)}{p_n(x)} = (x+1)\times\frac{\text{no. of terms } X_1,\ldots,X_n \text{ which are equal to } x+1}{\text{no. of terms } X_1,\ldots,X_n \text{ which are equal to } x}$$

(*ibidem*). Then $\phi_n(x) \xrightarrow{P} \phi_G(x)$ and Λ_n can be estimated by $\phi_{n-1}(X_n)$ in the above, that is,

$$\hat{\Lambda}_n = (X_n+1)\times\frac{\text{no. of terms } X_1,\ldots,X_{n-1} \text{ which are equal to the value of } X_n+1}{\text{no. of terms } X_1,\ldots,X_{n-1} \text{ which are equal to the value of } X_n}$$

This is Robbins' empirical Bayes method: the parameter Λ_n (which has a prior distribution G) is estimated directly from previous data,* and its estimator $\hat{\Lambda}_n$ is called an empirical Bayes estimator.†

As another example, Robbins also considered the case of a binomial distribution:

$$p_r(x\,|\,\lambda) = \binom{r}{x}\lambda^x(1-\lambda)^{r-x}, \quad x = 0,1\ldots,r; 0 \le \lambda \le 1.$$

The unconditional distribution of X is then

$$p_{G,r}(x) = \binom{r}{x}\int_0^1 \lambda^x(1-\lambda)^{r-x}\, dG(\lambda)$$

*Previously, von Mises had considered a similar problem to Robbins', but his approach was different (von Mises, 1942).

†We serendipitously used such an estimator earlier: the James–Stein estimator in (5.69) turns out to be an empirical Bayes estimator.

and the posterior mean is

$$\phi_{G,r}(x)=\frac{\int_0^1 \lambda^{x+1}(1-\lambda)^{r-x}\,dG(\lambda)}{\int_0^1 \lambda^{x}(1-\lambda)^{r-x}\,dG(\lambda)}.$$

The last two equations give the "fundamental relation"

$$\phi_{G,r}(x)=\frac{x+1}{r+1}\frac{p_{G,r+1}(x+1)}{p_{G,r}(x)};\quad x=0,1,\ldots,r$$

(*ibid*., p. 161). Robbins next defined

$$p_{n,r}(x)=\frac{\text{no. of terms } X_1,\ldots,X_n \text{ which are equal to } x}{n}$$

where X_i is the number of successes in the previous ith binomial experiment. Now, consider the sequence $X'_i,\ldots,X'_n$, where X'_i is the number of successes in the first $r-1$ (out of r) trials in the previous ith binomial experiment. Let

$$p_{n,r-1}(x)=\frac{\text{no. of terms } X'_1,\ldots,X'_n \text{ which are equal to } x}{n}$$

and

$$\phi_{n,r}(x)=\frac{x+1}{r}\frac{p_{n,r}(x+1)}{p_{n,r-1}(x)}.$$

Then the empirical Bayes' estimator of Λ_n is $\phi_{n,r}(X'_n)$.

Note that Robbins' empirical Bayes is *nonparametric* in nature because no attempt is made to model the prior G itself. In the *parametric* version of the empirical Bayes method, as later put forward by Efron and Morris (1971; 1972a; 1972b; 1973a; 1973b), a particular distribution of G is assumed, and the parameters of the latter distribution are then estimated from previous data.* Thus, in the Poisson example, one possibility in a parametric empirical Bayes method would be to assume that G is a gamma distribution, that is,

$$\Lambda \sim \text{Gamma}(\alpha,\beta),\quad \alpha>0,\beta>0,$$

where α,β are *hyperparameters* of the model. The unconditional distribution of X in (6.47) then becomes

$$p_{\alpha,\beta}(x)=\frac{\Gamma(\alpha+x)}{x!\Gamma(\alpha)}\frac{\beta^x}{(1+\beta)^{\alpha+x}}.$$

*On the other hand, in a purely Bayesian approach, these parameters would be further assigned a prior distribution.

Applying the method of moments, we obtain

$$\hat{\alpha} = \frac{\bar{X}^2}{\overline{X^2} - \bar{X}(1+\bar{X})}, \quad \hat{\beta} = \frac{\overline{X^2}}{\bar{X}} - (1+\bar{X}),$$

where $\bar{X} = \left(\sum_{i=1}^{n-1} X_i\right)\Big/(n-1)$ and $\overline{X^2} = \left(\sum_{i=1}^{n-1} X_i^2\right)\Big/(n-1)$.

Now using the prior above for Λ and the Poisson likelihood function, the posterior distribution of Λ is

$$\Lambda \mid x \sim \text{Gamma}\left(\alpha + x, \frac{\beta}{1+\beta}\right).$$

From the latter, the posterior mean is $(\alpha + x)\beta / (1+\beta)$ and hence the parametric empirical Bayes estimator of Λ_n is

$$\hat{\Lambda}_n^* = \frac{(X_n + \hat{\alpha})\hat{\beta}}{1+\hat{\beta}},$$

with the estimators $\hat{\alpha}$ and $\hat{\beta}$ calculated as above.

The reader will have observed that, in the implementation of both the nonparametric and parametric empirical Bayes method, we have made use of the data twice, once through the likelihood and the second time in the estimation of the parameter Λ_n (or hyperparameters α and β for the parametric case). But this is contrary to a genuine Bayesian analysis, and it is generally accepted that empirical Bayes methods are not Bayesian per se. Thus, denoting the prior distribution of Λ by $f(\theta \mid \phi)$, O'Hagan has said:

> ...The use of apparently Bayesian methods to estimate θ disguises the non-Bayesian treatment of the fundamental parameters ϕ. Empirical Bayes is not Bayesian because it does not admit a distribution for ϕ. (O'Hagan, 1994, p. 132)

Notwithstanding O'Hagan's admonishment, the empirical Bayes method has been involved in countless applications and is perhaps the closest many non-Bayesians would ever come to accepting a method carrying the name of Bayes. We therefore agree with Neyman's conclusion that indeed a "breakthrough [was] accomplished in Robbins' paper of 1955" (Neyman, 1962).

REFERENCES

A.E.T., 1930 Personal notes. *Journal of the American Statistical Association* **25**(171): 356–360.

Abbe E., 1863 *Über die Gesetzmässigkeit in der Vertheilung der Fehler bei Beobachtungsreihen.* Dissertation, Jena (Also in *Gesammelte Abhandlungen*, Vol. **II**, pp. 55–81 (1906). Jena: Gustave Fisher Ferlag).

Adams W.J., 2009 *The Life and Times of the Central Limit Theorem*, 2nd edition. American Mathematical Society, Providence, RI.

Adrain R., 1808 Research concerning the probabilities of the error which happen in making observations. *The Analyst or Mathematical Museum* **1**(4): 93–109.

Agarwal R.P., Sen S.K., 2014 *Creators of Mathematical and Computational Sciences.* Springer, New York.

Agwu N., Smith L., Barry A., 2003 Dr. David Harold Blackwell, African American Pioneer. *Mathematics Magazine* **76**(1): 3–14.

Aigner M., Ziegler G.M., 2004 *Proofs from THE BOOK*, 3rd edition. Springer-Verlag, Berlin.

Airy G.B., 1861 *On the Algebraical and Numerical Theory of Errors of Observations and the Combination of Observations.* Macmillan, London.

Aitken A.C., 1931 Some applications of generating functions to normal frequency. *The Quarterly Journal of Mathematics* **2**(1): 130–135.

Aitken A.C., Silverstone H., 1942 On the Estimation of Statistical Parameters. *Proceedings of the Royal Society of Edinburgh Section A* **61**(2): 186–194.

Aitken A.C., Turnbull H.W., 1932 *An Introduction to the Theory of Canonical Matrices.* Blackie & Son, London.

Aldrich J., 1997 R.A. Fisher and the making of maximum likelihood 1912–1922. *Statistical Science* **12**(3): 162–176.

Aldrich J., 2000 Fisher's "Inverse Probability" of 1930. *International Statistical Review* **68**(2): 155–172.

Aldrich J., 2005 The statistical education of Harold Jeffreys. *International Statistical Review* **73**(3): 289–307.

Alexander P.: Pearson, Karl (1857–1936); in Borchert D.M., (ed): *Encyclopedia of Philosophy* (Vol. **7**). Thomson/Gale, Detroit, 2006, pp. 159–161.

Classic Topics on the History of Modern Mathematical Statistics: From Laplace to More Recent Times, First Edition. Prakash Gorroochurn.

Allamand J.N.S., 1774 *Œuvres Philosophiques et Mathématiques de Mr. G.J. 's Gravesande* (Vol. **2**). MM Rey.

Andoyer H., 1922 *L'Oeuvre Scientifique de Laplace*. Payot, Paris.

Arbuthnot J., 1692 *Of the Laws of Chance, Or, A Method of Calculation of the Hazards of Game*. Motte, London.

Arbuthnot J., 1710 An Argument for Divine Providence, taken from the constant Regularity observ'd in the Births of Both Sexes. *Philosophical Transactions of the Royal Society of London* **27**: 186–190 (Reprinted in Kendall and Plackett (1977), pp. 30–34).

Aspin A.A., 1948 An examination and further development of a formula arising in the problem of comparing two mean values. *Biometrika* **35**: 88–96.

Aspin A.A., Welch B.L., 1949 Tables for use in comparisons whose accuracy involves two variances, separately estimated. *Biometrika* **36**: 290–296.

Baird D., 1983 The Fisher/Pearson Chi-squared controversy: a turning point for inductive inference. *British Journal for the Philosophy of Science* **34**(2): 105–118.

Barbacki S., Fisher R.A., 1936 A test of the supposed precision of systematic arrangements. *Annals of Eugenics* **7**(2): 189–193.

Barlow R.E.: Introduction to de Finetti (1937) Foresight: its logical laws, its subjective sources; in Kotz S., Johnson N.L., (eds): *Breakthroughs in Statistics* (Vol. **I**). Springer, New York, 1992, pp. 127–133.

Barnard G.A., 1945 A new test for 2 x 2 tables. *Nature* **156**: 783–784.

Barnard G.A., 1949 Statistical inference. *Journal of the Royal Statistical Society Series B (Methodological)* **11**(2): 115–149.

Barnard G.A., 1984 Comparing the means of two independent samples. *Applied Statistics* **33**:266–271.

Barnard G.A., 1987 R.A. Fisher: a true Bayesian? *International Statistical Review* **55**: 183–189.

Barnard G.A.: Introduction to Pearson (1900) On the criterion that a given system of deviations from the probable in the case of a correlated system of variables is such that it can be reasonably supposed to have arisen from random sampling; in Kotz S., Johnson N.L., (eds): *Breakthroughs in Statistics* (Vol. **II**). Springer, New York, 1992, pp. 1–10.

Barnard G.A., 1995 Pivotal models and the fiducial argument. *International Statistical Review* **63**: 309–323.

Barnard G.A., Jenkins G.M., Winsten C.B., 1962 Likelihood inference and time series (with discussion). *Journal of the Royal Statistical Society Series A (General)* **125**: 321–372.

Barnard G.A., Sprott D.A.: A note on Basu's examples of anomalous ancillary statistics; in Godambe V.P., Sprott D.A., (eds): *Foundations of Statistical Inference*. Holt, Rinehart and Winston, Toronto, 1971, pp. 163–176.

Barnett V., 1999 *Comparative Statistical Inference*, 3rd edition. John Wiley & Sons, New York.

Bartlett M.S., 1933 Probability and chance in the theory of statistics. *Proceedings of the Royal Society of London Series A* **141**(845): 518–534.

Bartlett M.S., 1936 The information available in small samples. *Proceedings of the Cambridge Philosophical Society* **32**(4): 560–566.

Bartlett M.S., 1939 Complete simultaneous fiducial distributions. *The Annals of Mathematical Statistics* **10**(2): 129–138.

Bartlett M.S., 1955 *An Introduction to Stochastic Processes*. Cambridge University Press, Cambridge.

Bartlett M.S., 1981 Egon Sharpe Pearson, 1895–1980. *Biometrika* **68**(1): 1–12.

Bartlett M.S.: Chance and change; in Gani J., (ed): *The Making of Statisticians*. Springer, New York, 1982, pp. 41–60.

Basu D., 1955 On statistics independent of a complete sufficient statistic. *Sankhya* **15**(4): 377–380.

Basu D., 1964 Recovery of ancillary information. *Sankhya* **26**: 3–16.

Bateson W., 1901 Heredity, differentiation, and other conceptions of biology: a consideration of Professor Karl Pearson's paper "On the Principle of Homotyposis". *Proceedings of the Royal Society of London* **69**: 193–205.

Bayes T., 1764 An essay towards solving a problem in the doctrine of chances. *Philosophical Transactions of the Royal Society of London* **53**: 370–418 (Reprinted in Pearson and Kendall (1970), pp. 134–53).

Behrens W.V., 1929 Ein beitrag zur fehlerberechnung bei wenigen beobachtungen. *Landwirtschaftliche Jahrbücher* **68**: 807–837.

Bellhouse D.R., 2002 On some recently discovered manuscripts of Thomas Bayes. *Historia Mathematica* **29**: 383–394.

Bennett J.H., 1990 *Statistical Inference and Analysis: Selected Correspondence of R.A. Fisher*. Oxford University Press, Oxford.

Benzecri J.P., 1982 *Histoire et Préhistoire de l'Analyse des Données*. Dunod, Paris.

Bera A.K., 2003 The ET interview: Professor C.R. Rao. *Econometric Theory* **19**: 331–400.

Berger J.O., Wolpert R.L., 1988 *The Likelihood Principle*, 2nd edition. Institute of Mathematical Statistics, Hayward, CA.

Bernardo J.M., Smith A.F.M., 2000 *Bayesian Theory*. John Wiley & Sons, New York.

Bernoulli D., 1735 Quelle est la cause physique de l'inclinaison des plans des orbites des planètes par rapport au plan de l'équateur de la révolution du soleil autour de son axe? Et d'où vient que les inclinaisons de ces orbites sont différentes en elles? *Pièces qui ont remporté le prix de l'Académie Royale des Sciencese en 1734* **3**: 95–122 (Original Latin version on pp. 125–144).

Bernoulli D., 1753 Réflexions et éclaircissemens sur les nouvelles vibrations des cordes. *Les Mémoires de l'Académie Royale des Sciences et des Belles-Lettres de Berlin de 1747 et 1748* **9**: 147–172.

Bernoulli D., 1778 Dijudicatio maxime probabilis plurium obseruationum discrepantium atque verisimillima inductio inde formanda. *Acta Academiae Scientiarum Imperialis Petropolitanea for 1777, pars prior 3–23* **1**: 3–23 (Translated by C.G. Allen (1961) as "The most probable choice between several discrepant observations and the formation therefrom of the most likely induction." *Biometrika*, **48**, 3–13. Commentary by M.G. Kendall. Also reprinted in Pearson and Kendall (1970), pp. 157–167).

Bernoulli J., 1713 *Ars Conjectandi*. Brothers Thurneisen, Basel (Translated in English as *The Art of Conjecturing, together with Letter to a Friend on Sets in Court Tennis* by E. Dudley Sylla (2006). Published by Johns Hopkins University Press).

Bernstein S.N., 1917 *Theory of Probability (in Russian)*, 4th edition. Gostekhizdat, Moscow-Liningrad.

Berry K.J., Johnston J.E., Mielke Jr P.W., 2014 *A Chronicle of Permutation Statistical Methods*. Springer, New York.

Bertrand J., 1888 Calcul des probabilities: seconde note sur la probabilite du tir a la cible. *Comptes Rendus, Académie des Sciences* **106**: 387–392.

Bertrand J., 1889 *Calcul des Probabilités*. Gauthier-Villars, Paris.

Bhattacharyya A., 1946 On some analogues of the amount of information and their use in statistical estimation. *Sankhya* **8**(1): 1–14.

Bickel P.J., Doksum K.A., 1977 *Mathematical Statistics: Basic Ideas and Selected Topics*. Holden-Day, San Francisco.

Bienaymé I.J., 1838 Mémoire sur la probabilité des résultats moyens des observations; démonstration directe de la règle de Laplace. *Mémoires de l'Académie des sciences de l'Institut de France* **5**: 513–558.

Bienaymé I.J., 1852 Sur la probabilité des erreurs d'aprés la méthode des moindres carrés. *Liouville's Journal de mathématiques pures et appliquées, Séries 1* **17**: 33–78. Also in *Mémoire de l'Académie Royale des Sciences de Paris* (2) **15**: 615–663

Biot J.B., 1804 Mémoire sur la propagation de la chaleur, lu à la Classe des Sciences Mathématiques et Physiques de l'Institut National. *Bibliothèque Britannique* **27**: 310–329.

Birembaut A.: Bravais, Auguste; in Gillispie C.C., (ed): *Dictionary of Scientific Biography* (Vol. **2**). Charles Scribner's Sons, New York, 1970, pp. 430–432.

Birnbaum A., 1962 On the foundations of statistical inference. *Journal of the American Statistical Association* **57**(298): 269–306.

Black M.: Induction; in Borchert D.M., (ed): *Encyclopedia of Philosophy* (Vol. **4**). Thomson/Gale, Detroit, 2006, pp. 635–650.

Blackwell D., 1947 Conditional expectation and unbiased sequential estimation. *The Annals of Mathematical Statistics* **18**(1): 105–110.

Blackwell D., Girshick M.A., 1954 *Theory of Games and Statistical Decisions*. John Wiley & Sons, New York.

Blom G., 1987 Harald Cramér 1893–1985. *The Annals of Statistics* **15**(4): 1335–1350.

BonJour L., 2009 *Epistemology: Classic Problems and Contemporary Responses*. Rowman & Littlefield, Lanham, MD.

Bose R.C., 1944 The fundamental theorem of linear estimation. *Proceedings of the 31st Indian Scientific Congress*: 2–3 (abstract).

Bowley A.L., 1928 *F.Y. Edgeworth's Contributions to Mathematical Statistics*. Royal Statistical Society, London.

Bowley A.L., 1934 Francis Ysidro Edgeworth. *Econometrica: Journal of the Econometric Society* **2**(2): 113–124.

Bowley A.L., Connor L.R., 1923 Tests of correspondence between statistical grouping and formulae. *Economica* **7**: 1–9.

Box J.F., 1978 *R.A. Fisher: The Life of a Scientist*. John Wiley & Sons, New York.

Box J.F., 1980 R.A. Fisher and the design of experiments, 1922–1926. *The American Statistician* **34**(1): 1–7.

Bradley L., 1971 *Smallpox Inoculation: An Eighteenth Century Mathematical Controversy*. University of Nottingham, Nottingham.

Braithwaite R.B., 1955 *Scientific Explanation: A Study of the Function of Theory, Probability and Law in Science*. Cambridge University Press, Cambridge.

Bravais A., 1846 Analyse mathématique sur les probabilités des erreurs de situation d'un point. *Mémoires de l'Institute de France* **IX**: 255–332.

Bravais A., 1866 *Études Cristallographiques*. Gauthier-Villars, Paris.

Breny H., 1955 L'état actuel du problème de Behrens-Fisher. *Trabajos de Estadística* **6**(2): 111–131.

Brillinger D.R., 1978 Henry Scheffé, 1907–1977. *Journal of the Royal Statistical Society A* **141**: 406–407.

Brillinger D.R., 2010 Erich Leo Lehmann, 1917–2009. *Journal of the Royal Statistical Society A* **173**: 683–686.

Brillinger D.R., 2011 David Harold Blackwell, 1919–2010 Obituary. *Journal of the Royal Statistical Society A* **174**(Part 1): 227–238.

Brown L.D., 1967 The conditional level of Student's *t*-test. *The Annals of Mathematical Statistics* **38**: 1068–1071.

Brownlee J., 1924 Some experiments to test the theory of goodness of fit. *Journal of the Royal Statistical Society* **87**(1): 76–82.

Brownlee K.A., 1965 *Statistical Theory and Methodology in Science and Engineering*, 2nd edition. John Wiley & Sons, New York.

Bru B.: Georges Darmois; in Heyde C.C., Seneta E., (eds): *Statisticians of the Centuries*. Springer, New York, 2001a, pp. 382–385.

Bru B.: Laplace, Pierre-Simon, Marquis de (1749–1827); in Smelser N.J., Baltes P.B., (eds): *International Encyclopedia of the Social & Behavioral Sciences*. Elsevier, Amsterdam/New York, 2001b, pp. 8378–8392.

Bru B., 2005 Poisson, the probability calculus and public education. *Journal Électronique d'Histoire des Probabilités et de la Statistique* **1**(2): 1–25.

Bru B., Jongmans F.: Joseph Bertrand; in Heyde C.C., Seneta E., (eds): *Statisticians of the Centuries*. Springer, New York, 2001, pp. 185–189.

Buehler R.: Fiducial inference; in Fienberg S.E., Hinkley D.V., (eds): *R.A. Fisher: An Appreciation*. Springer, New York, 1980, pp. 109–118.

Buehler R.J., 1959 Some validity criteria for statistical inferences. *The Annals of Mathematical Statistics* **30**: 845–863.

Buehler R.J., Feddersen A.P., 1963 Note on a conditional property of Student's *t*. *The Annals of Mathematical Statistics* **34**(3): 1098–1100.

Bühler W.K., 1987 *Gauss. A Biographical Study*. Springer, Berlin.

Bulmer M., 2003 *Francis Galton: Pioneer of Heredity and Biometry*. Johns Hopkins Press, Baltimore, MD.

Cajori F., 1928 *A History of Mathematical Notations* (Vol. **2**). Open Court, Chicago.

Cardano G., 1663 *Liber de ludo aleae* (in Opera Omnia, Vol. **1**, pp. 262–276).

Casella G., Berger R.L., 2002 *Statistical Inference*, 2nd edition. Duxbury Press, Pacific Grove, CA.

Cauchy A.L., 1853 Mémoire sur les résultats moyens d'un très grand nombre d'observations. *Comptes Rendus Hebdomadaires des Séances de l'Académie des Sciences* **37**: 381–385.

Chabert J.-L., 1989 Gauss et la méthode des moindres carrés. *Revue d'Histoire des Sciences* **42**: 5–26.

Chapman D.G., Robbins H., 1951 Minimum variance estimation without regularity assumptions. *The Annals of Mathematical Statistics* **22**(4): 581–586.

Chapman R.N., Gortner R.A., Scammon R.E., Walter F.K., Rosendahl C.O., 1930 J. Arthur Harris. *Science* **71**: 528–529.

Chatterjee S.K., 2003 *Statistical Thought: A Perspective and History*. Oxford University Press, Oxford.

Chrystal G., 1889 *Algebra: An Elementary Text-Book for the Higher Classes of Secondary Schools and for Colleges (Part II)*. Adam and Charles Black, Edinburgh.

Chrystal G., 1891 On some fundamental principles in the theory of probability. *Transactions of the Actuarial Society of Edinburgh* **2**: 420–439.

Cifarelli D.M., Regazzini E., 1996 De Finetti's contribution to probability and statistics. *Statistical Science* **11**(4): 253–282.

Cochran W.G., 1952 The χ^2 test of goodness of fit. *The Annals of Mathematical Statistics* **23**: 315–345.

Cochran W.G.: Fisher and the analysis of variance; in Fienberg S.E., Hinkley D.V., (eds): *R.A. Fisher: An Appreciation*. Springer, New York, 1980, pp. 17–34.

Condorcet M.-J.A.N.C.d., 1781 Sur les probabilités (Summary of Laplace's "Mémoire sur les probabilités"). *Histoire de l'Académie Royale des Sciences* **6**: 43–46 (1778).

Cook A., 1990 Sir Harold Jeffreys. 2 April 1891–18 March 1989. *Biographical Memoirs of Fellows of the Royal Society* **36**: 303–333.

Cooke R.M., 1991 *Experts in Uncertainty: Opinion and Subjective Probability in Science*. Oxford University Press, Oxford.

Coolidge J.L., 1925 *An Introduction to Mathematical Probability*. Oxford University Press, Oxford.

Coolidge J.L., 1926 Robert Adrain, and the beginnings of American mathematics. *American Mathematical Monthly* **33**: 61–76.

Costabel P.: Poisson, Siméon-Denis; in Gillispie C.C., (ed): *Dictionary of Scientific Biography* (Vol. **XV**, Supplement I). Charles Scribner's Sons, New York, 1978, pp. 480–490.

Courant R., Robbins H., 1941 *What is Mathematics?* Oxford University Press, London.

Cowan R.S., 1972 Francis Galton's statistical ideas: the influence of eugenics. *Isis* **63**: 509–528.

Cowan R.S.: Walter Frank Raphael Weldon; in Gillispie C.C., (ed): *Dictionary of Scientific Biography* (Vol. **14**). Charles Scribner's Sons, New York, 1981, pp. 251–252.

Cowles M., 2001 *Statistics in Psychology: An Historical Perspective*, 2nd edition. Lawrence Erlbaum, Hillsdale, NJ.

Cowles M., Davis C., 1982 On the origins of the .05 level of statistical significance. *American Psychologist* **37**(5): 553–558.

Cox D.R., 1958 Some problems connected with statistical inference. *The Annals of Mathematical Statistics* **29**(2): 357–372.

Cox D.R., 1971 The choice between alternative ancillary statistics. *Journal of the Royal Statistical Society Series B (Methodological)* **33**(2): 251–255.

Cox D.R., 1972 Regression and life tables. *Journal of the Royal Statistical Society Series B* **34**(2): 187–220.

Cox D.R., Hinkley D.V., 1979 *Theoretical Statistics*. Chapman & Hall, London.

Cramér H., 1937 *Random Variables and Probability Distributions*. Cambridge University Press, Cambridge.

Cramér H., 1946 *Mathematical Methods of Statistics*. Princeton University Press, Princeton.

Cushny A.R., Peebles A.R., 1905 The action of optical isomers: II. Hyoscines. *Journal of Physiology* **32**: 501–510.

Czuber E., 1914 *Wahrscheinlichkeitsrechnung* (Vol. **I**). Teubner, Leipzig.

d'Alembert J.L.R., 1747 Recherches sur la courbe que forme une corde tendue mise en vibrations. *Histoire de l'Académie Royale des Sciences et des Belles-Lettres de Berlin* **3**: 214–249.

d'Alembert J.L.R., 1761 *Opuscules Mathématiques* (Vol. **2**). David, Paris.

d'Alembert J.L.R., 1767 *Mélanges de Littérature, d'Histoire et de Philosophie (Tome V)*. Zacharie Chatelain & Fils, Amsterdam.

d'Alembert J.L.R., 1780 *Opuscules Mathematiques* (Vol. **7**). Chez Claude Antoine Jombert, fils ainé, Librairie du Roi, près le Pont-Neuf, Paris.

Dale A.I., 1985 A study of some early investigations into exchangeability. *Historia Mathematica* **12**(4): 323–336.

Dale A.I., 1999 *A History of Inverse Probability: From Thomas Bayes to Karl Pearson*, 2nd edition. Springer, New York.

Dale A.I.: Francis Ysidro Edgeworth; in Heyde C.C., Seneta E., (eds): *Statisticians of the Centuries*. Springer, New York, 2001, pp. 227–231.

Darmois G., 1928 *Statistique Mathématique*. Drouin, Paris.

Darmois G., 1935 Sur les lois de probabilité à estimation exhaustive. *Comptes rendus de l'Académie des Sciences, Paris* **260**: 1265–1266.

Darmois G., 1945 Sur les limites de la dispersion de certaines estimations. *Revue de l'Institut International de Statistique* **13**: 9–15.

Darwin C., 1876 *The Effects of Cross and Self Fertilisation in the Vegetable Kingdom*. John Murray, London.

Das Gupta S.: Distribution of the correlation coefficient; in Fienberg S.E., Hinkley D.V., (eds): *R.A. Fisher: An Appreciation*. Springer, New York, 1980, pp. 9–16.

DasGupta A. ed., 2011 *Selected Works of Debabrata Basu*. Springer, New York.

Daston L., 1988 *Classical Probability in the Enlightment*. Princeton University Press, Princeton.

Davenport C.B., 1930 James Arthur Harris. *Science* **71**: 474–475.

David H.A., 1995 First (?) occurrence of common terms in mathematical statistics. *The American Statistician* **49**(2): 121–133.

David H.A., 1998 First (?) occurrence of common terms in probability and statistics—a second list, with corrections. *The American Statistician* **52**(1): 36–40.

David H.A., 2005 Tables related to the normal distribution: a short history. *The American Statistician* **59**(4): 309–311.

David H.A., 2006 The introduction of matrix algebra into statistics. *The American Statistician* **60**(2): 162.

Davis P.J., Albers D.J., Alexanderson G.L., 2011 *Fascinating Mathematical People: Interviews and Memoirs*. Princeton University Press, Princeton.

Daw R.H., 1980 Johann Heinrich Lambert (1728–1777). *Journal of the Institute of Actuaries* **107**(3): 345–363.

de Beaumont É., 1865 Éloge historique d'Auguste Bravais. *Mémoires de l'Académie des Sciences* **35**: XXIII–XCIX.

de Finetti B., 1930a Fondamenti logici del ragionamento probabilistico. *Bollettino dell' Unione Matematica Italiana* **9**(5): 258–261.

de Finetti B., 1930b Funzione caratteristica di un fenomeno aleatorio. *Memorie Academia Nazionale dei Lincei* **4**(Series 6): 86–133.

de Finetti B., 1930c Problemi determinati e indeterminati nel calcolo delle probabilità. *Rendiconti della Reale Accademia Nazionale dei Lincei* **12**(Series 6 (facs. 9)): 367–373.

de Finetti B., 1937 La prévision: ses lois logiques, ses sources subjectives. *Annales de l'institut Henri Poincaré* **7**(1): 1–68.

de Finetti B.: Foresight: its logical laws in subjective sources; in Kyburg H.E., Smokler H.E., (eds): *Studies in Subjective Probability*. John Wiley & Sons, New York, 1964, pp. 93–158.

de Finetti B., 1972 *Probability, Induction and Statistics: The Art of Guessing*. John Wiley & Sons, New York.

de Finetti B., 1974 *Theory of Probability: A Critical Introductory Treatment*. John Wiley & Sons, New York (First published in 1970 as "Teoria Delle Probabilità", Giulio Einaudi s.p.a., Torino).

de Finetti B.: Probability and my life; in Gani J., (ed): *The Making of Statisticians*. Springer, New York, 1982, pp. 3–12.

de Moivre, A., 1711 De Mensura Sortis, seu, de Probabilitate Eventuum in Ludis a Casu Fortuito Pendentibus. Phil. Trans. **27**: 213–264.

de Moivre A., 1730 *Miscellanea Analytica de Seriebus et Quadraturis*. Touson & Watts, London.

de Moivre A., 1733 *Approximatio ad Summam Tenninorum Binomii* $(a + b)^n$ *in Seriem Expansi*. Printed for private circulation.

de Moivre A., 1738 *The Doctrine of Chances, or a Method of Calculating the Probabilities of Events in Play*, 2nd edition. Millar, London.

de Moivre A., 1756 *The Doctrine of Chances, or a Method of Calculating the Probabilities of Events in Play*, 3rd edition. Millar, London.

De Morgan A., 1837 Review of Laplace's Théorie Analytique des Probabilités (3rd edition). *Dublin Review* **2**: 338–354 & **3**: 237–248.

De Morgan A., 1838 *An Essay on Probabilities, and on their Application to Life Contingencies and Insurance Offices*. Longman, Orme, Brown, Green & Longmans, London.

De Morgan A.: Laplace, Pierre Simon; in Knight C., (ed): *The Penny Cyclopaedia of the Society for the Diffusion of Useful Knowledge* (London, 1833–1846). Charles Knight and Co., London, 1839, pp. 325–328.

DeGroot M.H., 1988 A conversation with George A. Barnard. *Statistical Science* **3**(2): 196–212.

Deming W.E., 1943 *Statistical Adjustment of Data*. John Wiley & Sons, New York.

Deming W.E., 1950 *Some Theory of Sampling*. Dover, New York.

Dempster A.P., 1964 On the difficulties inherent in Fisher's fiducial argument. *Journal of the American Statistical Association* **59**(305): 56–66.

Denis D., 2001 The origins of correlation and regression: Francis Galton or Auguste Bravais and the error theorists? *History and Philosophy of Psychology Bulletin* **13**: 36–44.

Diaconis P., 1977 Finite forms of de Finetti's theorem on exchangeability. *Synthese* **36**(2): 271–281.

Dirichlet P.G.L., 1829 Sur la convergence des séries trigonométriques qui servent á représenter une fonction arbitraire entre des limites données. *Journal für die Reine und Angewandte Mathematik* **4**: 157–169.

Dokic J., Engel P., 2003 *Frank Ramsey: Truth and Success*. Routledge, New York.

Doob J.L., 1936 Statistical estimation. *Transactions of the American Mathematical Society* **39**(3): 410–421.

Du Pré A.M., 1938 The first table of the normal probability integral: its use by Kramp, who constructed it. *Isis* **29**(1): 43–48.

Dugué D., 1937 Application des propriétés de la limite au sens du calcul des probabilités à l'étude des diverses questions d'estimation. *Journal de l'École Polytechnique III* **4**: 305–373.

Dugué D., 1960 Georges Darmois (1888–1960). *Revue de l'Institut International de Statistique/ Review of the International Statistical Institute* **28**(1/2): 122–123.

Dugué D., 1961 Georges Darmois, 1888–1960. *The Annals of Mathematical Statistics* **32**(2): 357–360.

Dugué D., 1974 Maurice Fréchet, 1878–1973. *International Statistical Review/Revue Internationale de Statistique* **42**(1): 113–114.

Dugué D.: Bienaymé, Jules; in Kruskal W.H., Tanur J.M., (eds): *International Encyclopedia of Statistics*. Free Press, New York, 1978, pp. 73–74.

Dunnington G.W., 1955 *Gauss: Titan of Science*. Mathematical Association of America, New York.

Dutka J., 1990 Robert Adrain and the method of least squares. *Archive for History of Exact Sciences* **41**(2): 171–184.

Dutka J., 1996 On Gauss' priority in the discovery of the method of least squares. *Archive for History of Exact Sciences* **49**(4): 355–370.

Eddington A.S., 1914 *Stellar Movements and the Structure of the Universe*. MacMillan, London.

Edgeworth F.Y., 1877 *New and Old Methods of Ethics: Or "Physical Ethics" and "Methods of Ethics"*. Parker, Oxford.

Edgeworth F.Y., 1881 *Mathematical Psychics: An Essay on the Application of Mathematics to the Moral Sciences*. Kegan Paul, London.

Edgeworth F.Y., 1883 The method of least squares. *The London, Edinburgh, and Dublin Philosophical Magazine and Journal of Science, 5th Series* **16**: 360–375.

Edgeworth F.Y., 1885 Methods of statistics. *Journal of the Statistical Society of London* (Jubilee Volume): 181–217.

Edgeworth F.Y., 1892 Correlated averages. *Philosophical Magazine, 5th Series* **34**: 190–204.

Edgeworth F.Y., 1908 On the probable errors of frequency-constants. *Journal of the Royal Statistical Society* **71**: 381–397, 499–512, 651–678.

Edgeworth F.Y., 1909 Addendum on "Probable errors of frequency-constants". *Journal of the Royal Statistical Society* **72**: 81–90.

Edwards A.W.F., 1974 The history of likelihood. *International Statistical Review* **42**: 9–15.

Edwards A.W.F., 1976 Fiducial probability. *The Statistician* **25**(1): 15–35.

Edwards A.W.F., 1978 Commentary on the arguments of Thomas Bayes. *Scandinavian Journal of Statistics* **5**(2): 116–118.

Edwards A.W.F., 1982 Pascal and the problem of points. *International Statistical Review* **50**: 259–266.

Edwards A.W.F.: George Udny Yule; in Heyde C.C., Seneta E., (eds): *Statisticians of the Centuries*. Springer, New York, 2001, pp. 292–294.

Edwards A.W.F.: R.A. Fisher, Statistical methods for research workers; in Grattan-Guinness I., (ed): *Landmark Writings in Western Mathematics 1640–1940*. Elsevier, Amsterdam, 2005, pp. 856–870.

Eells E., 1982 *Rational Decision and Causality*. Cambridge University Press, Cambridge.

Efron B., 1998 R.A. Fisher in the 21st century. *Statistical Science* **13**(2): 95–114.

Efron B., Morris C., 1971 Limiting the risk of Bayes and empirical Bayes estimators-Part I: the Bayes case. *Journal of the American Statistical Association* **66**(336): 807–815.

Efron B., Morris C., 1972a Empirical Bayes on vector observations: an extension of Stein's method. *Biometrika* **59**(2): 335–347.

Efron B., Morris C., 1972b Limiting the risk of Bayes and empirical Bayes estimators-Part II: the empirical Bayes case. *Journal of the American Statistical Association* **67**(337): 130–139.

Efron B., Morris C., 1973a Combining possibly related estimation problems. *Journal of the Royal Statistical Society Series B (Methodological)* **35**(3): 379–421.

Efron B., Morris C., 1973b Stein's estimation rule and its competitors—an empirical Bayes approach. *Journal of the American Statistical Association* **68**(341): 117–130.

Eisenhart C., 1964 The meaning of "least" in least squares. *Journal of the Washington Academy of Sciences* **54**: 24–33. Reprinted in Ku (1969), pp. 265–274.

Eisenhart C.: Karl Pearson; in Gillispie C.C., (ed): *Dictionary of Scientific Biography* (Vol. **10**). Charles Scribner's Sons, New York, 1974, pp. 447–473.

Eisenhart C., 1979 On the transition from "Student's" z to "Student's" t. *The American Statistician* **33**(1): 6–12.

Elderton W.P., 1902 Tables for testing the goodness of fit of theory to observation. *Biometrika* **1**: 155–163.

Elderton W.P., Johnson N.L., 1969 *Systems of Frequency Curves*. Cambridge University Press, Cambridge.

Ellis R.L., 1850 Remarks on an alleged proof of the "method of least squares," contained in a late number of the Edinburgh review. *Philosophical Magazine* **37**: 321–328 (Reprinted in *The Mathematical and other Writings of Robert Leslie Ellis*, Cambridge University Press, Cambridge, 1863).

Epstein R.A., 2009 *The Theory of Gambling and Statistical Logic*, 2nd edition. Academic Press, New York.

Euler L., 1748 Sur la vibration des cordes. *Mémoire de l'Académie Royale des Sciences et Belles-Lettres de Berlin* **6**: 69–85.

Euler L., 1782 Recherches sur une nouvelle espece de quarres magiques. *Verh Zeeuw Gen Weten Vlissengen* **9**: 85–239.

Euler L., 1798 Methodus facilis inveniendi series per sinus cosinusve angulorum multiplorum procedentes, quarum usus in universa theoria astronomiae est amplissimus (1793). *Nova Acta Academiae Scientarum Imperialis Petropolitinae* (11): 94–113 (Reprinted in Opera Omnia: Series 1, Vol. 16, pp. 311–332).

Everitt P.F., 1910 Tables of the tetrachoric functions for fourfold correlation tables. *Biometrika* **7**(4): 437–451.

Farebrother R.W., 1999 *Fitting Linear Relationships: A History of the Calculus of Observations 1750–1900*. Springer, New York.

Farebrother R.W.: Adrien-Marie Legendre; in Heyde C.C., Seneta E., (eds): *Statisticians of the Centuries*. Springer, New York, 2001, pp. 101–104.

Feller W., 1935 Über den zentralen Grenzwertsatz der Wahrscheinlichkeitsrechnung. *Mathematische Zeitschrift* **40**: 521–559.

Ferguson T.S., 1996 *A Course in Large Sample Theory*. Chapman & Hall, London.

Féron R.: Poisson, Siméon-Poisson; in Kruskal W.H., Tanur J.M., (eds): *International Encyclopedia of Statistics*. Free Press, New York, 1978, pp. 704–706.

Fienberg S.E., 2006 When did Bayesian inference become "Bayesian"? *Bayesian Analysis* **1**(1): 1–40.

Fienberg S.E., Lazar N.: William Sealy Gosset; in Heyde C.C., Seneta E., (eds): *Statisticians of the Centuries*. Springer, New York, 2001, pp. 312–317.

Fienberg S.E., Tanur J.M.: Jerzy Neyman; in Heyde C.C., Seneta E., (eds): *Statisticians of the Centuries*. Springer, New York, 2001, pp. 444–448.

Fine T.A., 1973 *Theories of Probability: An Examination of Foundations*. Academic Press, New York.

Fischer H., 2010 *A History of the Central Limit Theorem: From Classical to Modern Probability Theory*. Springer, New York.

Fischer W.: Helmert, Friedich Robert; in Gillispie C.C., (ed): *Dictionary of Scientific Biography* (Vol. **6**). Charles Scribner's Sons, New York, 1973, pp. 239–241.

Fishburn P.C., 1970 *Utility Theory for Decision Making*. John Wiley & Sons, New York.

Fisher R.A., 1912 On an absolute criterion for fitting frequency curves. *Messenger of Mathematics* **41**: 155–160.

Fisher R.A., 1915 Frequency distribution of the values of the correlation coefficient in samples from an indefinitely large population. *Biometrika* **10**: 507–521.

Fisher R.A., 1916 Biometrika. *Eugenics Review* **8**: 62–64.

Fisher R.A., 1918 The correlation between relatives on the supposition of Mendelian Inheritance. *Transactions of the Royal Society of Edinburgh* **52**(2): 399–433.

Fisher R.A., 1919 The causes of human variability. *Eugenics Review* **10**(4): 213–220.

Fisher R.A., 1920 A mathematical examination of the methods of determining the accuracy of an observation by the mean error, and by the mean square error. *Monthly Notices of the Royal Astronomical Society* **80**: 758–770.

Fisher R.A., 1921a On the "probable error" of a coefficient of correlation deduced from a small sample. *Metron* **1**: 3–32.

Fisher R.A., 1921b Studies in crop variation. I. An examination of the yield of dressed grain from Broadbalk. *The Journal of Agricultural Science* **11**(2): 107–135.

Fisher R.A., 1922a On the dominance ratio. *Proceedings of the Royal Society of Edinburgh* **42**: 321–341.

Fisher R.A., 1922b On the interpretation of χ^2 from contingency tables, and the calculation of P. *Journal of the Royal Statistical Society* **85**: 87–94.

Fisher R.A., 1922c On the mathematical foundations of theoretical statistics. *Philosophical Transactions of the Royal Society of London A* **222**: 309–368.

Fisher R.A., 1922d The goodness of fit of regression formulae, and the distribution of regression coefficients. *Journal of the Royal Statistical Society* **85**(4): 597–612.

Fisher R.A., 1923a Note on Dr Burnside's recent paper on errors of observation. *Proceedings of the Cambridge Philosophical Society* **21**: 655–658.

Fisher R.A., 1923b Statistical tests of agreement between observation and hypothesis. *Economica* **8**: 139–147.

Fisher R.A., 1924a On a distribution yielding the error functions of several well known statistics. *Proceedings of the International Congress of Mathematics (Toronto)* **2**: 805–813.

Fisher R.A., 1924b The conditions under which χ^2 measures the discrepancy between observation and hypothesis. *Journal of the Royal Statistical Society* **87**: 442–450.

Fisher R.A., 1924c The distribution of the partial correlation coefficient. *Metron* **3**: 329–332.

Fisher R.A., 1925a Applications of "Student's" distribution. *Metron* **5**: 90–104.

Fisher R.A., 1925b *Statistical Methods for Research Workers*. Oliver and Boyd, Edinburgh.

Fisher R.A., 1925c Theory of statistical estimation. *Proceedings of the Cambridge Philosophical Society* **22**: 700–725.

Fisher R.A., 1926a Bayes' theorem and the fourfold table. *Eugenics Review* **18**: 32–33.

Fisher R.A., 1926b The arrangement of field experiments. *Journal of the Ministry of Agriculture of Great Britain* **33**: 503–513.

Fisher R.A., 1930 Inverse probability. *Mathematical Proceedings of the Cambridge Philosophical Society* **26**(4): 528–535.

Fisher R.A., 1932 *Statistical Methods for Research Workers*, 4th edition. Oliver and Boyd, Edinburgh.

Fisher R.A., 1933 The concepts of inverse probability and fiducial probability referring to unknown parameters. *Proceedings of the Royal Society of London Series A* **139**(838): 343–348.

Fisher R.A., 1934a Probability likelihood and quantity of information in the logic of uncertain inference. *Proceedings of the Royal Society of London Series A* **146**(856): 1–8.

Fisher R.A., 1934b *Statistical Methods for Research Workers*, 5th edition. Oliver and Boyd, Edinburgh.

Fisher R.A., 1934c Two new properties of maximum likelihood. *Proceedings of the Royal Society of London A* **144**(852): 285–307.

Fisher R.A., 1935a *The Design of Experiments*. Oliver and Boyd, Edinburgh.

Fisher R.A., 1935b The detection of linkage with "dominant" abnormalities. *Annals of Eugenics* **6**(2): 187–201.

Fisher R.A., 1935c The fiducial argument in statistical inference. *Annals of Eugenics* **6**(4): 391–398.

Fisher R.A., 1935d The logic of inductive inference. *Proceedings of the Royal Society* **98**(1): 39–82.

Fisher, R.A. 1936 Uncertain inference. *Proceedings of the American Academy of Arts and Sciences* **71**(4): 245–258.

Fisher R.A., 1936 The half–drill strip system agricultural experiments. *Nature* **138**: 1101.

Fisher R.A., 1937a On a point raised by M.S. Bartlett on fiducial probability. *Annals of Eugenics* **7**(4): 370–375.

Fisher R.A., 1937b *The Design of Experiments*, 2nd edition. Oliver and Boyd, Edinburgh.

Fisher R.A., 1938 Review of "Lectures and Conferences on Mathematical Statistics," by J. Neyman. *Science Progress* **33**: 577.

Fisher R.A., 1939a Student. *Annals of Eugenics* **9**: 1–9.

Fisher R.A., 1939b A note on fiducial inference. *The Annals of Mathematical Statistics* **10**(4): 383–388.

Fisher R.A., 1941 The asymptotic approach to Behrens's integral, with further tables for the *d* test of significance. *Annals of Eugenics* **11**(1): 141–172.

Fisher R.A., 1945 A new test for 2 x 2 tables. *Nature* **156**(3961): 388.

Fisher R.A., 1950 *Contributions to Mathematical Statistics*. John Wiley & Sons, New York.

Fisher R.A., 1955 Statistical methods and scientific induction. *Journal of the Royal Statistical Society B* **17**: 69–78.

Fisher R.A., 1956a On a test of significance in Pearson's Biometrika Tables (No. 11). *Journal of the Royal Statistical Society B* **18**(1): 56–60.

Fisher R.A., 1956b *Statistical Methods and Scientific Inference*. Hafner Publising Company, New York.

Fisher R.A., 1959 Mathematical probability in the natural sciences. *Technometrics* **1**(1): 21–29.

Fisher R.A., 1960 Scientific thought and the refinement of human reasoning. *Journal of the Operations Research Society of Japan* **3**: 1–10.

Fisher R.A., Mackenzie W.A., 1923 Studies in crop variation. II. The manurial response of different potato varieties. *The Journal of Agricultural Science* **13**(3): 311–320.

Fisz M., 1963 *Probability Theory and Mathematical Statistics*, 3rd edition. John Wiley & Sons, New York.

Forbes J.D., 1849 XVII. On the alleged evidence for a physical connexion between stars forming binary or multiple groups, arising from their proximity alone: to the editors of the Philosophical Magazine and Journal. *Philosophical Magazine Series 3* **35**(234): 132–133.

Forbes J.D., 1850 LIV. On the alleged evidence for a physical connexion between stars forming binary or multiple groups, deduced from the doctrine of chances. *The London, Edinburgh, and Dublin Philosophical Magazine and Journal of Science* **37**(252): 401–427.

Fourier J.B.J., 1808 Mémoire sur la propagation de la chaleur dans les corps solides. *Nouveau Bulletin des Sciences de la Société Philomathique de Paris* **6**: 112–116 (Oeuvres 2, pp. 215–221).

Fourier J.B.J., 1822 *Théorie Analytique de la Chaleur*. Didot, Paris (Translated as *The Analytical Theory of Heat* by A. Freeman (1882). Published by Cambridge University Press, Cambridge).

Frápolli M. ed., 2004 *F.P. Ramsey: Critical Reassessments*. Continuum, New York.

Fraser D.A.S., 1968 *The Structure of Inference*. John Wiley & Sons, New York.

Fraser D.A.S.: The elusive ancillary; in Kabe D.G., Gupta R.P., (eds): *Multivariate Statistical Inference*. North Holland, Amsterdam, 1973, pp. 41–48.

Fraser D.A.S., 2004 Ancillaries and conditional inference. *Statistical Science* **19**(2): 333–369.

Fréchet M., 1928 *Les Espaces Abstraits et Leur Théorie Considéré Comme Introduction à l'Analyse Générale*. Gauthier-Villars, Paris.

Fréchet M., 1938 *Recherches Théoriques Modernes sur la Théorie des Probabilités*. Gauthier-Villars, Paris.

Fréchet M., 1943 Sur l'extension de certaines evaluations statistiques au cas de petits echantillons. *Revue de l'Institut International de Statistique* **11**(3/4): 182–205.

Freudenthal H.: Arbuthnot, John; in Gillispie C.C., (ed): *Dictionary of Scientific Biography* (Vol. **1**). Charles Scribner's Sons, New York, 1970, pp. 208–209.

Fuss P.H., 1843 *Correspondance Mathématique et Physique de Quelques Célèbres Géomètres du XVIIIème Siècle*. St. Petersburg.

Galton F., 1865 Hereditary talent and character. *Macmillan's Magazine* **12**: 157–166 (Part I), 318–327 (Part II).

Galton F. 1869 *Hereditary Genius*. Macmillan, London (Reprinted in 1979, Friedmann, London).

Galton F., 1875 Statistics by intercomparison, with remarks on the law of frequency of error. *Philosophical Magazine* **49**: 33–46.

Galton F., 1877 Typical laws of heredity. *Nature* **15**: 492–495, 512–514, 532–533. (Also in *Proceedings of the Royal Institution* **8**: 282–301 (1877)).

Galton F., 1885a Presidential address, Section H, Anthropology. *Nature* **32**(830): 507–510 (Also published in *British Association Reports* **55**: 1206–1214 (1885)).

Galton F., 1885b Regression towards mediocrity in hereditary stature. *Journal of the Anthropological Institute* **15**: 246–263.

Galton F., 1886 Family likeness in stature. *Proceedings of the Royal Society of London* **40**: 42–73 (Appendix by J.D. Hamilton Dickson, pp. 63–66).

Galton F., 1888 Co-relations and their measurement, chiefly from anthropometric data. *Proceedings of the Royal Society* **45**: 135–145.

Galton F., 1889 *Natural Inheritance*. Macmillan, London.

Galton F., 1890 Kinship and correlation. *North American Review* **150**: 419–431 (Reprinted in *Statistical Science* **4**: 81–86 (1889)).

Gani J.: Daniel Bernoulli; in Heyde C.C., Seneta E., (eds): *Statisticians of the Centuries*. Springer, New York, 2001, pp. 64–67.

Garber E., 1999 *The Language of Physics: The Calculus and the Development of Theoretical Physics in Europe, 1750–1914*. Birkhaüser, Boston.

Gauss C.F., 1801 *Disquisitiones Arithmeticae*. Berlin.

Gauss C.F., 1809 *Theoria Motus Corporum Coelestium*. Perthes et Besser, Hamburg (Translated as *Theory of Motion of the Heavenly Bodies Moving About the Sun in Conic Sections* by C.H. Davis (1857). Published by Little, Brown, Boston. Reprinted in 1963, Dover, New York).

Gauss C.F., 1823 *Theoria Combinationis Observationum Erroribus Minimis Obnoxia*. Dieterich, Göttingen (Translated as *Theory of the Combination of Observations Least Subject to Errors* by G.W. Stewart (1995). Published by SIAM, Philadelphia).

Gauss C.F., Bessel F.W., 1880 *Briefwechsel zwischen Gauss und Bessel*. W. Engelmann, Leipzig.

Gayen A.K., 1951 The frequency distribution of the product-moment correlation coefficient in random samples of any size drawn from non-normal universes. *Biometrika* **38**(1/2): 219–247.

Gehman H.M., 1955 America's second mathematician: not Adrian, but Adrain. *The Mathematics Teacher* **48**(6): 409–410.

Geisser S.: Basic theory of the 1922 mathematical statistics paper; in Fienberg S.E., Hinkley D.V., (eds): *R.A. Fisher: An Appreciation*. Springer, New York, 1980, pp. 59–66.

Geisser S.: Introduction to Fisher (1922) On the mathematical foundations of theoretical statistics; in Kotz S., Johnson N.L., (eds): *Breakthroughs in Statistics* (Vol. **I**). Springer, New York, 1992, pp. 1–10.

Geisser S., 2006 *Modes of Parametric Statistical Inference*. John Wiley & Sons, New York.

Ghosh J.K., 1988 *Statistical Information and Likelihood: A Collection of Critical Essays by Dr D. Basu*. Springer, New York.

Ghosh J.K., 2002 Debabrata Basu: a brief life-sketch. *Sankhya, Series A (1961–2002)* **64**(3).

Ghosh M., Pathak P.K. eds, 1992 *Current Issues in Statistical Inference: Essays in Honor of D. Basu*, IMS Lecture Notes, Vol. **17**. Institute of Mathematical Statistics, Hayward, CA.

Ghosh M., Reid N., Fraser D.A.S., 2010 Ancillary statistics: a review. *Statistica Sinica* **20**(4): 1309.

Gigerenzer G., Swijtink Z., Porter T., Daston L., Beatty J., Krüger L., 1989 *The Empire of Chance: How Probability Changed Science and Everyday Life*. Cambridge University Press, Cambridge.

Gillispie C.C., 1979 Mémoires inédits ou anonymes de Laplace sur la théorie des erreurs, les polynômes de Legendre et la philosophie des probabilités. *Revue d'Histoire des Sciences* **32**(3): 223–279.

Gillispie C.C., 2000 *Pierre-Simon Laplace, 1749–1827: A Life in Exact Science*. Princeton University Press, Princeton.

Glaisher J.W.L., 1872 On the law of facility of errors of observations, and on the method of least squares. *Memoirs of the Royal Astronomical Society* **39**: 75–124.

Glatzer I., 1913 A successful social reformer, Ernst Abbe, 1840–1905. *The Economic Journal* **23**(91): 329–339.

Gnedenko B.V., Sheynin O.B.: The theory of probability; in Kolmogorov A.N., Yushkevich A.P., (eds): *Mathematics of the 19th Century: Mathematical Logic, Algebra, Number Theory, and Probability Theory*. Birkhäuser Verlag, Basel/Boston, 1992, pp. 531–532.

Goldstine H.H., 1977 *A History of Numerical Analysis from the 16th through the 19th Century*. Springer, New York.

Good I.J., 1988 Bayes's red billiard ball is also a herring, and why Bayes withheld publication. *Journal of Statistical Computing and Simulation* **29**: 335–340.

Good I.J.: Introduction to Robbins (1955) An empirical Bayes approach to statistics; in Kotz S., Johnson N.L., (eds): *Breakthroughs in Statistics* (Vol. **I**). Springer, New York, 1992, pp. 379–387.

Gorroochurn P., 2011 Errors of probability in historical context. *The American Statistician* **65**(4): 246–254 (Reprinted in *The Best Writing on Mathematics 2013* (ed. M. Pitici) 2014, pp. 191–212, Princeton University Press, Princeton).

Gorroochurn P., 2012a *Classic Problems of Probability*. John Wiley & Sons, Inc., Hoboken, NJ.

Gorroochurn P., 2012b Some laws and problems of classical probability and how Cardano anticipated them. *Chance* **25**(4): 13–20.

Gorroochurn P., 2014 Thirteen correct solutions to the "Problem of Points" and their histories. *The Mathematical Intelligencer* **36**(3): 56–64.

Gosset W.S., 1936 Co-operation in large-scale experiments. *Supplement to the Journal of the Royal Statistical Society* **3**: 115–136.

Gosset W.S., 1970 *Letters from W.S. Gosset to R.A. Fisher, 1915–1936*. Printed for private circulation by Arthur Guiness, Son and Co.

Gouraud C., 1848 *Histoire du Calcul des Probabilités Depuis ses Origines Jusqu'à nos Jours*. Librairie D'Auguste Durand, Paris.

Gower B., 1982 Astronomy and probability: Forbes versus Michell on the distribution of the stars. *Annals of Science* **39**(2): 145–160.

Gower B., 1987 Planets and probability: Daniel Bernouilli on the inclinations of the planetary orbits. *Studies in History and Philosophy of Science Part A* **18**(4): 441–454.

Grattan-Guinness I.: Joseph Fourier, Théorie Analytique de la Chaleur (1822); in Grattan-Guinness I., (ed): *Landmark Writings in Western Mathematics 1640–1940*. Elsevier, Amsterdam, 2005, pp. 354–365.

Gray J.J., 1978 Johann Heinrich Lambert, mathematician and scientist, 1728–1777. *Historia Mathematica* **5**(1): 13–41.

Greenwood M., Yule G.U., 1915 The statistics of anti-typhoid and anti-cholera inoculations, and the interpretation of such statistics in general. *Proceedings of the Royal Society of Medicine* **8**: 113–194.

Gridgeman N.T., 1959 The lady tasting tea, and allied topics. *Journal of the American Statistical Association* **54**: 776–783.

Gridgeman N.T., 1960 Geometric probability and the number π. *Scripta Mathematica* **25**: 183–195.

Gridgeman N.T.: Fisher, Ronald Aylmer; in Gillispie C.C., (ed): *Dictionary of Scientific Biography* (Vol. **7**). Charles Scribner's Sons, New York, 1972a, pp. 7–11.

Gridgeman N.T.: Galton, Francis; in Gillispie C.C., (ed): *Dictionary of Scientific Biography* (Vol. **5**). Charles Scribner's Sons, New York, 1972b, pp. 265–267.

Grigorian A.T.: Lyapunov, Aleksandr Mikhailovich; in Gillispie C.C., (ed): *Dictionary of Scientific Biography* (Vol. **8**). Charles Scribner's Sons, New York, 1974, pp. 559–563.

Günther N.: Abbé, Ernst; in Gillispie C.C., (ed): *Dictionary of Scientific Biography* (Vol. **1**). Charles Scribner's Sons, New York, 1970, pp. 6–9.

Hacking I., 1965 *Logic of Statistical Inference*. Cambridge University Press, Cambridge.

Hacking I.: Gosset, William Sealy; in Gillispie C.C., (ed): *Dictionary of Scientific Biography* (Vol. **5**). Charles Scribner's Sons, New York, 1972, pp. 476–477.

Hagen G.H.L., 1837 *Grundzüge der Wahrscheinlichkeits-Rechnung*. Dümmler, Berlin.

Haggard E.A., 1958 *Intraclass Correlation and the Analysis of Variance*. Dryden Press, New York.

Hahn R., 2005 *Pierre Simon Laplace, 1749–1827: A Determined Scientist*. Harvard University Press, Cambridge, MA.

Hailperin T., 1996 *Sentential Probability Logic: Origins, Development, Current Status, and Technical Applications*. Lehigh University Press, Bethlehem, PA.

Hald A., 1952 *Statistical Theory with Engineering Applications*. John Wiley & Sons, New York.

Hald A., 1984 Nicholas Bernoulli's theorem. *International Statistical Review* **52**: 93–99.

Hald A., 1990 *A History of Probability and Statistics and Their Applications Before 1750*. John Wiley & Sons, New York.

Hald A., 1998 *A History of Mathematical Statistics from 1750 to 1930*. John Wiley & Sons, New York.

Hald A., 2007 *A History of Parametric Statistical Inference from Bernoulli to Fisher, 1713–1935*. Springer, New York.

Hall H.S., Knight S.R., 1891 *Higher Algebra: A Sequel to Elementary Algebra for Schools*, 4th edition. Macmillan, London.

Hall N.S., 2007 R.A. Fisher and his advocacy of randomization. *Journal of the History of Biology* **40**(2): 295–325.

Harris J.A., 1913 On the calculation of intraclass and interclass coefficients of correlation from class moments when the number of possible combinations is large. *Biometrika* **9**: 446–472.

Healy M.J.R., 2003 R.A. Fisher the statistician. *Journal of the Royal Statistical Society Series D (The Statistician)* **52**(3): 303–310.

Helmert F.R., 1875 Über die Berechnung des wahrscheinlichen Fehlers aus einer endlichen Anzahl wahrer Beobachtungsfehler. *Zeitschrift für angewandte Mathematik und Physik* **20**: 300–303.

Helmert F.R., 1876a Die Genauigkeit der Formel von Peters zur Berechnung des wahrscheinlichen Beobachtungsfehler direkter Beobachtungen gleicher Genauigkeit. *Astronomische Nachrichten* **88**: 113–132.

Helmert F.R., 1876b Über die Wahrscheinlichkeit der Potenzsummen der Beobachtungsfehler und über einige damit im Zusammenhange stehende Fragen. *Zeitschrift für angewandte Mathematik und Physik* **21**: 192–218.

Heron D., 1911 The danger of certain formulae suggested as substitutes for the corrleation coefficient. *Biometrika* **8**: 109–122.

Herschel J.F.W., 1849 *Outlines of Astronomy*. Longman, London.

Herschel J.F.W., 1850 Quetelet on probabilities. *Edinburgh Review* **92**: 1–57.

Herschel W., 1803 Account of the changes that have happened, during the last twenty-five years, in the relative situation of double-stars; with an investigation of the cause to which they are owing. *Philosophical Transactions of the Royal Society of London* **93**: 339–382.

Heyde C.C.: Bienaymé, Irénée-Jules; in Gillispie C.C., (ed): *Dictionary of Scientific Biography* (Vol. **15**). Charles Scribner's Sons, New York, 1976, pp. 30–33.

Heyde C.C., Seneta E., 1977 *I.J. Bienaymé: Statistical Theory Anticipated.* Springer-Verlag, New York.

Hinkley D.V.: Fisher's development of conditinal inference; in Fienberg S.E., Hinkley D.V., (eds): *R.A. Fisher: An Appreciation*. Springer, New York, 1980, pp. 101–108.

Hogben L., 1957 *Statistical Theory: The Relationship of Probability, Credibility and Error.* W.W. Norton & Co., Inc., New York.

Hotelling H., 1927 Review of Fisher's statistical methods for research workers. *Journal of the American Statistical Association* **22**: 411–412.

Hotelling H., 1930 The consistency and ultimate distribution of optimum statistics. *Transactions of the American Mathematical Society* **32**(4): 847–859.

Howie D., 2002 *Interpreting Probability: Controversies and Development in the Early Twentieth Century*. Cambridge University Press, Cambridge.

Hubbard R., Bayarri M.J., 2003 Confusion over measures of evidence (p's) versus errors (α's) in classical statistical testing. *The American Statistician* **57**(3): 171–178.

Hume D., 1748 *An Enquiry Concerning Human Understanding*, London (2007 edition by P. Millican, Oxford University Press, Oxford).

Huygens C., 1657 *De Ratiociniis in ludo aleae*. Johannis Elsevirii, Leiden (pp. 517–534 in Frans van Schooten's *Exercitationum mathematicarum liber primus continens propositionum arithmeticarum et geometricarum centuriam*).

Irwin J.O., 1935 Tests of significance for differences between percentages based on small numbers. *Metron* **12**: 83–94.

Itard J.: Lagrange, Joseph Louis; in Gillispie C.C., (ed): *Dictionary of Scientific Biography* (Vol. **7**). Charles Scribner's Sons, New York, 1975a, pp. 559–573.

Itard J.: Legendre, Adrien-Maris; in Gillispie C.C., (ed): *Dictionary of Scientific Biography* (Vol. **8**). Charles Scribner's Sons, New York, 1975b, pp. 135–143.

Jackman S., 2009 *Bayesian Analysis for the Social Sciences*. John Wiley & Sons, Inc., Hoboken, NJ.

James W., Stein C., 1961 Estimation with quadratic loss. *Proceedings of the Fourth Berkeley Symposium on Mathematical Statistics and Probability* **1**: 361–379.

Jaynes E.T., 1973 The well-posed problem. *Foundations of Physics* **3**(4): 477–492.

Jeffrey R.C., 1983 *The Logic of Decision*, 2nd edition. University of Chicago Press, Chicago.

Jeffreys H., 1924 *The Earth: Its Origins, History and Physical Constitution*. Cambridge University Press, Cambridge.

Jeffreys H., 1931 *Scientific Inference*. Cambridge University Press, Cambridge.

Jeffreys H., 1932 On the theory of errors and least squares. *Proceedings of the Royal Society of London A* **138**: 48–55.

Jeffreys H., 1933 Probability, statistics, and the theory of errors. *Proceedings of the Royal Society of London Series A* **140**: 523–535.

Jeffreys H., 1934 Probability and scientific method. *Proceedings of the Royal Society of London A* **146**: 9–16.

Jeffreys H., 1939 *Theory of Probability*. Clarendon Press, Oxford.

Jeffreys H., 1961 *Theory of Probability*, 3rd edition. Clarendon Press, Oxford.

Johnson W.E., 1924 *Logic, Part III: The Logical Foundations of Science*. Cambridge University Press, Cambridge.

Joyce J.M., 1999 *The Foundations of Causal Decision Theory*. Cambridge University Press, Cambridge.

Kalbfleisch J.G., 1985 *Probability and Statistical Inference, Vol. 2: Statistical Inference*, 2nd edition. Springer, New York.

Kalbfleisch J.G., Sprott D.A.: On the logic of tests of significance with special reference to testing the significance of Poisson-distributed observations; in Menges G., (ed): *Information, Inference and Decision*. Reidel, Dordrecht, 1974, pp. 99–109.

Kamlah A.: The decline of the Laplacian theory of probability; in Krüger L., Daston L.J., Heidelberger M., (eds): *The Probabilistic Revolution* (Vol. **1**). MIT Press, Cambridge, MA, 1987, pp. 91–116.

Kant I., 1781 *Kritik der reinen Vernunft*. J.K. Hartknoch, Riga.

Kendall D., 1968 Obituary: A.C. Aitken, D. Sc, FRS. *Proceedings of the Edinburgh Mathematical Society (Series 2)* **16**(2): 151–176.

Kendall D., 1977 Maurice Fréchet, 1878–1973. *Journal of the Royal Statistical Society A* **140**(4): 566.

Kendall M.G., 1945 *The Advanced Theory of Statistics* (Vol. **1**), 2nd Revised edition. Charles Griffin & Co. Ltd, London.

Kendall M.G., 1948 Who discovered the Latin Square? *The American Statistician* **2**(4): 13.

Kendall M.G., 1971 Studies in the history of probability and statistics. XXVI. The work of Ernst Abbe. *Biometrika* **58**(2): 369–373 (Reprinted in Kendall and Plackett (1977), pp. 331–335).

Kendall M.G., Moran P.A.P., 1963 *Geometrical Probability*. Charles Griffin & Co. Ltd, London.

Kendall M.G., Plackett R.L. eds, 1977 *Studies in the History of Statistics and Probability* (Vol. **II**). Griffin, London.

Keuzenkamp H.A., 2004 *Probability, Econometrics and Truth: The Methodology of Econometrics*. Cambridge University Press, Cambridge.

Keuzenkamp H.A., McAleer M., 1995 Simplicity, scientific inference and econometric modelling. *The Economic Journal* **105**(428): 1–21.

Keynes J.M., 1921 *A Treatise on Probability*. Macmillan & Co, London.

Kiefer J.C., Moore D.S.: Large sample comparison of tests and empirical Bayes procedures; in Owen D.B., (ed): *On the History of Statistics and Probability: Proceedings of a Symposium on the American Mathematical Heritage to Celebrate the Bicentennial of the United States of America, held at Southern Methodist University, May 27–29,1974*. Marcel Dekker, New York, 1976, pp. 349–365.

Kline M., 1972 *Mathematical Thought From Ancient to Modern Times* (Vol. **II**). Oxford University Press, Oxford.

Kolmogorov A.N., 1950 Unbiased estimates. *Izvestiya Rossiiskoi Akademii Nauk Seriya Matematicheskaya* **14**(4): 303–326.

Koopman B.O., 1936 On distributions admitting a sufficient statistic. *Transactions of the American Mathematical Society* **39**(3): 399–409.

Kramp C., 1799 *Analyse des Réfractions Astronomiques et Terrestres*. Dannbach, Strasbourg.

Kreps D.M., 1988 *Notes on the Theory of Choice*. Westview Press, Boulder, CO/London.

Kruskal W., 1946 Helmert's distribution. *American Mathematical Monthly* **53**: 435–438.

Ku H.H., 1969 *Precision, Measurement and Calibration: Statistical Concepts and Procedures*. U.S. Government Printing Office, Washington, DC.

Lagrange J.-L., 1759 Recherches sur la nature et la propagation du son. *Miscellanea Taurinensis* **2**: 1–112 (Oeuvres 1, pp. 39–148).

Lagrange J.-L., 1774 Sur une nouvelle espèce de calcul relatif à la différentiation et à l'intégration des quantités variables. *Nouveaux Mémoires de l'Académie Royale des Sciences et Belles-Lettres (1772)*: 185–221 (Oeuvres 3, pp. 441–476).

Lagrange J.-L., 1776 Mémoire sur l'utilité de la méthode de prendre le milieu entre les résultats de plusieurs observations. *Miscellanea Taurinensis* **5**(1770–1773): 167–232 (Oeuvres 2, pp. 173–234).

Lagrange J.-L., 1788 *Méchanique Analytique*. Chez La Veuve Desaint, Paris.

Lai T.L., Siegmund D., 1986 The contributions of Herbert Robbins to mathematical statistics. *Statistical Science* **1**(2): 276–284.

Lambert J.H., 1760 *Photometria sive de Mensura et Gradibus Luminis, Colorum et Umbrae*. Detleffsen, Augsburg.

Lambert J.H., 1764 *Neues Organon oder Gedanken über die Erforschung und Bezeichnung des Wahren und dessen Unterscheidung vom Irrthum und Schein*. Wendler, Leipzig.

Lambert J.H., DiLaura D.L., 2001 *Photometry, or on the Measure and Gradations of Light, Colors and Shade*. Illuminating Engineering Society of North America, New York.

Lancaster H.O., 1966 Forerunners of the Pearson χ^2. *Australian Journal of Statistics* **8**: 117–126.

Lancaster H.O., 1972 Development of the notion of statistical dependence. *Mathematical Chronicle* **2**: 1–16.

Lane D.A.: Fisher, Jeffreys, and the nature of probability; in Fienberg S.E., Hinkley D.V., (eds): *R.A. Fisher: An Appreciation*. Springer, New York, 1980, pp. 148–160.

Laplace P.-S., 1774a Mémoire sur la probabilité des causes par les événements. *Mémoire de l'Académie Royale des Sciences de Paris (savants étrangers)* **6**: 621–656 (OC 8, 27–65).

Laplace P.-S., 1774b Mémoire sur les suites récurro-récurrentes et sur leurs usages dans la théorie des hasards. *Mémoire de l'Académie Royale des Sciences de Paris (savants étrangers)* **6**: 353–371 (OC 8, 5–24).

Laplace P.-S., 1776a Mémoire sur l'inclinaison moyenne des orbites des comètes, sur la figure de la terre, et sur les fonctions. *Mémoire de l'Académie Royale des Sciences de Paris (savants étrangers)* **7**: 503–540 (1773, OC 8, 279–321).

Laplace P.-S., 1776b Recherches sur l'intégration des équations différentielles aux différences finies et leur usage dans la théorie des hasards. *Mémoire de l'Académie Royale des Sciences de Paris (savants étrangers)* **7**: (1773, OC 8, 69–197).

Laplace P.-S., 1780 Mémoire sur l'usage du calcul aux différences partielles dans la théorie des suites. *Mémoire de l'Académie Royale des Sciences de Paris*: 99–122 (1777, OC 9, 313–335).

Laplace P.-S., 1781 Mémoire sur les probabilités. *Histoire de l'Académie Royale des Sciences* **6**: 227–323 (1778, OC 9, 383–485).

Laplace P.-S., 1782 Mémoire sur les suites. *Mémoire de l'Académie Royale des Sciences de Paris*: 207–309 (1779, OC 10, 1–89).

Laplace P.-S., 1785 Mémoire sur les approximations des formules qui sont fonctions de très grands nombres. *Mémoire de l'Académie Royale des Sciences*: 1–88 (1782, OC 10, 209–291).

Laplace P.-S., 1786a Mémoire sur les approximations des formules qui sont fonctions de très grands nombres (Suite). *Mémoire de l'Académie Royale des Sciences*: 423–467 (1783, OC 10, 295–338).

Laplace P.-S., 1786b Sur les naissances, les mariages et les morts à Paris, depuis 1771 jusqu'en 1784, et dans toute l'étendue de la France, pendant les années 1781 et 1782. *Mémoires de l'Académie Royale des Sciences de Paris*: 693–702 (OC 11, 35–46).

Laplace P.-S., 1793 Sur quelques points du système du monde. *Mémoires de l'Académie Royale des Sciences de Paris*: **553**.

Laplace P.-S., 1796 *Exposition du Système du Monde* (**2** vols), Paris. (Translated as *The System of the World* by J. Pound (1809). Published by R. Phillips, London).

Laplace P.-S., 1799 *Traité de Mécanique Céleste*. Duprat, Paris (Vols 1 and 2: 1799, Vol. 3: 1802, Vol. 4: 1805).

Laplace P.-S., 1809 Mémoire sur divers points d'analyse. *Journal de l'École Polytechnique* **8**: 229–265.

Laplace P.-S., 1810a Mémoire sur les approximations des formules qui sont fonctions de très grands nombres et sur leur application aux probabilités. *Mémoire de l'Académie Royale des Sciences de Paris*: 353–415 (OC 12, 301–345).

Laplace P.-S., 1810b Supplément au mémoire sur les approximations des formules qui sont fonctions de très grands nombres et sur leur application aux probabilités. *Mémoire de l'Académie des Sciences*: 559–565 (OC 12, 349–353).

Laplace P.-S., 1811a Mémoire sur les intégrales définies et leur application aux probabilités, et spécialement à la recherche du milieu qu'il faut choisir entre les résultats des observations. *Mémoires de l'Académie Royale des Sciences de Paris* **11**: 279–347 (OC 12, 357–412).

Laplace P.-S., 1811b Sur les intégrales définies. *Bulletin de la Société Philomatique* **43**: 262–266.

Laplace P.-S., 1812 *Théorie Analytique des Probabilités*. Courcier, Paris.

Laplace P.-S., 1814 *Essai Philosophique sur les Probabilités*. Courcier, Paris (Translated as *A Philososphical Essay on Probabilities*, 6th edition by F.W. Truscott and F.L. Emory (1902). Reprinted in 1951 by Dover, New York).

Laplace P.-S., 1818 *Deuxième Supplément à la Théorie Analytique des Probabilités*. Courcier, Paris.

Laplace P.-S., 1902 *A Philosophical Essay on Probabilities*. John Wiley & Sons, New York (Translated by F. Truscott and F. Emory).

Le Cam L.: On the asymptotic theory of estimation and testing hypotheses; in Neyman J., (ed): *Proceedings of the Third Berkeley Symposium on Mathematical Statistics and Probability, Volume 1: Contributions to the Theory of Statistics*. California University Press, Berkeley, 1956, pp. 129–156.

Le Cam L., 1986 The Central Limit Theorem around 1935. *Statistical Science* **1**(1): 78–91.

Leadbetter M.R.: Harald Cramér; in Heyde C.C., Seneta E., (eds): *Statisticians of the Centuries*. Springer, New York, 2001, pp. 439–443.

Lecat M., 1935 *Erreurs de Mathématiciens des Origines à Nos Jours*. Anc[ne] Librairie Castaigne, Bruxelles.

Lee A.F.S., Gurland J., 1975 Size and power of tests for equality of means of two normal populations with unequal variances. *Journal of the American Statistical Association* **70**(352): 933–941.

Legendre A.M., 1784 Recherches sur la figure des planètes. *Mémoire de l'Académie Royale des Sciences* **10**: 370–389.

Legendre A.M., 1805 *Nouvelles Méthodes Pour la Détermination des Orbites des Comètes*. Courcier, Paris.

Lehmann E.L., 1946 Une propriété optimale de certains ensembles critiques du type A1. *Comptes rendus de l'Académie des Sciences, Paris* **223**(16): 567–569.

Lehmann E.L., 1959 *Testing Statistical Hypotheses*. John Wiley & Sons, New York.

Lehmann E.L.: Introduction to Neyman and Pearson (1933) On the problem of the most efficient tests of statistical hypotheses; in Kotz S., Johnson N.L., (eds): *Breakthroughs in Statistics* (Vol. **I**). Springer, New York, 1992a, pp. 67–72.

Lehmann E.L.: Introduction to Student (1908) The probable error of a mean; in Kotz S., Johnson N.L., (eds): *Breakthroughs in Statistics* (Vol. **II**). Springer, New York, 1992b, pp. 29–32.

Lehmann E.L., 1993 The Fisher, Neyman-Pearson theories of testing hypotheses: one theory or two? *Journal of the American Statistical Association* **88**(424): 1242–1249.

Lehmann E.L., 1994 Jerzy Neyman. *Biographical Memoirs* **63**: 395–420.

Lehmann E.L.: Scheffé, Henry; in Kotz S., Read C.B., Balakrishnan N., Vidakovic B., (eds): *Encyclopedia of Statistical Sciences*. John Wiley & Sons, Inc., Hoboken, NJ, 2006, pp. 7472–7474.

Lehmann E.L., 2011 *Fisher, Neyman, and the Creation of Classical Statistics*. Springer, New York.

Lehmann E.L., Scheffé H., 1950 Completeness, similar regions, and unbiased estimation: Part I. *Sankhya* **10**(4): 305–340.

Leibniz G.W., 1710 *Théodicée*. Garnier-Flammarion, Paris (1969 edition).

Leibniz G.W., 1971 *Mathematische Schriften* (ed. C.I. Gerhardt, 7 vols., Asher and Schmidt, Berlin, 1849–63). Geor Olms, Hildesheim.

Lenhard J., 2006 Models and statistical inference: the controversy between Fisher and Neyman-Pearson. *The British Journal for the Philosophy of Science* **57**(1): 69–91.

Lévy P., 1925 *Calcul des Probabilités*. Gauthier-Villars, Paris.

Lillehammer H., Mellor D.H., 2005 *Ramsey's Legacy*. Oxford University Press, Oxford.

Lindeberg J.W., 1922 Eine neue Herleitung des Exponentialgesetzes in der Wahrscheinlichkeitsrechnung. *Mathematische Zeitschrift* **15**: 211–225.

Lindgren B.W., 1993 *Statistical Theory*, 4th edition. CRC Press, London.

Lindley D.V., 1980 L.J. Savage—His work in probability and statistics. *The Annals of Statistics* **8**(1): 1–24.

Lindley D.V., 1986 Bruno de Finetti, 1906–1985 (obituary). *Journal of the Royal Statistical Society A* **149**: 252.

Lindley D.V., 1989 Obituary: Harold Jeffreys, 1891–1989. *Journal of the Royal Statistical Society A* **152**(Part 3): 417–419.

Lindley D.V.: Savage, L.J.; in Kotz S., Read C.B., Balakrishnan N., Vidakovic B., (eds): *Encyclopedia of Statistical Sciences*. John Wiley & Sons, Inc., Hoboken, NJ, 2006, pp. 7443–7445.

Linnik Yu.V., 1968 *Statistical Problems with Nuisance Parameters*. Translations of mathematical monographs, no. 20 (from the 1966 Russian edition) American Mathematical Society, New York.

Liouville J., 1852 Rapport sur un Mémoire de M. Jules Bienaymé, Inspecteur général des Finances, concernant la probabilité des erreurs d'aprés la méthode des moindres carrés. *Louiville's Journal de Mathématiques Pures et Appliquées* **17**: 31–32.

Little R.J., 1989 Testing the equality of two independent binomial proportions. *The American Statistician* **43**(4): 283–288.

Loève M., 1950 Fundamental limit theorems of probability theory. *The Annals of Mathematical Statistics* **21**(3): 321–338.

Lubet J.-P.: Laplace et *la Théorie Analytique des Probabilités*: Itinéraires de Découverte; in Barbin É., Lamarche J.P. (eds): *Histoires de Probabilités et de Statistiques*. Ellipses, Paris, 2004, pp. 197–224.

Luce R.D., Raiffa H., 1957 *Games and Decisions*. John Wiley & Sons, New York.

Lüroth J., 1876 Vergleichung von zwei Werten des wahrscheinlichen Fehlers. *Astronomische Nachrichten* **87**: 209–220.

Lyapunov A.M., 1900 Sur une proposition de la théorie des probabilités. *Mémoires de l'Académie Impériale des Sciences de St. Petérsbourg* **5**(13): 359–386 (In Adams (2009), pp. 151–171).

Lyapunov A.M., 1901a Nouvelle forme du théorème sur la limite de probabilité. *Mémoires de l'Académie Impériale des Sciences de St. Petersburg* **12**: 1–24 (In Adams (2009), pp. 175–191).

Lyapunov A.M., 1901b Sur un théorème du calcul des probabilités. *Comptes Rendus Hebdomadaires des Séances de l'Académie des Sciences de Paris* **132**: 126–128 (In Adams (2009), pp. 149–150).

Lyapunov A.M., 1901c Une proposition générale du calcul des probabilités. *Comptes Rendus Hebdomadaires des Séances de l'Académie des Sciences de Paris* **132**: 814–815 (in Adams (2009), pp. 173–174).

MacKenzie D.A., 1978 *Statistics in Britain, 1865–1930*. Edinburgh University Press, Edinburgh.

Magnello E.: Karl Pearson; in Heyde C.C., Seneta E., (eds): *Statisticians of the Centuries*. Springer, New York, 2001a, pp. 248–256.

Magnello E.: Walter Frank Raphael Weldon; in Heyde C.C., Seneta E., (eds): *Statisticians of the Centuries*. Springer, New York, 2001b, pp. 261–264.

Magnello M.E.: Karl Pearson, paper on the chi-square goodness-of-fit test (1900); in Grattan-Guinness I., (ed): *Landmark Writings in Western Mathematics 1640–1940*. Elsevier, Amsterdam, 2005, pp. 724–731.

Mahalanobis P.C., 1932 Auxiliary tables for Fisher's Z-test in analysis of variance. *Indian Journal of Agricultural Science* **2**: 679–693.

Maistrov L.E., 1974 *Probability Theory: A Historical Sketch*. Academic Press, New York.

Maraun M.D., Gabriel S., Martin J., 2011 The mythologization of regression towards the mean. *Theory & Psychology* **21**: 762–784.

Mardia K.V., 1990 Obituary: Professor B.L. Welch. *Journal of the Royal Statistical Society: Series A (Statistics in Society)* **153**: 253–254.

Markov A.A., 1900 *Ischislenie Veroiatnostei*. St. Petersburg.

Maskell E.J., 1929 Experimental error: a survey of recent advances in statistical method (continued). *Journal of Tropical Agriculture* **6**: 5–11.

May K.O.: Gauss, Carl-Friedrich; in Gillispie C.C., (ed): *Dictionary of Scientific Biography* (Vol. **5**). Charles Scribner's Sons, New York, 1972, pp. 298–315.

Mayo D.G., 1996 *Error and the Growth of Experimental Knowledge*. University of Chicago Press, Chicago.

McCormmach R., 2012 *Weighing the World: The Reverend John Michell of Thornhill*. Springer, New York.

McGrayne S.B., 2011 *The Theory That Would Not Die: How Bayes' Rule Cracked the Enigma Code, Hunted Down Russian Submarines, & Emerged Triumphant from Two Centuries of Controversy*. Yale University Press, New Haven, CT.

Mercer W.B., Hall A.D., 1911 The experimental error of field trials. *The Journal of Agricultural Science* **4**(2): 107–132.

Merriman M., 1877 A list of writings relating to the method of least squares, with historical and critical notes. *Transactions of the Connecticut Academy of Arts and Sciences* **4**: 151–232.

Michell J., 1767 An Inquiry into the probable parallax, and magnitude of the fixed stars, from the quantity of light which they afford us, and the particular circumstances of their situation. *Philosophical Transactions (1683–1775)* **57**: 234–264.

Millar R.B., 2011 *Maximum Likelihood Estimation and Inference: With Examples in R, SAS and ADMB*. John Wiley & Sons, Inc., Hoboken, NJ.

Molina E.C., 1930 The theory of probability: some comments on Laplace's théorie analytique. *Bulletin of the American Mathematical Society* **36**(6): 369–392.

Montmort P.R.d., 1708 *Essay d'Analyse sur les Jeux de Hazard*. Quillau, Paris.

Montmort P.R.d., 1713 *Essay d'Analyse sur les Jeux de Hazard*, 2nd edition. Quillau, Paris.

Moore D.S., 1971 Maximum likelihood and sufficient statistics. *American Mathematical Monthly* **78**: 50–52.

Morse P.M., 1982 Bernard Osgood Koopman, 1900–1981. *Operations Research* **30**(3): 417–427.

Mukhopadhyay N., 2000 *Probability and Statistical Inference*. Marcel Dekker, New York.

Napoléon [Bonaparte], 1870 *Correspondance de Napoléon 1er Publiée par Ordre de l'Empéreur Napoléon III*. Tome 30, Oeuvres de Napoléon 1er à Saint-Hélène. Plon and Dumaine, Paris.

Neyman J., 1934 On the two different aspects of the representative method: the method of stratified sampling and the method of purposive selection. *Journal of the Royal Statistical Society* **97**(4): 558–625.

Neyman J., 1935 Sur un teorema concernente le cosiddette statistiche sufficienti. *Giornale dell Istituto Italiano degli Attuari* **6**: 320–334.

Neyman J., 1938a A historical note on Karl Pearson's deduction of the moments of the binomial. *Biometrika* **30**(1/2): 11–15.

Neyman J., 1938b L'estimation statistique, traitée comme un problème classique de probabilité. *Actualitées Scientifiques et Industrielles* **739**: 25–57.

Neyman J., 1938c *Lectures and Conferences on Mathematical Statistics*. Graduatre School, US Department of Agriculture, Washington, DC.

Neyman J., 1941 Fiducial argument and the theory of confidence intervals. *Biometrika* **32**: 128–150.

Neyman J., 1950 *First Course in Probability and Statistics*. Henry Holt, New York.

Neyman J., 1951 Fisher's collected papers. *The Scientific Monthly* **42**: 406–408.

Neyman J., 1956 Note on an article by Sir Ronald Fisher. *Journal of the Royal Statistical Society B* **18**: 288–294.

Neyman J., 1962 Two breakthroughs in the theory of statistical decision making. *Review of the International Statistical Institute* **30**(1): 11–27.

Neyman, J., 1967 *A Selection of Early Statistical Papers of J. Neyman*. Univ. California Press. Berkeley.

Neyman J., Iwaszkiewicz K., 1935 Statistical problems in agricultural experimentation. *Supplement to the Journal of the Royal Statistical Society* **2**(2): 107–180.

Neyman J., Pearson E.S., 1928a On the use and interpretation of certain test criteria for purposes of statistical inference Part I. *Biometrika* **20A**: 175–240.

Neyman J., Pearson E.S., 1928b On the use and interpretation of certain test criteria for purposes of statistical inference. Part II. *Biometrika* **20A**: 263–294.

Neyman J., Pearson E.S., 1933 On the problem of the most efficient tests of statistical hypotheses. *Philosophical Transactions of the Royal Society of London A* **231**: 289–337.

Neyman J., Pearson E.S., 1936 Contributions to the theory of testing statistical hypotheses. (I) Unbiased critical regions of Type A and Type A_1. *Statistical Research Memoirs* **1**: 1–37.

Neyman J., Scott E.L., 1948 Consistent estimates based on partially consistent observations. *Econometrica: Journal of the Econometric Society* **16**(1): 1–32.

Nieuwentyt B., 1715 *Het regt gebruik der Wereldbeschouwingen*. Amsterdam.

O'Beirne T.H., 1965 *Puzzles and Paradoxes*. Oxford University Press, Oxford.

O'Hagan A., 1994 *Kendall's Advanced Theory of Statistic 2B: Bayesian Inference*. Edward Arnold, London.

Olkin I., 1989 A conversation with Maurice Bartlett. *Statistical Science* **4**(2): 151–163.

Pace L., Salvan A., 1997 *Principles of Statistical Inference: From a Neo-Fisherian Perspective*. World Scientific, Singapore.

Page W., 1984 An interview with Herbert Robbins. *The College Mathematics Journal* **15**(1): 2–24.

Parmigiani G., Inoue L.Y.T., 2009 *Decision Theory: Principles and Approaches*. John Wiley & Sons, Ltd, Chichester.

Parzen E., 1960 *Modern Probability Theory and its Applications*. John Wiley & Sons, New York.

Paty M.: D'Alembert et les probabilités; in Rashed R., (ed): *Sciences à l'Époque de la Révolution Française: Recherches Historiques*. Blanchard, Paris, 1988, pp. 203–265.

Pawitan Y., 2001 *In All Likelihood: Statistical Modelling and Inference Using Likelihood*. Oxford University Press, Oxford.

Pearce S.C.: Introduction to Fisher (1925) Statistical methods for research workers; in Kotz S., Johnson N.L., (eds): *Breakthroughs in Statistics* (Vol. **II**). Springer, New York, 1992, pp. 59–65.

Pearson E.S., 1925 Bayes' theorem, examined in the light of experimental sampling. *Biometrika* **17**(3/4): 388–442.

Pearson E.S., 1938 *Karl Pearson: An Appreciation of Some Aspects of His Life and Work*. Cambridge University Press, Cambridge.

Pearson E.S., 1939 "Student" as statistician. *Biometrika* **30**(3/4): 210–250 (Reprinted in Pearson and Kendall (1970), pp. 360–403).

Pearson E.S., 1955 Statistical concepts in the relation to reality. *Journal of the Royal Statistical Society Series B (Methodological)* **17**(2): 204–207.

Pearson E.S.: The Neyman-Pearson story: 1926–34; in David F.N., (ed): *Research Papers in Statistics: Festschrift for J. Neyman*. John Wiley & Sons, New York, 1966, pp. 1–23.

Pearson E.S., 1968 Studies in the history of probability and statistics. XX. Some early correspondence between W. S. Gosset, R.A. Fisher and Karl Pearson, with notes and comments. *Biometrika* **55**(3): 445–457 (Reprinted in Pearson and Kendall (1970), pp. 405–417).

Pearson E.S., Gosset W.S., Plackett R.L., Barnard G.A., 1990 *Student: A Statistical Biography of William Sealy Gosset*. Oxford University Press, New York.

Pearson E.S., Hartley H., 1954 *Biometrika Tables for Statisticians*. Cambridge University Press, Cambridge.

Pearson E.S., Kendall M.G. eds, 1970 *Studies in the History of Statistics and Probability* (Vol. **I**). Griffin, London.

Pearson K., 1892 *The Grammar of Science*. Walter Scott, London.

Pearson K., 1894 Contributions to the mathematical theory of evolution. *Philosophical Transactions of the Royal Society of London Series A* **185**: 71–110.

Pearson K., 1897a Mathematical contributions to the theory of evolution. On the law of ancestral heredity. *Proceedings of the Royal Society of London* **62**: 386–412.

Pearson K., 1897b *The Chances of Death and Other Studies in Evolution* (Vol. **I**). E. Arnold, London.

Pearson K., 1900a Mathematical contributions to the theory of evolution. VII. On the correlation of characters not quantitatively measurable. *Philosophical Transactions of the Royal Society of London Series A* **195**: 1–47.

Pearson K., 1900b On the criterion that a given system of deviations from the probable in the case of a correlated system of variables is such that it can be reasonably supposed to have arisen from random sampling. *Philosophical Magazine Series 5* **50**(302): 157–175.

Pearson K., 1901a Mathematical contributions to the theory of evolution: IX. On the principle of homotyposis and its relation to heredity, to variability of the individual, and to that of race. Part I: homotyposis in the vegetable kingdom. *Philosophical Transactions of the Royal Society of London Series A* **197**: 285–379.

Pearson K., 1901b *The Ethic of Freethought and Other Addresses and Essays*. A. and C. Black, London.

Pearson K., 1904 Mathematical contributions to the theory of evolution: XIII. On the theory of contingency and its relation to association and normal correlation. *Draper's Company Research Memoirs, Biometric Series, no 1 (London: Dulau)*: 1–37.

Pearson K., 1907 Reply to certain criticisms of Mr G.U. Yule. *Biometrika* **5**: 470–476.

Pearson K., 1908 Obituary notices of fellows deceased. *Proceedings of the Royal Society of London B (Biological Sciences)* **80**(544): i–xxxviii.

Pearson K., 1911 On the probability that two independent distributions of frequency are really samples from the same population. *Biometrika* **8**: 250–254.

Pearson, K., 1916 On the general theory of multiple contingency with special reference to partial contingency. *Biometrika* **11**: 145–158.

Pearson K., 1920 Notes on the history of correlation. *Biometrika* **13**: 25–45 (Reprinted in Pearson and Kendall (1970), pp. 185–205).

Pearson K., 1922 On the χ^2 test of goodness of fit. *Biometrika* **14**: 186–191.

Pearson K., 1924 On the difference and the doublet tests for ascertaining whether two samples have been drawn from the same population. *Biometrika* **16**(3/4): 249–252.

Pearson K., 1928a On a method of ascertaining limits to the actual number of marked members in a population of given size from a sample. *Biometrika* **20A**(1/2): 149–174.

Pearson K., 1928b The contribution of Giovanni Plana to the normal bivariate frequency surface. *Biometrika* **20**(3–4): 295–298.

Pearson K., 1930 *The Life, Letters and Labours of Francis Galton (Part IIIA)*. Cambridge University Press, Cambridge.

Pearson K., 1931 Historical note on the distribution of the standard deviations of samples of any size drawn from an indefinitely large normal parent population. *Biometrika* **23**: 416–418.

Pearson K., 1932 Experimental discussion of the (χ^2, P) test for goodness of fit. *Biometrika* **24**(3/4): 351–381.

Pearson K., 1978 *The History of Statistics in the 17th and 18th Centuries, Against the Changing Background of Intellectual, Scientific and Religious Thought*. Edited by E.S. Pearson. Lectures by Karl Pearson given at University College London during academic sessions 1921–1933. Griffin, London.

Pearson K., Filon L.N.G., 1898 Mathematical contributions to the theory of evolution. IV. On the probable errors of frequency constants and on the influence of random selection on variation and correlation. *Philosophical Transactions of the Royal Society of London Series A* **191**: 229–311.

Pearson K., Heron D., 1913 On theories of association. *Biometrika* **9**: 159–315.

Pearson K., Lee A., Bramley-Moore L., 1895 Mathematical contributions to the theory of evolution. II. Skew variations in homogeneous material. *Philosophical Transactions of the Royal Society of London Series A* **186**: 343–414.

Pearson K., Lee A., Bramley-Moore L., 1896 Mathematical contributions to the theory of evolution. III. Regression, heredity and panmixia. *Philosophical Transactions of the Royal Society of London Series A* **187**: 253–318.

Pedersen J.G., 1978 Fiducial inference. *International Statistical Review* **46**: 147–170.

Pfanzagl J., Sheynin O., 1996 Studies in the history of probability and statistics. XLIV. A forerunner of the t-distribution. *Biometrika* **83**: 891–898.

Pfanzagl J., Sheynin O.: Lüroth, Jakob; in Kotz S., Read C.B., Balakrishnan N., Vidakovic B., (eds): *Encyclopedia of Statistical Sciences* (Vol. **7**). John Wiley & Sons, Inc., Hoboken, NJ, 2006, pp. 4433–4434.

Pfeiffer P.E., 1990 *Probability for Applications*. Springer-Verlag, New York.

Picard R.: Randomization and design: II; in Fienberg S.E., Hinkley D.V., (eds): *R.A. Fisher: An Appreciation*. Springer, New York, 1980, pp. 46–58.

Pitman E.J.G., 1936 Sufficient statistics and intrinsic accuracy. *Mathematical Proceedings of the Cambridge Philosophical Society* **32**(4): 567–579.

Pitman E.J.G., 1939 The estimation of the location and scale parameters of a continuous population of any given form. *Biometrika* **30**(3/4): 391–421.

Pitman E.J.G.: Reminiscences of a mathematician who strayed into statistics; in Gani J., (ed): *The Making of Statisticians*. Springer, New York, 1982, pp. 111–125.

Plackett R.L., 1949 A historical note on the method of least squares. *Biometrika* **36**(3/4): 458–460.

Plackett R.L., 1972 The discovery of the method of least squares. *Biometrika* **59**(2): 239–251 (Reprinted in Kendall and Plackett (1977), pp. 279–291).

Plackett R.L., 1977 The marginal totals of a 2 x 2 table. *Biometrika* **64**: 37–42.

Plackett R.L., 1983 Karl Pearson and the chi-squared test. *International Statistical Review* **51**(1): 59–72.

Plana G.A.A., 1813 Mémoire sur divers problèmes de probabilité. *Mémoires de l'Académie Impériale de Turin, pour les Années 1811–1812* **XX**(20): 355–408.

Plummer H.C., 1940 *Probability and Frequency*. Macmillan, London.

Poincaré H., 1912 *Calcul des Probabilités*, 2nd edition. Gauthier-Villars, Paris.

Poisson S.D., 1811 Sur les intégrales définies. *Bulletin de la Société Philomatique* **50**: 375–380.

Poisson S.D., 1813 Mémoire sur les intégrales définies. *Journal de l'École Polytechnique* **9**: 215–246.

Poisson S.D., 1824 Sur la probabilité des résultats moyens des observations. *Connaissance des Temps pour l'an 1827*: 273–302.

Poisson S.D., 1837 *Recherches sur la Probabilité des Jugements en Matière Criminelle et Matière Civile*. Bachelier, Paris.

Popper K.R., 1959 *The Logic of Scientific Discovery*. Basic Books, New York (Originally published by Springer in 1934 as *Logik der Forschung*).

Porter T.M., 1986 *The Rise of Statistical Thinking, 1820–1900*. Princeton University Press, Princeton.

Porter T.M., 2010 *Karl Pearson: The Scientific Life in a Statistical Age*. Princeton University Press, Princeton.

Pratt J.W., 1976 F.Y. Edgeworth and R.A. Fisher on the efficiency of maximum likelihood estimation. *Applied Statistics* **4**(3): 501–514.

Preece D.A., 1990 R.A. Fisher and experimental design: a review. *Biometrics* **46**(4): 925–935.

Press S.J., 2009 *Subjective and Objective Bayesian Statistics: Principles, Models, and Applications*, 2nd edition. John Wiley & Sons, Inc., Hoboken, NJ.

Provine W.B., 2001 *The Origins of Theoretical Population Genetics*. University of Chicago Press, Chicago.

Ramsey F.P., 1931 *The Foundations of Mathematics and Other Logical Essays*. Rouledge and Kegan Paul, London (ed. R.B. Braithwaite).

Ramsey P.F.: Truth and possibility; in Kyburg H.E., Smokler H.E., (eds): *Studies in Subjective Probability*. John Wiley & Sons, New York, 1964, pp. 61–92.

Rao C.R., 1945 Information and accuracy attainable in the estimation of statistical parameters. *Bulletin of the Calcutta Mathematical Society* **37**(3): 81–91 (Reprinted in *Breakthroughs in Statistics: Volume I* (eds. S. Kotz and N.L. Johnson) 1992, pp. 235–247, Springer, New York).

Rao C.R., 1973 *Linear Statistical Inference and its Applications*, 2nd edition. John Wiley & Sons, New York.

Rao C.R., 1992 R.A. Fisher: the founder of modern statistics. *Statistical Science* **7**(1): 34–48.

Rao C.R.: Three score years of research in statistics: 1941–2000; in Das Gupta S., Ghosh J.K., Mitra S.K., Mukhopadhyay A.C., Prakash Rao B.L.S., Rao P.S.S.N.V.P., Rao S.B., Sarma Y.R., (eds): *Selected Papers of C.R. Rao* (Vol. **V**). Indian Statistical Institute/New Age International (P) Limited, Kolkata/New Delhi, 2001, pp. 396–459.

Reid C., 1996 *Courant*. Springer-Verlag, New York

Reid C., 1998 *Neyman*. Springer-Verlag, New York.

Reid N., 1994 A conversation with Sir David Cox. *Statistical Science* **9**(3): 439–455.

Reid N., 1995 The roles of conditioning in inference. *Statistical Science* **10**(2): 138–157.

Rhodes E.C., 1924 On the problem whether two given samples can be supposed to have been drawn from the same population: Part I. *Biometrika* **16**(3/4): 239–248.

Robbins H., 1956 An empirical Bayes approach to statistics. *Proceedings of the Third Berkeley Symposium on Mathematical Statistics and Probability, Volume 1: Contributions to the Theory of Statistics*: 157–163.

Robbins H.E., 1944 On the measure of a random set. *The Annals of Mathematical Statistics* **15**(1): 70–74.

Robert C.P., 2012 *Reading Théorie Analytique des Probabilités*. Report no. arXiv:1203.6249.

Rohde C.A., 2014 *Introductory Statistical Inference with the Likelihood Function*. Springer, New York.

Rojo J., 2011 Erich Leo Lehmann-A glimpse into his life and work. *The Annals of Statistics* **39**(5): 2244–2265.

Romano J.P., Siegel A.F., 1986 *Counterexamples in Probability and Statistics*. Wadsworth and Brooks/Cole, Monterey, CA.

Rosendahl C.O., Gortner R.A., Burr G.O., 1936 *J. Arthur Harris, Botanist and Biometrician*. The University of Minnesota Press, Minneapolis.

Roy R., 1961 Darmois, George 1888–1960. *Econometrica* **29**(1): 80–83.

Russell B., 1921 *The Analysis of Mind*. G. Allen & Unwin, London.

Sabbaghi A., Rubin D.B., 2014 Comments on the Neyman-Fisher controversy and its consequences. *Statistical Science* **29**(2): 267–284.

Sahlin N.E., 1990 *The Philosophy of F.P. Ramsey*. Cambridge University Press, Cambridge.

Salmon G., 1866 *Introductory to the Modern Higher Algebra*, 2nd edition. Hodges, Smith & Co., Dublin.

Salmon W., 1967 *The Foundations of Scientific Inference*. University of Pittsburgh Press, Pittsburgh.

Salsburg D., 2001 *The Lady Tasting Tea: How Statistics Revolutionized Science in the Twentieth Century*. W. H. Freeman & Co., New York.

Samueli J.J., 2012 *Les Fausses Démonstrations de l'Existence de Dieu*. Ellipses, Paris.

Samueli J.J., Boudenot J.C., 2009 *Une Histoire des Probabilités des Origines à 1900*. Ellipses, Paris.

Sarkar S., Pfeifer J., 2006 *The Philosophy of Science: An Encyclopedia*. Routledge, New York.

Satterthwaite F.E., 1946 An approximate distribution of estimates of variance components. *Biometrics Bulletin* **2**(6): 110–114.

Savage L.J., 1954 *The Foundations of Statistics*. John Wiley & Sons, New York.

Savage L.J.: The foundations of statistics reconsidered; in Neyman J., (ed): *Proceedings of the Fourth Berkeley Symposium on Mathematical Statistics and Probability, Volume 1: Contributions to the Theory of Statistics*. University of California Press, Berkeley, 1961, pp. 575–586.

Savage L.J., 1976 On rereading R.A. Fisher. *The Annals of Statistics* **4**(3): 441–500.

Savage L.J. 1981 *The Writings of Leonard Jimmie Savage – A Memorial Selection*. American Statistical Association, Washington.

Schay G., 2007 *Introduction to Probability with Statistical Applications*. Birkhauser, Boston.

Scheffé H., 1936 Asymptotic solutions of certain linear differential equations in which the coefficient of the parameter may have a zero. *Transactions of the American Mathematical Society* **40**(1): 127–154.

Scheffé H., 1942 An inverse problem in correlation theory. *The American Mathematical Monthly* **49**(2): 99–104.

Scheffé H., 1959 *The Analysis of Variance*. John Wiley & Sons, New York.

Scheffé H., 1970 Practical solutions of the Behrens-Fisher problem. *Journal of the American Statistical Association* **65**(332): 1501–1508.

Schervish M.J., 1995 *Theory of Statistics*. Springer-Verlag, New York.

Schmetterer L., 1974 *Introduction to Mathematical Statistics*. Springer, Berlin.

Scriba C.J.: Lambert, Johann Heinrich; in Gillispie C.C., (ed): *Dictionary of Scientific Biography* (Vol. **7**). Charles Scribner's Sons, New York, 1973, pp. 595–600.

Seal H.L., 1949 The historical development of the use of generating functions in probability theory. *Bulletin de l'Association des Actuaires Suisses* **49**: 209–228 (Reprinted in Kendall and Plackett (1977), pp. 67–86).

Seal H.L., 1967 Studies in the history of probability and statistics. XV. The historical development of the Gauss linear model. *Biometrika* **54**(1/2): 1–24 (Reprinted in Pearson and Kendall (1970), pp. 207–230).

Seidenfeld T., 1979 *Philosophical Problems of Statistical Inference: Learning from R.A. Fisher*. Reidel, Dordrecht.

Seidenfeld T., 1992 R.A. Fisher's fiducial argument and Bayes' theorem. *Statistical Science* **7**: 358–368.

Seneta E.: Russian probability and statistics before Kolmogorov; in Grattan-Guinness I., (ed): *Companion Encyclopedia of the History and Philosophy of the Mathematical Sciences* (Vol. **2**). Routledge, London, 1994, pp. 1325–1334.

Seneta E.: Poisson, Siméon-Denis; in Kotz S., Read C.B., Balakrishnan N., Vidakovic B., (eds): *Encyclopedia of Statistical Sciences*. John Wiley & Sons, Inc., Hoboken, NJ, 2006, pp. 6218–6220.

Senn S., 2004 Controversies concerning randomization and additivity in clinical trials. *Statistics in Medicine* **23**(24): 3729–3753.

Senn S., Richardson W., 1994 The first *t*-test. *Statistics in Medicine* **13**(8): 785–803.

Severini T.A., 1995 [Inference Based on Estimating Functions in the Presence of Nuisance Parameters]: Comment. *Statistical Science* **100**(2): 187–189.

Shafer G., 1986 Savage revisited. *Statistical Science* **1**(4): 463–485.

Shafer G.: Lambert, Johann Heinrich; in Kotz S., Read C.B., Balakrishnan N., Vidakovic B., (eds): *Encyclopedia of Statistical Sciences* (Vol. **6**). John Wiley & Sons, Inc., Hoboken, NJ, 2006, pp. 3947–3948.

Sheynin O.B., 1966 Origin of the theory of errors. *Nature* **211**: 1003–1004.

Sheynin O.B., 1971 J.H. Lambert's work on probability. *Archive for History of Exact Sciences* **7**(3): 244–256.

Sheynin O.B., 1973 Finite random sums (a historical essay). *Archive for History of Exact Sciences* **9**(4–5): 275–305.

Sheynin O.B., 1977 Laplace's theory of errors. *Archive for History of Exact Sciences* **17**(1): 1–61.

Sheynin O.B., 1978 S.D. Poisson's work on probability. *Archive for History of Exact Sciences* **18**(3): 245–300.

Sheynin O.B., 1979 C.F. Gauss and the theory of errors. *Archive for History of Exact Sciences* **20**(1): 21–72.

Sheynin O.B., 1989 A.A. Markov's work on probability. *Archive for History of Exact Sciences* **39**(4): 337–377.

Shoesmith E., 1985 Nicholas Bernoulli and the argument for divine providence. *International Statistical Review* **53**(3): 255–259.

Shoesmith E., 1987 The continental controversy over Arbuthnot's argument for divine providence. *Historia Mathematica* **14**(2): 133–146.

Shoesmith E.: Arbuthnot, John; in Kotz S., Read C.B., Balakrishnan N., Vidakovic B., (eds): *Encyclopedia of Statistical Sciences*. John Wiley & Sons, Inc., Hoboken, NJ, 2006, pp. 203–207.

Sinai Y., 2003 *Russian Mathematicians in the 20th Century*. World Scientific, Singapore.

Singpurwalla N.D., 2006 *Reliability and Risk: A Bayesian Perspective*. John Wiley & Sons, Inc., Hoboken, NJ.

Skyrms B., 2000 *Choice and Chance: An Introduction to Inductive Logic*, 4th edition. Wadsworth/Thomson Learning, Belmont, CA.

Slutsky E., 1913 On the criterion of goodness of fit of the regression lines and on the best method of fitting them to the data. *Journal of the Royal Statistical Society* **77**(1): 78–84.

Small C.G., 2010 *Expansions and Asymptotics for Statistics*. CRC Press, London.

Smirnov V.I., 1992 Biography of A.M. Lyapunov. *International Journal of Control* **55**(3): 775–784.

Smith A.G., 1989 Obituary: Sir Harold Jeffreys, FRS 1891–1989. *The Geographical Journal* **155**(3): 447–448.

Smith H.F., 1936 The problem of comparing the results of two experiments with unequal errors. *Journal of Scientific and Industrial Research* **9**: 211–212.

Smorynski C., 2012 *Chapters in Probability*. College Publications, London.

Snedecor G.W., 1934 *Calculation and Interpretation of Analysis of Variance and Covariance*. Collegiate Press, Ames, IA.

Solomon H., 1978 *Geometric Probability*. SIAM, Philadelphia.

Soper H.E., 1913 On the probable error of a correlation coefficient to a second approximation. *Biometrika* **9**: 91–115.

Soper H.E., Young A.W., Cave B.M., Lee A., Pearson K., 1917 On the distribution of the correlation coefficient in small samples. Appendix II to the papers of "Student" and R.A. Fisher. A cooperative study. *Biometrika* **11**: 328–413.

Spanos A., 1999 *Probability Theory and Statistical Inference*. Cambridge University Press, Cambridge.

Sprent P., 1994 E.J.G. Pitman, 1897–1993. *Journal of the Royal Statistical Society A* **157**: 153–154.

Sprott D.A., 1978 Gauss's contributions to statistics. *Historia Mathematica* **5**(2): 183–203.

Stapleton J.H., 2008 *Models for Probability and Statistical Inference: Theory and Applications*. John Wiley & Sons, Inc., Hoboken, NJ.

Stein C., 1956 Inadmissibility of the usual estimator for the mean of a multivariate normal distribution. *Proceedings of the Third Berkeley Symposium on Mathematical Statistics and Probability* **1**: 197–206.

Stein C., 1961 *Estimation of Many Parameters*. Wald Lectures, Institute of Mathematical Statistics. Unpublished.

Stigler S.M., 1973 Studies in the history of probability and statistics. XXXII. Laplace, Fisher and the discovery of the concept of sufficiency. *Biometrika* **60**(3): 439–445 (Reprinted in Kendall and Plackett (1977), pp. 271–277).

Stigler S.M., 1974 Studies in the history of probability and statistics. XXXIII. Cauchy and the witch of Agnesi: an historical note on the Cauchy distribution. *Biometrika* **61**(2): 375–380.

Stigler S.M., 1975 Studies in the history of probability and statistics. XXXIV. Napoleonic statistics: the work of Laplace. *Biometrika* **62**(2): 503–517.

Stigler S.M., 1977 An attack on Gauss, published by Legendre in 1820. *Historia Mathematica* **4**(1): 31–35.

Stigler S.M., 1978 Laplace's early work: chronology and citations. *Isis* **69**: 234–254.

Stigler S.M., 1981 Gauss and the invention of least squares. *The Annals of Statistics* **9**(3): 465–474.

Stigler S.M., 1982a Poisson on the Poisson distribution. *Statistics & Probability Letters* **1**(1): 33–35.

Stigler S.M., 1982b Thomas Bayes's Bayesian inference. *Journal of the Royal Statistical Society Series A* **145**(2): 250–258.

Stigler S.M., 1986a Laplace's 1774 memoir on inverse probability. *Statistical Science* **1**(3): 359–363.

Stigler S.M., 1986b *The History of Statistics: The Measurement of Uncertainty Before 1900*. Harvard University Press, Cambridge, MA.

Stigler S.M., 1989 Francis Galton's account of the invention of correlation. *Statistical Science* **4**(2): 73–79.

Stigler S.M., 1997 Regression towards the mean, historically considered. *Statistical Methods in Medical Research* **6**(2): 103–114.

Stigler S.M.: Ancillary history; in de Gunst M., Klaassen C., van der Vaart A., (eds): *State of the Art in Probability and Statistics: Festschrift for Willem R. van Zwet* (Vol. **36**). Institute of Mathematical Statistics, Beachwood, OH, 2001, pp. 555–567.

Stigler S.M.: P.S. Laplace, *Théorie Analytique des Probabilités*, First Edition (1812); *Essai Philosophique sur les Probabilités*, First Edition (1814); in Grattan-Guinness I., (ed): *Landmark Writings in Western Mathematics 1640–1940*. Elsevier, Amsterdam, 2005, pp. 329–340.

Stigler S.M., 2006 How Ronald Fisher became a mathematical statistician. *Mathématiques et Sciences Humaines (Mathematics and Social Sciences)* **176**: 23–30.

Stigler S.M., 2007 The epic story of maximum likelihood. *Statistical Science* **22**(4): 598–620.

Stigler S.M., 2008 Karl Pearson's theoretical errors and the advances they inspired. *Statistical Science* **23**(2): 261–271.

Stigler S.M., 2010 Darwin, Galton, and the Statistical Enlightenment. *Journal of the Royal Statistical Society Series A* **173**(3): 469–482.

Stigler S.M., 2013 The true title of Bayes's essay. *Statistical Science* **28**(3): 283–288.

Stone M.: Fiducial probability; in Kotz S., Read C.B., Balakrishnan N., Vidakovic B., (eds): *Encyclopedia of Statistical Sciences*. John Wiley & Sons, Inc., Hoboken, NJ, 2006, pp. 2302–2306.

Straub H.: Bernoulli, Daniel; in Gillispie C.C., (ed): *Dictionary of Scientific Biography* (Vol. **2**). Charles Scribner's Sons, New York, 1970, pp. 36–46.

Street D.J., 1990 Fisher's contributions to agricultural statistics. *Biometrics* **46**(4): 937–945.

Struik D.J.: Bertrand, Joseph Louis François; in Gillispie C.C., (ed): *Dictionary of Scientific Biography* (Vol. **2**). Charles Scribner's Sons, New York, 1970a, pp. 87–89.

Struik D.J.: Robert Adrain; in Gillispie C.C., (ed): *Dictionary of Scientific Biography* (Vol. **1**). Charles Scribner's Sons, New York, 1970b, pp. 65–66.

Stuart A., Ord J.K., 1994 *Kendall's Advanced Theory of Statistics, Volume 1: Distribution Theory*, 6th edition. Arnold, London.

Stuart A., Ord J.K., Arnold A., 1999 *Kendall's Advanced Theory of Statistics, Volume 2A: Classical Inference and the Linear Model*, 6th edition. Arnold, London.

Student, 1908a The probable error of a correlation coefficient. *Biometrika* **6**: 302–310.

Student, 1908b The probable error of a mean. *Biometrika* **6**: 1–25.

Student, 1917 Tables for estimating the probability that the mean of a unique sample of observations lies between -∞ and any given distance of the mean of the population from which the sample is drawn. *Biometrika* **11**(4): 414–417.

Student, 1923 On testing varieties of cereals. *Biometrika* **15**(3/4): 271–293.

Student, 1925 New tables for testing the significance of observations. *Metron* **5**: 113–120.

Student, 1936 The half-drill strip system agricultural experiments. *Nature* **138**: 971–972.

Student, 1938 Comparison between balanced and random arrangements of field plots. *Biometrika* **29**(3/4): 363–378.

Sukhatme P.V., 1938 On Fisher and Behrens' test of significance for the difference in means of two normal samples. *Sankhya* **4**(1): 39–48.

Sverdrup E., 1967 *Laws and Chance Variations: Basic Concepts of Statistical Inference (Vol II: More Advanced Treatment)*. North-Holland, Amsterdam.

Swetz F.J., 2008 The mystery of Robert Adrain. *Mathematics Magazine* **81**(5): 332–344.

Swirles B., 1992 Harold Jeffreys from 1891 to 1940. *Notes and Records of the Royal Society of London* **46**(2): 301–308.

Székely G.J., 1986 *Paradoxes in Probability Theory and Mathematical Statistics*. Reidel, Dordrecht.

Taylor B., 1713 De motu nervi tensi. *Philosophical Transactions* **28**: 26–32.

Taylor B., 1715 *Methodus Incrementorum Directa et Inversa*. Innys, London.

Todhunter I., 1865 *A History of the Mathematical Theory of Probability From the Time of Pascal to That of Laplace*. Macmillan, London (Reprinted by Chelsea, New York, 1949, 1965).

Torretti R., 1990 *Creative Understanding*. University of Chicago Press, Chicago.

Tricomi F.G.: Giovanni Plana; in Gillispie C.C., (ed): *Dictionary of Scientific Biography* (Vol. **10**). Charles Scribner's Sons, New York, 1981, pp. 267–279.

Truesdell C., 1960 *The Rational Mechanics of Flexible or Elastic Bodies 1638–1788. Introduction to Leonhardi Euleri Opera Omnia* (2nd series, Vol. XI (2)). Orell Füssli Turici, Zürich.

Uspensky J.V., 1937 *Introduction to Mathematical Probability*. McGraw-Hill, New York.

Van Aarde M., 2009 *Epistemology of Statistical Science*. Sun Press, Stellenbosch.

van der Waerden B.L., 1968 *Mathematical Statistics*. Springer-Verlag, New York (Translated from *Mathematische Statistik*, 2nd edition (1965). Published by Springer-Verlag, Berlin).

van Fraassen B.C., 1980 *The Scientific Image*. Oxford University Press, Oxford.

van Fraassen B.C., 1989 *Laws and Symmetry*. Oxford University Press, Oxford.

van Zwet W.R., 2011 Remembering Erich Lehmann. *The Annals of Statistics* **39**(5): 2266–2279.

Venn J., 1866 *The Logic of Chance*. Macmillan, London.

von Kries J., 1886 *Die Principien der Wahrscheinlichkeitsrechnung*. J.C.B. Mohr, Tübingen.

von Mises R., 1931 *Wahrscheinlichkeitsrechnung und ihre Anwendung in der Statistik und theoretischen Physik*. Franz Deuticke, Leipzig.

von Mises R., 1942 On the correct use of Bayes' formula. *The Annals of Mathematical Statistics* **13**(2): 156–165.

von Mises R., 1957 *Probability, Statistics, and Truth*, 2nd revised English edition. George Allen & Unwin Ltd., London (First published in German by Springer in 1928, first English edition in 1939).

von Neumann J., Morgenstern O., 1944 *Theory of Games and Economic Behavior*. Princeton University Press, Princeton.

Wald A., 1939 Contributions to the theory of statistical estimation and testing hypotheses. *The Annals of Mathematical Statistics* **10**(4): 299–326.

Wald A., 1950 *Statistical Decision Functions*. John Wiley & Sons, New York.

Walker H.M., 1928 The relation of Plana and Bravais to theory of correlation. *Isis* **10**(2): 466–484.

Walker H.M., 1929 *Studies in the History of Statistical Method*. Williams & Wilkins, Baltimore, MD.

Wallace D.L.: The Behrens-Fisher and Fieller-Creasy problems; in Fienberg S.E., Hinkley D.V., (eds): *R.A. Fisher: An Appreciation*. Springer, New York, 1980, pp. 119–147.

Waterhouse W.C., 1990 Gauss's first argument for least squares. *Archive for History of Exact Sciences* **41**(1): 41–52.

Watson G.: Aitken, Alexander Craig; in Kotz S., Read C.B., Balakrishnan N., Vidakovic B., (eds):*Encyclopedia of Statistical Sciences*. John Wiley & Sons, Inc., Hoboken, NJ, 2006, pp. 101–103.

Weerahandi S., 2003 *Exact Statistical Methods for Data Analysis*. Springer, New York.

Weiss L.: Wald, Abraham; in Kotz S., Read C.B., Balakrishnan N., Vidakovic B., (eds): *Encyclopedia of Statistical Sciences* (Vol. **9**). John Wiley & Sons, Inc., Hoboken, NJ, 2006, pp. 9018–9020.

Welch B.L., 1938 The significance of the difference between two means when the population variances are unequal. *Biometrika* **29**(3/4): 350–362.

Welch B.L., 1947 The generalization of Student's problem when several different population variances are involved. *Biometrika* **34**: 28–35.

Welch B.L., 1956 Note on some criticisms made by Sir Ronald Fisher. *Journal of the Royal Statistical Society B* **18**: 297–302.

Welch B.L., 1958 "Student" and small sample theory. *Journal of the American Statistical Association* **53**(284): 777–788.

Weldon W.F.R., 1883 Memoirs: note on the early development of Lacerta Muralis. *Quarterly Journal of Microscopical Science* **23**: 134–144.

Weldon W.F.R., 1890 The variations occurring in certain *Decapod Crustacea*. I: *Crangon vulgaris*. *Proceedings of the Royal Society* **47**: 445–453.

Weldon W.F.R., 1892 Certain correlated variations in *Crangon vulgaris*. *Proceedings of the Royal Society* **51**: 2–21.

Weldon W.F.R., 1893 On certain correlated variations in *Carcinus moenas*. *Proceedings of the Royal Society* **54**: 318–329.

Welsh A.H., 1996 *Aspects of Statistical Inference*. John Wiley & Sons, New York.

Westergaard H., 1932 *Contributions to the History of Statistics*. King, London.

Whittaker E.T., 1921 On some disputed questions of probability. *Transactions of the Faculty of Actuaries* **8**: 163–206.

Whittaker E.T., Robinson G., 1944 *The Calculus of Observations: A Treatise on Numerical Mathematics*, 4th edition. Blackie & Son, London. (Reprinted in 1967 by Dover, New York)

Whittaker J.M., Bartlett M.S., 1968 Alexander Craig Aitken. 1895–1967. *Biographical Memoirs of Fellows of the Royal Society* **14**: 1–14.

Whittle P., 2004 Maurice Stevenson Bartlett. 18 June 1910–8 January 2002. *Biographical Memoirs of Fellows of the Royal Society* **50**: 15–33.

Wilks S.S., 1940 On the problem of two samples from normal populations with unequal variances (abstract). *Annals of Mathematical Statistics* **11**: 475–476.

Wilks S.S., 1962 *Mathematical Statistics*. John Wiley & Sons, New York.

Williams E.J., 1994 Edwin James George Pitman 1897–1993. *Historical Records of Australian Science* **10**(2): 163–185.

Williams E.J.: Edwin James George Pitman; in Heyde C.C., Seneta E., (eds): *Statisticians of the Centuries*. Springer, New York, 2001, pp. 468–471.

Wishart J., Bartlett M.S., 1932 The distribution of second order moment statistics in a normal system. *Mathematical Proceedings of the Cambridge Philosophical Society* **28**(4): 455–459.

Wishart J., Bartlett M.S., 1933 The generalised product moment distribution in a normal system. *Mathematical Proceedings of the Cambridge Philosophical Society* **29**(2): 260–270.

Wolfenden H.H., 1942 *The Fundamental Principles of Mathematical Statistics*. Macmillan, Toronto.

Wolfowitz J., 1952 Abraham Wald, 1902–1950. *The Annals of Mathematical Statistics* **23**(1): 1–13.

Wood T.B., Stratton F.J.M., 1910 The interpretation of experimental results. *The Journal of Agricultural Science* **3**(4): 417–440.

Wrinch D., Jeffreys H., 1919 On some aspects of the theory of probability. *The London, Edinburgh, and Dublin Philosophical Magazine and Journal of Science* **38**(228): 715–731.

Wrinch D., Jeffreys H., 1921 On certain fundamental principles of scientific inquiry. *The London, Edinburgh, and Dublin Philosophical Magazine and Journal of Science* **42**: 369–390.

Wrinch D., Jeffreys H., 1923 On certain fundamental principles of scientific inquiry. *The London, Edinburgh, and Dublin Philosophical Magazine and Journal of Science* **45**: 368–374.

Yates F., 1934 Contingency tables involving small numbers and the χ^2 test. *Supplement to the Journal of the Royal Statistical Society* **1**(2): 217–235.

Yates F., 1952 George Udny Yule. 1871–1951. *Obituary Notices of Fellows of the Royal Society* **8**(21): 309–323.

Yates F., 1964 Fiducial probability, recognisable sub-sets and Behrens' test. *Biometrics* **20**(2): 343–360.

Yates F., 1984 Tests of significance for 2 x 2 contingency tables. *Journal of the Royal Statistical Society A* **147**(3): 426–463.

Yule G.U., 1897 On the significance of Bravais' formulae for regression, &c., in the case of skew correlation. *Proceedings of the Royal Society of London Series A* **60**: 477–489.

Yule G.U., 1900 On the association of attributes in statistics: with illustrations from the material of the childhood society, &c. *Philosophical Transactions of the Royal Society of London A* **194**: 257–319.

Yule G.U., 1906a On a property which holds good for all groupings of a normal distribution of frequency for two variables, with applications to the study of contingency-tables for the inheritance of unmeasured qualities. *Proceedings of the Royal Society of London A* **77**(517): 324–336.

Yule G.U., 1906b On the influence of bias and of personal equation in statistics of ill-defined qualities: an experimental study. *Proceedings of the Royal Society of London A* **77**(517): 337–339.

Yule G.U., 1907 On the theory of correlation for any number of variables, treated by a new system of notation. *Proceedings of the Royal Society of London A* **79**(529): 182–193.

Yule G.U., 1911 *An Introduction to the Theory of Statistics*. Griffin, London.

Yule G.U., 1912 On the methods of measuring association between two attributes. *Journal of the Royal Statistical Society* **75**(6): 579–652.

Yule G.U., 1922 On the application of the χ^2 method to association and contingency tables, with experimental illustration. *Journal of the Royal Statistical Society* **85**(1): 95–104.

Yushkevich A.P., 1987 Nicholas Bernoulli and the publication of James Bernoulli's *Ars Conjectandi*. *Theory of Probability and its Applications* **31**(2): 286–303.

Zabell S.L., 1988 Buffon, Price, and Laplace: scientific attribution in the 18th century. *Archive for History of Exact Sciences* **39**(2): 173–181 (Also in Zabell (2005), pp. 74–83).

Zabell S.L., 1989a R.A. Fisher on the history of inverse probability. *Statistical Science* **4**(3): 247–256 (Also in Zabell (2005), pp. 142–160).

Zabell S.L., 1989b The rule of succession. *Erkenntnis* **31**: 283–321 (Also in Zabell (2005), pp. 38–73).

Zabell S.L., 1992 R.A. Fisher and the fiducial argument. *Statistical Science* **7**: 369–387 (Also in Zabell (2005), pp. 161–198).

Zabell S.L., 2005 *Symmetry and its Discontents*. Cambridge University Press, Cambridge.

Zehna P.W., 1966 Invariance of maximum likelihood estimators. *Annals of Mathematical Statistics* **37**: 744.

INDEX

Note: page numbers in bold refer to illustrations

Classic Topics on the History of Modern Mathematical Statistics: From Laplace to More Recent Times, First Edition. Prakash Gorroochurn

Printed and bound by CPI Group (UK) Ltd, Croydon, CR0 4YY

16/04/2025

14658517-0005